MANUFACTURING TECHNOLOGY

John R. Lindbeck, *Ph.D., Professor of Engineering Technology*

Molly W. Williams, *Ph.D., P.E., Professor of Mechanical Engineering*

Robert M. Wygant, *Ph.D., Professor of Industrial Engineering*

Western Michigan University

CONTRIBUTING AUTHORS

Paul V. Engelmann, *Ed.D., Assistant Professor of Engineering Technology*

Nancy H. Steinhaus, *Ph.D., Associate Professor of Consumer Resources Technology*

Western Michigan University

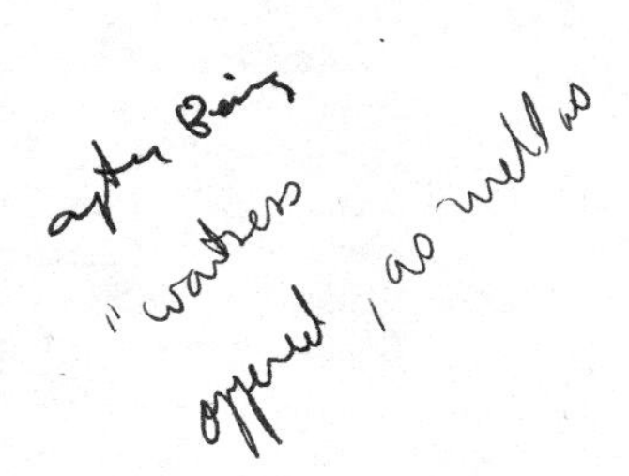

Prentice Hall
Englewood Cliffs, New Jersey 07632

Library of Congress Cataloging-in-Publication Data

LINDBECK, JOHN ROBERT.
Manufacturing technology.

Bibliography: p.
Includes index.
1. Production management. 2. Manufacturing processes. I. Williams, Molly W. II. Wygant, Robert M. III. Title.
TS155.L588 1990 658.5 88–31754
ISBN 0–13–487315–7

Editorial/production supervision and interior design: Patrice Fraccio
Cover design: Diane Saxe
Manufacturing buyer: Mary Noonan
Page layout: Marty Behan
Cover photograph: Courtesy of *Forest Machine Tool* (a subsidiary of *Brisard Machines Outils*)

Printed in the United States of America

10 9 8 7 6 5 4 3

ISBN 0-13-211690-1

Prentice-Hall International (UK) Limited, *London*
Prentice-Hall of Australia Pty. Limited, *Sydney*
Prentice-Hall Canada Inc., *Toronto*
Prentice-Hall Hispanoamericana, S.A., *Mexico*
Prentice-Hall of India Private Limited, *New Delhi*
Prentice-Hall of Japan, Inc., *Tokyo*
Simon & Schuster Asia Pte. Ltd., *Singapore*
Editora Prentice-Hall do Brasil, Ltda., *Rio de Janeiro*

CONTENTS

PREFACE

The authors have planned this volume as an introduction to the subject of contemporary manufacturing organization, methods, and processes, for those students who are embarking on careers in this field. The range of this subject makes it impossible to treat in one volume, so the authors must be selective. All effort was directed toward presenting basic information in an interesting fashion, and to including sufficient illustrations and explanations of manufacturing fundamentals. The intention was that with a grasp of this basic knowledge, and a comprehension of how things are made, students will be better prepared to deal with specialized course work in manufacturing process areas, materials, organizational structures, and advanced systems.

The many chapters on material processing are generally, but not totally, related to metallics. There are two main reasons for this:

1. Metal process information is important to manufacturing personnel, because the machines and tooling used to process materials such as wood and plastic are made of metal. Metal processing information is necessary in any manufacturing economy.
2. Many of these processes are basic to all material areas. The theories and practices of drilling, abrading, and coating, for example, are generally applicable to most materials. Students should bear this in mind as they make their way through the chapters.

Several special processes dealing with plastics, woods, and fibers are discussed in those chapters. This was done for the student's convenience and to facilitate an understanding of material process differences.

We dedicate this book to these students, so that they may become inquisitive, excited, and committed to producing better manufactured products.

ACKNOWLEDGMENTS

A textbook such as this evolves over a period of several years, and requires the dedication, effort, and forebearance of many people and institutions. No volume dealing with manufacturing practices can emerge without the cooperation of representative industries. Authors must turn to these industries for information, data, and illustrations so that their book reflects the current, state-of-the-art processes used in the manufacture of industrial products. The authors of this text are sincerely indebted to these industries, and we further identify their contributions as a courtesy line accompanying each of their illustrations appearing in this text.

Special mention must be made to the work of Dr. David H. Gregg who produced the numerous computer-aided drawings which so effectively portray the range of industrial processes and operations. He displayed not only a great self-taught talent, but tremendous patience in meeting tight deadlines and making necessary modifications called for by the authors.

Ms. Sharon Howes exhibited equal patience in her tireless work at the word processor, producing drafts, making endless changes, and always remaining understanding and capable.

Mr. David Wormmeester was invaluable in preparing the excellent sketches under great pressure to meet the final time constraints.

To all of these, and to their families, and to others too numerous to name, the authors are indeed grateful.

INTRODUCTION

Very simply, manufacturing is converting raw materials into usable products. A further refinement of this fundamental definition is that these products may be reproduced in quantity, at a quality which makes them functional, and at a cost which makes them competitive. There is a basic difference, therefore, between an artisan who designs and crafts a single chair, and a number of workers involved in building a quantity of identical chairs. This act of replication requires the organization of product design, materials, special tools and equipment, skilled people, and a management team into a productive system.

The processes involved in converting materials into products can be classified as cutting, forming, assembling, and finishing. Ultimately, these are the only acts one can impose upon a substance to change it into something which people can use. Furthermore, these changes imposed upon materials are designed to produce modifications in the mass or bulk of the material, so that a subsequent change in geometry or shape occurs. These modifications can be described as mass or bulk *reducing* (cutting), mass or bulk *conserving* (forming), mass or bulk *increasing* (assembling), and mass or bulk *conditioning* (finishing). Stated another way, the geometry or configuration of a material can be changed by removing unnecessary material, moving material from one area to another, joining materials together, or modifying the surface of a material. These primary changes are here described briefly in a somewhat theoretical, yet descriptive, fashion to introduce the contents of this book.

Cutting is a mass or bulk reduction process, where shape change is achieved by *removing* material particles or chips until the workpiece has the desired geometry. Cutting is also a *separation* process, where a part size is diminished by trimming or shearing to a specified dimension. Cutting therefore differs from hot or cold forming, where bulk deformation is employed to create a new shape, but where no gross part of the workpiece is removed.

Many shapes produced by cutting can be made by hot or cold forming or by casting. The greatest advantage of cutting is the ability to produce intricate shapes having extremely accurate dimensions. Material removal can be more precisely controlled than can plastic flow. Furthermore, the problems of shrinkage in hot-formed parts and the springback of cold-formed parts are eliminated. Because cutting does not rely on plastic flow to modify the shape of a workpiece, a cut part can have the same properties as the starting piece. For example, machined metal parts retain the benefits of prior cold work or heat treatment.

Cutting is usually more economical than forming for very complex parts, or when only a few pieces are to be produced. For most shapes, however, cutting is more expensive and must be used judiciously. In comparison with metal forming, for example, machining is slower, produces more waste material, and requires greater skill.

One of the greatest uses of cutting is to finish parts made by other processes. Many parts are produced by first making the rough shape by forming, and then completing the part by machining. Most turbine blades, for example, are milled from forged blanks. Some parts are produced almost entirely by forming, and then are machined only at areas where extremely accurate or complex shapes are required. For example, metal shafts for pumps or propellers are often hot- or cold-formed, with the bearing journals and threaded ends produced by machining.

There are a number of processes which logically fall under the category of cutting. Sawing and shearing are cutting by *separation* methods. Milling, shaping, planing, turning, drilling, boring, and reaming commonly are called *machining* processes, and involve material *removal* to produce shapes. *Abrading* is technically a material removal process, although abrasive saws are separation devices.

Several nontraditional cutting techniques are employed to meet special material processing requirements. Among these are thermal, chemical, and pressure methods such as lasers, photochemical etching, and abrasive jet cutting.

Forming is a mass or bulk conserving process, where shape change is achieved by manipulation or deformation. The bulk of the end product is equal, or almost equal, to the initial bulk of the workpiece or raw substance. In cutting, a shape was produced by material removal. In forming, shape occurs by moving the material about (except in the process of casting, to be explained later).

The fundamental forming processes include bending, drawing, rolling, forging, extruding, and casting. In all but casting, the geometry change is effected by deformation. Metal can be squeezed, stretched, twisted, turned, or pushed to produce various shapes. The casting of metals or ceramics involves the introduction of a fixed amount of material into a mold cavity. The material is conserved; only the precise amount needed goes into the finished product. Nothing is cut away (except gates, risers, etc.) to form the final part geometry. Similarly, plastic is injected into a die, with the same part configuration and mass conserving results. Forming becomes a significant method to produce pieces to near final shape and size at great economy, and with certain structural attributes. And, generally, it costs less.

Assembling involves joining together several components to create a final product. As such, it is a mass or bulk increasing process, because the total mass of that final product is the sum total of the masses of the individual parts of which it is made. Parts can be assembled by joining components with mechanical fasteners such as nuts and bolts, by fusing them together as weldments, or by adhering them with glues and solders. Some assemblies are meant to be permanent, while others are meant to be taken apart—disassembled—as required. See Figure INT-1.

Finishing involves workpiece surface modification for the purposes of protection or appearance. A fine piece of furniture is lacquered both to protect its surfaces and to make it more attractive. It is a mass or bulk conditioning process, inasmuch as the bulk of the end product remains essentially unchanged. A board may be sanded or a metal part surface-ground, processes whereby modest amounts of material are removed. Wood may be painted or a metal part chrome-plated, and in the process minute additive surface changes will occur. However, in both examples, the bulk change is insignificant. Process examples include polishing, buffing, plating, painting, staining, anodizing, and heat treating. In the cases of the last three examples, no surface additions or subtractions occur. The surface structure or color is the sole resultant change. See Figure INT-2.

Any discussion of material processing leads to the ultimate conclusion that process involves selection and option. For example, holes may be *generated* in a piece of metal in a variety of ways. They may be drilled, punched, milled, laser-cut, chiselled, chemically-etched, pierced, abrasive-cut, liquid jet-cut, or thermally-cut with a oxy-fuel torch. How do you cut the hole? The decision must be based on so many factors—the quality and accuracy of the hole, the reason for having a hole in the first place, the kind of metal being cut, the available equipment, the shape of the hole, the matters of automated systems, numbers of parts, and speed, as well as others. The many people involved in the manufacture of the part must make the final decision and select the most appropriate process.

The same factors must be studied for the categories of forming, assembling, and finishing processes: What selection is made, and why?

Figure INT-1 Assembly processes involve the options of mechanical, adhesion, or cohesion systems.

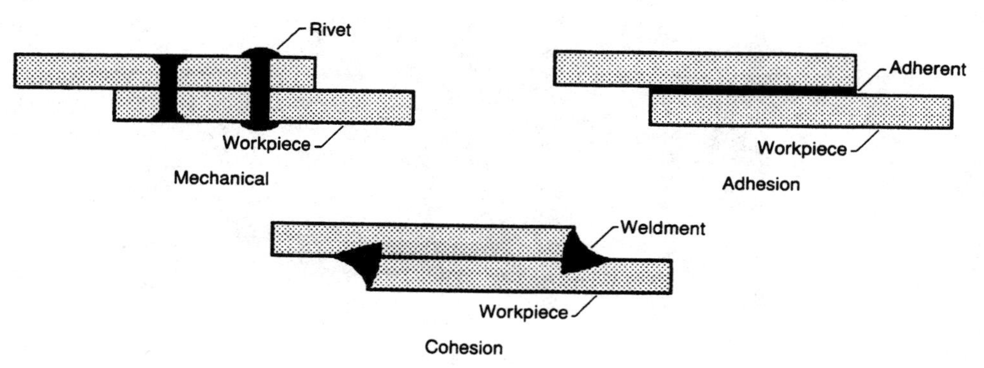

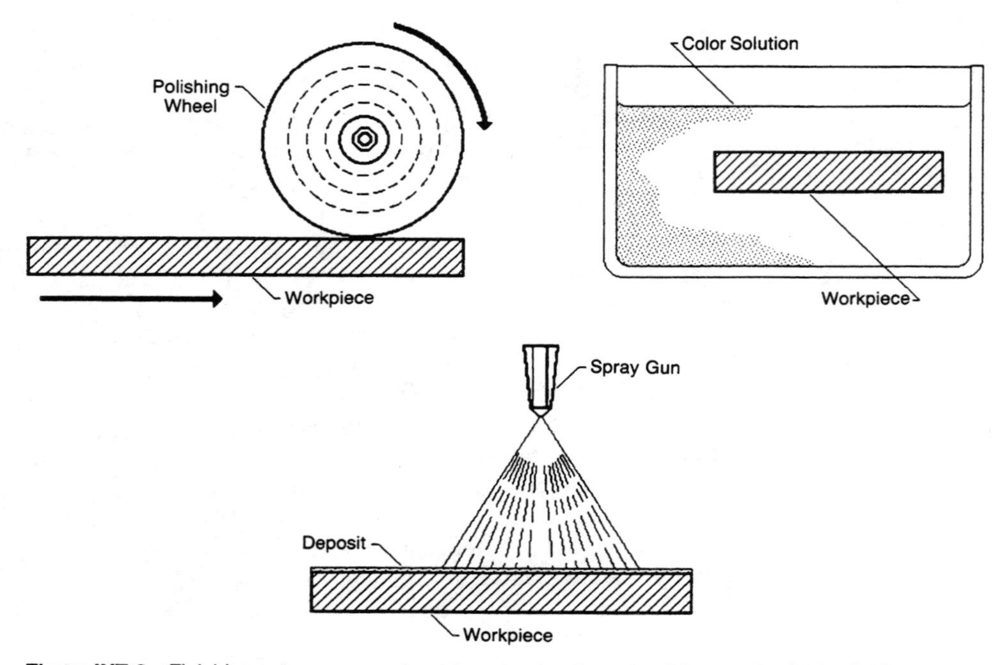

Figure INT-2 Finishing processes may be categorized as those involving mechanical polishing, staining or coloring, or coating.

In modern manufacturing, process independent of systematic organization counts for very little. Random, independent material changes may lead to a product, but seldom one which meets the requirements of reproducibility, consistency, measurable quality, or cost-effectiveness. Similarly, the various supporting units in a manufacturing enterprise must be linked together with process to fulfil their missions.

Manufacturing is generally considered to be a series of integral *functions* dedicated to some product mission. First, there is a *planning* or design function, typically called research and development, and charged with the responsibility for product design. A *tooling* function is necessary for preparing the dies, equipment, and machines to produce the object. A *production control* function provides the vital forecasting-scheduling-dispatching to set the manufacturing wheels in motion. *Quality control* is a function which monitors production to assure end products or uniform high

Figure INT-3 Simple, common aluminum cans are the result of accurate and carefully monitored production controls. (Courtesy of Reynolds Metals Company)

quality. *Human resource management* functions to direct the selection, hiring, training, and caring of persons needed in the production scheme. The attendant planning and monitoring functions of these elements lead to the production of quality goods.

An excellent example of the need for a systematic production management scheme is the manufacture of common aluminum beverage cans. See Figure INT-3. Beginning with the can design according to customer specifications, followed by a range of forming processes, through a series of conditioning operations and final testing, can manufacture demands precise control and monitoring. See Figure INT-4.

Advanced technology plays an increasingly important role in making available the equipment and systems for unattended machine operations and automatic inspection, storage, and inventory. Increasingly, this technology becomes necessary if an industry is to remain competitive and keep abreast of rapid change.

Figure INT-4 The production sequence for aluminum beverage cans. The ends or can lids are joined to the can body after filling. Adhesive and mechanical joints are employed. (Courtesy of Reynolds Metals Company)

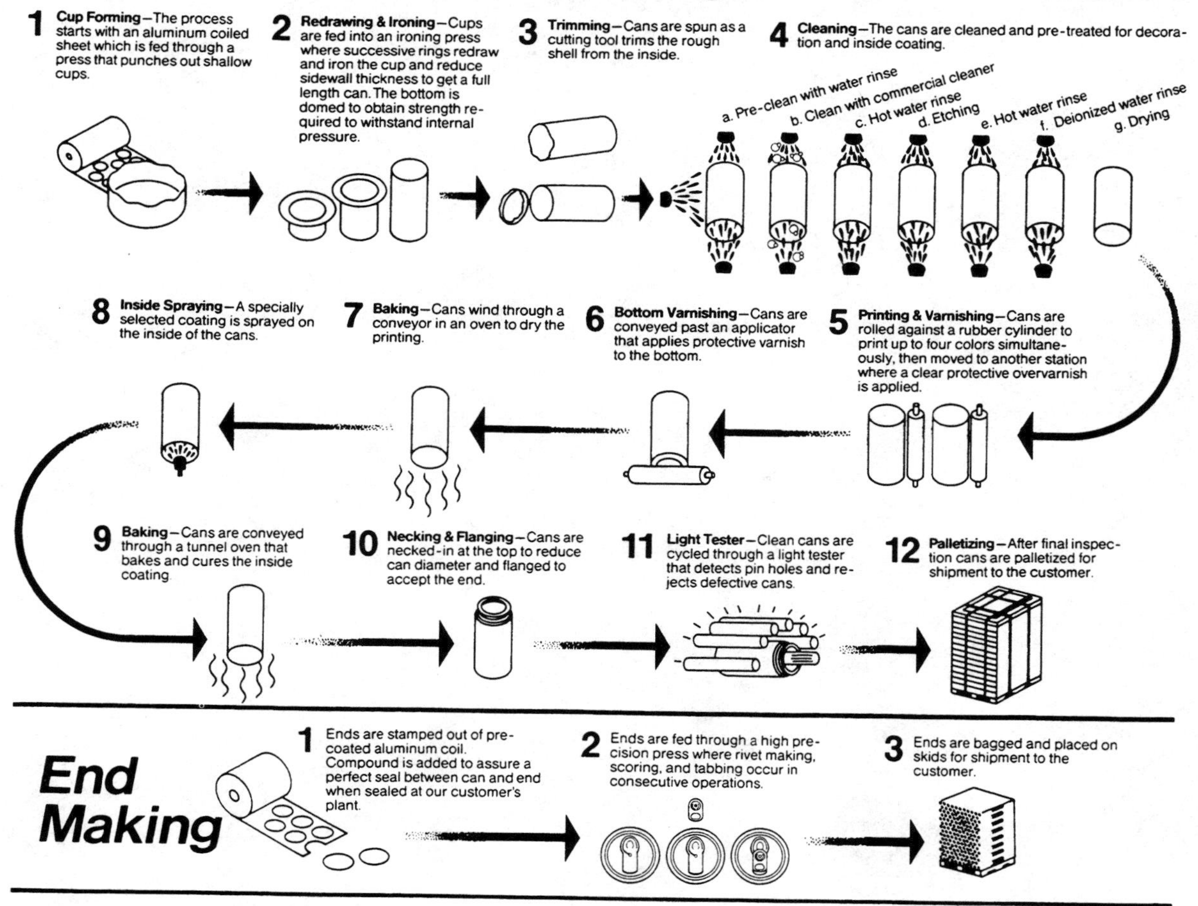

part 1 PRINCIPLES OF MANUFACTURING

1 EVOLUTION OF THE MANUFACTURING INDUSTRY

Manufacturing is the process of transforming raw materials into a finished product. Originally manufacturing was a human activity that described the making of articles by hand. As the demand for more goods increased, new inventions provided the machinery and power to assist the human in producing goods in larger quantities. With developments in electronic controls came the technology that enabled industries to automate processes so they could turn out products by mechanical power with a minimum of human control.

This chapter reviews the development of manufacturing and the trends in productivity in the United States, and shows how some of the major technological developments in manufacturing have contributed to economic growth.

HISTORICAL OVERVIEW

The early settlers in the United States were primarily engaged in agriculture. European emigrants brought with them crops, tools, and farming methods from their native countries. The original settlers, by necessity, were highly self-sufficient. In addition to growing a major portion of their own food supplies, most households provided many of their own clothes, tools, and manufactured articles. With 90 percent of the population involved in agricultural production, it was only natural that supporting businesses and trades, such as storekeepers and blacksmiths, catered to the needs of the farmers.

The stream of immigration into Virginia, Maryland, and New England gave rise to the growth of settlements on the East Coast. As a result of this expanding population, the needs for supporting crafts and merchants increased.

Cottage industries. Until the late 1700s, manufacturing in the United States was dominated by work done by highly skilled individuals and small "cottage" industries that were maintained by the crafters and apprentices. The household manufacturers made goods primarily for local consumption and utilized the skill of the craft workers, rather than elaborate equipment.

Each community had its local craft workshops. The art of spinning, as shown in Figure 1-1, was carried on by women and children. Spinning was applied to the reducing of silk, flax, hemp, and wool into thread.

Another important craft was blacksmithing, still being practiced by smiths, Figure 1-2, people who work with metals to manufacture a vast variety of articles used both in the home and in the production of goods. Most common of the smiths are blacksmiths. Others are called whitesmiths, or brightsmiths; these polish their work to improve the appearance.

The potter, as shown in Figure 1-3, formed many different items such as cups, plates, and vessels of baked earth. The art of pottery is one of the oldest trades, dating back to the ancient Greeks and Etruscans.

Figure 1-1 Spinning. (Courtesy of Dover Publications)

Figure 1-2 A Smith. (Courtesy of Dover Publications)

Figure 1-3 Potter. (Courtesy of Dover Publications)

Figure 1-4 Wool-comber. (Courtesy of Dover Publications)

Some of the other "cottage" crafts that emerged as a result of growing demand included the wool-comber, to provide wool for blankets and coarse cloth; the jeweller; the cooper for the manufacture of casks, tubs, pails, and containers for domestic use; and the currier, who prepared hides for use by shoemakers, coachmakers, saddlers, bookbinders, etc. See Figures 1-4 through 1-7.

Because of the shortage of skilled labor, workers often used their skills to perform more than one trade: A tanner often did currier work and served as the village shoemaker; blacksmiths repaired all types of equipment; and wool-combers and spinners frequently made clothing for sale to local merchants. As expected, the output from these operations was limited and was generally inefficient.

Using machines. Although Leonardo da Vinci designed many machines such as the change gear lathe and the internal grinding machine, the actual development of these machines came much later. It wasn't until basic machines were capable of making other machines and parts to reproduce themselves that the industrial revolution began. The first basic machine tool was the boring mill, developed by John Wilkinson in Bersham, England in 1775. This mill was used to accurately bore the cylinders for James Watt's steam engines. Thus, the machine age and the power age were born together.

Figure 1-5 Jeweller. (Courtesy of Dover Publications)

Figure 1-6 Cooper. (Courtesy of Dover Publications)

Figure 1-7 Currier. (Courtesy of Dover Publications)

In 1800, another engineer from England, Henry Maudslay, developed the screw cutting lathe with lead screw, change gears, and compound slide rest. Not only did a new age of accurate machining begin but, perhaps just as important, the mass production of accurate screw threads of varied pitch was now possible. The profile or copying lathe, developed by the American Thomas Blanchard, provided the capability to duplicate irregular forms. A treadle-powered metalcutting lathe is shown in Figure 1-8.

The first metal planer to produce duplicate flat surfaces was developed by Richard Roberts in Manchester, England, about 1817. It was patterned after an earlier woodworking model that was first made and used by Sir Mark Brumel in France about 1793. Figure 1-9 shows a planing machine that was made by the Merian Company in 1870. It is interesting to note that the bed and ways of the first machine were made by hand with chisel and file. Once this handmade planer was completed, other flat surfaces could now be machined with accuracy.

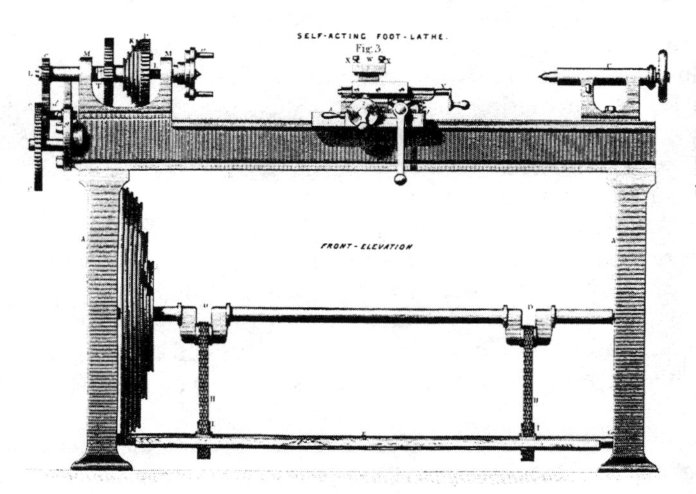

Figure 1-8 Engraving of a treadle-powered metalcutting lathe of 1873 built by Jos. Whitworth & Co. (Courtesy of The National Museum of American History, Smithsonian Institution)

Figure 1-9 Daniels planing machine, Richardson Merian and Company, Worcester, MA., Ca 1870. (Courtesy of the National Museum of American History, Smithsonian Institution)

Eli Whitney developed the first milling machine in New Haven, Connecticut, in approximately 1818. This machine performed the first truly accurate duplication of parts, in many configurations of material, by mechanically controlling the course and cutting motion of a rotating, multiple-edged cutting tool. With this invention the operation and control functions were transferred from skilled hand craftsmen to machine tools.

The first automatic drilling machine with power feed was developed in approximately 1840 by J. Nasmyth of Manchester, England. This power feed feature made metal drilling and reaming practical by controlling the rate and direction of the cutting action to produce accurate holes.

Measuring techniques. It wasn't until late in the eighteenth century that instruments and standards of measurement were developed that enabled manufacturers to make interchangeable parts. (James Watt's micrometer, 1772, made possible the accurate measurement of parts produced by new machinery such as Wilkinson's boring mill.)

In addition to his work with developing new machinery, Maudslay designed and built a measuring instrument, called the "Lord Chancellor," in 1800. This new instrument, shown in Figure 1-10, could detect minute differences in length and exactness, which provided much greater accuracy than the requirements of that period in manufacturing.

In 1850 the Brown & Sharpe Company in Rhode Island began to manufacture steel rules and vernier calipers which provided machinists with the tools that were necessary for making accurate measurements. This system of line measurement is still widely used in manufacturing processes.

Expanding on the idea of Jean Laurent Palmer's "screw caliper," Joseph Brown and Lucian Sharpe introduced the micrometer in America in the late 1860s. See Figure 1-11.

A standard of measure was made available to all machine shops with the development of gauge blocks. In 1896 Carl Johansson of Sweden started to make a combination of gauge blocks that have universally been called "jo-blocks" and are synonymous with precise measurement in machining operations throughout the world. A set of Whitworth plug and ring gauges are shown in Figure 1-12.

Figure 1-10 Henry Maudslay's "Lord Chancellor" measuring instrument. (Courtesy of The National Museum of American History, Smithsonian Institution)

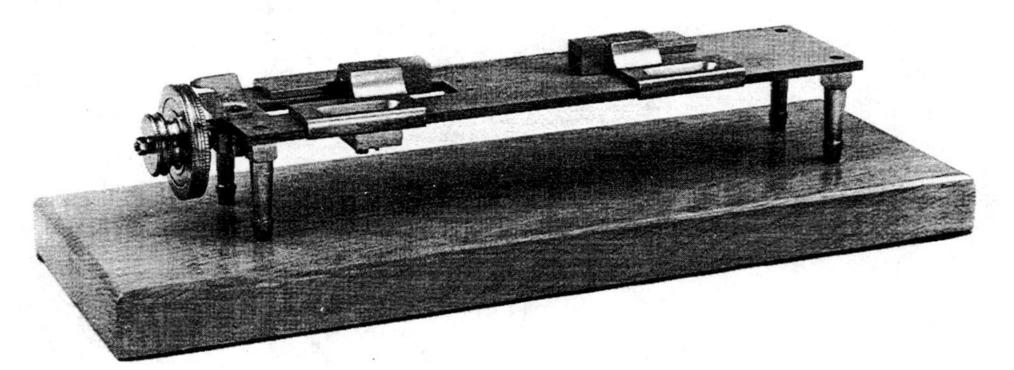

POCKET
SHEET METAL GAUGE,

For the use of Machinists, Jewelers, Silversmiths, Sheet Brass Rollers and Workers, Sheet Iron Rollers, Wire Drawers, Rubber Manufacturers, Paper Makers, Type Founders, &c.

(Full Size.)

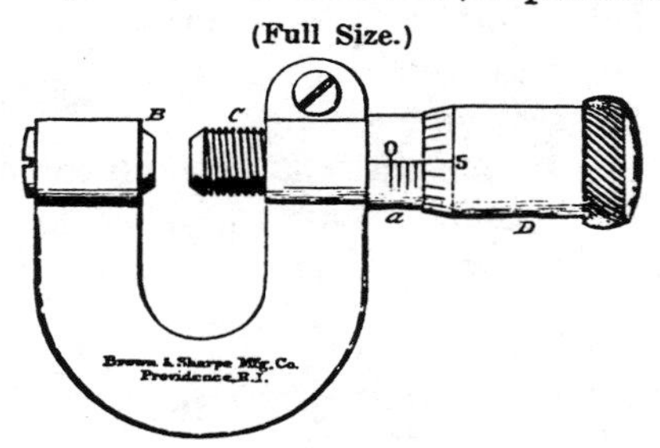

MADE BY

BROWN & SHARPE MF'G. CO., PROVIDENCE, R. I.

Price, $5. In Morocco Case, $5.50.

Sent per Mail on receipt of Price.

This gauge will measure the thickness of sheet metal or other material, by thousandths of an inch up to three tenths of an inch at any point within half an inch from the edge and can be applied as easily as the common gauge. It will also answer to measure the diameter of wire. Means of adjustment are provided in case of wear by continued use.

Figure 1-11 Brown & Sharpe "Pocket Sheet Metal Gage." (Courtesy of The National Museum of American History, Smithsonian Institution)

Figure 1-12 A set of Whitworth plug and ring gauges. (Courtesy of The National Museum of American History, Smithsonian Institution)

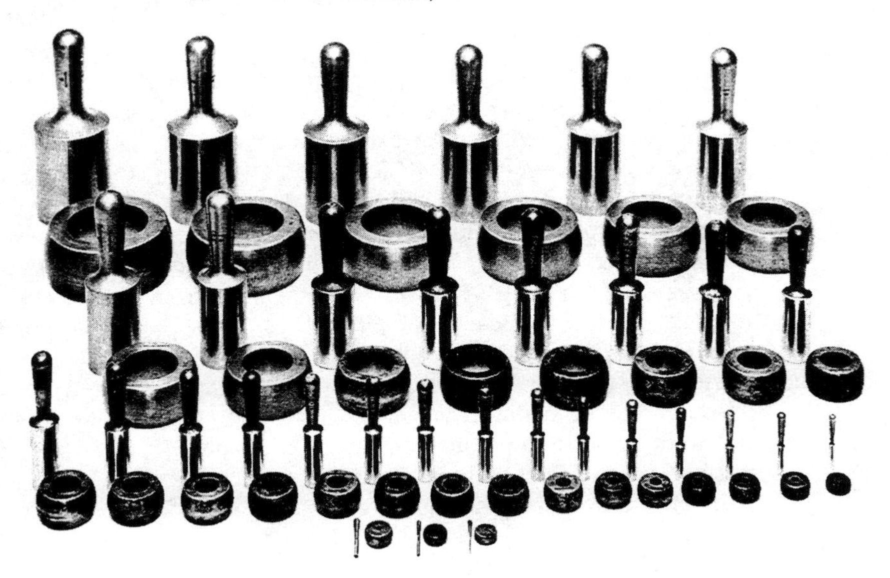

Factory systems. In 1832 Charles Babbage published *On the Economy of Machinery and Manufacture*, in which he described two distinct systems of production—the "making" system and the "manufacturing" system. In the "making" system each of the products are made one at a time, or in small quantities, similar to what is now referred to as "job shop" operations. The "manufacturing" system referred to making a large quantity of the same product, where the manufacturing methods varied depending on the quantity to be produced. This method is often called "mass production."

In the nineteenth century the predominate factory system in European countries was the "making" system, while the "manufacturing" system is considered to have originated in American arsenals in the manufacture of munitions.

EMERGENCE OF MODERN MANUFACTURING

The beginning of manufacturing in the United States can be traced to the early 1800s when Eli Whitney contracted to deliver 10,000 muskets to the U.S. government and organized the factory manufacturing system. Although widely recognized for the use of interchangeable parts, Whitney was instrumental in developing the concepts of jigs, fixtures, and specialized machinery to be used in a progressive work flow within the factory. Other innovators were able to duplicate and refine these concepts, particularly the arms makers in the New England states.

A second stage of manufacturing involved the division of operations into specialized components and separate factory work units. An example of this type of arrangement was noted by Chandler (cited by Abernathy and others, 1983, p. 34).

> [manufacturers] turned the day-to-day operations of the new factories over to the foremen of the several departments. As in the case of iron and steel mills, these foremen controlled; they hired, fired, and promoted their working force. In those departments requiring the most intricate processing techniques in grinding, polishing, and other finishing of metal components, the foremen were responsible for the profitability as well as the production of their departments.

Improvements in transportation, communications, and finance paved the way for new methods of manufacturing, which were necessary because of the mass market for new products. An example of a typical machine shop in the late 1800s is shown in Figure 1-13.

During the period from 1860 to 1930 a second industrial revolution began. While the previous period mainly focused on iron-making, steam power, and the cotton textile industry, this second phase of industrial development was comprised of four major areas (Ayres, 1984, p. 110): 1) the shift from iron to steel as an engineering material, 2) the application of electricity, 3) the internal combustion engine, and 4) the mass production of consumer goods.

As the Bessemer steel-making process was implemented, the quantity and price of steel showed a dramatic change. In 1867, only 2,500 tons of Bessemer steel rail were made in the United States, at $170 per ton. By 1884, production of steel rail was up to 1,500,000 tons, and prices were down to $32 per ton (Ayres, 1984, p. 112). Figure 1-14 indicates how a boiler drum was forged on a mandrel.

The application of electricity had a slower rate of acceptance, beginning with the first electric telegraph in the 1840s. It was not until the 1890s that hydroelectric

Figure 1-13 Machine shop—c. 1880. (Courtesy of The National Museum of American History, Smithsonian Institution)

Figure 1-14 Hollow forging boiler drum on mandrel—Midvale Steel Works, 1930. (Courtesy of The National Museum of American History, Smithsonian Institution)

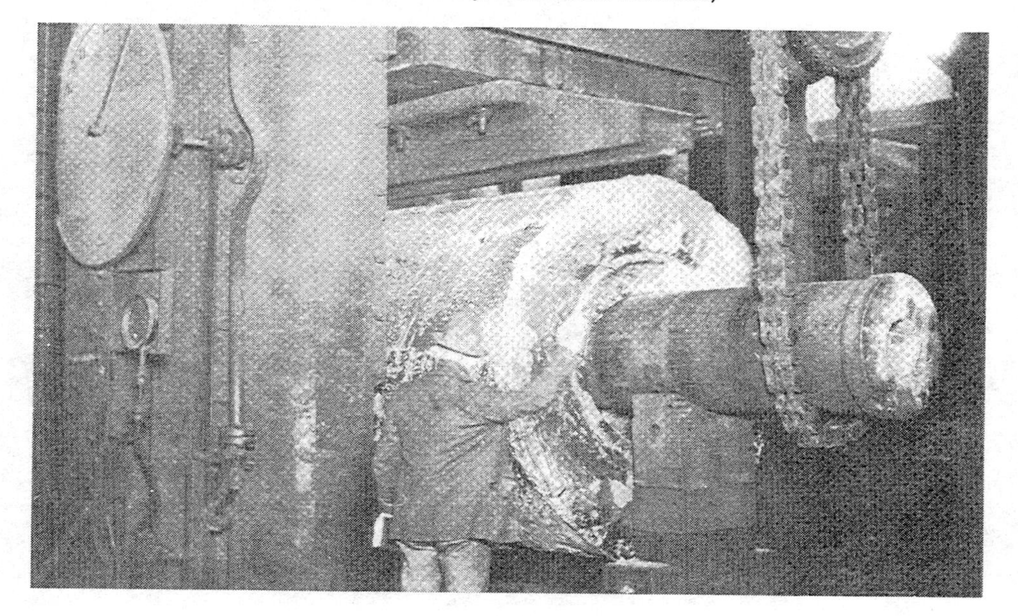

sources were used to supply electric power for industrial motors and individual homes. By the 1920s, the use of electric washing machines, sewing machines, refrigerators, vacuum sweepers, and other electrical appliances was common in many urban homes.

The development of the internal combustion engine opened up a whole new method of individual travel, and with it came a vast new market for manufactured goods. See Figure 1-15.

In 1899 R.E. Olds established the first factory in the United States which was devoted exclusively to manufacturing automobiles. The first cars produced in the new Olds Motor Works were entirely assembled vehicles using a hand-to-hand assembly line. As reported in Kettering and Orth (1955, p. 53), "The result was an unprecedented production—18,500 cars had been built by the end of 1905. When considering that there were only about 4,000 cars in America in 1900, and that those were owned by wealthy people and sportsmen, imagine what the result might be of putting over 18,000 new, inexpensive cars in the hands of average people."

In 1909 Henry Ford sold over 10,000 Model Ts. Within a period of only six years, his sales had reached a quarter of a million cars a year. The relationship of this dramatic growth and the cost per auto is shown in Figure 1-16. With the increase in volume came a drop in price from $950 in 1909 to $490 in 1914.

Figure 1-15 Assembly of small power pumps—Dean Works, c. 1918. (Courtesy of The National Museum of American History, Smithsonian Institution)

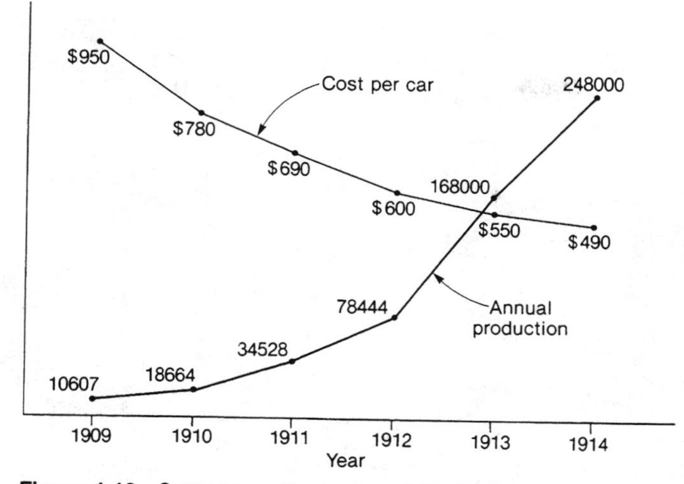

Figure 1-16 Cost versus Production of Model T Automobile from 1909 to 1914. (Source: *American Battle for Abundance*, p. 68)

The popularity of the new automobile started a new phase in U.S. manufacturing—the mass production of technologically complex parts. See Figures 1-17 and 1-18. The rapid growth within the automobile industry required new and improved methods of assembly, greater control of quality, standardization of parts, and new systems of management control.

The next major breakthrough in industrial technology was in the areas of electronics and automation. A large part of these advances resulted from the development of computers and their applications in processing information and controlling manufacturing machines and equipment.

Figure 1-17 Flywheel Magneto Assembly—Ford Motor Co., 1913. (Courtesy of The National Museum of American History, Smithsonian Institution)

Figure 1-18 Gear Cutting—Dean Works, 1917. (Courtesy of The National Museum of American History, Smithsonian Institution)

Early efforts to automate manufacturing processes can be traced to the programmable loom designed by Joseph Jacquard in the early 1800s. Charles Babbage's work on a mechanical calculating machine in the 1820s was the first step in the design of a computer with stored memory. The first electronic computer, the ENIAC, was built by J.P. Eckert and J.W. Mauchly at the University of Pennsylvania during the period of 1943 to 1946.

The first automatic machines were controlled by mechanical stops, cams, and limit switches. The invention of the transistor (1947) and the silicon chip (1960) made it possible for the wide application of microprocessors in the control of manufacturing equipment. By 1952 the first numerically controlled machine was developed at Massachusetts Institute of Technology. In 1962 General Motors installed its first robot. Direct numerical control and computer numerical control of production machinery soon followed.

In discussing the factory of the future, Ayres (1984, p. 160) states that:

> it is only in the last few years that the new generation of computer-controlled machine tools is substantially replacing the older types in most applications. As compared to the manual/mechanical control systems of less than two decades earlier, output per machine has risen by a factor of 3 to 5. Thus, although only around 4 percent of the machine tools in the United States are numerically controlled, these tools may account for as much as 50 percent of the value added by the metalworking industries. This alone might justify the notion of a coming qualitative change in manufacturing technology.

MANUFACTURING AND PRODUCTIVITY

After centuries of tribal life in America, the real income of the typical Indian family did not exceed $1 a month, in addition to the food and shelter provided within the family. This tends to raise questions concerning one common theory, that American economic growth was created by the abundance of land, natural resources, and a favorable climate.

With the Indian societies maintaining a steady-state standard of living, the rate of economic growth in America was near zero until 1700. It wasn't until the arrival of the early settlers in North America that there was a dramatic increase in economic growth.

The change in the economy of the United States was influenced by two major factors. One factor was the method in which these settlers earned their livings. Most of the colonists worked as independent farmers, artisans, or merchants. A second economic force was that virtually every white family owned its own land. By 1774, three out of every four free families owned their own farm. As a result of this new settlement by people with different values, there was a rapid rate of growth in the United States per capita national product, as shown in Table 1-1. These data indicate that the rate of growth from independence to 1860 was almost three times that of the colonial period.

TABLE 1-1
U.S. NATIONAL PRODUCT PER PERSON ANNUAL GROWTH RATE

Year	Percent
1700–1800	+0.4
1800–1860	+1.1
1860–1967	+1.7

Source: Lebergott, 1984, p. 60.

Abramowitz and David (1973) have examined the forces that contributed to economic growth, and they conclude: "Over the course of the nineteenth century the pace of increase of the real gross domestic product was accounted for largely by the traditional, conventionally defined factors of production: labor, land, the tangible reproducible capital."

As the demand for more products increased, it soon became apparent that the individual crafters and those working at home could no longer produce the quantities of goods that were required to meet the expanding markets. This new growth was answered with factory production. Goods made in factories could be produced in greater volume and more efficiently than "home manufacturers." The comparative cost of factory man-hours versus home man-hours for various products is shown in Table 1-2.

With the advent of factories came advances resulting from new sources of technology. New methods of production increased the output and improved operating efficiency in many plants. The demand for trained engineers, scientists, and technicians increased as their worth became evident.

Some of the original ideas for improved manufacturing methods came from Frederick W. Taylor and his "scientific management" approach to factory operations. Not only was he able to increase output and profits, but the workers also received higher earnings. Taylor looked upon human work in the same way that he analyzed machinery and mechanical devices for improvements in methods of manufacturing.

TABLE 1-2
FACTORY MAN-HOURS AS A PERCENT OF HOME MAN-HOURS IN THE NINETEENTH CENTURY

Product	Percent
Cotton cloth	1
Hosiery	2
Soap	5
Overalls	10
Men's shirts	14
Underwear	19
Men's suits	36

Source: Lebergott, 1984, p. 135.

The Midvale-Heppenstall Steel Works where Taylor did some of his early work is shown in Figure 1-19.

Another force which acted on the need for expanded manufacturing was the gain in the standard of living in the United States. From 1900 to 1930 there was a dramatic rise in the ownership of household facilities and durable goods. Along with the increase in demand came an unprecedented gain in productivity during the period of 1920 to 1929. Beginning in the 1930s and continuing through 1948, productivity in the United States showed a decline from the previous growth experienced in the 1920s.

After World War II productivity began another upswing that carried into the 1960s. A study by Denison (1962) indicates that about half of overall productivity growth between 1948 and 1966 was due to advances in productivity knowledge, as cost-reducing technological innovations were incorporated into the organizations and instruments of production.

Between 1966 and 1976, the rate of growth in total productivity slowed by half to a 1.4 percent average rate. Much of this slowdown is attributed to volume-related factors reflecting the incomplete recovery in 1976 from the 1974–75 recession.

In addition to advances in knowledge, the rate of capital investment is also an important source of growth in productivity. A study by the Productivity Policy Board indicated the United States trailing other major industrialized nations in capital investment as a percentage of output in both total economy and growth in output per employee hour. Manufacturing investment in Japan is almost three times the rate in the United States. As a result, the United States has an aging industrial base: For instance, only 31 percent of machine tools were under 10 years old in 1979 in the United States, as compared to 61 percent in Japan.

The loss of U.S. competitiveness is no more evident than in the automotive industry. Abernathy, Clark, and Kantro (1983, p. 60) have made a comparison of the costs and labor productivity in selected United States and Japanese automobile companies. The Japanese have a true labor advantage of between $1,100 and $1,400 per vehicle. The difference in cost per hour worked ($8.72) accounts for $700 of the difference. The remaining advantage is explained by the differences in labor-hours per car.

Figure 1-19 Midvale—Heppenstall Steel Works, Pittsburgh—Steam hammer gang, c. 1930. (Courtesy of The National Museum of American History, Smithsonian Institution)

Another major area of U.S. industrial decline has been in the machine tool industry. The United States has been a leader in developing, building, and exporting machine tools since the late nineteenth century. As reported by Ayres (1984, p. 24):

> In the early 1970s U.S. machine-tool builders still exported twice as much as they imported. Exports exceeded imports until 1977, when imports of computer-numerically controlled machine tools from Japan soared. By 1982 the balance of trade in machine tools had reversed: Imports exceeded exports by more than 2 to 1, and over 50 percent of the machine tools purchased in the United States were manufactured elsewhere, mainly in Japan. The U.S. machine tool industry has lost half its traditional market in the past five years.

The impact of the decline in U.S. competitiveness is most evident in the job base of skilled factory workers. Since 1979 the United States has lost 288,000 jobs in machinery production, 439,000 in primary metal-making, 241,000 in motor-vehicle production, and 186,000 jobs in textiles. In total, American factories have lost 1,834,000 jobs in a period of only six years.

THE FUTURE OF MANUFACTURING

Automation and productivity. An editorial from *Automation News* (Dec. 3, 1984) discusses some of the reasons for automation:

> We have all heard a thousand good reasons why an industry or a nation should automate: increased productivity, greater efficiency, reduced labor costs, greater product flexibility,

and increased competitiveness are commonly cited. There is one reason, however, nearly overlooked by the analysts. It is best described as maintaining a profit at reduced capacity. As we all know, any plant can be figured for its minimum capacity that still generates profit. However, factory automation, for the very reasons stated above, can make it possible to derive profits from a factory operating at below 50 percent of capacity.

Flexible automation in this case means more than just the ability to make a fast turnaround in product mix or design. It means, as well, the ability to respond successfully to the financial and operating conditions of a tight marketplace—in short, a greater chance at profits. If there is a better reason for jumping in and using automation, we have yet to hear of it.

The automotive industry is perhaps the leader in attempting to meet the competition from overseas manufacturers by utilizing the latest advances in technology. By 1990, General Motors Corporation each year will be teaching 14,000 robots new places to weld, paint, drill, and inspect (Lehner and Marcom, 1984).

The computer will play a major role in the push for new and better ways to improve productivity. GMC has developed a computer program that will assist their engineers in placing robots on the assembly lines. It also has initiated a Manufacturing Automation Protocol (MAP), which is an attempt to form a network between the various elements of automation so that they may communicate with each other. The set of standard communication rules that have been established by GM with their MAP is sure to force other manufacturers into standardization, which is necessary if the efficiencies that will result from automation are to be maximized.

The introduction of programmable automation has generally been associated with batch production because of its ability to perform a variety of tasks. However, more companies are beginning to realize the benefits of this type of automation for mass production. Programmable automation can be divided into three general categories: 1) computer-aided design; 2) computer-aided manufacturing (for example, robots, computerized machine tools, flexible manufacturing systems); and 3) computer-aided techniques for management (for example, management information systems and computer-aided planning). Together, operating as a system, these tools are known as computer-integrated manufacturing.

Robots and automation. Industrial robots were developed and sold in the United States in 1959, but U.S. managers were slow in accepting the new technology. The Japanese received their first license agreement for robots in 1969, but in contrast to the U.S. they immediately started a vigorous program to produce and use robots throughout their manufacturing operations. By 1980, there were at least 6,000 units in Japan as compared to 3,500 in the United States.

As more manufacturing organizations integrate robots into their production operations, the robot population in the U.S. will continue to grow. The advantages of using robots to improve productivity are being realized throughout every phase of manufacturing. Walter Weisel, when president of the Robotic Industries Association, was quoted as saying "It is not uncommon for the industrial robot, along with its supporting equipment, to yield productivity increases from 20 to 30 percent."

Part of the increased productivity is because robots can operate continuously over a 24-hour period. This longer utilization time, as compared to a conventional eight-hour shift, generally increases operating efficiencies even when robot cycle times are longer than traditional methods.

Because the robotic industry in the United States is new compared to other technologies, forecasting robotic growth is difficult. It wasn't until 1984 that any confirmed statistics were available on the size of the U.S. robotics industry. For the first time, the Robotic Industries Association (RIA) began collecting and reporting statistics on robot sales and orders. Based on the results of the statistical collection program, RIA estimates that there were 30,000 industrial robots at work in American industries in 1988. If the present growth rate is maintained, more than 90,000 robots will be installed in America by 1995. An article in the *Harvard Business Review* (Foulkes and Hirsch, 1984) compares the cost of an industrial robot performing the same task as human labor in the automotive industry. They show that a robot will cost $6 per hour compared to the average cost of $24 per hour for a worker in the typical U.S. automotive plant.

SUMMARY

During this century, technological improvements have been one of the major sources of economic growth in this country. The introduction of new products and more efficient methods of manufacturing have increased productivity, which has allowed this nation to experience a rising standard of living.

Beginning in the 1970s, the United States has shown signs of losing its technological lead. If this trend continues, the standard of living along with many of the social gains that have been made will decline. Without technological gains, business and industry will not be able to create new jobs, productivity will continue to decline, and competition from other countries will force a slowdown in the overall economy.

QUESTIONS

1. What was the major factor that started the industrial revolution?
2. Name three of the first machine tools and the men that developed them.
3. Who organized the factory system of manufacture in the United States?
4. When did the second industrial revolution begin?
5. What were the four major areas of the second industrial revolution?
6. Discuss the relationship between the cost and production of the Model T automobile.
7. How were the first automatic machines controlled?
8. What factors accounted for the economic growth during the nineteenth century?
9. Describe the rate of productivity in the United States from 1900 to 1975.
10. How did the loss in U.S. competitiveness affect U.S. industry?
11. What part did the computer play in productivity growth?
12. Discuss the use of robots in the United States.
13. What is the difference between the "making" system and the "manufacturing" system?

2 MANUFACTURING OPERATIONS

The cottage or household industries described in the preceding chapter were small enough so that a single individual could manage these operations without any formal organization. As the demand for products grew, the factory system of manufacturing began to replace the small craft operations.

During the nineteenth century there were no schools for managerial training, and most entrepreneurs in the United States relied on trial and error in the administration of large-scale operations. Fortunately for the new managers, most of the businesses, with the exception of transportation and mining, remained small and were generally located in one geographical area.

In this chapter the changes that took place as a result of the expansion of basic industries are presented. These changes include organizational structure and the various approaches to managing industrial production. Also discussed are staff and line functions that are found in typical manufacturing organizations in the United States.

BASIC INDUSTRIES

The first industries were actually small crafts, which supplied goods for an immediate local market. As the demand for manufactured goods increased, it was necessary to find new ways to expand the output of the crafts. As a result, many of the early American inventions were aimed at reducing the need for scarce labor and capital by utilizing abundant natural resources, principally forest and agricultural products.

The first major manufactured goods for export, colonial shipbuilding and the production of naval stores, resulted from the vast supply of virgin forests which provided the lumber that supported this growth.

During the first half of the nineteenth century, cotton manufacturing was the largest industry in the United States. An Englishman named Samuel Slater was an apprentice to Richard Arkwright, inventor of cotton-spinning machinery. Slater familiarized himself with this machinery and came to America, where he contracted with the firm of Almy and Brown, Providence, RI, to reproduce the cotton machinery; this he did from memory.

Another inventor and manufacturer who is often considered the single most important individual in the expansion of the textile industry in the United States was Francis Lowell. He duplicated and improved English machinery for carding, spinning, and weaving. Lowell also pioneered improvements in factories that became known as the "Waltham System of Manufacturing." He advocated good, clean, and safe factories that employed young women—a paternalistic approach to employment.

The mass-production revolution began when industries in the United States started using the technological principles of interchangeable parts, division of labor, and special-purpose machinery to produce multiple units of the same product. This method of production became known as "The American System of Manufactures." One of the first factories for mass-production of arms was the Springfield Armory, shown in Figure 2-1.

Some of the first products manufactured in the United States that received attention in England and the European countries included Colt revolvers, McCormick farm equipment, locks made by Alfred Hobbs, and Singer sewing machines. See Figure 2-2. It was the Singer sewing machine that became the best known U.S. product overseas. Singer also built the first U.S. factory overseas in 1868.

Figure 2-1 Springfield Armory 1830. (Courtesy of The National Museum of American History, Smithsonian Institution)

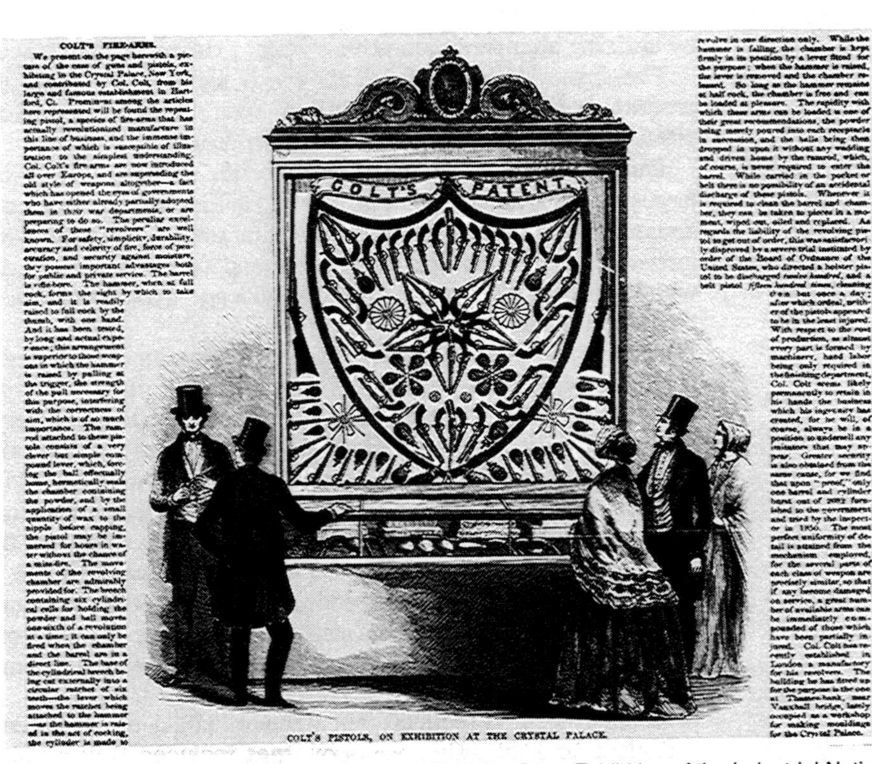

COLT'S FIRE-ARMS.

We present on the page herewith a picture of the case of guns and pistols, exhibiting in the Crystal Palace, New York, and contributed by Col. Colt, from his large and famous establishment in Hartford, Ct. Prominent among the articles here represented will be found the repeating pistol, a species of fire-arms that has actually revolutionized manufacture in this line of business, and the immense importance of which is susceptible of illustration to the simplest understanding. Col. Colt's fire-arms are now introduced all over Europe, and are superseding the old style of weapons altogether—a fact which has opened the eyes of governments who have either already partially adopted them in their war departments, or are preparing to do so. The peculiar excellences of these "revolvers" are well known. For safety, simplicity, durability, accuracy and celerity of fire, force of penetration, and security against moisture, they possess important advantages both for public and private service. The barrel is rifle-bore. The hammer, when at full cock, forms the sight by which to take aim, and it is readily raised to full cock by the thumb, with one hand. And it has been tested, by long and actual experience; this arrangement is superior to those weapons in which the hammer is raised by pulling at the trigger, the strength ...

COLT'S PISTOLS, ON EXHIBITION AT THE CRYSTAL PALACE.

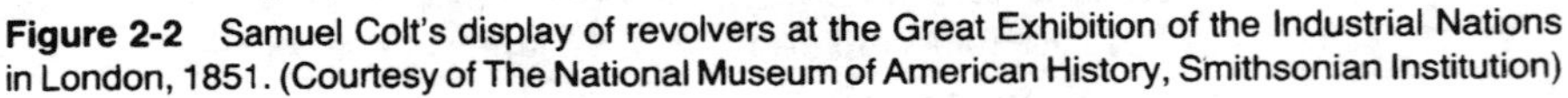

Figure 2-2 Samuel Colt's display of revolvers at the Great Exhibition of the Industrial Nations in London, 1851. (Courtesy of The National Museum of American History, Smithsonian Institution)

With the introduction and spread of steam power, machinery production, and the factory system, a new group of industries was born. Often referred to as "heavy industries," these new manufacturing companies produced goods such as iron and steel, railroad equipment, large machinery, and many of the raw materials, such as ores and coal.

One of the basic heavy industries was the manufacture of iron and steel. Even on a small scale, large production equipment was required in the smelting and refining process. Handling and storing of the raw materials presented new problems which did not exist in the manufacture of smaller products that were common in "light industries." A rapid increase in the demand for heavy metals resulted from the application of steam engines, rail transportation, and industrial machinery.

MANUFACTURING ORGANIZATIONS

Organizational structure. During the early days of manufacturing there were very few American books on manufacturing administration and structure. Before 1880, the foreman in many shops was an independent contractor who had the responsibility to produce a specified quantity of a given product. The supervisor received an agreed amount of money for his services and was responsible for hiring and paying his own group of workers.

As factories became larger and more complex, a new level of management was introduced that reduced the authority of the line foreman. The basic organization structure can be defined as patterns of work and responsibility, and the hierarchical arrangement of these functions.

In 1916 a French engineer and industrialist by the name of Henri Fayol wrote a book, *General and Industrial Management,* which presented his famous 14 points or "General Principles of Management." These principles include the following topics:

1. Division of work
2. Authority and responsibility
3. Discipline
4. Unity of command
5. Unity of direction
6. Subordination of individual interest to the general interest
7. Enumeration
8. Centralization
9. Scalar chain line of authority
10. Order
11. Equity throughout all levels
12. Stability of tenure of personnel
13. Initiative
14. *Esprit de corps* = a sense of union and common interests

Fayol believed that managers can not be effective without principles of organization and operation. He stressed that the soundness and good working order of any organization depends on certain principles that must be flexible as the objective of the organization changes.

The new mass producers of manufactured goods could no longer function with some of the traditional organizational structures. Many enterprises integrated both vertically and horizontally. This created new areas of specialization.

As markets expanded, specialized skills were required for marketing, sales, and distribution of the goods. Many companies produced their own raw and semifinished materials that were managed by purchasing, warehousing, and traffic departments. Smaller departments handled such functions as research, personnel, finance, and legal affairs.

Since many individuals work within a given organization, a method must be developed that involves dividing job activities in such a way that the goals of the company will be satisfied. Establishing patterns of work, responsibility, and hierarchical arrangements is referred to as organizational structure.

APPROACHES TO ORGANIZATION

Organizations are structured in a number of different ways: Organizing by function, by geographical area, and by product or customers are common forms of organizational structure found in companies engaged in manufacturing operations.

Functional. One of the most common forms of organization is grouping according to the tasks or functions being performed by the people in the work group. A typical organization by function appears in Figure 2-3 and shows separate departments for finance, sales and marketing, personnel, manufacturing, and engineering and research.

A major advantage of this type of structure is that it allows the use of specialists to perform the functions that they are most qualified to do efficiently. A disadvantage of functional organization is that the specialists may focus their attentions on their individual areas of responsibility without considering the overall goals of the company.

Geographical. Another common form of organization is by geographical area. For large companies the basic structure may be by country or international regions such as South America, Europe, and the Far East. In the domestic market a manager may be assigned all activities in a region, such as eastern or western operations. An organization based on geographical area is shown in Figure 2-4.

By locating manufacturing facilities in different locations, companies are able to benefit from the availability of raw materials and trained labor, develop better relationships with local customers, and operate as independent units. A major disadvantage in this structure is the limited availability or duplication of specialized staffs.

Product. In some companies departments are grouped by the type of product they manufacture. See Figure 2-5. Personnel in this form of organizational structure are able to concentrate their skills and expertise on a single product or group of similar products. Manufacturing, engineering, and sales personnel are able to coordinate their efforts toward meeting the needs of the customer. Automotive companies and conglomerates are typically organized by product line.

One of the difficulties with the product form of organization is the need to duplicate many activities, since each division must have its own personnel and facilities necessary to be in business.

Some guidelines for structuring an organization include the following basic principles of organization:

1. The Hierarchial Principle or Scalar Principle
   a. Ascending and descending levels of authority
2. The Unity of Command Principle
   a. Each employee has a single immediate person he or she is responsible to.
3. The Line and Staff Principle
   a. The line performs directly. The staff advises the line.
   b. Occasionally the staff will act in a coordinative role.
4. The Principle of Delegation
   a. The making of decisions should be delegated to the lowest possible level in the hierarchy.
5. The Principle of Control
   a. The use of internal checks and balances
   b. The use of information (facts) to establish if the goals of the organization are being accomplished
6. The Principle of Span of Control
   a. The number of people one individual can effectively supervise
   b. Technical staff vs. line operators

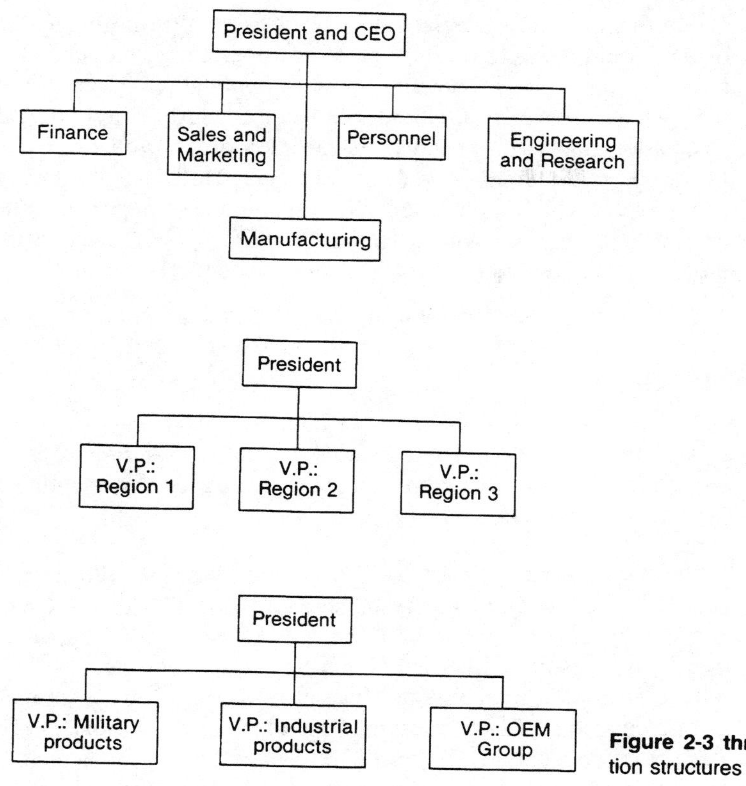

Figure 2-3 through 2-5 Organization structures

7. The Principle of Coordination
   a. Units of an organization tend to grow apart unless held together.
   b. Getting people to work together
8. The Principle of Grouping or Division of Labor
   a. People work most efficiently when grouped according to their specialities.
   b. Labor often divides itself logically.
      1) Purpose
      2) People served
      3) Place (regions)
      4) Time (shifts)

APPROACHES TO MANAGEMENT

Division of labor. In 1776 Adam Smith wrote the first economics text, *Wealth of Nations,* in which he presented his now famous theory on the division of labor. Using the example of the manufacturing of pins, Smith demonstrated how the greatest improvement in production can be made through applying the division of labor. In the processes of pin manufacturing one worker draws the wire, one straightens the wire, another cuts it to length, one makes the head, etc. In all, 18 separate operations are performed in making a single pin.

Smith shows how specialization leads to savings in time by the increased dexterity of individual workers performing single operations, reductions in time lost changing from one process to another, and the application of proper tools and machinery whenever a worker's attention is directed toward a single operation.

Charles Babbage expanded on the idea of division of labor in his book *On the Economy of Machinery and Manufacturers*. He wrote, "perhaps the most important principle on which the economy of a manufacture depends, is the division of labour amongst the persons who perform the work." He discussed the various principles on which the advantages of this system depend. These include:

1. The time required for learning—the greater the number of distinct processes, the longer the time to learn the task.
2. Time lost in changing from one operation to another.
3. Skill acquired by repetition of the same processes.
4. Contrivance of tools and machinery to execute processes are much more likely to occur in the operator's mind when his or her attention is devoted to a very few operations.

Scientific management. It wasn't until the late 1800s that a systematic organization of factory work received increased attention in the United States. Frederick W. Taylor, shown in Figure 2-6, became known as the "Father of Scientific Management" as a result of his theories on time and motion study, functional supervision, standard times for each task, and incentive pay based on the amount of work performed.

In June 1895, Taylor presented his ideas at the annual meeting of the American Society of Mechanical Engineers and later before the United States Congress. In his famous paper, "Shop Management," he describes the duties of management in the development of scientific management:

1. The development of a science of management for each element of a man's work which replaces the old rule-of-thumb method.

Figure 2-6 Frederick W. Taylor (1856–1915). (Courtesy of The National Museum of American History, Smithsonian Institution)

2. Scientific selection of the best worker for each job, and then training, teaching, and developing the workman in place of the former practice of allowing the worker to select his own task and train himself.

3. The development of cooperation between the management and the workers to ensure that all work being performed conforms to the principles of the scientific method that was developed.

4. The equal division of work and responsibility between management and the workers, each department taking over the work for which it is better fitted instead of the former condition in which most of the work and responsibility were thrown on the men.

Taylor explained that the prosperity of the employer is not possible without the prosperity of the employees. He believed that the greatest prosperity for both the employer and employee can only be obtained when the worker attains the highest level of productivity. Taylor is one of the first to recognize the importance of the individual by stressing that the most important factor affecting productivity is the training and development of each worker.

Another pioneer in scientific management was Frank B. Gilbreth. Among his many contributions, his original work in the study of body motions used in performing tasks is the most widely accepted. Gilbreth was the founder of modern motion study techniques, which are used to improve operations by eliminating unnecessary motions and simplifying necessary motions.

Frank Gilbreth, working with his wife Lillian, held that for the greatest results the best methods of work must be recorded. The methods must be constantly improved, and causes and reasons for inefficiencies must be noted and corrected.

In an address to scientists and engineers on "One Best Way To Do Work," Frank stressed that the science of management is assigning measurement to management. This includes units of measurement, methods of measurement, and measuring devices. By this method, standards of efficiency can be determined to discover and conserve those methods that have proven to be the most efficient.

The Gilbreths believed that everything should be standardized down to the minute detail before starting a job. Motion study, skill study, time study, and methods study were all tools of scientific management that they used to find the one best way to do work. By using these techniques production can be increased, fatigue reduced, and operators can be instructed in the best method for performing an operation.

Behavioral science. The behavioral approach to management emphasizes increased production through a better understanding of people and how they act in an organization. This behavioral approach began with the classic studies performed at the Hawthorne Works of the Western Electric Company in the late 1920s and early 1930s. Experiments were conducted to determine the factors in the work situation which affected the morale and productive efficiency of workers.

The first study was in the Relay Test Room. Its purpose was to find the relationship between lighting levels and worker output. After a series of tests in which the intensity of the lighting was changed, it became apparent that no matter what light level employees were exposed to, productivity increased. After the individual workers were interviewed it was determined that the major reasons for the increase in output was due to factors other than the physical environment of the workplace.

The researchers found that members of the Relay Test Room group had developed a high level of morale as a result of the interpersonal relationships with each other and their supervisor, the perception that they were part of a special group, and the participation and freedom they had in determining work assignments and work pace.

Based on information from the first experiment, the research team decided to conduct new studies involving the social relationships among members of the group, and the effect their supervisor had on motivating them to higher levels of production and quality. An experiment was set up to analyze the effect of rest periods, shorter work days, and employee attitudes toward their work and the company.

The second series of experiments lasted almost two years, during which time rest periods were changed, length of the work day was varied, and methods of pay were changed. Once again, it became obvious that the output of the subjects was affected by more than the mechanical changes attributed directly to the experiments.

Elton Mayo, one of the chief investigators, concluded that "The supervisors took a personal interest in the employees and their achievements; he showed pride in the record of the group. He helped the group to feel that its duty was to set its own conditions of work; he helped the workers find the 'freedom' of which they so frequently spoke in the course of the experiment."

There are numerous questions concerning the findings of the Hawthorne experiments. The results of these studies, however, gave wide support for a human relations approach to management, which holds that each worker must be treated as an individual rather than a tool of the organization.

Quantitative. In the 1930s mathematics and statistics were applied in the management of industrial operations. These techniques were an extension of scientific management, with a greater emphasis on quantitative approaches to decision making.

During World War II mathematical models were used to determine optimal solutions for military strategies. These techniques are generally referred to as operations research and are now widely used to arrive at the best or optimal solution to industrial problems.

LINE AND STAFF FUNCTIONS

Staff functions include special technical skills to support the efforts of line personnel, whose activities make a direct contribution to the production of the organization's basic products. Staff responsibilities are supportive; they assist line personnel in the performance and improvement of production operations. The distinction between staff and line authority is not always clear-cut. Problems often occur when the relationships and responsibilities of line vs. staff are not clarified. Conflict is created when staff personnel assume line authority, credit for success is not shared, and when line and staff personnel do not work together on the objectives of the organization as a whole. An example of line and staff organization is shown in Figure 2-7.

Research and engineering. The primary responsibility of this department is the formulation and recommendation of research programs to meet the needs of

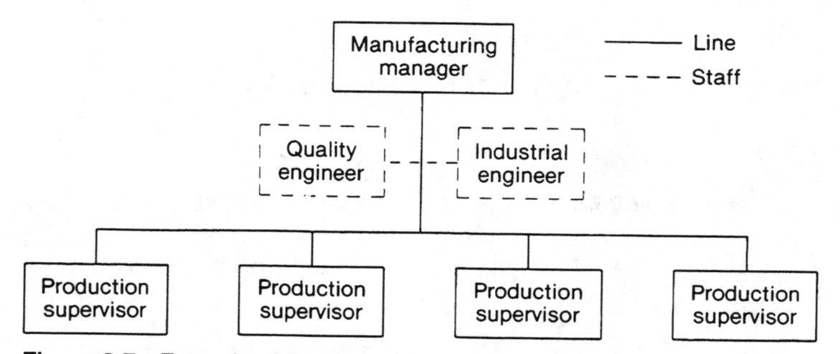

Figure 2-7 Example of line/staff organization chart.

the company for the improvement and cost reduction of existing products and processes, as well as the introduction of new products and processes.

Industrial engineering. The principal duties of the Industrial Engineering department are broad and varied, depending on the company. Almost all industrial engineers are involved in planning, developing, maintaining, and administering a continuous program to reduce cost, improve quality, and increase operating efficiency throughout the manufacturing operations.

Specific responsibilities include methods and labor standards programs, plant layout and material handling, and equipment selection and replacement. The industrial engineers coordinate cost-reduction activities resulting from technological improvements, methods changes, product design, and equipment replacement. Frequently the industrial engineering staff is responsible for preparation of the annual capital plan forecast, which includes preparation of data necessary for justification.

A job description that outlines the basic functions and specific duties of a typical industrial engineer is shown in Figure 2-8.

Manufacturing engineering. This group is sometimes called process engineering. Their functions include the selection of methods of manufacturing; selection of production equipment; and the design, manufacture, and control of the tool room. In some organizations the functions of Industrial Engineering and Maintenance are also under the control of Manufacturing Engineering. An organization chart for manufacturing engineering in a machining and fabrication plant is shown in Figure 2-9.

Material control. The overall responsibility of material control includes coordinating the activities of purchasing, materials requirements planning, production scheduling, inventory control, warehousing, shipping, and receiving.

The purchasing group is responsible for the procurement of raw materials, supplies, and equipment at the lowest cost consistent with approved standards.

Material requirements planning involves the interpretation of sales forecasts to determine the quantities of raw materials and other production items which will be required to meet master production schedules.

Quality assurance. The quality personnel are involved in both product design and manufacturing to ensure conformity to product specifications. During the design

POSITION: Industrial Engineer

REPORTS TO: Industrial Engineering Chief

BASIC FUNCTIONS:

The primary functions of this position involve coordinating activities among Manufacturing, Production, Engineering, Accounting, and Product Engineering in establishing work simplification, improvement of product standards, and cost-reduction programs.

SUPERVISES THE WORK OF: No one

SPECIFIC DUTIES:

1. Establishes programs to influence design or design changes on production parts which result in reduced finished product costs.
2. Coordinates cost-reduction programs among various departments.
3. Makes cost analyses of existing processes and recommends changes in design, specifications, methods, tooling, and processes to reduce cost and improve quality.
4. Recommends the selection of material handling equipment, handling techniques, and material flow.
5. Prepares and coordinates plant and office layouts.
6. Assists Manufacturing Departments in the preparation of their annual requests for capital expenditures, analysis of cost-control problems and the review and evaluation of cost-reduction suggestions.
7. Summarizes estimates for processing, tooling, equipment, and labor costs. Studies capability of new equipment to justify purchase items designed to reduce manufacturing costs.
8. Develops manufacturing performance and cost controls.
9. Issues status and progress reports for all major manufacturing projects.
10. Maintains complete familiarity with the work measurement program and provides assistance in maintaining this program as required. Assists in discussions with labor Relations and/or the Union concerning labor standards.
11. Performs related work as assigned by the Industrial Engineering Chief.

Figure 2-8 Job description for Industrial Engineer.

phase they promote and participate in design reviews. They offer suggestions for the design until all involved are satisfied that it will accomplish its purpose and can be manufactured within the proposed cost constraints.

Quality functions include inspection of the product to make sure it conforms to the engineering drawings, tolerance limits, and performance specifications. Inspection data identifies causes of defects and provides information for corrective action for both purchased and manufactured parts and materials.

Finance. The finance function includes the activities of obtaining, allocating, and controlling the resources of the organization. The activities of the finance group include:

1. Arranging for the necessary funding to operate both on a daily basis and for long-term financing.
2. Preparation of budgets to plan for the financial requirements of the company. Budgets are usually made for each year and generally a five- or ten-year plan.
3. Accounting functions such as monetary controls, payroll, accounts receivable, and accounts payable.
4. Preparation of economic evaluations of investment alternatives such as new products, equipment, and facilities.

Sales and marketing. This group's activities involve the selling of the manufactured products. Marketing is responsible for advertising, customer relations, sales forecasting, and the development of new product ideas. In some companies the marketing group has the responsibility for distribution of the product and has a major influence on production scheduling and finished inventory levels.

Industrial relations. The industrial relations or personnel department is charged with the recruitment of personnel, labor relations and contract negotiations (when the company has a union), training, job evaluation, wage and salary administration, and in most plants, employee health and safety programs.

Safety professional. The responsibility for maintaining a safe and healthy environment should be clearly assigned to a specific individual with support from each of the operating departments. Usually this function comes under the direct supervision of the Industrial Relations or Personnel Department.

Supervision. A Manufacturing Manager or Production Superintendent is generally assigned to oversee the manufacturing activities of the plant through the direct management of supervisors, with the objectives being continuous and efficient production in desired quantity and quality.

Specific duties of production supervision are:

1. Direct the activities of individual production departments with respect to production volume, cost and quality of production, meeting production schedules, and delivery dates.
2. Motivate production workers in order to receive maximum efficiency and productivity.

Figure 2-9 Manufacturing Engineering organization chart.

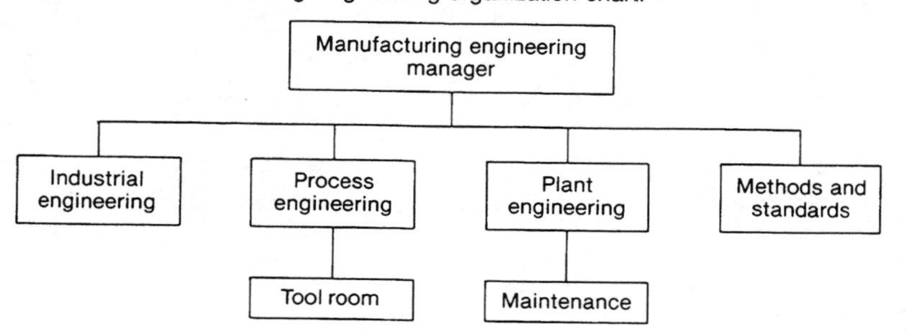

3. Encourage the efficient utilization of equipment, facilities, and personnel to maintain conformity to budgetary limitations.
4. Interpret and administer compliance with division policies, safety standards, and good housekeeping practices.
5. Supervise the timely preparation of required reporting relative to production and inventory and recommend such changes as may be advisable.
6. When there is a union, participate in collective bargaining grievance procedures.

Plant engineering and maintenance. Plant Engineering is charged with the design and specification of new facilities, the design and operation of utility systems, and the maintenance and repair of existing plant equipment, buildings, and grounds. This responsibility includes the development of drawings, specifications, bills of materials, and cost estimates.

The Plant Engineering department arranges for the procurement of outside contractors to carry out specific approved projects and becomes responsible for satisfactory performance in the best interest of the company.

TABLE 2-1
HISTORICAL SUMMARY OF OPERATIONS MANAGEMENT

Date (Approximate)	Contribution	Contributor
1776	Specialization of labor in manufacturing	Adam Smith
1799	Interchangeable parts, cost accounting	Eli Whitney and others
1832	Division of labor by skill; assignment of jobs by skill; basics of time study	Charles Babbage
1900	Scientific management; time study and work study developed; dividing planning and doing of work	Frederick W. Taylor
1900	Motion study of jobs	Frank B. Gilbreth
1901	Scheduling technique for employees, machines, jobs in manufacturing	Henry L. Gantt
1915	Economic lot sizes for inventory control	F.W. Harris
1931	Statistical inference applied to product quality; quality control charts	Walter A. Shewhart
1935	Statistical sampling applied to quality control; inspection sampling plans	H.F. Dodge and H.G. Romig
1940	Operations research applications in World War II	P.M.S. Blacket and others
1946	Digital computer	John Mauchly, J.P. Eckert
1947	Linear programming	George B. Dantzig, William Orchard-Hays, and others
1950	Mathematical programming, nonlinear and stochastic processes	A. Charnes, W.W. Cooper, H. Raiffa, and others
1951	Commercial digital computer; large-scale computations available	Sperry Univac
1960	Organizational behavior; continued study of people at work	L. Cummings, L. Porter, and others

From Adam, Everett E. Jr. and Ebert, Ronald J. *Production and Operations Management*, 2nd ed. Englewood Cliffs, NJ: Prentice Hall, 1982, p. 19.

SUMMARY

Management theories and organizational methods of companies engaged in producing a product are continuously changing. Managers are constantly challenged to find better ways to adapt to the ever-increasing technological improvements in manufacturing. Table 2-1 shows a historical summary of operations management.

QUESTIONS

1. What were the first major industries in the United States?
2. Describe three common forms of organizational structures.
3. List five guidelines for designing or structuring an organization.
4. Who was Frederick W. Taylor?
5. What effect did lighting levels have on worker output during the "Hawthorne studies"?
6. What kind of job functions does a Manufacturing Engineering department perform?
7. What are the basic principles of organization?
8. Draw an organization chart for a typical manufacturing company.

3 DESIGN AND CONTROL OF MANUFACTURING SYSTEMS

To be successful in any manufacturing endeavor a great deal of attention must be given to the design and control of the manufacturing facilities and operations. These activities include the selection of a manufacturing site, whether it be a new plant or part of an existing facility. After the product has been designed, the method of manufacture must be determined and the equipment purchased. Finally, provisions must be made to ensure that a quality product is manufactured in the most efficient and safe manner possible.

The functions that are necessary to satisfy the requirements of the manufacturing organization are covered in this chapter. An overview of the major methods of design and control introduces the principal factors that are involved in manufacturing operations. Additional coverage of each of these topics may be found in the references at the end of the text.

LOCATION AND CAPACITY PLANNING

After a product has been selected for manufacture, one of the next decisions facing management is to determine where the product will be produced. For some, this may be a problem of determining the best location within an existing manufacturing facility. When a new or expanded plant is required, the complexity of the decision is increased because of the numerous factors that must be considered in selecting the best geographical location.

Location decisions. The location of a new plant may be infrequent, and in some cases may never be required during the career of a manufacturing professional. However, when decisions on a new location must be made they may have a major effect on the future operations and profitability of the company.

Location alternatives should always be evaluated when a manufacturing organization is first formed. A new location should also be considered to determine if multiplant operations may be advantageous when expansion of existing operations requires additional space. Economic factors may indicate that certain advantages in customer service and/or operating costs can be improved by investment in facilities in different locations.

Location decisions are generally made in two stages. The first step involves the selection of an area or region. The second step includes consideration of factors concerning the actual site within a community.

METHODS ENGINEERING

Methods engineering is the industrial science which is chiefly concerned with increasing the effectiveness of labor. It has greatly influenced our lives since the industrial revolution. No other factor has had a greater impact on low production costs and high standards of living.

Methods engineering may be defined as:

1. A specific combination of layout and working conditions, materials, equipment, tools, and motion patterns involved in accomplishing a given operation or task.
2. The sequence of operations and/or processes used to produce a given product or accomplish a given job.
3. The procedure or sequence of motions used by one or more individuals to accomplish a given operation or work task.

When applying methods engineering techniques to improve productivity, the analysts must a) take all factors into account when planning for greater effectiveness—consider all alternatives; b) understand that the nature of many jobs is continually changing; c) recognize the importance of standards for estimating, scheduling, and controlling operations; and d) help employees understand what productivity means, why it is important, and how it can be increased.

The six steps in methods improvement are shown in Figure 3-1. During the conceptual model development phase, the proposed or present method must be carefully analyzed and documented. This is often done by using some of the graphic tools that are described in the following sections.

GRAPHIC TOOLS OF METHODS ENGINEERING

When the methods analyst is designing a new work area or improving an existing operation, it is often necessary to obtain factual information relating to the process. Some of the more common graphic tools for analysis are discussed following.

1. *Select and Define the Problem*
   a) recognize that a problem exists
   b) determine the scope of the problem
   c) define the goal or objective
2. *Visualize Problem and Record Information*
   a) obtain data
   b) establish criteria for evaluation
   c) list any restrictions
   d) use graphic tools for analysis
3. *Critical Examination for Solutions*
   a) eliminate where possible
   b) select several alternatives
4. *Develop and Improve Alternatives*
   a) no single correct answer
   b) consider
      1. time—labor and process
      2. cost of equipment
      3. quality of product
      4. human aspects
      5. product quantity and life cycle
5. *Install New Method*
   a) communicate with all individuals
   b) seek participation
6. *Follow-up—Evaluate Results*

Figure 3-1 Six steps in methods improvement.

Operation process chart. This chart is a graphic representation of the points at which materials are introduced into a process, and of the sequence of inspections and all other operations except those involving material handling. It includes information considered desirable for analysis, such as time required and location. Two symbols are used in constructing the operation process chart. The circle denotes an operation—where a part is intentionally transformed or is being studied or planned before performing productive work on it. The square denotes an inspection—the part is examined to determine its conformity to a standard. An example of an Operation Process Chart is shown in Figure 3-2.

Flow process chart. A somewhat different type of process chart which is useful in studying methods is the Flow Process Chart. The official ASME definition is:

> A flow process chart is a graphic representation of the sequence of all operations, transportations, inspections, delays, and storages occurring during a process or procedure, and includes information considered desirable for analysis such as time required and distance moved.

Five symbols are used in constructing flow process charts. They are:

operation ○ delay D
transportation ⇨ storage ▽
inspection □

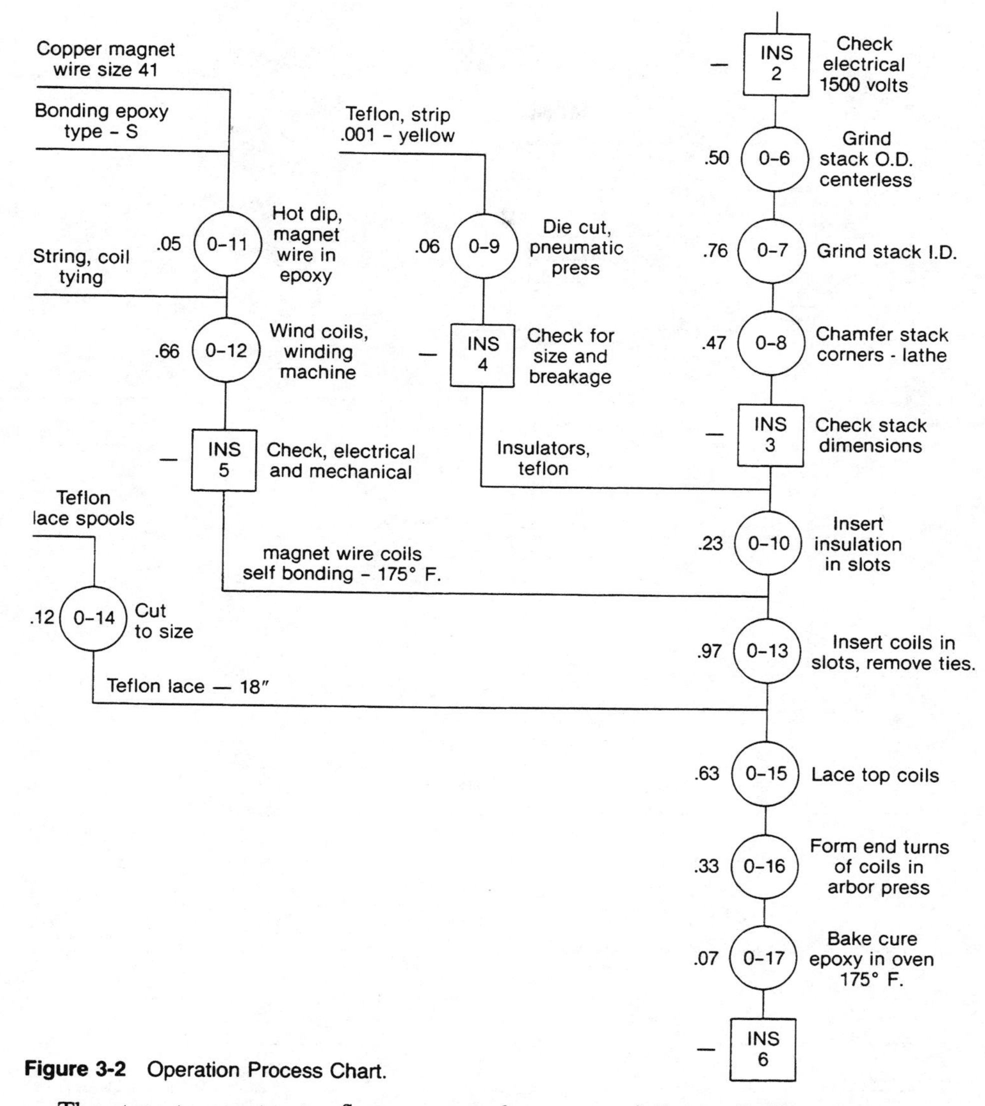

Figure 3-2 Operation Process Chart.

The steps to construct a flow process chart are as follows: 1) Choose the subject to be followed—select either the operator or material, 2) write a brief description of each detail, 3) classify the steps using one of the five elements of work, 4) apply the symbols, 5) enter distance and time, and 6) summarize the data and record it in the summary block on the chart. An example of a flow process chart for an office task is shown in Figure 3-3.

During the analysis of the data, the content of individual jobs and job elements may be analyzed using the "Questioning Technique" that was developed from Rudyard Kipling's "Six Honest Men":

I have six honest serving men
They taught me all I know
Their names are who and what and when
And why and where and how.

ANALYSIS — WHY? — WHAT? WHERE? WHEN? WHO? HOW? — QUESTION EACH DETAIL

20-762 3/62

FLOW PROCESS CHART

NO. 1

PAGE 1 OF 1

SUMMARY

	PRESENT NO.	PRESENT TIME	PROPOSED NO.	PROPOSED TIME	DIFFERENCE NO.	DIFFERENCE TIME
○ OPERATIONS	13					
TRANSPORTATIONS	4					
□ INSPECTIONS	0					
D DELAYS	5					
▽ STORAGES	0					
DISTANCE TRAVELED	120 FT.		FT.		FT.	

JOB Pricing and Posting Orders

☐ MAN OR ☒ MATERIAL Unpriced Orders

CHART BEGINS On A's desk

CHART ENDS On A's desk

CHARTED BY John Smith DATE 4/15/

	DETAILS OF (PRESENT ~~PROPOSED~~) METHOD	OPERATION / TRANSPORT / INSPECTION / DELAY / STORAGE	DISTANCE IN FEET	QUANTITY	TIME	CHANGE: ELIMINATE	COMBINE	SEQUE.	PLACE	PERSON	IMPROVE	NOTES
1.	Placed on A's desk	○										By messenger
2.	Time Stamped	○										
3.	Placed in OUT box	○										
4.	Waits	D										
5.	Picked up by C	○										At least every 15 min.
6.	To desk	⇨	20									
7.	Sorted into Priced & Unpriced	○										
8.	To B's desk	⇨	40									By C
9.	Placed on desk	○										
10.	Waits	D										
11.	Priced	○										
12.	To C's desk	⇨	40									By B
13.	Placed on desk	○										
14.	Waits	D										
15.	Posted	○										
16.	To A's desk	⇨	20									By C
17.	Placed on A's desk	○										
18.	Waits	D										
19.	Sorted	○										For inside and outside hospital
20.	Placed in envelopes	○										
21.	Placed in OUT box	○										
22.	Waits	D										
23.												

Figure 3-3 Flow Process Chart.

All of the facts are examined to find out the following information:

WHAT	is done?
HOW	is it done?
WHEN	is it done?
WHERE	is it done?
WHO	does it?
WHY	is it done?
WHY	that way?
WHY	then?
WHY	there?
WHY	that person?

Operations analysis. A study which encompasses all those procedures concerned with the design or improvement of production, the purpose of the operation or other operations, inspection requirements, material used, and the manner of handling material, setup, tool equipment, working conditions, and methods used is called an operation analysis. A form that lists all the important factors influencing the effectiveness of an operation and which is often used to guide the analyst in the study is shown in Figure 3-4.

Human-machine chart. This chart is used to study the coordinated synchronous or simultaneous activities of a work or system of one or more machines and/or one or more operators. Each machine and/or operator is shown in a separate, parallel column indicating the activities as related to the rest of the system. The primary benefit of this chart is to identify and eliminate idle time from an operation cycle. By determining exactly when the idle period occurs, how long it lasts, and its relation to the part of the cycle during which the operator is occupied, undesirable idle time may be minimized.

The chart is drawn to scale to show the relationship between the busy and idle time of the operators and the machines. See Figure 3-5. Plotting the information to scale enables the analyst to determine at a glance the relative time consumed by each part of the job.

Workplace charts. A workplace or right- and left-hand chart is a graphic representation of the coordinated activities of the right and left hands.

Where the job is sufficiently repetitive to warrant a detailed study of the right and left hands, a workplace study may be made. Moves, operations, holds, or delays performed by each hand are charted, using flow process chart symbols.

A workplace chart or RH-LH chart is shown in Figure 3-6. In parallel vertical columns, the elements are charted to show the simultaneous activity of each hand. A brief description is provided alongside the regular charting symbols. Whenever a change occurs to either or both hands, it is shown on the next line of the chart, regardless of how short or long the time element.

OPERATION ANALYSIS SHEET

DATE STARTED ________ PART NAME ________ PART NO. ________

DATE COMPLETED ________ DEPARTMENT ________ DRG. NO. ________

OPERATION ________ MATERIAL ________

CHECK	ANALYSIS	DESCRIPTION
____ ____ ____ ____ ____ ____	IS IT NECESSARY AT ALL? CAN IT BE ELIMINATED BY PRIOR PLANNING? DOES IT ACCOMPLISH ITS INTENDED PURPOSE? CAN IT BE COMBINED WITH ANOTHER OPERATION? CAN THE SUPPLIER DO IT FOR LESS? CAN THIS OPERATION BE MADE TO ACCOMPLISH MORE TO SIMPLIFY SUCCEEDING OPERATIONS?	NO. 1 PURPOSE ________ ________ ________ ________
____ ____ ____ ____ ____ ____ ____	HAVE THE PRINCIPLES OF VALUE ANALYSIS BEEN APPLIED? ARE ALL OF THE PARTS NECESSARY? CAN ONE PART BE RE-DESIGNED TO FUNCTION FOR TWO? CAN STANDARD PARTS BE USED? DOES DESIGN PERMIT THE BEST ASSEMBLY AND PROCESSING TECHNIQUES? WILL DESIGN LEND ITSELF TO NUMERICAL CONTROL MACHINING, SEMI-AUTOMATION, OR AUTOMATION PROCESSING? CAN ANYTHING BE LEARNED FROM COMPETITOR'S DESIGN?	NO. 2 PART DESIGN (SKETCH)
____ ____ ____ ____ ____	ARE MATERIAL REQUIREMENTS OUT OF DATE? CAN LESS EXPENSIVE MATERIAL BE USED? CAN SCRAP BE REDUCED? WOULD BETTER MATERIAL REDUCE PROCESSING OR MACHINING COSTS? CAN PACKAGING BE IMPROVED?	NO. 3 MATERIAL A. OTHER MATERIAL POSSIBILITIES ________ B. SCRAP COST REDUCTION ________ C. PROCESS MATERIALS ________
____ ____ ____ ____ ____ ____ ____	CAN MATERIALS GO DIRECTLY TO THE JOB? CAN GRAVITY BE USED TO MOVE MATERIAL? CAN HANDLING BE REDUCED FURTHER? HAS DISTANCE MOVED BEEN CONSIDERED? CAN SIGNALS (BELLS, LIGHTS, ETC.) BE USED TO ALERT MATERIAL HANDLERS WHEN MATERIAL IS READY TO BE MOVED? ARE PROPER CONTAINERS BEING UTILIZED? CAN CRANES, TRUCKS, OR CONVEYORS HELP?	NO. 4 MATERIAL HANDLING A. THE MATERIAL IS BROUGHT TO THE WORKPLACE BY: ________ B. THE MATERIAL IS HANDLED AT THE WORKPLACE BY: ________ C. THE MATERIAL IS TAKEN FROM THE WORKPLACE BY: ________
____ ____ ____ ____ ____ ____	IS STATISTICAL QUALITY CONTROL IN USE? A. IF NOT, CAN IT BE USED? ARE PRESENT INSPECTION METHODS SUFFICIENT? ARE THE PRESENT INSPECTION REQUIREMENTS: A. TOO EXPENSIVE? B. NECESSARY? C. TO ELABORATE?	NO. 5 INSPECTION A. SPECIFICATIONS & TOLERANCES ________ ________ B. INSPECTION METHODS ________ ________

Figure 3-4 Operation Analysis Sheet.

CHECK	ANALYSIS	DESCRIPTION
	CAN ANY OF THE OPERATIONS BE: A. COMBINED? B. ELIMINATED? C. PERFORMED INTERNAL TO MACHINE TIME? D. REDUCED IN CONTENT? CAN OPERATION SEQUENCE BE IMPROVED? CAN IT BE PERFORMED IN ANOTHER DEPARTMENT AT LESS COST? ARE QUANTITIES OR LOT SIZES ECONOMICAL?	NO. 6 ANALYSIS OF PROCESS (LIST ALL OPERATIONS) SEQ / DEPT / OPERATION
	LAYOUT: A. ARE LAWS OF MOTION ECONOMY USED? B. IS WORK AREA UNCLUTTERED AND FREE? C. ARE MATERIALS & SUPPLIES EASILY OBTAINED? D. CAN SET-UP BE IMPROVED? MACHINE: A. ARE CONTROLS EASILY ACCESSIBLE? B. IS OPERATOR COMFORTABLE AT MACHINE? C. CAN OPERATOR RUN TWO MACHINES? TOOLS: A. ARE TOOLS SUITABLE FOR THE JOB? B. ARE JIGS & FIXTURES USED PROPERLY? C. CAN QUICK ACTING CLAMPS BE USED?	NO. 7 WORKPLACE LAYOUT (SKETCH IMPROVEMENTS)
	CONSIDER LIGHTING, AIR CONDITIONING, HEAT, VENTILATION, FUMES, REST ROOMS, COFFEE BREAKS, ETC. OTHER:	NO. 8 WORKING CONDITIONS (SUGGEST IMPROVEMENTS)
	COMPARE METHODS OF SEVERAL OPERATORS USE IMPROVEMENTS FROM OTHER JOBS USE FOOT OPERATED DEVICES USE SIMULTANEOUS 2 HANDED METHODS CONSIDER MOTION ECONOMY DEVICES PROVIDE CORRECT CHAIRS & BENCHES.	NO. 9 OTHER POSSIBILITIES FOR IMPROVEMENT

DATE	PROPOSED IMPROVEMENT	ACTION TAKEN	BY

COMMENTS: ____________________

STUDY TAKEN BY: __________ APPROVED BY: __________ ESTIMATED SAVINGS: __________

Figure 3-4 Cont'd.

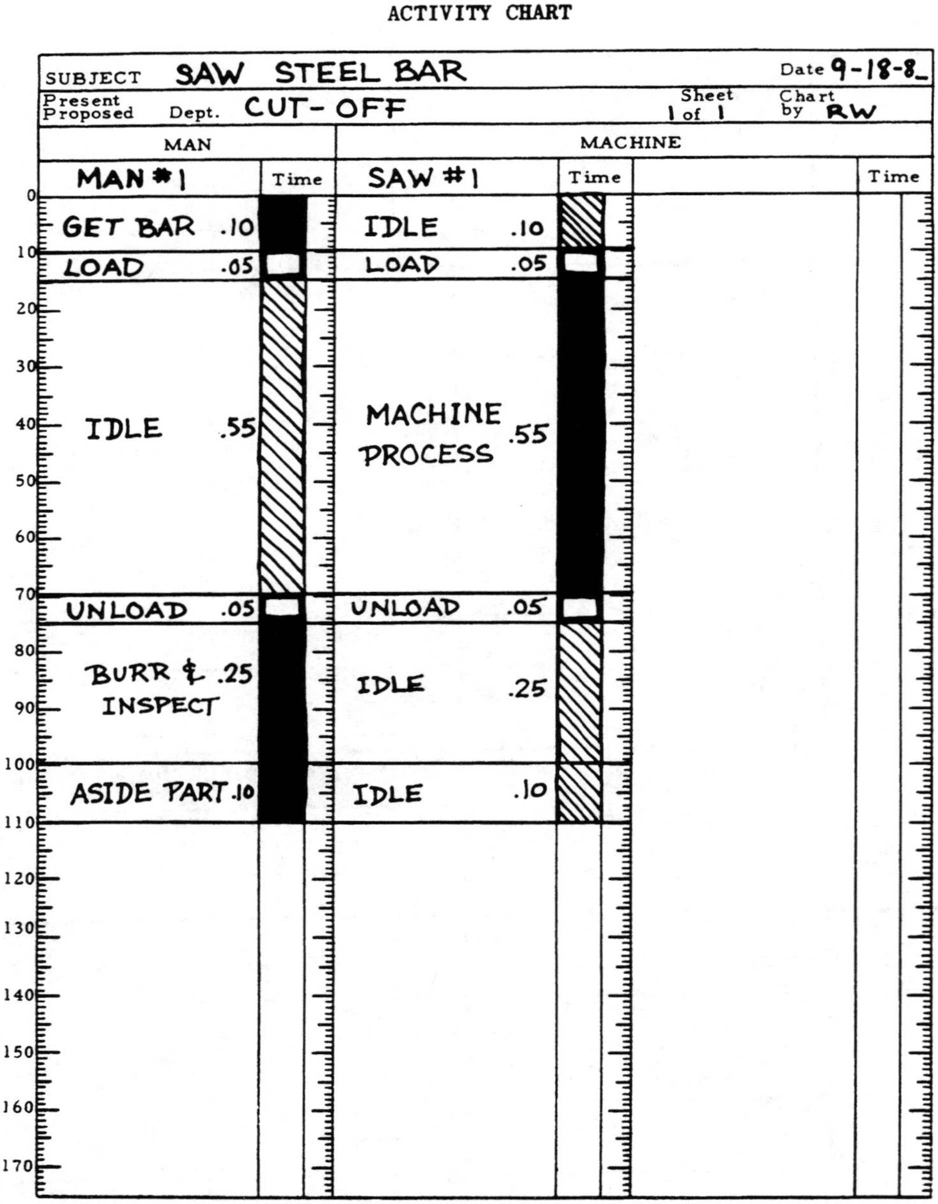

ACTIVITY CHART

SUBJECT SAW STEEL BAR						Date 9-18-8_
Present / Proposed Dept. CUT-OFF				Sheet 1 of 1		Chart by RW
MAN		MACHINE				
MAN #1	Time	SAW #1	Time		Time	
GET BAR .10		IDLE .10				
LOAD .05		LOAD .05				
IDLE .55		MACHINE PROCESS .55				
UNLOAD .05		UNLOAD .05				
BURR & INSPECT .25		IDLE .25				
ASIDE PART .10		IDLE .10				

Time scale: 0, 10, 20, 30, 40, 50, 60, 70, 80, 90, 100, 110, 120, 130, 140, 150, 160, 170

Figure 3-5 Operator-Machine (Activity) Chart.

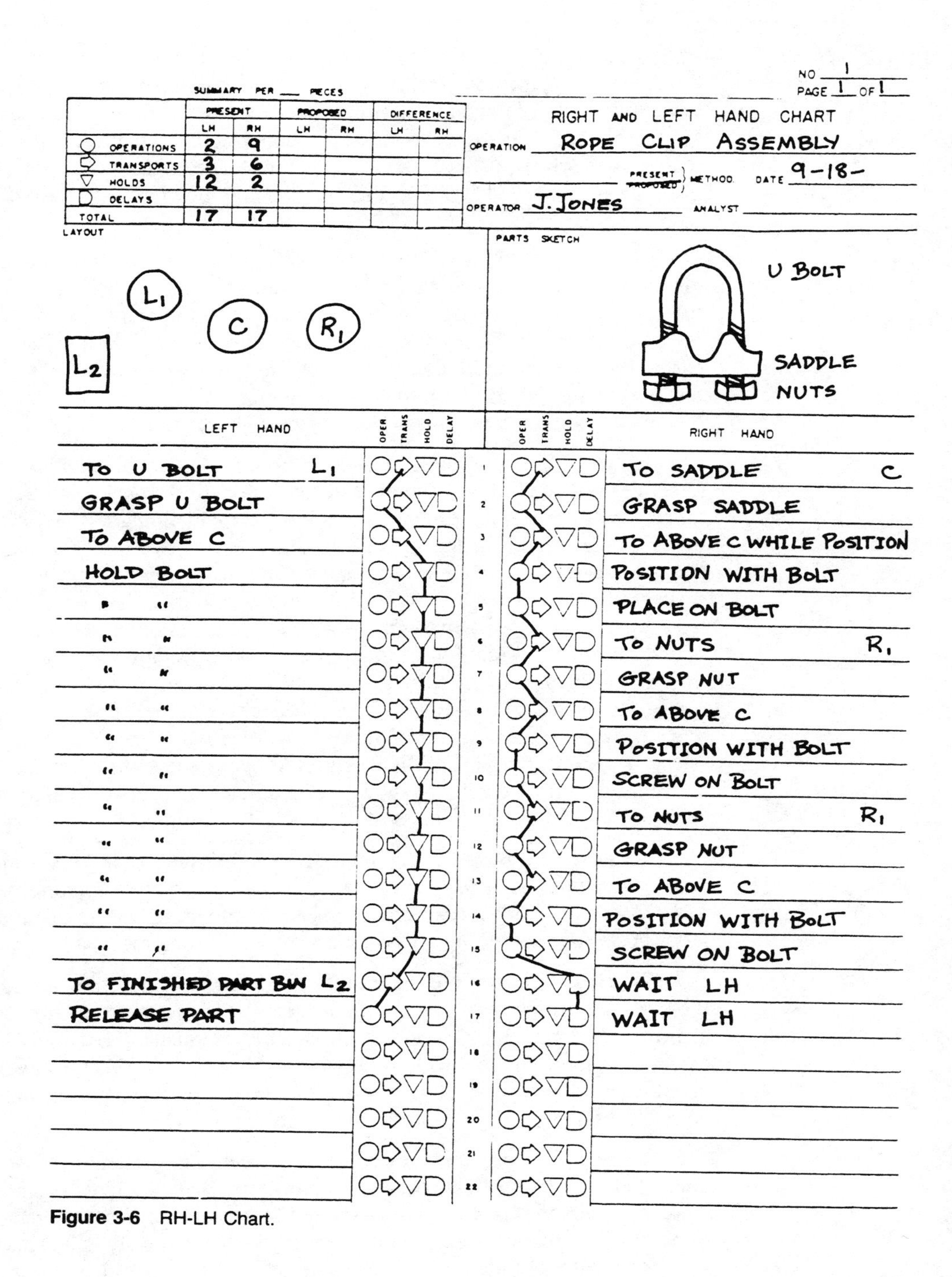

NO 1
PAGE 1 OF 1

RIGHT AND LEFT HAND CHART

OPERATION ROPE CLIP ASSEMBLY

PRESENT / PROPOSED METHOD DATE 9-18-

OPERATOR J. JONES ANALYST

SUMMARY PER ___ PIECES

	PRESENT LH	PRESENT RH	PROPOSED LH	PROPOSED RH	DIFFERENCE LH	DIFFERENCE RH
OPERATIONS	2	9				
TRANSPORTS	3	6				
HOLDS	12	2				
DELAYS						
TOTAL	17	17				

LAYOUT

PARTS SKETCH

LEFT HAND	OPER TRANS HOLD DELAY		OPER TRANS HOLD DELAY	RIGHT HAND
TO U BOLT L1		1		TO SADDLE C
GRASP U BOLT		2		GRASP SADDLE
TO ABOVE C		3		TO ABOVE C WHILE POSITION
HOLD BOLT		4		POSITION WITH BOLT
" "		5		PLACE ON BOLT
" "		6		TO NUTS R1
" "		7		GRASP NUT
" "		8		TO ABOVE C
" "		9		POSITION WITH BOLT
" "		10		SCREW ON BOLT
" "		11		TO NUTS R1
" "		12		GRASP NUT
" "		13		TO ABOVE C
" "		14		POSITION WITH BOLT
" "		15		SCREW ON BOLT
TO FINISHED PART BIN L2		16		WAIT LH
RELEASE PART		17		WAIT LH
		18		
		19		
		20		
		21		
		22		

Figure 3-6 RH-LH Chart.

WORK MEASUREMENT

In this section the question of "Why work measurement?" will be answered before discussing the different techniques of work measurement. When considering the problems involved in administering a work measurement program—extra paper work, the grievances, the expense of a staff department to establish the standards—is it really worthwhile to have such a program? The management of most companies gives an unqualified "yes" to this question. Following are some of the reasons for their support:

1. To maintain a more consistent work pace in the shop, reducing the burden on the supervisor to achieve production requirements. Without work measurement, the production output often depends on what the operator desires to produce. His or her goal may be one that is set by the supervisor, by fellow workers, or in most cases, by what the individual feels is a reasonable day's work. One can readily see the problems with this. Output between operators can vary greatly depending on the person who is determining the expected production.

2. To provide a systematic means of providing a planning and scheduling program. One does not need to be a member of the Production Control and Scheduling department to be aware of the problems involved in maintaining a flow of material through the different phases of the manufacturing process, even with a work measurement program to serve as a guide in planning the flow. Without production standards to serve as a basis for machine loading and operator assignment, material could be held up at one operation while a machine or operator performing a subsequent operation might be idle for an indefinite period waiting for the material. This results in poor utilization of machinery and other resources, which is very costly. If the same jobs were done over and over with the same machinery, the same methods, and the same employees, eventually schedules would become fairly accurate. But one can see the problems when new employees, new methods, new jobs, or new equipment might be employed. The scheduler would have to start from scratch to develop production guidelines.

3. To provide a means for estimating new jobs and/or processes more accurately. Any company must depend upon new jobs and new processes to remain competitive and profitable. Work measurement plays an important part in accurately estimating labor costs involved in new jobs and/or revised processes.

4. To provide a cost measurement. Work measurement has an important role in the cost accounting system. By establishing a production standard for an operation, a direct labor cost for that operation can be established. By applying standards to each direct labor operation, any segment of the operation that exceeds the estimated or anticipated costs can be pinpointed. If standards were not established, it would be most difficult to locate the trouble spots.

5. To provide the employee with a goal—what is expected. Most employees are interested in what their supervisors expect of them and in return want to fulfill these expectations as well as possible. Most of the operators in the shop are the same in this respect. Work measurement is the company's means of telling the operator on the production job what his or her goal should be—or in other words, what the company expects. Even though the operator may not always agree with

the goal set by the company, all employees appreciate knowing what the company expects of them.

Advantages to the employee. There are advantages and satisfaction for the employee in a work measurement system. Some of these are:

1. Increased job security through more effective utilization of time and reduction of downtime and delay costs.
2. In some instances, work measurement may make the employee's job easier. This may be brought about by improved equipment maintenance, methods, and better tools.
3. Provides personal satisfaction in recording and displaying good individual performance.
4. Will eliminate any blame to the employees for conditions affecting their productivity which are beyond their control.
5. Will allow employees to know how management is appraising their performance.
6. Will allow employees to feel that they are receiving fair and equitable treatment in relation to other employees.
7. In the case of incentive plans, the employee is afforded the opportunity to earn a higher wage.

Advantages for the company. There are definite advantages for the company in a work measurement plan. Some of these advantages are:

1. Will provide better control of labor cost and equipment utilization.
2. Will accurately indicate the performance of all employees and indicate those employees who need assistance in attaining productivity goals.
3. Will allow the company to accurately determine setup costs, which are vital for estimating, pricing, and economical manufacturing.
4. Will record and identify the causes of downtime and delay time, allowing management to minimize excess costs in this area.
5. Will allow improved scheduling, which in turn will reduce production costs and improve customer deliveries.
6. In total, work measurement will reduce labor and operating costs per unit of product, which in turn should allow the company to increase its sales and profits.

Methods of measurement. The primary methods for measuring work include 1) estimates or historical data, 2) stopwatch time study, 3) standard data, 4) predetermined time standards, and 5) work sampling.

Many companies have found that standards based on estimates or historical data are inconsistent and in most cases are not accurate. For this reason most companies use some form of "engineered work standards."

The most widely used method for measuring work is stopwatch time study. This technique was introduced by Frederick W. Taylor in the late 1800s. The stopwatch method is relatively simple to learn and explain to operators. The analyst is able to observe the complete cycle and to record the actual time the operator takes to perform a given task. During the observation the analyst is also recording the method and has the opportunity to make any improvements that will help to reduce the amount of work that is required to complete the task.

A major disadvantage with stopwatch time study is that the analyst must use a subjective evaluation of the worker's skill and effort to determine a performance rating factor. The rating factor is applied to the recorded time to arrive at a "normal" time that is considered fair and attainable by all employees.

Standard data is a synthesis of operations or elements of work analyzed and arranged for ready application to a range of equipment, parts, or operations. This method eliminates the need for performance rating and provides consistent time standards. By using standard data tables, methods and standards may be established in advance of actual production.

Predetermined time standards are a system of elemental manual motion times covering the principal body and body extremity activities. Predetermined time values have been determined from experimentation and measurement for each significantly different variation of each motion. Some of the more frequently used systems are Methods-Time Measurement (MTM), Maynard Operational Sequence Technique (MOST), and Work Factor (WOFAC).

A work sampling study consists of a number of observations taken at random intervals; in taking the observations, the state or condition of the object of the study is noted, and this state is classified into predefined categories of activity pertinent to the particular work situation. From the proportions of observations in each category, inferences are drawn concerning the total work activity under study. The work sampling technique has several advantages that the other methods do not offer. Information is obtained at a fraction of the cost and is more representative of the true conditions. In many cases work sampling is more accurate than other methods of work measurement.

FACILITIES LAYOUT

The location of various departments, work stations, and equipment affects operating costs and manufacturing effectiveness. There are numerous theories and methods for determining the layout that will provide the greatest amount of flexibility and at the same time economical manufacturing methods. The design of facilities layouts are generally categorized by three different work flow patterns that are described in this section.

Product layout. The typical product layout would be an assembly line where large volumes of a single product require the same sequence of operations for production. The machines and equipment are arranged so that each of the operations can be completed in sequence with minimum travel between work stations. Specialized tasks are performed at each of the work stations along the line or flow between stations. See Figure 3-7.

Figure 3-7 Product layout.

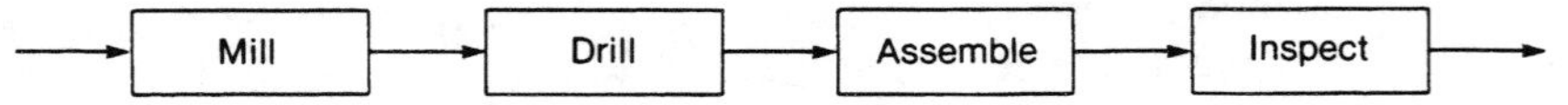

Process layout. In a process layout similar types of machines or equipment are grouped together and the product is moved from one work area to another. In this type of layout a variety of products are produced in small quantities with an intermittent flow. See Figure 3-8.

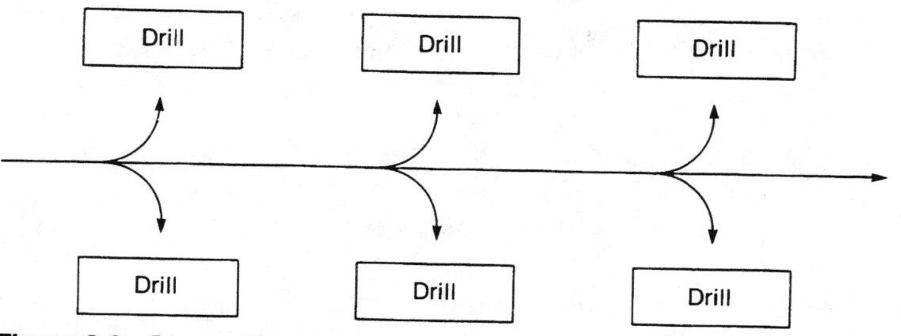

Figure 3-8 Process layout.

Fixed-position layout. When the characteristics of the product require that it remain in a single location, the tools, materials, and labor are moved to the product as required for the manufacturing functions to be performed on the product. Examples of this type of layout are the manufacture of aircraft, large ships, and specialized machinery.

Table 3-1 summarizes some of the ways in which basic layouts differ from one another.

FUNDAMENTALS OF PRODUCTION AND INVENTORY CONTROL

Inventory management is important to the success of all business and manufacturing organizations. For many years industry in the United States has treated inventory as a necessary requirement of doing business. It was not uncommon to find companies that had inventory levels as high as six months to a year of annual sales volume on certain products. As a result of the success of Japanese companies in the application of Just-in-Time inventory, a greater emphasis has been placed on controlling and reducing inventory levels.

This section discusses the reasons for holding inventory, costs of inventory, and methods for controlling inventory.

Reasons for holding inventory. One of the major problems encountered with controlling inventory is the conflict between the basic objectives of running an efficient manufacturing operation, minimizing the investment in inventory, and providing the best possible service to the customer. The sales department would like to maintain a large inventory of finished goods and alter manufacturing schedules to enable them to respond to the demands of the consumer. Both of these objectives increase investments in new materials, supplies, and finished products. Managers of manufacturing operations want to staff and run the processes at a constant level

TABLE 3-1
CHARACTERISTICS OF LAYOUT DESIGNS

Aspect of the Conversion Process	Product-oriented	Process-oriented	Fixed-position
Product characteristics	Layout geared to producing a standardized product, in large volume, at stable rates of output	Layout for diversified products requiring common fundamental operations, in varying volume, at varying rates of output	Low volume, each unit often unique
Product flow pattern	Straight line flow of product; same sequence of standard operations on each unit	Diversified flow pattern; each order (product) may require unique sequence of operations	Little or no product flow; equipment and human resources brought to site as needed
Human skills requirement	Tolerance for performing routine, repetitive tasks at imposed pace; highly specialized work content	Primary skilled craftsmen; can perform without close supervision and with moderate degree of adaptability	High degree of task flexibility often required; specific work assignments and location vary
Supporting staff	Large administrative and indirect support staff for scheduling materials and people, work analysis, and maintenance	Must possess skills for scheduling, materials handling, and production and inventory control	High degree of scheduling and coordinating skills required
Material handling	Material flows predictable, systematized, and often automated	Type and volume of handling required is variable; duplication of handling often occurs	Type and volume of handling required is variable, often low; may require heavy-duty general-purpose handling equipment
Inventory requirements	High turnover of raw material and work-in-process inventories	Low turnover of raw material and work-in-process inventories; high raw materials inventories	Variable inventories due to lengthy production cycle can result in inventory tie-ups for long periods
Space utilization	Efficient utilization of space, high rate of product output per unit of space	Relatively low rate of output per unit of facility space; large work-in-process requirements	For conversion within the facility, a low rate of space utilization per unit of output may occur
Capital requirements	High capital investment in equipment and processes that perform very specialized functions	Equipment and processes are general purpose and feature flexibility	General-purpose equipment and processes that are mobile
Product cost components	Relatively high fixed costs; low unit direct labor and materials costs	Relatively low fixed costs; high unit costs for direct labor, materials (inventory), and materials handling	High labor and materials costs; relatively low fixed costs

Source: *Production and Operations Management*, 2nd ed. by Everett E. Adam, Jr. and Ronald J. Ebert 1983, p. 241.

of production with as few changes as possible. The financial group would like to keep investment costs low by reducing inventory of both raw materials and finished goods.

Costs of inventory. The costs of maintaining inventories are a combination of ordering costs, carrying costs, out-of-stock costs, and capacity-related costs.

The costs of ordering may be either a result of purchasing raw materials and parts or manufacturing ordering costs. When a company purchases material, orders must be placed with the supplier, incoming goods must be inspected and processed, and invoices must be written to cover the cost of the goods. When an order is scheduled in a manufacturing facility, the lot size must be determined, releases issued, and machines and equipment must be set up for each production lot.

Inventory carrying costs include the expense of storing the product, insurance and taxes on the investment, obsolescence, and the capital that is invested in the inventory.

When inventory is not available to satisfy the demands of the customer, additional costs may be incurred in scheduling overtime, expediting orders, and possibly in lost business. Manufacturing efficiencies may also be affected by interrupting planned schedules to run small lots of priority items. This not only increases the production costs but also extends the manufacturing cycle, which leads to delinquent deliveries of the existing products.

Planning and control. The various techniques of production and inventory control are shown in Figure 3-9.

The Economic Lot—Size or Economic Order Quantity (EOQ) minimizes the combined costs of placing inventory replenishment orders and investments in holding costs. The frequently used formula for EOQ is:

$$\text{EOQ} = \sqrt{\frac{2\,DS}{IC}}$$

where

D = annual demand in units
S = setup or procurement cost per order
I = carrying cost, expressed as a percentage rate
C = cost per unit

Example A company that manufactures pumps has an annual demand for 2,000 housings. To set up the machinery to make the housings costs $800 per order. Each completed housing has a value of $40. The carrying cost is 20 percent of the unit cost.

$$\text{EOQ} = \sqrt{\frac{2 \times 2000 \times 800}{0.20 \times 40}}$$

$$= 632 \text{ Units}$$

Another function of inventory planning and control is the classification of inventory items in order to determine what items are to be controlled. For any group of items in inventory, a small percentage of the items account for a large percentage

Economic Lot-Sizing:

Square root	Part-period balancing	Least total cost
Welch families	Least unit cost	Lot-for-lot
Major/minor setups	Period order quantity	LIMIT

Forecasting:

Delphi	First-order exponential smoothing
Judgment	Second-order exponential smoothing
Running average	Base index (seasonal)
Weighted average	S-curves
Focus	

Materials Planning:

General:
ABC Classification, aggregate trade-off curves.

Order point:

Classical	Reserve stock (normal distribution)
Time-phased	Reserve stock (Poisson distribution)
Two-bin	Periodic review
	Visual review

Material requirements planning:

Explosion chart	Structuring bills of material
Conventional	Engineering change control
Regeneration	Allocations
Net change	Pegging

Capacity Planning:

Production planning	Detailed CRP
Rough-cut CRP	Graphical

Input Control:

Selecting orders	Block scheduling	CPM
Forward scheduling	Infinite loading	CPS
Backward scheduling	Finite loading	PERT

Output Control:

Input/output control	Expediting	Critical Ratio
Dispatching	Feedback	Flow control

Figure 3-9 Popular techniques of production and inventory control (Source: Production and Inventory Control, George W. Plossl, 1985, p. 11).

of the total value. Analysis of this type is called ABC classification. Using this method, inventory can be separated into three classes:

- A items—high value items that account for 70 to 80 percent of the total cost of the inventory. These units make up only 10 to 20 percent of the total number of items.
- B items—intermediate value and usage; approximately 30 to 40 percent of the items with a value of 15 to 20 percent.
- C items—low value items, accounting for 5 to 10 percent of the value, but including 40 to 50 percent of the total items. See Figure 3-10.

Inventory analysis such as the ABC classification determines the items that need to be controlled. The degree of control for each classification is as follows:

- A items—tight control, regular review by top-level supervision, close follow-up, accurate records.
- B items—normal control, regular attention, good records.
- C items—simple control, low priority, large inventories.

ABC classification enables companies to apply their resources (dollars and personnel) to maximize their return. This technique can be used for establishing ordering

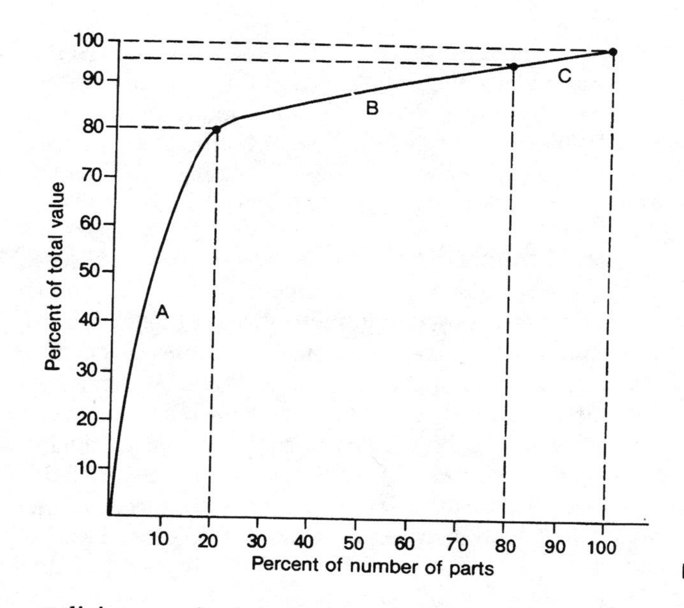

Figure 3-10 ABC classification.

policies, purchasing, setting priorities for scheduling, determining order quantities, and making safety stock decisions.

The largest percentage of manufacturing products are a combination of several components and/or raw materials. Material Requirements Planning (MRP) is being used by many companies to plan and control the purchase and scheduling of these materials. The implementation of MRP requires an accurate data base that includes:

1. Master Schedule—what is to be made, how many, and when the finished goods are required.
2. Bills of Material—a complete breakdown of all components and materials that are required for the finished product.
3. Inventory Status—current information on the quantities that are in stock.
4. In-process Status—quantities and locations of material that has been released for production.
5. Process Information—cycle times, machinery capacity, lead times, and material flow.

EOQ, ABC, and MRP are not solutions to all production and inventory control problems. These are simply some of the techniques that can assist in the management of manufacturing operations.

SAFETY

The importance of designing and maintaining a safe workplace is a major concern for the welfare of the employee and the success of the organization. Management has both a moral and legal obligation for the work-related safety and health of employees. Most managers also recognize that the advantages of maintaining a safe and clean work environment far exceed the costs associated with accidents and health hazards at the work site.

Federal and state regulations. In 1971 the United States government passed the Occupational Safety and Health Act (OSHA). This act established safety and health regulations that covered the majority of industrial operations. The act provided the authority to inspect workplaces, issue citations for standards violations, assess penalties in the form of fines, and in extreme cases, to obtain court orders to close down operations until unsafe practices are corrected.

Planning and organization. Some safety experts estimate that for every major injury, there are over 300 noninjury accidents. By identifying and eliminating these noninjury accidents it is possible to prevent major injuries.

For the employer, the costs associated with unsafe operating methods include medical and insurance payments, a cost of property damage, loss of output, and disrupted production. From the standpoint of the worker, accidents may mean physical suffering, permanent disabilities, and potential loss of earnings.

Traditionally used indexes for accident measurement include accident and injury frequency rates and severity rates. Frequency measured the numbers of cases per standard number of workhours and included only those cases in which the worker missed at least one day of work. Severity was a measure of the total lost workdays per standard quantity of workhours.

The current method of accident measurement uses a total injury/illness incident rate that includes all injuries or illnesses which require medical treatment. The method by which the incident rates is computed is

$$\text{total injury/illness incident rate} = \frac{\text{number of injuries and illnesses} \times 200{,}000}{\text{total hours worked by all employees during the period covered}}$$

For example, a plant that had 12 injuries and job-related illnesses during a period in which all employees worked a total of 800,000 hours would have a total injury/illness incident rate of

$$\frac{12 \times 200{,}000}{800{,}000} = 3.0 \text{ per } 200{,}000 \text{ hours worked}$$

The current method of calculating incident rates includes all cases involving medical treatment, not just lost injuries that resulted in lost workdays.

The safety professional may also find it beneficial to track other incident rates such as:

1. separate injury rates
2. separate illness rates
3. lost workday cases rates (similar to old frequency rates)
4. number of lost workdays rate (similar to old severity rates)

Under the current method of calculating statistical measures for safety performance, all rates use a factor of 200,000. This number represents the number of hours worked per year by 100 employees:

$$40 \text{ hours/week} \times 50 \text{ weeks/year} \times 100 \text{ employees} = 200{,}000$$

One of the disadvantages of using accident statistics is that they are collected after the fact. To be of any value the data must be collected over a long period of time. Often by the time the accident statistics are established, the work conditions may have changed. However, accident statistics have proven valuable to the safety professional, safety insurance companies, and regulatory agencies for comparative purposes.

Willie Hammer, in *Occupational Safety Management and Engineering*, lists 21 specific functions of safety personnel:

1. Ensures that federal, state, and local safety laws, regulations, codes, and rules are observed.

2. Ensures that OSHA record-keeping and reporting requirements are met.

3. Assists top managers in preparing safety policies, and then ensuring that they are carried out. Any noncompliance with those policies should be reported to the manager, or managers, capable of directing that corrective action be taken.

4. Monitors all activities where accidents could occur that would cause injury to personnel, damage to equipment or facilities, or losses of material.

5. Halts any operation or activity that constitutes an imminent hazard to personnel or could result in loss of equipment or facilities.

6. Establishes suitable liaison and working arrangements with other activities concerned with, or which might be involved in, accident prevention, such as plant security, fire prevention and fire-fighting department, plant engineering office, and the medical department. Coordinates safety activities in which these organizations might be involved, such as development of emergency plans.

7. Assists in the formation of safety committees and assists them in carrying out their activities.

8. Establishes and monitors programs for detecting and correcting hazardous conditions. Ensures that use of toxic or other hazardous materials is properly controlled and that proper safeguards, such as personnel protective equipment, are available, used, and in good working order prior to possible use.

9. Reviews and approves the safety aspects of plant facility layouts and designs, and of equipment being procured. Either checks or has qualified personnel check new equipment being installed to ensure that it is safe and is provided with prescribed safeguards.

10. Makes certain that hazardous areas, entrances and exits, and dangerous equipment are posted in accordance with prescribed standards.

11. Controls selection, acquisition, and use of hazard monitoring, detection, and warning equipment; of personal protective equipment and materials; and of emergency equipment.

12. Conducts safety training of personnel at all levels. Where supervisors are to conduct training of employees under their jurisdiction, ensures that supervisors are knowledgeable in the subjects being presented and that training is accomplished as programmed.

13. Conducts investigations of accidents, near-misses, and hazardous conditions. Prepares reports and takes action to prevent recurrences.

14. Measures and records any tests made of environmental hazards, such as noise levels or the presence of toxic gases or radiation.

15. Disseminates information on safety to all activities to alert them to specific hazards, or to maintain general interest in accident prevention.

16. Accompanies inspectors from governmental agencies and insurance companies on surveys and audits of the plant. Reviews any reports of discrepancies and initiates action for their correction.

17. Establishes or approves procedures for hazardous operations, such as tank entry, and ensures that personnel who will engage in these operations know the hazards involved, that proper protective equipment is available and will be used, and that all concerned are familiar with actions to be taken in an emergency.

18. Makes on-site reviews of activities and determines whether any procedures or methods could lead to accidents. Recommends changes if a possibility exists that a human or material failure could result in injury or damage.

19. Periodically inspects emergency supplies located in the plant to ensure that nothing has been removed without authorization, and that they are ready for their intended purposes.

20. Maintains all records that relate to safety activities.

21. Keeps informed on latest developments in activities that relate to his or her job, such as new equipment; hazardous materials that might be used in the plant; changes in laws, regulations, codes, and rules; new methods of accident prevention; worker's compensation rulings; or accidents that have occurred at similar facilities or in related industries.

SUMMARY

Managers are continually changing the ways in which they are designing and controlling manufacturing systems. This chapter covers an introduction into some of the more important functions of design and control. For a more detailed explanation of these techniques the reader should consult some of the books listed at the end of this text.

QUESTIONS

1. What factors affect the decision on where to locate a new manufacturing facility?
2. Describe the six steps in methods improvement.
3. What things should a methods analyst consider when he or she is designing a new workplace for an assembly operation?
4. What is the difference between an Operation Process Chart and a Flow Process Chart?
5. What are the advantages and disadvantages of work measurement?
6. List five methods for measuring work. Which of these methods would you recommend? Why?

7. The annual usage of an electronic part is 30,000 units per year. It costs $400 to process an order for the part. Each part costs $10 and the cost to keep this part in inventory is $1. What is the Economic Order Quantity?
8. Describe the ABC system of classification. Give an example where this may be used.
9. What is OSHA?
10. List the functions of a safety engineer.

4 METROLOGY AND QUALITY CONTROL

Many scholars have dealt with the topic of interchangeable manufacture and have emphasized primarily machines and processes. In addition to precision machine tools, it is necessary to understand the importance of other elements in the overall system of manufacturing: (1) units of measure, (2) precision gauges and measuring instruments, (3) uniformly accepted standards of measurement, and (4) methods to ensure that the desired levels of accuracy and uniformity are maintained.

THE METRIC SYSTEM OF MEASURES

The international language of technical measurement is the SI Metric System, and all persons in science, engineering, and technology must be fluent in that language. The symbol "SI" denotes the International System of Units, and is predicated upon the seven base units and two supplementary units shown in Figure 4-1. Units for all other quantities are derived, with special names which relate to the base and supplementary units in a coherent manner; in brief, they are expressed as products and ratios of the nine base and supplementary units without numerical factors. All other SI derived units are similarly derived in a coherent manner from the twenty-seven base, supplementary, and special-name SI units. For use with the SI units, there is a set of 16 prefixes to form multiples and submultiples of these units. See Figure 4-2. For mass the prefixes are to be applied to the gram instead of to the SI unit, the kilogram.

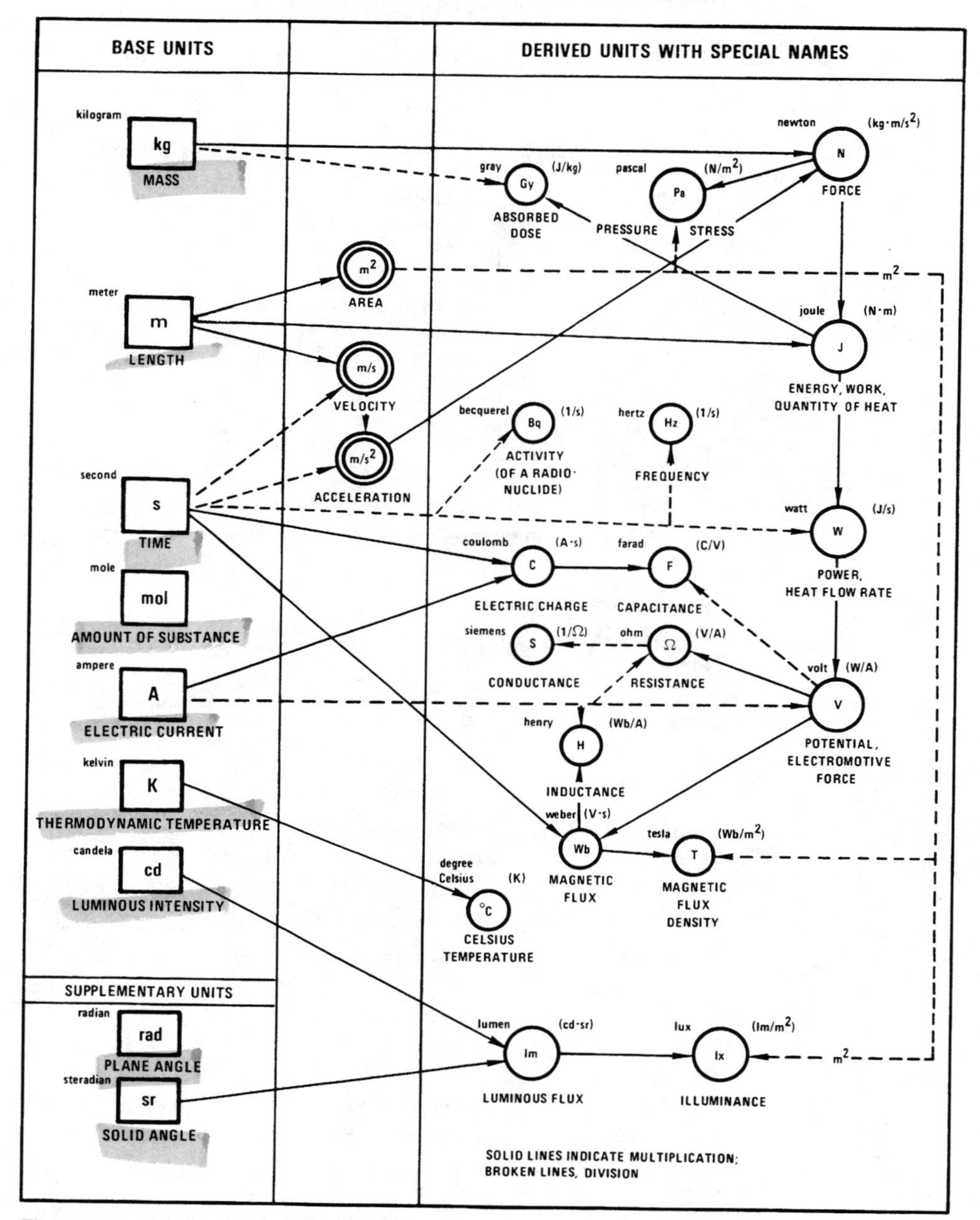

Figure 4-1 Relationships of SI units with names. (Courtesy U.S. Department of Commerce)

The SI units together with the SI prefixes provide a logical and interconnected framework for measurement in science, industry, and commerce. Leading national and international professional and standardized bodies encourage familiarity with, and diffusion of, SI units throughout all sectors of United States activities. Further metric tabular data are found in Tables 4-1 through 4-4.

SI UNIT PREFIXES

Multiplication Factor	Prefix	Symbol	Pronunciation (USA)*	Term (USA)	Term (Other Countries)
1 000 000 000 000 000 000 = 10^{18}	exa	E	as in Texas	one quintillion	one trillion
1 000 000 000 000 000 = 10^{15}	peta	P	as in petal	one quadrillion	one thousand billion
1 000 000 000 000 = 10^{12}	tera	T	as in terrace	one trillion	one billion
1 000 000 000 = 10^{9}	giga	G	jig ' a (a as in about)	one billion	one milliard
1 000 000 = 10^{6}	mega	M	as in megaphone	one million	
1 000 = 10^{3}	kilo	k	as in kilowatt	one thousand	
100 = 10^{2}	hecto	h	heck 'toe	one hundred	
10 = 10	deka	da	deck ' a (a as in about)	ten	
0.1 = 10^{-1}	deci	d	as in decimal	one tenth	
0.01 = 10^{-2}	centi	c	as in sentiment	one hundredth	
0.001 = 10^{-3}	milli	m	as in military	one thousandth	
0.000 001 = 10^{-6}	micro	μ	as in microphone	one millionth	
0.000 000 001 = 10^{-9}	nano	n	nan ' oh (nan as in Nancy)	one billionth	one milliardth
0.000 000 000 001 = 10^{-12}	pico	p	peek ' oh	one trillionth	one billionth
0.000 000 000 000 001 = 10^{-15}	femto	f	fem ' toe (fem as in feminine)	one quadrillionth	one thousand billionth
0.000 000 000 000 000 001 = 10^{-18}	atto	a	as in anatomy	one quintillionth	one trillionth

*The first syllable of every prefix is accented to assure that the prefix will retain its identity.

Figure 4-2 SI unit prefixes. (Courtesy American National Metric Council)

TABLE 4-1
SOME SI DERIVED UNITS EXPRESSED IN TERMS OF BASE UNITS

Quantity	SI Unit	Unit Symbol
area	square meter	m^2
volume	cubic meter	m^3
speed, velocity	meter per second	m/s
acceleration	meter per second squared	m/s^2
wave number	1 per meter	m^{-1}
density, mass density	kilogram per cubic meter	kg/m^3
current density	ampere per square meter	A/m^2
magnetic field strength	ampere per meter	A/m
concentration (of amount of substance)	mole per cubic meter	mol/m^3
specific volume	cubic meter per kilogram	m^3/kg
luminance	candela per square meter	cd/m^2

DIMENSIONING AND TOLERANCES

The use of interchangeable parts had a great impact upon manufacturing methods. Since mating parts were manufactured independently of each other, some scheme had to be developed to ensure the proper fit of the components at assembly. At first a scheme was adopted to make a ''master'' part and to produce parts just like

TABLE 4-2
SOME SI DERIVED UNITS EXPRESSED BY MEANS OF SPECIAL NAMES

Quantity	SI Unit		
	Name	Symbol	Expression in Terms of SI Base Units
dynamic viscosity	pascal second	Pa·s	$m^{-1}\cdot kg\cdot s^{-1}$
moment of force	newton meter	N·m	$m^2\cdot kg\cdot s^{-2}$
surface tension	newton per meter	N/m	$kg\cdot s^{-2}$
power density, heat flux density, irradiance	watt per square meter	W/m^2	$kg\cdot s^{-3}$
heat capacity, entropy	joule per kelvin	J/K	$m^3\cdot kg\cdot s^{-2}\cdot K^{-1}$
specific heat capacity, specific entropy	joule per kilogram kelvin	J/(kg·K)	$m^2\cdot s^{-2}\cdot K^{-1}$
specific energy	joule per kilogram	J/kg	$m^2\cdot s^{-2}$
thermal conductivity	watt per meter kelvin	W/(m·K)	$m\cdot kg\cdot s^{-3}\cdot K^{-1}$
energy density	joule per cubic meter	J/m^3	$m^{-1}\cdot kg\cdot s^{-2}$
electric field strength	volt per meter	V/m	$m\cdot kg\cdot s^{-3}\cdot A^{-1}$
electric charge density	coulomb per cubic meter	C/m^3	$m^{-3}\cdot s\cdot A$
electric flux density	coulomb per square meter	C/m^2	$m^{-2}\cdot s\cdot A$
permittivity	farad per meter	F/m	$m^{-3}\cdot kg^{-1}\cdot s^4\cdot A^2$
permeability	henry per meter	H/m	$m\cdot kg\cdot s^{-2}A^{-2}$
molar energy	joule per mole	J/mol	$m^2\cdot kg\cdot s^{-2}\cdot mol^{-1}$
molar entropy, molar heat capacity	joule per mole kelvin	J/(mol·K)	$m^2\cdot kg\cdot s^{-2}\cdot K^{-1}\cdot mol^{-1}$
exposure (x and γ rays)	coulomb per kilogram	C/kg	$kg^{-1}\cdot s\cdot A$
absorbed dose rate	gray per second	Gy/s	$m^2\cdot s^{-3}$

it. Since the parts produced actually varied, this procedure soon caused much consternation and concern. How much could the parts vary from the master before they were defective? What happened if the distance between the holes was slightly greater than on the master? These and many more questions required answers.

Many solutions were attempted, templates and mating parts were used for checks, and many tempers were taxed to the breaking point. Out of the many procedures described came one which has persisted throughout the years. This procedure entailed the drawing of each part showing the desired dimensions and other quality characteristics. These drawings are known as blueprints.

Originally each dimension on the blueprint was shown by including the maximum and minimum size which would permit proper function. The difference between the maximum and minimum size was termed the tolerance. As long as the dimension or quality characteristic was encompassed by these limits the part was acceptable; if it wasn't, the part was to be considered defective.

The task of determining these limits was and still is tremendous. It is nearly impossible for any one person to envision the complete picture of function and production. However, these tolerances are the backbone of present-day production methods. They are necessary to establish the uniformity of parts required to ensure proper assembly and function.

TABLE 4-3
EXAMPLES OF CONVERSION FACTORS FROM NON-SI UNITS TO SI

Physical Quantity	Name of Unit	Symbol for Unit	Definition in SI Units
length	inch	in	2.54×10^{-2} m
length	nautical mile*	nmi	1852 m
length	angstrom*	Å	10^{-10} m
velocity	knot*	kn	(1852/3600) m/s
area	barn*	b	10^{-28} m^2
acceleration	gal*	Gal	10^{-2} m/s^2
mass	pound (avoirdupois)	lb	0.453 592 37 kg
force	kilogram-force	kgf	9.806 65 N
pressure	conventional millimeter of mercury	mmHg	$13.5951 \times 9.806\ 65$ N·m^{-2}
pressure	atmosphere*	atm	101 325 N·m^{-2}
pressure	torr	Torr	(101 325/760) N·m^{-2}
pressure	bar*	bar	10^5 Pa
stress	pound-force per sq. in	lbf/in^2	6 894.757 Pa
energy	British thermal unit	Btu	1055.056 J
energy	kilowatt hour	kWh	3.6×10^6 J
energy	calorie (thermochemical)	cal	4.184 J
activity (of a radio-nuclide)	curie*	Ci	3.7×10^{10} Bq
exposure (x or γ rays)	röntgen*	R	2.58×10^{-4} C·kg^{-1}
absorbed dose	rad*	rd	1×10^{-2} Gy

* The General Conference on Weights and Measures has sanctioned the temporary use of these units.

TABLE 4-4
UNITS IN USE WITH THE INTERNATIONAL SYSTEM

Name	Symbol	Value in SI Unit
minute	min	1 min = 60 s
hour	h	1 h = 60 min = 3 600 s
day	d	1 d = 24 h = 86 400 s
degree	°	1° = (π/180) rad
minute	′	1′ = (1/60)° = (π/10 800) rad
second	″	1″ = (1/60)′ = (π/648 000) rad
liter	L*	1 L = 1 dm^3 = 10^{-3} m^3
metric ton (tonne)	t	1 t = 10^3 kg
hectare	ha	1 ha = 10^4 m^2

* The international symbol for liter is "l," which can be easily confused with the numeral "1." Accordingly, the symbol "L" is recommended for United States use.

During the nineteenth century interchangeable manufacturing was achieved through the use of fixtures and common tooling—not accurate measurement. As manufacturing operations expanded into new locations and parts were exchanged between different companies, it became necessary to develop measuring instruments to obtain absolute measurement of parts.

Rules. One of the most basic of all measuring instruments is the steel rule, which is the tool from which many other tools have been developed. See Figure 4-3. Rules are so essential and so frequently used on a variety of work that they are available in a truly amazing selection to suit the needs of the precision worker. Steel rules are graduated in customary or metric units, and sometimes scales for both systems are provided on a single rule. They can be graduated on each edge of both sides and even on the ends. Customary graduations are commonly as fine as one-hundredth (0.010) inch. Metric graduations are usually one-half millimeter (0.5 mm).

The hook rule feature, shown in Figure 4-4, is simpler to use. Not only does it provide an accurate stop at the end of the rule for setting calipers, but it also can be used for taking measurements where it is not possible to be sure that the end of the rule is even with the edge of the work.

The basic combination square consists of a hardened steel graduated rule and moveable combination square and miter head with spirit level and scriber (see Figure 4-5). In itself, it is a most versatile and useful layout and measuring tool which can be used as a try square, miter, depth gauge, height gauge, and level. The addition of a center head provides an easy means of locating the center of cylindrical work.

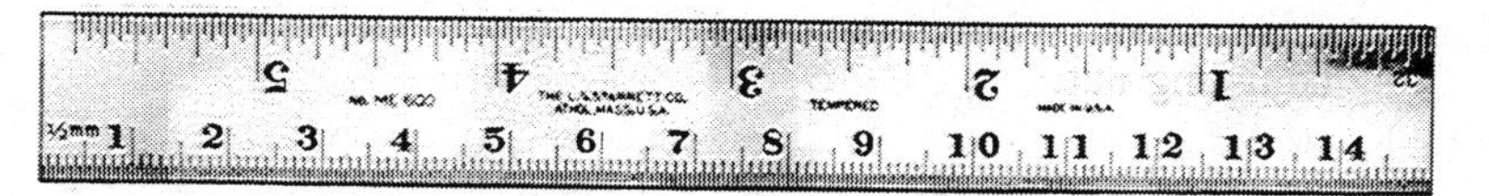

Figure 4-3 A common steel rule with both millimeter and inch graduations. (Courtesy of The L.S. Starrett Company)

Figure 4-4 The hook rules. (Courtesy of The L.S. Starrett Company)

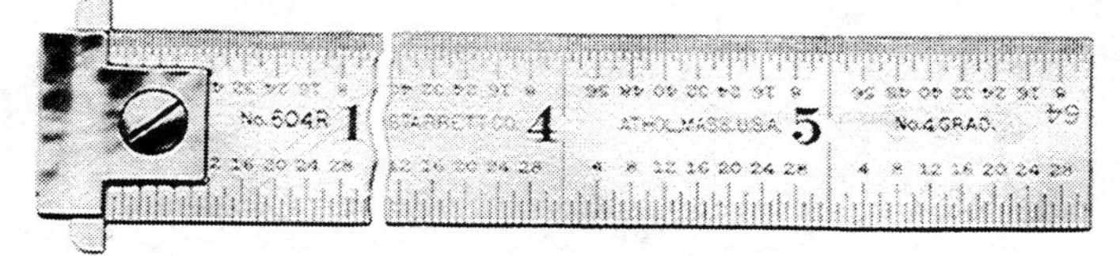

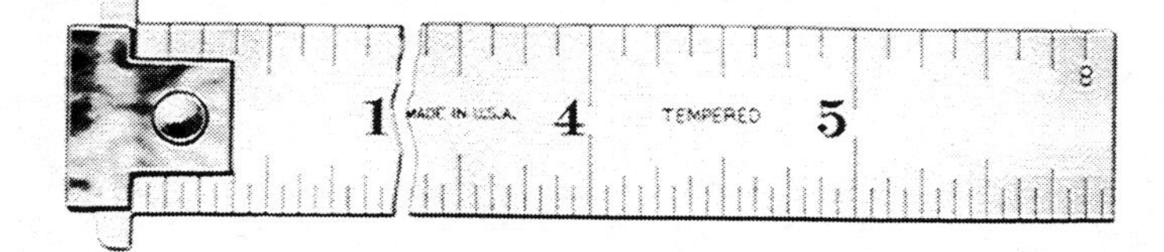

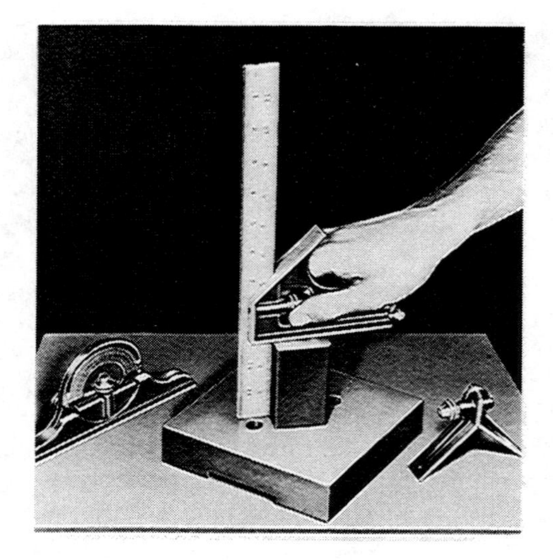

Figure 4-5 Various elements of the combination square employ a 300 mm (12 inch) steel rule with heads to measure linear angles, shaft centers, and, as shown, to determine heights of parts. (Courtesy of The L.S. Starrett Company)

Calipers. Calipers are used to measure diameters or thicknesses of parts with greater ease and accuracy than can be done with the steel rule. Several types of these instruments are used. The simple caliper has two legs which can be adjusted to the size of a workpiece, which is then measured with a steel rule. See Figure 4-6. There are separate types for inside and outside diameter measures. Slide calipers are a refinement of the steel rule, which insures greater accuracy in aligning the graduated scale with the edges or points to be measured. Figure 4-7 shows a typical slide caliper. They are useful in rough measures of diameters, thicknesses, and holes. The Vernier caliper, invented by the sixteenth century French mathematician, Pierre Vernier, consists basically of a stationary bar and a movable slide assembly. The stationary rule is a graduated bar with a fixed measuring jaw. The movable Vernier slide assembly combines a movable jaw, Vernier plate, clamp screws, and adjusting nut.

Figure 4-6 Simple calipers are used to measure the diameters of objects, as shown here. (Courtesy of The L.S. Starrett Company)

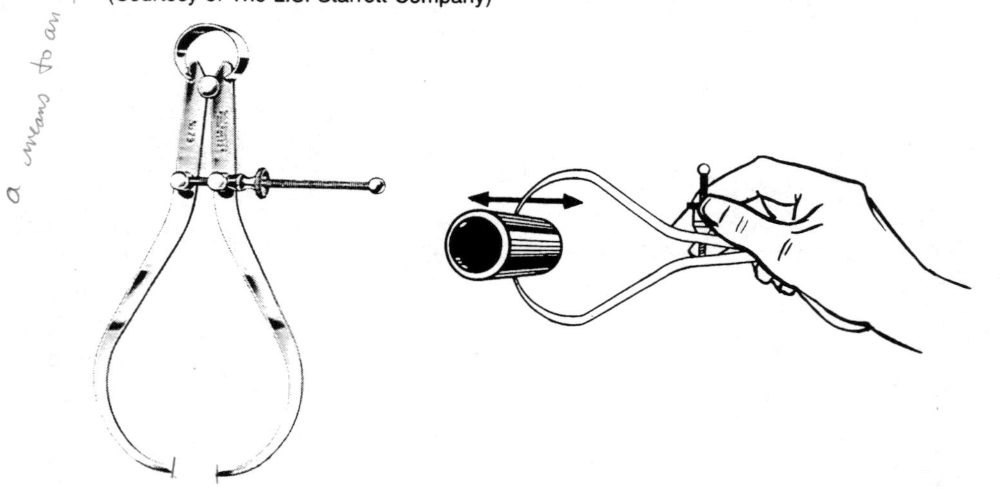

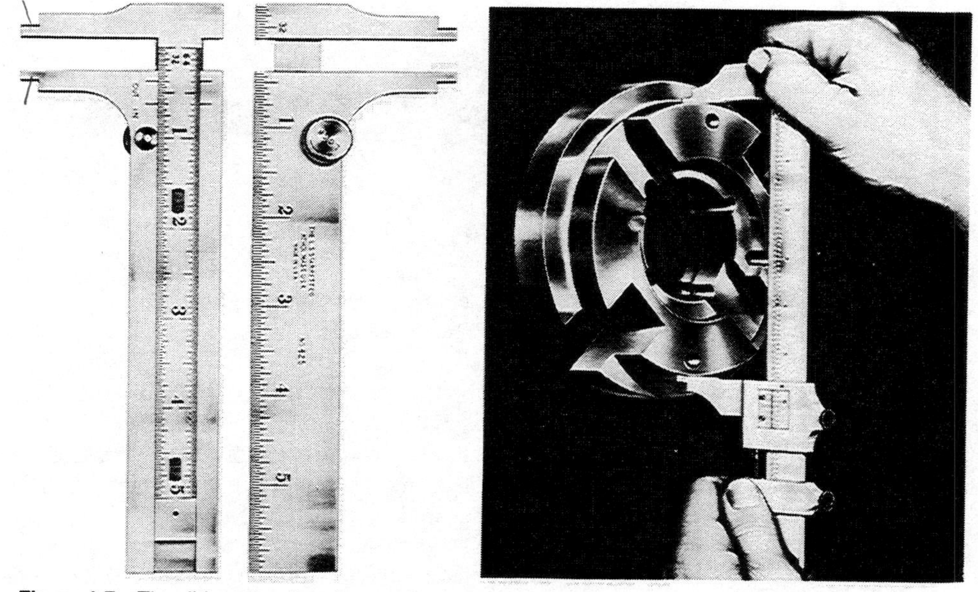

Figure 4-7 The slide caliper is an improvement of the steel rule for inside-outside basic measurements. (Courtesy of The L.S. Starrett Company)

Figure 4-8 Using the Vernier caliper to measure the diameter of a metal part. (Courtesy of The L.S. Starrett Company)

The Vernier slide assembly moves as a unit along the graduations of the bar to bring both jaws in contact with the work. Readings are taken in hundredths of a millimeter (thousandths of an inch) by reading the position of the Vernier plate in relation to the graduations on the stationary bar. See Figure 4-8. Though accurate, this basic tool can be difficult to read and interpret because it requires the visual alignment of the graduations. A refinement of the basic tool is the dial caliper, shown in Figure 4-9, which provides a direct reading. The electronic digital caliper

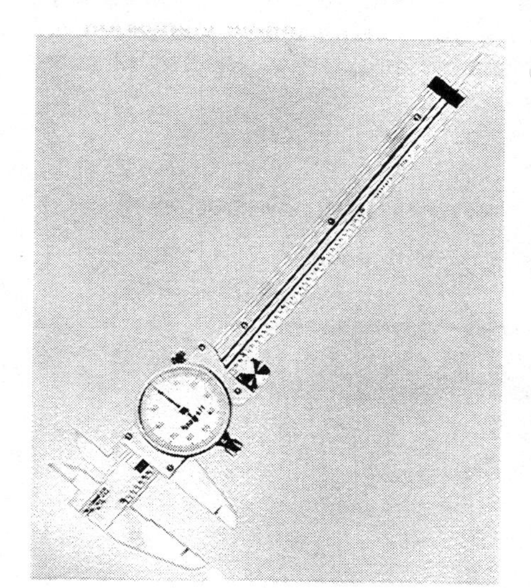

Figure 4-9 The dial caliper provides a quick-reading visual measurement. (Courtesy of The L.S. Starrett Company)

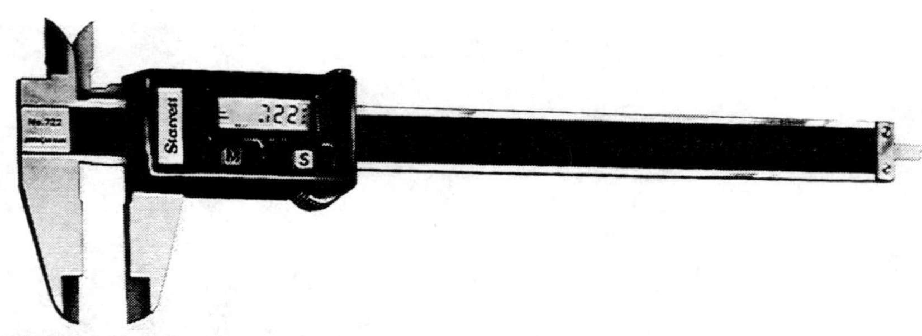

Figure 4-10 The electronic digital caliper can be linked to a computer for data analysis. (Courtesy of The L.S. Starrett Company)

provides a resolution of 0.001 mm (0.0001 inch) and an outlet jack to link it to a computer for printouts and data analysis. See Figure 4-10.

The micrometer caliper (or more simply, "micrometer"), is used to make a variety of accurate measurements without the possibility of error through misreading which exists when other finely graduated tools are used. An example of a micrometer is illustrated in Figure 4-11.

In effect, a micrometer caliper combines the double contact of a slide caliper with a precision screw adjustment which may be read with great accuracy. It operates on the principle of screw thread pitches. A typical application of this instrument is shown in Figure 4-12.

The tool functional theory is as follows. Since the pitch of the spindle screw is one-half millimeter (0.5 mm), one revolution of the thimble advances the spindle toward or away from the anvil the same 0.5 mm distance. See Figure 4-13.

The reading line on the sleeve is graduated in millimeters (1.0 mm) with every fifth millimeter being numbered from 0 to 25. Each millimeter is also divided in half (0.5 mm), and it requires two revolutions of the thimble to advance the spindle 1.0 mm.

The beveled edge of the thimble is graduated in 50 divisions, every fifth line being numbered from 0 to 50. Since one revolution of the thimble advances or withdraws the spindle 0.5 mm, each thimble graduation equals 1/50 of 0.5 mm or 0.01 mm. Thus, two thimble graduations equal 0.02 mm; three graduations 0.03 mm, etc.

Figure 4-11 Micrometer caliper nomenclature. (Courtesy of U.S. Navy)

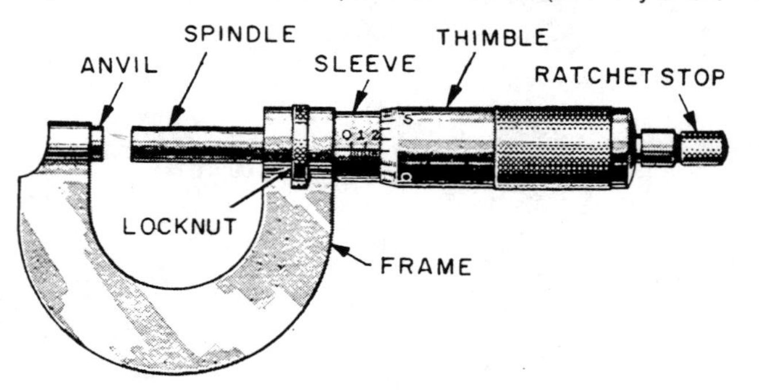

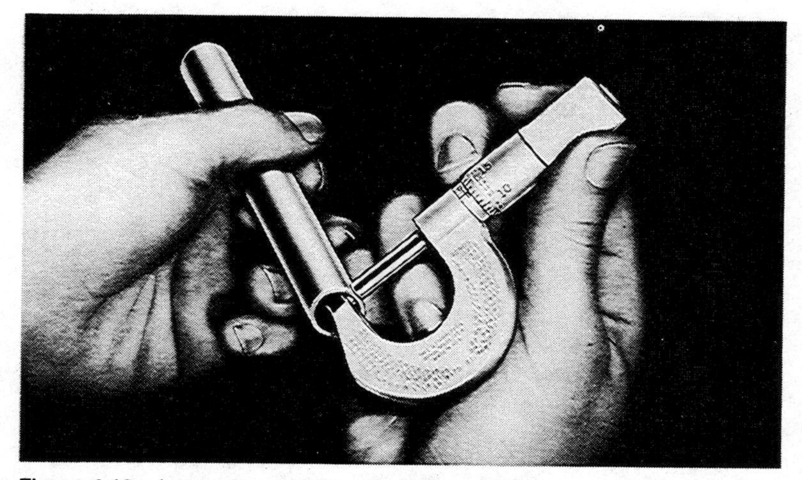

Figure 4-12 A typical use of the micrometer caliper is shown here. (Courtesy of The L.S. Starrett Company)

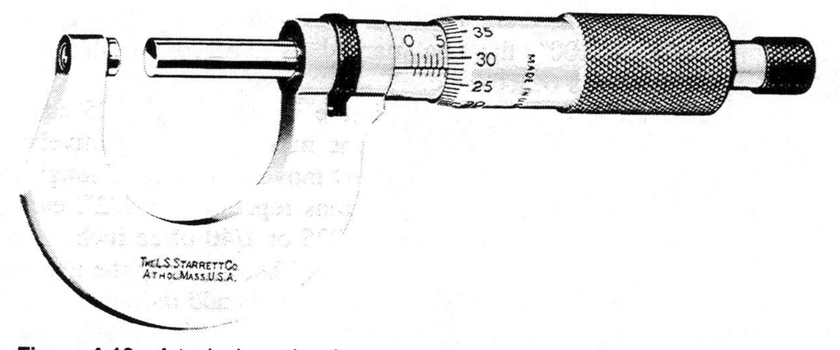

Figure 4-13 A typical metric micrometer reading. (Courtesy of The L.S. Starrett Company)

To read the micrometer, add the number of millimeters and half-millimeters visible on the sleeve to the number of hundredths of a millimeter indicated by the thimble graduation which coincides with the reading line on the sleeve.

Example Referring to the picture and drawing:

The 5 mm sleeve graduation is visible.	5.00 mm
One additional 0.5 mm line is visible in the sleeve	0.50 mm
Line 28 on the thimble coincides with the reading line on the sleeve, so 28 × 0.01 mm.	= 0.28 mm
The micrometer reading is	5.78 mm

When using an inch micrometer, the pitch of the screw on the spindle is 1/40″ or 40 threads per inch, and one complete revolution of the thimble advances the spindle face toward or away from the anvil face precisely 1/40 or 0.025 inches. See Figure 4-14.

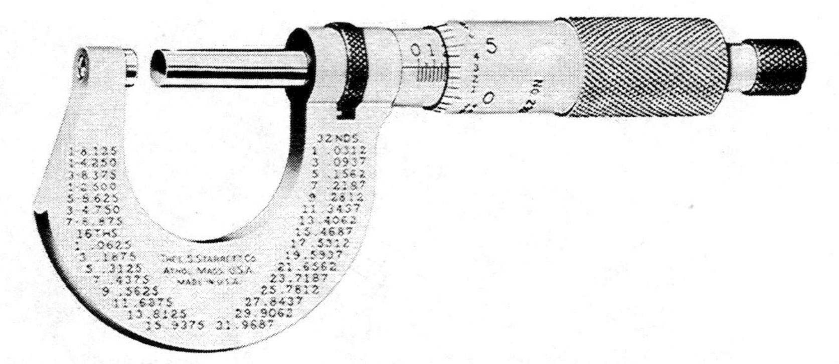

Figure 4-14 A typical inch micrometer reading. (Courtesy of The L.S. Starrett Company)

The reading line on the sleeve is divided into 40 equal parts by vertical lines that correspond to the number of threads on the spindle. Therefore, each vertical line designates 1/40 or 0.025 inches and every fourth line which is longer than the others designates hundredths of thousandths. For example: The line marked "1" represents 0.100″, the line marked "2" represents 0.200″, and the line marked "3" represents 0.300″, etc.

The beveled edge of the thimble is divided into 25 equal parts with each line representing 0.001″ and every line numbered consecutively. Rotating the thimble from one of these lines to the next moves the spindle longitudinally 1/25 of 0.025″ or 0.001 inch; rotating two divisions represents 0.002″, etc. Twenty-five divisions indicate a complete revolution, 0.025 or 1/40 of an inch.

To read the micrometer in thousandths, multiply the number of vertical divisions visible on the sleeve by 0.025″, and to this add the number of thousandths indicated by the line on the thimble which coincides with the reading line on the sleeve.

Example Refer to the illustration in Figure 4-14.

The "1" line on the sleeve is visible, representing . . . . 0.100″
There are three additional lines visible,
each representing 0.025″ . . . . 3 × 0.025″ = 0.075″
Line "3" on the thimble coincides with the reading
line on the sleeve, each line representing 0.001″ . . . 3 × 0.001″ = 0.003″
The micrometer reading is . . . . 0.178″

An easy way to remember is to think of the various units as if you were making change from a 10 dollar bill. Count the figures on the sleeve as dollars, the vertical lines on the sleeve as quarters, and the divisions on the thimble as cents. Add up your count and put a decimal point instead of a dollar sign in front of the figures.

Digital micrometers, shown in Figure 4-15, make readings faster and easier for every machinist, regardless of experience. The frame-mounted counter saves handling time since it can be read without removing fingers from the thimble or the micrometer from the work.

A number of special micrometers are used to measure threads, hole depths, and inside dimensions.

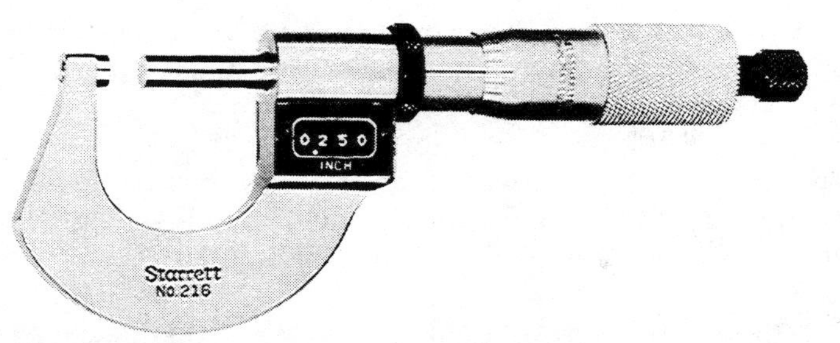

Figure 4-15 The digital micrometer simplifies the calculation of precision measures. (Courtesy of The L.S. Starrett Company)

Dial indicators. One of the instruments widely used in layout, inspection, and quality control operations is the dial indicator. Specially designed and manufactured to fine watch-making standards with either jeweled or plain bearings, it has precisely finished gears, pinions, and other working parts that make possible measurements as accurate as 0.00025 mm (0.000001 inch). See Figure 4-16.

In this tool any movement up against the probe is transmitted to the dial pointer, where a reading is made. The instrument can be used in conjunction with a lathe attachment to check the accuracy or runout of a hole bored in a chuck-mounted workpiece. They also are used to check the surfaces of flat parts, tool alignment, and machine setups.

Surface gauge. A surface gauge is a measuring tool generally used to transfer measurements to work by scribing a line, and to indicate the accuracy or parallelism of surfaces. This type of gauge is shown in Figure 4-17. It consists of a base with an adjustable spindle, to which may be clamped a scriber or an indicator. Surface gauges are made in several sizes and are classified by the length of the spindle, generally ranging from 100 to 300 mm (4 to 12 inches). The scriber is fastened to

Figure 4-16 A dial indicator. (Courtesy of The L.S. Starrett Company)

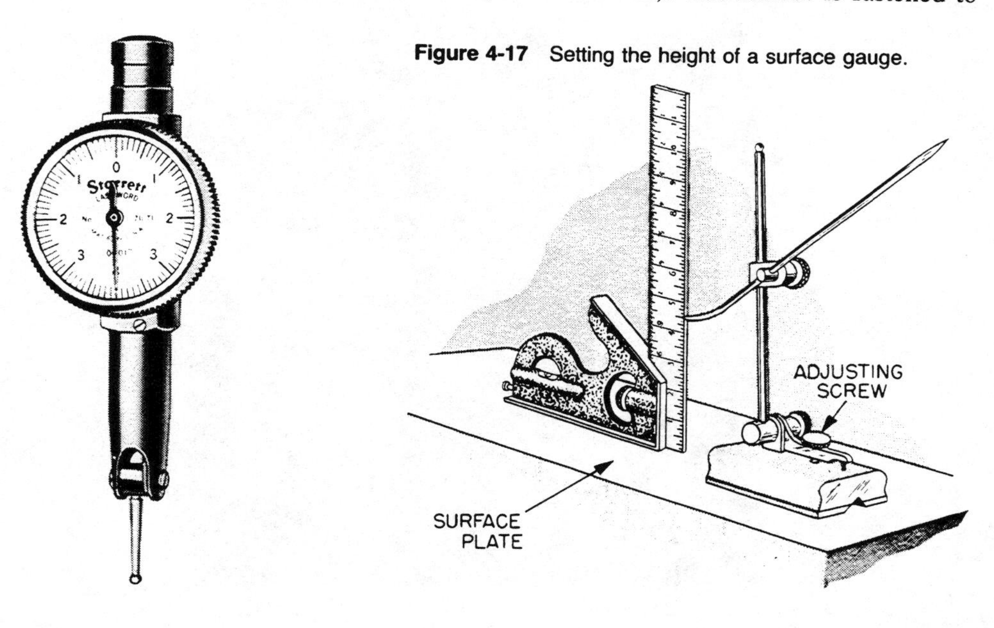

Figure 4-17 Setting the height of a surface gauge.

the spindle with a clamp. The bottom and the front ends of the base of the surface gauge have deep V-grooves cut in them, which allow the gauge to be seated on a cylindrical surface.

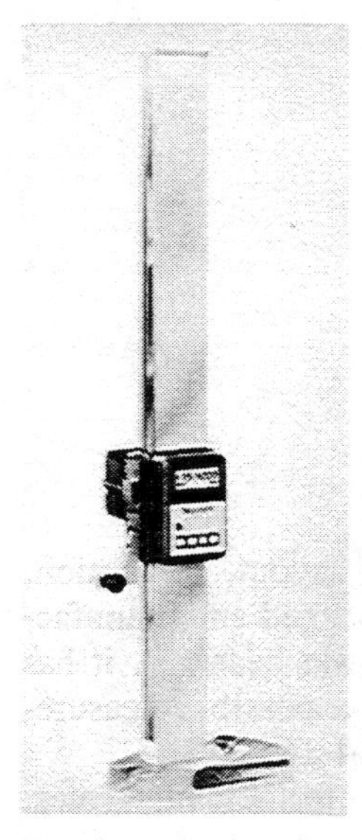

Figure 4-18 An electronic height gauge. (Courtesy of The L.S. Starrett Company)

Electronic height gauge. These gauges are extremely sensitive and are accurate to plus or minus 0.005 mm (0.0002 inch). See Figure 4-18. They are similar in use to the surface gauge for height position transfers.

Coordinate measuring machine (CMM). This state-of-the-art measuring unit employs four basic elements to determine the sizes of three-dimensional parts. They are machine structure, the probes, the computer and software, and custom accessories. See Figure 4-19. The machine features a bar fixture into which are fit a variety of probes. The bar, as illustrated in Figure 4-20, permits the probe to be moved within a specified cubic measuring area, defined by X-Y-Z axes. The part to be measured is fixed to a granite surface plate, ready for probe orientation. When the probe contacts a point on the part, the point is defined by its relationship to the axes. Reader heads travelling on each axis transmit the point location data to a communications interface with a succession of point readings. The computer then calculates precise geometric and dimensional part information. These machines feature speed, ease, and accuracy, and a concomitant reduction of inspection time. CMM can be used with forgings, castings, stampings, and machined pieces for in-process and final inspection.

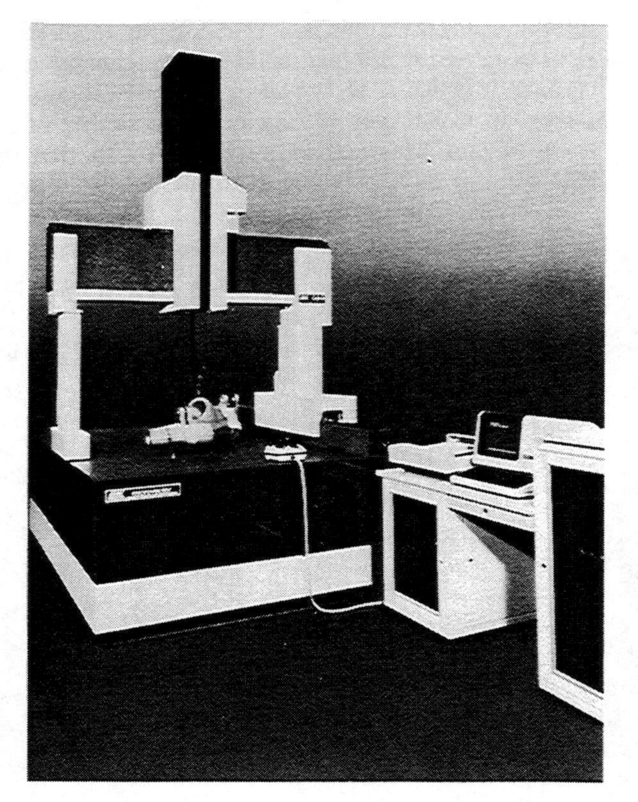

Figure 4-19 A coordinate measuring machine. (Courtesy of LK Tool U.S.A., Inc.)

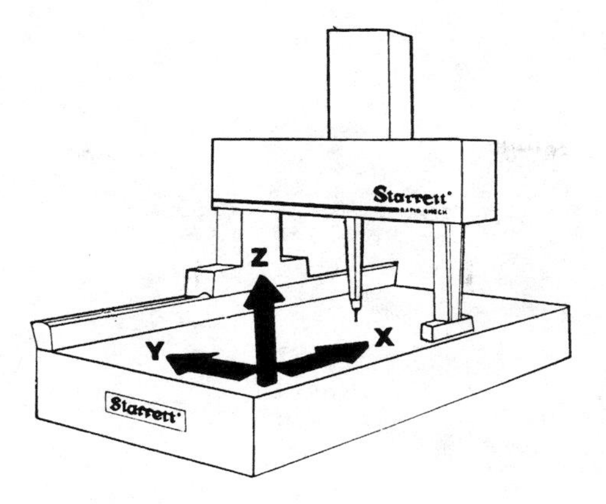

Figure 4-20 The CMM determines coordinate measures on the axes shown. (Courtesy of The L.S. Starrett Company)

Laser bench micrometer. This laser tool provides a high level of precision, noncontact measuring and gauging. See Figure 4-21.

When a part is placed in the V-block, the laser beam scans the item 120 times per second and measures the precise shadow cast by the object. The receiver detects the shadow's edges and sends the data to the microprocessor, where the dimension is calculated. The calculation is based on the precise timing of the shadow's edge positions and the scanning speed of the laser beam. This calculated dimension then appears on the digital display. See Figure 4-22.

When the gauge is first turned on and no measurements have been taken, all that appears in the display area is a group of large asterisks. As soon as measurements are taken, alphanumeric digits display the gauged dimension and statistical data. To the right of the digits is a group of display status indicators which inform the operator whether the gauge is displaying millimeters or inches, is set for single or continuous measurements, is displaying the dimension of the part being measured

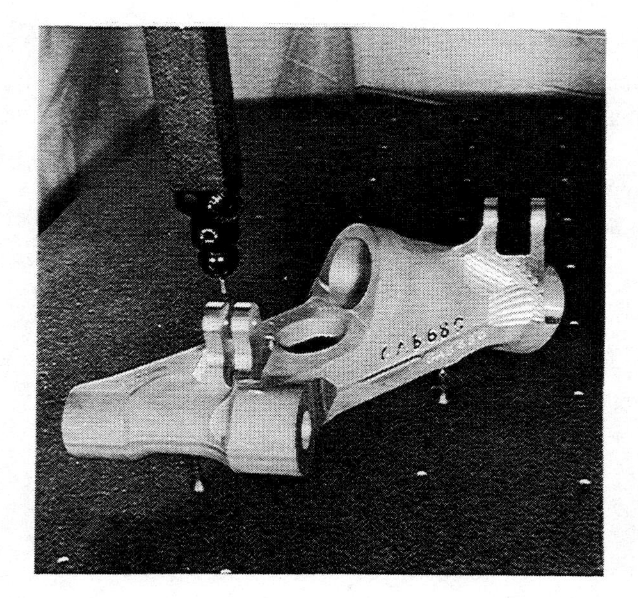

Figure 4-21 Part inspection using the coordinate measuring machine. (Courtesy of The L.S. Starrett Company)

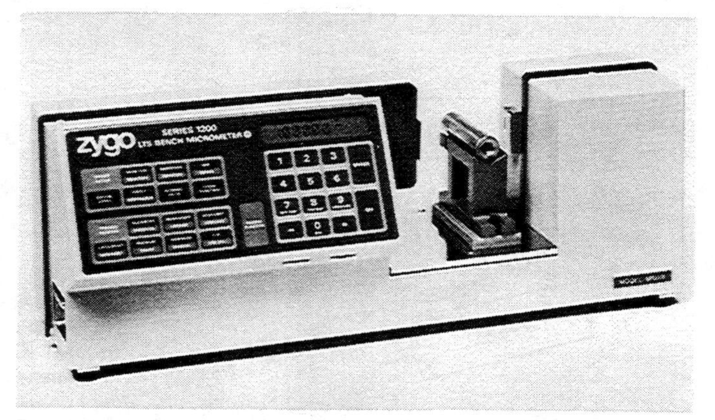

Figure 4-22 Bench laser micrometer. (Courtesy of Zygo Corporation)

or the measured deviation from a nominal value, and if the gauged dimension is within preset high and low limit dimensions.

The machine can be used to measure the in-process diameter of rolled rod, check the positions of holes and slots in razor blades, monitor part shapes, and perform off-line gauging. Some typical part measure arrangements are shown in Figures 4-23 and 4-24.

CONTROL AND QUALITY CONTROL

Control refers to the process employed in order to meet standards. This process consists of observing the actual performance, comparing the performance with some standard, and then taking action if the observed performance is significantly different from the standard.

The steps in a control process include:

1. Selecting what is to be controlled.
2. Setting a standard or goal for the control.
3. Measuring actual performance.
4. Identifying the difference between actual and standard.
5. Taking action to correct the difference.

Figure 4-23 Diagram of laser micrometer operation. (Courtesy of Zygo Corporation)

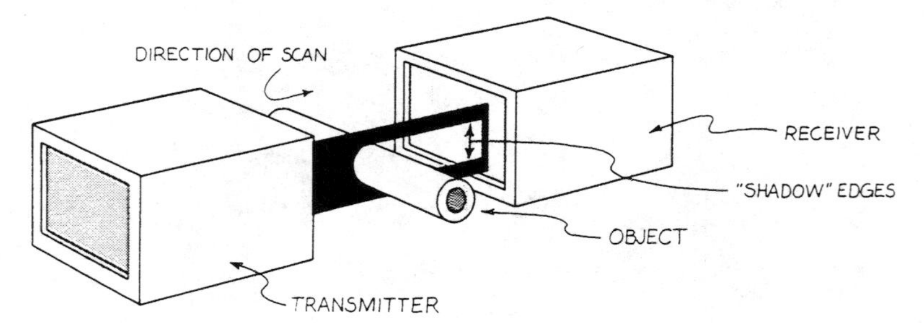

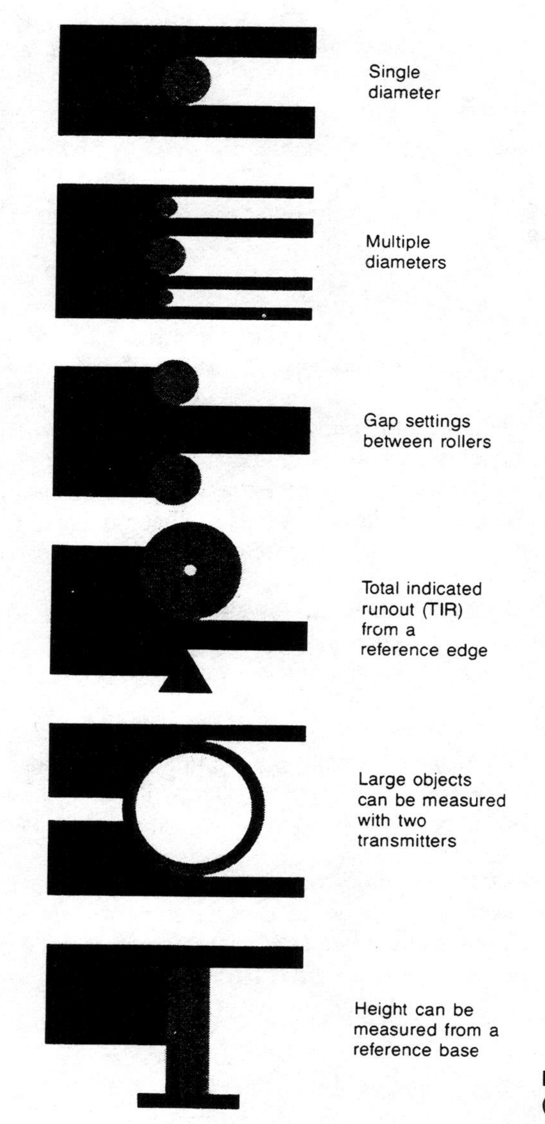

Figure 4-24 Typical laser micrometer setups. (Courtesy of Zygo Corporation)

QUALITY CONTROL

The term quality control is used in a very broad sense. It encompasses all those activities of a manufacturing enterprise that attempt to improve the conformance of parts to specifications, and to the review and revision of existing specifications to provide for a more realistic production goal. This activity has existed for many years, actually since the beginning of factories. It involves, among other things, the following activities:

1. An inspection plan to determine what inspections to make, how to inspect, and where to perform the inspections.

2. A gauging or testing policy to maintain control over gauges and testing equipment.
3. Scrap and salvage control to provide for disposal of substandard material.
4. Acceptance procedures for checking material delivered by outside or interdepartmental suppliers.
5. Process capability studies to ensure that the process is capable of meeting the desired specifications.

Quality control by statistical methods. When referring to quality control by statistical methods, or to statistical quality control, this implies those quality control techniques that specifically involve the use of statistical methods.

The term "statistics" has caused a great deal of consternation among students and practitioners in business and industry. It is generally associated with highly involved theories and mathematics. It should be emphasized that although the derivation of many statistical techniques may be extremely complicated and may involve advanced mathematics, the use of these methods has been reduced to very elementary arithmetic and may be substantiated by simple logical deduction.

Tools of statistical quality control. There are four fields to which statistical quality control techniques are generally applied. They are: (1) process capability; (2) process control; (3) acceptance techniques; and (4) special job studies.

Process capability. Every stable process has a capability which is defined as a six standard deviation (or six sigma) spread. This means that if the process is behaving normally, 99.73 percent of the product will have characteristics within plus or minus three standard deviations. This is what the process is capable of producing.

The standard deviation is an indicator of how values in a distribution differ on the average from the mean of the distribution. Calculation of the standard deviation involves determining how much a value deviates from the mean. The standard deviation, shown by the Greek letter sigma, is found as follows:

1. Calculate the average size or mean, called x-bar, of the measurements in the sample.
2. Calculate the standard deviation by

$$\sigma = \sqrt{\frac{\Sigma(x_i - \bar{x})^2}{n - 1}}$$

where σ = standard deviation
x_i = each measurement in the sample
$\bar{x}$ = average size or mean of all measurements in the sample
n = number of parts in each sample

Example If the length of six rods is 10.1, 11.0, 10.0, 10.5, 10.7, and 10.4 inches, the mean would be

$$\frac{10.1 + 11.0 + 10.0 + 10.5 + 10.7 + 10.4}{6} = 10.45$$

The standard deviation would be

$$\sqrt{\frac{(10.1 - 10.45)^2 + (11.0 - 10.45)^2 + \cdots + (10.4 - 10.45)^2}{6 - 1}} = .037$$

The standard deviation will show how far away a value in a distribution lies from the mean in relationship to the rest of the values. When referring to process capability the question is, "Is this process capable of meeting the product specifications?" The difference between the upper specification limit and the lower specification limit gives the tolerance range.

Some general rules for determining if a process is capable of meeting specifications with characteristics that have two-sided limits are:

1. If six standard deviations is less than the tolerance range and the process is well centered, the process is capable of meeting specifications.
2. If six standard deviations is equal to the tolerance range, the process is capable if the process mean is perfectly centered and never changes.
3. If six standard deviations is greater than the tolerance range, the process is not capable of meeting specifications.

Some companies may have their own rules to define process capability, but the preceding rules can be used as general guidelines in determining whether or not the process is capable of meeting the product specifications.

If the product characteristic is a one-sided specification, different rules for process capability apply. A simple approach is to add the mean to three standard deviations and compare this to the tolerance range:

1. If the mean plus three standard deviations is less than the tolerance range, then the process is capable.
2. If the mean plus three standard deviations equals the tolerance range, the process is capable if the mean does not increase.
3. If the mean plus three standard deviations is greater than the tolerance range, the process is not capable.

Process control. As material is processed, samples are taken and inspected. Based upon the results of these inspections, the process is either judged to be operating in a normal fashion or subject to unusual conditions. This decision is based upon the interpretation of a quality control chart. The control charts may be divided into two categories; those for variable data and those for attribute data.

Whenever the inspections involve a numerical classification of a quality characteristic, the data is said to involve variables. When a diameter is recorded in inches, hardness in Rockwell units, chemical contents in percentages, or an angle in degrees, the data are recorded as variables. If a part is simply judged good or bad, the data are recorded in terms of attributes. The material either is the proper hardness or it isn't, the chemical content is either acceptable or it isn't, the weld either exists or it doesn't, the finish is either smooth enough or it's rough. These classifications provide data in attribute form.

Control charts called $\overline{X}$–R charts are used for variable data. Charts for attribute data include p, np, c, and c charts.

Quality control charts provide their greatest benefit by detecting, at the earliest possible time, conditions which cause defective material. Thus a process may be corrected before it produces large quantities of borderline or defective material.

Acceptance sampling. Acceptance sampling techniques are designed to permit valid decisions on material to be based upon a relatively small sample. This permits companies to inspect a sample of the parts and decide, based on this evidence, whether or not the entire group of parts is satisfactory. Most of the acceptance sampling plans are based upon attribute inspection. The sampling plans which have the widest acceptance are the *Military Standard Sampling Procedures and Tables for Inspecting by Attributes*, or MIL-STD-105D.

Special job studies. Very often certain quality information is desired which is not available from the control charts and acceptance techniques in use. Special statistical techniques may be used to analyze the data collected. This may involve special applications of the control chart techniques or may involve other more advanced statistical techniques. Some of the more common studies include the review of existing specifications, causes for process variability, the investigation of alternative processes, and product reliability studies.

CONSTRUCTION OF $\bar{X}$–R CHARTS

No other tool of statistical quality control has received the recognition which has been accorded the Shewhart control chart for variables. This chart has found wide usage and continues to be used in nearly every process quality control application. Dr. Walter D. Shewhart of the Bell Telephone Laboratories has been generally recognized as the "father of statistical quality control."

The control chart for variables has been known popularly as the $\bar{X}$–R chart. As its name implies, it is used where the inspection data is collected in actual measurements. This type of chart may be used whenever a quality characteristic can be reduced to a numerical form. The field of application for this type of chart is extremely broad. The following list is included to illustrate a few of the applications:

1. Temperature of iron, furnace, or air
2. Strength tests on products
3. Volumes of containers
4. Weights of packaged goods
5. Moisture contents of lumber
6. Sizes of formed dimensions

Preliminary considerations. Before an $\bar{X}$–R chart can be put to use, certain decisions must be made. The success of the entire program may depend upon how carefully these decisions are reached.

1. What characteristic is to be studied? Very often it is somewhat difficult to determine just what characteristic should be analyzed by $\bar{X}$–R charts. For instance, in turning a shaft to a certain size, other characteristics may be important, such as

concentricity of the turned surface, the parallelism of the turned surface with some other surface, the finish of the surface, and the run-out of the outside diameter with regard to centers. The quality control person must determine which characteristics are important to the specific shaft and set up a separate control chart for each one to be studied.

2. How often should the checks be made? Depending upon the characteristic to be studied, the frequency of checks must be determined. Should a check be made every 15 minutes, every 4 hours, every 100 parts? How frequently is it necessary that checks be made? This problem basically becomes one of economics, balancing the cost of frequent checks against the losses which may occur if checks are not made often enough. No single formula can possibly be determined to analyze the conditions peculiar to each application. This decision must be based upon a knowledge of the process and with an eye to ultimate product quality and economy.

3. How are items to be checked selected? Whenever the inspector has a group of items from which to select those to be checked, a plan must be established to eliminate any bias. The sample items must be selected at random from the group. This means that each and every item should have the same chance of being selected. This often presents a problem which, however, may be overcome by proper planning.

4. How many observations should be made at each check? When a check is made, how many observations should be involved? How large should the sample be? From a practical standpoint, this problem has been simplified by commonly setting the sample size at four, five, or six, and then adjusting the frequency of check to obtain the desired control. Most $\bar{X}$–R charts are based on a sample size of five. Only rarely are other sample sizes used, even though the size of the group from which the sample was taken may vary greatly.

5. What gauges or equipment will be used to obtain the measurements? The gauging equipment or testing apparatus must be reviewed to insure that it accurately measures the characteristic desired. The user of the equipment must be certain that it is correctly calibrated and properly used. The decisions reached from an analysis of an $\bar{X}$–R chart can be no better than the data collected. If the data is subject to bias or to errors of measurement, the data is worse than useless—it is misleading. The success of the entire quality control program is dependent upon the care and consideration given the collection of the data.

$\bar{X}$–R data sheet. It is difficult to overemphasize the need for adequate information on the data sheet. The headings on the sample data sheet shown in Figure 4-25 describe the part, the operation performed, the location of the operation, the specification, and other pertinent inspection information. Careful study of the headings provides the information necessary to intelligently interpret the data. The data sheet does not indicate the frequency of sampling. However, by dividing the lot size by the pieces per hour, it will give an approximate inspection interval.

Referring to the data sheet, the first measurement in the first sample is 2.05. The average of the five values in the first sample is 2.08 and is shown in the $\bar{X}$ column.

The range of values in the first sample is from 2.05 to 2.11 or a total of .06. This value is inserted in the R column.

VARIABLES CONTROL CHART ($\bar{X}$ & R)

PART NO.	CHART NO.
165A	1

PART NAME (PRODUCT)	OPERATION (PROCESS)	SPECIFICATION LIMITS
FACE PLATE	10-BORE HOLES	2.08 ± 0.05

OPERATOR	MACHINE	GAGE	UNIT OF MEASURE	ZERO EQUALS
	1610	PLUG # 615	INCH	

		1	2	3	4	5	6	7	8	9	10
DATE											
TIME											
SAMPLE MEASUREMENTS	1	2.05	2.13	2.03	2.09	2.10	2.03	2.10	2.07	2.07	2.07
	2	2.08	2.08	2.12	2.07	2.06	2.13	2.06	2.07	2.09	2.06
	3	2.08	2.05	2.06	2.10	2.06	2.08	2.06	2.11	2.08	2.11
	4	2.11	2.09	2.09	2.09	2.07	2.06	2.08	2.07	2.09	2.09
	5	2.10	2.10	2.11	2.07	2.08	2.10	2.08	2.08	2.09	2.07
SUM		10.42	10.45	10.41	10.42	10.37	10.40	10.38	10.40	10.42	10.40
AVERAGE, $\bar{X}$		2.08	2.09	2.08	2.08	2.07	2.08	2.08	2.08	2.08	2.08
RANGE, R		.06	.08	.09	.03	.04	.10	.04	.04	.02	.05
NOTES											

AVERAGES

RANGES

Figure 4-25 $\bar{X}$-R Data.

VARIABLES CONTROL CHART ($\bar{X}$ & R)

PART NAME (PRODUCT) FACE PLATE	OPERATION (PROCESS) 10 - BORE HOLES	PART NO. 165A	CHART NO. 1
		SPECIFICATION LIMITS 2.08 ± 0.05	

OPERATOR	MACHINE 1610	GAGE PLUG # 615	UNIT OF MEASURE INCH	ZERO EQUALS

		1	2	3	4	5	6	7	8	9	10	11	12	13	14	15	16	17	18	19	20	21	22	23	24	25
DATE																										
TIME																										
SAMPLE MEASUREMENTS	1	2.05	2.13	2.03	2.09	2.10	2.03	2.10	2.07	2.07	2.07															
	2	2.08	2.08	2.12	2.07	2.06	2.13	2.06	2.07	2.09	2.06															
	3	2.08	2.05	2.06	2.10	2.06	2.08	2.06	2.11	2.08	2.11															
	4	2.11	2.09	2.09	2.09	2.07	2.06	2.08	2.07	2.09	2.09															
	5	2.10	2.10	2.11	2.07	2.08	2.10	2.08	2.08	2.09	2.07															
SUM		10.42	10.45	10.41	10.42	10.37	10.40	10.38	10.40	10.42	10.40															
AVERAGE, $\bar{X}$		2.08	2.09	2.08	2.08	2.07	2.08	2.08	2.08	2.08	2.08															
RANGE, R		.06	.08	.09	.03	.04	.10	.04	.04	.02	.05															
NOTES																										

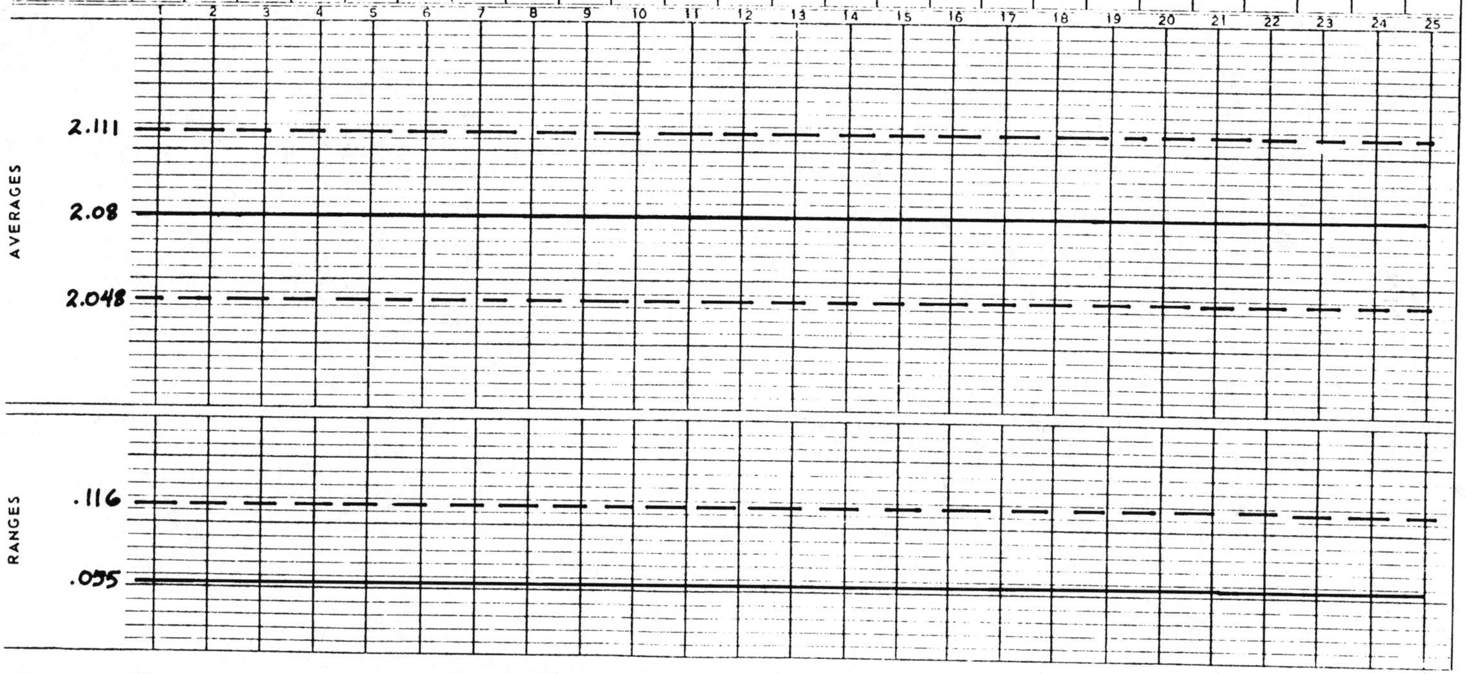

AVERAGES: 2.111 (dashed), 2.08 (solid), 2.048 (dashed)

RANGES: .116 (dashed), .055 (solid)

Figure 4-26 $\bar{X}$-R Control Chart.

The procedure for collecting samples and determining the average and range values for each sample is continued until the average, $\overline{X}$, and range, R, have been collected and recorded for all 20 samples.

$\overline{X}$–R chart. After the initial samples have been taken, the next step is to calculate center lines and control limits for $\overline{X}$ and R values.

The first center line of interest is the center line of the $\overline{X}$ values. This line represents the average of the sample averages or the grand average, $\overline{\overline{X}}$. The sum of the $\overline{X}$'s may be obtained from the data sheet. Next divide the sum of the $\overline{X}$'s by the number of samples to find the grand average. In the example shown in Figure 4-25, the grand average is

$$\overline{\overline{X}} = 2.08$$

The grand average should be plotted as a solid line, as indicated on the $\overline{X}$–R chart in Figure 4-26.

The average range, R, is found by adding together all of the R's from the samples and dividing this total by the number of R's in the total. Using the data from Figure 4-25, the average range is

$$R = 0.055$$

The average range should be plotted on the bottom half of the control chart as shown in Figure 4-26.

The following formulas show how to calculate the control limits for averages:

$$\text{Upper Control Limit for Averages } UCL_{\bar{x}} = \overline{X} + A_2R$$

$$\text{Lower Control Limit for Averages, } LCL_{\bar{x}} = \overline{X} - A_2R$$

The control limit formulas are based on plus or minus three standard deviations and use the central line equal to the average of the data used in calculating the control limits. Values for the A factors used in the formulas are given in Table 4-5. The $UCL_{\bar{x}}$ and $LCL_{\bar{x}}$ should be shown on the $\overline{X}$ chart as broken lines extending through the length of the chart.

TABLE 4-5
FACTORS FOR $\overline{X}$ AND R CONTROL CHARTS

Number of Observations in Sample	Factors	
	A_2	D_4
2	1.880	3.268
3	1.023	2.574
4	0.729	2.282
5	0.577	2.114
6	0.483	2.004

The following formula shows how to calculate the control limit for the range:

$$\text{Upper Control Limit for Ranges, } UCL_R = D_4R$$

The factor for D_4 can be found in Table 4-1. When the sample size is less than seven, the lower control limit is zero. When this is the case, the LCL_R is usually omitted from the chart. After the UCL_R has been calculated, it should be plotted on the R chart as a broken line extending through the length of the chart.

Interpretation of $\bar{X}$–R charts. After the control charts have been prepared and the control limits established, the $\bar{X}$ and R values for each of the new samples are plotted on the charts.

A certain amount of variation exists in the human, machine, and material at all times. This produces the variation which is due to inherent variability. The control limits for both the $\bar{X}$ and R are located so that they will encompass nearly all the variation due to inherent causes. Therefore, if a plotting on either the $\bar{X}$ chart or on the R chart falls outside the control limits, an assignable cause exists and may be located by proper investigation. Occasionally, about three times out of a thousand, a plotting subject only to inherent variation will fall outside the limits. This occurs so infrequently that any plotting outside the limits should be adequate reason for an investigation.

CONSTRUCTION OF THE *p*-CHART

The most popular and widely used of all attribute charts is the p-chart. This chart has been referred to as either a fraction defective or percent defective chart. The number of defective items in the sample is divided by the sample size and multiplied by 100 to find the percent defective.

Data sheet for the *p*-chart. In addition to the information necessary to completely identify the product, the operation, and characteristics to be checked, the data sheet should provide a suitable means for recording the inspection information and transforming it into usable form. A sample data sheet to record p-chart information is shown in Figure 4-27.

***p*-Chart.** The vertical scale should be calibrated to permit the plotting of all the information which might be anticipated. A plotting should be made on the chart to correspond with the inspection results for each sample. The plots are commonly joined by straight lines to avoid the loss of any data.

The first step in constructing the p-chart is to calculate the center line or average line. This line is labeled p and is the average fraction or percent defective.

The average fraction defective may be calculated by the following formula:

$$p = \frac{\text{Total number of defective parts observed}}{\text{Total number of parts inspected}}$$

After the center line has been calculated, the control limits should be determined. The formulas for the fraction defective are

$$\text{Upper Control Limit, } UCL_p = p + 3\sqrt{\frac{p(1-p)}{n}}$$

$$\text{Lower Control Limit, } LCL_p = p - 3\sqrt{\frac{p(1-p)}{n}}$$

average percent of defective parts

P DATA SHEET

Plant

Department No.	67	Part Name Stud	Part No. O-115
Machine No.	1966	Oper. No. & Description #10 Turn Thread one end + cut Off	
Pcs. per Hr.	300		
Subgroup Size	300		
Sample Size	50		

			Reasons for Reject					
NO.	Number Defective	Fraction Defective	P.D. Oversize	P.D. Undersize	Stud length	Thread length	1/16" × 45° chamfer	Remarks
1	6		6				2	No chamfer
2	7			7		7		Thread length US
3	4			4				
4	3		2				1	No chamfer
5	3			3		3		Thread length US
6	6			6				
7	12		2	10	4			Stud length OS
8	13			13			2	No chamfer
9	2			2				
10	12		10	2			4	No chamfer
11	0							
12	2			2				
13	6			6				
14	0							
15	7			7		4		Thread length US
16	13			13	4			Stud length OS
17	6			6				
18	8			8				
19	12		12					
20	8		8					
21	11			11				
22	2			2				
23	6			6				
24	4			4				
25	5			5				

Figure 4-27 p Chart Data Sheet.

For the sample information recorded on the data sheet in Figure 4-27, the average fraction defective is

$$p = \frac{158}{1250} = .1264$$

and the control limits are

$$\text{UCL}_p = .1264 + 3\sqrt{\frac{(.1264)(1 - .1264)}{1250}} = .1546$$

$$\text{LCL}_p = .1264 - 3\sqrt{\frac{(.1264)(1 - .1264)}{1250}} = .0982$$

P CHART

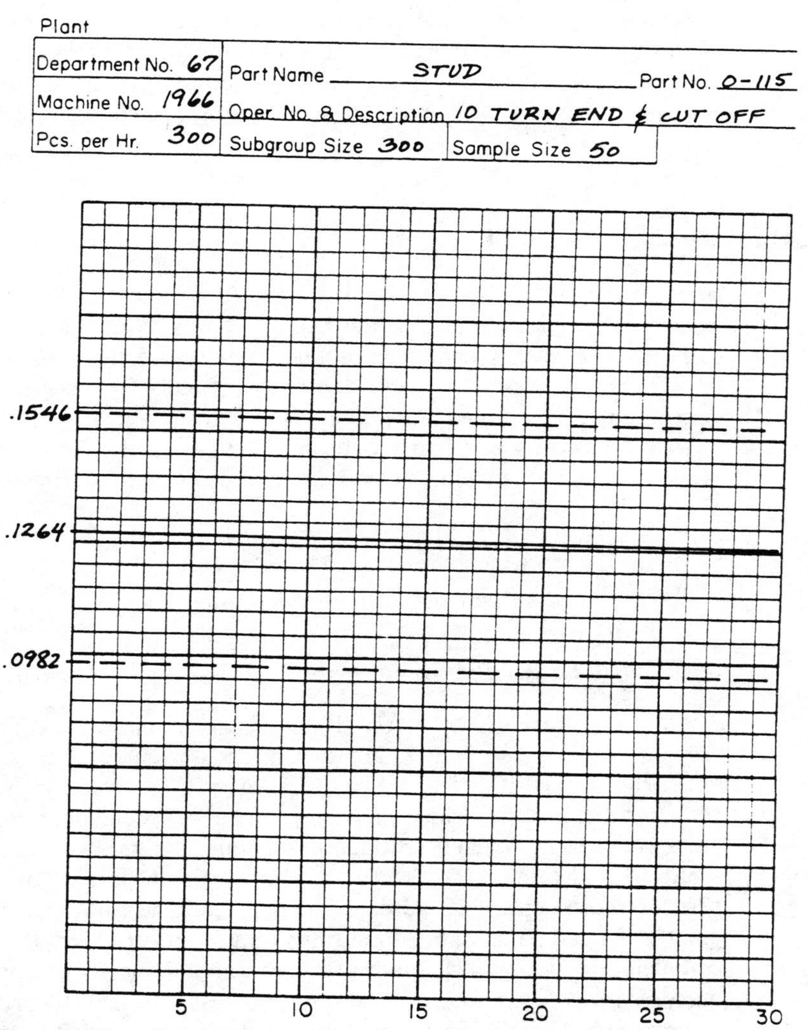

Figure 4-28 p Control Chart.

The average fraction defective and control limits are shown on the *p*-chart in Figure 4-28.

Interpretation of the *p*-chart. There are two questions which require an answer when a *p*-chart is analyzed. Is the process in control? Is the process satisfactory?

If all the plots are within the control limits and dispersed about the center line, the process is said to be in a state of control. In more general terms, the variation among the plots is such that it could be expected to occur if all the plots were based on samples taken from the same lot of material. The variation is apparently due only to sampling fluctuations. There is no indication of a process change.

If the process is in control, the average percent defective may be considered the expected quality level of the process. It has been producing material of a certain percent defective in the past, and unless some process change is made it can be expected to continue at approximately the same level. No simple rule exists for setting an acceptable quality level. The level must be selected on the basis of knowledge concerning the function of the part, assembly difficulties, further processing, and the possibility of a defective part actually getting into an assembly or finished part.

ACCEPTANCE SAMPLING

Acceptance sampling is used when material is presented for use and it must either be accepted or rejected, depending upon its quality characteristics. This is the type of problem generally encountered when material is received from an outside vendor or from some preceding fabrication or processing department, for which a certified control chart concerning the process is not available.

In the past this problem has been met in a number of different ways. Some of the more obvious advantages and shortcomings of the following techniques will be discussed:

1. No testing whatsoever.
2. One hundred percent sorting or detailing.
3. Spot-checking.
4. Scientific sampling.

Occasionally parts or materials are received into a manufacturing plant which are not checked but, rather, are sent directly to storage, processing, or assembly. When this occurs, it may result in increased costs for reworking the defective material, for substitute material which must be ordered for special delivery, and for the possible costs which arise when a production schedule is retarded.

At the other extreme, some manufacturers inspect all incoming material 100 percent. Sometimes 100 percent checking is necessary, particularly where life or great risks are involved. More often, however, it is exorbitantly expensive and conducive to sloppy inspection practices. The efficiency of any checking technique decreases as the number of checks is increased. When 100 percent of the product is inspected, the individuals performing the checking are subject to inspection fatigue. Experience has shown that their efforts are rarely better than 90 percent efficient.

In an attempt to arrive at a compromise between no checking and 100 percent checking, some people have resorted to spotchecking random shipments. This, of course, will stop some defective material from entering production, but can only be partially effective, since many of the lots are accepted with no check.

Scientific sampling plans possess many of the same characteristics of the arbitrary sampling plans, in that they provide a compromise between 100 percent inspection and no check. They have the added virtue of distinguishing between critical and less critical characteristics, the risks of making a wrong decision are known, and all material of the same classification is subjected to a test of similar discriminatory power (the large lots are not discriminated against, nor are the small lots under-inspected).

The most frequently used acceptance sampling plans employ MIL-STD-105D tables when maximum acceptance is desired at a predetermined quality level (AQL). These tables have provisions for classifying defects as critical, major, or minor and using different AQL values for each classification. The following will summarize the information required for their use:

1. Select the AQL to be used from the preferred values given. AQLs less than or equal to 10.0 may be expressed either in percent defective or defects per hundred units. Those greater than 10.0 are expressed only in defects per hundred units.

2. Determine the lot size. Refer to Table 4-6 to find sample size code letter. Generally, inspection level II is usually specified. Level I is less discriminating, and level III is more discriminating.

3. Select the appropriate sampling plan.
   a. Use Table 4-7 for single sample.
   b. Separate tables are required for double and multiple sampling.

The preceding sampling plans are referred to as normal inspection plans. MIL-STD-105D has provisions for switching from normal to tightened inspection, from tightened to normal inspection, from normal to reduced inspection, and from reduced to normal inspection. Tightened inspection essentially reduces the sampling risks when the process average is at an unacceptable level. Reduced inspection decreases the number in the sample when the process average is consistently better than the desired AQL.

TABLE 4-6
MIL-STD-105D SAMPLE SIZE CODE LETTERS

Lot or Batch Size			Special Inspection Levels				General Inspection Levels		
			S-1	S-2	S-3	S-4	I	II	III
2	to	8	A	A	A	A	A	A	B
9	to	15	A	A	A	A	A	B	C
16	to	25	A	A	B	B	B	C	D
26	to	50	A	B	B	C	C	D	E
51	to	90	B	B	C	C	C	E	F
91	to	150	B	B	C	D	D	F	G
151	to	280	B	C	D	E	E	G	H
281	to	500	B	C	D	E	F	H	J
501	to	1200	C	C	E	F	G	J	K
1201	to	3200	C	D	E	G	H	K	L
3201	to	10000	C	D	F	G	J	L	M
10001	to	35000	C	D	F	H	K	M	N
35001	to	150000	D	E	G	J	L	N	P
150001	to	500000	D	E	G	J	M	P	Q
500001	and	over	D	E	H	K	N	Q	R

TABLE 4-7
MIL-STD-105D *SINGLE SAMPLING PLANS FOR* NORMAL *INSPECTION (MASTER TABLE)*

Sample size code letter	Sample size	Acceptable Quality Levels (normal inspection)																									
		0.010	0.015	0.025	0.040	0.065	0.10	0.15	0.25	0.40	0.65	1.0	1.5	2.5	4.0	6.5	10	15	25	40	65	100	150	250	400	650	1000
		Ac Re	Ac Re	Ac Re	Ac Re	Ac Re	Ac Re	Ac Re	Ac Re	Ac Re	Ac Re	Ac Re	Ac Re	Ac Re	Ac Re	Ac Re	Ac Re	Ac Re	Ac Re	Ac Re	Ac Re	Ac Re	Ac Re	Ac Re	Ac Re	Ac Re	Ac Re
A	2	↓	↓	↓	↓	↓	↓	↓	↓	↓	↓	↓	↓	↓	↓	0 1	↓	↓	1 2	2 3	3 4	5 6	7 8	10 11	14 15	21 22	30 31
B	3	↓	↓	↓	↓	↓	↓	↓	↓	↓	↓	↓	↓	↓	0 1	↑	↓	1 2	2 3	3 4	5 6	7 8	10 11	14 15	21 22	30 31	44 45
C	5	↓	↓	↓	↓	↓	↓	↓	↓	↓	↓	↓	↓	0 1	↑	↓	1 2	2 3	3 4	5 6	7 8	10 11	14 15	21 22	30 31	44 45	↑
D	8	↓	↓	↓	↓	↓	↓	↓	↓	↓	↓	↓	0 1	↑	↓	1 2	2 3	3 4	5 6	7 8	10 11	14 15	21 22	30 31	44 45	↑	↑
E	13	↓	↓	↓	↓	↓	↓	↓	↓	↓	↓	0 1	↑	↓	1 2	2 3	3 4	5 6	7 8	10 11	14 15	21 22	30 31	44 45	↑	↑	↑
F	20	↓	↓	↓	↓	↓	↓	↓	↓	↓	0 1	↑	↓	1 2	2 3	3 4	5 6	7 8	10 11	14 15	21 22	↑	↑	↑	↑	↑	↑
G	32	↓	↓	↓	↓	↓	↓	↓	↓	0 1	↑	↓	1 2	2 3	3 4	5 6	7 8	10 11	14 15	21 22	↑	↑	↑	↑	↑	↑	↑
H	50	↓	↓	↓	↓	↓	↓	↓	0 1	↑	↓	1 2	2 3	3 4	5 6	7 8	10 11	14 15	21 22	↑	↑	↑	↑	↑	↑	↑	↑
J	80	↓	↓	↓	↓	↓	↓	0 1	↑	↓	1 2	2 3	3 4	5 6	7 8	10 11	14 15	21 22	↑	↑	↑	↑	↑	↑	↑	↑	↑
K	125	↓	↓	↓	↓	↓	0 1	↑	↓	1 2	2 3	3 4	5 6	7 8	10 11	14 15	21 22	↑	↑	↑	↑	↑	↑	↑	↑	↑	↑
L	200	↓	↓	↓	↓	0 1	↑	↓	1 2	2 3	3 4	5 6	7 8	10 11	14 15	21 22	↑	↑	↑	↑	↑	↑	↑	↑	↑	↑	↑
M	315	↓	↓	↓	0 1	↑	↓	1 2	2 3	3 4	5 6	7 8	10 11	14 15	21 22	↑	↑	↑	↑	↑	↑	↑	↑	↑	↑	↑	↑
N	500	↓	↓	0 1	↑	↓	1 2	2 3	3 4	5 6	7 8	10 11	14 15	21 22	↑	↑	↑	↑	↑	↑	↑	↑	↑	↑	↑	↑	↑
P	800	↓	0 1	↑	↓	1 2	2 3	3 4	5 6	7 8	10 11	14 15	21 22	↑	↑	↑	↑	↑	↑	↑	↑	↑	↑	↑	↑	↑	↑
Q	1250	0 1	↑	↓	1 2	2 3	3 4	5 6	7 8	10 11	14 15	21 22	↑	↑	↑	↑	↑	↑	↑	↑	↑	↑	↑	↑	↑	↑	↑
R	2000	↑	↑	1 2	2 3	3 4	5 6	7 8	10 11	14 15	21 22	↑	↑	↑	↑	↑	↑	↑	↑	↑	↑	↑	↑	↑	↑	↑	↑

↓ = Use first sampling plan below arrow. If sample size equals, or exceeds, lot or batch size, do 100 percent inspection.

↑ = Use first sampling plan above arrow.

Ac = Acceptance number.
Re = Rejection number.

TABLE 4-8
MIL-STD-105D *SINGLE SAMPLING PLANS FOR TIGHTENED INSPECTION (MASTER TABLE)*

Sample size code letter	Sample size	Acceptable Quality Levels (tightened inspection)																									
		0.010	0.015	0.025	0.040	0.065	0.10	0.15	0.25	0.40	0.65	1.0	1.5	2.5	4.0	6.5	10	15	25	40	65	100	150	250	400	650	1000
		Ac Re	Ac Re	Ac Re	Ac Re	Ac Re	Ac Re	Ac Re	Ac Re	Ac Re	Ac Re	Ac Re	Ac Re	Ac Re	Ac Re	Ac Re	Ac Re	Ac Re	Ac Re	Ac Re	Ac Re	Ac Re	Ac Re	Ac Re	Ac Re	Ac Re	Ac Re
A	2	↓	↓	↓	↓	↓	↓	↓	↓	↓	↓	↓	↓	↓	↓	↓	↓	↓	↓	1 2	2 3	3 4	5 6	8 9	12 13	18 19	27 28
B	3	↓	↓	↓	↓	↓	↓	↓	↓	↓	↓	↓	↓	↓	↓	0 1	↓	↓	1 2	2 3	3 4	5 6	8 9	12 13	18 19	27 28	41 42
C	5	↓	↓	↓	↓	↓	↓	↓	↓	↓	↓	↓	↓	↓	0 1	↓	↓	1 2	2 3	3 4	5 6	8 9	12 13	18 19	27 28	41 42	↑
D	8	↓	↓	↓	↓	↓	↓	↓	↓	↓	↓	↓	↓	0 1	↓	↓	1 2	2 3	3 4	5 6	8 9	12 13	18 19	27 28	41 42	↑	↑
E	13	↓	↓	↓	↓	↓	↓	↓	↓	↓	↓	↓	0 1	↓	↓	1 2	2 3	3 4	5 6	8 9	12 13	18 19	27 28	41 42	↑	↑	↑
F	20	↓	↓	↓	↓	↓	↓	↓	↓	↓	↓	0 1	↓	↓	1 2	2 3	3 4	5 6	8 9	12 13	18 19	↑	↑	↑	↑	↑	↑
G	32	↓	↓	↓	↓	↓	↓	↓	↓	↓	0 1	↓	↓	1 2	2 3	3 4	5 6	8 9	12 13	18 19	↑	↑	↑	↑	↑	↑	↑
H	50	↓	↓	↓	↓	↓	↓	↓	↓	0 1	↓	↓	1 2	2 3	3 4	5 6	8 9	12 13	18 19	↑	↑	↑	↑	↑	↑	↑	↑
J	80	↓	↓	↓	↓	↓	↓	↓	0 1	↓	↓	1 2	2 3	3 4	5 6	8 9	12 13	18 19	↑	↑	↑	↑	↑	↑	↑	↑	↑
K	125	↓	↓	↓	↓	↓	↓	0 1	↓	↓	1 2	2 3	3 4	5 6	8 9	12 13	18 19	↑	↑	↑	↑	↑	↑	↑	↑	↑	↑
L	200	↓	↓	↓	↓	↓	0 1	↓	↓	1 2	2 3	3 4	5 6	8 9	12 13	18 19	↑	↑	↑	↑	↑	↑	↑	↑	↑	↑	↑
M	315	↓	↓	↓	↓	0 1	↓	↓	1 2	2 3	3 4	5 6	8 9	12 13	18 19	↑	↑	↑	↑	↑	↑	↑	↑	↑	↑	↑	↑
N	500	↓	↓	↓	0 1	↓	↓	1 2	2 3	3 4	5 6	8 9	12 13	18 19	↑	↑	↑	↑	↑	↑	↑	↑	↑	↑	↑	↑	↑
P	800	↓	↓	0 1	↓	↓	1 2	2 3	3 4	5 6	8 9	12 13	18 19	↑	↑	↑	↑	↑	↑	↑	↑	↑	↑	↑	↑	↑	↑
Q	1250	↓	0 1	↓	↓	1 2	2 3	3 4	5 6	8 9	12 13	18 19	↑	↑	↑	↑	↑	↑	↑	↑	↑	↑	↑	↑	↑	↑	↑
R	2000	0 1	↑	↓	1 2	2 3	3 4	5 6	8 9	12 13	18 19	↑	↑	↑	↑	↑	↑	↑	↑	↑	↑	↑	↑	↑	↑	↑	↑
S	3150			1 2																							

↓ = Use first sampling plan below arrow. If sample size equals or exceeds lot or batch size, do 100 percent inspection.

↑ = Use first sampling plan above arrow.

Ac = Acceptance number.
Re = Rejection number.

TABLE 4-9
MIL-STD-105D *SINGLE SAMPLING PLANS FOR REDUCED INSPECTION (MASTER TABLE)*

Sample size code letter	Sample size	Acceptable Quality Levels (reduced inspection)†																									
		0.010	0.015	0.025	0.040	0.065	0.10	0.15	0.25	0.40	0.65	1.0	1.5	2.5	4.0	6.5	10	15	25	40	65	100	150	250	400	650	1000
		Ac Re	Ac Re	Ac Re	Ac Re	Ac Re	Ac Re	Ac Re	Ac Re	Ac Re	Ac Re	Ac Re	Ac Re	Ac Re	Ac Re	Ac Re	Ac Re	Ac Re	Ac Re	Ac Re	Ac Re	Ac Re	Ac Re	Ac Re	Ac Re	Ac Re	Ac Re
A	2	⇩	⇩	⇩	⇩	⇩	⇩	⇩	⇩	⇩	⇩	⇩	⇩	⇩	⇩	0 1	⇩	⇩	1 2	2 3	3 4	5 6	7 8	10 11	14 15	21 22	30 31
B	2	⇩	⇩	⇩	⇩	⇩	⇩	⇩	⇩	⇩	⇩	⇩	⇩	⇩	0 1	⇧	⇩	0 2	1 3	2 4	3 5	5 6	7 8	10 11	14 15	21 22	30 31
C	2	⇩	⇩	⇩	⇩	⇩	⇩	⇩	⇩	⇩	⇩	⇩	⇩	0 1	⇧	⇩	0 2	1 3	1 4	2 5	3 6	5 8	7 10	10 13	14 17	21 24	⇧
D	3	⇩	⇩	⇩	⇩	⇩	⇩	⇩	⇩	⇩	⇩	⇩	0 1	⇧	⇩	0 2	1 3	1 4	2 5	3 6	5 8	7 10	10 13	14 17	21 24	⇧	⇧
E	5	⇩	⇩	⇩	⇩	⇩	⇩	⇩	⇩	⇩	⇩	0 1	⇧	⇩	0 2	1 3	1 4	2 5	3 6	5 8	7 10	10 13	14 17	21 24	⇧	⇧	⇧
F	8	⇩	⇩	⇩	⇩	⇩	⇩	⇩	⇩	⇩	0 1	⇧	⇩	0 2	1 3	1 4	2 5	3 6	5 8	7 10	10 13	⇧	⇧	⇧	⇧	⇧	⇧
G	13	⇩	⇩	⇩	⇩	⇩	⇩	⇩	⇩	0 1	⇧	⇩	0 2	1 3	1 4	2 5	3 6	5 8	7 10	10 13	⇧	⇧	⇧	⇧	⇧	⇧	⇧
H	20	⇩	⇩	⇩	⇩	⇩	⇩	⇩	0 1	⇧	⇩	0 2	1 3	1 4	2 5	3 6	5 8	7 10	10 13	⇧	⇧	⇧	⇧	⇧	⇧	⇧	⇧
J	32	⇩	⇩	⇩	⇩	⇩	⇩	0 1	⇧	⇩	0 2	1 3	1 4	2 5	3 6	5 8	7 10	10 13	⇧	⇧	⇧	⇧	⇧	⇧	⇧	⇧	⇧
K	50	⇩	⇩	⇩	⇩	⇩	0 1	⇧	⇩	0 2	1 3	1 4	2 5	3 6	5 8	7 10	10 13	⇧	⇧	⇧	⇧	⇧	⇧	⇧	⇧	⇧	⇧
L	80	⇩	⇩	⇩	⇩	0 1	⇧	⇩	0 2	1 3	1 4	2 5	3 6	5 8	7 10	10 13	⇧	⇧	⇧	⇧	⇧	⇧	⇧	⇧	⇧	⇧	⇧
M	125	⇩	⇩	⇩	0 1	⇧	⇩	0 2	1 3	1 4	2 5	3 6	5 8	7 10	10 13	⇧	⇧	⇧	⇧	⇧	⇧	⇧	⇧	⇧	⇧	⇧	⇧
N	200	⇩	⇩	0 1	⇧	⇩	0 2	1 3	1 4	2 5	3 6	5 8	7 10	10 13	⇧	⇧	⇧	⇧	⇧	⇧	⇧	⇧	⇧	⇧	⇧	⇧	⇧
P	315	⇩	0 1	⇧	⇩	0 2	1 3	1 4	2 5	3 6	5 8	7 10	10 13	⇧	⇧	⇧	⇧	⇧	⇧	⇧	⇧	⇧	⇧	⇧	⇧	⇧	⇧
Q	500	0 1	⇧	⇩	0 2	1 3	1 4	2 5	3 6	5 8	7 10	10 13	⇧	⇧	⇧	⇧	⇧	⇧	⇧	⇧	⇧	⇧	⇧	⇧	⇧	⇧	⇧
R	800	⇧	⇧	0 2	1 3	1 4	2 5	3 6	5 8	7 10	10 13	⇧	⇧	⇧	⇧	⇧	⇧	⇧	⇧	⇧	⇧	⇧	⇧	⇧	⇧	⇧	⇧

⇩ = Use first sampling plan below arrow. If sample size equals or exceeds lot or batch size, do 100 percent inspection.

⇧ = Use first sampling plan above arrow.

Ac = Acceptance number.
Re = Rejection number.
† = If the acceptance number has been exceeded, but the rejection number has not been reached, accept the lot, but reinstate normal inspection.

Example For a production lot size or 2,000 parts, inspection level II, and a desired AQL of 1.0 percent:

1. Use Table 4-6 to find the lot size code = K.
2. From Table 4-7 for single sample, normal inspection, the sample size is 125. Accept the lot on three or less defects and reject on four or more.
3. From Table 4-8 with tightened inspection the sample size is 125. Accept the lot with two or less defects and reject the lot on three or more defects.
4. From Table 4-9 with reduced inspection the sample size is 50. Accept when the lot has one or no defects and reject if the lot has four or more defects. If two or three defectives are found, the lot is accepted, but the type of inspection changes from reduced to normal. A change to normal inspection is also required when a lot is rejected.

If a vertical arrow is encountered, the first sampling plan above or below the arrow is used. When this occurs, the sample-size code letter and the sample size change. For example, if a single sample normal plan (Table 4-7) is used with AQL of 0.25 and a code letter of E, the code letter changes to H and the sample size changes from 13 to 50.

QUESTIONS

1. What is the difference between a steel rule and a caliper?
2. Describe three measurements that can be made with a dial indicator.
3. What is a Coordinate Measuring Machine?
4. List five activities of Quality Control. Give an example of how each of these functions may be used in the manufacture of wood furniture.
5. The measurements from five parts are 3.1, 3.0, 2.9, 3.0, and 3.2 cm. Calculate the mean and standard deviation of this sample.
6. What is the difference between an $\overline{X}$–R chart and a p-chart?
7. Take a sample of 50 parts and construct an $\overline{X}$–R chart including the Upper and Lower control limits.
8. Inspection of 500 parts indicated that there were 40 defective parts in the lot. What are the Upper and Lower control limits for this part?
9. The Inspection Department wants to sample a lot of 1,500 parts. The inspection level is II and the desired AQL is 1.5. How many parts should be inspected?
10. If the sample of parts inspected in question nine had six bad parts, should the lot be accepted or rejected?

Columbus Casting

part 2 MATERIALS OF MANUFACTURING

5 PROPERTIES OF MATERIALS

When materials are selected for use in a manufacturing process, there are many properties which may be important. Strength, which is the ability to support a load without deforming or breaking, is often a significant consideration. Other important properties may include hardness and toughness. Hardness is related to the ability to withstand wear, and toughness is the ability to withstand impact.

Materials must be selected so that their properties are suited for their particular application. Structural members in buildings are required to maintain nearly the same shape at all times. On the other hand, springs in automobile suspensions are expected to tolerate large deformations and then to return to their original shape. Sometimes parts are selected intentionally so that they will fail. Rupture disks in pressure vessels or shear pins in some machinery serve as inexpensive, easily replaceable weakest links whose failure protects against much more expensive accidents.

A designer might also consider thermal stability or corrosion resistance. Those properties which are the most critical for a particular application, along with considerations such as weight, volume, cost, and availability determine which material is selected.

It is important to recognize that the same properties that make a material durable in service may also make it difficult to process or shape into its final form. It is often desirable to have a material which is soft during the forming operations and then to harden it afterwards. This situation is possible with some metal alloys and with some polymers. In this chapter, we will examine the types of materials that are available and show why their properties are different.

PHYSICAL PROPERTIES

Physical properties are those which can be measured without changing the chemical nature of the material.

Melting and boiling points. For example, the melting or boiling point of a material is considered a physical property because the molecules do not change form during melting or boiling. In the three phases, gas, liquid, or solid, the molecules are all the same, differing only in their degree of freedom and separation from each other. Cooling a liquid below its freezing point will cause it to solidify and return to its solid form. Although we usually refer to the melting point of a solid and the freezing point of a liquid, these two terms are simply different names for the same point—that temperature at which it is possible for liquid and solid forms to coexist. Table 5-1 shows the wide range of melting and boiling points for some metals.

TABLE 5-1
MELTING AND BOILING POINTS OF SOME METALS

Metal	Melting Point (°C)	Boiling Point (°C)
Mercury	−38.9	356.6
Tin	231.9	2,270
Lead	327.4	1,620
Aluminum	660	2,057
Silver	961	1,950
Copper	1,083	2,336
Iron	1,535	3,000
Tungsten	3,370	5,900

Generally, the higher the melting point of a substance, the better it will retain its strength in high temperature service environments. Figure 5-1 illustrates the maximum use temperatures for a variety of available materials. However, there are

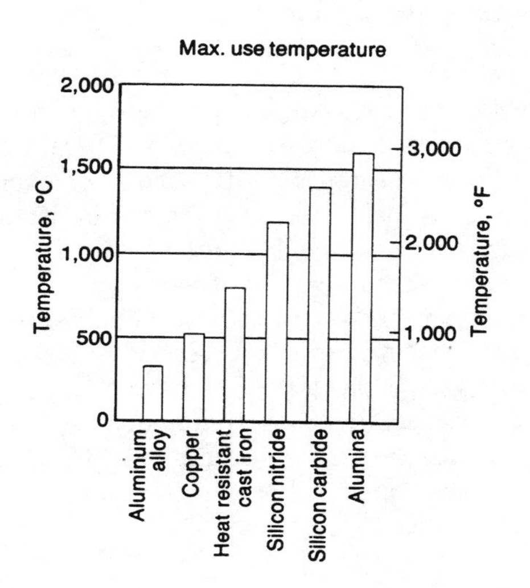

Figure 5-1 Maximum practical use temperatures for selected engineering materials. (From *Guide to Engineered Materials*, ASM International [formerly American Society for Metals], 1986. Reprinted with permission.)

certainly situations in which low melting materials are desirable. For example, solders and fuse materials used in electrical circuits and in heat-activated automatic sprinkler systems are chosen to melt predictably at selected temperatures.

Thermal expansion. Another physical property that may be important in some applications is thermal expansion. Generally, solids expand when they are heated. The thermal expansion coefficient, often designated by the Greek letter alpha, α, is used in the following relationship.

$$\delta L = \alpha L \delta T$$

In this equation, δL represents the change in length, L, that occurs when a substance is heated through a temperature change, δT. Larger values of α imply greater dimensional changes for the same change in temperature. Representative values of thermal expansion coefficients for a few commercial materials are listed in Table 5-2.

TABLE 5-2
THERMAL EXPANSION COEFFICIENTS

Material	$\alpha(°C^{-1})$
Aluminum	24.1×10^{-6}
Brass	19×10^{-6}
Copper	$14 - 18 \times 10^{-6}$
Steel	$11 - 13 \times 10^{-6}$
Invar (Nickel Steel)	0.9×10^{-6}
Cement and Concrete	$10 - 14 \times 10^{-6}$
Glass	7.2×10^{-6}
Fused Quartz	0.05×10^{-6}
Low-Density Polyethylene	180×10^{-6}
Natural Rubber	670×10^{-6}
Silicone Rubber	1200×10^{-6}

Source: Flinn & Trojan, *Engineering Materials and Their Applications*, 2nd ed., Houghton Mifflin Co., Boston, 1981.

The wide range of thermal expansion coefficients is readily apparent. Some polymers expand nearly 1,000 times as much as do metals. There are a few materials that exhibit very low thermal expansion. Fused quartz, for example, is used in high temperature applications in which other glass materials would break because of differences in thermal expansion caused by inhomogeneous temperatures. The steel alloy, Invar, is used to manufacture very high sensitivity gauges whose dimensions can be relied upon to remain nearly constant even if environmental temperatures change. Figure 5-2 compares the thermal expansion rates of several different materials.

Occasionally, materials are selected because they have a large thermal expansion coefficient. For example, the bimetallic strips commonly used in thermostats are made of two metals whose thermal expansion coefficients are very different. Thus small changes in temperature cause large changes in shape, making the device very sensitive to temperature fluctuations.

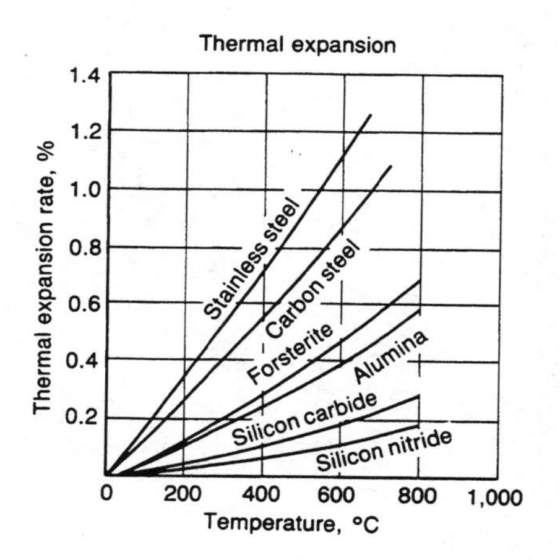

Figure 5-2 Thermal expansion rates for selected materials. (From *Guide to Engineered Materials*, ASM International [formerly American Society for Metals], 1986. Reprinted with permission.)

CHEMICAL PROPERTIES

In contrast to physical changes, *chemical changes* are those which alter the molecules in a substance. For example, many corrosion processes change a metal into an oxide. The tendency of a material to undergo oxidation is critical in its resistance to corrosion. Most metals will form oxides. Consequently, it is important to protect the most active metals from corrosive environments. Ceramics, on the other hand, are already metal oxides or other compounds which can often be used in high temperature oxidizing environments without further chemical reaction causing degradation of their properties.

Chemical changes involve the alteration of the molecular structure of a material. Polymerization reactions in which many small molecules are bonded together to form much longer chains are used to form plastics. Sometimes heat or light can destroy chemical bonds in these materials, causing a degradation of their strength. In all of these examples, the chemical changes alter the types of molecules present.

MECHANICAL PROPERTIES

Mechanical Properties are frequently considered in the contexts of manufacturing and materials processing. Generally, mechanical properties are those which influence the ease with which a material can be shaped. These properties, such as strength or hardness, are very much like physical properties in that the bulk material does not change its average chemical composition under mechanical processing or heating. However, there are often microscopic chemical reactions taking place, such as the formation of small precipitate particles, which may drastically alter the mechanical characteristics of a material.

Tensile testing. The mechanical properties of a material are often determined by a tensile test. The configuration for such a test is shown in Figure 5-3. When a specimen is subjected to a load, its initial deformation is *elastic*. This term, elastic,

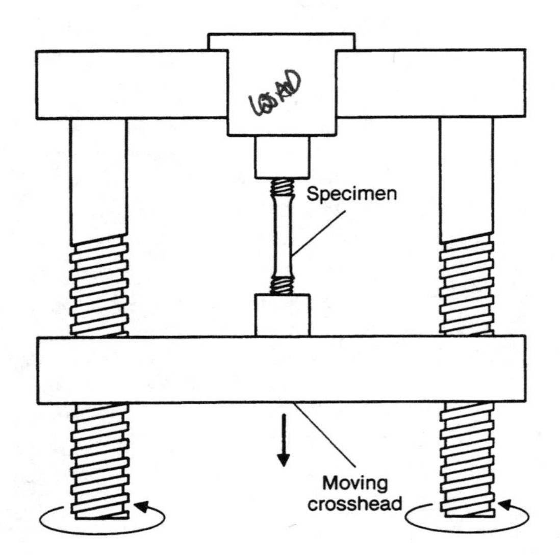

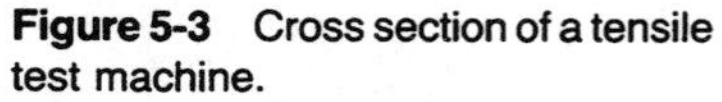
Figure 5-3 Cross section of a tensile test machine.

means that the deformation is reversible. That is, the object changes shape when a load is applied, but returns to its original shape when the load is removed. Usually, such deformation is linear with load. In other words, twice the load produces twice the deformation.

Elastic modulus. In order to define the property known as *elastic modulus*, or, alternatively, as Young's modulus, or modulus of elasticity, we must express the applied load as a stress, the force per unit cross-sectional area.

$$\sigma = F/A$$

Here, the greek letter sigma, σ, represents stress. Usually stress is in units of pounds per square inch (psi) or in Newtons per square meter, also known as a Pascal. Because the Pascal is too small a stress to be of practical significance, engineering stresses expressed in SI units are given in megapascals, MPa. The applied force, F, is given in either pounds or Newtons. The cross-sectional area, A, is usually in either square inches or square meters.

Strain. Instead of expressing the deformation of a specimen simply in terms of elongation, we use the parameter known as strain. Strain is the elongation per unit length.

$$\epsilon = \delta L/L$$

In this equation, δL is the change in length, and L is the original length of the specimen. Because the units for δL and L are the same, strain is a dimensionless quantity. However, it is sometimes expressed as inches per inch.

The advantage to using stress and strain rather than force and elongation to characterize the response of a specimen to load is that the effects of the shape of the specimen have been eliminated. Consequently, the ratio between stress and strain is characteristic of the material itself and not of the specimen dimensions.

For a material that behaves elastically, a graph of stress versus strain is a straight line whose slope is defined as the elastic modulus, E.

$$E = \sigma/\epsilon$$

Typical values for the elastic modulus of various materials are given in Table 5-3. Note the wide range of values for the materials. Plastics have much lower elastic moduli and, consequently, they deform to a much greater extent than do metals. The range of available elastic moduli is shown in the bar graph in Figure 5-4.

TABLE 5-3
ELASTIC MODULI

Material	E (psi)	E (MPa)
Aluminum	10×10^6	70,000
Brass	16×10^6	110,000
Steel	30×10^6	205,000
Concrete	2×10^6	14,000
Glass	10×10^6	70,000
Rubber (synthetic)	600 – 11,000	4 – 75
Polystyrene	0.4×10^6	2,800

Source: Van Vlack, *Elements of Materials Science and Engineering*, 5th edition, Addison Wesley, Reading, Massachusetts, 1985.

True stress versus engineering stress. A brief qualification should be made of the preceding definitions of stress and strain. When a specimen is stretched, its length and cross-sectional area change slightly. In the equations defining stress and strain, the question arises as to whether we should use the area and length measured before or after deformation. For small deformations, the differences are so slight as to cause no practical difference in the values calculated by the two methods. However, in more extreme cases, the distinction must be made between what is known as engineering stress and true stress. The engineering stress is defined as applied load divided by the original area. True stress is the applied load divided by the actual cross-sectional area. For specimens in which the deformation significantly reduces the cross-sectional area, there may be an appreciable difference between engineering and true stress.

A similar distinction is made between engineering strain and true strain. Here, as before, engineering strain is based on the original dimensions. It is the change in length divided by the original length. In contrast, true strain is given by the natural logarithm of the ratio of the original length, L_0, to the final length, L, as shown in the following equation.

$$\epsilon_{true} = ln\ (L_0/L)$$

In this equation, L_0 is the original length, and L is the length after deformation. For small strains, there is no practical difference between the two ways of calculating strain. In most practical service situations, the distinction is unnecessary. However, the distinction can become important in the severe deformations that can occur during some forming operations.

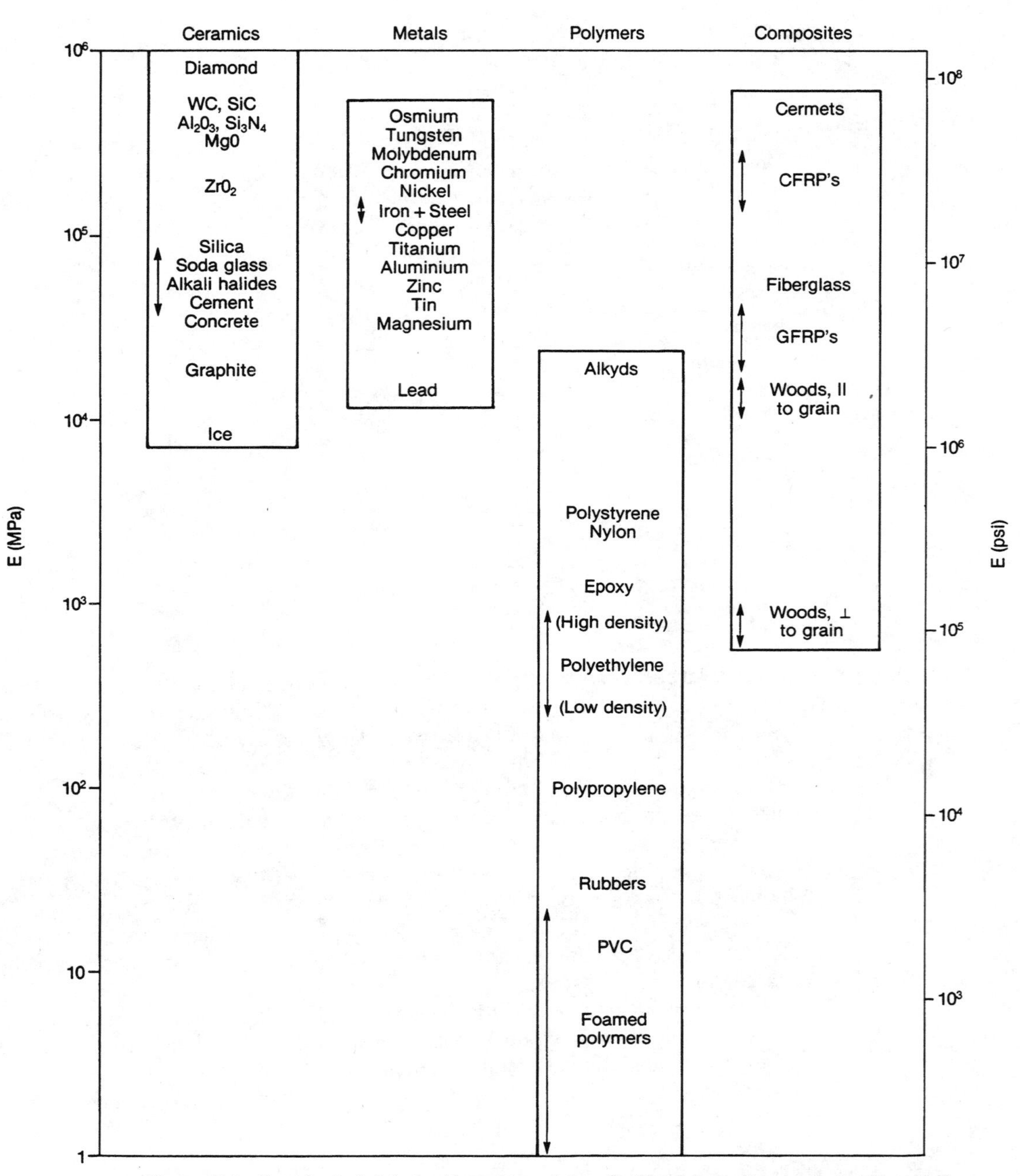

Figure 5-4 Bar chart of data for Young's modulus, E. (Reprinted with permission from M.F. Ashby and D.R. Jones, *Engineering Materials 1, An Introduction to their Properties and Applications*, 1980, Pergamon Books Limited.)

Plastic deformation. When a material is stretched beyond the elastic range, irreversible deformations occur. Deformation which occurs when the material does not return to its original shape after removal of the load is known as plastic deformation. The stress necessary to cause plastic deformation is known as the yield stress. When an applied stress is greater than the yield stress of the material, then plastic deformation, or permanent offset, occurs.

A complete stress-strain curve typically has an elastic region, in which stress is directly proportional to strain. This region is followed by a plastic region, in which the linear stress-strain relationship is no longer valid. Eventually, failure or breakage of the specimen occurs at the stress known as the ultimate strength. Typical stress-strain curves for different materials are shown in Figure 5-5.

Generally, for steel and for impure iron, there is a very clear stress at which plastic yielding occurs. For other materials, there is no such abrupt change; the

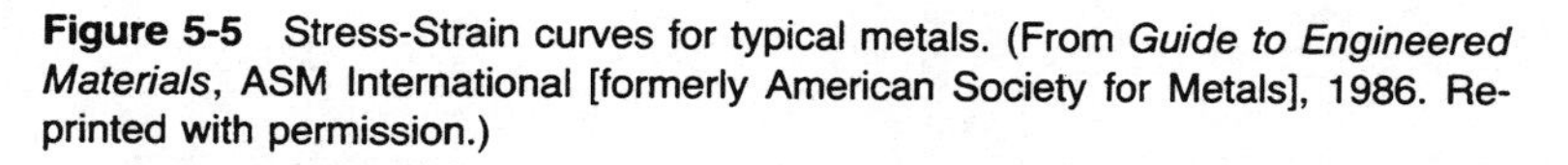

Figure 5-5 Stress-Strain curves for typical metals. (From *Guide to Engineered Materials*, ASM International [formerly American Society for Metals], 1986. Reprinted with permission.)

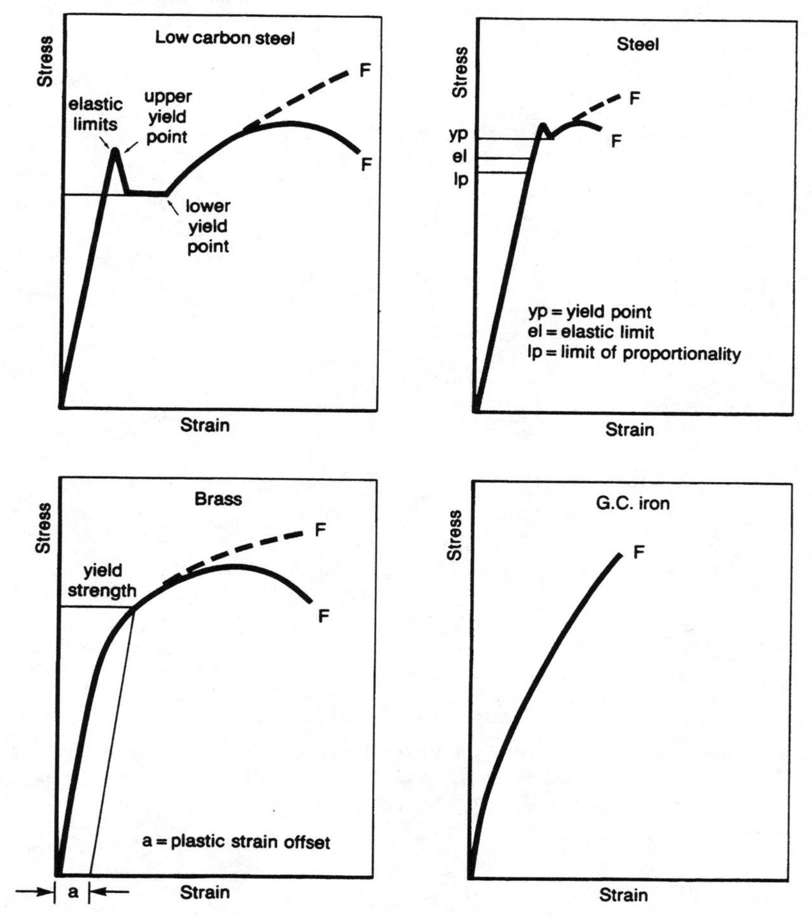

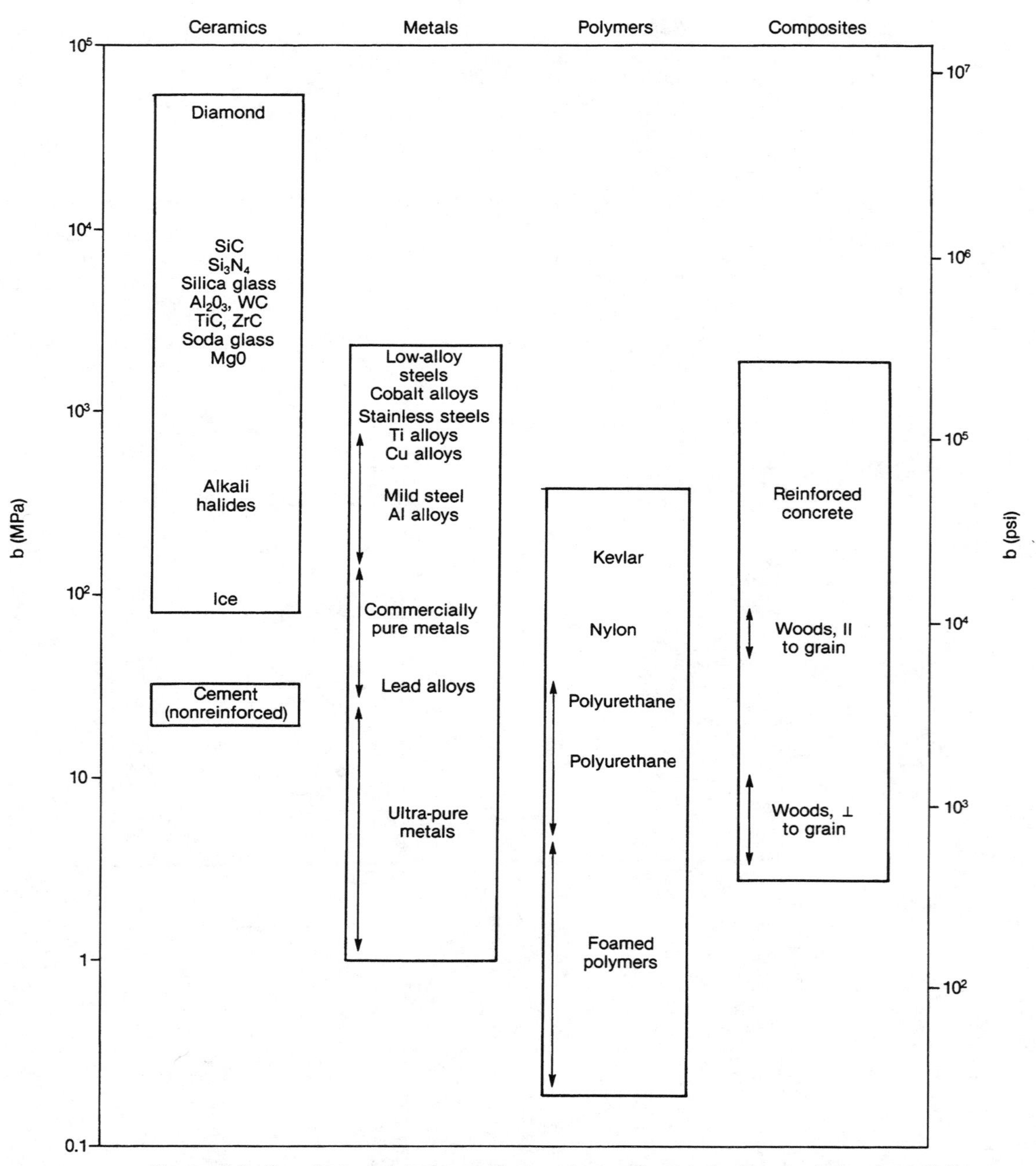

Figure 5-6 Bar chart of data for yield strength, σ. (Reprinted with permission from M.F. Ashley and D.R. Jones, *Engineering Materials 1, An Introduction to their Properties and Applications*, 1980, Pergamon Books Limited.)

stress-strain curve gradually decreases in slope. This behavior is illustrated by the curve shown for brass. In this situation, the yield point is arbitrarily defined as that stress which will produce a permanent offset of 0.2 percent. Yield points for a variety of engineering materials are shown in Figure 5-6. The wide range of values for different materials should be noted.

Brittle failures. The relative lengths of the elastic and plastic regions of a stress-strain curve are an indication of the type of failure that occurs when a material finally breaks. Failures which occur with very little plastic deformation of the specimen are known as brittle failures. The stress-strain curve for such materials is a nearly straight line ending in failure. Substances which behave this way are many ceramics, glass, and cast iron. Alternatively, those materials in which a large plastic deformation occurs undergo what is known as a ductile failure.

Toughness. The area underneath a stress-strain curve is a measure of the energy per unit volume which will cause failure. Ductile failures generally require much more energy than do brittle failures. Materials which can undergo some plastic deformation are much less likely to break under impact than are brittle materials. This property of being able to withstand impact loading is known as toughness. Tests of toughness generally involve impact upon a notched specimen. In Figure 5-7 the test specimen for the Charpy V-notch impact test is illustrated. This test is commonly used for metals. It further shows how the toughness of a low carbon steel increases with increasing temperature. Many materials can become quite brittle at low temperatures.

Toughness should be considered along with yield strength in determining the suitability of a material for a particular application. The yield strength of a material is the maximum stress that a material can withstand without plastic deformation,

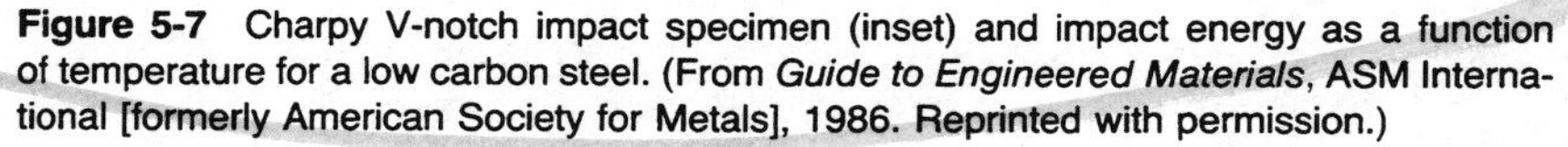

Figure 5-7 Charpy V-notch impact specimen (inset) and impact energy as a function of temperature for a low carbon steel. (From *Guide to Engineered Materials*, ASM International [formerly American Society for Metals], 1986. Reprinted with permission.)

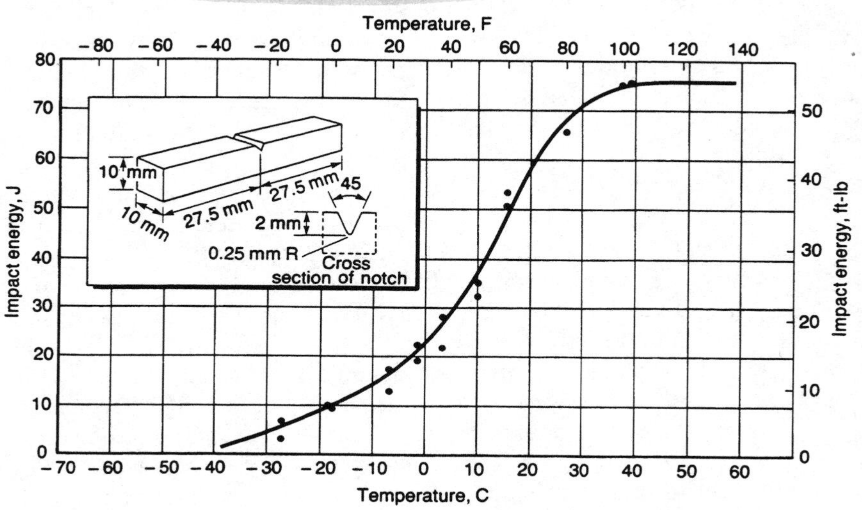

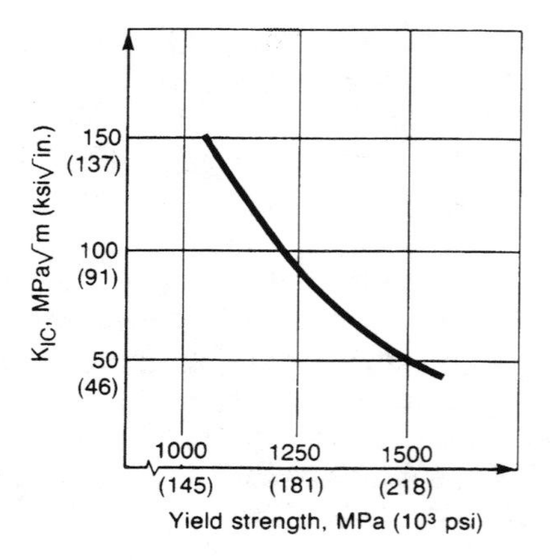

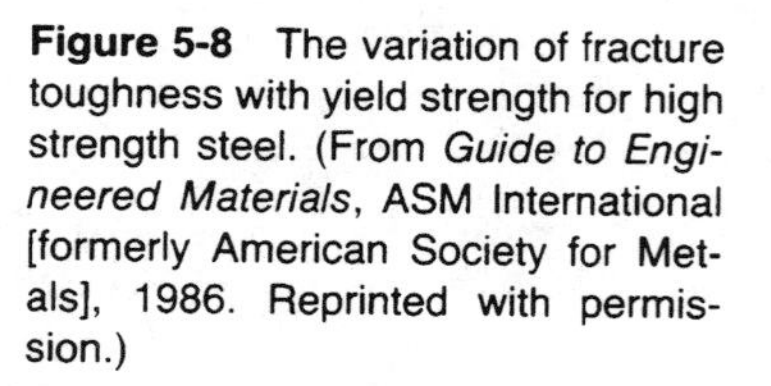

Figure 5-8 The variation of fracture toughness with yield strength for high strength steel. (From *Guide to Engineered Materials*, ASM International [formerly American Society for Metals], 1986. Reprinted with permission.)

while toughness is a measure of a material's ability to withstand impact. It is generally true that increasing the yield strength of a material can decrease its toughness. Figure 5-8 shows this phenomenon for high strength steel.

Anisotropy. In some materials, properties such as strength or elastic modulus are dependent on direction within the material. This characteristic, known as anisotropy, is observed in many materials in which processing creates directional properties within the material. A common example of anisotropy is seen in the paper used as newsprint. Anyone who has ever tried to tear an article out of a newspaper has observed that it is easy to tear the paper vertically between columns, but that it is much more difficult to make a straight horizontal tear across columns. This behavior is due to the orientation of long fibers which are aligned as the paper is produced and are oriented vertically in the final printed newspaper. The direction in which tears propagate easily is along the fibers rather than across them.

Other examples of anisotrophy are seen in the solidification direction in metal castings and in the axes along which sheet metal is rolled, wire is drawn, or polymers extruded. Even the direction in which a surface is ground affects its wear resistance.

Perhaps the most obviously anisotrophic material is wood. Long fibers of cellulose lie along the direction of the grain. The strength of wood in compression is typically 10 times as great along the grain as it is transverse to the grain. Similarly, resistance to shear is much greater transverse to the grain than along the grain.

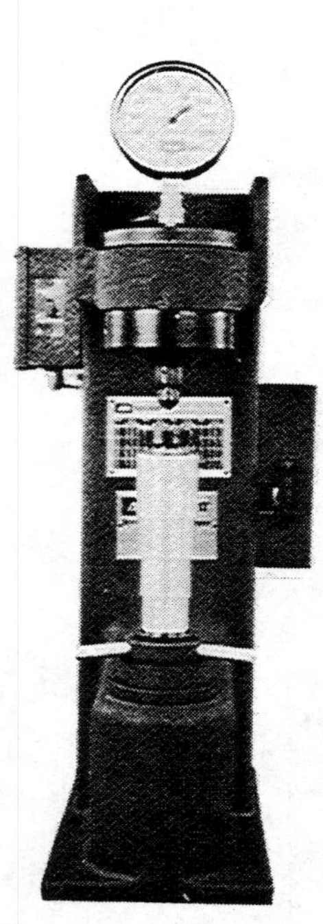

Figure 5-9 A Brinell hardness tester. (From *Metals Handbook, Desk Edition*, ASM International [formerly American Society for Metals], 1985. Reprinted with permission.)

Hardness. Another mechanical property that must be mentioned is hardness. Traditionally, the concept of hardness was related to what materials could scratch others. The historical Mohs hardness scale utilized several common materials as standards against which other materials could be measured. The scale ranged from talc, the softest substance used, to diamond, the hardest. Because of the very qualitative nature of the Mohs scale, it is seldom used now. Modern investigators use other testing methods, from which quantitative hardness values can be obtained.

Hardness is very closely related to yield strength, because the modern methods for measuring hardness are related to the amount of plastic deformation which occurs when an indenter is pressed into the surface with a standard load. There are several common types of hardness testers. In a Brinell tester, a hardened steel ball is pressed into the surface with a specific load. See Figures 5-9 and 5-10. The size of the

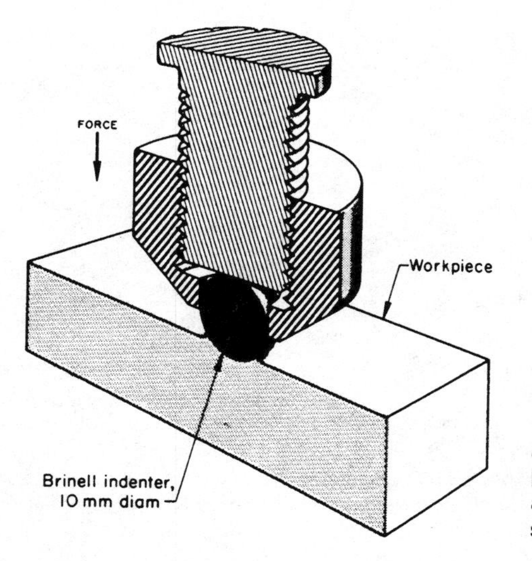

Figure 5-10 Sectional view of the Brinell indenter. (From *Metals Handbook, Desk Edition*, ASM International [formerly American Society for Metals], 1985. Reprinted with permission.)

resultant depression is then measured. The diameter is used to calculate the Brinell hardness according to the equation:

$$HB = L/\{(\pi D/2)[D - (D^2 - d^2)^{1/2}]\}$$

where L is the load in kilograms, D is the diameter of the ball in millimeters, and d is the diameter of the indentation in millimeters.

A Rockwell hardness tester has interchangeable indenters, either a diamond brale or a hardened steel sphere. Figure 5-11 shows the principal components of the Rockwell hardness tester. The indenter, shown in Figure 5-12, is applied to the specimen surface with a nominal small load while an indicator dial is zeroed. Then

Figure 5-11 Principal components of a Rockwell hardness tester. (From *Metals Handbook, Desk Edition*, ASM International [formerly American Society for Metals], 1985. Reprinted with permission.)

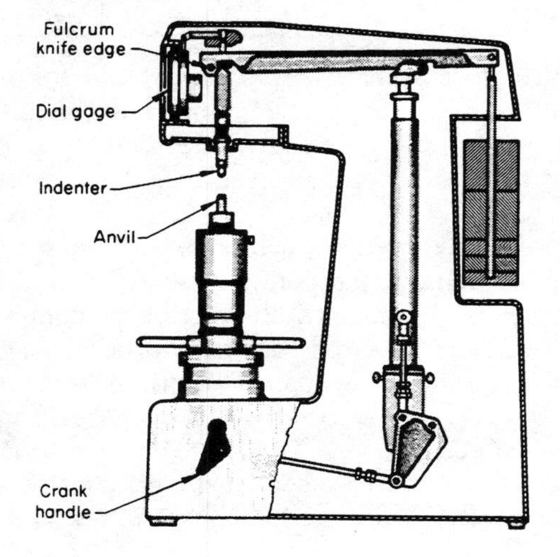

Figure 5-12 Diamond brale indenter used in Rockwell hardness test. (From *Metals Handbook, Desk Edition*, ASM International [formerly American Society for Metals], 1985. Reprinted with permission.)

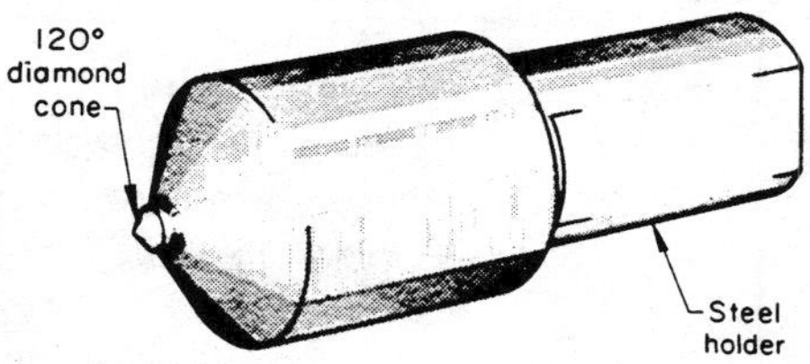

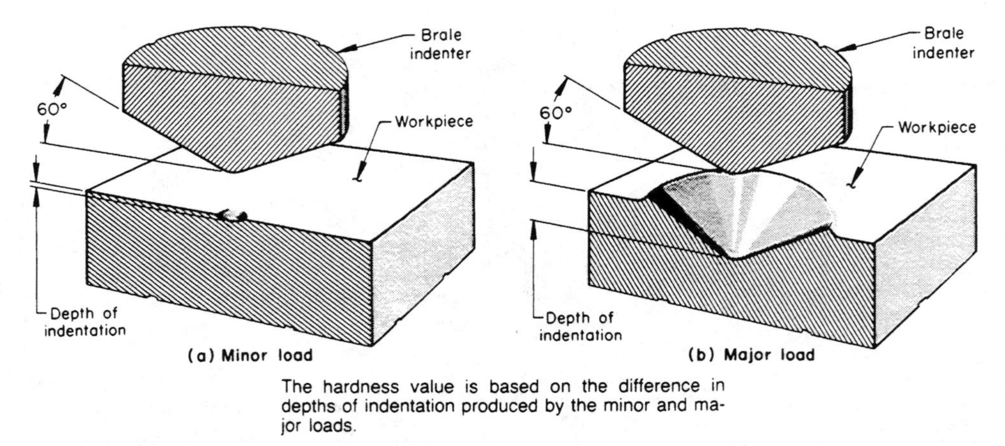

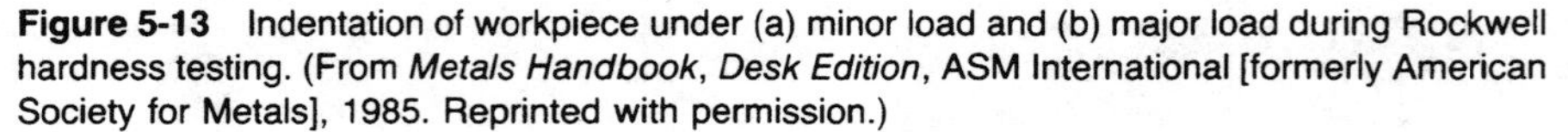

Figure 5-13 Indentation of workpiece under (a) minor load and (b) major load during Rockwell hardness testing. (From *Metals Handbook*, *Desk Edition*, ASM International [formerly American Society for Metals], 1985. Reprinted with permission.)

a larger load is applied and removed (Figure 5-13). Depth of penetration of the indenter, as indicated on the dial, is related to Rockwell hardness. Different Rockwell hardness scales call for different indenters and loads.

Other hardness tests, known as Knoop and Vickers, use a diamond pyramid indenter. As with the previously described means of hardness testing, the size or depth of the indentation is used to indicate the hardness of the material. Of course, for any hardness reading to be valid, the hardness of the indenter must be greater than that of the material to be tested. Also, it should be noted that these indenters measure only plastic deformation. A material that easily deforms elastically, such as rubber, will nevertheless appear quite hard by these testing methods because there will be very little plastic deformation. Consequently, it is important not to confuse elastic modulus with hardness.

TYPES OF MATERIALS

Now that we have defined several material properties, we can look at the types of materials that are available and explain why their properties differ. Although materials used in manufacturing are seldom in their elemental states, a periodic table of the elements is nevertheless useful in describing their behavior (Figure 5-14).

Metallic bonding. Those elements which are metals are in a region occupying the left side and lower portions of the periodic table. Metal atoms have one or more outer electrons which can be shared with other metal atoms to form a metallic solid. In solid metals, the outer electrons are free to occupy any position in the material, and they move easily under an applied electric field. Consequently, metals are usually very good conductors of electricity and of heat. In both cases, the electrons are the principal carriers.

The forces that bond metal atoms together usually act equally well in any spatial direction. Consequently, metal atoms in the solid phase will often have the maximum

Periodic table of the elements

Metals — Nonmetals

Key to chart: Atomic Number → 50; Symbol → Sn; Atomic Weight → 118.69; Oxidation States → +2 +4; Electron Configuration → -18-18-4

Transition Elements: groups IIIb to VIII

Iª	IIª	IIIᵇ	IVᵇ	Vᵇ	VIᵇ	VIIᵇ	VIII	VIII	VIII	Iᵇ	IIᵇ	IIIª	IVª	Vª	VIª	VIIª	0	Orbit
1 +1 −1 H 1.0079 1																	2 0 He 4.00260 2	K
3 +1 Li 6.939 2-1	4 +2 Be 9.0122 2-2											5 +3 B 10.81 2-3	6 +2 +4 −4 C 12.011 2-4	7 +1 +2 +3 +4 +5 −1 −2 −3 N 14.0067 2-5	8 −2 O 15.9994 2-6	9 −1 F 18.998403 2-7	10 0 Ne 20.17$_9$ 2-8	K-L
11 +1 Na 22.9898 2-8-1	12 +2 Mg 24.312 2-8-2											13 +3 Al 26.98154 2-8-3	14 +2 +4 −4 Si 28.08 2-8-4	15 +3 +5 −3 P 30.97376 2-8-5	16 +4 +6 −2 S 32.06 2-8-6	17 +1 +5 +7 −1 Cl 35.453 2-8-7	18 0 Ar 39.948 2-8-8	K-L-M
19 +1 K 39.09 8-8-1	20 +2 Ca 40.08 8-8-2	21 +3 Sc 44.9559 8-9-2	22 +2 +3 +4 Ti 47.9 8-10-2	23 +2 +3 +4 +5 V 50.941 8-11-2	24 +2 +3 +6 Cr 51.996 8-13-1	25 +2 +3 +4 +7 Mn 54.9380 8-13-2	26 +2 +3 Fe 55.847 8-14-2	27 +2 +3 Co 58.9332 8-15-2	28 +2 +3 Ni 58.71 8-16-2	29 +1 +2 Cu 63.54 8-18-1	30 +2 Zn 65.38 8-18-2	31 +3 Ga 69.72 8-18-3	32 +2 +4 Ge 72.59 8-18-4	33 +3 +5 −3 As 74.9216 8-18-5	34 +4 +6 −2 Se 78.96 8-18-5	35 +1 +5 −1 Br 79.904 8-18-7	36 0 Kr 83.80 8-18-8	-L-M-N
37 +1 Rb 85.467 18-8-1	38 +2 Sr 87.62 18-8-2	39 +3 Y 88.9059 18-9-2	40 +4 Zr 91.22 -18-10-2	41 +3 +5 Nb 92.9064 -18-12-1	42 +6 Mo 95.94 -18-13-1	43 +4 +6 +7 Tc 98.9062 -18-13-2	44 +3 Ru 101.07 -18-15-1	45 +3 Rh 102.905 18-16-1	46 +2 +4 Pd 106.4 -18-18-0	47 +1 Ag 107.868 18-18-1	48 +2 Cd 112.40 -18-18-2	49 +3 In 114.82 -18-18-3	50 +2 +4 Sn 118.69 -18-18-4	51 +3 +5 −3 Sb 121.75 -18-18-5	52 +4 +6 −2 Te 127.60 -18-18-6	53 +1 +5 +7 −1 I 126.9045 -18-18-7	54 0 Xe 131.30 -18-18-8	-M-N-O
55 +1 Cs 132.9054 -18-8-1	56 +2 Ba 137.3 -18-8-2	57* +3 La 138.9055 -18-9-2	72 +4 Hf 178.49 -32-10-2	73 +5 Ta 180.948 -32-11-2	74 +6 W 183.85 -32-12-2	75 +4 +6 +7 Re 186.207 -32-13-2	76 +3 +4 Os 190.2 -32-14-2	77 +3 +4 Ir 192.9 -32-15-2	78 +2 +4 Pt 195.09 -32-16-2	79 +1 +3 Au 196.9665 -32-18-1	80 +1 +2 Hg 200.59 -32-18-2	81 +1 +3 Tl 204.37 -32-19-3	82 +2 +4 Pb 207.19 -32-18-4	83 +3 +5 Bi 208.980 -32-18-5	84 +2 +4 Po (209) -32-18-6	85 At (210) -32-18-7	86 0 Rn (222) -32-18-8	N-O-P
87 +1 Fr (223) -18-8-1	88 +2 Ra 226.0254 -18-8-2	89** +3 Ac (227) -18-9-2	104 +4 Rf (261) -32-10-2	105 Ha (262) -32-11-2	106 (263) -32-12-2													-O-P-Q

															Orbit
*Lanthanides	58 +3 +4 Ce 140.12 -20-8-2	59 +3 Pr 140.9077 -21-8-2	60 +3 Nd 144.24 -22-8-2	61 +3 Pm 147 -23-8-2	62 +2 +3 Sm 150.4 -24-8-2	63 +2 +3 Eu 151.96 -25-8-2	64 +3 Gd 157.25 -25-9-2	65 +3 Tb 158.925 -27-8-2	66 +3 Dy 162.50 -28-8-2	67 +3 Ho 164.9304 -29-8-2	68 +3 Er 167.26 -30-8-2	69 +3 Tm 168.9342 -31-8-2	70 +2 +3 Yb 173.04 -32-8-2	71 +3 Lu 174.967 -32-9-2	-N-O-P
**Actinides	90 +4 Th 232.038 -18-10-2	91 +5 +4 Pa 231.0359 -20-9-2	92 +3 +4 +5 +6 U 238.029 -21-9-2	93 +3 +4 +5 +6 Np 237.0482 -22-9-2	94 +3 +4 +5 +6 Pu 239.052 -24-8-2	95 +3 +4 +5 +6 Am (243) -25-8-2	96 +3 Cm (247) -25-9-2	97 +3 +4 Bk (247) -27-8-2	98 +3 Cf (251) -28-8-2	99 +3 Es (254) -29-8-2	100 +3 Fm (257) -30-8-2	101 +2 +3 Md (258) -31-8-2	102 +2 +3 No (259) -32-8-2	103 +3 Lr (260) -32-9-2	-O-P-Q

Numbers in parentheses are mass numbers of most stable isotope of that element

Figure 5-14 Periodic table of the elements. (From *Metals Handbook, Desk Edition*, ASM International [formerly American Society for Metals], 1985. Reprinted with permission.)

number of nearest neighbors that can be accommodated. Some experimentation with a handful of marbles should show that a sphere can have a maximum of 12 identically sized spheres for neighbors. Many metals adopt this close-packed arrangement in the solid phase. Many others have 8 instead of 12 nearest neighbors. However, in either case, metals are among the most dense elements that exist. Figure 5-15 shows the three most common metallic crystal structures.

The other implication of nondirectional bonding of metals is that the atoms can roll over each other without serious disruption of the interatomic bonding. This characteristic implies that metals can be deformed relatively easily without causing

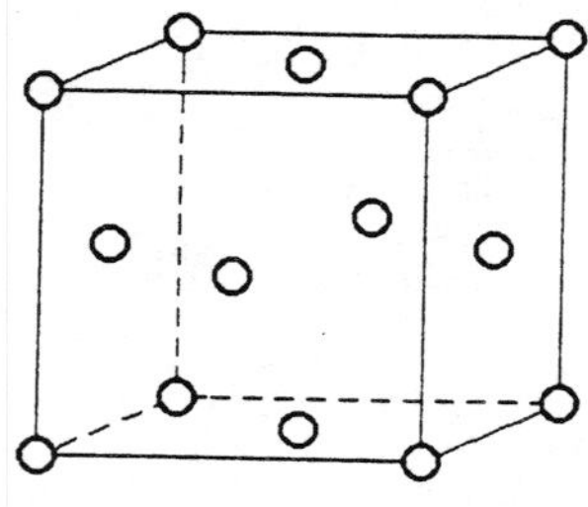

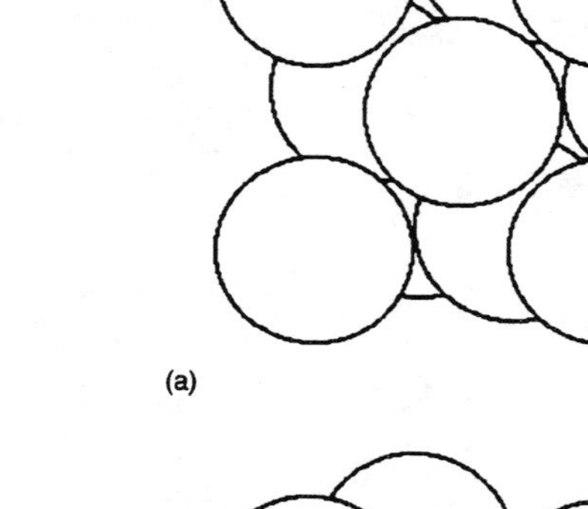

Figure 5-15 The three most common metallic crystal structures, (a) face centered cubic, (b) body centered cubic, and (c) close-packed hexagonal. Structures on the left illustrate atomic positions. Those on the right are "hard sphere" models.

(a)

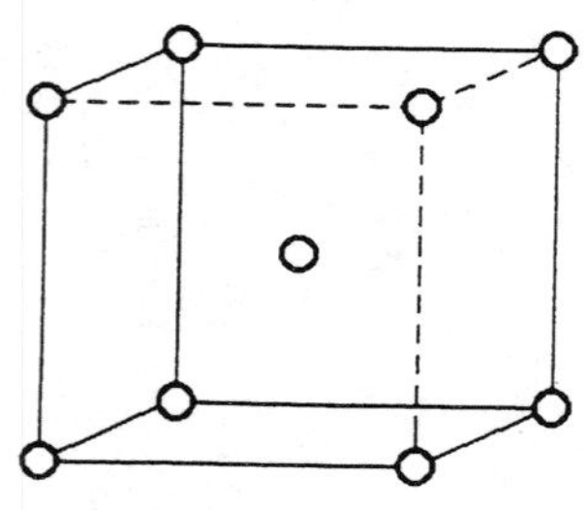

(b)

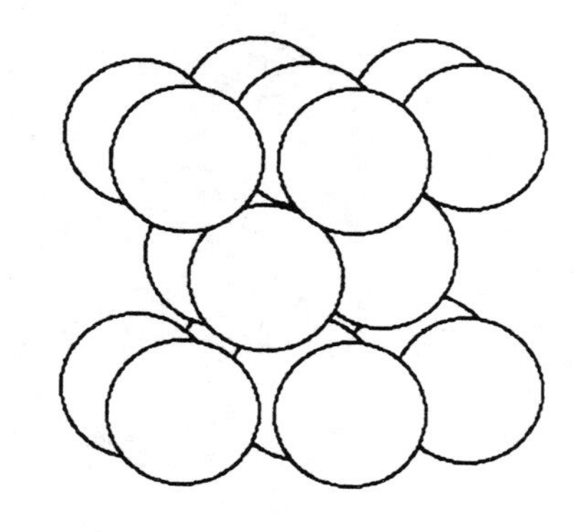

(c)

cleavage between planes of atoms. The ductile behavior of metals is a direct consequence of the metallic bond.

Ionic bonding. The elements on the right side of the periodic table are nonmetals. Their atoms are one or more electrons short of having a complete outer electronic shell. With the addition of the needed electrons, these atoms become negatively charged species known as ions. Examples are the oxide ion, 0^{-2}, or chloride ion, $C1^{-}$. Those extra electrons can come from electron-donating atoms such as metals. Metal atoms can lose their outer electrons to become positive ions such as $A1^{+3}$ or Mg^{+2}. Once this charge transfer has taken place, the resultant species are a positive and a negative ion. Because of their opposite charges the two ions attract each other, forming what is known as an ionic bond. An ionic bond results because of the electrostatic attraction of oppositely charged objects for each other.

Metal oxides are included in the class of materials known as ceramics. In these materials, the interatomic or, more accurately, the interionic bonding is very different from that seen in metals. In metal oxides, the solid is held together by electrostatic forces. Each cation (positive ion) is surrounded by anions (negative ions).

The number of nearest neighbors is limited by the relative sizes of the two ions. A very small cation might be able to accommodate only four surrounding anions. Other common nearest neighbor numbers are six or eight. Two common ceramic crystal structures are shown in Figure 5-16. In any case, the structures are usually less closely packed than are the metals. Consequently, ceramics are generally less dense than metals.

The ionic bonding of metal oxides makes them more brittle than the metals. Attempts at moving one layer of ions over another generally requires such a large force that cleavage results rather than plastic deformation. Although some ceramics exhibit plastic behavior at high temperatures, under ambient conditions they are almost always very hard, brittle materials.

Covalent bonding. There is yet another type of chemical bonding which can occur when the two atoms forming a compound are very similar. Covalent bonding generally occurs between two atoms occupying positions near each other in the periodic table. In this situation, the atoms share their electrons. This type of bonding

Figure 5-16 Two common ceramic crystal structures, (a) the cesium chloride structure, (b) the sodium chloride structure

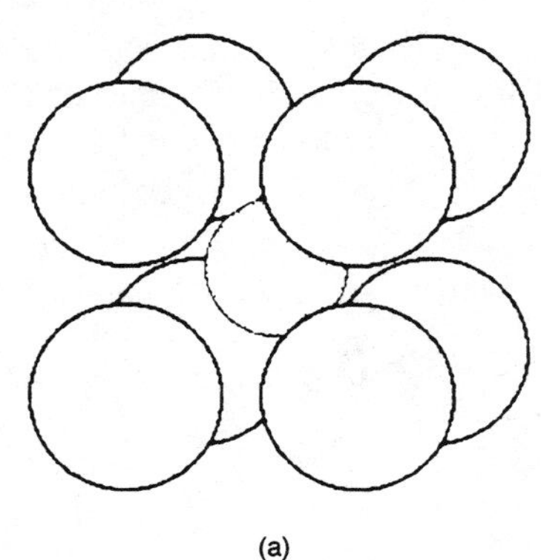

(a) (b)

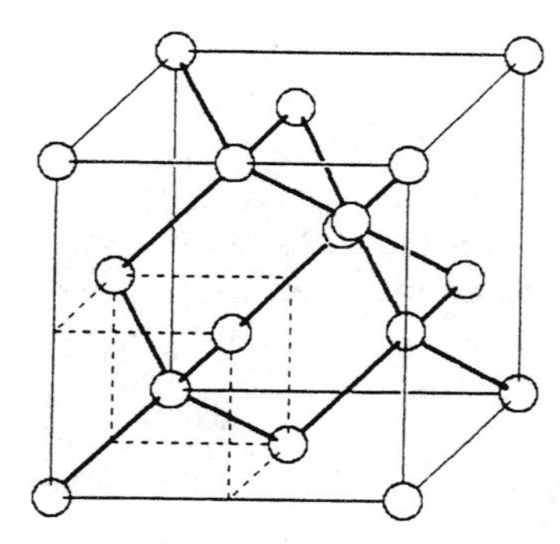

Figure 5-17 The diamond crystal structure.

results in a very strong, localized attachment of one atom to another. Compounds such as SiC (silicon carbide), SiO_2 (silica), and BN (boron nitride) can form very large networks of covalent bonds so that the entire solid is essentially one giant molecule. Silicon carbide adopts the diamond structure shown in Figure 5-17. Such materials are very hard and also very brittle.

Furthermore, because covalent bonds have preferred directions in space, the atoms have a few (usually one to four) nearest neighbors positioned with very specific angular separations. This geometrical constraint implies that covalently bonded structures are more likely to form open network structures rather than the more dense close-packed structures associated with metals. As a consequence, the covalently bonded network solids are less dense than are the metals.

Another class of covalently bonded materials is the polymers. These materials are made of long chain molecules hundreds or thousands of atoms long (Figure 5-18). The chains themselves are usually carbon and oxygen atoms with side groups which may include carbon, oxygen, nitrogen, hydrogen, fluorine, or chlorine atoms.

Although the polymer molecule itself is strongly held together by covalent bonds, the forces between two of these long chains are much weaker. These weaker intermolecular forces are known as VanderWaals forces. It is relatively easy to cause the chains to slide over each other producing permanent changes in shape. Consequently, polymers are often low melting, soft materials with low elastic moduli and low yield strengths. It is possible, however, to harden plastics by chemical reactions that cause cross-linking of the polymer chains. These reactions effectively freeze the structure, making the material much harder and more rigid and, at the same time, more brittle.

Composite materials. A class of materials that has recently received much attention is that of composites. They are replacing metals in applications such as car bodies and sports equipment, in which strength combined with lighter weight is desirable. These are materials in which there are at least two materials combined.

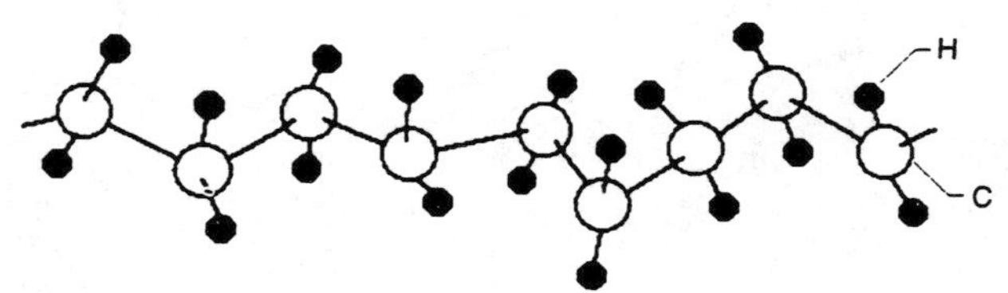

Figure 5-18 Schematic drawing of a polyethylene molecule.

For example, fibers of glass or graphite might be dispersed in a polymer matrix, creating a material which is much stronger than the polymer alone and much tougher than the glass or graphite. The mechanical properties of composite materials can be varied widely depending on the relative amounts of the two component phases and on the shape of the dispersed phase.

SUMMARY

We have seen in this chapter some of the properties that are significant in selecting materials for a given application. These may include chemical or physical or mechanical properties. In addition, we have seen how the chemical nature of the material, the position of the atoms in the periodic table and how they are bonded together, affects the bulk properties of the material.

QUESTIONS

1. Define the following terms:
   a. physical properties
   b. chemical properties
   c. mechanical properties
   d. elastic strain
   e. plastic strain
   f. elastic modulus
   g. anisotropy
   h. ionic bonding
   i. covalent bonding
   j. metallic bonding
2. Sketch the types of stress-strain curves that would be typical of
   a. ductile material
   b. brittle materials
3. Compare the mechanical behavior of
   a. metals
   b. ceramics
   c. polymers
4. Calculate how much the length of a steel bridge will change due to temperature changes during the year. The bridge is 100 meters long. Annual temperatures range from $-20°C$ to $+40°C$.

   Solution: Use the value for thermal expansion coefficient in Table 5.2. For steel, $\alpha = 12 \times 10^{-6}\ °C^{-1}$. We know that $\delta L = \alpha\ L\ \delta T$. Therefore, $\delta L = 12 \times 10^{-6}\ °C^{-1} \times 100\ m \times (60°C) = 0.072$ m or 7.2 cm.
5. What is the stress which occurs when a load of 150 pounds is suspended from a wire whose diameter is 1/16 in.?

   Solution: We know that $\sigma = F/A$. The area of the wire is related to diameter by the equation $A = \pi\ D^2/4$.

$$\sigma = 150\ \text{lb}/(\pi\ (1/16)^2/4)\ \text{in}^2 = 150/0.00307 = 48{,}900\ \text{psi}$$

6. What is the strain when a 50 cm rod is elongated by 3 mm?

Solution: $\epsilon = \delta L/L = 3$ mm/ 50 cm = 3 mm/ 500 mm = 0.006

7. What strain is produced by a stress of 5000 psi on an aluminum tensile specimen?

Solution: $E = \sigma/\epsilon$ Therefore, $\epsilon = \sigma/E = 5000$ psi/ 10^7 psi = 0.0005

8. What stress (in MPa) will produce a strain of 0.001 in steel?

Solution: $\sigma = E\epsilon = 205{,}000$ MPa $\times$ .001 = 205 MPa

9. A specimen originally 0.5 inches in diameter is deformed until the diameter is 0.3 inches under an applied load of 5,000 lb. Find the true stress and the engineering stress.

Solution: Engineering Stress = F/original area = 5,000 lb/ (π $0.5^2/4$)in^2 = 25,500 psi

True stress = F/actual area = 5,000 lb/(π $0.3^2/4$)in^2 = 70,700 psi

10. A specimen originally 10.0 cm in length is deformed by applying a tensile load. Compare engineering strain to true strain for deformations of 0.1 mm, 1.0 cm, and 10 cm.

Solution: engineering strain = $\delta L/L$, true strain = $ln\ (L_0/L)$, and $L_0 = \delta L + L$

Deformation	True Strain	Engineering Strain
0.1 mm	0.0010	0.0010
1.0 cm	0.0953	0.100
10 cm	0.693	2.0

6 METALS

Metals, like most solid materials, are crystalline. That is, the atoms are not randomly arranged, as they would be in an amorphous substance, but are in regularly repeating unit cells which extend in three dimensions for distances encompassing many atoms. The most common metallic crystal structures were shown in the preceding chapter. Figure 5-15 showed the face centered cubic, body centered cubic, and close-packed hexagonal structures. These unit cells are repeated many times in all directions to form a grain. Solid metals are made up of many grains separated by disordered regions known as grain boundaries (Figure 6-1). Usually, grain sizes are so small that individual grains cannot be resolved without a microscope. However, in some cases it is possible for single grains to be quite obvious. The mottled appearance of galvanized iron is due to grains of zinc whose diameter may be as large as a centimeter. Individual grains are also visible in large castings. Although recently developed technology can produce large single crystals for applications such as turbine blades, most metals used in manufacturing are polycrystalline.

CRYSTALLINE STRUCTURES

Body centered cubic

Within the metal grains, atoms are arranged in repeating unit cells. One of the common unit cells, the body centered cubic crystal structure, is exemplified by room temperature iron, by tungsten, and by the alkali metals. Each atom is surrounded

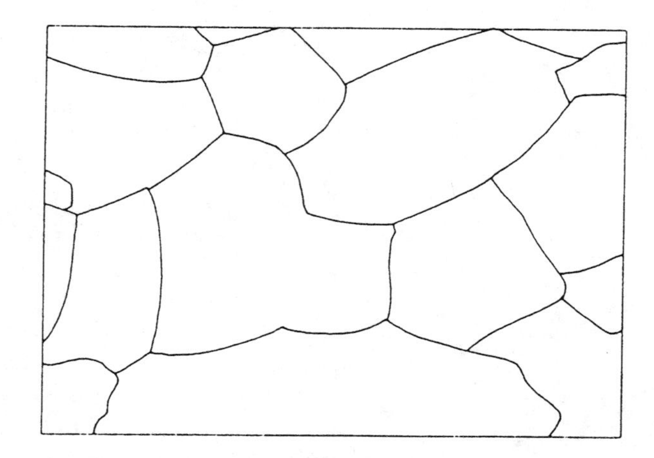

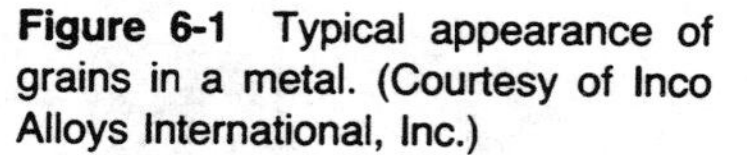
Figure 6-1 Typical appearance of grains in a metal. (Courtesy of Inco Alloys International, Inc.)

by eight nearest neighbors. When this structure is extended over many unit cells, it becomes clear that the atoms occupy positions on large extended planes. For example, the plane which includes the face diagonals on two opposite sides and the cube edges joining them looks like that shown in Figure 6-2. Planes of this orientation have the most atoms per unit area of any in the body centered cubic structure. Furthermore, the separation between adjacent planes is greater for these planes than for any other. When plastic deformation occurs, it is these planes which slide over each other.

When a metal undergoes plastic deformation, the shape of each grain within that metal is altered. This process takes place when one plane of atoms slides over another. Therefore, it is useful to know what planes are the farthest apart. These maximum density planes are the ones on which slip most often occurs.

Face centered cubic

Another very common metallic crystal structure is the face centered cubic structure, characteristic of copper, aluminum, gold, and silver. This structure is somewhat more densely packed than is the body centered cubic structure. In face centered structures, the most densely packed plane is the one which includes the face diagonals of three sides. The atoms are arranged as shown in Figure 6-3. In this structure,

Figure 6-2 The body centered cubic unit cell, showing the principal slip plane.

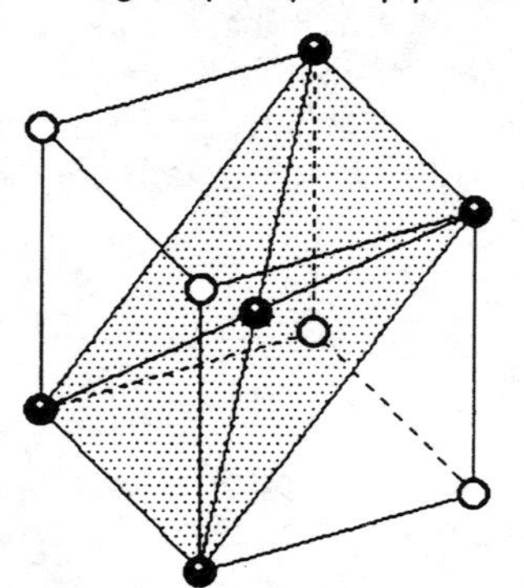

Figure 6-3 The face centered cubic unit cell, showing the principal slip plane.

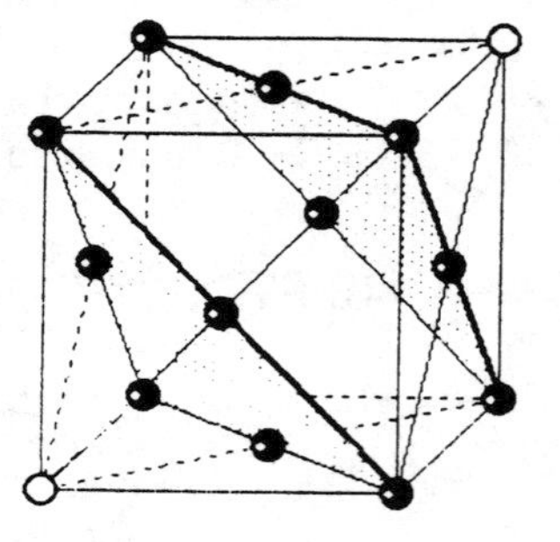

Figure 6-4 Atoms arranged in the face centered cubic structure. Some atoms have been removed to show the slip plane at the top left corner. (Courtesy of Inco Alloys International, Inc.)

there are four different orientations for these planes, providing a wide variety of possible slip directions. Consequently, these metals are usually very ductile. Figure 6-4 shows a collection of atoms arranged in the face centered cubic structure. Some of the atoms have been removed to show a slip plane.

Close-packed hexagonal

The third common metallic crystal structure, shown in Figure 6-5, is known as the close-packed hexagonal structure. It is exemplified by zinc. Its close-packed planes are the basal planes. Note that the atomic arrangement on these planes is the same as that for the face centered cubic structure. The way in which these planes are stacked distinguishes between the two crystal forms. There is only one orientation at which these close-packed planes exist in the hexagonal close-packed structure. There are fewer possible slip orientations. Therefore, hexagonal close-packed metals are less easily deformed than are the cubic metals.

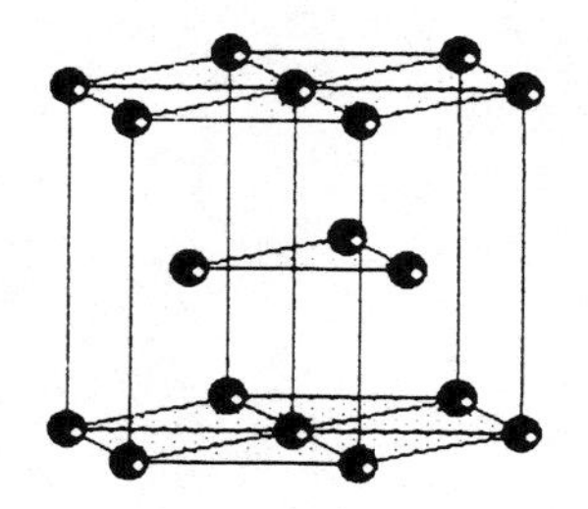

Figure 6-5 The close-packed hexagonal unit cell, showing the basal planes.

DISLOCATIONS

The mechanism by which plastic deformation occurs involves the movement of defects within the metal grains. These defects, known as dislocations, are local perturbations in the periodic organization of the atoms. A dislocation may be considered as an extra plane of atoms that terminates somewhere within the grain (Figure 6-6). Very small atomic movements can cause the dislocation to move laterally within the grain. When a dislocation moves entirely across the grain, an offset of one interatomic spacing occurs, causing a permanent change in the shape of the grain (Figure 6-7).

The motion of a dislocation takes much less force than would be needed to move an entire plane of atoms simultaneously. An analogy can be made between this mechanism of motion by dislocations, and a relatively easy way to reposition a carpet on a floor. One can put a small bump near one edge of the carpet and then push the bump across the carpet and off the far side. This technique requires much less force than dragging the entire carpet all at once.

Metals can accommodate dislocations more easily than can many other types of materials because the metallic bond is not very directional; neighboring atoms can be at any orientation about the central atom, so that the energy associated with an extra plane of atoms is relatively small. Also, because all the metal atoms are the same (or nearly so, in the case of some alloys) they are all attracted to each other. This circumstance is in contrast to the situation in ionic solids, such as many ceramics. In the latter case, an extra plane of atoms would be likely to juxtapose ions of the same charge, an occurrence requiring very high energy, and consequently very improbable.

Work hardening

We have seen that plastic deformation of metals is associated with motion of dislocations within grains. Dislocations can also interact among themselves. Because an extra plane of atoms causes a local disruption in the perfect order of a crystal, there will be a local stress field associated with the dislocation, causing compression where the extra plane of atoms is and tension where the plane is absent. This stress field repels other similar dislocations. If many dislocations are within a grain, they can repel each other so strongly that further deformation becomes more and more difficult. This phenomenon is clearly observable in the bulk behavior of metals. It is known as work hardening.

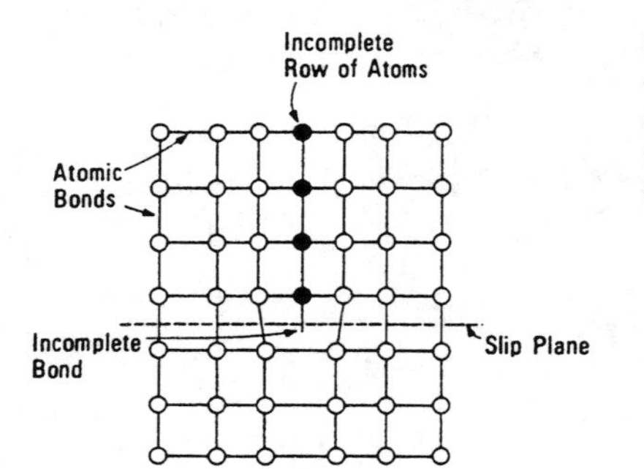

Figure 6-6 An expanded view of a crystal showing an extra plane of atoms resulting in an edge dislocation. (Courtesy of Inco Alloys International, Inc.)

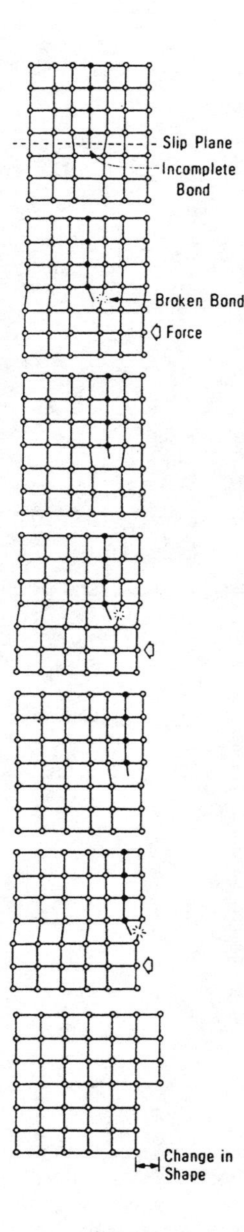

A dislocation caused by an incomplete row of atoms above a slip plane. The incomplete row does not have a matching row of atoms on the opposite side of the slip plane, resulting in an incomplete bond.

A shear force applied to the structure breaks the bond in the row of atoms next to the dislocation row.

The row of atoms that was incomplete has captured the bottom row lost by its neighbor. The row that lost its bond now has no matching row below the slip plane. The dislocation has moved from one row to another.

Continued application of force repeats the process. The row next to the dislocation loses its grip on its matching row below the slip plane.

The row that was incomplete grabs the row lost by its neighbor, moving the dislocation another row to the right.

Continued application of force breaks the last bond, freeing the dislocation from the structure.

The dislocation has reached the surface, where the extra row of atoms becomes a step or offset in the structure. As the dislocation traveled along the slip plane, it helped the atoms on one side of the plane to slip over the atoms on the opposite side.

Figure 6-7 Slip caused by dislocation movement. (Courtesy of Inco Alloys International, Inc.)

A specimen containing very few dislocations is initially very soft. When it is plastically deformed, more dislocations are introduced into the grains making each successive strain increment more difficult. During this process, the grains are elongated and distorted from their original equiaxed shape (Figure 6-8). The stress-strain curve for a softened material looks like that in Figure 6-9a. When the material is plastically deformed and then unloaded, the stress-strain curve follows that in Figure 6-9b. Strain caused by subsequent loading will follow the last curve back up again. Therefore, it can be seen that work hardening causes the elastic range of the specimen to increase.

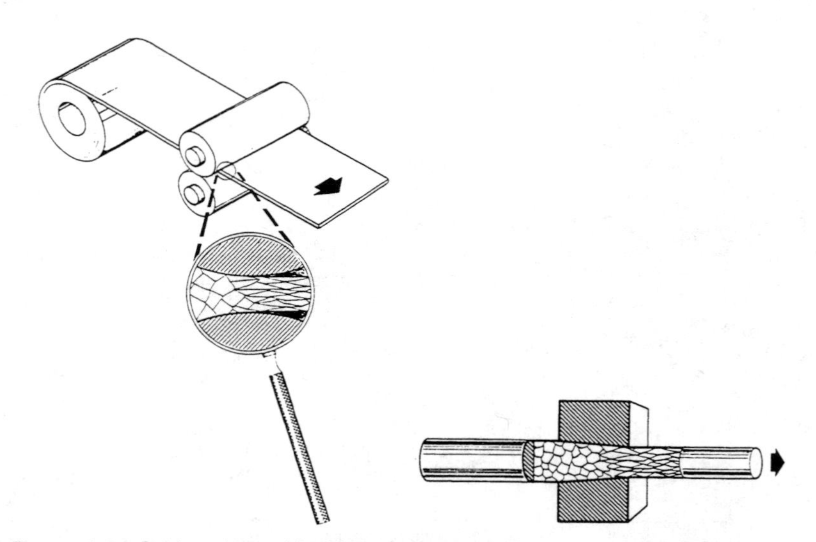

Figure 6-8 Cold working either by drawing or rolling squeezes and elongates the grains. (Courtesy of Inco Alloys International, Inc.)

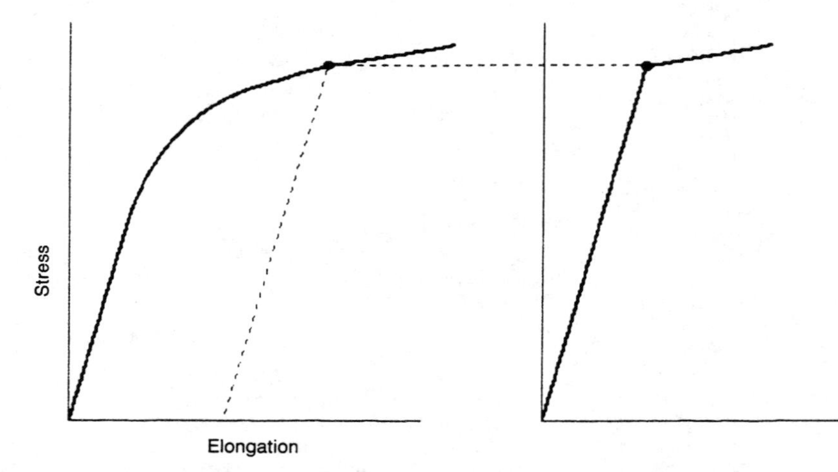

Figure 6-9 Correlation of work-hardening with the stress-strain curve. The graph on the left shows a typical stress-strain curve for an annealed metal. If the material is stressed beyond its yield point and then unloaded, the stress-strain relationship will follow the dotted line back to the horizontal axis leaving some permanent offset. If the material is again loaded, its stress-strain curve will follow the curve on the right, which has a longer linear elastic range than the curve on the left. Thus, straining the material has raised its yield point.

This behavior leads to several very significant conclusions. First, plastic deformation raises the yield point of a material. That is, cold working makes a material harder. Second, this same process decreases the ductility of the material. The amount of further deformation that can occur without fracture is reduced. Third, although the elastic range of a material is increased by cold working, the elastic modulus is not changed. The slope of the stress-strain curve is not affected.

Stress relief

The dislocations introduced by cold working remain permanently in the metal grains until some energy source is provided. Heat or thermal energy is sufficient in some cases to cause some stress relief. Heating, also known as annealing, of a cold-worked metal can cause several events to occur. First, the dislocations may become sufficiently mobile to allow some of them to move out of the grains and into the grain boundaries. This process, called strain relief, softens the material slightly and allows further deformation to take place without as much likelihood of fracture.

Recrystallization

At higher temperatures and at longer times, a second stage occurs. This process is called recrystallization (Figure 6-10). At the grain boundaries, new crystals form which are nearly strain free. Atoms in the old, highly strained grains find lower energy positions in the new grains. The new strain-free grains grow over the old strained ones. The metal does melt during this process. The material remains solid throughout the recrystallization process.

The recrystallization process can be used to produce grain refinement. All that is needed is for a material to be slightly strained and then heated. The grain refinement results from the formation of new grains originating at the grain boundaries. The old grains completely disappear as the new grains grow over the old ones. After recrystallization, the material has a smaller grain size, a lower yield point, and higher ductility.

Grain growth

If the recrystallization process is allowed to continue further, the new grains will grow to larger sizes. The larger grains will grow at the expense of smaller ones, with the small ones eventually disappearing completely. This third stage, known as grain growth, results in progressively larger grain sizes (Figure 6-11).

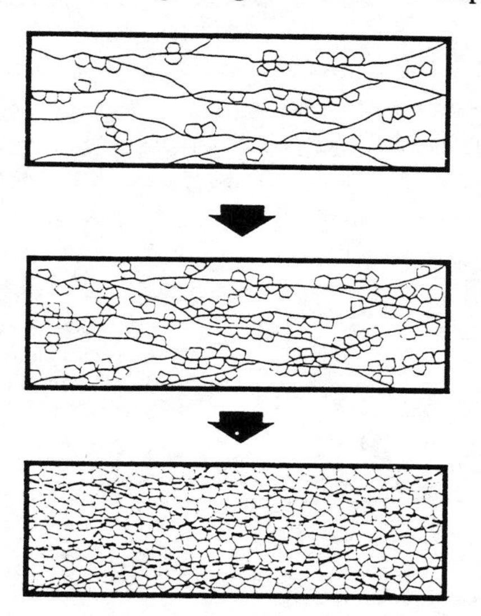

Figure 6-10 When a cold worked metal is heated, small grains begin to grow at the grain boundaries. Eventually, they completely replace the old grains. (Courtesy of Inco Alloys International, Inc.)

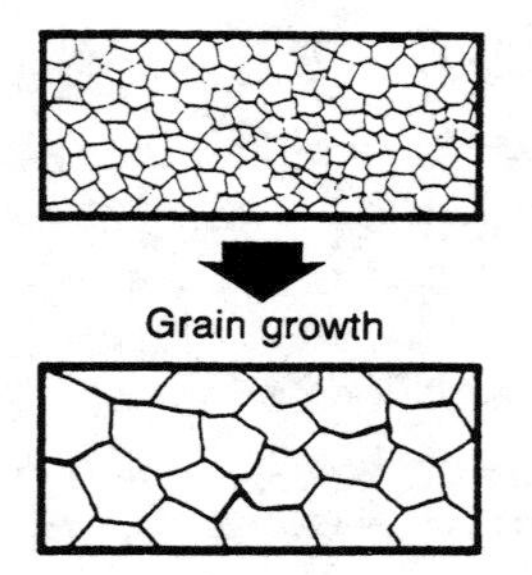

Figure 6-11 When a metal is held at annealing temperature, the grains become larger. (Courtesy of Inco Alloys International, Inc.)

The temperatures at which these processes occur readily are dependent on the amount of cold working and on the material itself. Generally, the lower the melting point of a material, the lower will be its recrystallization temperatures. Furthermore, heavily cold-worked metals recrystallize more readily than do lightly cold-worked metals. Heavily cold-worked materials also will have a finer grain size than will lightly cold-worked materials (Figure 6-12).

Now we can understand the difference between cold- and hot-working of a metal. If a material is to be formed at low temperatures, it is often necessary to perform the forming operations in several stages with intermediate annealing steps. Otherwise the part may fracture during the later steps. If a material is hot-worked, the temperature is high enough for recrystallization to occur at the same time as the deformation. At elevated temperatures, materials can be deformed to a much greater extent without the risk of fracture.

A simple demonstration can illustrate the difference between hot and cold working. A metal paper clip can be broken rather easily by repeated bending. Usually, bending and then straightening the wire half a dozen times is sufficient to cause fracture. A similar effort with a piece of lead foil will be unsuccessful. Lead has such a low melting point that it recrystallizes at room temperature. Consequently, it cannot be work hardened at normal room temperature. Ambient temperature for lead is sufficiently high that it is, in effect, hot working temperature. Lead retains its ductility indefinitely at room temperature.

Figure 6-13 shows cross sections of two screws whose threads have been formed by two different methods. In one case, the threads were cut, so that metal was removed during the forming operation. In the other case, the threads were formed by rolling. No metal was removed, but the surface was reformed from a smooth cylinder to the final threaded shape. The deformed grains produced within the threads

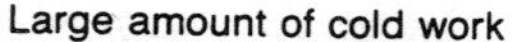

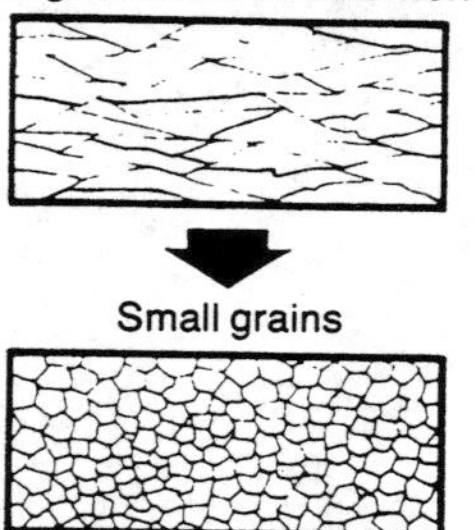

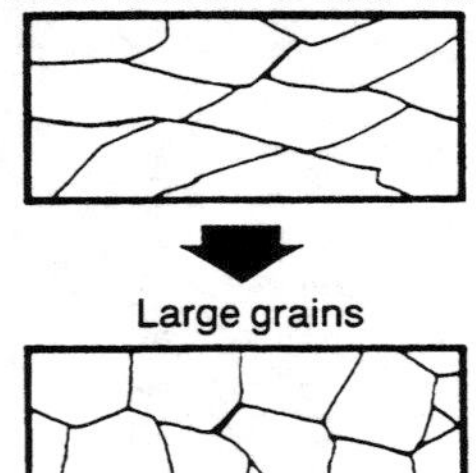

Figure 6-12 Small amounts of cold work produce large grains, while large amounts of cold work produce small grains. (Courtesy of Inco Alloys International, Inc.)

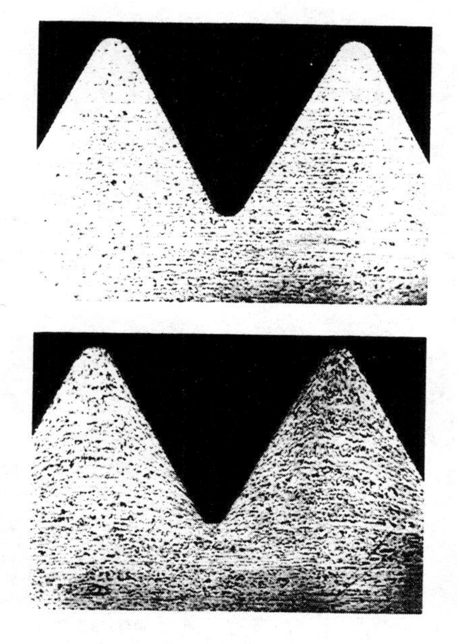

Figure 6-13 Cut screw threads (above), and rolled threads (below). (Courtesy of Reed Rolled Thread Die Company)

by the rolling operation are clearly visible. The screw with the rolled threads would be expected to be harder than the one with the cut threads because of the hardening effects of the cold working.

ALLOY SYSTEMS

While work hardening will harden almost any metal below its recrystallization temperature, there are two other means available to harden alloys. These other methods are possible only if another element is present. They can be used only in alloy (multiple component) systems.

When two metallic elements are melted together, it is sometimes possible to make a solid solution. In a solid solution the atoms are in a regular crystalline array, as they were in the case of an elemental metal, but the two different atoms are randomly distributed over the atomic sites. The solubility of one metal in another is influenced by whether or not the two atoms are similar in size and position in the periodic table. Two elements that have nearly the same atomic size, the same crystal structure, and similar chemical properties are likely to be very soluble in each other. In some cases, it is possible to make a solid solution of any composition, all the way from zero to 100 percent of one element. The more dissimilar the two atoms are, the less one will dissolve in the other.

Solution hardening

It is well established that a solid solution is harder than the pure elements from which it is made. This occurrence is known as solution hardening. An example of a solution hardened alloy is α-brass, copper with a small amount of zinc. Although it has the same crystal structure as pure copper, the solid solution alloy has much greater hardness.

Precipitation hardening

Many alloy systems rely on yet another means for hardening. When the limit of solubility has been exceeded, then it is not possible to form a solid solution. A second phase forms. It is desirable for this second phase to be in the form of small, widely distributed, hard particles whose presence prevents the motion of dislocations through the grains. The iron carbides present in steels are very important in determining the mechanical properties. Many aluminum alloys are also precipitation hardened.

BINARY SYSTEMS

Before discussing individual alloy systems, it is desirable to present some general information about what happens when two elements are combined. In order to understand the behavior of multicomponent systems, it is necessary to describe what happens when two different elements are mixed together. There are several possibilities. If the two elements are very different from each other, it is possible that they will form a chemical compound having properties very different from either element. An example of this result is the combination of iron and oxygen to form iron oxide. Iron and oxygen are very far apart on the periodic table. Iron, being a metal, is more likely to lose electrons than is oxygen, a more electronegative element which is more likely to gain electrons. Even when both elements are metals, if they have very different properties then it is possible for an intermetallic compound to form. Such compounds have very narrow composition ranges.

Another possible outcome is that the elements may form a solid solution. If the elements are very similar in chemical properties and in atomic size, they can blend into a solution in which the two types of atoms are randomly distributed over the sites in the crystal lattice. This situation can occur when the elements are very near each other in the periodic table. Solid solutions can also be observed to form when a smaller atom occupies the spaces between larger atoms. This type of solution, known as an interstitial solid solution, is exemplified by the solution of carbon in iron.

The third possible outcome is that the elements may be soluble only up to some solubility limit. Beyond that limit, a two-phase mixture develops. There will be two identifiably different crystal structures that can be seen when the solid is sectioned, polished, and etched. The tin-lead system is a good example of this behavior. A small amount of lead can be dissolved in tin, and a small amount of tin can be dissolved in lead. However, alloys of intermediate composition contain two phases, a tin-rich phase and a lead-rich phase.

PHASE DIAGRAMS

A convenient means of illustrating the properties of a two component system is known as a phase diagram. The horizontal axis represents composition, usually in weight percent, although atomic percents are sometimes used. The vertical axis is temperature. Regions within the diagram indicate whether one or two phases are present.

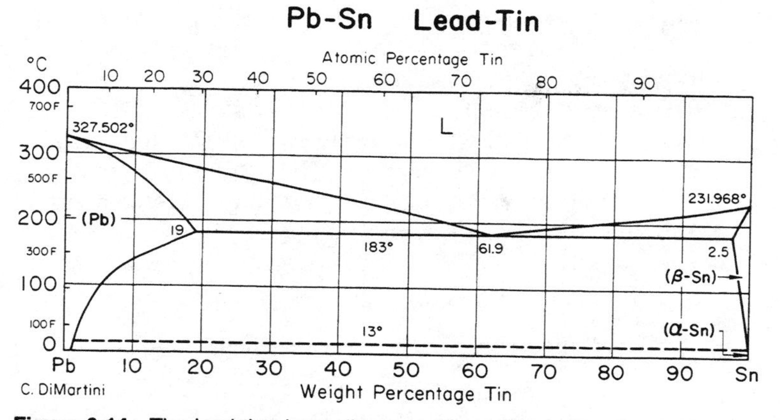

Figure 6-14 The lead-tin phase diagram. (From *Metals Handbook*, Vol. 8, 8th ed., ASM International [formerly American Society for Metals], 1973. Reprinted with permission.)

As an example, observe the lead-tin phase diagram shown in Figure 6-14. This binary system is selected because of its simplicity in comparison with other more complex systems. Several important features can be determined from the phase diagram:

1. What phases are present at any temperature and average composition?
2. What are the compositions of these phases?
3. What are the relative amounts of these phases?

For any temperature and composition, the phases present are indicated on the phase diagram. In a two-component (binary) system, there can be only one or two phases present, except in very restricted circumstances. Usually, only the single phase fields are identified. In the lead-tin diagram they are identified as (β-Sn), the tin-rich solid having the same form as pure tin; (Pb), the lead-rich solid having the same crystal form as pure lead; and *L*, the liquid phase. Parentheses are used to enclose the designations of the metal phases to avoid confusion with the pure elements. That is, "Pb" implies pure lead, while "(Pb)" implies a solution of lead with some tin dissolved in it. Incidentally, the (α-Sn) phase in the lower right corner of the diagram is a very low temperature form of tin which is much less ductile than the usual ambient form. Historically, the formation of (α-Sn) was responsible for the deterioration of tin organ pipes during the winter in unheated European cathedrals. For the purposes of the following discussion, we will deal only with the beta form of tin.

Regions between the single phase fields are two-phase, made up of the two single components on the two sides of the field. The additional designations could be added to the figure to indicate the two-phase regions, *L*+(Pb), *L*+(β-Sn), (Pb)+(β-Sn).

At a temperature of 100°C, tin-rich solutions, indicated on the diagram by (β-Sn), can be formed for compositions from 98 to 100 percent tin. Lead-rich solutions, indicated by (Pb) are possible for compositions containing up to approximately 5 percent tin. At compositions from 5 to 98 percent tin there will be two phases formed. Within the two-phase region, the compositions of the two phases are those indicated by the point where the temperature line intersects the boundaries between the single and two-phase regions. For example, at 100°C, at an average composition of 40 percent tin, the (β-Sn) phase will be 98 percent tin, 2 percent lead, and the (Pb) phase will be 95 percent lead, 5 percent tin.

The relative amounts of the two phases can be determined by a "lever rule." The percentages of the two phases are given by the proportion into which the horizontal line within the two-phase region is divided when the bulk composition is used as a fulcrum. In the preceding example, for an average composition of 40 percent tin at 100°C, the proportion of the tin-rich phase is given by the distance on the opposite side of the lever (98%–40%) divided by the total length of the lever (98%–5%).

$$(98–40)/(98–5) = 0.62 \text{ or } 62\%$$

Similarly, the proportion of lead-rich phase is given by the following equation:

$$(40–5)/(98–5) = 0.38 \text{ or } 38\%$$

As expected, the closer the average composition is to a single phase region, the higher is the proportion of that phase.

Now that the rudiments of phase diagrams have been established, it is possible to understand what happens during the solidification of liquid. Again, in the same binary system, there is a special composition at 62 percent tin. When liquid of this composition is cooled, it remains uniform in composition until a temperature of 183°C is reached. At that temperature, the liquid freezes into two phases in the reaction:

$$L \rightarrow (\beta\text{-Sn}) + (\text{Pb})$$

This composition is known as the eutectic composition. The temperature at which this reaction occurs is the only temperature at which three phases can coexist in a binary alloy system. The solid phases of eutectic compositions often have a very characteristic microstructure. Because the two phases must grow simultaneously, they are often in very fine alternating regions of plates or rods.

When liquids having other than the eutectic composition are cooled, a solid phase, known as a pro-eutectic phase, solidifies first, beginning at a temperature above the eutectic temperature. For example, a 40 percent tin alloy begins to solidify at 240°C with the formation of (Pb). Just above the eutectic temperature of 183°C, the system consists of (Pb) and liquid of the eutectic composition. With further cooling, the liquid transforms to the solid eutectic microstructure while the pro-eutectic phase remains unchanged. Consequently, the final solid material will have regions of (Pb), which solidified at temperatures between 240°C and 183°C, and regions of the eutectic mixture (Pb) and (β-Sn) (Figure 6-15).

Other alloy systems can be much more complex. However, the principles used to determine phases, compositions, and amounts are the same as in the preceding example. In our discussion of other alloy systems, we will refer to phase diagrams in order to explain the properties of these materials.

IRON AND STEEL

Production of iron from ore

Iron, as it exists naturally in iron ore, is chemically combined with oxygen, sulfur, and silicon. These compounds must be reduced chemically to produce elemental

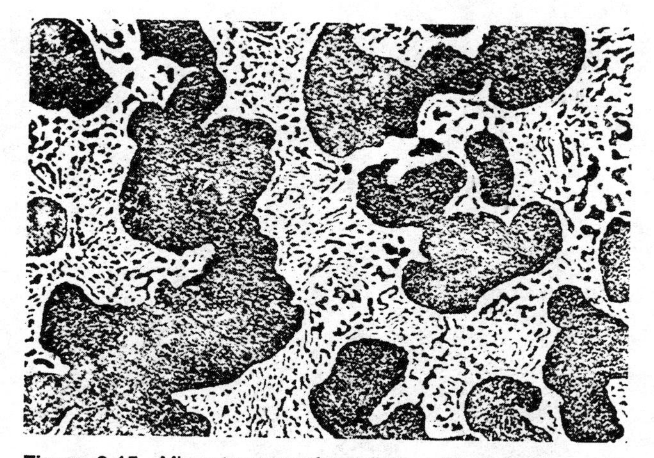

Figure 6-15 Microstructure of a solidified 60 percent lead, 40 percent tin alloy. The dark grains are the pro-eutectic (Pb) phase, and the surrounding lighter mottled areas are the eutectic mixture of (β-Sn) and (Pb). (From *Metals Handbook*, Vol. 7, 8th ed., ASM International [formerly American Society for Metals], 1972. Reprinted with permission.)

iron. The refining process occurs in a blast furnace (Figure 6.16). There, crushed iron ore is mixed with coke and limestone, and heated air is forced through the furnace to promote the chemical reaction.

Coke, which is prepared from coal, is a porous form of carbon. Its functions are to provide combustion heat for the refining process and to combine with the oxygen in the iron ore to produce oxides of carbon and to reduce the iron oxide to elemental iron. The limestone is important to the chemistry of the flux, that layer of oxides and silicates which floats on top of the purified liquid iron. The molten metal from a blast furnace is transferred into molds in which it solidifies. This product is commonly known as pig iron.

Cast iron

Pig iron is the material used to make cast iron. Often, scrap iron is added during the melting process. The term "cast iron" actually refers to family of materials. Typically, they all contain 2 to 4 weight percent carbon and often some silicon. The various types of cast irons are distinguished by the form which is taken by the carbon (Figure 6-17).

The iron-carbon phase diagram shows that carbon is not very soluble in solid iron (Figure 6-18). At room temperature, less than 0.02 weight percent carbon can be dissolved in the iron. In gray iron, the most common form of cast iron which is widely used in applications such as automobile engine blocks, the undissolved carbon is in the form of large graphite flakes. Graphite is quite soft and is often used as a solid lubricant. Cast iron is quite brittle. Consequently, it cannot be shaped by any operations requiring bending. However, it is easily machined. The graphite flakes lubricate the cutting tool and also cause chips to break off easily.

Other forms of cast iron known as white iron, ductile iron, and malleable iron can be produced. In these other materials, the graphite may be in spheroidal shapes

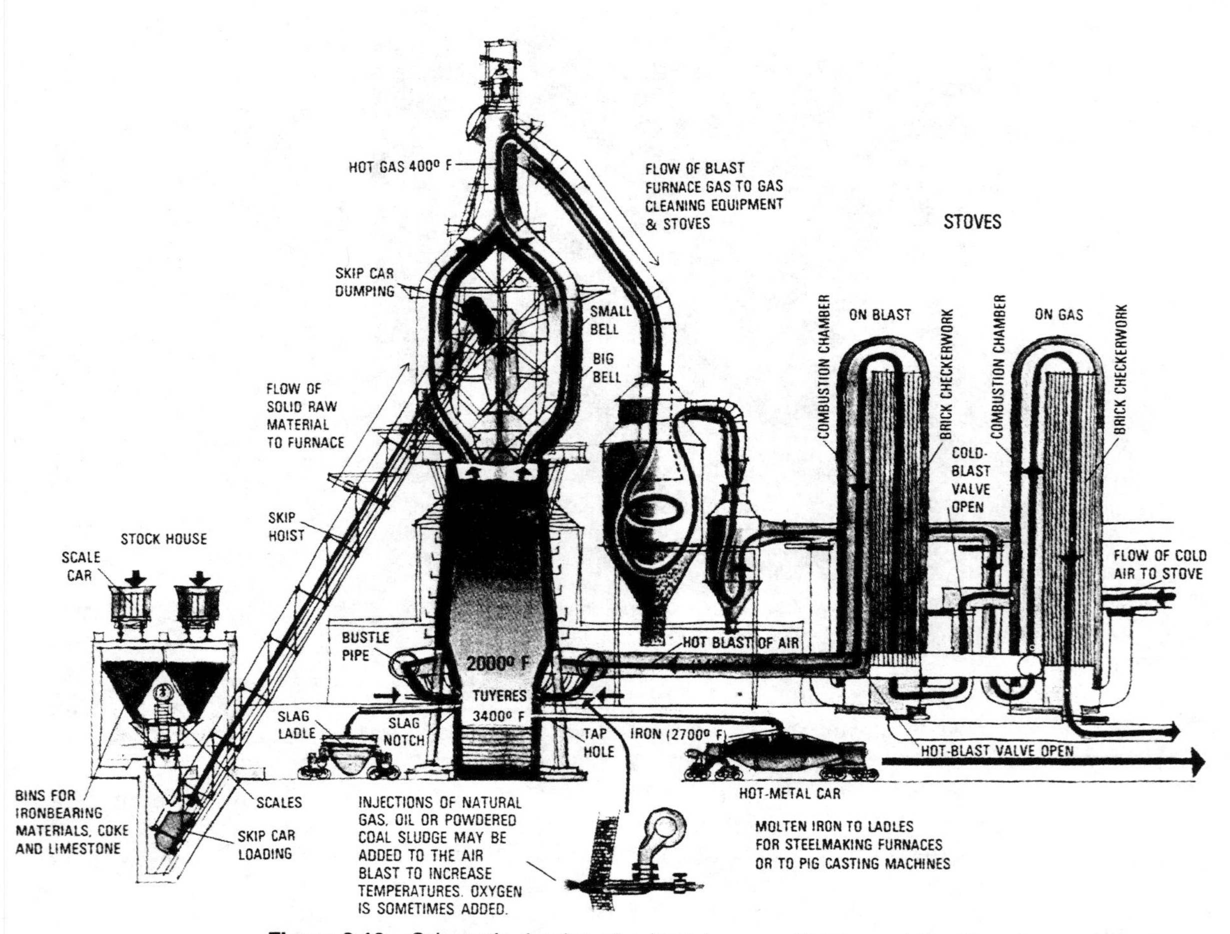

Figure 6-16. Schematic drawing of a blast furnace. (Courtesy of American Iron and Steel Institute)

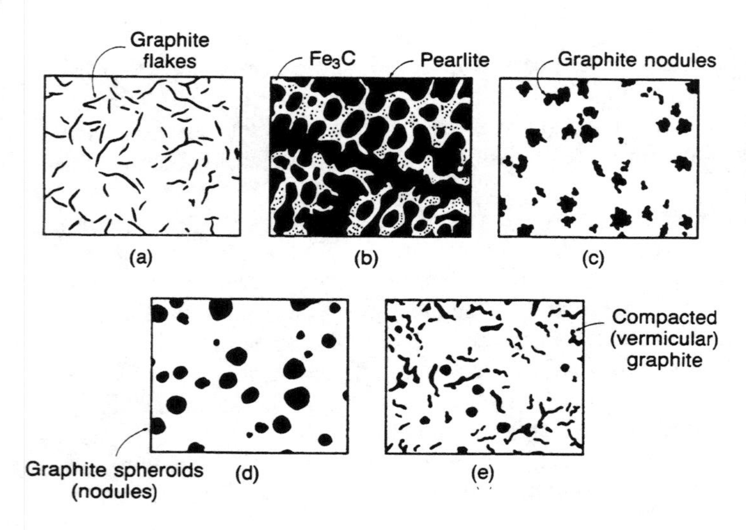

Figure 6-17 Schematic drawings of the five types of cast irons (a) gray iron, (b) white iron, (c) malleable iron, (d) ductile iron, and (e) compacted graphite iron. Source: Askeland, *The Science and Engineering of Materials*, Brooks/Cole Engineering Division, Monterey, 1984 (Courtesy of PWS-Kent Publishing Company)

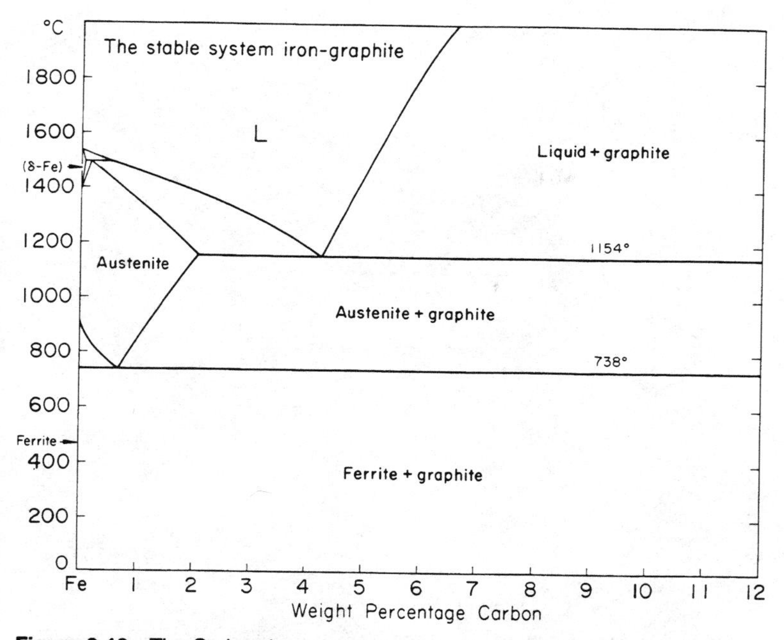

Figure 6-18 The Carbon-Iron phase diagram. (From *Metals Handbook*, Vol. 8, 8th ed., ASM International [formerly American Society for Metals], 1973. Reprinted with permission.)

rather than in flakes, or the carbon may be in iron-carbon compounds known as carbides. These other materials have higher strength than does gray iron, but more complex heat treatment cycles are needed to produce them.

Steel

Steels are also iron-carbon alloys, but the amount of carbon is lower than that found in cast iron. Steel is commonly made in a basic oxygen furnace or in an open hearth furnace (Figures 6-19 and 6-20). In both of these processes, the amount of carbon is reduced by burning it out with oxygen.

Steel is based on an iron-iron carbide binary system. When metals cool quickly, there is often not time for the equilibrium structure to form. This is the case with steel. Although the equilibrium state at room temperature would be iron and graphite, the precipitation of graphite is a very slow process. Faster cooling produces Fe_3C, an iron carbide, also known as cementite. It is quite stable and can serve as one of the components in a pseudobinary system. The phase diagram for the iron-iron carbide binary system can be used to explain the properties of plain carbon steel (Figure 6-21).

Ferrite and austenite

There are two different crystal forms of pure iron. At room temperature, α-iron, also known as ferrite, is the stable form. Ferrite has the body centered cubic structure. At higher temperatures, γ-iron, also known as austenite, having the face centered cubic structure, is the stable form. Austenite can dissolve much more carbon than

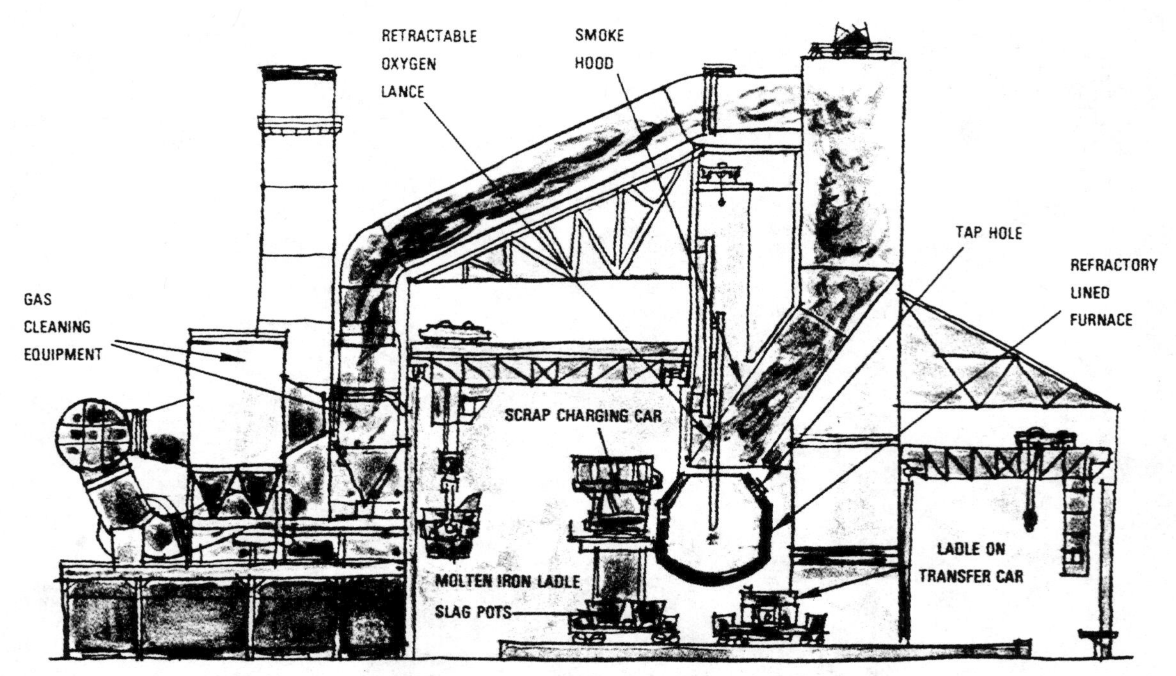

Figure 6-19 Schematic of a basic oxygen furnace. (Courtesy of American Iron and Steel Institute)

Figure 6-20 Schematic of an open hearth furnace. (Courtesy of American Iron and Steel Institute)

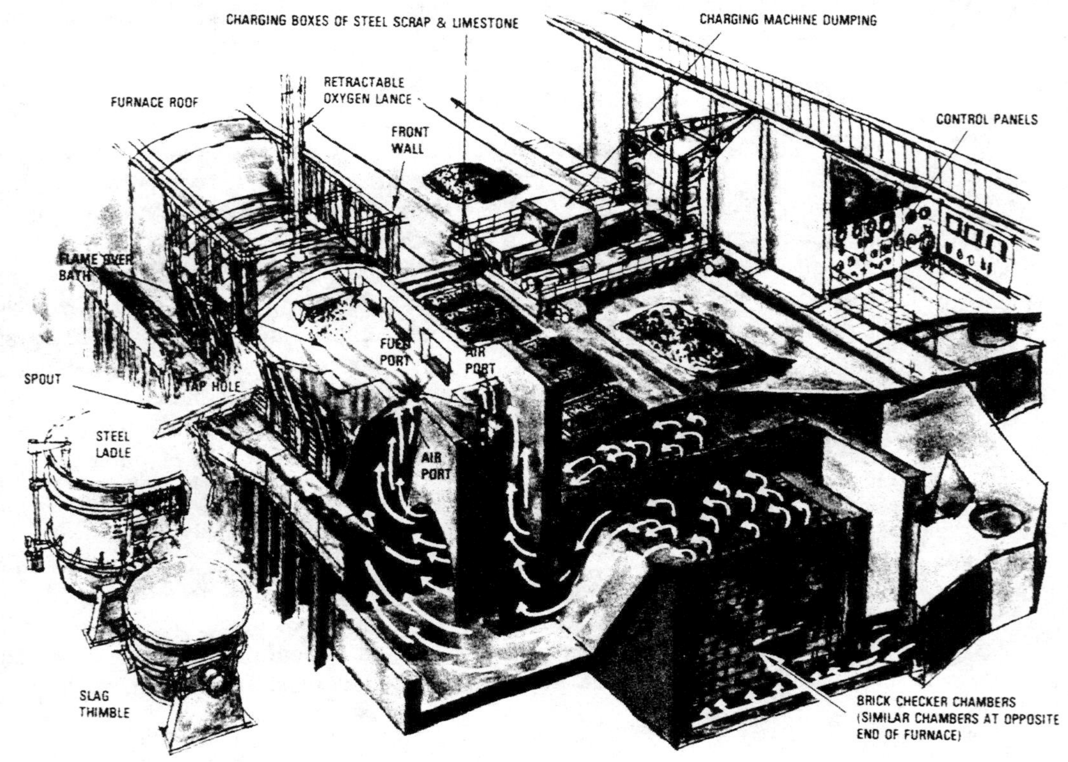

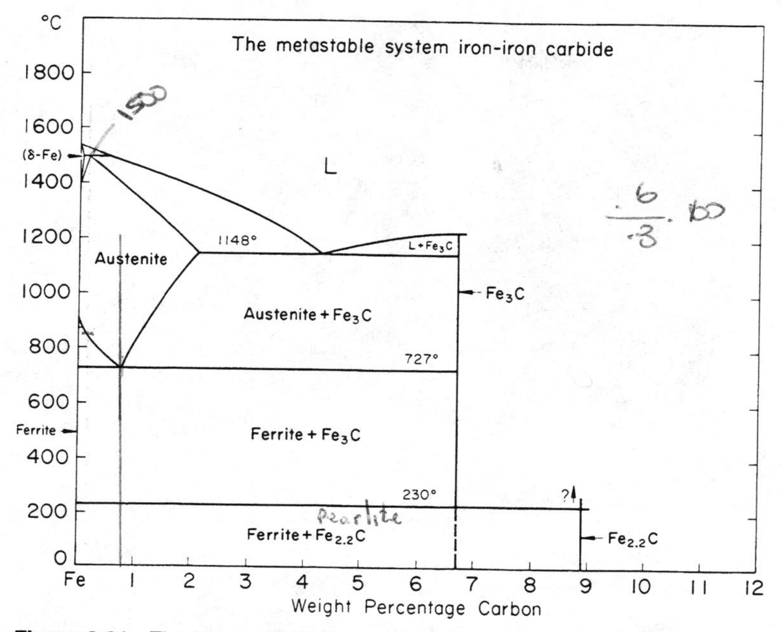

Figure 6-21 The iron-iron carbide system. (From *Metals Handbook*, Vol. 8, 8th ed., ASM International [formerly American Society for Metals], 1973. Reprinted with permission.)

can ferrite, up to 2 weight percent at 1148°C and lesser amounts at lower temperatures. Plain carbon steels have carbon contents such that at some temperature all of the carbon can be dissolved in the austenite.

Medium carbon steels

Consider what happens when a medium carbon steel containing 0.4 weight percent carbon is heated to 1,000°C and then cooled to room temperature. At 1,000°C all of the carbon is dissolved in the austenite. As the iron cools, ferrite begins to form at approximately 780°C. The ferrite grains grow larger as the system cools further, until just above 727°C there are grains of both ferrite and austenite in amounts which can be determined by the lever rule. For a carbon content of 0.4 weight percent, the amounts of ferrite and austenite are approximately equal. Cooling below 727°C causes a solid-state reaction in which the austenite decomposes into ferrite and cementite. This reaction is known as a eutectoid decomposition. The term, eutectoid, is used when the parent phase is a solid, in contrast to the previously mentioned eutectic reaction, in which the parent phase was a liquid. The resulting structure consists of very fine alternating plates of ferrite and cementite. This two-phase structure was given the name pearlite because of its iridescent appearance. The structure is so fine that many optical microscopes cannot resolve the lamellar structure. Figure 6-22 was prepared by means of electron microscopy.

When steel is fully cooled, it consists of regions of ferrite, which formed above the eutectoid temperature, and regions of pearlite, which formed from the eutectoid decomposition of the austenite. Figure 6-23 shows a typical ferrite/pearlite mixture.

Figure 6-22 Lamellar pearlite (magnification 500x). (From *Metals Handbook*, Vol. 7, 8th ed., ASM International [formerly American Society for Metals], 1972. Reprinted with permission.)

Figure 6-23 Micrograph (magnification = 500x) for an alloy steel containing 0.3 wt% carbon. (From *Metals Handbook*, Vol. 7, 8th ed., ASM International [formerly American Society for Metals], 1972. Reprinted with permission.)

Low and high carbon steels

Plain carbon steels can all have this microstructure. The carbides are very hard, while the ferrite is relatively soft and ductile. The properties of the steel depend on the relative amounts of the two phases. Low carbon steels, having carbon contents of less than 0.2 weight percent are used in applications such as rolled sheet for use in car bodies. It can be bent or formed by stamping. Higher carbon steels, having carbon contents near 0.8 weight percent, have much higher proportions of carbides and are, consequently, very much harder and durable in service. However, they are less easily formed. They are used in applications such as railroad rails.

Martensite

There is yet another aspect of steel alloys which is not apparent from the equilibrium diagram. When austenite with dissolved carbon is cooled very rapidly, by immersion in water or oil, there is insufficient time for the pearlite to form. However, the face centered cubic structure of the austenite is unstable at room temperature. It undergoes a transformation to a distorted body centered structure in which the carbon atoms remain in solution. This phase is known as martensite (Figure 6-24). It is extremely hard but also very brittle. It has very little impact resistance.

Figure 6-24 Micrograph (magnification = 500x) of alloy steel containing 0.4 wt% carbon, heated and quenched to produce martensite. (From *Metals Handbook*, Vol. 7, 8th ed., ASM International [formerly American Society for Metals], 1972. Reprinted with permission.)

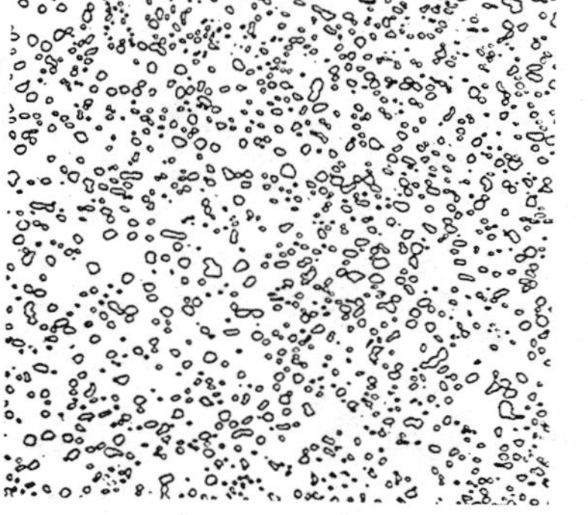

Figure 6-25 Micrograph (magnification = 500x) showing spheroidal graphite particles in steel. (From *Metals Handbook*, Vol. 7, 8th ed., ASM International [formerly American Society for Metals], 1972. Reprinted with permission.)

When martensite is tempered by heating at temperatures below those that would produce austenite, the carbon can come out of solution and form carbides. The carbides will be finely dispersed spherical particles, not like the plates that were in pearlite (Figure 6-25). Growth of the carbides increases the toughness of the steel while also decreasing its hardness. By appropriate choice of tempering time and temperature, one can obtain the desired combination of hardness and impact strength.

Hardening of steel can be carried out after most of the forming operations have occurred. A heat/quench/temper cycle may introduce some warping in large parts. However, heat treatment can often be done when the part is near completion and only finishing operations remain.

Furthermore, hardening can be done locally on some pieces. Playing a flame over the surface followed by a water spray can harden one area but leave the remaining areas unchanged. Such a procedure can be used to harden tools such as chisels or wedges. The leading edge must be very hard, but impact resistance must be preserved so that the tool does not shatter.

Alloy steels

Other alloying elements are often used in steel to obtain a better set of properties. Some elements promote the formation of carbides in alloys in which high hardness is desired. Other elements slow the eutectoid decomposition so that martensite can be formed at greater depths where the cooling rate during quenching is slower than at the surface. The effect produces deeper hardening materials.

Corrosion resistance may be affected by the addition of alloying elements. Very large amounts (10 to 20 weight percent) of nickel and chromium are used to produce stainless steels, known for their excellent corrosion resistance. These large amounts of alloying elements can stabilize the austenite phase so that it remains even at room temperature. Austenitic stainless steels are very resistant to corrosion. However, they are also very tough and ductile, making machining and forming operations very difficult.

Aluminum

Nonferrous metals are included in many other widely used alloys. Aluminum refining is an example of a relatively recent technology. Aluminum could not be refined economically until the 1940s. The energy required to reduce aluminum ore to aluminum metal is too great to be provided by the oxidation of another element. Consequently, electrochemical means must be employed. In the Hall-Heroult process, an electric current is forced through molten aluminum salts at temperatures of approximately 1,000°C (Figure 6-26). Pure aluminum is generated at one electrode.

Aluminum has several characteristics which make it a valuable engineering material. Under most conditions it is very resistant to corrosion. Although aluminum is chemically a very active metal, the oxide, which forms readily in air, becomes a protective layer which inhibits further oxidation. This behavior contrasts with that of iron, whose oxide, commonly known as rust, flakes off, continually exposing the metal to the air.

Furthermore, aluminum is much lighter than other commonly used engineering materials. Aluminum alloys are used in applications such as in aircraft, where weight reduction is critical.

As is the case with most pure metals, aluminum is too soft to be used without some means of hardening. Aluminum alloys often contain some silicon, copper, or

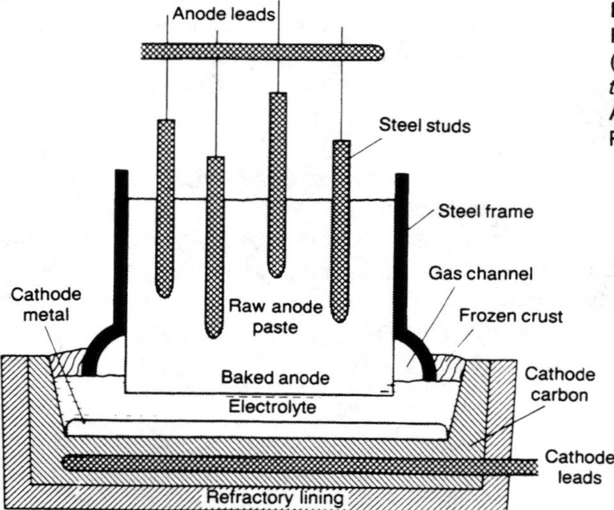

Figure 6-26 Schematic of the Hall Heroult process for refining aluminum. (From *Metals Handbook, Desk Edition*, ASM International [formerly American Society for Metals], 1985. Reprinted with permission.)

magnesium. These elements form finely dispersed precipitates which are critical in the hardening of aluminum alloys.

Age hardening

The process known as age hardening is best described by reference to the aluminum-magnesium phase diagram (Figure 6-27). Observe that the solubility of magnesium in aluminum changes drastically with temperature. At room temperature, less than 0.5 weight percent magnesium can be dissolved, while at temperatures near 450°C, nearly 15 weight percent can be dissolved. An alloy of 4 weight percent magnesium would be a single phase solid solution at temperatures above 250°C, but would be a two-phase mixture of aluminum and the intermetallic compound, β, at room temperature. Such an alloy is typically solution treated by heating it to a temperature above 250°C. The higher the temperature, the faster the solution process occurs. However, too high a temperature would cause melting or undesirable grain growth. After the alloy is solution treated, it is quenched by cooling it rapidly to room temperature. Rapid cooling does not permit the precipitate to form immediately. Rather, the precipitate develops slowly, typically over a few days at ambient temperatures. Consequently, a newly solution-treated and quenched alloy is in a softened condition, permitting forming operations to be accomplished readily. A few days later, the alloy will have become precipitation hardened. The rate of hardening is very sensitive to temperature. Boiling water temperatures can produce hardening in less than an hour, yet the process can be postponed indefinitely at temperatures below freezing. Solution treated rivets can be stored for many weeks in a freezer and then be removed on the day they will be used.

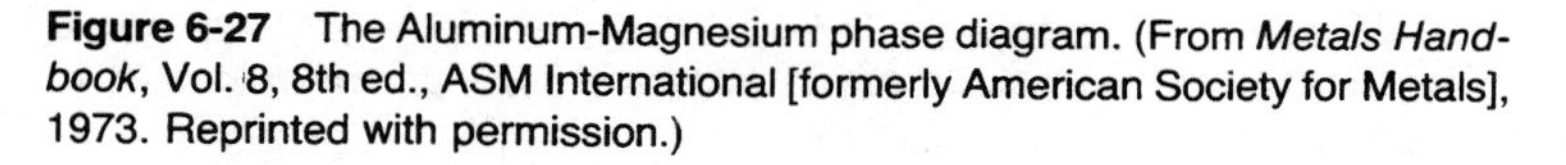

Figure 6-27 The Aluminum-Magnesium phase diagram. (From *Metals Handbook*, Vol. 8, 8th ed., ASM International [formerly American Society for Metals], 1973. Reprinted with permission.)

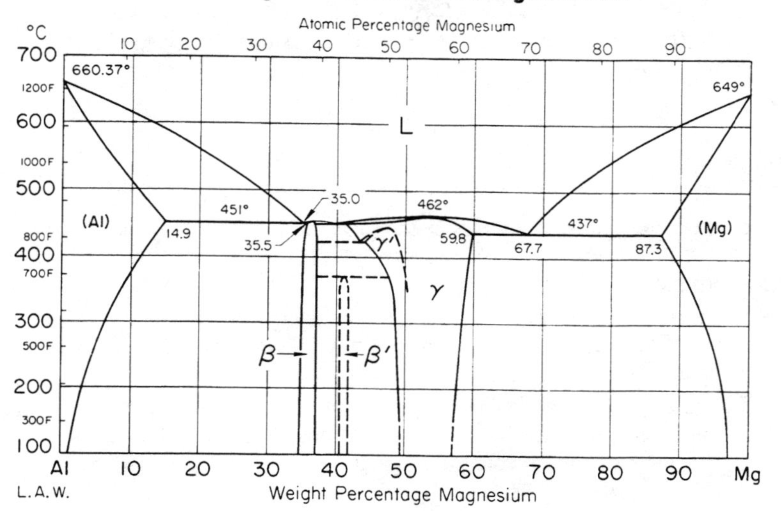

Copper and brass

Copper, another important nonferrous metal, is one of the most easily refined metals. Its early use thousands of years ago during the Bronze Age attests to the ease with which it can be refined from its ore. In some areas of the world, copper can be found in elemental form. Usually, however, copper is found in ores composed mostly of copper oxides and sulfides.

Copper is refined by a thermochemical process. Heating the ore drives off the oxygen and some sulfur oxides, which must be removed by scrubbers to decrease air pollution. Further heating with additional oxygen drives off the rest of the sulfur, leaving ''blister copper'' typically of 98 percent purity. It is so named because of the blistering effect of residual bubbles of SO_2. Additional refining can be accomplished by further fire refining or by electrowinning.

Pure refined copper is very ductile. That feature, along with its high thermal and electrical conductivity and excellent corrosion resistance, makes it a material widely used in electrical wiring and in heat transfer applications. It is often used in flexible tubing.

Copper is also a principal component in bronze and brass alloys. Both of these alloys are quite hard and resistant to corrosion, making them suitable for bearings and for other parts which might be subject to heavy wear or corrosive environments. Copper and its alloys are widely used in domestic and industrial environments.

SUMMARY

Metals are crystalline materials. Pure metals are generally soft and easily deformed. They can be hardened by three processes: cold-working, solution hardening, or precipitation hardening. The behavior of major alloy groups such as steels, aluminum alloys, and brasses can be explained by reference to phase diagrams, which illustrate the phases that exist for any composition and temperature. The mechanical properties of these materials can be modified through appropriate heat treatment processes. Metals can be shaped while they are relatively soft. Then they can be hardened to provide greater durability in service. This capability explains the wide use of metals in manufacturing.

QUESTIONS

1. Examine the micrographs in the volume of the *ASM Metals Handbook* ''Metallography and Microstructures'' to find examples of (a) eutectic microstructures, and (b) precipitate formation.
2. Examine the binary phase diagrams in the appropriate volume of the *ASM Metals Handbook* to determine with which metals aluminum could form an alloy which could be precipitation hardened.
3. Use any available spheres (marbles, styrofoam balls, candies) to construct models of the three principal metallic crystal structures. Make the models several unit cells in size so that the extended planes of atoms are visible. Show that the planes with the most atoms are the farthest apart.

4. Over what range of temperatures will the following plain carbon steels be pure austenite? (a) 0.2 weight percent C, (b) 0.4 weight percent C, (c) 0.8 weight percent C, (d) 1.0 weight percent C.

Answers: (a) 830–1,450°C, (b) 780–1,450°C, (c) 727–1,380°C, (d) 820–1,350°C

5. Describe what phases form as the steels listed in question 4 are cooled from the austenite range to room temperature.

Answers: (a) At 830°C ferrite begins to precipitate. Just above the eutectoid temperature (727°C), the alloy is approximately 75 percent ferrite and 25 percent austenite. At 727°C, the remaining austenite undergoes the eutectoid decomposition to form pearlite resulting a final microstructure which is 75 percent ferrite, 25 percent pearlite.

(b) Ferrite precipitation begins at 780°C. Just above 727°C, the alloy is 50 percent ferrite, 50 percent austenite. At 727°C the austenite decomposes to form pearlite, resulting in a final microstructure of 50 percent ferrite, 50 percent pearlite.

(c) This alloy has the eutectoid composition. It will remain 100 percent austenite until it has cooled to 727°C, at which temperature it will form pearlite.

(d) At 820°C iron carbide, Fe_3C, will begin to form. At temperatures just above 727°C the alloy will be 97 percent austenite, 3 percent Fe_3C. After the decomposition of the austenite to form pearlite, the alloy will consist of 97 percent pearlite, 3 percent carbides.

6. What is the temperature range over which aluminum alloys containing the given amounts of magnesium will be a solid solution? (a) 4 weight percent, (b) 6 weight percent, (c) 8 weight percent.

Answers: (a) 220–600°C, (b) 300–570°C, (c) 340–540°C.

7 PLASTICS MATERIALS, APPLICATIONS, AND PROCESSING

Plastics, once regarded as an inexpensive substitute for more expensive traditional materials, are now regarded as the preferred manufacturing materials for a wide variety of products. The correct application of plastics promises improved efficiencies in transportation, construction, telecommunications, manufacturing, farming, retailing, and medicine. One of the great challenges for those in engineering and manufacturing will be to correctly apply the ever-increasing diversity of polymers to the wide range of products produced.

Plastics, as defined by The Society of the Plastics Industry (SPI), are as follows:

> A large and varied group of materials which consist of or contain as an essential ingredient a substance of high molecular weight which, while solid in the finished state, at some state of its manufacture is soft enough to be formed into various shapes—most usually through the application (either singly or together) of heat and pressure.

It is important to note that in addition to the polymer itself, plastics may incorporate a series of additives to enhance their properties. These additives may include plasticizers to increase the flexibility of the polymer, flame retardants to decrease the rate at which a polymer burns, and colorants to enhance the appearance of a product. In addition, some additives may change the weight or strength of a polymer. Examples of this include fillers, which generally change the weight and reduce the cost of the polymer, and reinforcements, which improve the mechanical properties of the polymer, principally tensile and compressive strengths. Furthermore, blowing agents which enable a polymer to foam may decrease a product's weight and improve its rigidity through the structure of the foam. Finally, additives may be used to improve

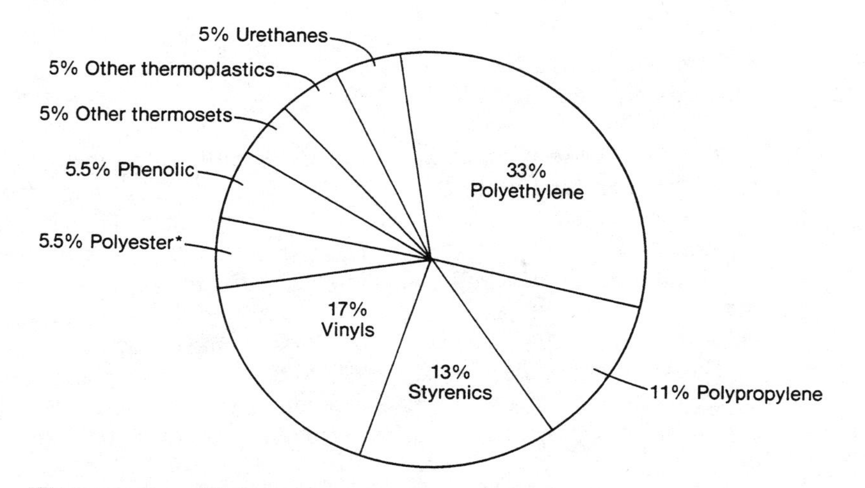

*Thermoplastic and thermoset percentages based on SPI figures.

Figure 7-1 U.S. resin consumption in 1985.

the processability of the polymer. In some cases, lubricants may be used in this capacity to decrease the friction between the resin and the machine. Stabilizers may be used to improve a liquid polymer's shelf life, and catalysts and promoters may be used to increase the speed of polymerization or "setting up."

Polymers may be divided into two broad categories: thermoplastics, those polymers capable of being remelted and reused, and thermosets, polymers which once set or polymerized may never be brought to a liquid state again. A good analogy of a thermoplastic is the behavior of wax. When wax is exposed to heat it becomes a liquid. When the heat source is removed, it becomes a solid. Similarly, when a thermoplastic is exposed to heat it becomes a liquid or achieves a plastic state. When the heat is removed, the polymer returns to a solid state. In the case of a thermoset, a good analogy is that of concrete. Once water is added to the ingredients of concrete, it attains a liquid state. Following an exothermic or heat-producing chemical reaction, the concrete sets up. Once hard, the concrete can never be reliquified and may only have its shape changed through some type of material removal such as breaking, sawing, or chiselling. Similarly, if the shape of a thermoset part needs to be altered, the only option available is generally to machine or abrade the surface of the existing part so that it conforms to the new desired shape.

A majority of the polymers processed are thermoplastics because of their wide range of capabilities and ability to be reprocessed. In Figure 7-1 it should be noted that over three-quarters of the polymers processed in the U.S. are thermoplastics. In addition, it is noteworthy to recognize that polyethylene alone constitutes roughly one-third of all plastics processed in the U.S. This is due principally to polyethylene's wide application in the areas of film and food packaging production.

THERMOPLASTICS

Thermoplastics are divided up into a number of groups, often referred to as families. Families are grouped together because of their close chemical structure. In fact, in many cases members of the family are minor variations of each other. Thus, this section will examine the general characteristics of certain families of plastics as well as significant individual polymers.

Acrylics. Acrylics, or more precisely, polymethyl methacrylate (PMMA) are noted for their high optical qualities. See Figure 7-2. In fact, they are second only to glass in their ability to transmit light. They have a fair chemical resistance but are attacked by many solvents. Their most notable resistance is that to outdoor weathering, as acrylics will maintain their clarity for decades in some cases. Acrylics are available in clear and a wide range of transparent and opaque colors. Domestic consumption in 1987 totaled 665 million pounds. Areas of application included acrylic sheet commonly found under the trade name Plexiglas. Furthermore, lenses of all description, from precision-ground magnification lenses to those found on the tail lights of every automobile, are commonly produced from acrylic.

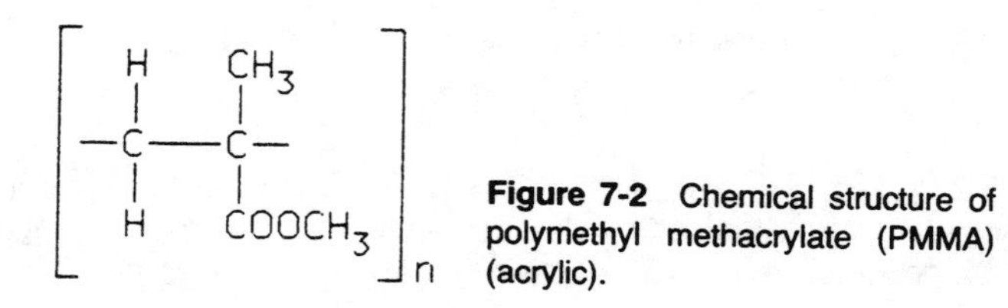

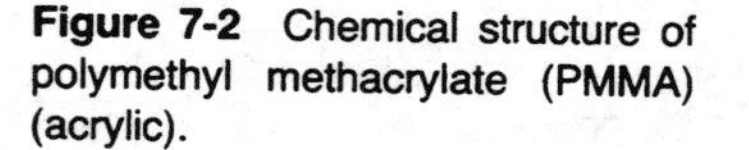

Figure 7-2 Chemical structure of polymethyl methacrylate (PMMA) (acrylic).

Cellulosics. Cellulosics include materials such as cellulose acetate (CA), cellulose acetate propionate (CAP), cellulose acetate butyrate (CAB), and ethyl cellulose (EC). See Figure 7-3. Cellulosics are noted for being rigid, tough, and durable. Their chemical resistance is fair although not outstanding. Their domestic consumption in 1987 equaled 88.4 million pounds. The largest areas of applications were in the sheet and film areas, although a considerable amount of cellulosics are used in the production of hand tools, optics, and personal items such as brushes and pens.

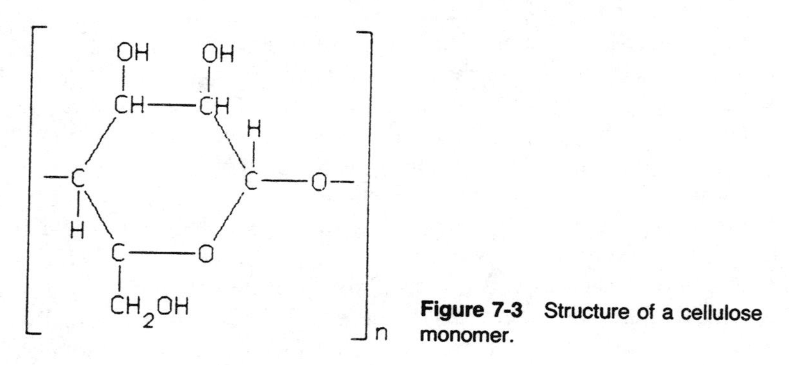

Figure 7-3 Structure of a cellulose monomer.

Polyolefins. The polyolefin family consists of a number of very important high volume resins, including polyethylene (PE), polypropylene (PP), and the stryenics.

Polyethylene. The polyethylenes consist of a number of major grades based upon the density and molecular weight of the resins, See Figure 7-4. These grades

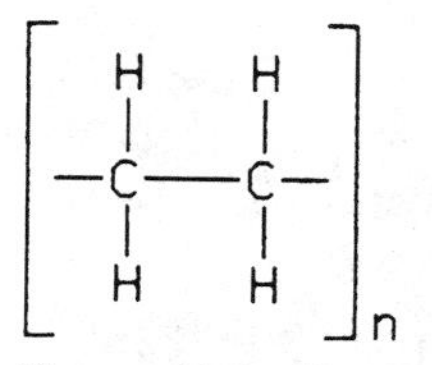

Figure 7-4 Chemical composition of polyethylene (PE).

include low density polyethylene (LDPE), high density polyethylene (HDPE), ultra high molecular weight polyethylene (UHMWPE), and linear low density polyethylene (LLDPE). Generally speaking, the lower the density the greater the flexibility of the resin. Polyethylene is known for its superior chemical resistance and low cost. In fact, it is one of the preferred packaging materials for very strong acids and bases. Polyethylene is available in a wide range of translucent and opaque colors. Domestic consumption of polyethylene in 1987 topped 25.94 billion pounds. Major markets included clear films, shrink wraps, coatings, containers, and bags, with the major consumers in the areas of packaging and film and sheet production.

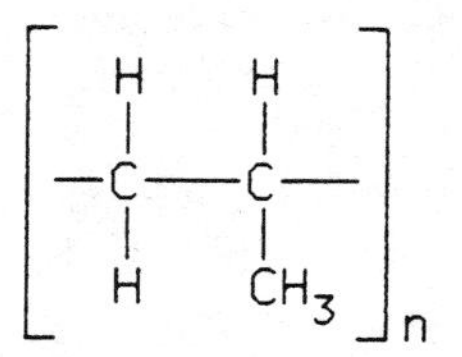

Figure 7-5 Structure of a polypropylene (PP) molecule.

Polypropylene. Polypropylene is a dense, somewhat flexible material and may easily be mistaken for ultra-high molecular weight polyethylene. See Figure 7-5. It is strong and has excellent chemical resistance; however, it may pose coloration problems in some areas. Domestic consumption of polypropylene in 1987 topped 6.47 billion pounds. Major application areas included fiber and filaments, film, containers and caps, automotive batteries, and appliance parts.

Styrenics. The styrenics or styrene group includes such materials as polystyrene (PS), high impact polystyrene (HIPS), styrene-acryonitrile (SAN), and acrylonitrile-butadiene-styrene (ABS).

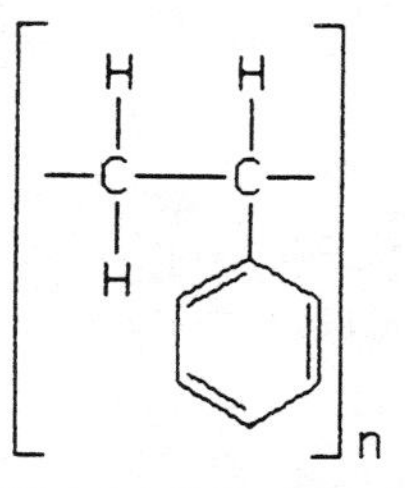

Figure 7-6 Chemical makeup of polystyrene (PS).

Polystyrene. Polystyrene is a rigid, low cost, and reasonably brittle material with fairly low heat and chemical resistance. See Figure 7-6. Domestic consumption in 1987 topped 4.86 billion pounds. Major applications include all types of expandable polystyrene bead and Dow styrofoam products as well as disposable packaging, flatware and cutlery and housewares.

Acrylonitrile-butadiene-styrene. Due to ABS's impact modification through the addition of butadiene, it may be applied to a much broader range of durable items than polystyrene. See Figure 7-7. ABS has fair chemical resistance and is attacked by numerous solvents. However, it has very good dimensional stability

Figure 7-7 Chemical structure of acrylonitrile-butadiene-styrene (ABS). Impact resistance is varied by changing the proportion of butadiene (upper right).

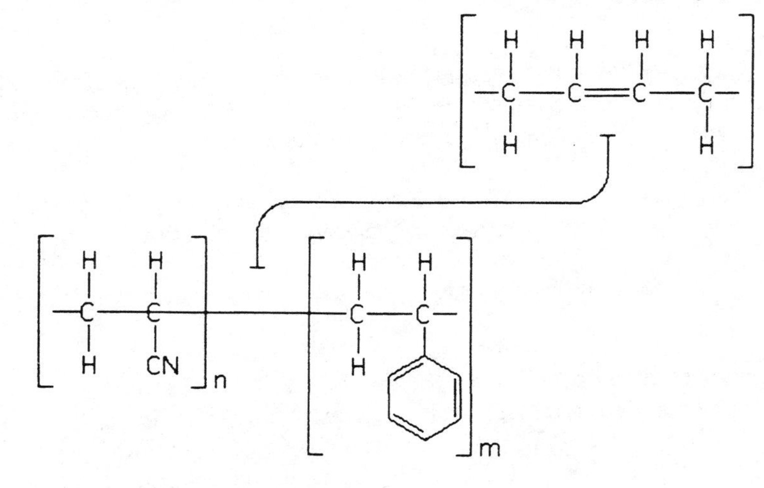

and is one of the leading resins used for electroplating. Domestic consumption of ABS was 1.19 billion pounds in 1987. Major applications include automotive grillwork and trim, water pipe used in building construction, and appliance parts.

Vinyls. The two principal members of the family of vinyls include polyvinyl chloride (PVC) and polyvinylidene chloride (PVDC).

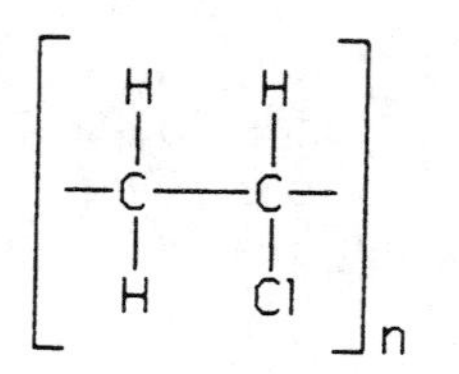

Figure 7-8 Chemical structure of polyvinyl chloride (PVC).

Polyvinyl chloride. Polyvinyl chloride is available in powder, granular, and liquid forms. See Figure 7-8. In addition, it is also available in a wide range of durometers or "flexibilities." Those grades that are rigid tend to have excellent electrical properties and are fairly resistant to exterior weathering and moisture. PVC has fairly good chemical resistance and is only attacked by certain solvents such as toluene. Domestic consumption of PVC in 1987 approached 8.06 billion pounds. Major markets included floor coverings, rigid vinyl siding, gutters and windows, water and sewer pipes, electrical wire and cable sheating, and upholstery materials.

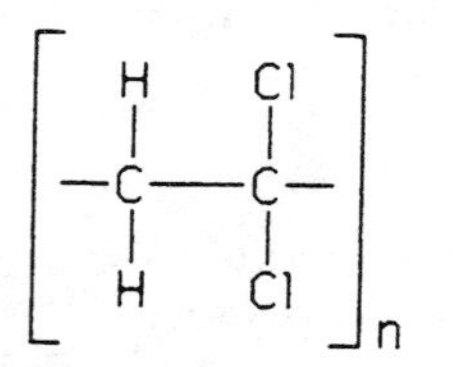

Figure 7-9 Chemical composition of polyvinylidene chloride (PVDC).

Polyvinylidene chloride. Polyvinylidene chloride is noted for its flexibility and its superior resistance to both water and oxygen. See Figure 7-9. Its principal use is for film and sheet material used in packaging of food products. In fact, polyvinylidene chloride is the principal ingredient in Dow's Saran Wrap.

Fluoroplastics. The fluoroplastics are a large family of fluorine based resins including polytetrafluorethylene (PTFE), fluorinated ethylenepropylene (FEP), ethylene tetrafluoroethylene copolymer (ETFE) polyvinyl fluoride (PVF), and perfluoroaikoxytetrafluoroethylene (PFA).

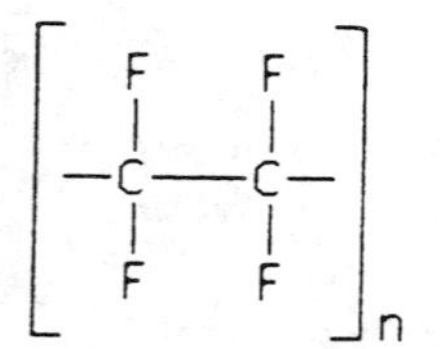

Figure 7-10 Structure of polytetrafluorethylene (PTFE).

Polytetrafluorethylene. This resin is noted for its superior chemical resistance, excellent heat resistance, low coefficient of friction, and extremely good wear qualities. See Figure 7-10. However, the cost for the resin is fairly high, in the neighborhood of $6.00 per pound (1988). High cost is typical of all fluoroplastics and, in fact, some resins may run as high as $30.00 plus per pound. Major uses for PTFE include chemical resistant linings, antiwear surfaces, and some nonstick surfaces used on cooking utensils.

Engineering polymers. Unlike most of the families of plastics, the title engineering polymers does not necessarily refer to the chemical similarities between the types of resins, but rather to their capacity to endure large structural loads.

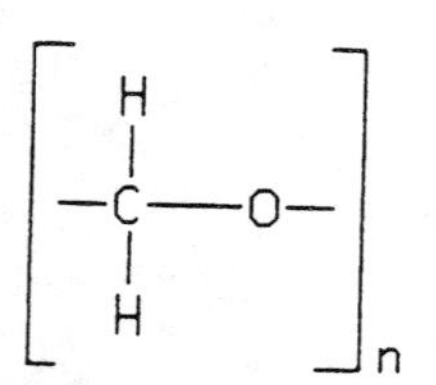

Figure 7-11 Chemical structure of polyacetal (acetal).

Polyacetal. Polyacetal usually just referred to as acetal is noted for being very strong and rigid with good resistance to vibration, chemicals, and abrasion. See Figure 7-11. Its U.S. consumption in 1987 amounted to 122 million pounds. Applications included power tools, industrial components, plumbing, and transportation parts.

Polycarbonate. Polycarbonate (PC) is noted for being extremely rigid with some of the highest impact strengths available in an engineering polymer. See Figure 7-12. Domestic consumption for 1987 totaled 387 million pounds. Polycarbonate has fairly good chemical resistance, especially where acids are concerned. Excellent

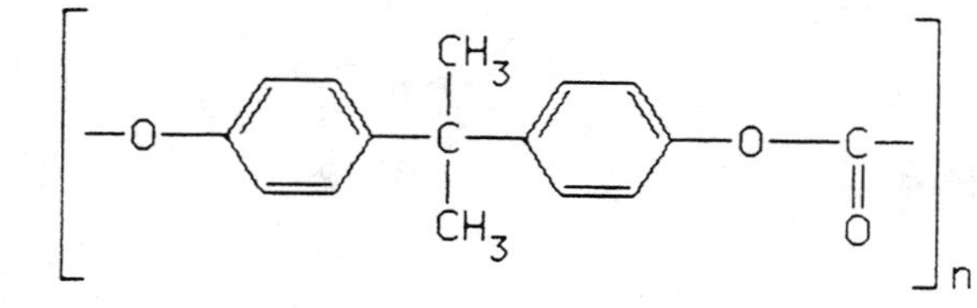

Figure 7-12 Structure of a polycarbonate (PC) molecule.

applications for polycarbonate include sporting goods such as shoulder pads and helmets, high impact lenses and brackets for the transportation industry, high impact clear sheet, and appliance and electrical parts. Polycarbonate is so impact resistant that it is now used for many bullet-proof vests.

Polyamides "nylon." Polyamides (PA) represent the oldest group of "engineering polymers," with the introduction of nylon in 1938. See Figure 7-13. Nylon is characterized by a rigid, glassy surface, high processing temperatures, and high strength. The major drawback of all types of nylons is that they are hygroscopic or water absorbent. In addition to nylons, other polyamides are represented by Dupont's Nomex and Kevlar aramid fibers. These fibers have played a key role in the development of polymer composites, since they are less brittle, lighter, and stronger than traditional fiberglass. U.S. consumption of nylon consisted of 471 million pounds in 1987. Nylon is well known for its excellent electrical properties and chemical resistance. Typical applications include film, electronics, and industrial applications such as gears and bearings, with the largest single market being the transportation industry. It is noteworthy that with the introduction of nylon components in vehicle transmissions that the noise emissions from those vehicles has dropped significantly.

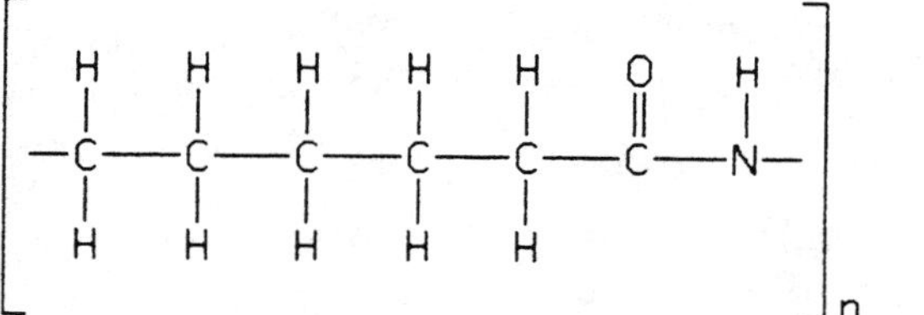

Figure 7-13 Chemical composition of polyamide (PA) (nylon 6).

Polyamid imides. Polyamid imides (PI) constitute a tiny portion of the plastics produced. See Figure 7-14. However, because of their unique properties, they lend themselves to very harsh and severe mechanical applications. An example of this material is Amoco's Torlon. These polymers tend to have surface temperatures

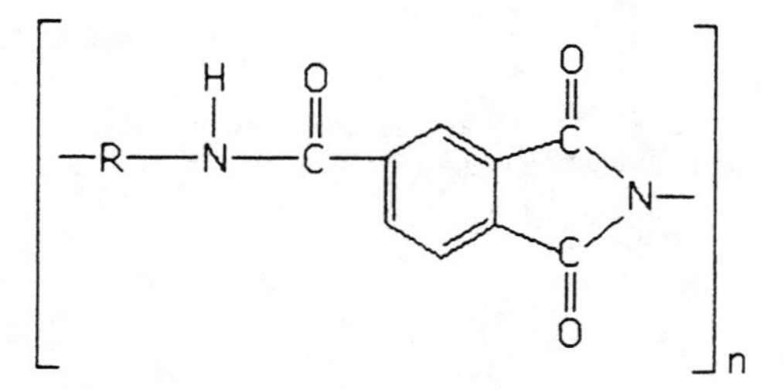

Figure 7-14 Structure of polyamid imide (PAI).

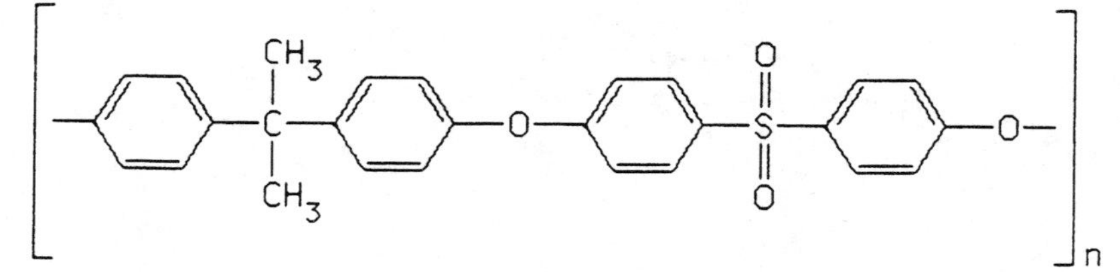

Figure 7-15 Chemical structure of polysulfone (PSF).

around 500° F, with peak tolerance approaching 900° F. It is very significant to note that these are indeed thermoplastic materials. Typically, these polymers have both high impact strength and high abrasion resistance and may in fact offer self-lubrication with the addition of a graphite composite. Typical costs for these materials may range from $5.00 to $20.00 per pound. Typical applications include power transmission and bearing parts, engine, and other mechanical components.

Polysulfone. Polysulfone (PSF) is a rigid, high temperature polymer generally used in conjunction with mineral or glass fillers and reinforcements. See Figure 7-15. It is a prime candidate for metal replacement due to its high impact strength and rigidity at elevated temperatures. Polysulphone occupies a rather tiny portion of the domestic market due to its high cost, approximately $4.00 per pound in 1988. Polysulphone exhibits good chemical resistance in some cases but may be dissolved or attacked by benzene, chlorine, methylene chloride, and acetone. Typical applications include small appliance parts, mechanical components, electrical connectors, and some medical apparatus.

Polyphenylene-based alloys. Polyphenylene-based alloys include polyphenylene oxide (PPO), polyphenylene sulfide (PPS), and polyphenylene ether. Domestic consumption of all polyphenylene-based alloys amounted to 175 million pounds in 1987.

Polyphenylene oxide. Polyphenylene oxide is a rigid polymer that feels much like polystyrene or ABS. See Figure 7-16. An example of this material is GE's Noryl. Polyphenylene oxide is noted for its dimensional stability and superior mechanical and electrical properties. In addition, PPO is a good candidate for structural foam. Many PPO components are formed to improve their rigidity and reduce their weight. PPO has good chemical resistance, with the exception of certain hydrocarbons. Typical applications include appliances, transportation, and business machines. In fact, the majority of all computer housings are made of polyphenylene oxide structural foam.

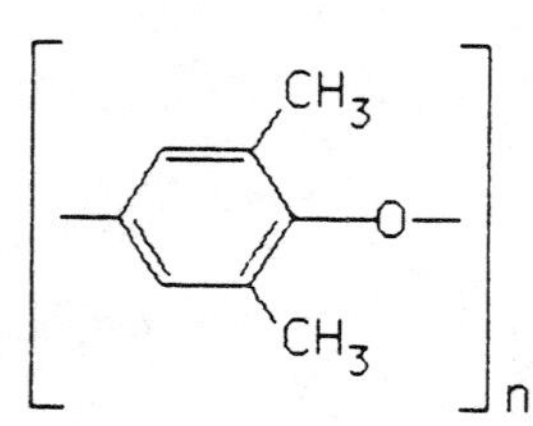

Figure 7-16 Chemical make-up of polyphenylene oxide (PPO).

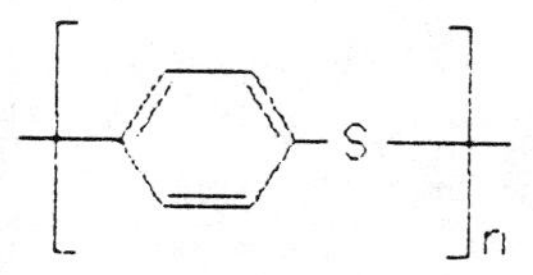

Figure 7-17 Structure of a polyphenylene sulfide (PPS) molecule.

Polyphenylene sulfide. Like PPO, PPS is a rigid material. See Figure 7-17. However, it has improved heat stability and outstanding chemical resistance. Like PPO, primary areas of application include electronics and certain mechanical parts.

Thermoplastic polyesters. Thermoplastic polyesters include polyethylene terephthalate (PET) and polybutylene terephthalate (PBT). The use of both engineering grades of PBT and PET and nonengineering grades of PET have increased sharply in recent years. Engineering grades alone account for 133 million pounds of resin per year as of 1987.

Polyethylene terephthalate. PET may be found in both engineering and nonengineering grades. See Figure 7-18. Engineering grades of PET generally have higher crystallinity and are coupled with glass or some other reinforcing material. Engineering grades of PET are noted for their extremely high mechanical strengths and the stability of these properties at elevated temperatures. Nonengineering grades of PET account for 1.67 billion pounds in 1987. Primary uses of PET include fibers used for fabrics and tire cord, film used for freezer-to-microwave food pouches, blow molded beverage bottles, and mechanical components for the transportation industry.

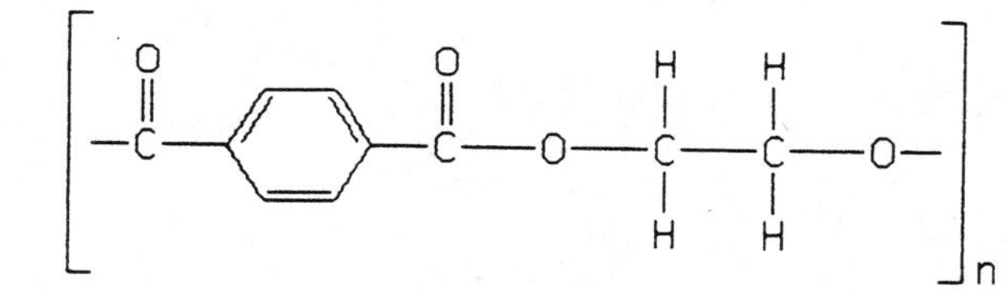

Figure 7-18 Chemical structure of polyethylene terephthalate (PET).

Polybutylene terephthalate. Like PET, PBT is rigid, with excellent electrical and mechanical properties and low moisture absorption. See Figure 7-19. Major uses include electrical appliances, industrial components, and exterior and mechanical automotive parts including such things as radiator top and bottom tanks.

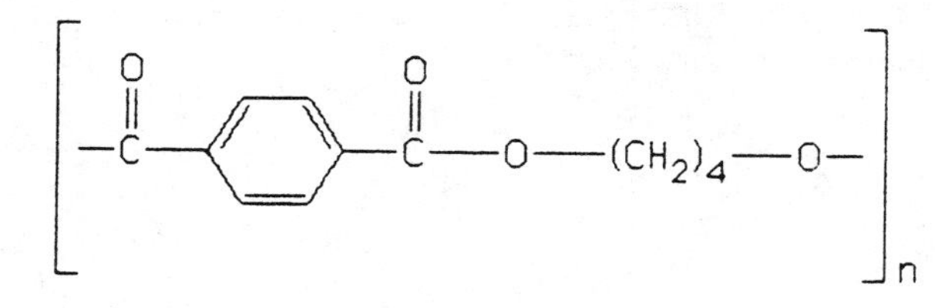

Figure 7-19 Chemical composition of polybutylene terephthalate (PBT).

THERMOSETS

When considering thermosetting resins, it is important to remember that any scrap generated cannot be remelted and reused. Thus, rejected thermosetting parts will pose a disposal problem unless they can be ground and used as a filler for some other process. The most serious disposal problems exist in large, sheet molded compound polyester and reaction injection molded urethane components. At this writing, creative solutions have not been widely employed and the majority of rejected parts are placed in sanitary landfills. It is the author's opinion that this is both economically and ecologically unsound and the employment of creative solutions poses a major challenge for the plastics industry.

Epoxy. Epoxy is a two-part liquid or semiliquid resin. When the resins are combined, they produce a hard, extremely strong, high temperature polymer. See Figure 7-20. Dependent upon the grade used, epoxies may range from a very brittle to semiflexible material. Domestic consumption of epoxy in 1987 amounted to 404 million pounds. Along with epoxy's high temperature resistance and superior mechanical properties, it is noted for its excellent chemical resistance and electrical insulation.

Due to epoxy's impervious nature, nearly 45 percent of all epoxies used are employed as surface coatings in one form or another. These coatings may be used to protect everything from concrete floors to chemical storage tanks. Other major markets include the electronics industry, adhesives, and tubing. However, one of the most fascinating areas of application is the use of graphite epoxy composites to form the structural members and outer skins of both aircraft and spacecraft. These graphite epoxy composites represent some of the highest strength-to-weight ratios of any material. Although the number of tons used in this area is relatively small at the present time, this technology is projected to revolutionize the aerospace industry.

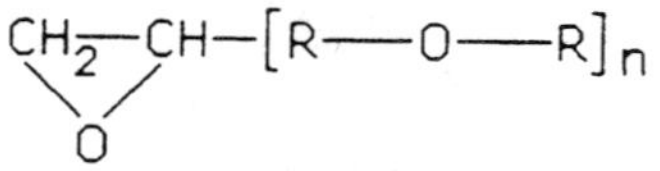

$$CH_2 - CH - [R - O - R]_n$$
$$O$$

Figure 7-20 Structure of epoxy.

Formaldehyde-based resins. Formaldehyde-based resins include melamine formaldehyde (MF), phenol formaldehyde (PF), and urea formaldehyde (UF).

Melamine formaldehyde. Melamine is a very hard, low impact resin with moderate chemical resistance. See Figure 7-21. It performs well under elevated temperatures, and unlike many thermosetting resins, it accepts colorants well. Major uses of melamine include high pressure laminants for countertops, adhesives, protective coatings, and some housewares.

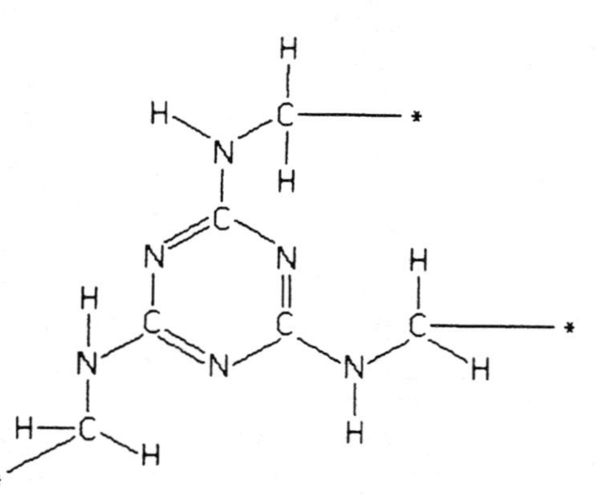

Figure 7-21 Chemical structure of melamine formaldehyde (MF).

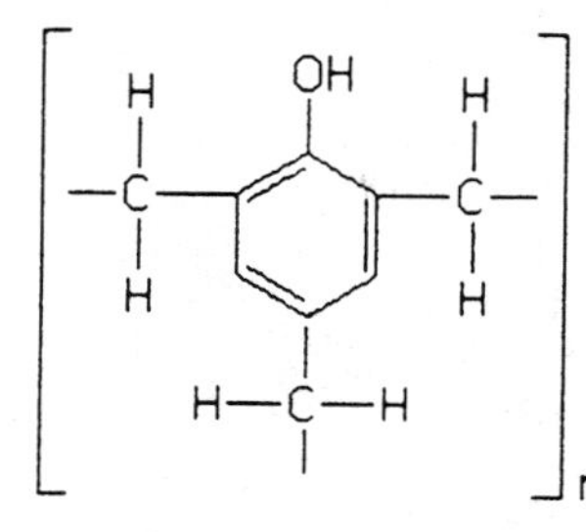

Figure 7-22 Chemical makeup of phenol formaldehyde (PF) (phenolic).

Phenol formaldehyde. Phenol formaldehyde as shown in Figure 7-22, commonly known as phenolic resin or under the trade name "bakelite," is like melamine in that it is a hard and brittle resin. Phenolic's principle attributes are that it has superior chemical resistance and electrical insulation properties. Although phenolic was once the major resin in the plastics industry, its market share has steadily fallen to a level of 197 million pounds in 1987. Major markets for phenolic components are in the electronics and telecommunications area. Furthermore, a considerable amount of the resin is used in the production of appliances, housewares, and to some extent in the transportation industry.

Polyesters. Unsaturated polyesters are known for their hardness and chemical resistance. See Figure 7-23. However, without the addition of a filler or reinforcement, polyester tends to be extremely brittle in nature. For this reason, the vast majority of all thermosetting polyester is used in conjunction with glass fibers and/or mineral fillers. Domestic consumption of thermosetting polyester has stabilized at about 1.62 billion pounds in 1987.

Major markets for polyester include building construction, corrosion resistant mechanical products, and transportation, including aircraft, aerospace, surface transportation, and marine vessels. The combined transportation market accounts for nearly 40 percent of all polyester used. In fact, the vast majority of all privately owned marine vessels from canoes and kayaks to large yachts have polyester-glass composite or "fiberglass" hulls.

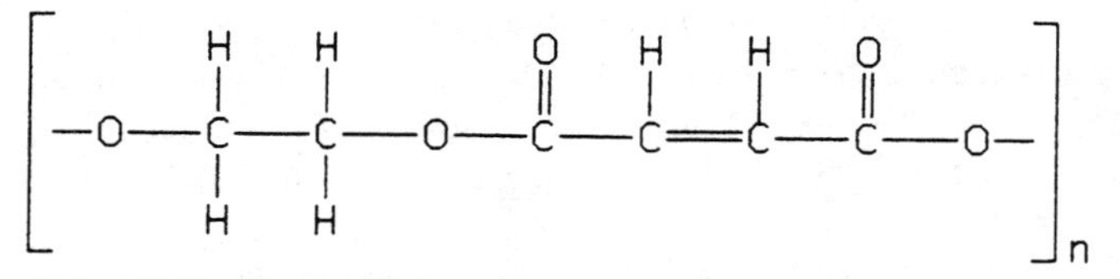

Figure 7-23 Structure of a polyester molecule.

Polyurethanes. Polyurethanes (PUR) may be found as either a rigid or flexible thermosetting foam or as a thermosetting or thermoplastic elastomer. See Figure 7-24. Elastomers are flexible, solid resins and may be found in a wide range of durometers from extremely flexible to very dense and rigid. Rigid urethane elastomers provide a tough, impact and abrasion resistant material that may be applied where rubbers are not able to function. Rigid thermosetting foams provide superior insulation qualities as well as low flammability. Thermosetting flexible urethane foams provide

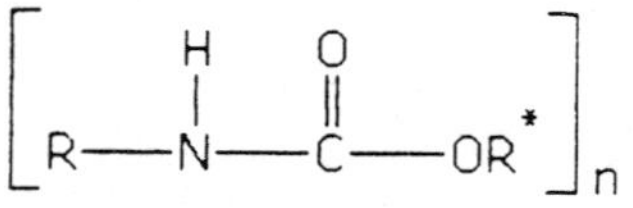

Figure 7-24 Chemical structure of polyurethane (PUR).

the majority of cushioning materials used in the world today. Total consumption of elastomers, flexible and rigid foams, reached 2.68 billion pounds in 1987.

Major applications of polyurethanes include flexible foam cushions used in furniture in the transportation industry, and flexible foam pads for carpets and bedding. Rigid PUR foam is used as insulation for buildings, refrigeration, and other industrial products. Furthermore, reaction injection molded elastomers are used principally by the transportation industry for fenders, door panels, and bumpers.

Silicones. Silicones (SI) are tough and extremely flexible materials which are resistant to virtually all chemicals. See Figure 7-25. Silicones may be found as either a one part air-curing resin or a two-part air or heat-curing resin. Major uses for silicones include adhesives, sealants, caulks, and flexible tooling components.

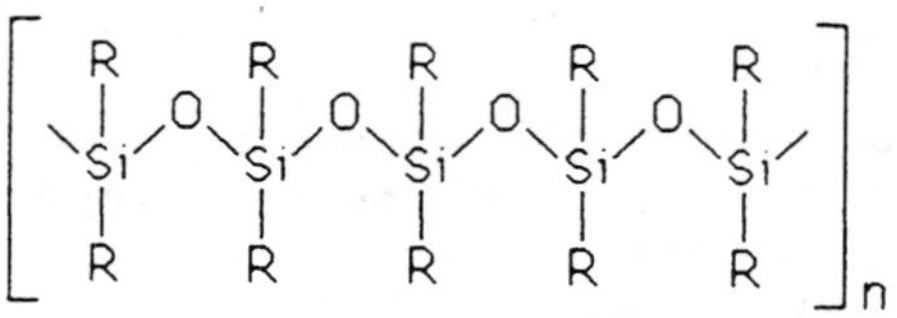

Figure 7-25 Structure of silicone (SI).

APPLICATION OF PLASTICS

Product packaging has become the largest single market for the plastics industry. Some 23.5% of all plastics produced are used for some sort of packaging. The reasons for plastics success in the packaging industry has been due to two major factors. First, with the wide range of polymers available virtually any complex group of packaging requirements can be met. Secondly, through high production plastics processing, a low unit cost may be achieved. That is to say, if the requirements for a product dictate that a great number of products be produced the cost of production falls sharply. This is especially true in molding and extrusion operations. Thus, plastics have offered the packaging industry a wide range of packaging options at a low cost per unit. Major production methods used in the packaging industry include extrusion, injection molding, blow molding, and thermoforming. Special resin properties such as the ability to insulate against shock and abrasion and to produce clear light-weight packages which may be frozen and then cooked make plastics the material of choice for the packaging industry. See Figure 7-26.

The second largest user of plastics is the building construction industry. One-fifth of all plastics produced will end up in some form of building construction. The number of pounds found in the average house rises with each year. New applications of plastics have also revolutionized commercial construction. Major areas of importance include foam insulation to reduce building heat loss and conserve valuable energy resources. Molded plastic electrical components and the polymer coating of

Figure 7-26 This tray is designed to be frozen, then cooked, and finally used to serve the fully prepared meal. (Courtesy of Shaefer Advertising, Inc.)

wires serves to prevent electrical shock and produce maximum electrical insulation. In addition, plastic plumbing pipes, fixture, and fittings provide fast installation and in some cases resistance to rupture due to freezing. Furthermore, the majority of all nonwood floor coverings, including carpet, tile, and coatings, are polymeric in nature. This is true of most construction adhesives, paints, and coatings. In addition, an increasing amount of buildings are constructed with fiberglass shingles and vinyl siding, windows, gutters, and downspouts. The list of potential improvements in construction is nearly limitless. The issue of product flammability is of major importance in the building construction industry. If plastics are to be correctly applied and accepted, they must not only resist burning but fail to give off toxic fumes when heated. Major strides have been made in the area of polymer flammability. However, this is an area of continued research.

Housewares, appliances, furniture, and toys account for about 8 percent of all plastics in use today. Unlike building construction where the majority of products are extruded, the majority of home and office items are molded. There is a high probability that if 100 home and office items were picked up at random, the majority of them would be injection molded. In the housewares area, everything from cookware and laundry baskets to coat hangers and novelty items are made of plastics. The modern household could barely function without the use of plastics in the kitchen. Plastics enable us to store food for extended periods of time both at and below room temperature. The impact of plastics on appliances has been equally dramatic. Consider for a moment a refrigerator with all plastics removed. The liners, drawers, rack supports, handles, name plates, insulation, wiring, and even the paint on the outside would disappear without the use of plastics. This same scenario is true for the majority of business machines, telecommunications equipment, home appliances, hand power tools, and even lawn and garden tools.

Furnishings are another area where plastics have been heavily applied. Unfortunately, wide usage of molded panels to replace traditionally wooden components was tried as a cost-cutting measure. Many poorly designed and poorly constructed products were turned out on the open market. The net effect was that plastics in the furniture industry became synonymous with something inexpensive and of low quality. Only within the last decade have we begun to apply plastics to the furniture industry in a logical way. Many plastic components no longer try to imitate other materials. A primary example of this is the wide use of plastics in the office systems area. A $1,000 office chair, regardless of what it is made out of, is neither inexpensive nor of low quality. The use of plastics in the office environment is able to provide a soft, pleasing, sound absorbent, light and space efficient workplace which would be hard to reproduce without the use of the polymer. See Figure 7-27.

As with furniture, toys suffer from many of the early stigmas about low quality products. Here again, we are finally seeing products designed to be made from plastics in the first place, rather than converted to a low cost resin to save money. The results are durable and attractive toys that may be passed from generation to generation as traditional toys have always been.

By far, one of the areas of greatest potential growth is the transportation industry. Today, the transportation industry only uses about 13.3 percent of all plastics produced. Yet, with gross vehicle weight becoming an increasing concern for the automotive, aircraft, and marine industries, the introduction of increasingly lightweight components is a paramount consideration. Major improvements in the fuel efficiencies of most automobiles and trucks may be attributed in part, to the increased use of plastics throughout the vehicles as shown in Figure 7-28. It is projected that in the very near future the entire automotive frame and floor plan subassemblies will be reaction injection molded in a single operation. Since the polymers used do not

Figure 7-27 An all injection molded work station, fully adjustable so that work surface heights and angles provide maximum comfort for personnel. (Courtesy of Rubbermaid Office Products)

Figure 7-28 Pontiac Fiero showing all of the outer body panels made of plastics. (Courtesy of CPC Division, General Motors)

corrode and are extremely shock absorbent, vehicle damage due to the corrosion and collision should be dramatically reduced. In addition, overall quality should improve since frame and floor plan unibody subassemblies are the critical component in the alignment of a vehicle. With a single molding operation, the potential for error caused during assembly is greatly reduced.

Work is also progressing on polymeric engine and transmission components which could drastically reduce the frictional and inertial losses in a vehicle's drive train. Similarly, even though a relatively small number of external graphite epoxy composite aircraft components are in use today, by the end of the century composites should comprise a major portion of aircraft exterior panels and structural members. See Figure 7-29. In any event, the use of carbon/graphite fiber epoxy composite will inevitably change many of the design characteristics now accepted as standard on aircraft.

Figure 7-29 Polymer composites currently in use on the outer surfaces of the Boeing 767. (Courtesy of E.I. DuPont DeNemours and Company)

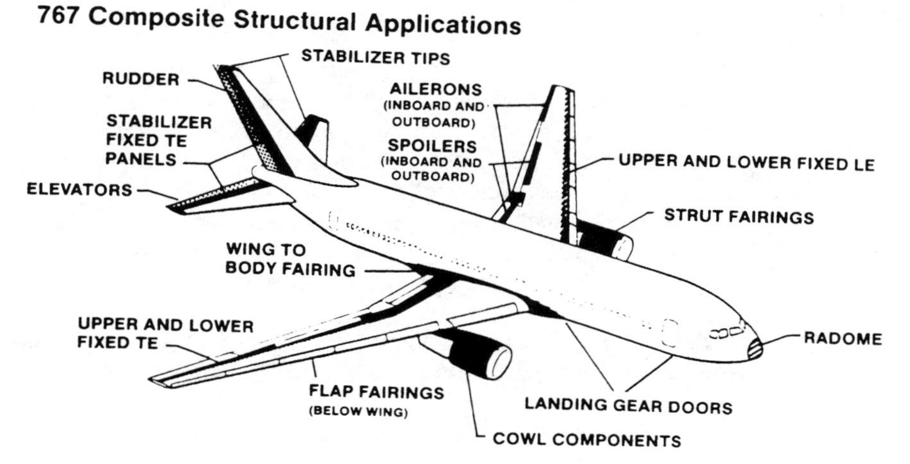

Similarly, the days of linen sails and wooden hulls have been replaced by a lightweight nylon and kevlar® sails and fiberglass polyester hulls. These new craft are lighter, sleeker, faster, and easier to manage than craft made of traditional materials. In fact, the majority of personal pleasure craft have more components made from plastics than from any other types of materials. Everything from the deck cushions and galley countertops to the hull, mast, and sails are made from some type of plastics.

A great deal of medical equipment has been developed through the use of plastics. Everything from suturing staplers to custom made orthopedic braces are made out of plastics. Plastics have also been responsible for several very sophisticated improvements in medical care. Most notable are the introduction of the artificial heart and new developments in artificial polymeric skin to be used on burn patients. The importance of artificial skin development is that it promotes healing and regeneration after severe damage to the body. Finally, we are seeing an increasing number of polymeric drugs, something unheard of only a few decades ago.

The scope of plastics is by no means limited to the few major categories listed here. Plastics are widely applied in agriculture, textiles, personal care products, military hardware, and the list is almost endless. Two fundamental questions that must be asked when applying plastics to products of any variety are: "What are the important design parameters needed to build a successful product?" and "What kind of condition is this product going to be subjected to over its life?" Sometimes the final answers to these questions are that the polymers that will fulfill all of the design requirements are by no means cost effective, either due to material cost or the volume of products needed. The fundamental mistake made early in the plastics industry's history was the over application of certain polymers for the sole purpose of cutting costs. Correctly applied, plastics are superior to any material. Incorrect application can lead to disastrous results and will serve as a liability to anyone connected with the project. Manufacturing must always strive to employ the many positive characteristics of polymers to improve products so that they function more reliably, more efficiently, more conveniently, and more cost-effectively than products made from other traditional materials.

PLASTICS PROCESSING

The range of plastics processes is nearly as vast as the range of polymers themselves. This is due to the wide variety of polymer processing requirements, product design specifications, and part quantities required. Plastics processing may be divided up into several major categories: extrusion, molding, thermoforming, reinforced composites, casting, expansion, coating, decorating, and assembly. Each of these processes has specific strengths and weaknesses and is applicable given specific parameters. The challenge of plastics manufacturing is to specify the best and most efficient process or groups of processes to produce a given product.

Extrusion. Presently, about 33 percent of all plastics are processed into semifinished or finished parts through extrusion. In simple terms, extrusion is a method of continuously melting a polymer and forcing it through a die. The die has an orifice similar to the cross section of the desired finished product. The size and shape of the orifice are dependent upon the melt flow and shrinkage characteristics of the specific resin. The part or ''extrudate'' must be cooled, drawn, and cut to the desired length as it leaves the orifice of the die. A simple illustration of the principle of extrusion is the act of squirting tooth paste from a tube. Commonly, a resin must be extruded in some specific shape and size before it may be molded or formed into finished parts. To effectively handle the large and varied requirements of processing, four types of extruders have been developed: single screw, twin screw, plunger, and gear pump.

Single screw extruders. Single screw extruders as shown in Figure 7-30, are the most versatile and widely used type of extruders found in industry today. They have an extruder screw driven by a motor and reduction gears. The screw rides in a closely fitting barrel which provides a good seal between the top of the screw flights and barrel walls. The screw acts to both auger the polymer forward and add frictional melting heat. In the screw process, the majority of the melting heat is generated through friction. In fact, once a screw extruder is up to heat, 100 percent of the melting heat required can be generated through polymer friction. This is known as ''adiabatic extrusion,'' but is not practical in most production situations since the majority of temperature control is lost. To overcome the control problem, frictional heat is moderated by both electric heating and air or water cooling of the barrel.

Twin screw extrusion. Twin screw extrusion employs the same principles as single screw extrusion. The major difference is that a twin screw extruder has two intermeshing screws. The screws may be designed for either corotation or counterrotation. The action of the two screws greatly improves both mixing and output capacity. However, the machinery and maintenance cost tend to be considerably greater than single screw machines. Twin screw extrusion is generally used for compounding of polymers or on very high production long-running jobs.

Figure 7-30 Cross-section of a typical single screw extruder for plastics. (Courtesy of the Society of The Plastics Industry, Inc.)

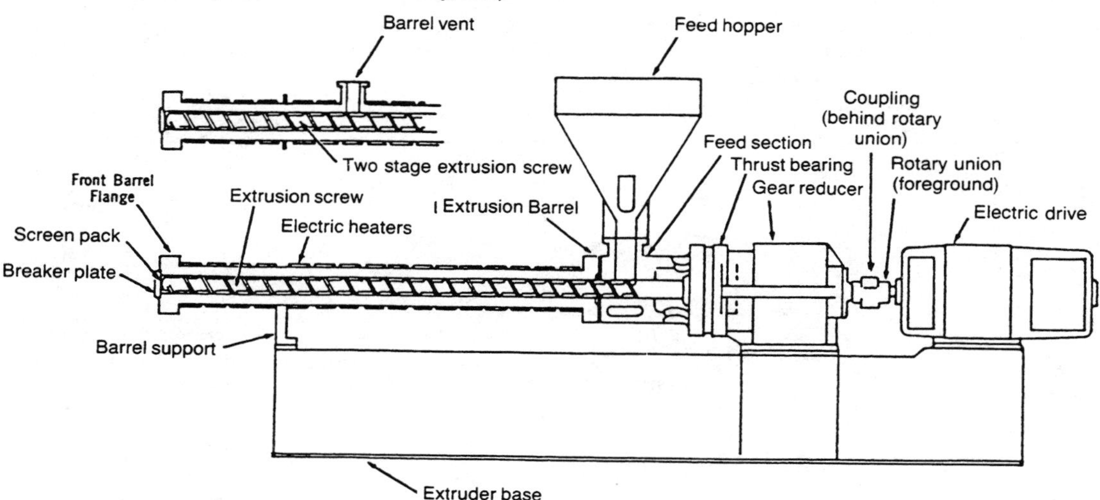

Minor types of extrusion. Ram or plunger type extruders have been developed for both continuous and intermittent operations. In these extruders, a ram or plunger is substituted for an extrusion screw. The major disadvantage is that the melt tends to have poor mixing and thus a nonuniform thermal profile. A second minor category is that of gear pump extrusion. A gear pump is characterized by several sets of intermeshing gears that squeeze the resin forward. A gear pump is generally considered a high constant pressure, low-output machine.

Plastification mechanisms. A screw extruder is divided into zones and sections and characterized by its length to diameter ratio shown in Figure 7-31. Zones refer to the areas of external heating and cooling of the barrel, whereas sections refer to the functional areas along the length of a screw. Generally, the back heat zone of an extruder is cooler to allow the resin pellets or powder to enter the barrel without prematurely melting together and blocking the throat of the machine. In a single stage extruder screw there are three sections: feed, compression, and metering. The feed section forces the unmelted resin forward into the compression section. The compression or transition section compresses, shears, and melts or plastifies the resin. Finally, the metering section thoroughly mixes the melt and insures a relatively even pressure at the breaker plate. The screen pack, breaker plates, and die serve to straighten the flow and further improve the overall pressure characteristics of the melt.

Heat, pressure, and time are the most critical factors in a melt. Polymers as a group exhibit non-Newtonian flow characteristics. Unlike water and most other liquids, a plastics' viscosity is nonuniform in nature. A Newtonian fluid such as water has a uniform viscosity throughout its liquid temperature range. Plastics, on the other hand, become less viscous as their temperature increases. Furthermore, since an increase in pressure will cause an increase in temperature, the potential viscosity changes govern all other aspects of polymer extrusion. The viscosity of the melt is generally a result of polymer shearing. Shearing occurs as the resin is slid along the screw and sticks to the barrel walls. Layers of the melt are slid over one another

Figure 7-31 Cross-section of an extruder barrel showing the screw. (Courtesy of the Society of The Plastics Industry, Inc.)

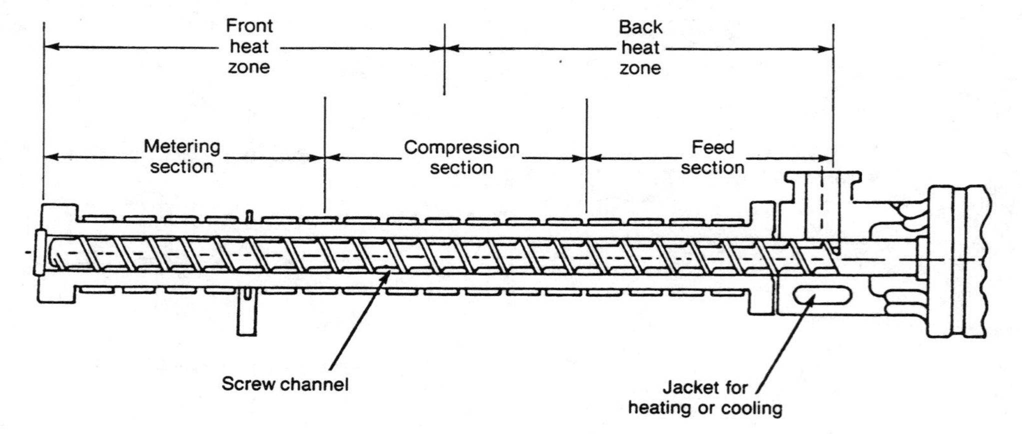

causing the tearing and alignment of the polymer chains. Since shearing is a major producer of plastification heat, both heat history and shearing serve as predictors of polymer behavior, property stability, and potential failure.

If a thermoplastic melt is nonuniform in nature, several major problems may occur. First, the melt may exhibit poor additive mixture. For instance, a poor colorant mixture may result in anything from uneven color intensity to stripes, streaks, and blotches. The second major problem is a nonuniform temperature profile. This may mean that some resin pellets are not melted while others are to the point of degradation. The problem of nonuniform melt is the main reason that plungers are considered very inferior melting tools for polymers.

Another problem that may occur in a polymer melt is that of excess moisture in the resin. If allowed to stay in the melt, moisture will cause bubbles and streaks. Moisture problems are most evident in hygroscopic or water absorbing resins. The majority of engineering thermoplastics exhibit at least some level of hygroscopic behavior. The two major solutions for hygroscopic behavior are drying the resin prior to barrel entry and venting of the barrel.

Elements of the extrusion line. An extrusion line can be divided into up-stream and down-stream equipment. See Figure 7-32. Up-stream equipment includes all of the equipment involved in melting and pressurizing the polymer, whereas downstream equipment is involved in cooling and sizing the extrudate. Thus, everything after the die is considered to be down-stream or down-line equipment.

It is common to have several times the cost of an extruder tied up in down-stream equipment. Cooling and pulling equipment are the most important components

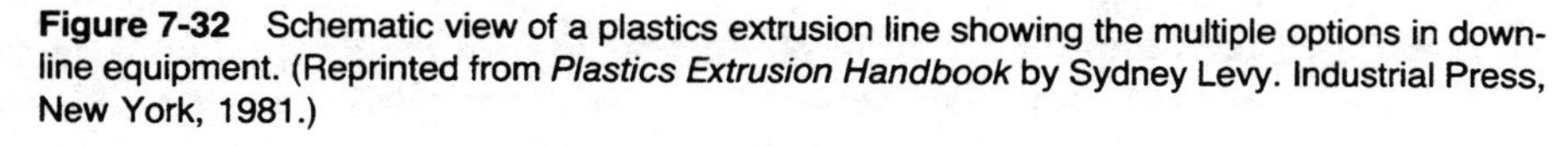

Figure 7-32 Schematic view of a plastics extrusion line showing the multiple options in down-line equipment. (Reprinted from *Plastics Extrusion Handbook* by Sydney Levy. Industrial Press, New York, 1981.)

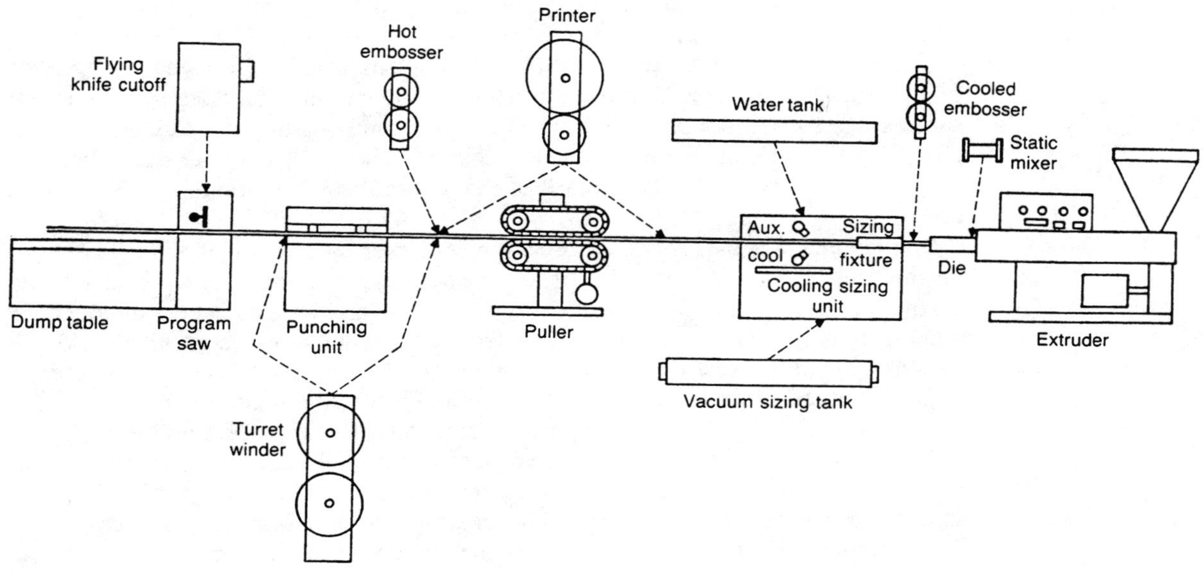

of down-stream equipment; an extruder cannot function without them any more than it could function without a die. However, down-stream equipment may include a wide range of cooling, sizing, pulling, cutting, winding, and on-line measurement apparatus.

Cooling and sizing equipment may in fact be one and the same. Cooling methods include the use of water baths, chilled air, and refrigerated sleeves. Sizing equipment may act to both cool and shape the extrudate at the same time. Examples of this include chilled sizing rollers for sheet, vacuum sizing tanks for pipe and tubing, and chilled fixtures for profile extrusion. Sizing and post forming operations may also include noncooled embossing rolls. A majority of all sizing equipment is built for a specific job and may range in cost from a few dollars to several tens of thousands of dollars.

Delivery equipment, on the other hand, tends to have a greater range of application. This includes pullers, cutters, and winders. Pulling is generally done by either a set of pinch rollers or caterpillar treads. Pinch rolls come in many configurations and serve as the major method of providing steady tension on the extrudate. Without steady tension, proper sizing would be impossible. Caterpillar treads or belt pullers are used on extrudates which might be crushed or distorted by pinch rolls. Often a cutting unit may be directly attached to the puller. Cutting may involve either partial or complete cut-off and may incorporate the use of high speed knives, shears, saws, or non-conventional methods. Finally, it may be necessary to spool or wind coated wire, tubing, or film, whereas sheet stock or pipe are traditionally boxed or palleted.

A more recent development in extrusion has been the employment of on-line measuring devices. It is possible to measure virtually any critical dimension or aspect of the extrudate during the process. The range of measurement methods is quite vast and the creativity with which the methods are employed is the key to future improvement of product quality. In the most refined cases, the part variations are fed to a computerized process controller on the extruder which in turn adjusts elements of the process to compensate for variations in the part.

Types of extrusion processing. Extrusion is usually classified by the type of the products produced. These products fall into eight basic categories with numerous subcategories. Major extrusion categories include compounding, pipe/tubing, profile, rod, filament, sheet film, and extrusion coating. The same extruder could be used for all types of products. The area that changes is the configuration of the die and down-line equipment.

Compounding is the process of combining polymers and additives and then extruding multiple strands of the resin. The strands are then chopped or pelletized into small granules (Figure 7-33) to be used in blow molders, injection molders, or other extruders. Pipe and tubing dies look quite similar to each other, and the most notable difference is that pipe is defined as being generally $\frac{1}{2}''$ in diameter or larger. Tubing is smaller than $\frac{1}{2}''$ in diameter and is usually flexible in nature. Profile extrusions are generally nonuniform in nature, and the range of possible shapes is only limited by the creativity of the product designer. Examples of profile shapes include vinyl house siding and window frames, automotive door gaskets and moldings, stair treads, table edge moldings, and retainers of all descriptions. See Figure 7-34. Frequently, rods are classified as a part profile extrusion. However, round

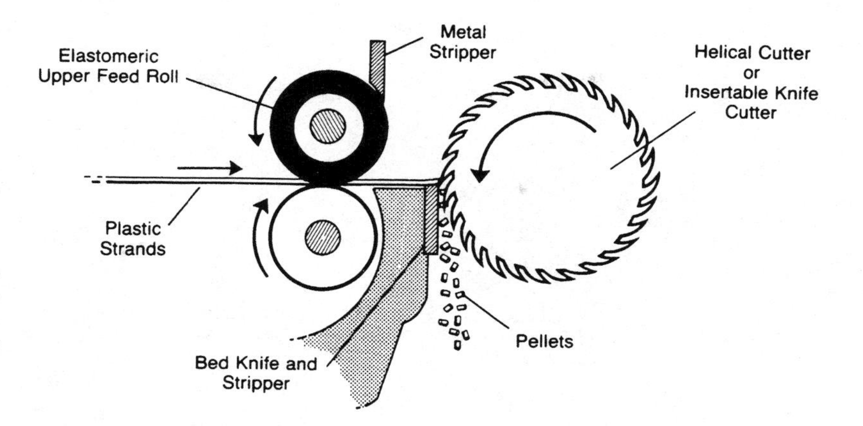

Figure 7-33 Cross-section of a strand pelletizer showing its operation. (Courtesy of American Roller Company)

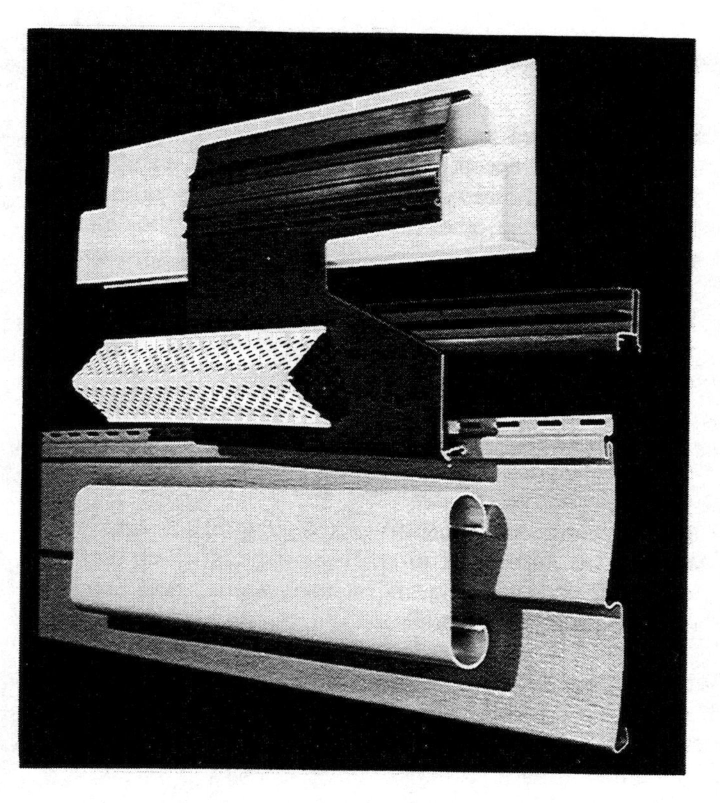

Figure 7-34 Products produced through profile extrusion. (Courtesy of Crane Plastics)

rods are more accurately an extension of filament extrusion. Rods are generally round and larger than .060″, whereas filaments are classified as strands smaller than .060″ in diameter. Filaments of less than .005″ make up the majority of all textile fibers.

By far the most capital intensive lines to set up are sheet and film operations. Sheet is extruded through large flat dies into a three roll mill. See Figure 7-35. Each roll may require an optical quality finish to produce smooth flawless sheet. The rolls are sequentially cooled and adjusted to provide the desired sheet thickness. It is common for a large mill to cost several hundred thousand dollars or more.

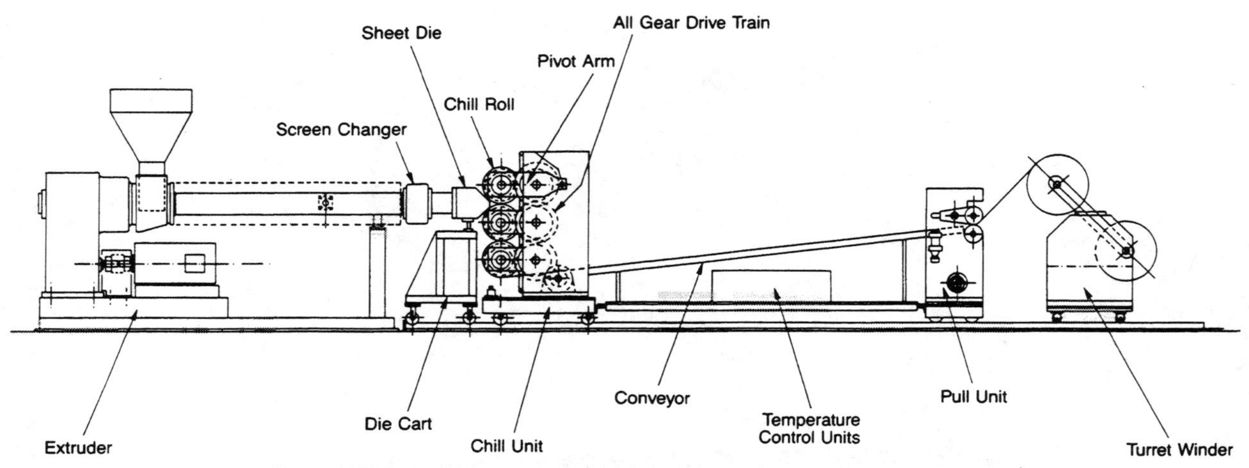

Figure 7-35 A typical sheet extrusion line. (Courtesy of Sterling Extruder Corp.)

Sheet is classified as being larger than .010″ thick. Cast film is quite similar to sheet production except that it uses a smaller die and a single roll mill. Film quality is not superior; however, it is the least expensive method of producing flat stock. Blown film, on the other hand, provides a superior quality product that may be as thin as .00025″. Blown film employs the same principel as blowing up a balloon. In Figure 7-36, a large tube is extruded, usually vertically, and is then blown up, which decreases the wall thickness and increases the diameter of the bubble. Vertical lines generally require a film tower several stories tall and the cost of take-off equipment generally runs in the hundreds of thousands of dollars.

Coextrusion. It should also be noted that any method of extrusion may also be performed with coextruded plastics. Coextrusion is a method of extruding several layers of resin into the same die. The layers do not mix but stay segregated retaining their original properties. This process requires a separate extruder for each resin or color, and is capable of producing products which employ the varied properties of many different polymers. It is little wonder that extrusion is the largest plastics processing method with all of the versatility it has to offer.

Calendaring. Calendaring is another method of producing thermoplastic sheet stock. It is like sheet extrusion in that it requires a series of rollers to achieve the correct stock thickness by squeezing the plastic sheet. See Figure 7-37. However, at this point the similarity to extrusion stops. The melting mechanism in the calendaring process consists of one or more sets of rollers which melt and squeeze a powdered resin in sheet stock. Thus, the calendaring rolls replace the extruder screw in this process.

Molding Processes. Although molding is not unique to plastics, it certainly is the group of processes plastics is best known for. Generally, molding is characterized by three things: a plastification chamber, a clamping unit (usually high pressure), and a unique mold for each new product. There are five major molding processes for plastics: injection, reaction injection, blow, compression, and transfer.

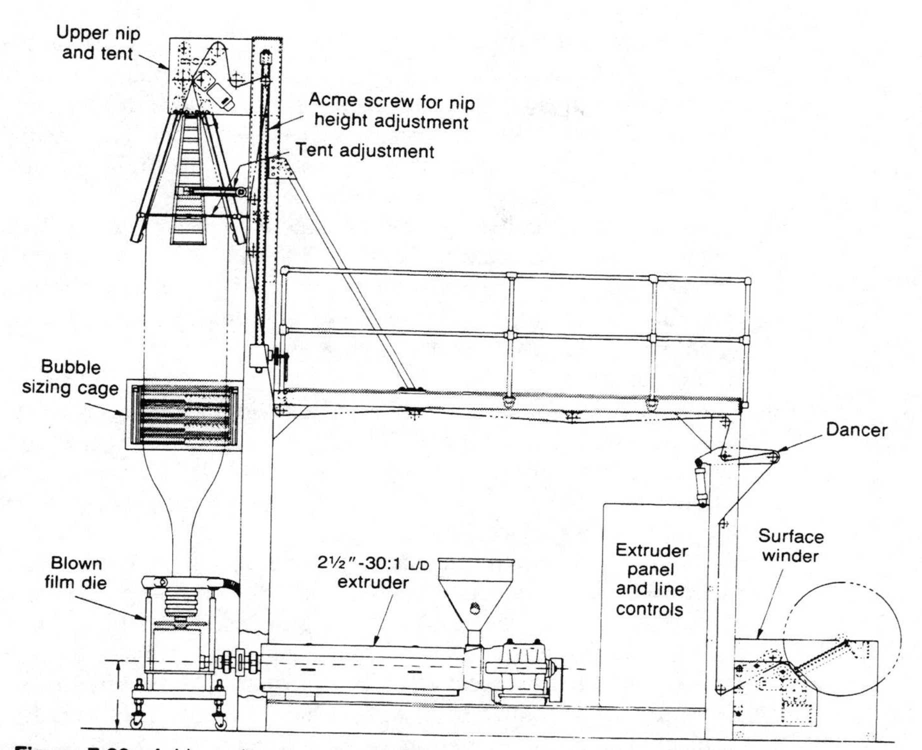

Figure 7-36 A blown film line showing a vertical bubble, overhead take-off, and winder located behind the extruder. (Courtesy of Sterling Extruder Corp.)

Figure 7-37 Diagram showing the plastics path through inverted "L" calendar rolls. (Courtesy of Society of Plastics Engineers)

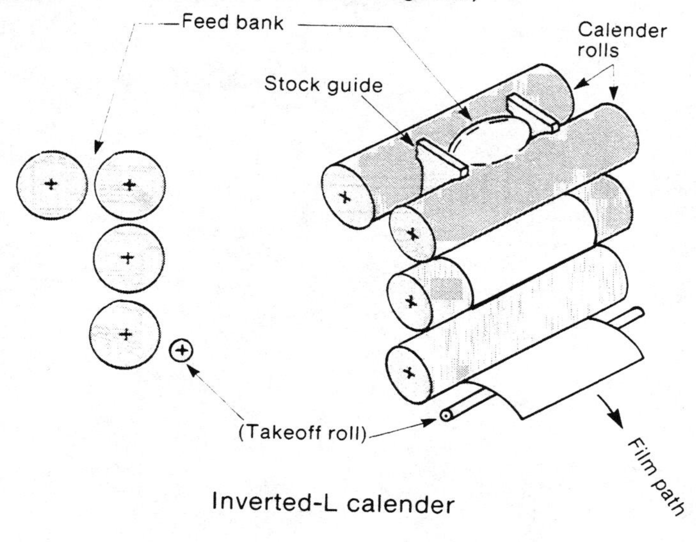

Injection molding. Injection molding is one of the most important processes in use for plastics today. Injection molding is capable of producing finished parts in huge quantities with great detail and precision and at very low unit cost. The key to injection molding's success is the minimal amount of labor per part after the tooling has been produced. In fact, it is possible to produce over 100 finished parts in a single four-second cycle.

An injection molder is very similar to a die caster, with the main difference being in the plastification unit. See Figure 7-38. There are three major types of injection molders: reciprocating screw, plunger, and two-stage. For each machine type the configuration of the plastification unit changes. However, all of them rely upon a precision mold and a high pressure clamping unit to hold the mold closed during injection.

By far the most common type of injection molder uses a reciprocating screw. A reciprocating screw (Figure 7-39) is similar to the screw found in an extruder, with the major difference being at the tip or delivery end of the screw. Unlike an extruder, the injection molder injects parts with resin on an intermittent basis, which serves as only part of the machine cycle. Proper filling of the mold requires that all of the resin be injected at high pressure in a short period of time. Injection often takes only several seconds, and may involve under 20 percent of the total cycle time. During the rest of the cyle the screw plasticizes and holds resin for the next part. In order for the machine to plasticize and hold resin, the screw slides backwards to permit resin to collect in front of the screw tip. A special tip called a nonreturn valve allows resin to flow ahead of the screw until enough resin for the next shot is melted. At the start of injection, the screw moves forward or reciprocates like a piston. The nonreturn valve closes as soon as the screw starts forward preventing resin from flowing back up the screw flights during injection. Once injection is completed, the screw starts to plasticize a new shot of resin. While the screw plasticizes resin for the next shot, the previously injected parts are cooling in the mold. When

Figure 7-38 A typical reciprocating screw injection molder. (Courtesy of Society of Plastics Engineers)

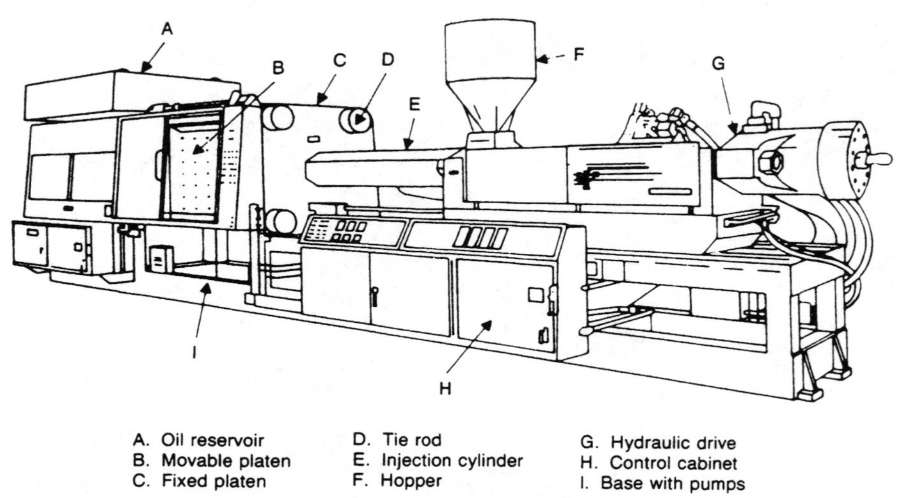

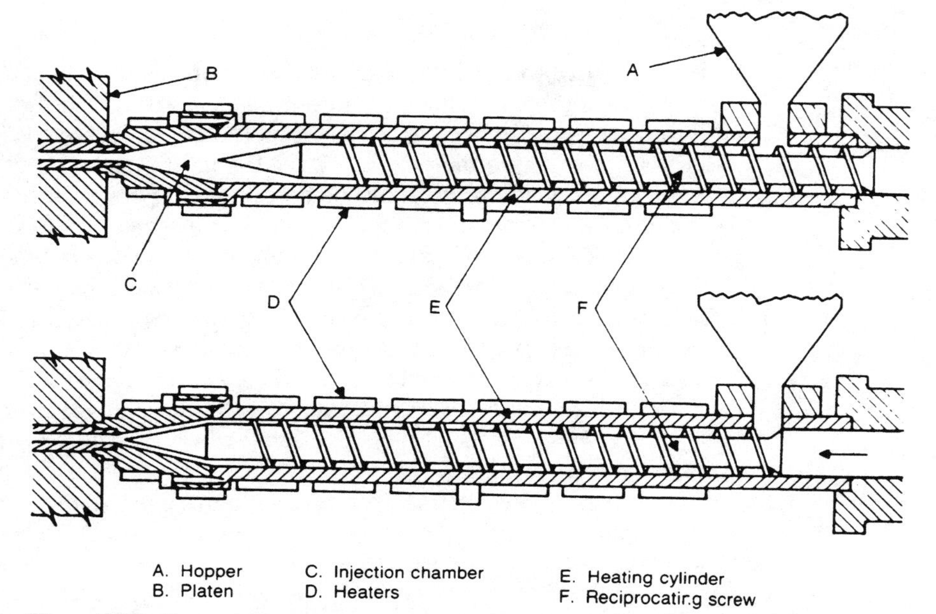

A. Hopper
B. Platen
C. Injection chamber
D. Heaters
E. Heating cylinder
F. Reciprocating screw

Figure 7-39 Cross-section of a reciprocal screw injection molder showing the screw retracted prior to injection (top), screw reciprocated forward during injection (bottom). (Courtesy of Society of Plastics Engineers)

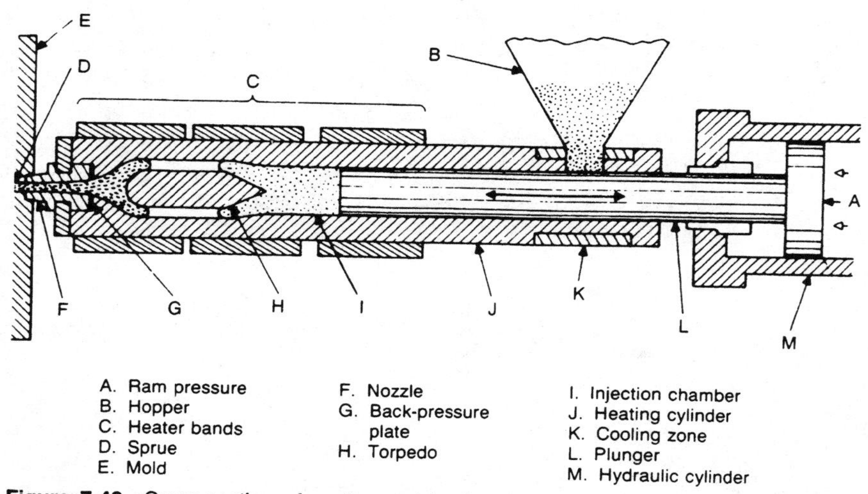

A. Ram pressure
B. Hopper
C. Heater bands
D. Sprue
E. Mold
F. Nozzle
G. Back-pressure plate
H. Torpedo
I. Injection chamber
J. Heating cylinder
K. Cooling zone
L. Plunger
M. Hydraulic cylinder

Figure 7-40 Cross-section of a plunger injection molder. (Courtesy of Society of Plastics Engineers)

the parts are sufficiently cooled, the mold opens and the parts are ejected. The mold then recloses and a new set of parts are injected.

The cycle of a plunger injection molder is quite similar to that of a reciprocating screw injection molder. The difference is that a plunger or ram both plasticizes and injects at the same time. See Figure 7-40. There is also a greater reliance on heater bands to soften the resin pellets. Plunger injection molding is the oldest type of injection molding and represents less than 1 percent of all injection molders built in the U.S. today. Like ram type extrusion, plunger injection molding suffers from all the problems of a nonuniform melt and heat history. The loss of part uniformity, especially in complex parts, makes this process generally uneconomical.

Two-stage injection molding is by far the most complex and expensive process in terms of the machine cost and maintenance. However, it has the capability for the highest production levels and may deliver the most uniform melt of all of the injection molders. A two-stage screw plunger machine acts like an extruder plus a plunger injection molder. See Figure 7-41. The extruder may run all the time which increases the machine's capability to plasticize resin. During the majority of the cycle, the screw extrudes resin into a second barrel in front of a plunger. The plunger moves backwards as the barrel fills. At the time of injection, a valve stops the flow of resin from the extruder and allows the plunger to inject into the mold cavity. Following injection, the valve moves back to allow resin to pass from the extruder into the plunger again.

For all types of injection molders there are two major groups of clamps. Mechanical or toggle clamps are generally found on smaller machines. Toggle clamps are actuated by a small hydraulic cylinder and apply leverage pressure to keep the mold closed. The second group of clamps are fully hydraulic units. Hydraulic clamps are generally found in most sizes of injection molders. They rely on one or more hydraulic cylinders to close and hold the mold during injection. Clamping is critical since the force of injection may result in several thousand tons of internal pressure in very large mold cavities.

The majority of injection molds are complex and expensive to build. They are usually built of special high strength steels and require precise machining. Thermoplastics require a cold mold in order for the resin to solidify. For this reason, injection molds are fluid cooled. In order for a part to be of good quality, the temperature and internal pressure distribution of the part must be fairly uniform at a given point in time. Parts molded under differing conditions exhibit nonuniform physical characteristics and behavior. Thus, the design and construction of the mold is critical to part performance.

Figure 7-41 Cross-section of a two-stage screw plunger injection molder. (Courtesy of Society of Plastics Engineers)

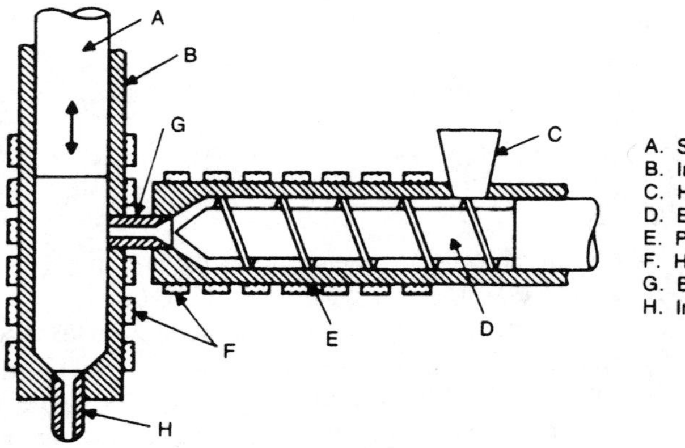

A. Shooting piston
B. Injection cylinder
C. Hopper
D. Extrusion screw
E. Preplasitcizer cylinder
F. Heaters
G. Ball-check nozzle
H. Injection nozzle

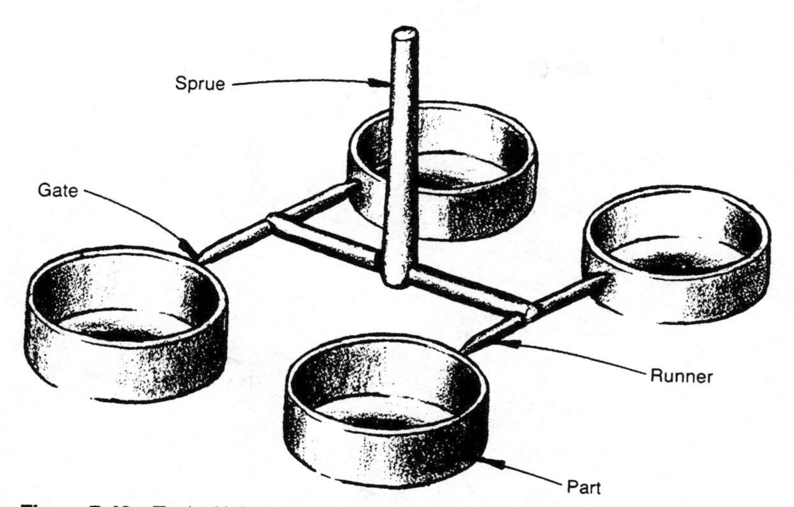

Figure 7-42 Typical injection molded shot showing the sprue, runners, gates, and parts.

As in metal casting, there are four parts of a mold that the molten plastics fill: the sprue, runner, gate, and part cavity. See Figure 7-42. The sprue conducts plastics from the nozzle of the machine to the runner system. The runner system distributes resin to each of the gates. Finally, the gates are the points of entry between the runner system and the part cavities. In cold runner molds the gates, runners, and sprues are cooled and ejected with the parts. However, in a hot runner mold the sprues, runners, and gates are kept fluid so that only the part is cooled for ejection. Resins which lend themselves to hot runner processing can save the cost of runner reprocessing.

Reaction injection molding. Reaction injection molding (RIM) differs from injection molding in several ways. The principle difference is that RIM uses a two-part liquid resin, commonly thermosetting urethane. The resin is mixed by spraying very high pressure streams of each liquid directly at one another in a mixing chamber. See Figure 7-43. This process is known as high pressure impingement mixing. The mixture is then injected into the mold at a considerably lower pressure than in injection molding. Since injection is relatively low pressure, far less clamping pressure is required to hold the mold closed. In addition, molds may be made out of aluminum since they have lower pressure requirements. Generally, RIM parts have large surface areas that make injection molding tooling far too costly. Automotive bumpers, valance panels, and front spoilers have been good candidates for RIM.

Blow molding. Extrusion blow molding is a cross between plastics extrusion and glass blowing. Blow molding is primarily a thermoplastic process and, therefore, the rules that govern thermoplastics extrusion apply here too. In extrusion blow, a tube or parison is extruded at right angles to the floor. This is accomplished either by using a cross-head die to turn the orifice of the die 90 degrees to the floor on a horizontal extruder, or by placing the extruder in the vertical down position. In Figure 7-44, the hot parison is extruded down between two mold halves. The mold

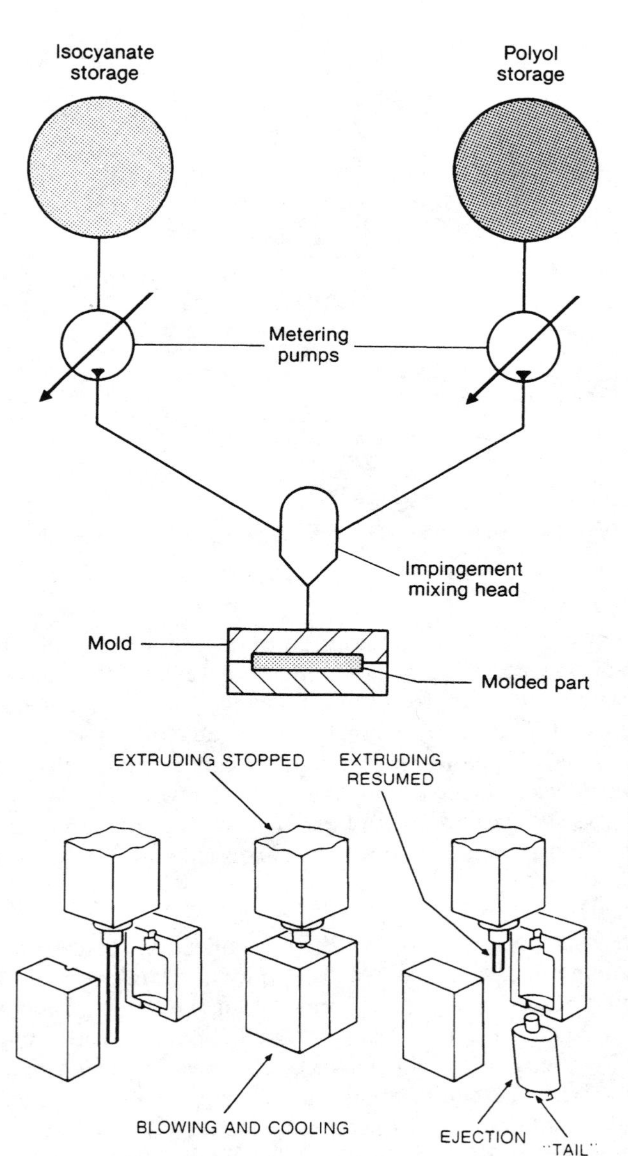

Figure 7-43 Schematic of the reaction injection molding (RIM) process.

Figure 7-44 Schematic showing the intermittent extrusion blow molding process. (Courtesy of USI Chemical Co., Division of National Distillers and Chemical Corporation)

halves close on the hot parison, which seals one or both ends of the tube. Either the parison extrusion stops or the mold is moved to a separate blowing station. Air is then injected into the hot parison through a blow needle which expands the tube in the mold similar to blowing up a balloon. As the hot plastics contact the water cooled mold walls, they cool and the finished part is ejected. Extrusion blow molding is capable of producing containers from less than one ounce to over 100 gallons of capacity. With the introduction of UHMWPE automotive gas tanks and 55 gallon drums, the concept of blow molding as "bottle blowing" has begun to change.

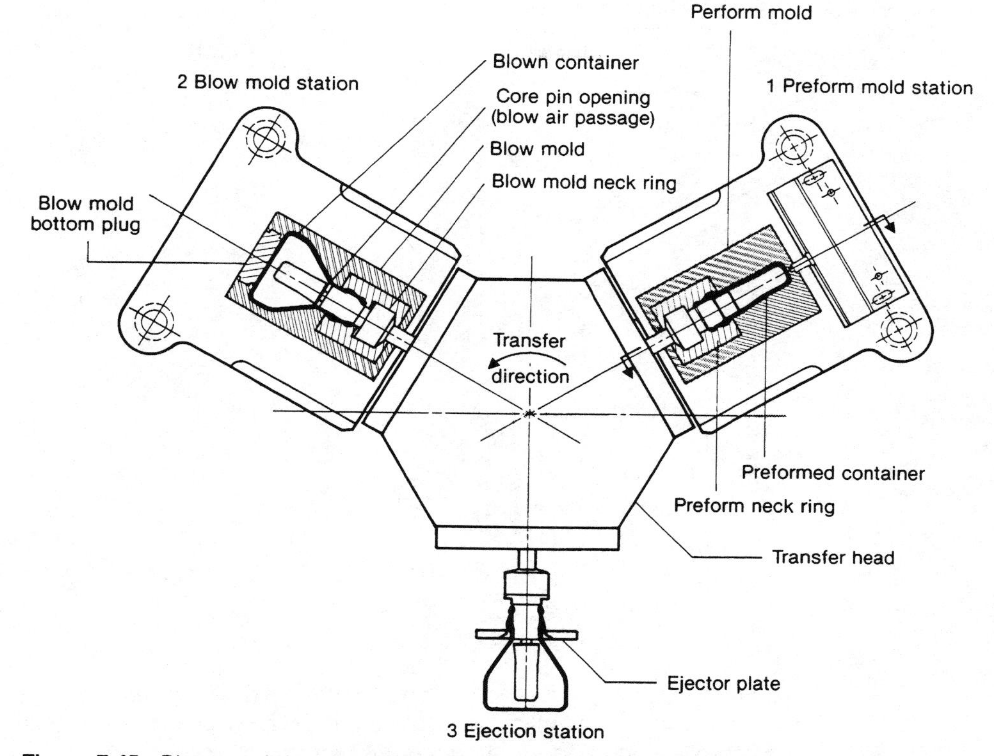

Figure 7-45 Diagram of the three positions found in the injection blow molding process. (Courtesy of Jomar Company)

Hollow objects requiring complex shapes, molded-in inserts, and very large capacities are now fair game for blow molding product development.

The second type of blow molding is the injection blow process. It is characterized by multiple cavity molds, high production, and small precision parts. See Figure 7-45. In injection blow, the parisons are injection molded much like a test tube made of plastics. The parisons stay on core rods and are swung into the blowing station while still hot. Blow molds then close over the parisons and air is injected through the ends of the core rods. The parts (usually bottles less than 4 liters in size), are then swung into an ejection station as soon as they are cool. Injection blow molding is a more complex process and requires about three times the tooling expenditures. However, production rates may reach several hundred parts per minute and the part quality is hard to match with the most sophisticated extrusion blow operation.

Compression molding. Compression molding is the oldest plastics molding process. Most of the early work in plastics was done with compression molding. In fact, compression was the major process in the industry until after World War II. The majority of compression molded parts are made from thermoset resins. Compression is the only molding process that does not require a separate plastification chamber. Since resin melting takes place in the mold cavity itself, plastification

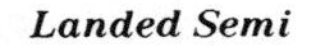

Landed Semi

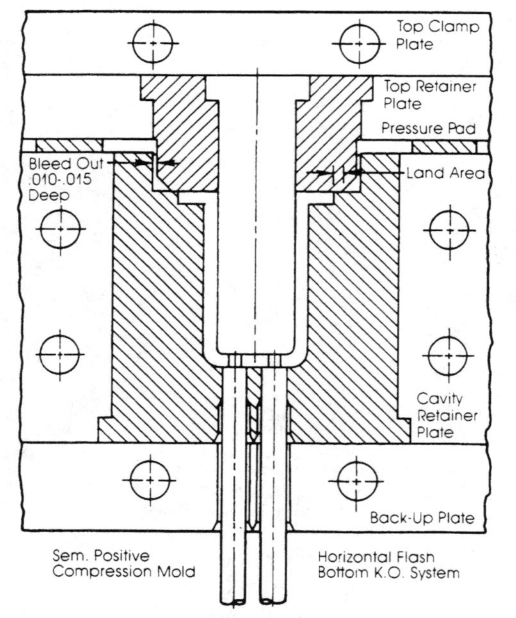

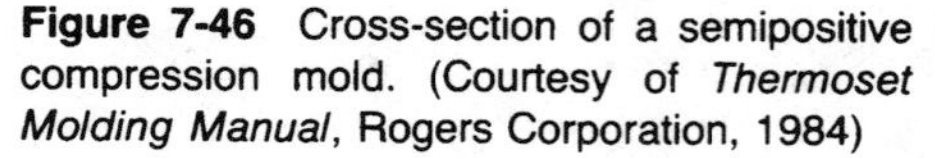

Figure 7-46 Cross-section of a semipositive compression mold. (Courtesy of *Thermoset Molding Manual*, Rogers Corporation, 1984)

relies on clamping pressure and the heating of the mold and platens. A compression molder is basically a large hydraulic press with a heated mold and platens. The platens support and align the mold halves.

There are three types of compression molds: flash, semi-positive, and fully positive. A flash type mold relies on excess material flashing or squeezing out of the cavity into a flash well. Flash type molds demand less accuracy in measuring the charge of resin but require the waste of resin in the flash well. In a fully positive mold, the flash well is not present and the parting line of the mold telescopes toward the part. The parting line seals at the moment of final compression. This provides no waste through flash and requires precision measurement of the resin charge. Fully positive molds tend to be the most costly to build and are used for long production runs. Semipositive molds are a compromise between flash and positive molds. See Figure 7-46.

Probably the most important development in recent years has been the increased usage of Sheet Molding Compound (SMC) for large panels. SMC is the production method used for hoods, doors, and fenders in the transportation industry. SMC is a method of producing sheets of polyester and chopped glass that can then be placed into large molds. The finished products have the same basic characteristics of fiberglass laminated panels. However, the several hours required to lay-up a laminated panel are reduced to a few minutes for SMC processing.

Transfer molding. Transfer molding is an offshoot of compression molding. Due to the problems associated with resin plastification within the mold cavity, many types of inserts and detail are impractical in compression molding. For these reasons, two types of transfer molding were developed. Plunger transfer molding consists of a compression press with a ram or plunger that can operate through the

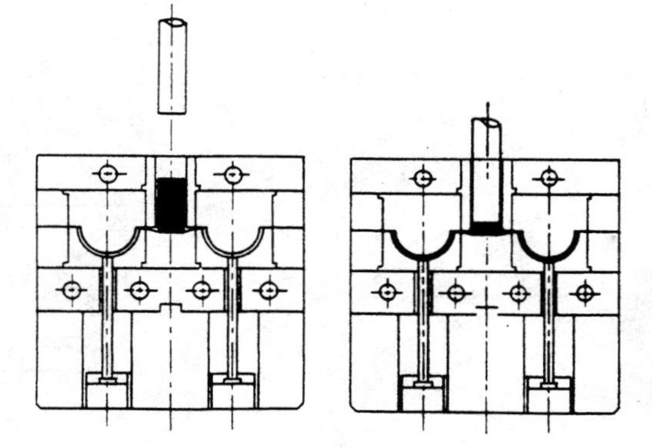

Figure 7-47 Cross-section of a plunger-type transfer mold. (Left) preform of resin with transfer plunger retracted. (Right) plunger transfers resin into the part cavities. (Courtesy of McGraw-Hill)

center of the top platen. See Figure 7-47. The plunger melts resin in the transfer chamber and forces it into the mold cavity through a runner system. Since the mold is fully closed during resin transfer, there tends to be more uniform melt characteristics than with compression molding. Pot type transfer molding employs a standard compression press. A transfer pot and plunger are built into the mold itself. See Figure 7-48. Upon full compression, the plunger plasticizes the resin and forces it through a sprue and runner system. The cavity, gate, runner, and sprue construction is similar to that of injection molding, and in fact many former transfer parts are now being produced through thermoset injection molding. Traditionally transfer molding has produced some of the most dimensionally accurate parts in the entire industry. However, today product quality is being matched by thermoset injection and many electronic and computer components are no longer transfer molded.

Thermoforming. Thermoforming is a group of predominantly thermoplastic processes which are used to form sheet, rods, and tube. Unlike the majority of other plastics processes, thermoforming does not require that the polymer reach a liquid state at any time during processing. Forming generally takes place during

Figure 7-48 Cross-section of a pot-type transfer mold. (Left) preform prior to transfer. (Center) resin transferred to part cavity. (Right) mold opening and part ejection. (Courtesy of McGraw-Hill)

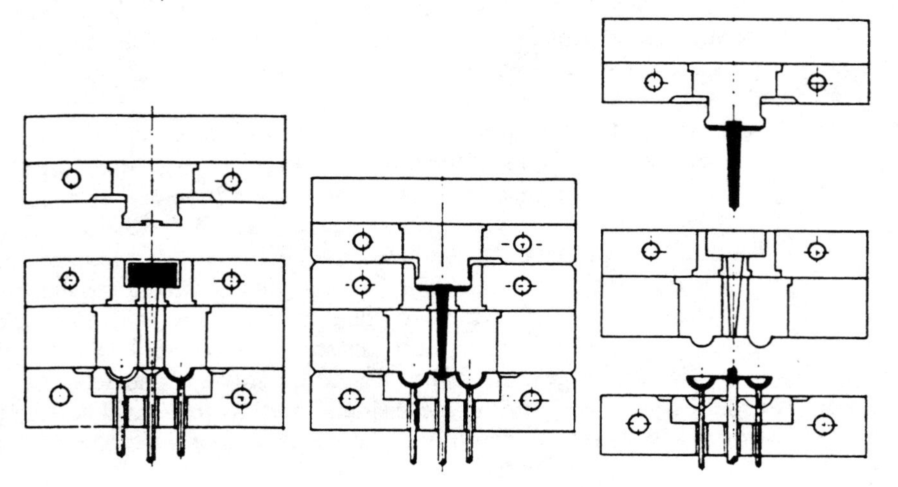

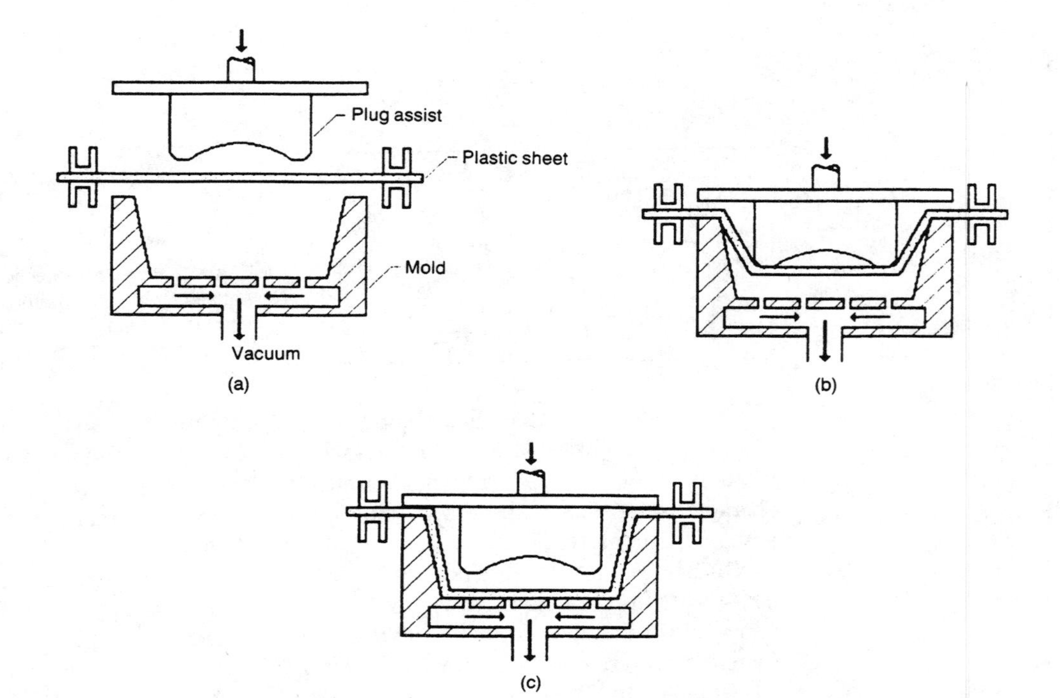

Figure 7-49 Diagram of straight thermoforming with a plug assist. (1) Heated sheet is brought into position. (2) Plug assist stretches the heated sheet into the mold. (3) Vacuum draws the heated sheet against the mold surface.

the plastic state of the material. There are three basic methods of thermoforming: straight, drape, and mechanical forming.

Straight or vacuum forming employs a female or cavity mold. A heated plastic sheet is brought into position over the mold as shown in Figure 7-49 and vacuum is applied to draw the sheet into the mold cavity. In some instances, air pressure may be applied to the opposite side of the sheet to help force the sheet into the mold. In addition, a mechanical plug assist may force the sheet into the mold cavity before vacuum or air pressure are applied. The plug assist acts to rapidly position the sheet inside the cavity and also changes the thinning characteristics of the sheet.

Drape forming is similar to vacuum forming, except that it uses a male mold. See Figure 7-50. A mechanical or ring assist is sometimes used in place of a plug assist. As with straight forming, thinning is a problem, although in different areas of the part. Generally speaking, the part wall will be the thickest in those areas of the mold which the sheet contacted first and thinnest in those contacted last.

The third type of thermoforming, mechanical forming, requires neither air pressure nor vacuum. Instead, a mold more like a compression mold is used with both a male and female half. The mold halves are closed on the plastic sheet, thus compressing the sheet to the desired shape. Since the sheet is contacted on both sides, good detail may be produced on the inside as well as the outside of the part. In addition, since the distance between the mold halves regulates the part thickness, the problem of thinning is eliminated.

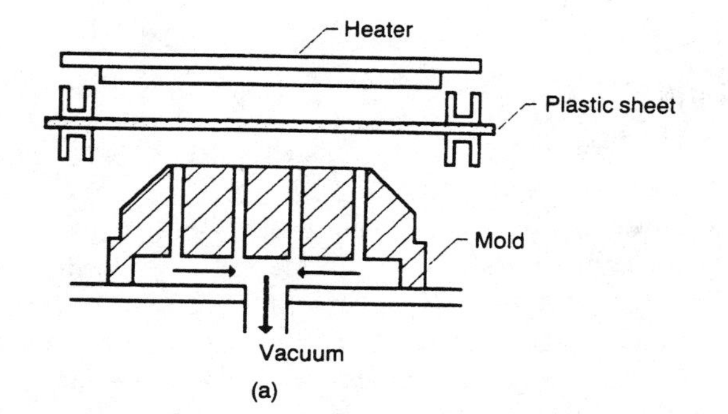

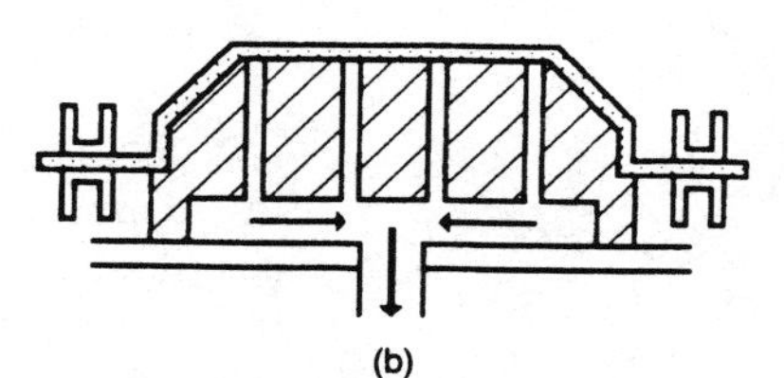

Figure 7-50 Diagram of drape thermoforming. (1) Heated sheet is brought into position above the mold. (2) The sheet is pulled down over the mold, vacuum draws the sheet against the mold surface.

The greatest advantage of thermoforming is that the tooling costs are far lower than injection and compression. In straight and drape forming, a single-sided tool is used and the low pressure requirements allow tools to be made from cast aluminum, epoxy, and even wood in some cases. Thermoforming provides an excellent source of prototype and short-run parts. Typical products include many packaging items such as cottage cheese containers and disposable drinking glasses, as well as aircraft fairings, hot tubs, and appliance door inserts.

Special Processes for Polymer Composites. Reinforced polymer composites are certainly a major development area within the plastics industry over the next several decades. In many cases polymer composites may be processed by injection, compression, RIM, extrusion, or even a modified type of thermoforming. However, special processes are required, for the use of long or uniformly aligned fibers. The majority of special reinforced polymer processing is done by three processes: lamination, filament winding, and pultrusion.

Lamination/lay-up. Lay-up is a method of saturating the fibers of a reinforcing material with a liquid resin. The mixture is applied to either a male or female mold. In most cases the resin used is either thermosetting polyester or epoxy. The layer of resin-saturated fibers shown in Figure 7-51 may also be enhanced by the

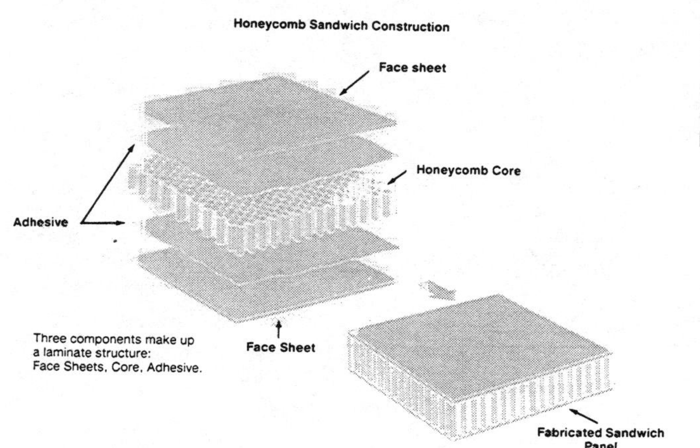

Figure 7-51 Lamination of a honeycomb sandwich panel used in the aircraft industry. (Courtesy of E.I. DuPont De-Nemours and Company)

addition of honeycomb structural sheets and preimpregnated (prepreg) cores. The fibers may consist of fiberglass, thermosetting, or thermoplastic polymers, metal, or graphite. The lay-up creates a shell over the mold which is in turn stripped from the mold to produce the desired finished part. Lay-up may be done by hand or mechanically or by a combination of both.

In many operations a gun is employed to spray either resin or a mixture of resin and chopped fibers. This process is often referred to as "spray-up" and allows for fast deposition of the composite. Following the deposition process, the composite is rolled, pressed, or evacuated in a vacuum bag to eliminate the majority of air and unwanted internal voids. This process is currently being used to produce boat hulls and a wide variety of aerospace components such as wing and fuselage sections.

Filament winding. Filament winding is an outgrowth of lay-up processing. Resin-saturated fiber is wound over a mandrel or form to produce a hollow vessel. See Figure 7-52. In some cases the mandrel is hollow and becomes the inner lining of the part produced. In other cases, the mandrel is either deflated, collapsed, or dissolved to remove it once the finished part has cured. Because of the fiber and or prepreg alignment, a maximum amount of part strength may be achieved. Thus filament wound tanks and vessels achieve some of the highest strength-to-weight ratios of parts made of polymers and polymer composites.

Pultrusion. Pultrusion is a cross between extrusion and filament winding. Fibers are saturated with liquid resin drawn through a heated die and cured. See Figure 7-53. Thus, the resulting part may achieve almost any shape achieved through extrusion. However, the part has a great deal of strength, due to the alignment of the reinforcing fibers. Pultrusions are commonly used for fishing poles and archery bows and may soon replace many flat springs and beams in current aluminum and steel structures. The first U.S. space station will probably be made from pultruded beams, now under development by NASA.

Casting Processes. Casting is a group of processes which do not require the application of much pressure, if any, in order to form the resin. Casting of plastics may involve either powdered or liquid resins used in male or female molds. In cases where thermoplastic resins are used, the resin is heated at some point in the process and thermofusion takes place. Thermofusion is simply the melting and coalescence of the resin into a single homogeneous mass. The resin is then cooled and the finished part may be removed from the mold. In the case of thermoset resins, two-part liquid resins are used. The two prepolymers are mixed prior to

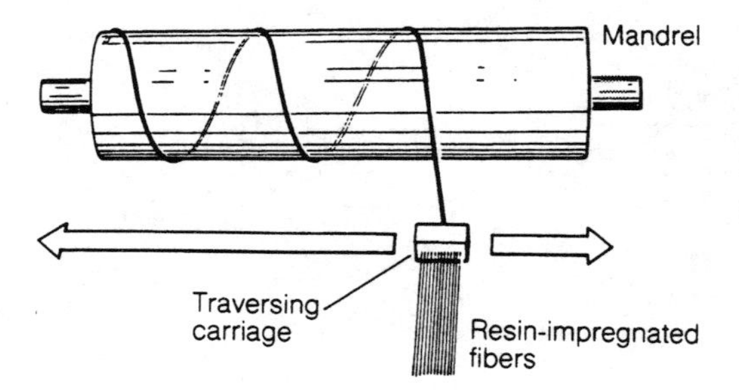

Figure 7-52 Diagram of lathe type filament winding. (Reprinted with permission, *High Technology* magazine. October 1983. Copyright © 1983 by Info Technology Publishing Corporation, 214 Lewis Wharf, Boston, MA 02110.)

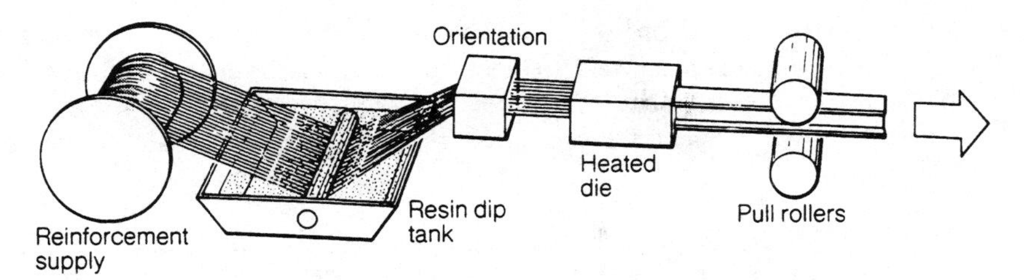

Figure 7-53 Diagram showing the pultrusion process. (Reprinted with permission, *High Technology* magazine, October 1983. Copyright © 1983 by Info Technology Publishing Corporation, 214 Lewis Wharf, Boston, MA 02110.)

entering the mold. In the mold the resin undergoes polymerization and may then be removed from the mold as a finished part.

Rotational casting. By far the most important single polymer casting process is rotational casting. In rotational casting a cold two-piece mold is partially filled with resin and the mold halves are then closed to provide a sealed vessel. In Figure 7-54, the mold is then rotated on two axes and brought to an appropriate temperature for thermofusion. The resin is spread evenly over the inside of the mold during thermofusion. The mold is then cooled during further rotation to produce a hollow, balanced, and air-tight parts with great potential detail. Rotational casting is presently able to produce vessels with internal capacities over 1,000 gallons, such as large fertilizer tanks. In addition, most balls and other balanced parts are made by rotational casting.

Full mold and slush casting. Full mold and slush casting are variations on the same principle. During the casting cycle a substraight or component may be submerged in the liquid. When the resin hardens, the component is permanently sealed. This is known as potting or encapcilation and is generally performed on

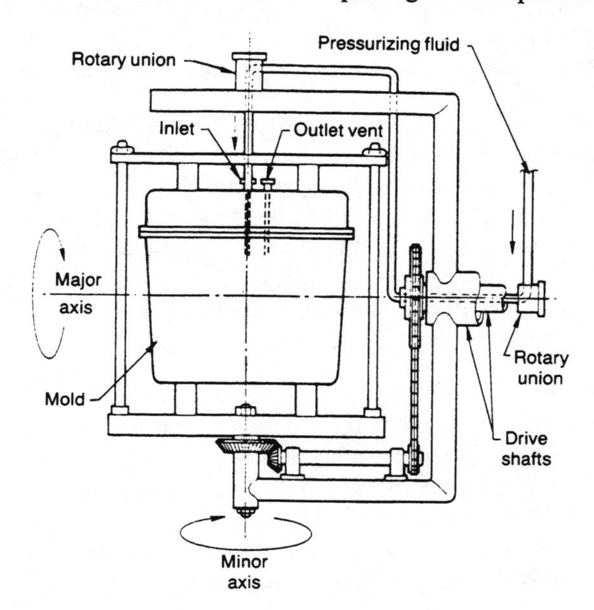

Figure 7-54 Diagram of the rotational casting process. (Courtesy of McNeil Akron)

electrical components unable to withstand exposure to the elements. Slush casting is generally reserved for thermoplastics and involves filling a heated mold with resin. The majority of the resin is then poured out of the mold after thermofusion and has produced the desired part wall thickness along the inner surface of the mold.

Dip casting. Dip casting is exactly like slush casting, with the exception that a male mold is coated with resin in a tank. In most cases, molds are sequenced through a bath of liquid plastisol. Many hollow flexible parts such as chemical resistant gloves and automotive shifting boots are produced by dip casting.

Expansion Processes. Expansion processing is a unique contribution of plastics to manufacturing as a whole. The ability to foam polymers enables us to provide flotation and insulation to other manufactured products. Foamed plastics fall into two groups: liquid foams and bead foams. Both types of foam rely on a blowing or gassing agent introduced into the resin at some point in the process. The blowing agent in turn causes the resin to form under a prescribed set of conditions. Liquid foaming may be used during extrusion, injection molding, RIM, and casting operations.

Structural foam. In the case of injection molded structural foam, a thermoplastic resin is plasticized in the barrel of the machine as in a normal injection molding operation. A blowing agent may either be added in solid form to the resin in the hopper of the machine or a gas may be introduced in the injection barrel. See Figure 7-55. The resulting foam may both lighten and strengthen the molded parts, and may also be extruded for similar results. Structural foam is used widely in the production of computer housings and other similar items.

Liquid foam. Liquid foam is most often associated with RIM. Either flexible or rigid thermosetting urethane foam is generally used. Flexible RIM is frequently used to produce seat cushions, padded automotive dashes, and handles. Ridged RIM may be used to insulate refrigeration components or other hollow components.

Bead foam. Bead foam, once restricted to polystyrene, is now found in a wide range of thermoplastics from polyethylene to PMMA. The bead foam process relies on the introduction of a blowing agent into tiny rigid beads of a resin. Figure 7-56 shows the process steps involved in bead foam molding. When the beads are heated through the use of steam, the blowing agent expands to form tiny foam balls or beads, which fuse together through the use of steam heat. Expandable bead foam is used in a wide range of products, from disposable drinking cups and insulated food carriers to marine floatation devices and evaporative or "lost" foam casting patterns.

Coating Processes. One of the unique properties of polymers is that they form excellent surface coatings on other types of materials. To this end, there is a wide range of plastics suitable for use in coatings as well as a number of different coating processes. The reason for the diversity in process types is that there are often limitations due to either the molecular structure of the polymer, the substrate being coated, or the design parameters of the finished product.

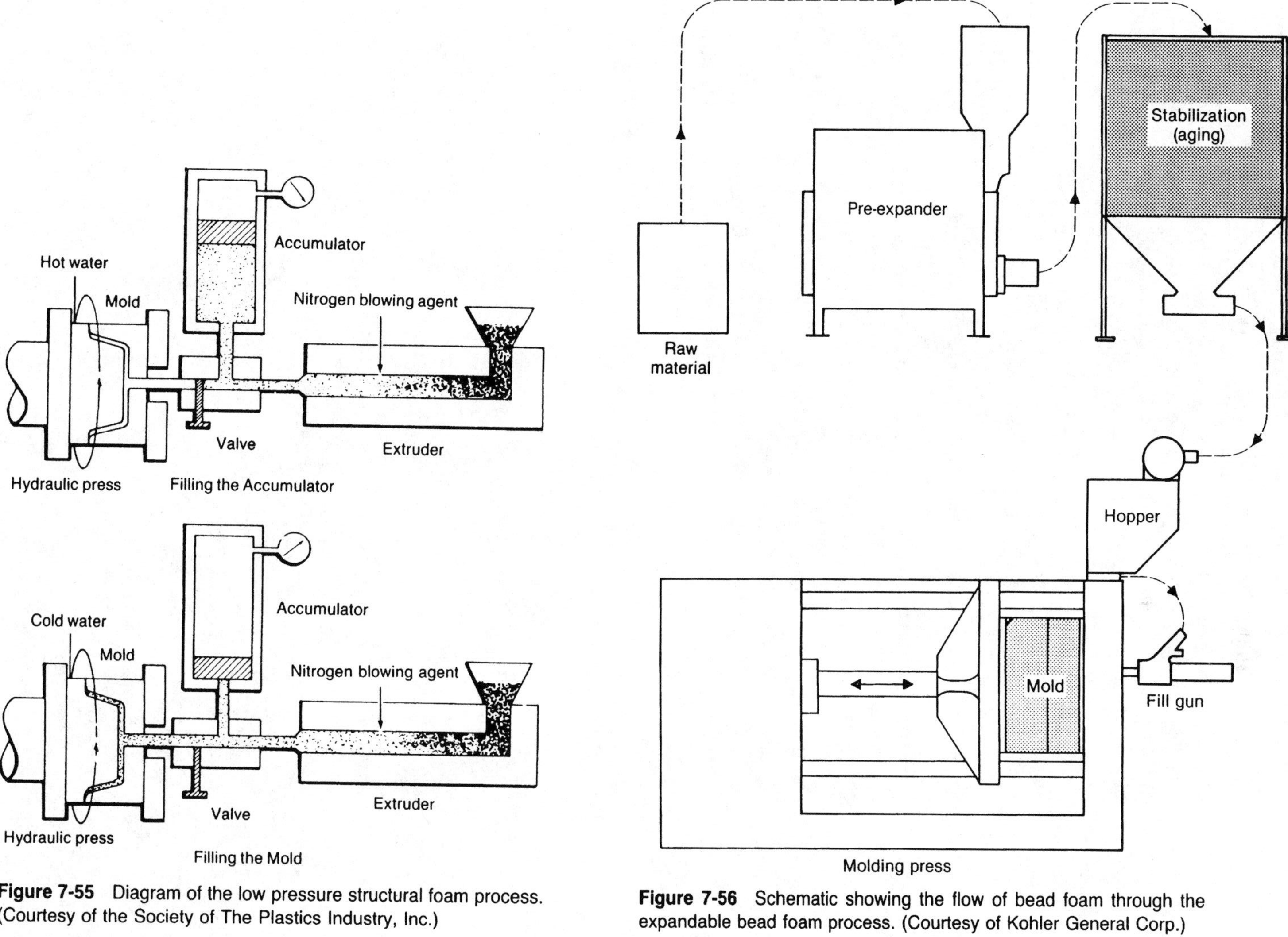

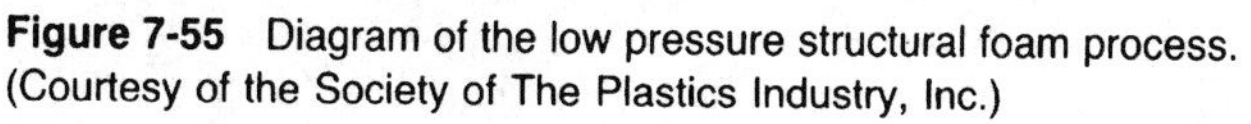
Figure 7-55 Diagram of the low pressure structural foam process. (Courtesy of the Society of The Plastics Industry, Inc.)

Figure 7-56 Schematic showing the flow of bead foam through the expandable bead foam process. (Courtesy of Kohler General Corp.)

Extrusion coating. In the extrusion coating process, a screw-type extruder is used to melt a resin and apply it to a substrate. This process is used for wire coating, in which the wire substrate is passed through the back of the extrusion die in order to be completely coated, by passing through the flow of molten plastics. See Figure 7-57. Much of extrusion coating is done in the form of profile extrusion and may result in products such as vinyl coated wood cored window casements and metal cored edge protectors for automobile interior trim.

Dip coating. Dip coating is very similar to dip casting, with the primary difference that the coated substrate remains an integral part of the product. The major limitation of dip coating is that the polymer must be maintained in a liquid state during the process. An excellent example of dip coating are the vinyl coated plier and wrench handles which insulate against electrical shock.

Powder coating. Unlike dip coating, in the powder coating process the polymer is brought to a semiliquid state after it is applied to the substrate itself. The two major variations of this process are fluidized and electrostatic coating.

A fluidized coating generally requires a bed of powdered resin to be charged with compressed air until the powder begins to behave like a liquid. This is due to the capillary movement of the air as it passes through the powder. The substrate is heated above the fusion temperature of the powder and is submerged into the bed. The surface temperature of the substrate melts the powder that comes in contact

Figure 7-57 Schematic showing the extrusion coating process with the substraight to be coated moving from left to right. (Courtesy of USI Chemicals Co., Division of National Distillers and Chemicals Corporation)

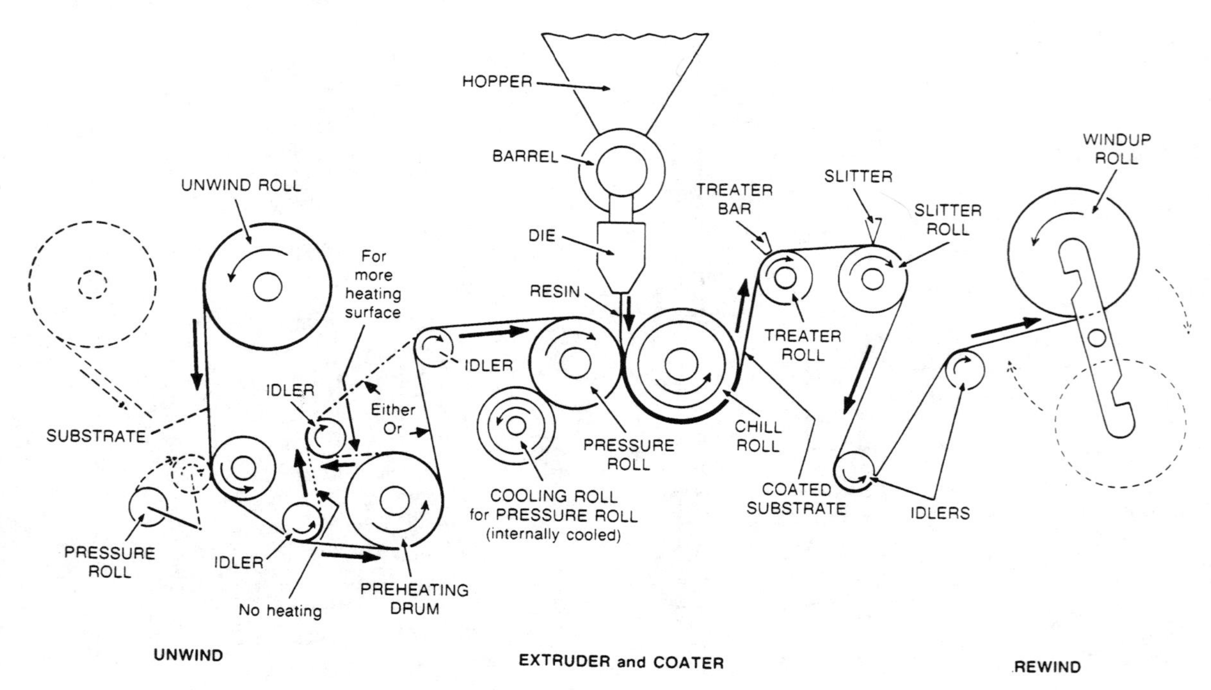

with it. The longer the substrate is in the bed, the thicker the coating becomes, to the point that thermal insulation prevents any further powder from sticking to the product. Following the application of the bed, the product is then reheated for final and complete fusing of the coating. The major disadvantages of this method are that the substrate must be raised above the fusion temperature of the polymer, and the heavier sections of a part tend to coat more heavily because of their greater thermal energy.

The second major variation of powder coating is electrostatic coating. In electrostatic coating, a positive charge is applied to the part and a negative charge is applied to either the powder bed or a powder spraying gun. When the negatively charged powder comes in contact with the positively charged part, the two stick together due to electrical attraction. As more powder accumulates on the part, the amount of electrical insulation increases. When the positively charged part is completely insulated by negatively charged powder, no more powder will stick to the part. Because of this phenomenon, an even coating thickness is applied to all surfaces of the part. The part is then heated until the surface temperature reaches the fusion temperature of the polymer. The part is then cooled and the positive charge shut off. An excellent example of powder coating is the coating found on most dishwasher racks.

Decorating Processes. In addition to molded in color and mold in metallic finishes, many products require additional surface finishing to enhance both their appearance and usefulness. Many of these processes are unchanged from those used to finish metal, ceramic, or paper products.

Painting. In many cases, a layer of paint is applied to ensure a uniform surface luster or color to an assembly or group of products. Products may be dipped or sprayed to produce a variety of desired results. When painting plastics, it is necessary to select a paint whose chemical base will bond to the polymer's surface without attacking or weakening structure or integrity of the product. Conversely, a lack of bonding will result in a peeling condition.

Plating. At present, it is possible to plate a wide range of thermoplastics and thermosets either through the use of electroless plating or by electroplating. Electroless plating is similar to a painting or coating process, and is primarily used as the base coat for electroplating, as radio frequency interference (RFI), or as electromagnetic interference (EMI) shielding. It is used for an electroplating base coat because it will adhere to the surface of the polymer and provide a conductive surface which can then be electroplated. Electroless plating is used for RFI/EMI shielding for things such as computer housings, radio cabinets, satellite dishes, and audiovisual to provide an effective shield against interference.

Once a coating of an electroless copper or nickel has been applied to the surface of a polymer, it may be electroplated as if the part were made of metal. Through electroless plating, it is possible to create a durable chrome finish on automotive and appliance parts comparable to any steel stamping or die cast part traditionally found in these applications. Good examples of this would include automotive grill work, appliance handles, and mixer bodies. The major problems with electroplating are that it is an expensive and time-consuming process and requires the use of a number of toxic chemicals.

Vacuum metalizing. Vacuum metalizing is a process used in place of traditional electroplating because of its low cost and reduced toxic waste. In the vacuum metalizing process, parts are placed into a vacuum chamber. After a vacuum is drawn, a small amount of the desired plating metal is vaporized through the use of an electric arc. In the presence of the vacuum, the vaporized metal distributes itself throughout the vacuum chamber and adheres to the cooler surfaces of the parts in the chamber. The parts must be rotated or multiple platings must be applied, since the vapor adheres to the surfaces in a direct line of sight from its point of vaporization. This process is similar to a suntan; something that is shadowed from the source of the vapor will not receive plating. Because vacuum metalizing has a lower surface adhesion than electroplating, it is generally applied to areas of relatively low wear. Examples of vacuum metalized parts may often be found in the interiors of automobiles and low wear areas of consumer goods.

Printing. Many traditional printing methods are used on plastics. Leading methods of printing on thermoformed containers include both offset and flexography. In addition, many blow molded containers may also be printed on by the use of offset, flexographic, or screen printing methods. See Figure 7-58.

However, one printing method has gained tremendous acceptance in the printing of injection molded components. This technique is referred to as transfer pad or pad printing. Transfer pad printing uses an engraved plate quite similar to a gravure plate. Ink is wiped into the depressions in the plate or "cliche" and a silicone pad transfers the ink from the depressions in the engraved plate to the surface of the molded product as shown in Figure 7-59. The advantage of pad printing is that the surface of the product does not need to be uniform, since the pad will conform to

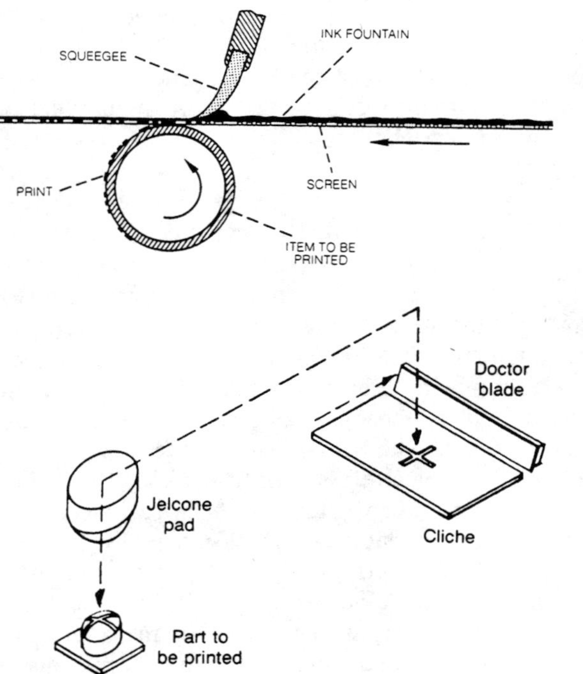

Figure 7-58 Diagram of the screen printing process showing a round container or bottle being printed. (Courtesy of USI Chemicals Co., Division of National Distillers and Chemicals Corporation)

Figure 7-59 Diagram of the pad printing process. (Courtesy of Service Tectronics, Inc.)

the surface. In addition, pad printing does not require foils or films like hot stamping, and there is no surface deformation due to the heated dies.

Hot stamping. Hot stamping is a method of transferring a pigment or foil to the surface of the molded product by means of a hot die. See Figure 7-60. The foil or pigment is temporarily bonded to a backing of carrier film. In addition, a layer of heat and pressure-sensitive adhesive is applied over the pigment or foil. When the hot die is applied to the carrier film, the adhesive is liquified at those points contacted by both the die and the part beneath. After the die has been removed, the adhesive rapidly cools and the carrier film is stripped off. The result is a foil or pigment impression left where the die met the part. The advantage of hot stamping is that it may be used to selectively plate areas of a part allowing for reveal lines, lettering, and details. The disadvantages are that the die may leave unwanted surface deformation, and the film is more expensive than printing operations.

Welding and Assembly. When assembling components made out of plastics, there are three major groups of processes that may be used. These processes include adhesion, cohesion, and mechanical assembly. Simply defined, adhesion is the bonding of two or more parts to one another with a dissimilar material or glue. Cohesion is the fusion of two or more similar parts to one another, either through the application of a solvent or heat to melt the polymers. Mechanical assembly is the fastening of two or more parts together by the use of some mechanical holding device such as a screw, rivet, or through the interference between the two parts such as a snap fit.

Adhesion. Adhesion includes a wide range of gluing applications and adhesion materials. Adhesion bonding of parts is an increasingly accepted and sophisticated assembly method. There is a far greater amount of adhesion bonding of nonplastics than of plastics. However, almost without exception, the adhesives used in industry today are made up of some type of polymeric material. Common adhesives include urethanes, alkides, silicones, epoxies, ethylenes, and ureas.

Figure 7-60 Diagram of the hot stamping process showing a round container or bottle being hot stamped. (Courtesy of USI Chemicals Co., Division of National Distillers and Chemicals Corporation)

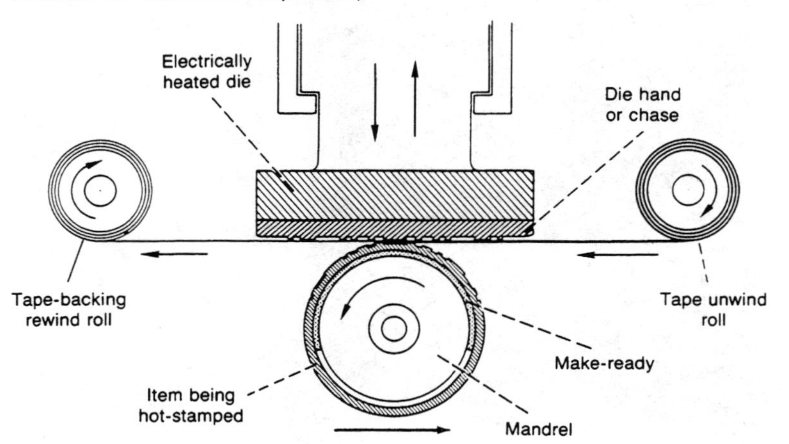

There are three major types of adhesive application: solvent carrier, chemical reaction, and hot melt. A solvent carrier adhesive is one that drives or allows the solvents to evaporate out of the adhesive. Solvents may range from the water we find in common wood glues to the more toxic ketones found in construction adhesives. Reaction type adhesives include things like epoxies, where a catalyst is added to the adhesive to cause polymerization of the glue. The majority of hot melt glues fall into the ethylene group and are preferred in many applications because of their rapid set up time. With the advances in adhesive technology, the days of multipurpose glues in manufacturing are rapidly coming to an end. There is a wide range of durabilities, flexibilities, and atmospheric and chemical resistances found in specific adhesives. When considering an adhesive for a given manufacturing process, a specific list of requirements should be formulated so that adhesives suppliers are able to specify the correct adhesive for the given job.

Cohesion through solvent cementing. Cohesion through solvent cementing has long been a viable manufacturing option, dependent upon the areas to be bonded and the type of polymer used in the process. By far the most common polymers to receive solvent cementing assembly are acrylics, styrenes, ABS, and some polycarbonates, and cellulosics. Solvent cementing requires a very closely fitting joint, so that the solvent causes the opposing surfaces to melt and intermingle. When the solvent evaporates, the resulting joint is formed by chemical bonds between the two parts. The major drawbacks to solvent cementing include the evaporation of the solvents and the length of time required to form the chemical bonds. For this reason, solvent cementing is often used on limited production and prototyping applications. Generally, for a long-run production, a welding operation is more efficient because of the reduced cycle time during cohesion through welding.

Ultrasonic welding. Ultrasonic welding relies on the ability of high frequency sound waves to excite the molecular structure of a polymer. See Figure 7-61. In a typical ultrasonic welding application, 20,000 hertz of high frequency vibration are passed from the welder's transducer unit through a horn. By varying the configuration

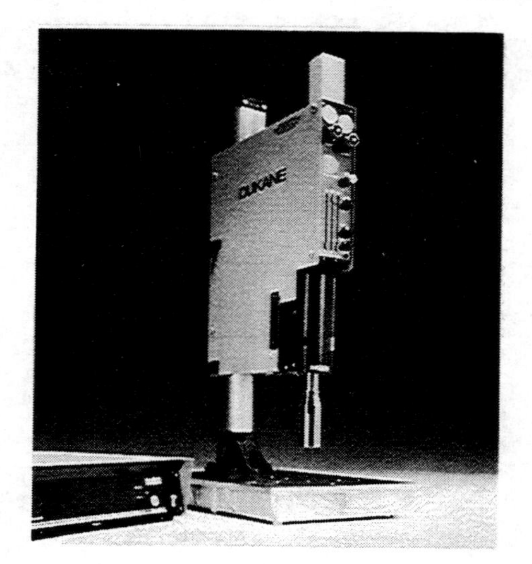

Figure 7-61 An ultrasonic welder. (Courtesy of DuKane)

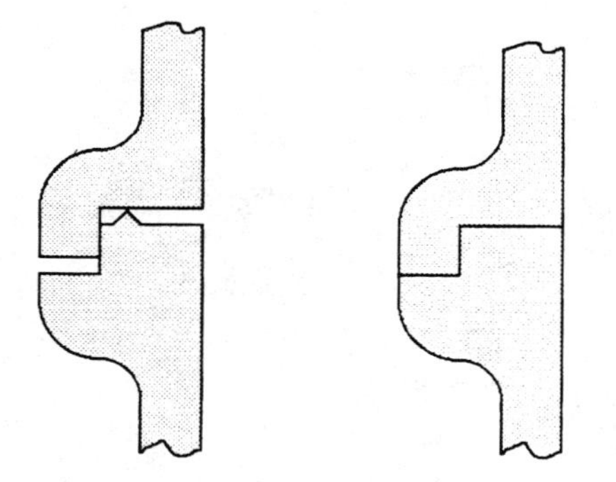

Figure 7-62 Cross-section of an ultrasonic welding joint showing the energy director (left) and the completed joint after welding (right).

of the horn, it is possible to increase the amplitude and thus the effective welding energy to the part. The two parts to be welded together must be designed so that there is a closely fitting joint with an energy director along the length of the weld. The energy director serves to concentrate the welding vibration at the center of the joint to be fused. See Figure 7-62. With a correctly designed horn and part fixture, the only area to vibrate is the point of weld between the two parts. With this process, a housing and cover may be welded together without the welder ever touching the welded joint. The high frequency vibration would pass through the welding joint and fuse the cover to the housing in a single several-second operation. With ultrasonic welding, welding heat distortion is not a problem, and cosmetically attractive and hermetically sealed joints may be created at high speed with uniform high quality.

Fusion bonding/hot plate welding. In the fusion bonding process, both sides of the joint to be welded must be heated to the melting temperature of the polymer and then brought in contact with each other. The process steps for fusion bonding are illustrated in Figure 7-63. This is done by nesting each of the parts to be welded in opposing fixtures. The parts are then brought to contact with a double-sided heat plate. The heat plate is removed and the parts are forced together. The resulting weld has excellent structural integrity but tends to develop a flash line along the entire weld. For this reason, it is recommended that a flash well be designed so that the flash ends up in a nonvisible area. Hot plate welding is especially suitable to weld parts which do not have uniform surfaces. It is possible to level the opposing joint sides through the application of the heat plate. The heat plate must have a nonstick surface applied to it, and its temperature is critical in order to avoid sticking, warping, or cold weld conditions. One application of fusion bonding is in the manufacture of automotive and commercial batteries, which tend to have nonuniform cases and covers due to differential molding shrinkage. With a fusion bonding application, the joints may be surface leveled to produce a leak-proof, acid-resistant seal around the entire top of the battery.

Hot air welding. Hot air welding is very similar to oxyacetylene torch welding of metals. In hot air welding, a stream of superheated air is applied to the welding joint. In addition, a filler rod is fed into the joint at a prescribed rate to ensure closure. Hot air welding has long been employed in the repair of thermoplastics which do not lend themselves to repairs with adhesives or solvents. Care must be taken to ensure that the filler rod is compatible with both surfaces being joined. As with other welding processes, the addition of colorants, lubricants, and fillers to the polymer may cause serious complications in the welding process.

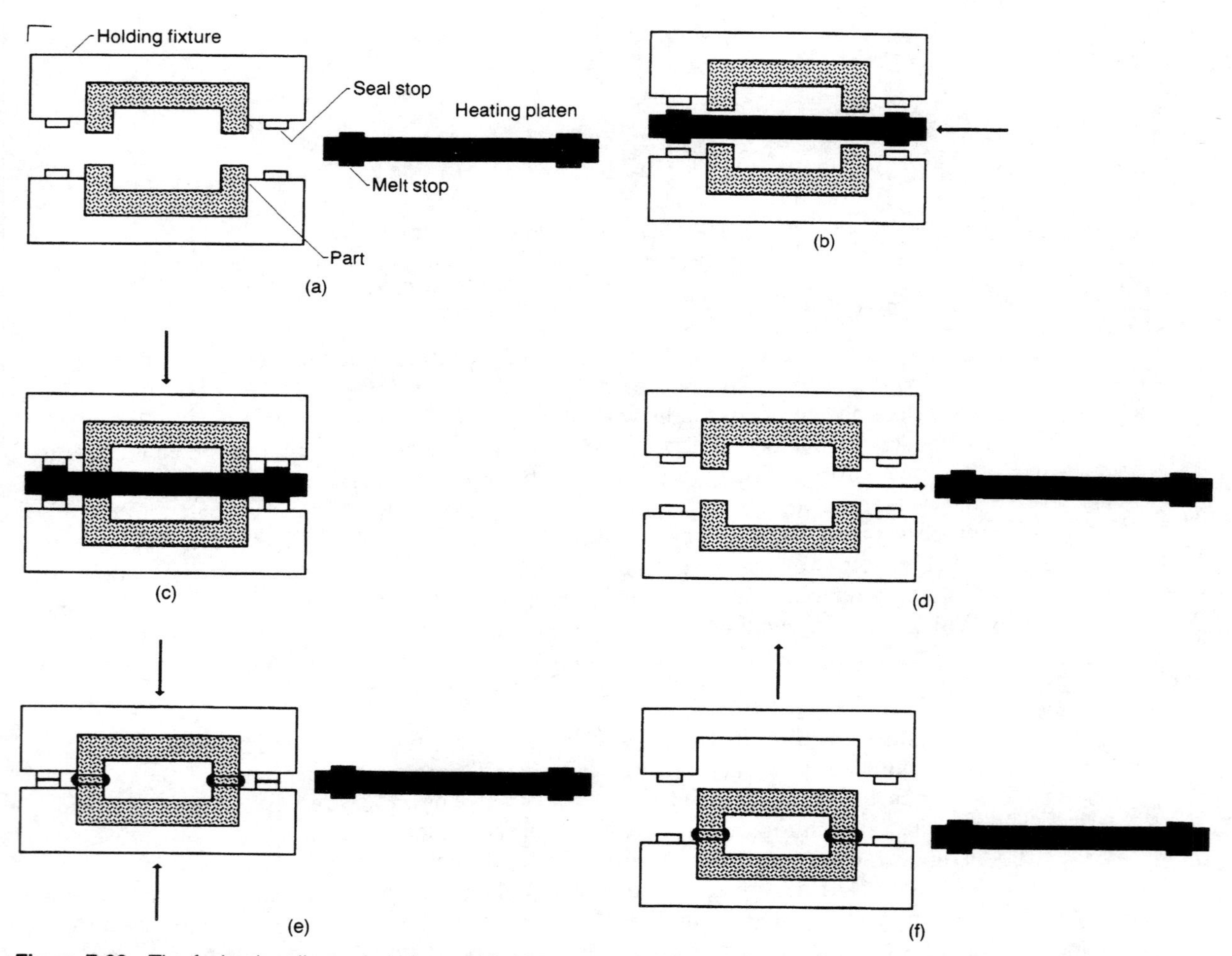

Figure 7-63 The fusion bonding or hot plate welding process. (a) Parts are held in a fixture. (b) Heating platen is brought into position. (c) Platen melts the surfaces to be welded. (d) Platen is retracted. (e) Heated joints fuse together. (f) Assembly is ejected.

Spin welding. Spin welding of plastics is very similar to the spin welding of metals used in the manufacture of valves for internal combustion engines. In the spin welding process, the two parts to be welded are each held in separate fixtures. One part is held stationary while the other is rotated at extremely high speed. See Figure 7-64. The areas of the two parts that come in contact with each other experiences tremendous friction, which results in the melting and fusion of the adjoining surfaces. When melting takes place, rotation is abruptly stopped and the two parts are allowed to cool while being held in alignment. Thus, by the nature of the process, cylindrical and spherical objects become logical candidates for spin welding. Spin welding has been used for many types of packaging containers, as well as consumer goods and toys such as whiffle balls.

Figure 7-64 Diagram of the spin welding process. (Courtesy of Packaging Systems Corporation)

Impulse welding. Impulse welding is generally used to seal layers of film and sheet thermoplastics together. The impulse process relies on a nichrome wire's ability to be rapidly heated and cooled. Both sheets to be welded are clamped together in close contact with a nichrome wire. In most cases, the wire is shielded so that it does not directly touch the weld seam. Current is passed through the nichrome wire causing it to heat up and fuse the two sheets together. Typical applications of impulse welding may be found in a variety of packaging applications, from garbage bags to sealed pouches of food for use in the microwave. On a larger scale, impulse welding may be used for things as extensive as pool liners and large canopies and awnings.

Mechanical assembly. There are several methods of mechanical assembly which may be applied to the field of plastics. Quite frequently fasteners are applied to plastics, especially where disassembly and reassembly are required. Fasteners may include self-tapping screws of either a thread cutting or thread forming style, customary and metric pitch threaded fasteners, rivets, heat stakes, and velcro.

When considering a threaded fastener, the load applied to the component is critical. In some instances, a self-tapping screw will not hold and a standard threaded fastener will be required. Standard thread fasteners may be screwed into either molded-in threaded holes or molded-in threaded inserts. In many cases, threaded inserts tend to be more durable. however, they tend to be more expensive in long production runs. It is also very important to consider the effect of the threads on the polymer. In some polymers, self-tapping screws will apply so much stress to the part that it will rapidly fail in the field. For this reason, care must be exercised when selecting threaded fasteners.

Rivets and heat stakes are often applied, especially when sheet metal and plastics are being assembled together. In a heat staking operation, a small plastic peg or stake is molded on the underside of a part. Following the installation of the sheet metal part, the peg is staked or headed over similar to a riveting operation. See Figure 7-65. This may be done through cold forming, ultrasonic staking, or hot/cold staking.

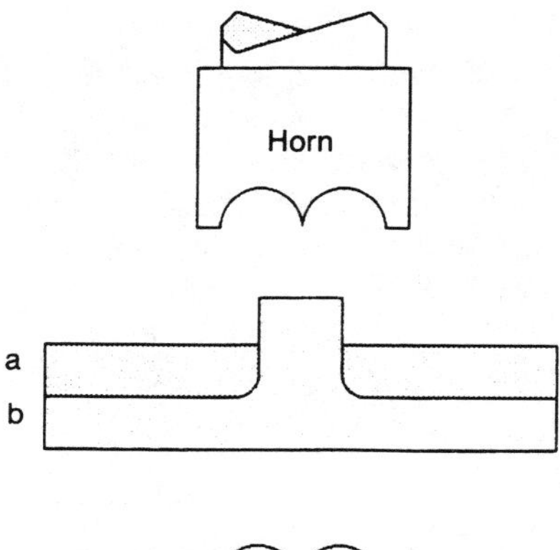

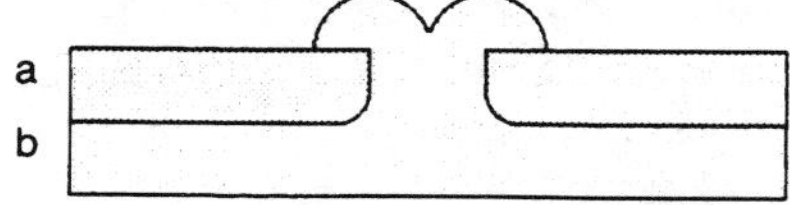

Figure 7-65 Diagram of the ultrasonic staking process. (Top) horn is positioned over stake. (Bottom) stake is "headed over" to form a secure assembly.

Figure 7-66 Velcro hook and loop assembly system as applied to an automotive door panel. (Courtesy of Velcro USA, Inc.)

A very recent alternative in plastics assembly is the application of velcro stripes, which may be molded directly into the parts. Velcro's hook and loop design allows for a wide range of bonding strengths, dependent upon the type of material used and the configuration of hooks and loops. Bonding strengths may range from a few pounds to many hundreds of pounds per square inch. In addition to being molded in place, velcro may be produced from the same material as the parts being molded from, which allows for welding of the velcro strips to the polymer components. This type of assembly method is being widely applied in the transportation industry. See Figure 7-66.

The final major type of mechanical assembly comes in the form of snap-fits. A snap-fit is produced when two parts are molded as opposing pieces of a locking assembly. The tab or lip on one part must be able to slide into the recess or undercut of another part. As the two parts are seated, the plastic memory cause the tab to snap into the recess holding the two components firmly together. Snap-fits may be designed for permanent assembly or for open and closure. Good examples of closure snap-fits include many bottle caps and household containers, especially in the area of food storage. Correctly applied, the snap-fit will provide rapid assembly and excellent part durability.

SUMMARY

The key factor in applying plastics to any manufactured product is that a correct selection of both process and resin must be made. Thus, the diversity of both polymers and processes must be taken into account. The design and manufacturing personnel should weigh the benefits of several prospective resins and processes. Careful selection will afford both a durable and cost effective product. However, if consideration is not given to the full range of possibilities, the manufacturer stands to miss either structural or financial opportunities. By keeping track of new developments in plastics, the manufacturing industry will be able to maintain a competitive edge.

QUESTIONS

1. Explain the major differences between thermoplastics and thermosets.
2. List some characteristics that engineering polymers possess which nonengineering or general purpose polymers do not.
3. Make a list of the advantages and disadvantages of thermoplastics. Also make a list of the advantages and disadvantages of thermosets. Determine which advantages and disadvantages thermoplastics and thermosets have in common and which they do not.
4. Select an object of your choice and specify three potential resins from which the object could be produced. Explain why each resin would be well-suited and what further information you would need to make a final decision.
5. Explain the word "plastification."
6. Compare and contrast extrusion, injection molding, blow molding, and thermoforming.
7. List the factors you would use to determine whether a product should be injection blow molded, extrusion blow molded, or rotationally cast.
8. Explain the difference between coating and decorating.
9. Explain why ultrasonic welding is such a versatile welding process.
10. Choose an assembled product made of plastics. This object should be reasonably simple and have a minimum number of pieces. Good examples would include a mechanical drafting pencil, disposable lighter, cassette tape, broken calculator, and so forth. Disassemble the object and lay out the parts on a table or bench top. Based on the knowledge you have gained from this chapter, attempt to determine the resins and processes used to produce each part. Write a summary report discussing how you reached your conclusions and what additional information you would need to be certain of your results.
11. Explain the difference between liquid foam, either rigid or flexible and, bead foam.
12. Define a "polymer composite" as it applies to manufacturing.

8 WOOD TECHNOLOGY

Wood is a hard, fibrous, organic material derived from trees, and is a valuable engineering material used in building construction, furniture manufacturing, and in an array of other industrial products.

Wood is one of the oldest construction materials. Early humans used it for building shelters, boats, tools, utensils, weapons, and craft objects. In many product applications, wood has been supplanted by metals and plastics, but it still maintains a prominent position in the construction and furniture industries. There are hundreds of wood species in use throughout the world. Some are tough and pliable, others are valued for their beauty or aroma, and still others are notable because of an ability to resist decay. Wood is unique in its physical and mechanical characteristics, is a renewable resource, and because of these attributes will surely continue to be a valuable substance in modern technology. See Figure 8-1.

WOOD STRUCTURE

All wood is comprised of cellulose, lignin, ash-forming minerals, and extractives formed in a cellular matrix. Variations in the composition of these four components and in cellular structure result in woods with different properties and characteristics. Wood species therefore can be light or heavy, stiff or flexible, hard or soft, open or closed grain, richly patterned, or colorless.

Figure 8-1 Custom manufacture of wooden structural windows. (Courtesy of Pella/Rolscreen Company)

The fibrous nature of wood strongly influences how it is used. Specifically, wood is composed primarily of elongated, hollow, reed-shaped cells which are arranged in a parallel orientation along the trunk of a tree. The nature of this orientation, and the manner in which wood structure occurs, can be seen by examining the cross section of a tree. See Figure 8-2. The primary elements of this section are as follows:

1. The *cambium* is a microscopic layer of living cells lying between the wood and the inner bark where tree growth takes place, and all such new growth occurs with the addition of new cells, not the further development of old ones. The new wood cells are formed on the inside, and the new bark cells on the outside, of the cambium.

2. The *phloem*, or inner bark, is a layer of moist and soft tissue featuring sieve tubes which carry natural nutrients from the leaves to all growing parts of the tree.

3. The *outer bark* is a rough, coarse corklike layer of dry, dead tissues which provides general protection against injury and the elements. The inner and outer bark are separated by a bark cambium as a demarcation line.

4. The *sapwood* is a pale-colored wood beneath the inner bark which contains both dead and living tissue. This region serves to conduct water and nutrients from the roots to the leaves. It is this region which is most susceptible to damage and decay.

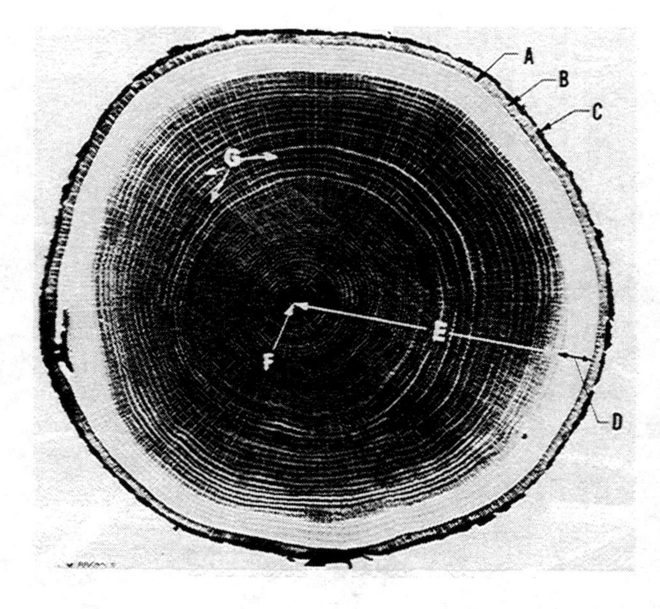

Figure 8-2 A cross-sectional cut of a white oak log. A) Cambium. B) Phloem. C) Outer bark. D) Sapwood. E) Heartwood. F) Pith. G) Wood rays. Light bands are latewood, dark bands are earlywood. (Courtesy of Forest Products Laboratory, Forest Service, U.S. Department of Agriculture)

5. The *heartwood* is that part of the tree extending from the pith to the sapwood, the cells of which are dead and no longer participate in the life processes of the tree. It is here that gums and resins accumulate to make the heartwood darker and more decay resistant than sapwood. Heartwood, formed by the gradual change in the softwood, is usually harder and stronger than sapwood, and as such provides the structural support necessary to the living tree.

6. *Pith* is the small, soft core located in the structural center of the tree trunk. This is the tissue region about which the first wood growth takes place in the newly formed twig.

7. *Wood rays* are strips of cells extending radially within a tree to connect the several wood tissue layers, from pith to bark. Their purpose is to transfer and store food, and they range in size from a few cells in height in some species to a hundred millimeters (4 inches) or more in oak. It is these rays which give the characteristic pattern or fleck in quartersawn oak boards.

Growth rings. With most species of trees growing in temperate climates, there is sufficient difference between the wood formed early and that formed late in a growing season to produce well-marked annual *growth rings*. The age of a tree at any cross section of the trunk may be determined by counting these rings. See Figure 8-3. Growth rings, or annual rings, are most readily seen in hardwood species such as ash and oak, and among softwoods in the yellow pine group, owing to the sharp contrast between early wood and late wood in these species. In some other species, such as water tupelo, sweetgum, and soft maple, differential of early and late growth is slight, and the annual growth rings are difficult to recognize. In some tropical regions, growth may be practically continuous throughout the year, and no well-defined annual rings are formed.

The inner part of the growth ring formed first in the growing season is called *early wood*, and the outer part formed later in the growing season, *late wood*.

Figure 8-3 A cross-section of a ponderosa pine log. Light bands are earlywood, dark bands are latewood. An annual ring is composed of the earlywood ring and the latewood ring outside it. (Courtesy of Forest Products Laboratory, Forest Service, U.S. Department of Agriculture)

Actual time of formation of these two parts of a ring may vary with environmental and weather conditions. Early wood is characterized by cells having relatively large cavities and thin walls and late wood cells have smaller cavities and thicker walls. The transition from early wood to late wood may be gradual or abrupt, depending on the kind of wood and the growing conditions at the time it was formed. In some species, such as the maples, gums, and yellow-poplar, there is little difference in the appearance of the inner and outer parts of a growth ring.

When growth rings are prominent, as in the southern yellow pines and ring-porous hardwoods, early wood differs markedly from summerwood in physical properties. Early wood is lighter in weight, softer, and weaker than summerwood; it shrinks less across and more lengthwise along the grain of the wood. Because of the greater density of late wood, the proportion of late wood is sometimes used to judge the quality or strength of wood. This method is useful with such species as the southern yellow pines, Douglas fir, and the hardwoods, ash, hickory, and oak.

Wood cells. Specifically, wood is comprised of covalently bonded glucose units which join to form giant cellulose polymers (chains) 5,000 to 10,000 units long. In addition to the bonds within the chains, there are also secondary bonds between the chains themselves, such as the van der Waals forces. The formation of these wood cells occurs when some hundred or so of these long-chain cellulose molecules achieve a nearly parallel orientation into subunits called *crystallites*. In subsequent formative phases, from 10 to 100 crystallites link together in a bundle called a *microfibril*. Following in the cell wall, about a hundred microfibrils join to form a ropelike *lamella* (sometimes called a fibril).

The arrangement of the lamellae in the cell wall is shown in Figure 8-4. In the *primary wall* they form a loose, irregular network, and passing into the first layer of the *secondary wall*, *S*1, the network becomes more precise. In the *S*2 layer of the secondary wall, the lamellae run almost parallel to each other in a spiral around the cells. This (S2) layer is the thickest of the several in the cell wall and has the

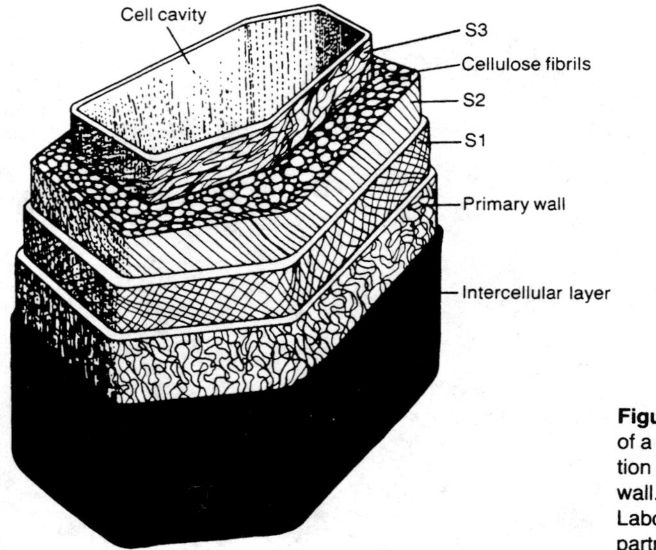

Figure 8-4 Schematic cross-section of a woody cell illustrating the orientation of the several layers in the cell wall. (Courtesy of Forest Products Laboratory, Forest Service, U.S. Department of Agriculture)

greatest effect on how the cell behaves. The smaller the angle which the fibrils make with the long direction of the cell, the stronger is the cell. Finally, in the innermost layer of the cell wall, *S*3, the lamellae are once again in a netlike arrangement.

The *intercellular layer* lies between the walls of adjacent cells, and is comprised essentially of a gluelike *lignin* to bind the cells together. While cellulose gives toughness to the cell wall, the lignin provides its rigidity. Lignin also is dispersed among the lamellae to give the unit its characteristic substantive structure. Cells are extremely small. It is estimated that a cubic meter contains over 270 billion individual cells.

Three types of cells are formed by the above arrangement of cellular elements:

1. Parenchyma, the short, thin-walled cells which function primarily for the storage and distribution of nutrients.
2. Tracheids, the long (up to 7 mm or 0.28 inch) fibrous, lignified cells which are primarily supportive.
3. Vessels, the wide-nutrient-conductive cells.

It was stated earlier that the four major components of wood are extractives, ash-forming, inorganic minerals, lignin, and cellulose. The extractives are not part of the wood structure, but they contribute to the wood such properties as color, odor, taste, and resistance to decay. They include tannins, starch, coloring matter, oils, resins, fats, and waxes, and can be removed from the wood by neutral solvents, such as water, alcohol, acetone, benzene, and ether. The inorganic constituents typically are calcium, magnesium, potassium, and lesser quantities of other elements. This inorganic matter makes up the ash when wood is burned.

These cells develop and arrange themselves into two distinct classes, the *hardwoods*, or the broad-leaved trees, and the *softwoods*, or the cone-bearing trees, which have needles or scalelike leaves. Typical hardwoods, such as the oaks, maples, and birches, characteristicslly drop their leaves in the autumn. Softwoods, such as pines, firs, and spruces, are evergreen.

Hardwoods and softwoods. The separation into these two groups has no direct application to the hardness or the softness of the wood. The true differences lie in cellular structure. An examination of these two illustrations reveals a number of interesting differences and similarities. (See Figures 8-5 and 8-6.) For example, there is a greater variety of cells in the hardwood, and these cells are not arranged in orderly, radial rows characteristic of the softwood. Another important difference is that hardwoods have vessel segments, whereas softwoods do not. When viewed in cross section, these vessels are called *pores*. Hardwoods are therefore called porous woods and softwoods are said to be *nonporous*, in that they do not contain vessels. Furthermore, in hardwoods such as chestnut, oak, and ash, the vessels are of the same diameter in both the early wood and late wood. Such woods are called *diffuse-porous*, because the vessels are scattered throughout the entire growth ring. In hardwoods such as birch, maple, and yellow poplar, the vessels are narrower in the late wood than the early wood. Since these differences in vessels appear as rings in a log cross section, such woods are called *ring-porous*. Diffuse-porous woods usually require a wood filler in a finishing operation.

Vessel segments develop from single cells. These segments are joined end-to-end to form the vessels which serve as the main passageway for liquids moving from the roots to the crown (leaves and branches) of a tree. In softwoods, resin

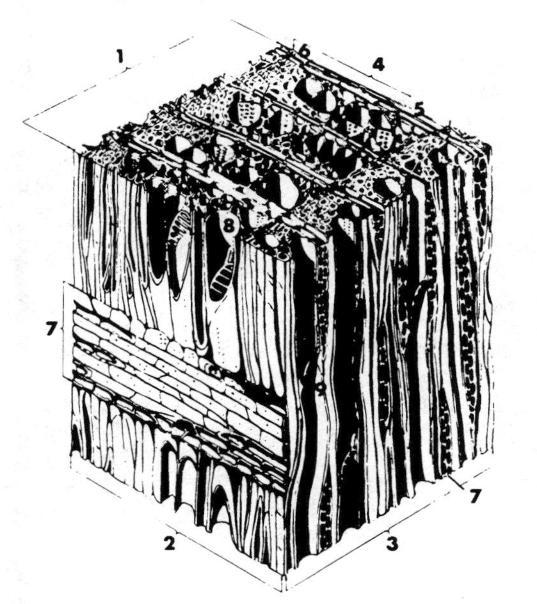

Figure 8-5 Wood cell structure of a hardwood. Note the smaller, tighter pore construction. 1) Cross-sectional face. 2) Radial face. 3) Tangential face. 4) Annual ring. 5) Earlywood. 6) Latewood. 7) Wood ray. 8) Vessel. 9) Sieve plate. (Courtesy of Forest Products Laboratory, Forest Service, U.S. Department of Agriculture)

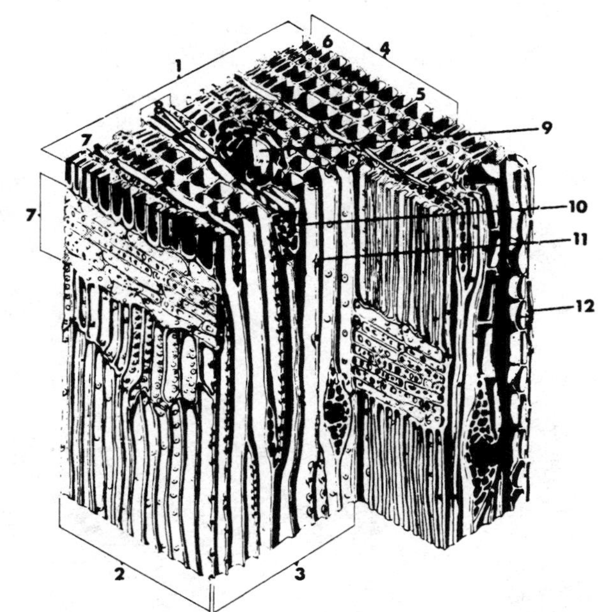

Figure 8-6 Wood cell structure of a softwood. Note the open pore construction. 1) Cross-sectional face. 2) Radial face. 3) Tangential face. 4) Annual ring. 5) Early-wood. 6) Latewood. 7) Wood ray. 8) Fusiform ray. 9) Vertical resin duct. 10) Horizontal resin duct. 11) Bordered pit. 12) Simple pit. (Courtesy of Forest Products Laboratory, Forest Service, U.S. Department of Agriculture)

ducts are formed when a space among several cells expands to form an enlarged opening in the wood. Sticky resin is released from the cells lining the duct.

In summary, it is the structure, variety, and organizational pattern of the wood cells which give hardwoods and softwoods their characteristic appearance and properties.

PROPERTIES OF WOOD

Wood properties provide a basis for selecting a suitable wood for a specific design application. The property factors determine the use of walnut in the construction of a fine chair, birch for cabinet doors, fir or pine for house construction, and spruce for pulpwood. In all such situations, the decision is based upon a knowledge of the properties of the wood material, in comparison with the specifications for a product design. Certain woods have outstanding physical appearance, while others are tough, rot-resistant, or aromatic. Some important properties are considered following.

Moisture content. Wood materials have an affinity for water in both vapor and liquid forms, absorbing and losing moisture according to variations in temperature and humidity. Moisture content is the term which describes the total amount of water in a given wood sample, and is expressed as a percentage of the ovendry weight of the wood. These values are determined according to the following formula:

$$\text{MC} = \frac{\text{weight of water in wood}}{\text{ovendry weight of wood}} \times 100$$

Measures of moisture content are important because of the effects of moisture on both the mechanical and nonmechanical properties in wood materials.

Of special importance is that wood, like many other materials, shrinks as it loses moisture and swells as moisture is absorbed. As wood dries, it shrinks most in the direction of the annual rings (tangentially), somewhat less across these rings (radially), and very little along the grain (longitudinally). The combined effects of tangential and radial shrinkage are shown in Figure 8-7. Wood which contains grain irregularities will frequently shrink in a distorted fashion longitudinally, resulting in warped or bowed boards. It will be noted that quartersawn lumber has the most desirable shrinkage pattern.

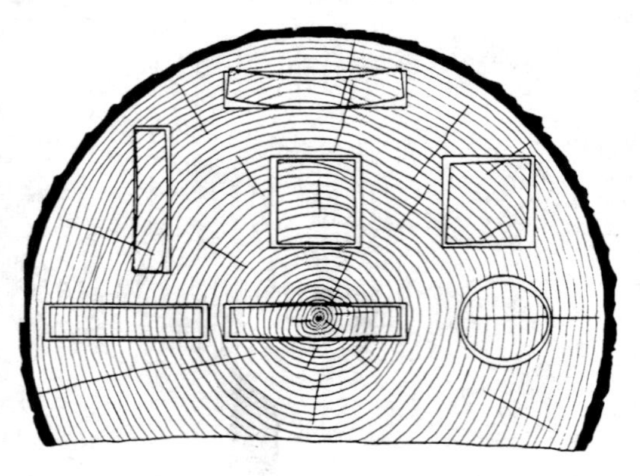

Figure 8-7 Characteristic shrinkage and distortion of flat, square, and round lumber cuts, as affected by the direction of the annual rings. Tangential shrinkage is about twice as great as radial. Note that the two flat and square pieces with the least amount of shrinkage are quartersawn lumber. (Courtesy of Forest Products Laboratory, Forest Service, U.S. Department of Agriculture)

Specific gravity. This measure is the ratio of the weight of a given volume of wood to that of an equal volume of water at a standard temperature. Because the weight of wood varies with changes in moisture content (due to shrinking and swelling), specific gravity has to be measured under controlled conditions. This measurement is obtained from the formula:

$$G \text{ (specific gravity)} = \frac{\text{ovendry weight of wood}}{\text{weight of displaced volume of water}}$$

This value varies for different woods according to the amount of cell wall material and the sizes of the cells. The term density is distinguished from specific gravity, as density refers to the weight per unit volume, as expressed in pounds per cubic foot. It is found by the equation:

$$\text{Density} = \frac{\text{weight of wood with moisture}}{\text{volume of wood with moisture}}$$

Both specific gravity and density influence the strength properties of wood and the suitability of various woods for specific applications.

Physical properties. As described earlier, wood consists primarily of relatively long, hollow cells running lengthwise in the trunk and branches. These cells, as well as those which radiate from the central axis, are cemented together by lignin. This structure, resembling a bundle of hollow tubes joined together, accounts for many of the unique physical characteristics of wood materials.

Cell structure, beside influencing the mechanical properties, also affects the appearance of wood. The varied arrangement in cellular structure which is visible on the cross-sectional face of a log is what gives the wood its grain and texture. These terms are often used interchangeably, but grain actually refers to both the annual rings and the direction of fibers longintudinally—that is, straight or spiral grain. The terms open and closed grain indicate the relative size of the pores. Texture refers more specifically to the finer structure of wood.

The grain pattern, which greatly affects the decorative features of wood materials, is itself influenced by the method in which the log is sawed into boards. Quartersawn and plainsawn boards are shown in Figure 8-8. Note the unique grain pattern for each. In quartersawn (or slash-cut) lumber, the annual rings intersect the surface at angles between 45 and 90 degrees. In plainsawed pieces, the rings intersect at angles less than 45 degrees.

The color of wood and possibly the presence of knots combine with the grain structure to provide the familiar beauty of wood. Perhaps the most fascinating visual aspect of wood is the tremendous range of possible patterns. Additionally, it can be said that wood is thermally a good insulator, and electrically a poor conductor.

Mechanical properties. Mechanical characteristics of wood behavior are essentially the same as with other solid materials, in that they are expressions of the ability to resist applied forces. Some of the more important forces or stresses related to wood are discussed here.

1. Shear Strength. Shearing strength is a measure of the ability of wood to resist the slipping of one segment in relation to another along the grain of the workpiece.

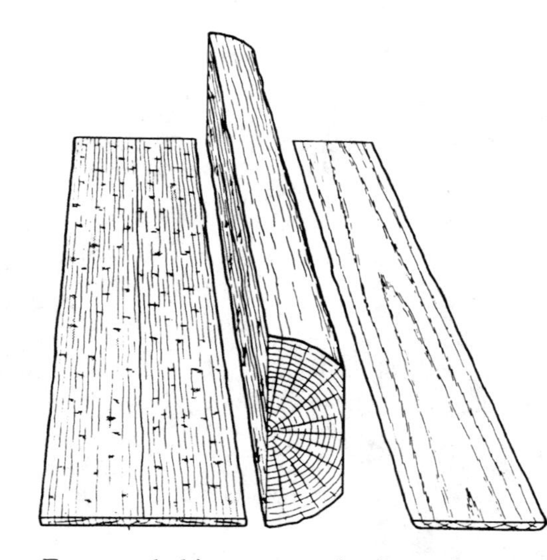

Figure 8-8 Note the different grain patterns in a quartersawn board (left) and a plainsawn board (right). The quartersawn board also is less subject to shrinkage, distortion, and warping. (Courtesy of Forest Products Laboratory, Forest Service, U.S. Department of Agriculture)

For wood this property is determined only in a direction running parallel to the grain, since the shear strength of wood is weakest in this axis.

2. Compressive Strength. The ability of a wood member to resist a force acting to shorten its dimensions is called its compressive strength. Values are determined for loading both along and across the grain of wooden members. The compressive failure of short columns is caused by shear along the grain, or by a collapse of the compressed fibers. While wood reacts favorably to loading parallel to the grain, its compressive strength across the grain is low.

3. Tensile Stress. Forces acting to increase the dimensions of a wooden member are tensile stresses, and the ability to resist such loading is referred to as tensile strength. Tensile strength tests are made perpendicular to the grain, and they measure the resistance of wood to those tension forces tending to tear it apart. Tensile loading along the grain of wood is more difficult to measure because it also involves shearing forces at the end connections.

4. Bending Stress. Such stress is a combination of shear, compressive, and tensile loading forces and causes wood to bend or flex. Generally the strain induced in a wooden member is proportional to the applied stress. If the duration of stress is short, the member will usually return to original shape. The term used to describe this constant proportion between stress and strain is the elastic modulus, or *modulus of elasticity*. In wood this modulus is a measure of stiffness or rigidity. For a beam, this measure indicates resistance to deflection, and as such the modulus of elasticity is a measure of the ability of a material to return to its original size and shape after a load is removed.

Besides the modulus of elasticity, a wood is tested and measured in several ways related to bending. The other static bending tests include a measure of the energy absorbed by a beam when it is stressed to the property which indicates the ability of the wood to absorb shock without permanent damage. Another test involves bending to maximum load and represents a measure of the ability of wood to absorb shock up to the point of complete failure. This value indicates the combined strength and toughness of wood under bending stresses. A workpiece exhibiting considerable flex, and gradually breaking with the absorption of much shock, is considered tough. Brittleness is a term used to describe workpieces which break abruptly with little shock absorption. In the impact bending test, a weight is dropped on a beam from successively increased heights until total rupture occurs. This measures the ability of wood to absorb shocks causing stresses beyond the proportional limit, that is, beyond the point at which strain is proportional to stress.

Hardness. Hardness testing is described at length in Chapter 5. However, it seems appropriate here to mention certain facts related specifically to wood. The hardness of wood refers to its resistance to wear, marring, and denting. Woods vary greatly in these respects, and the terms "hardwood" and "softwood," as used conventionally, are no indication of the property of hardness. Values are different for end grain and side grain surfaces and are measured by the load required to embed a 0.444 inch ball to one-half its diameter in these surfaces.

WOOD STANDARDS AND CLASSIFICATIONS

Sawmill operations employed to produce lumber include debarking logs and cutting them into boards of standard sizes. Standard lengths of rough hardwood lumber range from 4′ to 16′, by increments of 1′. Standard thicknesses include $\frac{3}{8}''$ to $\frac{3}{4}''$, by $\frac{1}{8}''$ increments; $\frac{3}{4}''$ to $1\text{–}\frac{1}{2}''$ to 6″, by increments of $\frac{1}{2}''$. The minimum standard width for hardwood is 3″. Softwood standards are essentially the same as for hardwoods, except that in practice, lengths are produced in increments of 2′, and the minimum rough width is 2″. Metric measures of these standard sizes will be used in the future.

The term *board measure* refers to a system of volume lumber measure. In this system the *board foot* is the basic unit of measurement. A board foot is 1″ thick and 12″ square. The number of board feet in a piece of lumber is obtained by multiplying the nominal thickness in inches by the nominal width in inches by the length in feet, and dividing by 12:

$$\frac{(T \times W \times L)}{12}$$

Lumber less than 1″ in thickness is figured as 1″. The metric wood volume measure is the cubic meter.

A log sawed into marketable lumber generally yields boards of varying quality. To provide the purchaser with the material best suited to his or her needs, the lumber is assigned grade classifications according to the uniform standards of national associations. The grade of a piece of lumber is based upon the presence or absence of defects such as checks, knots, pitch pockets, and stains, which may affect the strength, durability, or utility value of the lumber. The best grades are practically free of such defects. However, the presence of defects does not necessarily prevent lumber from being acceptable for many uses.

Most hardwood boards are not used in their entirety, but are cut into smaller pieces suitable for furniture and other products. The highest grade is termed Firsts, and the next grade Seconds. First and Seconds, (generally written "FAS"), are in practice combined in one grade. The third grade is termed Selects, followed by No. 1 Common, No. 2 Common, No. 3 Common, and No. 3B Common. Hardwood flooring is graded under separate standards, according to species.

Softwood lumber is graded according to the rules of a number of different associations. In order to eliminate grading differences among these several associations, softwood grades are based upon a set of simplified rules known as the American Lumber Standards. These standards include three main classes: yard lumber; structural lumber (or timber); and factory and shop lumber. Generally the standard grades A and B are sold together as B or better. Grade A is practically clear wood, while Grade B permits some imperfections such as pin knots and small stains. Grade C allows a limited number of similar imperfections which can be covered by paints. Grade D permits any number of surface imperfections which do not detract from the appearance of the finish when painted. Boards in Grade 1 through Grade 5 contain defects which detract from appearance, but which are suitable for utility purposes. Grades 1 and 2 are for use without waste, whereas Grades 3, 4, and 5 permit limited amounts of waste.

OTHER WOOD FORMS

Rough lumber is dried and planed to ready it for use in manufacturing. Some is flat-planed on both surfaces and edges (four square), while other pieces are planed to a variety of useful shapes. See Figure 8-9.

Aside from lumber, wood is also processed into plywoods and fiberboards.

Plywood is a term which designates glued wood panels made up of several layers or plies, generally lying at right angles to one another. As compared with solid wood, plywood has these important advantages:

- Equalized strength properties along the length and width of the panel.
- Greater resistance to splitting and checking.
- Dimensionally more stable with changes in moisture content.

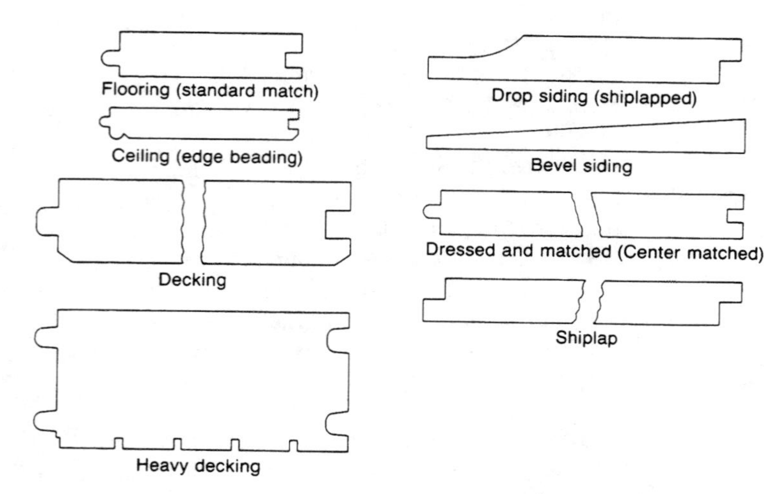

Figure 8-9 Common shapes of lumber produced with a planer. (Courtesy of Forest Products Laboratory, Forest Service, U.S. Department of Agriculture)

Fiberboard, hardboard, waferboard, and particle board are materials made from vegetable fibers or chips and binding agents (such as glue) and pressed into sheets of varying densities. These sheets are used for insulation, and as paneling covered with veneers. The properties of these sheet materials are similar to those of plywood.

Modified wood. Modern technology has led to the development of a number of ways of treating wood to change its structure, improve its properties, and increase its range of applications. One of the most important of these treatments is wood plastic composition (WPC). In this process, a wood workpiece is placed in a vacuum chamber where air is removed from the cells of the workpiece. A plastic monomer such as methacrylate is then introduced into the chamber, where it enters the wood cells. Next the wood is bombarded with radioactive isotopes, causing polymerization to take place. This process in effect converts the wood into plastic. The antimoisture, strength, and dimensional stability properties are improved greatly.

Materials with properties improved or altered, when compared to the original wood, can also be obtained by applying chemical and compression treatments to wood. Wood treated with a thermosetting resin, such as a phenolic, and cured without compression is known as an impreg. Through this process a penetrating, bulking agent is deposited in the fibers of veneers, which improves the dimensional stability of the veneer. This process, though not really common, has had some use in plywood paneling, especially for exterior applications where the necessity to eliminate checking is present.

Compreg is similar to impreg except that the wood is compressed prior to curing the resin within the wood structure. The resultant properties are similar to those of impreg, with the additional advantages of improved appearance and moldability. Typical applications include antenna masts, small airplane propellers, and products for which dielectric characteristics are important. A surface densification or ''case hardening'' can also be achieved by applying compreg techniques to solid wood.

Resin-treated wood in both the uncompressed (impreg) and compressed (compreg) forms is generally more brittle than the original wood. To meet the demand for a tougher compressed product, a compressed wood containing no resin has been developed. It will not lose its compression under swelling conditions as will ordinary compressed, untreated wood. This material is made by modifying the compressing conditions so that the lignin will flow enough to eliminate internal stresses. (Lignin is the cementing material between the cellulose fibers.) The resulting product is not as water resistant as compreg, but is about twice as tough and has higher tensile and flexural strength properties.

This material can be used in the same way as compreg, where extremely high water resistance is not needed. Likely applications are for use in propellers, tool handles, forming dies, and picker sticks and shuttles for weaving, where high impact strength is needed.

The wood modification processes described above do not involve a pretreatment of the wood to soften it. Such a process, called plasticization, temporarily acts on the lignin-cellulose structure to loosen it. Plasticization is followed by the simultaneous application of heat and pressure, generally between forming dies, to produce a desired contour. The resultant product is excellent for such applications as forming rounded corners on desk legs.

A purely chemical treatment that gives wood a high degree of dimensional stability is the polyethylene glycol (PEG) process. The treatment consists of soaking rough-sized green wood in the PEG solution for several weeks, thus preventing the cell walls from shrinking during the drying process.

PROCESSING WOOD MATERIALS

Wood materials are transformed into usable products by cutting, forming, fastening, and finishing operations or processes. In this section, some of the more significant machine woodworking operations are illustrated and described.

Cutting. Wood materials are subjected to a range of cutting operations in order to prepare them for subsequent processing, or to produce a final shape or form. Two major types of cutting can be identified. Wood materials are sawed in order to secure stock of proper length for further processing. They may also be planed to thickness, a process involving the *removal* of unwanted material to give a desired dimension to a workpiece. Usually these processes involve multiple operations. For example, a rough board is jointed and planed, then sawed to desired width and length. The primary cutting operations include jointing, planing, routing, shaping, turning, and sawing. Additional processes include drilling and abrading.

Jointing is a basic woodworking process for the purpose of trueing or smoothing one surface of a workpiece by means of a single peripheral cutting head. See Figure 8-10. Accordingly, this process does not involve the machining of a workpiece to a specified thickness. The main function of jointing is to prepare the workpiece for subsequent processing. Manual and semi-automatic jointing machines are used.

Planing is a process used to smooth one or more surfaces of a workpiece, thereby bringing it to some specified dimension. See Figure 8-11. The process is also called milling, and commonly is employed to smooth and size the two surfaces and two edges of a piece of lumber, such as a two-by-four (which is in fact 1.5 × 3.5 inches). A *single-surfacer* (or planer, or mill) planes one surface; a *double-surfacer*

Figure 8-10 Schematic diagram of a wood jointing operation.

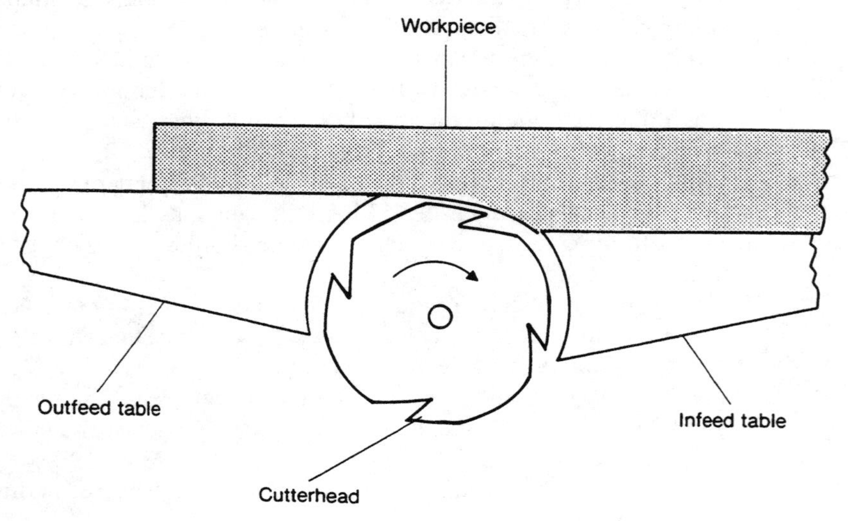

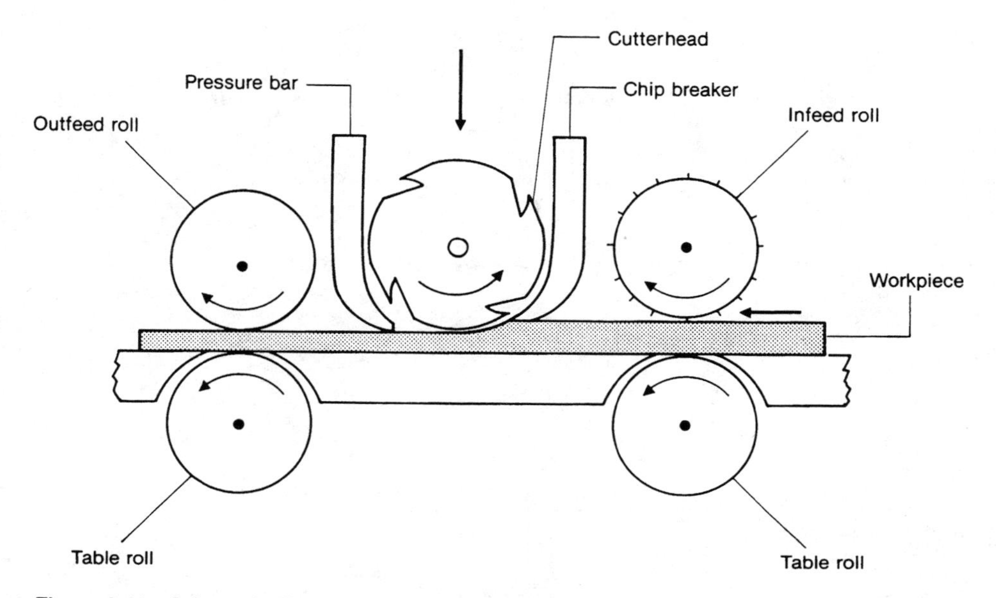

Figure 8-11 Schematic diagram of a wood planing operation.

planes two surfaces; and the *planer and matcher* planes both surfaces and both edges as described previously. See Figure 8-12. Abrasive drum planers are also used.

Moulding is quite similar to planing, except that the moulder has staggered cutting heads, and can produce contoured as well as flat surfaces.

High-speed industrial *routing* is one of the most versatile processes using the volume production of wooden parts. It is an overhead cutting tool which can enter the workpiece as a drill and then proceed to cut a recess, or an interior shape as desired. See Figure 8-13. It can also function as a mortising, shaping, boring, and drilling machine to meet the varied requirements of a particular part piece. Hand held routers also are available for custom and limited production applications. Some routers have floating spindlers which can move up and down automatically to perform surface carving operations. See Figure 8-14.

Computer controlled routers are typically used in the woodworking industry for accurate, precision, high production work. See Figure 8-15. An infinite number of cutters are available for routing work. These are secured to the spindle chuck, and they rotate at speeds between 10,000 and 20,000 rpm. Special air-actuated fixtures can be mounted on the table for the speedy positioning and securing of workpieces.

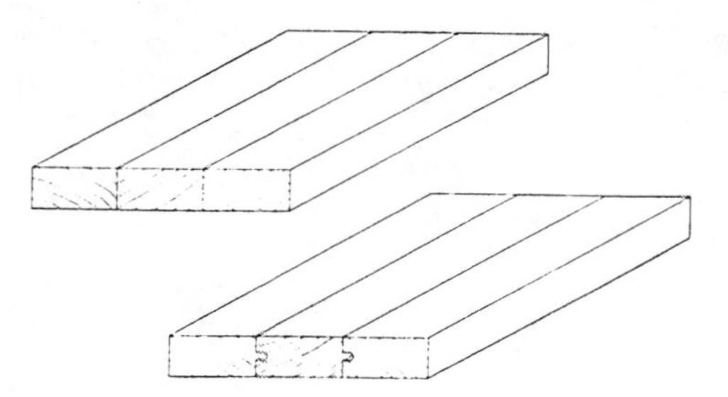

Figure 8-12 Planer/matcher planing operations are used to produce "four-square" boards such as these. (Courtesy of Forest Products Laboratory, Forest Service, U.S. Department of Agriculture)

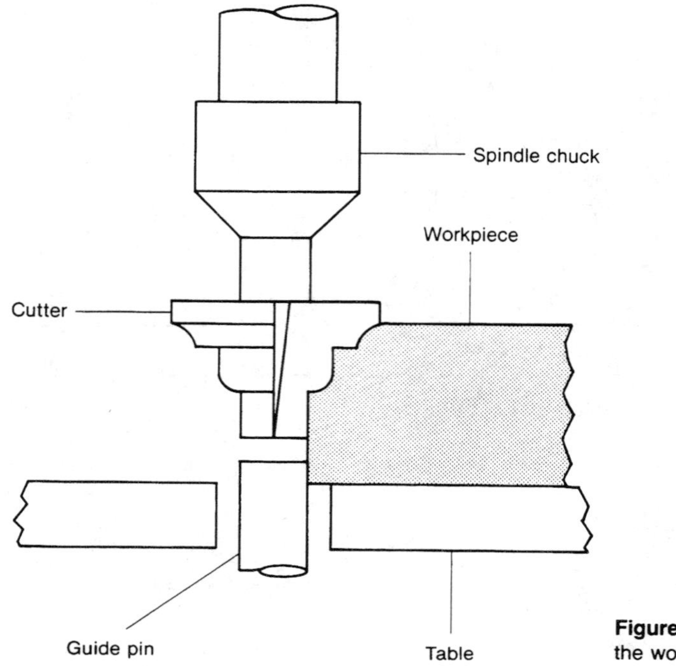

Figure 8-13 Schematic diagram of the wood router operation.

Figure 8-14 Typical parts produced on the routing center. (Courtesy of Powermatic, A Division of Stanwich Industries, Inc.)

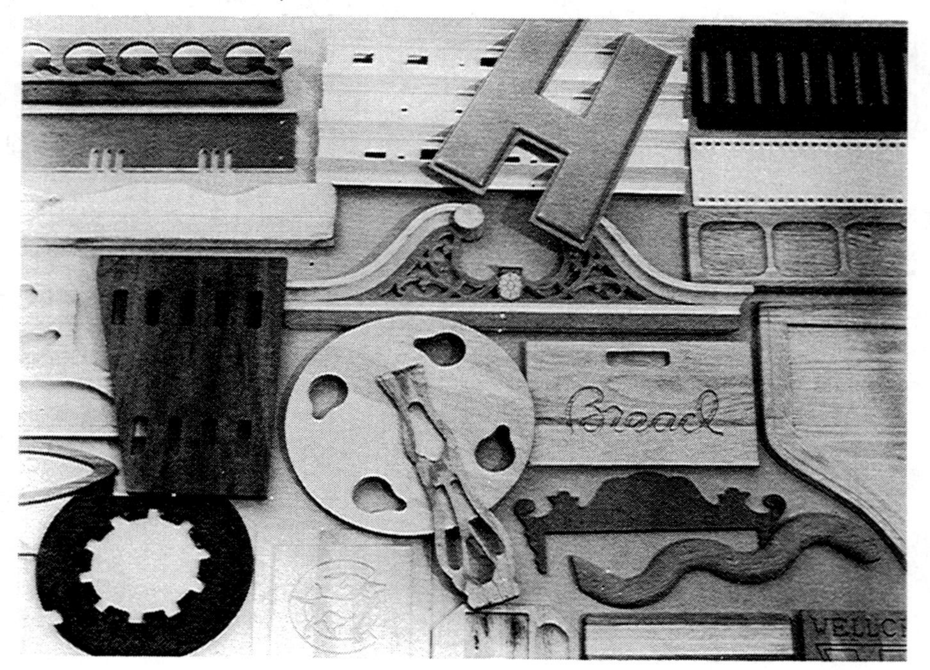

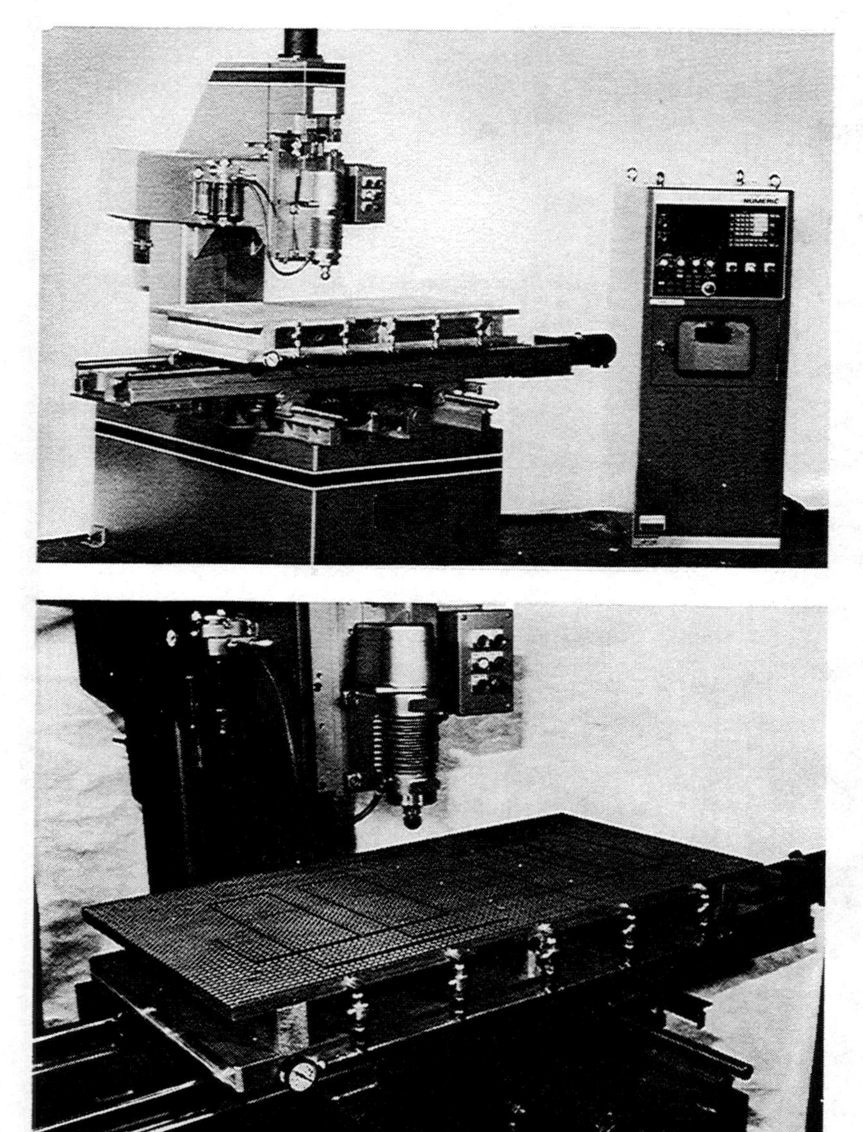

Figure 8-15 A computer controlled routing center for cutting wood workpieces. (Courtesy of Powermatic, A Division of Stanwich Industries, Inc.)

Figure 8-16 This vacuum chuck is constructed of 25 mm (one inch) thick aluminum jig plate. The gasket or vacuum seal material positioned on the chuck is necessary to generate a vacuum in specific areas of the chuck, and to firmly hold the subplate. The vacuum seal pattern is determined by the geometry of the parts to be machined. (Courtesy of Powermatic, A Division of Stanwich Industries, Inc.)

Vacuum chuck tables are also used. See Figure 8-16. Such tables feature grid patterns which can accept vacuum seal strips to hold large workpieces, or special subplates for multiple workpiece holding. See Figure 8-17. Vacuum ports provide the necessary access to an air system. Subplates are prepared by drilling vacuum parts and routing grooves to hold the gasket seal strips. The specific configuration is determined by the size and number of workpieces to be positioned on the plate. Workpieces are then placed over the work areas, secured by the vacuum control, and are ready for the CNC routing operation. The table moves to index the workpiece under the cutter head. See Figure 8-18. Multiple spindle heads also are used. See Figure 8-19.

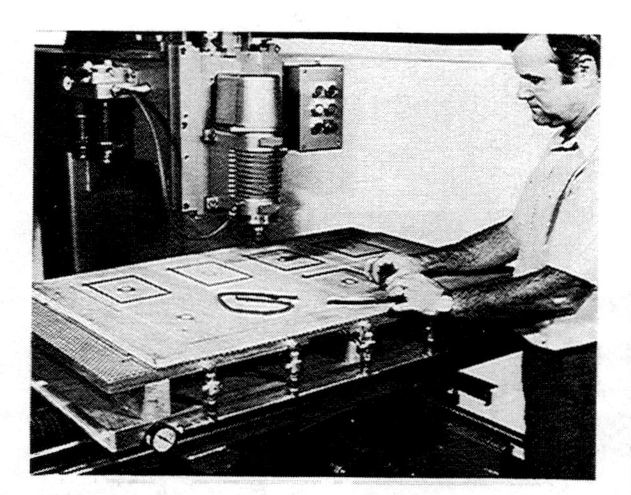

Figure 8-17 A plywood subplate is set on the vacuum chuck to hold the pieces in place to help assure accuracy during a production run. A single subplate can often be used for multiple operations. Gasket material is applied in grooves to assure a positive workpiece seal. (Courtesy of Powermatic, A Division of Stanwich Industries, Inc.)

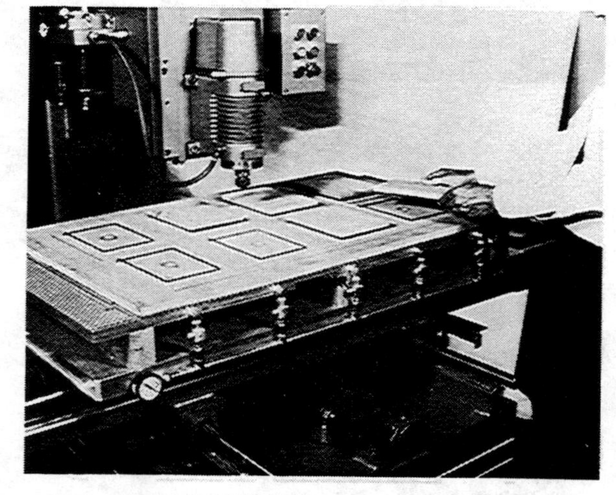

Figure 8-18 Precut blanks to be machined are placed on the subplate. The vacuum is activated by engaging control valves and the blanks are now ready to be machined. When the cycle is complete, the vacuum is released to allow quick removal of the finished parts from the table. (Courtesy of Powermatic, A Division of Stanwich Industries, Inc.)

Figure 8-19 Dual spindles may be used on the routing center for the efficient production of wooden parts. (Courtesy of Powermatic, A Division of Stanwich Industries, Inc.)

Shaping is similar to routing except that the cutter spindle is located at table level, and not overhead as with the router. See Figure 8-20. The shaper is used primarily for cutting edge profiles on the side, end, or periphery of a workpiece. Additionally, inside contours and scooping cuts for shaping bowl interiors are done with this machine. Workpieces may be fed manually, or multiple station automatic systems may be employed.

Turning operations in woodwork are similar to those metal turning processes described in Chapter 11. Workpieces can be chuck-mounted, or secured for off-hand, manual turning, or for template contour workpiece cutting. *Back-knife* lathes feature an s-shaped pattern cutter for producing furniture legs and baseball bats. See Figure 8-21. *Rotary-knife* lathes are used to thin parts such as drumsticks. Multiple-spindle *carving* is a special turning operation where many workpieces rotate simultaneously under router spindles which profile-turn objects to a desired shape.

Sawing operations in wood processing employ the same types of band and circular sawing machines as discussed in Chapter 11. Variations of these basic types include vertical panel saws, sliding table dimension saws, and double end tenoning saws, among others. Such saws can perform precision edging, mitering, parallel cutting, scoring, squaring, and crosscutting with manual and power feeds. The primary

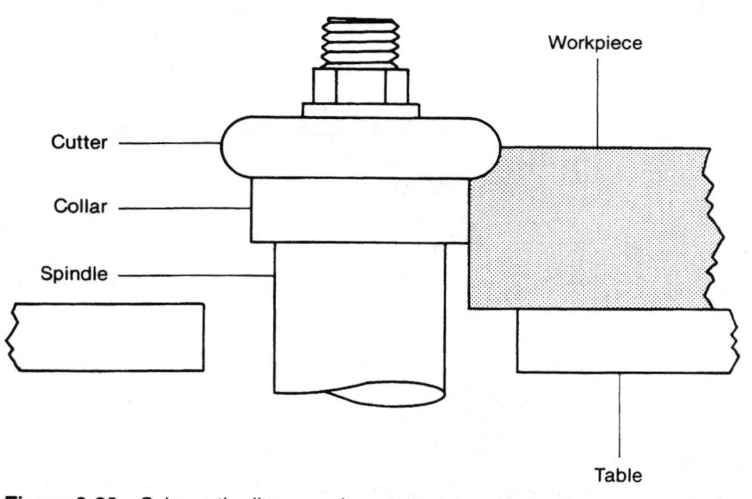

Figure 8-20 Schematic diagram of a wood shaper operation.

Figure 8-21 This turned workpiece is mounted between centers, and produced by a shaped pattern back knife.

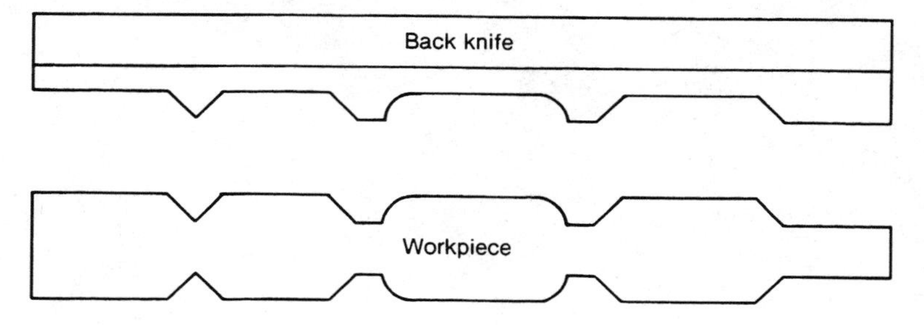

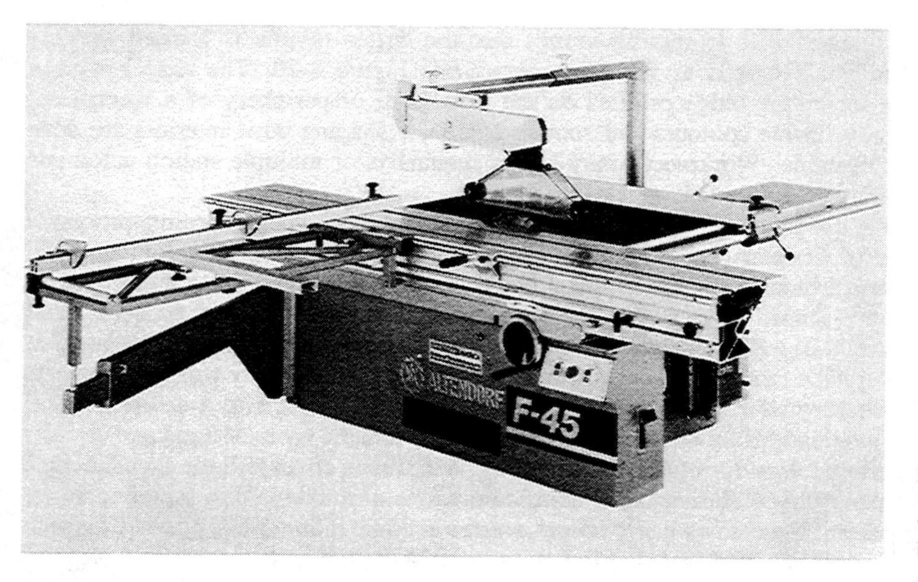

Figure 8-22 A sliding dimension table saw for accurate production work. (Courtesy of Altendorf America)

attributes of these machines are high volume repeatability, accuracy, and smooth cuts. See Figures 8-22 and 8-23.

Drilling wood is similar to drilling as described in Chapter 12. Drilling in wood technology is generally called boring, and utilizes many of the common types of drilling tools. The equipment ranges from the common vertical, single spindle floor drill press, to vertical multiple spindle boring machines, to horizontal single and multiple head borers. Additionally, these machine heads may be tilted to bore holes at desired angles. A typical machine is shown in Figure 8-24. These versatile machines permit the simultaneous, precision drilling of multiple holes. Such operations are especially important in modular cabinet and casegood furniture construction using the 32 mm system. (See following.)

Figure 8-23 This electric power feed miter saw is used to make angular cuts on wood panels. (Courtesy of Evans Rotork, Inc.)

Figure 8-24 This boring machine is equipped with one horizontal head with 21 spindles and two vertical heads with 15 spindles each, located beneath the work area for bottom drilling. The three drilling heads in combination with five pneumatic fence stops preclude costly change-over time and result in accurate, volume production. (Courtesy of Altendorf America)

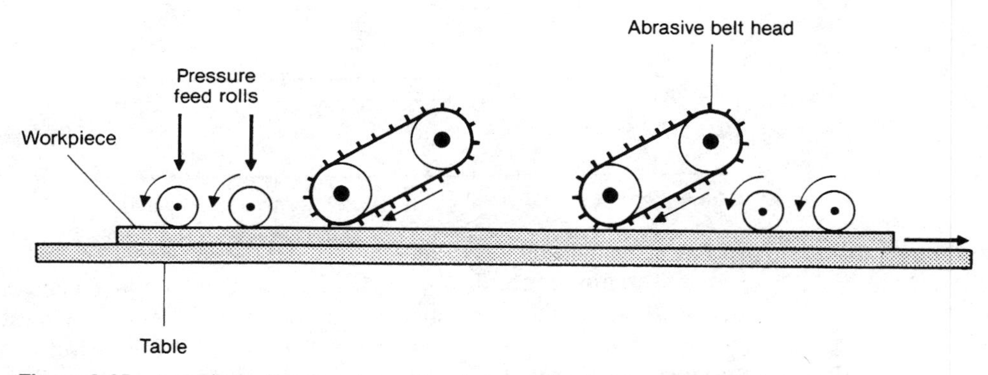

Figure 8-25 A single surface panel sanding diagram. The belt heads generally hold abrasive belts of different grit values for progressive sanding operations.

Abrading, or sanding as the process is known in wood product manufacture, is fully described in Chapters 13 and 21. Typical sanding machines are shown in Figures 8-25, 8-26, and 8-27.

Forming. Wooden parts can be steam-bent, and pressed to create contoured chair seats, laminated beams, and objects such as tennis rackets. Forming technology is therefore often applied to wooden parts, but by far the greater product volume is in casegoods and joined components.

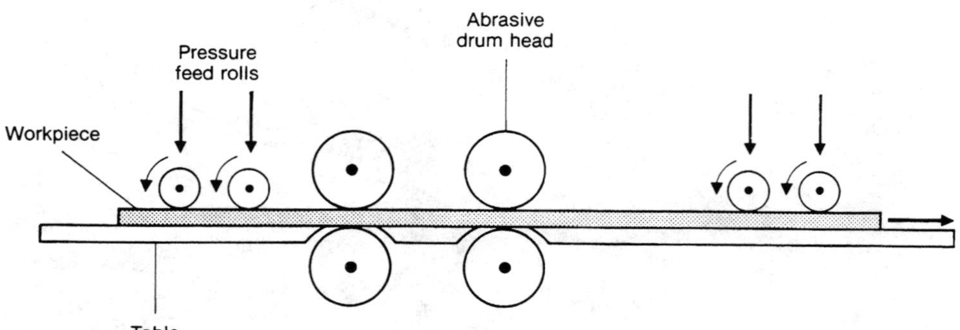

Figure 8-26 A double surface panel sanding diagram.

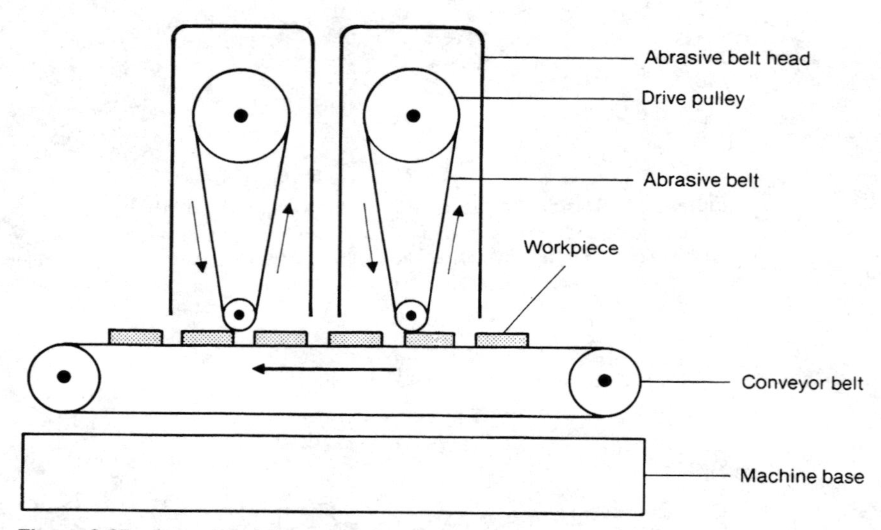

Figure 8-27 A two-head flat sander is used for small part pieces.

Joining and assembling. As occurs with all product manufacture, once components for a product have been processed they must be assembled into the final product configuration. All wood assembly involves the careful preparation of joint systems, appropriately called joinery. Joint design imposes certain problems for the designer, such as strength, appearance, elimination of coarse end grain, ease of preparation, and ease of assembly. For example, an array of corner joints is shown in Figure 8-28. These are typically used in drawer case and cabinet construction. With the exception of the rebate (or rabbet) joint, all are of a special locking variety which require only glue to complete the construction. They are so designed to preclude the use of nails, staples, or screws, to both speed assembly and to improve finish appearance.

Methods of joining shelves, case dividers, and furniture legs involve yet another series of joint designs. See Figure 8-29. With the exception of butt joint, all are self-locking, can be glued, need no mechanical fasteners, and are relatively easy to assemble. The butt joint normally would need glue, and screws, nails, or staples.

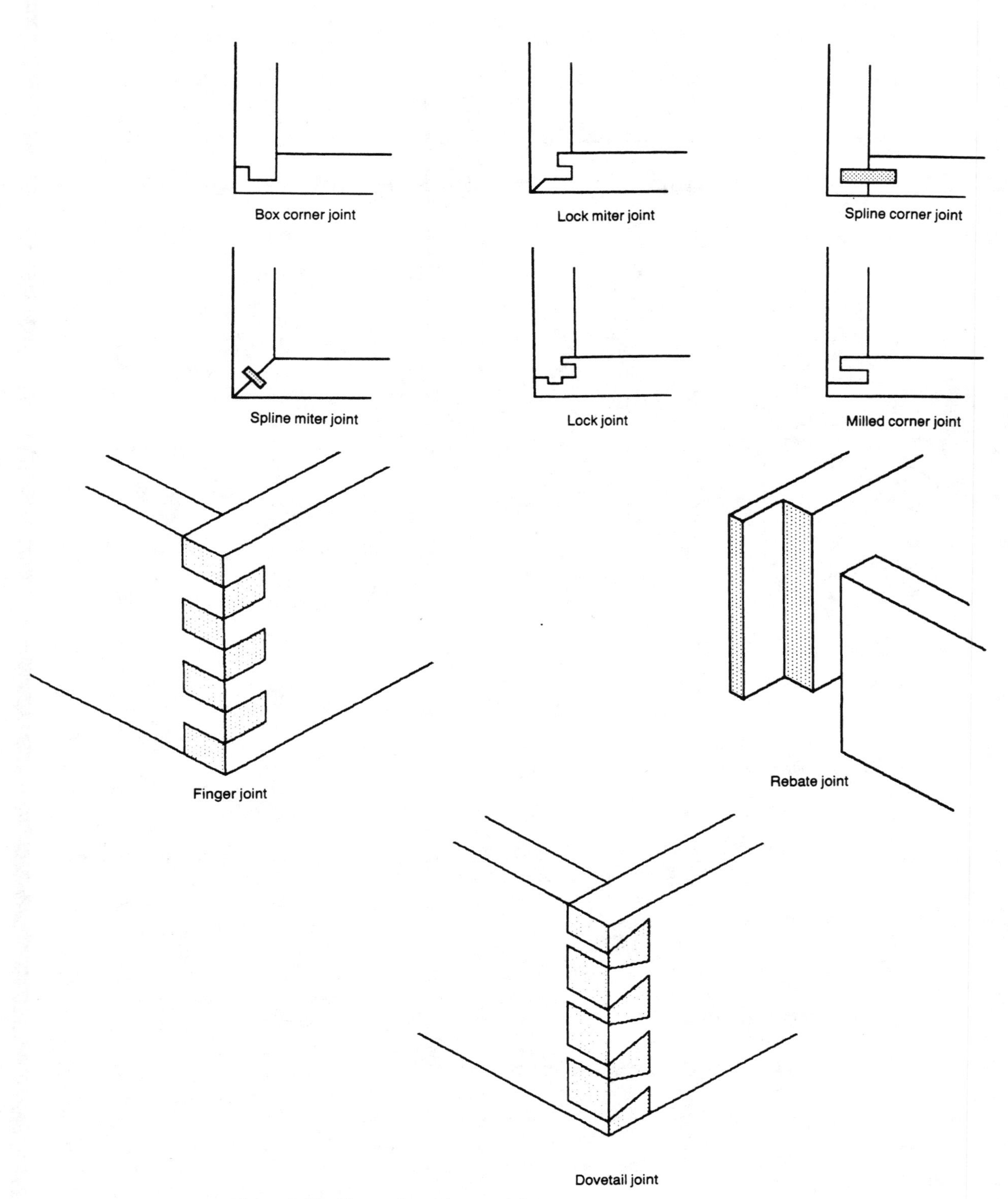

Figure 8-28 Types of wood corner joints.

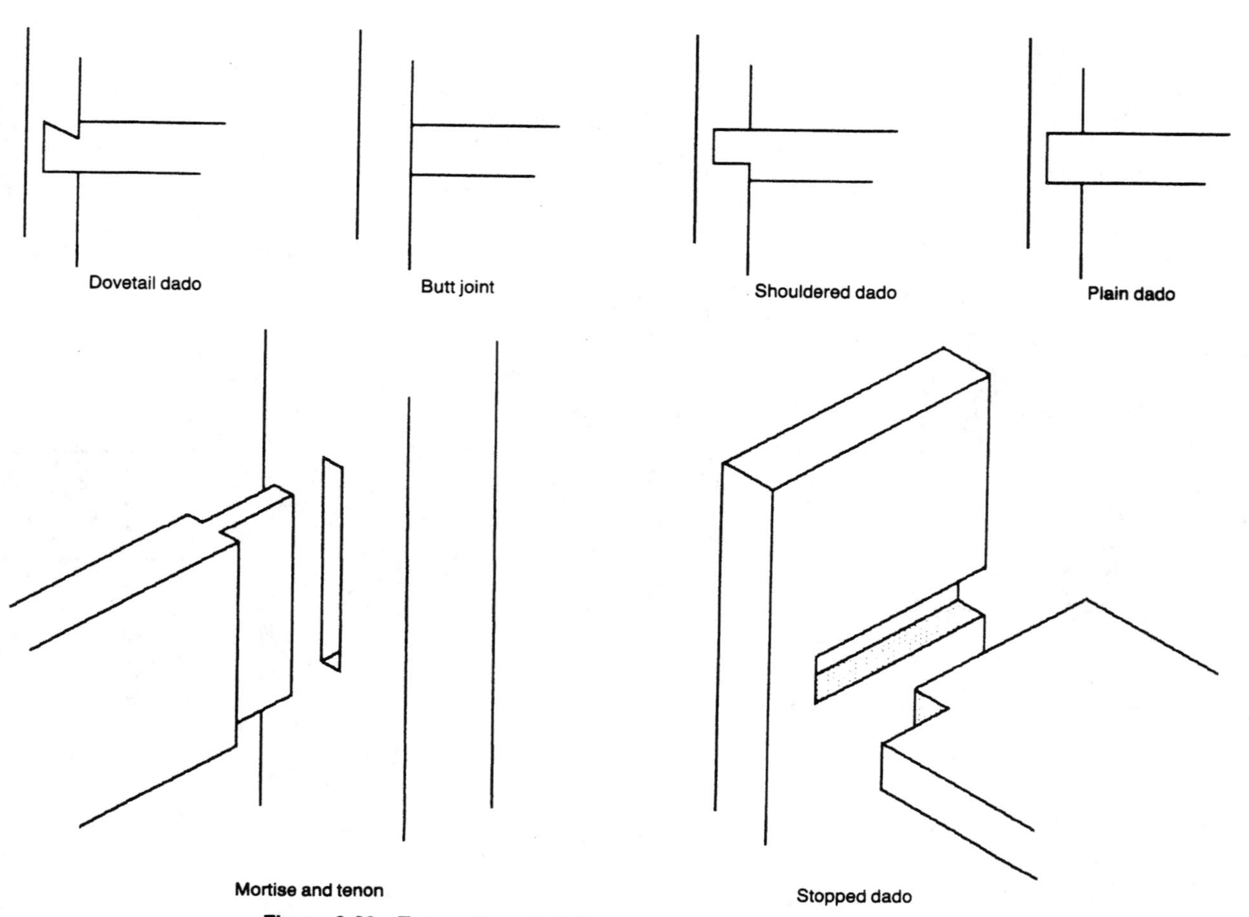

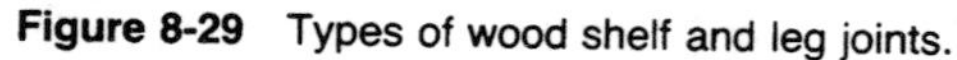

Figure 8-29 Types of wood shelf and leg joints.

Lap joints are employed to join lengths of molding and trim boards, which can be coat finished to hide any objectionable joint line. See Figure 8-30.

A more extended treatment of mechanical and adhesive joining techniques are described in Chapters 18 and 19. Many of these principles are applicable to wood products and should be examined by the reader.

A number of mechanical wood joining systems have been developed to speed up component assembly. The efficacy of such systems is that they are entirely mechanical and provide a sure means of joining parts without the use of glue. The problems associated with gluing components together in mass production operations are the need for special clamping devices, the necessary attention to clean, smooth joints, proper glues and gluing conditions, and the matter of removing excess glue from the joint. These special mechanical connections assist to avoid these problems. Some examples are shown in Figure 8-31. These connectors find wide use in the European 32 mm system used in making kitchen cabinets.

The European system is based upon a 32 mm (1.25 inches) grid employed to achieve maximum standardization and variation in cabinet design and manufacture.

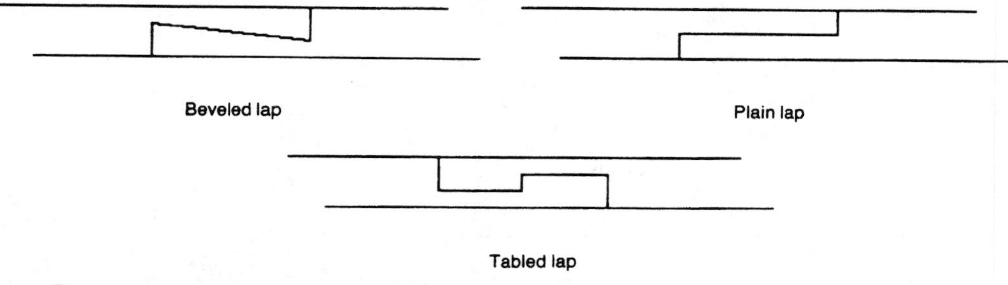

Figure 8-30 Types of lap joints.

Figure 8-31 Applications of special mechanical connectors to join wooden parts. Note the nut inserts which are screwed or pressed into one member, and the threaded connector which passes through a hole in the adjoining member, and into the fixed insert. This provides a strong, durable, and tight fastening arrangement. (Courtesy of Murakoshi)

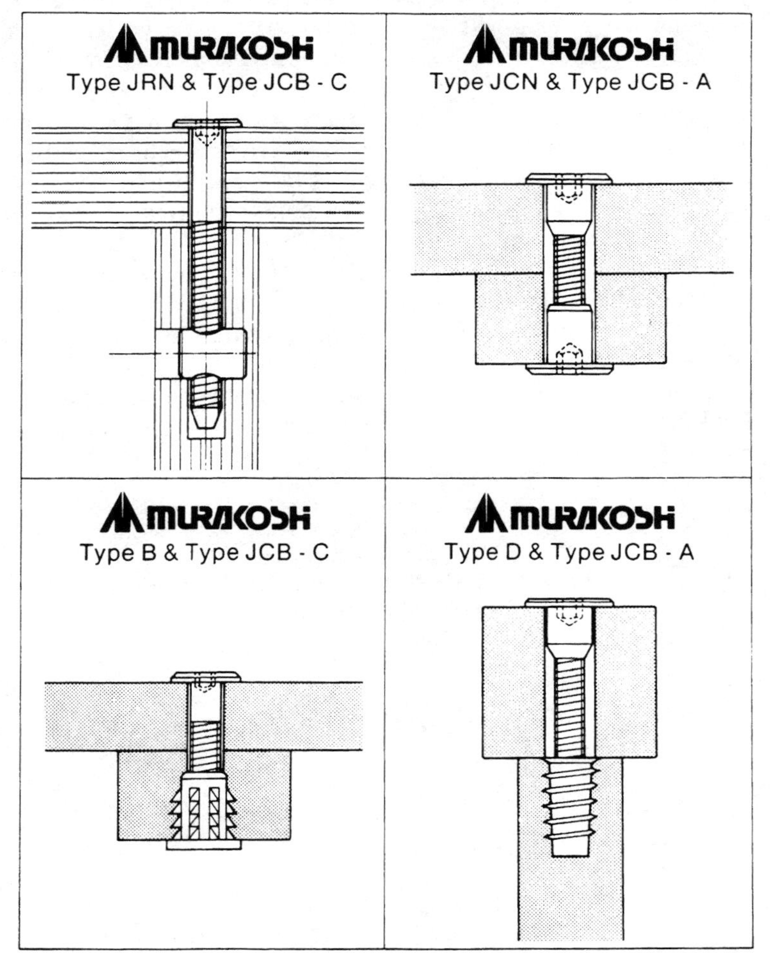

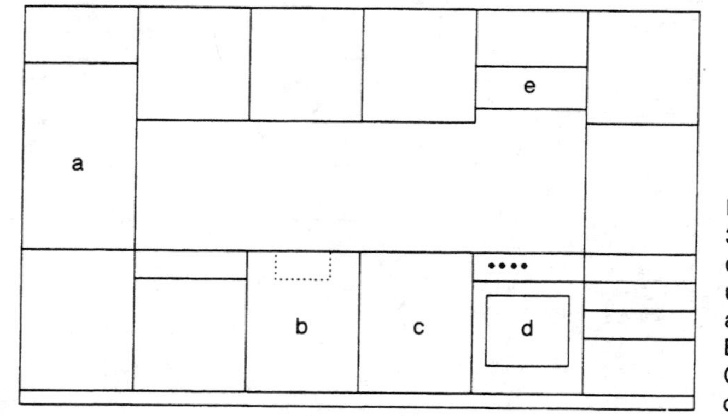

Figure 8-32 An example of the 32 mm grid system used in kitchen cabinet design. The standard modules make possible an unlimited arrangement of components. A) Built-in refrigeration. B) Sink unit. C) Dishwasher unit. D) Stove and oven unit. E) Exhaust vent.

See Figure 8-32. The function of the grid is to locate the positions of drilled holes to accommodate concealed hinges, drawer guides, dowels, mechanical connectors, shelf supports, construction fittings, and drawer/door pulls. Once the design of a cabinet series has been completed, the panels may be cut to size and holes drilled, utilizing the sawing and automatic, multiple hole boring machines described and illustrated earlier. The flexibility of the system is such that any combination of doors, drawers, and appliance openings is possible. See Figure 8-33. The use of this system enhances both the automatic manufacturing of the units, and their assembly. Some machinery which is used in producing these units is shown in Figures 8-34, 8-35, and 8-36. These systems do have their limitations, and fine furniture continues to be assembled with glue, joints, and fixtured with automatic clamps.

Finishing. Wood products are subjected to the same range of preservation and presentation finishing techniques as described for other materials in Chapters 21 and 22. There are, however, some basic demands which are obtained with such products.

Being an organic material, wood must "breathe" in order that moisture may be transmitted and absorbed through its surfaces. The wood trim or furniture in a

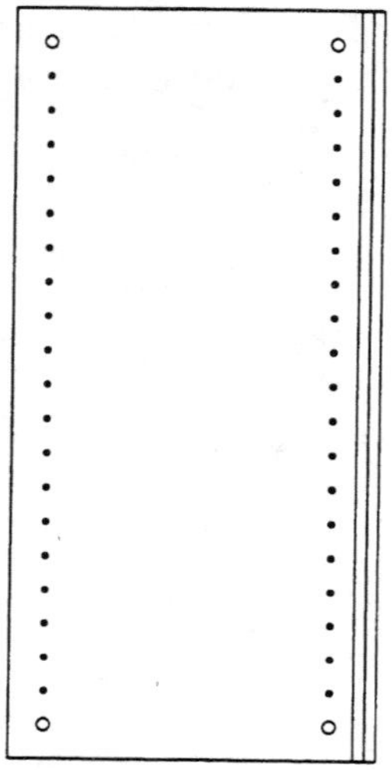

Figure 8-33 A drilling pattern for a cabinet side panel. All holes are 32 mm on center to accommodate shelf supports, drawer guides, fasteners, and other hardware. All holes are drilled in one operation.

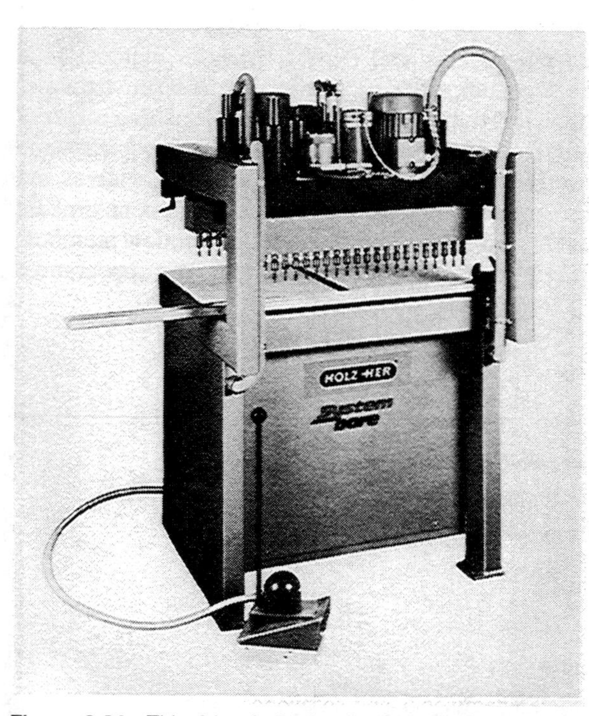

Figure 8-34 This 64 spindle line boring machine is used to drill the holes for the sample panel in Fig. 8-33. (Courtesy of HOLZ-HER U.S., Inc.)

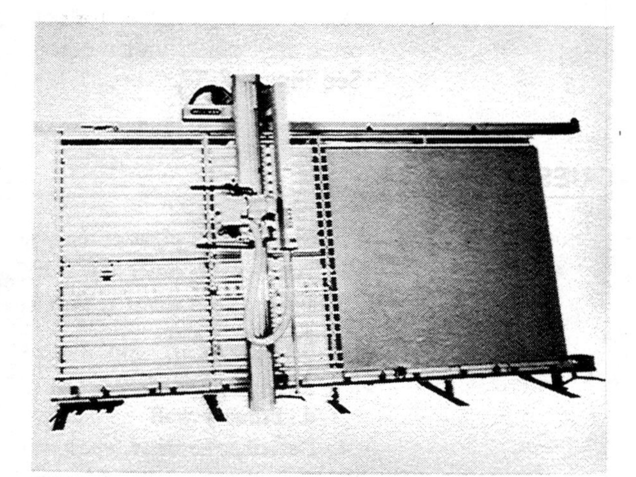

Figure 8-35 This panel saw is used in the precision cutting of 32 mm system cabinet panels. (Courtesy of HOLZ-HER U.S., Inc.)

Figure 8-36 This device assures the accurate and sure positioning of drawer units as they are assembled. The air-actuated clamping cylinders speed up assembly with quick-set glues. (Courtesy of Altendorf America)

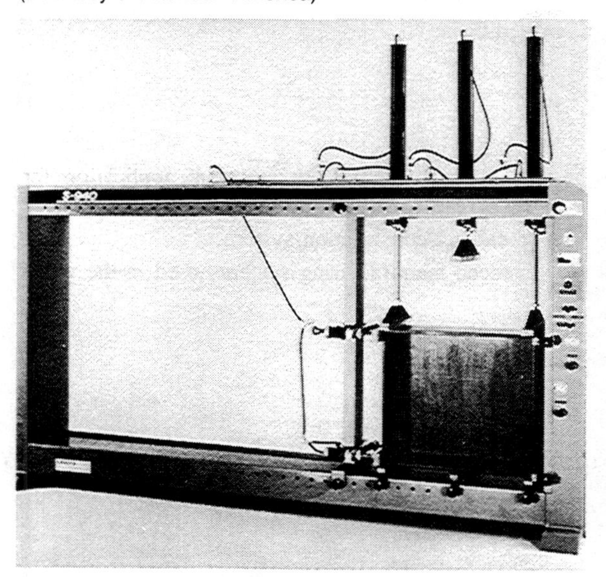

Figure 8-37 Aluminum cladding on windows offers protection from the weather and eliminates exterior painting. (Courtesy of Pella/Rolscreen Company)

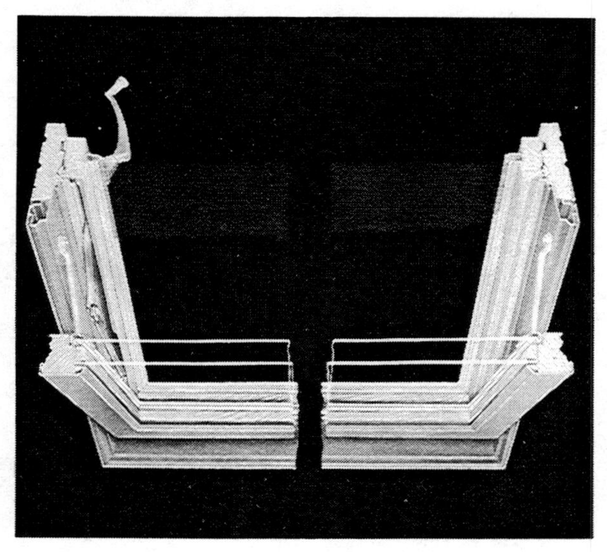

home absorbs moisture in the summer months and expels it in the winter. As a result, drawers and doors mysteriously stick or bind sometimes and run freely at others. Obviously, a wood surface which is tightly coated with some finishing material and exposed to moisture possibly may peel, warp, or check. This is especially true for exterior wood applications, in which cases exterior stains or preservatives are preferable. Wooden structural members such as dock, deck, or retainment timbers can be pressure-dipped for permanent preservation. Exterior wooden window members often are "clad" with a metal or plastic extrusion to protect and maintain appearance. See Figure 8-37.

QUESTIONS

1. Describe the eight primary elements of the cross section of a tree.
2. Describe the process which occurs to produce growth or annual rings in trees.
3. Define these terms as they relate to wood fibers:
   a. Crystallites
   b. Microfibrils
   c. Lamella
   d. Primary wall
4. Describe the three types of wood cells.
5. Describe the major differences between hardwoods and softwoods.
6. Describe the meaning and importance of the following wood properties:
   a. Moisture content
   b. Specific gravity
7. Differentiate between the physical and mechanical properties of wood.
8. Define the following wood classification terms:
   a. Board foot
   b. Grade
   c. Modified wood
9. Describe the following wood cutting operations;
   a. Jointing
   b. Planing
   c. Routing
   d. Shaping
10. Sketch some of the common wood joints and describe an important application for each.
11. Describe the theory of the 32 mm wood cabinet construction system.
12. Prepare a library research report on advanced manufacturing systems used in the wood products industry.

9 APPAREL TECHNOLOGY

A *fiber* is a particle of inorganic or organic matter in which the length is at least 100 times the diameter. Many fibers exist naturally, but only those which are pliable, strong, and at least 5 mm (0.02 inch) long are usable commercially. Some fibers, such as cotton and wool, are short and must be spun together to form yarns. These are called *staple fibers*. Fibers of an indefinite length, such as silk and the manufactured fibers (nylon, polyester, etc.) are called *filaments*.

A variety of natural and manufactured fibers are ultimately made into yarns which are woven, knitted, or otherwise made into fabrics for use in apparel, home furnishings, or household textiles. But fibers are also useful in industry when mixed with binders to make fiber construction panels and insulation, when twisted into rope, and when used to produce paint or cleaning brushes. Strands of metal wire (fibers) are even woven into screen cloth and filter baskets for a broad range of industrial applications. See Figure 9-1.

THE NATURE OF FIBERS

Some fibers, such as linen and cotton, are composed of natural plant products whereas other fibers, such as silk and wool, are of animal origin. Fibers emerging from plant and animal sources are governed by their respective biological growth processes and possess different properties and characteristics.

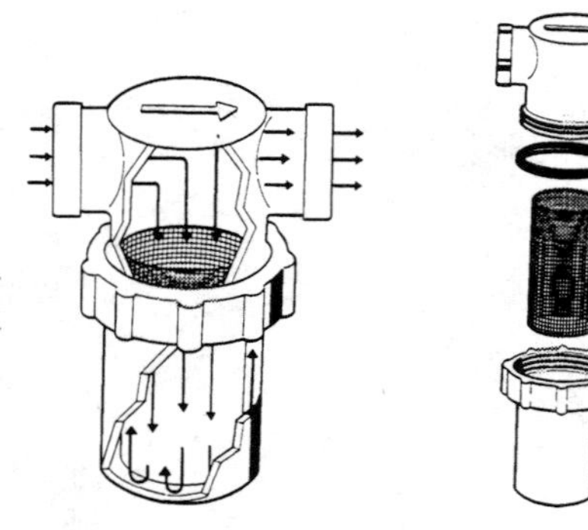

Figure 9-1 Typical metal strainer basket application assembly and disassembly. (Courtesy of RON-VIK Inc., Minneapolis, MN.)

With few exceptions, manufactured fibers are polymeric materials, characterized by molecules of great length. Examples include nylon (Antron), polyester (Dacron), and acrylics (Orlon). There are also inorganic fibers such as asbestos and fiberglass; however, this class of fibers is more important in ceramics and plastics.

Organic fibrous materials consist of large molecules in which covalently bonded atoms of carbon, hydrogen, and oxygen predominate. Molecules of the polymer are attracted to each other by weak van der Waals forces. The individual polymer molecules may also be cross-linked to each other through covalent bonds. This results in stronger and more rigid fibers.*

Man-made fibers can be engineered to possess specific characteristics to fulfill a particular need in end-use. Manmade fibers are classified as *cellulosic* and *noncellulosic*. Rayon, acetate, and triacetate are classified as cellulosic fibers. Noncellulosic fibers include nylon, polyester, and acrylic and are made from chemical derivatives of petroleum. Man-made fibers are made from polymers, which must be either dissolved in a solvent or in a molten form. The liquid polymer material is then extruded through the tiny holes of a spinneret to form continuous filaments. These filaments solidify as they pass through either air or a chemical bath. The long filaments may be processed into yarns or cut into staple lengths and blended with natural fibers. Blends are important in the production of textiles because they capitalize on the best qualities of each fiber, combining the look and feel of the natural fiber with the easy-care properties of the synthetics.

FABRIC MANUFACTURING

Fibers are the basis of fabric manufacturing. Any fabric made from fibers by any method can be broadly defined as a *textile*. Textiles include nonwoven fabrics, woven fabrics, and knitted fabrics. Fibers used to make nonwoven materials, such as felt, are arranged and processed directly into a fabric by moisture, heat, and pressure. Bonded fabrics are formed from a web or sheet of fibers held together by adhesives or treated with heat or solvents, which soften the fibers and bond them into position. Most textile fibers, however, are formed into long strands called *yarns* before being made into fabrics.

* See Chapter 5, Properties of Materials.

Spinning. Staple fibers, whether natural or man-made, are formed into yarns by the *cotton system* and the *woolen* and *worsted system* or by a variety of newer methods. In the cotton system, yarns are made by carding and/or combing the fibers to separate them, provide a parallel orientation, and remove short unmanageable fibers. During subsequent processes, the fibers are drawn together, and sufficient twist is introduced to bind the yarn firmly together.

Yarns spun from wool and man-made staple fibers employ the woolen or worsted system. These systems are similar to the cotton system, except that in the woolen system yarns are only carded, resulting in a more random fiber arrangement, whereas worsted yarns are carded and combed, producing a highly oriented fiber arrangement. The worsted system produces smoother and finer quality yarns. Yarns produced from short staple fibers are referred to as *spun yarns* or *staple fiber yarns*.

Filament yarns are made from natural or man-made fibers of indefinite length. The manufacturing process is simple and direct. Individual fibers are brought together, with or without a twist, and formed into a yarn. In the *continuous process*, yarns are made immediately after the fiber has been extruded. Other filament fibers are wound onto cones immediately after extrusion and stored until needed. When fibers are made into yarns at a later time, the process is referred to as a *discontinuous process*. Yarns made from filament fibers are called *filament yarns* or *thrown yarns*.

Weaving. *Weaving* is the process of making fabrics from sets of yarns which interlace at right angles. Those running the length of the fabric are called the *warp*, and those lying across the width of the piece are called *filling yarns* or *weft*.

There are three basic weaves: plain, twill, and satin. A *plain weave* is formed when warp and filling yarns alternately pass over and under one another. This weave is the simplest, most common weave and is relatively durable and inexpensive. Muslin, cheesecloth, and handkerchief linen are examples of fabrics made from a plain weave.

Twill weaves are characterized by a diagonal pattern on the fabric. Filling yarns pass over a number of warp yarns before going under a warp yarn. This sequence is repeated again and again; however, it is moved over one step each row creating the diagonal line. Fabrics made from a twill weave are strong and durable. They are also more expensive than plain weave fabrics. Examples of twill weave fabrics include denim, gabardine, and serge.

In the *satin weave*, the warp yarn crosses over several filling yarns resulting in "floats" on the fabric face. If filament yarns are used, the long float yarns reflect light, giving the fabric a lustrous and smooth surface. Satin weaves are not as durable as plain and twill weaves, because the float yarns are easily snagged. Antique satin and bridal satin are examples of satin weave fabrics.

These weave systems and their many variations, along with the type of yarns employed, produce fabrics which are loose or close, soft or hard. They also provide a broad range of interesting patterns and textures, such as hopsacking and herringbone. Weaving is an old art, but the basics remain unchanged. Modern technology has introduced sophisticated equipment to produce a broad spectrum of woven fabrics at an affordable price and in a reasonable length of time.

Knitting. Knitted fabrics are especially desirable in applications requiring flexibility and porosity, such as in underwear, sweaters, and stockings. In knitting, the loop is the basic structural unit. As each new loop is formed, it is drawn through the one previously made. Many rows of interlocking loops result in a knit fabric. In *filling* or *weft knitting*, loops run across the width of the fabric or around a circle. This knitting method is inexpensive and produces a fabric with good stretch. When loops are formed in a vertical direction, the method of construction is known as *warp knitting*. These fabrics are generally flat and can be manufactured quickly and easily. Through the use of machinery, interesting variations and novelty knits can be created. Hand knitting is still a popular art, but modern machines are used for commercial production.

FINISHING MANUFACTURED TEXTILES

Fabrics are finished in order to change or improve their characteristics, appearance, or feel before being made into consumer products. The term *greige goods* is used to refer to fabrics that have not received any finishing process. Depending on the fiber content, certain finishes are routinely applied in preparing a fabric for end use. Cotton, for example, would typically be singed, desized, scoured, bleached, mercerized, tentered, calendered, and inspected. A different sequence of procedures would be utilized for other fabrics. Additional finishes might be applied for specific purposes. Some common finishes include singeing, mercerization, compressive shrinkage, and water resistance.

Singeing eliminates projecting fiber ends which cause roughness, pilling, and dullness if left on the fabric surface. Fiber ends are brushed up and subjected to open flames or heated plates to remove the unwanted fiber ends. Fabrics of natural and blended fibers may be singed.

Mercerization imparts a sheen to fabrics made of cellulosic fibers, such as cotton and linen. The fabric is treated with a cold, caustic chemical solution which swells the fibers allowing them to reflect light and thus achieve a lustrous, silky appearance. Additionally, mercerization increases strength, improves dyeability, and partially reduces residual shrinkage of the fabric.

Compressive shrinkage produces cotton and linen fabrics that will have less than 1 or 2 percent shrinkage during laundering. Slightly damp fabric placed over a thick blanket is fed over a roller and against a heated cylinder causing the fabric to retract, which compresses or shortens the fabric. A fabric with the trade name Sanforized is an example of compressive shrinkage.

Water-resistant finishes utilize agents such as silicones, fluorochemicals, and ammonium compounds to coat and adhere to individual fibers, causing them to resist wetting. Water-resistant finishes do not alter the appearance of the fabric, are comfortable to wear, and repel stains.

Some finishes are applied to fabrics with mechanical equipment; others utilize chemical solutions to produce the desired result. Some finishes are considered durable if they can withstand "normal" wear; others come off during wear or cleaning and can be reapplied. Whatever the requirement or specific characteristic, finishes offer the consumer many wear, care, and comfort benefits.

APPAREL MANUFACTURING

Once the spinning, weaving, knitting, and finishing processes are completed, the resultant fabrics are manufactured into a variety of industrial and consumer products. The manufacturing of apparel is one example of how textile materials are converted into finished products.

Apparel manufacturing consists of three major segments: preproduction, production, and postproduction. Regardless of factory size, these processes are generally further divided according to tasks. *Preproduction* includes designing, patternmaking, grading and marking, and cutting. *Production* involves small parts stitching and garment assembling, while *postproduction* consists of operations such as finishing, pressing, quality control, inspection, packaging, and shipping.

PREPRODUCTION

Design. Design inspiration comes from a variety of sources and, traditionally, designs have been sketched in two-dimensional form on paper. The designer needs an understanding of construction techniques and must be able to illustrate exact style and construction details so that the pattern maker can work from the drawing. See Figure 9-2. Designers traditionally have sketched their ideas free-hand; however, contemporary technology provides today's designer with a tool that eliminates the

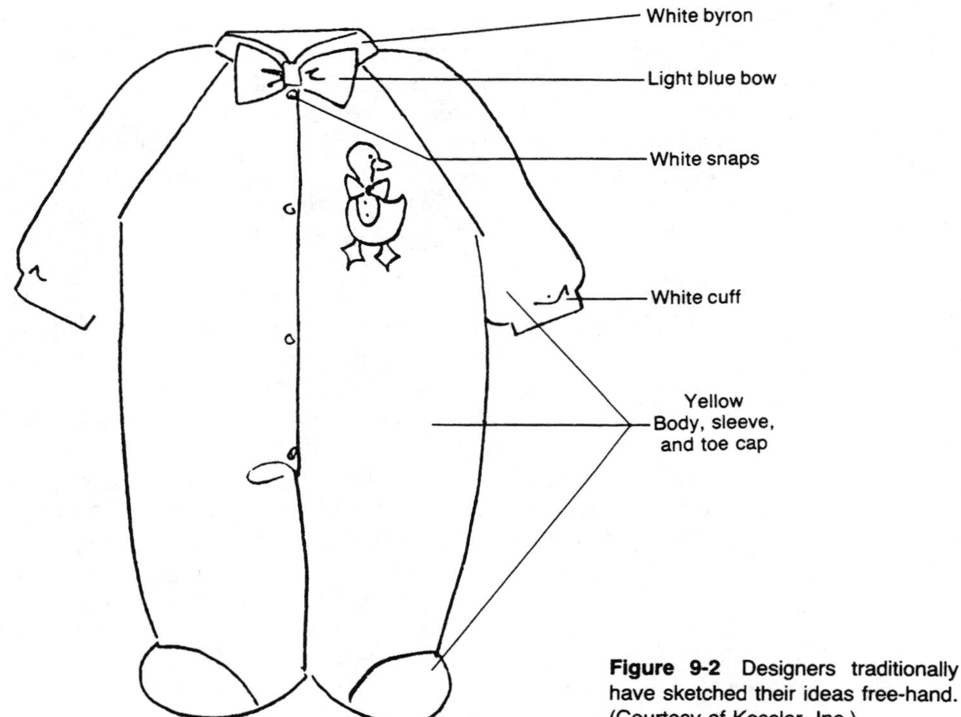

Figure 9-2 Designers traditionally have sketched their ideas free-hand. (Courtesy of Kessler, Inc.)

Figure 9-3 Computerized fashion design systems allow the designer to replace the sketch pad, pencil, and pen with a data tablet, pen, and display screen. (Courtesy of Gerber Garment Technology)

necessity for sketching. Computerized fashion design systems allow the designer to replace the sketch pad, pencil, and pen with a data tablet, pen, and display screen. See Figure 9-3. Sketches, style modifications, color combinations, texture, and fabric pattern can be quickly displayed, assessed, and changed until the best design is achieved. This process is quicker than hand-drawn illustrations and shortens the time required to turn designs into finished products.

Computer technology is making great inroads in all phases of the apparel industry. Benefits resulting from computer-aided design (CAD) and computer-aided manufacturing (CAM) include greater flexibility, increased productivity, reduced throughput time, lower production cost, and consistent product quality. Incorporating computer technology requires a large initial outlay in cost, however, and a company must be convinced that the benefits will justify the expenditure. As imports threaten American-made products, the textile and apparel industry has developed a ''Quick Response'' technology that narrows the cost differential.

Patternmaking. After a design has been created, the first pattern must be made. First patterns may be developed by *draping*, *drafting*, or a combination of the two. When a first pattern is draped, the pattern maker cuts and shapes muslin on a three-dimensional dress form until the desired look is achieved. All style lines and shaping techniques must be carefully placed. Finally, the muslin is removed from the dress form, marked to indicate all construction and seam details, perfected with French curves and rulers, and copied onto heavy paper.

Utilizing pencil, paper, L-square, and a set of prescribed measurements, the pattern maker may draft the full-size first pattern by hand at a drawing board or graphically display a scaled down basic flat pattern on a computer screen. In either case, the basic pattern is manipulated by the pattern maker to assume the style of the new design. See Figures 9-4 and 9-5. The computer pattern drafting system eliminates the time-consuming task of tracing, measuring, and drawing patterns and interacts with a plotter to produce a hard copy of the first pattern within minutes. Once the pattern is in the computerized system, it may be recalled at any time for redesign or adjustment.

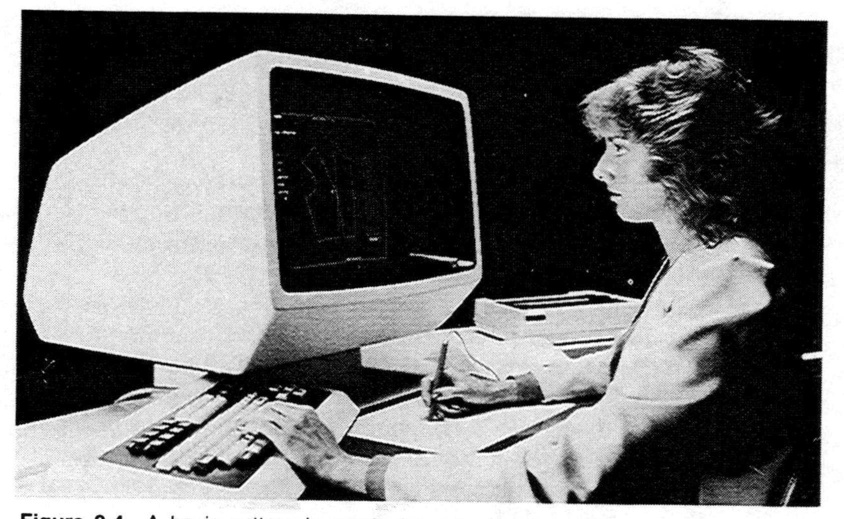

Figure 9-4 A basic pattern is manipulated by the pattern maker to assume the style of the new design. (Courtesy of Gerber Garment Technology)

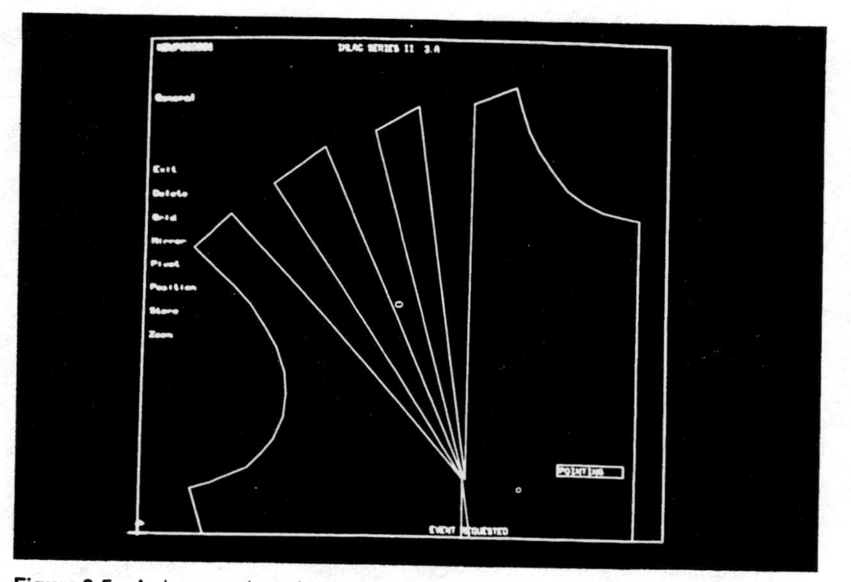

Figure 9-5 A close-up view of a pattern displayed on the computer screen. (Courtesy of Gerber Garment Technology)

The first pattern is made in a "sample size" and is used to construct a sample garment. The sample, made by a sample maker, is a test garment. Factory construction methods are employed as the garment is sewn to determine any construction problems that may result in production. The sample garment is then evaluated on a mannequin as well as on a live model whose measurements correspond to the sample size. Problems in fit, comfort, or appearance that do not show up on a mannequin may be apparent when a live model puts the garment in motion. An evaluation team may make changes in the style, ease requirements (which allow for body movement), and fit of the sample. The sample garment is corrected or, if necessary, a new pattern is made, thereby requiring the construction of a new sample garment. This procedure is repeated until all aspects of fit, comfort, and appearance of the garment are perfected.

A retailer who has contracted for a private label for the design may request an opportunity to evaluate the sample garment and has the right to make additional adjustments to meet company specifications. When all parties are satisfied with the design, final approval is granted.

Some very sophisticated computer design systems allow the designer to produce on a display screen a three-dimensional rather than a two-dimensional view of the garment on an image of an actual model. The designer can see exactly what the garment will look like—front, back, and side—without going through the expense and time of making a sample garment. Photographs of the design may be generated in various colors, patterns, and textures and may be sent to potential buyers in a fraction of the time required when working with traditional samples.

The production pattern. The production pattern maker may adjust the sample pattern or make an entirely new pattern, either by hand or computer, based on the corrected sample pattern. The pattern maker assures that all style lines are accurate for mass production and that all measurements conform to standardized size specifications of the company, or the retailer if the design is a private label garment. These specifications include room for body movement and style fullness.

All pattern pieces must be made to fit precisely the piece to which it ultimately will be sewn. Identifying marks, known as *notches*, are placed on adjacent pattern pieces by the pattern maker so that assembly line operators can accurately match garment parts during construction. The completed production pattern is then made either by hand or computer and is referred to as a *master block* or *sloper*. Perfection of the production pattern is essential to the ultimate fit, appearance, and success of the completed garment. The cost of mass producing garments that are not acceptable when completed is exorbitant and can create serious problems for a firm's profits and reputation.

Precosting and costing. The designer keeps accurate records on all designs as they develop. These records include information essential to determining garment costs. Called the work sheet, it is sent to the costing department where *precost* is estimated from the various components of the design such as fabric, trim, closures, etc. Precosting occurs before the design is actually adopted into the line. If precosts exceed the manufacturer's usual price structure, the design is returned to the designer for modification until production costs are consistent with the rest of the manufacturer's line.

After the garment has been accepted as one of the season's styles, *final cost* is calculated. Final costing involves a detailed analysis of the materials, sample garment, production pattern, step-by-step construction operations, and manufacturer's overhead to establish an accurate cost figure for the garment. In addition to material and production costs, a detailed cost analysis includes material handling, freight, sales commission, and profits. Generally, cost is determined per dozen garments.

Grading the pattern. After the production pattern is perfected, the pattern must be made proportionally smaller and larger to produce the manufacturer's complete range of sizes. This procedure is known as *grading*. In children's wear, for example, a sample-size garment might be a 3T (3 Toddler). The pattern grader would reduce the sample-size production pattern by predetermined amounts in width, length, and circumference to create smaller sizes. The 3T would also be increased to provide larger sizes. Each company determines its own specifications, or grade rules, for grading increments. Manufacturers of very expensive garments may choose to have each size test-fit after grading.

The grader may use a grading machine in producing the various sizes from the production pattern, or a digitizer may be used to enter pattern grade points in a computer. When the grading machine is used, the master block is placed in the machine over tagboard and grading is done one size at a time. The grading machine has knobs which are turned by hand to move the pattern either horizontally or vertically to make the specified increments at each pattern point. Each piece of a pattern is shifted and traced until all parts have been graded for the entire size range. The pattern grader is responsible for determining the amount to be added, since the grading machine does not automatically do this. For example, if a company's specifications called for a 1-inch width increment for a given size at the waist, the grader would have to know that half of the amount of the increment must be added at the waist level on the front pattern piece, and half of the increment must be added on the back pattern piece. This grading increment must be divided by the total number of seams in each pattern piece, and an equal amount of the increment must be added to each seam of each pattern piece making up the waistline of the garment. In a design that has many pattern pieces comprising the front and back sections, the grading procedure is even more complex. Competence in mathematics is essential for successful grading. The grader must also have a good understanding of pattern structure, since pattern lines and curves must be redrawn or blended without altering the original style lines of the pattern.

Computer technology has enabled the pattern grader to accomplish this task with greater ease and efficiency at a graphic workstation, by entering pattern grade points into a computer with a digitizer. A digitizer resembles a drafting table that contains numerous closely spaced horizontal and vertical wires acting as X and Y coordinates. The master pattern is placed on the digitizer and a cursor is moved around the outer edges of the pattern recording the location of critical grade points such as the side waist, armhole, shoulder, etc. See Figure 9-6. Software on the computer can

Figure 9-6 The master p tern is placed on the digitizer and a cursor is moved around the outer edges of the pattern recording the location of critical grade points. (Courtesy of Gerber Garment Technology)

automatically increase and decrease each grade point a specified amount in a specified direction for each size within the range according to the company's grade rules. See Figure 9-7. When grading is completed, all pattern pieces for the size range of the design may be printed out on a plotter. Companies not wishing to invest in computerized equipment may receive the efficiency and cost benefits by subscribing to a grading service.

Making the marker. When grading is completed, a marker is made. *Markers* are cutting guides or pattern layouts placed on top of the fabric prior to cutting. All pattern pieces of varying sizes are drawn on paper equal in width to the garment fabric and as long as required to accommodate the cutting order. Grain direction, one-way fabric design, and fabric nap must be considered when making a marker. Markers may be made either by hand or by computer.

In the hand operation, the marker maker traces graded tagboard pattern pieces onto the proper size paper. The marker maker experiments with pattern placements until the required pieces are placed as economically as possible. Substantial fabric

Figure 9-7 Software on the computer can automatically increase and decrease each grade point a specified amount in a specified direction for each size within the range according to the company's grade rules. (Courtesy of Gerber Garment Technology)

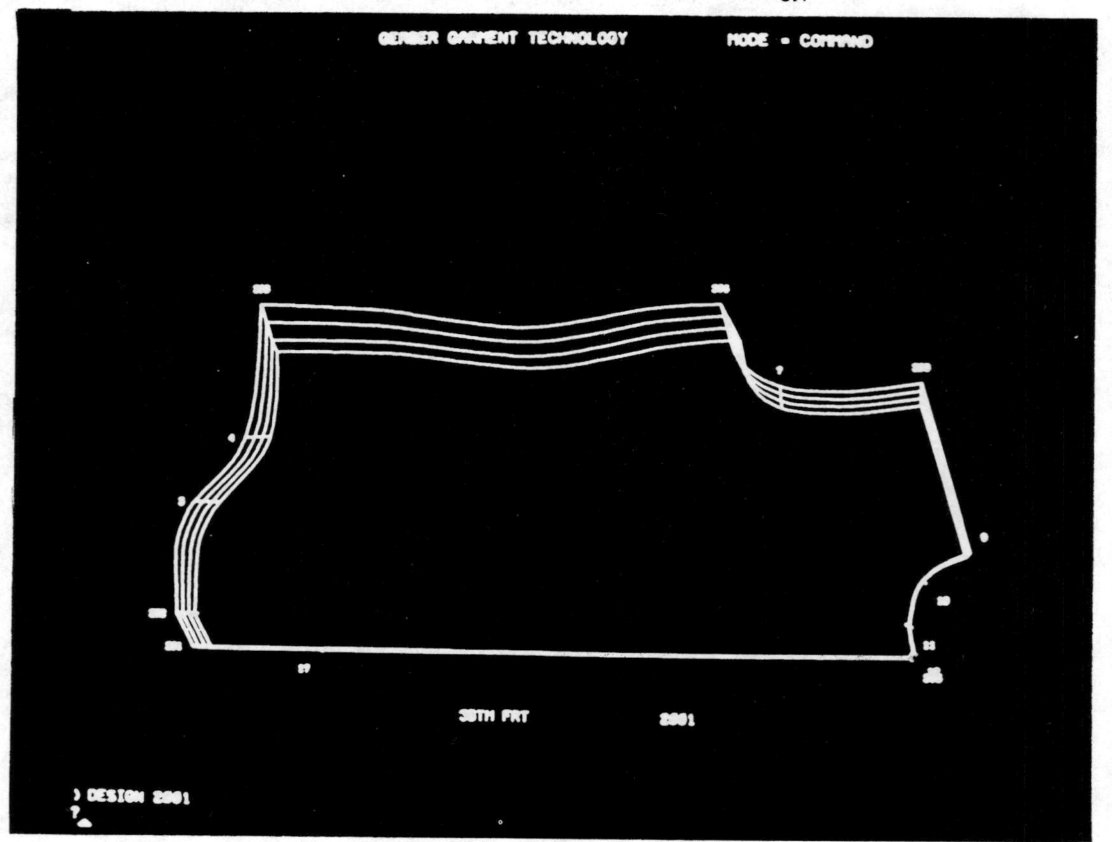

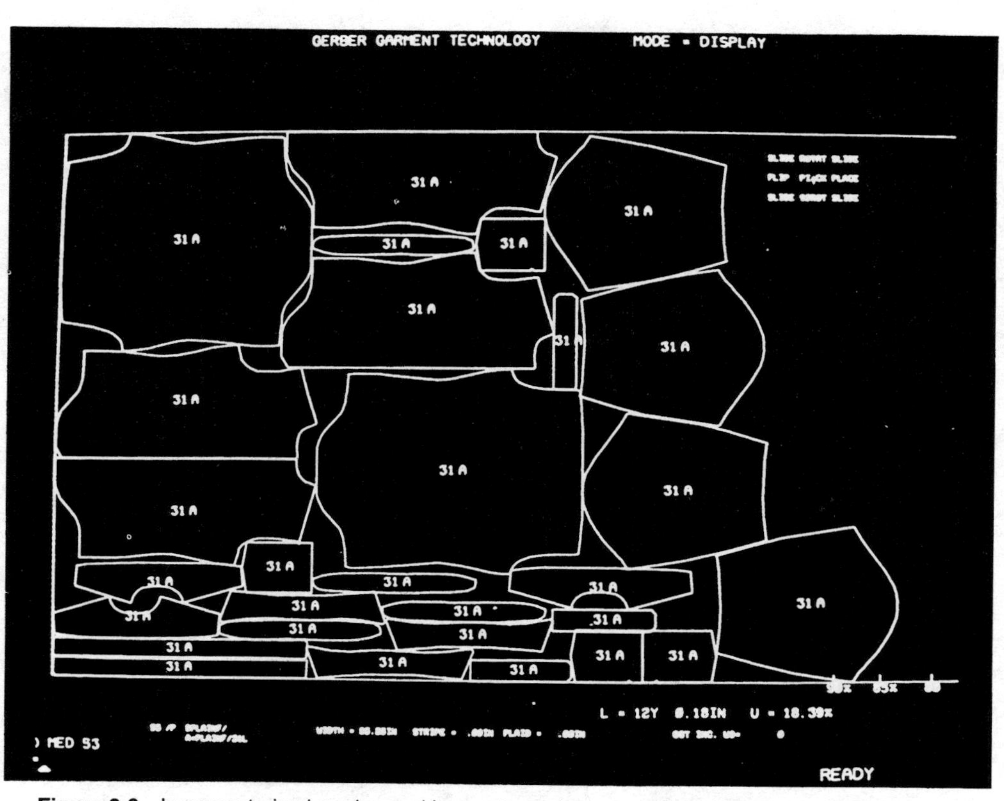

Figure 9-8 In computerized marker making, pattern pieces are nested into position to fill the scaled down width and length of the fabric layout. The computer is programmed to estimate layout efficiency. (Courtesy of Gerber Garment Technology)

savings can be realized by conserving mere inches per garment cut. A marker is finalized and considered ''tight'' when fabric efficiency is maximized and waste is minimized. Duplicate markers are made and sent to the cutting room while the original is rolled and stored for future reference.

A computer can be used to make a marker in a fraction of the time required for the manual operation. The graded pattern pieces are reduced in size and displayed graphically on a monitor eliminating the handling of full-scale pattern pieces. A stylus is used to manuever (or nest) pattern pieces into position, filling the scaled down width and length of the anticipated fabric layout. See Figure 9-8. Computerized markers are easily adjusted to material variation. The computer is programmed to estimate the fabric efficiency of the pattern layout, and the operator continues to make adjustments until maximum efficiency is obtained. The computer interacts with a plotter, which draws full-scale markers for production. The marker is stored in the computer's data base for future reference, and the full-scale marker is transferred to the cutting department. Depending on the garment style and fabric type, a tight marker may have close to or better than 90 percent material utilization. Computer-aided marking can improve material utilization by as much as 5 percent. (see Chapter 9 References, *Cuttings*).

Fabric inspection. Manufacturers generally inspect incoming fabrics after they arrive at the warehouse. See Figure 9-9. Fabric rolls are placed on an inspection frame where they are checked to assure yardage length and width accuracy, locate and mark visible defects such as streaks, flaws, and holes, and determine whether garment pieces can be cut together and mixed so that no shade differences will be perceived.

Once on the inspection frame, fabric yardage is automatically determined as the fabric is transferred by rollers from one bolt to another. Fabric yardage length can be a source of disagreement between fabric suppliers and manufacturers. Accurate measurement can be affected by the tension of the inspection frame relative to the type of fabric (woven, stretch woven, or knit) and by the percent of relative humidity in the air when fabrics containing wool, cotton, and rayon are measured. In checking fabric width, the traditional procedure is to manually measure the width at three locations on the roll and record the narrowest dimension for use in the cutting order.

To assure that no color differences occur from one fabric roll to another, apparel manufacturers take swatches from each roll and match them either visually or by instrument against a standard swatch under several different sources of light. Color matching is always important to the manufacturer, but is especially important when the production of two- and three-piece coordinate outfits occurs at different manufacturing locations.

While on the inspection frame, fabrics pass over lighted viewing tables to be checked for defects. See Figure 9-10. Imperfections are marked at the side of the fabric so that flaws are easily detected during the spreading operation prior to cutting. To assure a quality finished product, defects must be avoided or removed. Although fabric suppliers have significantly reduced the number of defects in shipped goods, occasionally an unacceptable roll of fabric must be returned. Inspected fabrics are either sent to the cutting room or placed in a warehouse until needed for production.

Fabric inspection significantly delays production and adds to the total manufacturing cost. With continuing competition from imports, both apparel manufacturers and fabric suppliers recognize the necessity to reduce the time and cost of getting the fabric into production. Consequently, the Textile Apparel Linkage Council (TALC), under the sponsorship of the American Apparel Manufacturers Association (AAMA),

Figure 9-9 Fabric rolls are transferred from the receiving area to be placed on the inspection frame. (Courtesy of (c) 1987 Gerber Products Company)

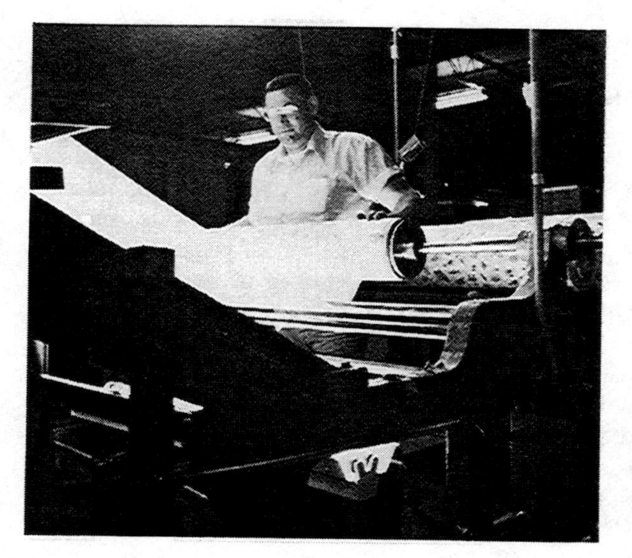

Figure 9-10 While on the inspection frame, fabrics pass over lighted viewing tables to be checked for defects. (Courtesy of (c) 1987 Gerber Products Company)

has been established. Its objectives are to improve competitive position and profitability by eliminating the duplication of inspection and its subsequent costs currently incurred by both the fabric producer and user, to improve overall response time, and to establish voluntary standardization of data and communication systems.

Specific areas have been identified as having the highest priority for voluntary standardization and production improvement. It is suggested that fabric suppliers be responsible for accurately marking on each fabric roll the minimum width and length, percent of relative humidity under which lengths were measured, shade description, identification information, and defect locations, thus limiting the need for apparel manufacturers to duplicate these tasks. Further, TALC encourages the development of inspection frames capable of measuring all of these factors simultaneously to provide reliability and consistency in fabric quality. Industry-wide compliance with TALC recommendations would benefit both textile and apparel companies through more efficient utilization of time and money and greater retailer service and satisfaction (see Chapter 9 References, *Fortess*).

Laying up the cut. Prior to cutting, multiple layers or plies of fabric must be spread on a long lay-up table. The actual length of the lay-up depends on the marker size. Large bolts of fabric are loaded on a moveable frame called a *spreader*. The spreader is guided along the lay-up table by two operators who smooth out fabric wrinkles when necessary. Depending on fabric thickness and the cutting order, fabric can be spread up to 300 plies. Fabrics that have some degree of elasticity, such as jerseys and stretch terry cloths, can create problems during spreading and must be left to relax overnight to assure accurate pattern size after cutting.

A computer controlled spreading machine requires little or no human involvement. It has the capability to load the fabric bolt, spread the fabric tension-free to a predetermined ply height, identify fabric defect locations, and unload the empty bolt according to information supplied by the computerized cut plan in the main system computer. Upon completion, the entire spread can be easily moved by air flotation to a moveable table, which will then transfer the spread to the cutting table.

Cutting. Several cutting systems are available to the manufacturer. The most commonly used include single-layer, die, straight knife, and computerized cutting systems. Selection is determined by fashion and production needs.

In *single-layer cutting*, the operator cuts fabric one layer at a time. This method is employed in some better suit and coat houses, where quantity is not a primary concern, or in making sample garments. Single-layer cutting is completely accurate. The pattern and fabric are held in place by weights, and the cutter uses shears or a round blade knife to cut around individual pattern pieces.

Mass production requires the cutting of multiples of the same pattern piece at one time. Die cutting may be selected for garments or garment parts that do not change style from one season to the next. A die is like a large cookie cutter which takes the shape of the outer edges of the pattern piece. In some cases, more than one die may be joined together, resulting in a "gang die." After the material has been spread to the desired height, the die is positioned and forced through the plies by a hydraulic press. Die edges are quite sharp and result in a fast and accurate cutting method. See Figure 9-11.

When style changes occur more frequently, *straight-knife cutting* or *computerized cutting systems* are employed. A marker, placed on top of the fabric plies, must be used as a cutting guide for the straight-knife procedure. The straight-knife machine can cut up to 10 or more inches of fabric thickness successfully. This hand-held tool has a long, thin blade that moves rapidly in a vertical direction as it is guided through the fabric plies. See Figure 9-12. The cutter has a choice of speeds and blades depending on fabric and pattern shape requirements.

The computerized cutting system is designed to cut high plies of material quickly and more accurately than any other method, and is an economical answer to the problems found in a quick turnaround, high fashion, and fast changing manufacturing environment. After the fabric has been spread to the desired height, it is covered by a plastic film and compressed by a vacuum hold-down system to a fraction of its original height. See Figure 9-13. The fabric is held securely on a work surface

Figure 9-11 Pattern pieces to be made into vinyl baby pants were die cut since style does not change frequently. (Courtesy of (c) 1987 Gerber Products Company)

Figure 9-12 The straight knife cutter has a long thin blade that moves rapidly in a vertical direction as it is guided through the fabric. (Courtesy of DeLong Sportswear)

Figure 9-13 After the fabric has been spread to the desired height, it is covered by a plastic film and compressed by a vacuum hold down system to a fraction of its original height. (Courtesy of Gerber Garment Technology)

comprised of bristle squares, which allow for precision cutting by a high speed reciprocating knife. Utilizing computer integrated manufacturing, data stored in the computer directs the cutting head to follow the "tight" marker previously made by the marker maker. See Figure 9-14.

Although laser and water-jet cutting techniques have been successful in the leather and plastic industries, they have limited success in the apparel industry. The laser cuts perfectly; however, it tends to fuse plies together. In water-jet cutting a small stream of water under very high pressure shears the fabric as it follows the pattern outline. Although plies are not actually fused, as in laser cutting, they are difficult to separate. Both of these methods are generally restricted to low-ply or single-ply cutting. Until these problems are eliminated, laser and water-jet cutting are unlikely to gain widespread acceptance.

Figure 9-14 Data stored in the computer directs the cutting head to follow the tight marker previously made by the marker maker. (Courtesy of Gerber Garment Technology)

As cuts are completed, stacks of identical garment pieces are moved from the cutting table, plymarked so that pattern pieces from the same fabric layer are assembled together in the sewing operation, and tied into *bundles*. Plymarking is an important step, since it is essential that all parts of a garment are from the same fabric ply to assure color and shading match in the finished garment.

PRODUCTION

Production involves the actual assembling of the garment pieces. If garments are to be assembled on-site, bundles of garment pieces are transported to the sewing department. On-site assembly is referred to as an *inside shop*, where all operations necessary to produce finished garments occur at the same location where preproduction operations are performed.

A contractor, or *outside shop*, may be hired to cut garment pieces as well as sew and assemble finished garments according to a manufacturer's specifications. Outside shops may be owned by the manufacturer and may be located far from the central facility where designing, patternmaking, and marker making occur. If sewing and assembly are to be done elsewhere, bundles must be shipped to the off-site location.

Garment assembly involves many operations. The number required depends on the apparel type and style. A simple child's crewneck top or pant may require 7 different sewing operations, whereas a man's suit may require 200. The operations performed and the type and complexity of equipment utilized also vary with the garment. In spite of these differences, the basic assembly order is similar. The assembly system utilized depends on the size of the manufacturer and the type of apparel.

Assembly systems. Three assembly systems are available to the manufacturer: the tailor or whole garment system, the progressive bundle system, and the unit production system. The *tailor or whole garment system* may be used in manufacturing garments such as women's better dresses, suits, and coats. One operator is responsible for assembling one complete garment. Operator skill requirements are high and productivity is low with this assembly system. Garment production cost is understandably higher as well.

In the *progressive bundle system*, operators repeatedly perform the same task on dozens of the same or similar style garments in a given day. One operator might stitch a cuff to the sleeve, another attach the sleeve to the shirt body, while the third closes the side seam. This type of assembly-line operation is generally laid out so that the operators performing the same task are grouped together and bundles of garment parts move from one workstation to another in sequential construction order. Operators become quite skilled and efficient at one particular procedure, thus increasing productivity.

In the apparel industry, assembly-line sewing machine operators frequently are paid by the number of garment pieces they sew in a day. This is known as *piecework*. Bundles of pieces are transported to each work station. Operators untie a bundle of garment parts, perform a specific task on each part, retie the bundle and place it in a cart or conveyor tote box to be moved to the next workstation. Upon completion of each bundle, operators attach a bundle pay ticket to their worksheet as proof of their productivity. Pay rates vary with the time and level of skill required to perform a given task. Operators who are highly skilled, work fast, and make few mistakes can increase their earnings substantially.

The *unit production system* (UPS) is a computer controlled transport system utilizing overhead carriers to move cut garment pieces from the cutting room to individual workstations for assembly. Garment pieces are suspended on carriers singly or in multiples and offered to the worker in the proper position for sewing. The system increases productivity by reducing the amount of time sewers spend handling bundles, garment parts, and bundle pay tickets. Consequently, the manufacturer gains in total throughput. UPS consists of a network of computers constantly exchanging information about the manufacturing process. In addition to the central computer at a supervisory workstation, computers are located at each sewing station, which allows the operator to communicate any problems or needs with the central system, determine individual productivity rates, and make work adjustments as desired. Each operator's speed and productivity is analyzed and work is supplied as needed creating a balanced work flow. The UPS is programmed to know the correct sequence of sewing operations as well as the operations that can be performed out of sequence. When all sequential stations are full, the UPS seeks out an alternative operation that is ready to receive. This feature eliminates production back-up and reestablishes balance on the sewing floor.

The unit production system reduces in-process inventory, moves garments through the production line sooner, and helps cut down on delivery time. Consequently, overall manufacturing time and production costs are reduced. UPS is an excellent example of a Quick Response technology giving manufacturers an important advantage in a highly competitive market.

Joining methods. In order to produce a finished item, the individual cut pieces must be joined together and the various garment parts assembled. Depending on the design, material, and utility requirements of an item, two joining methods available are stitching and heat sealing.

Stitching. Most items made from textile or nontextile materials are joined by passing a strand(s) or loop(s) of thread through the material. Done repeatedly, this process is known as *stitching* and is the basis of sewing. Stitches are introduced into material to join two or more pieces of material together, to finish a raw edge, or to decorate the material. A series of identical stitches joining two or more pieces of material is called a *seam*. Seams are used to assemble parts in the production of sewn items. The stitch selected for use with each seam type varies with individual requirements for the item being sewn.

Government specifications for the fabrication of sewn items state certain standards for seams and stitches. These federal standards define and illustrate four seam classifications: superimposed, lapped, bound, and flat. See Figure 9-15. A *superimposed seam* is formed by placing two or more layers of material together and seaming them near the edge with one or more rows of stitches. The superimposed seam has many variations and is utilized for all-purpose apparel applications such as closing seams, attaching waistbands, or applying decorative tape trims to action wear.

When two or more plies of material are overlapped and seamed with one or

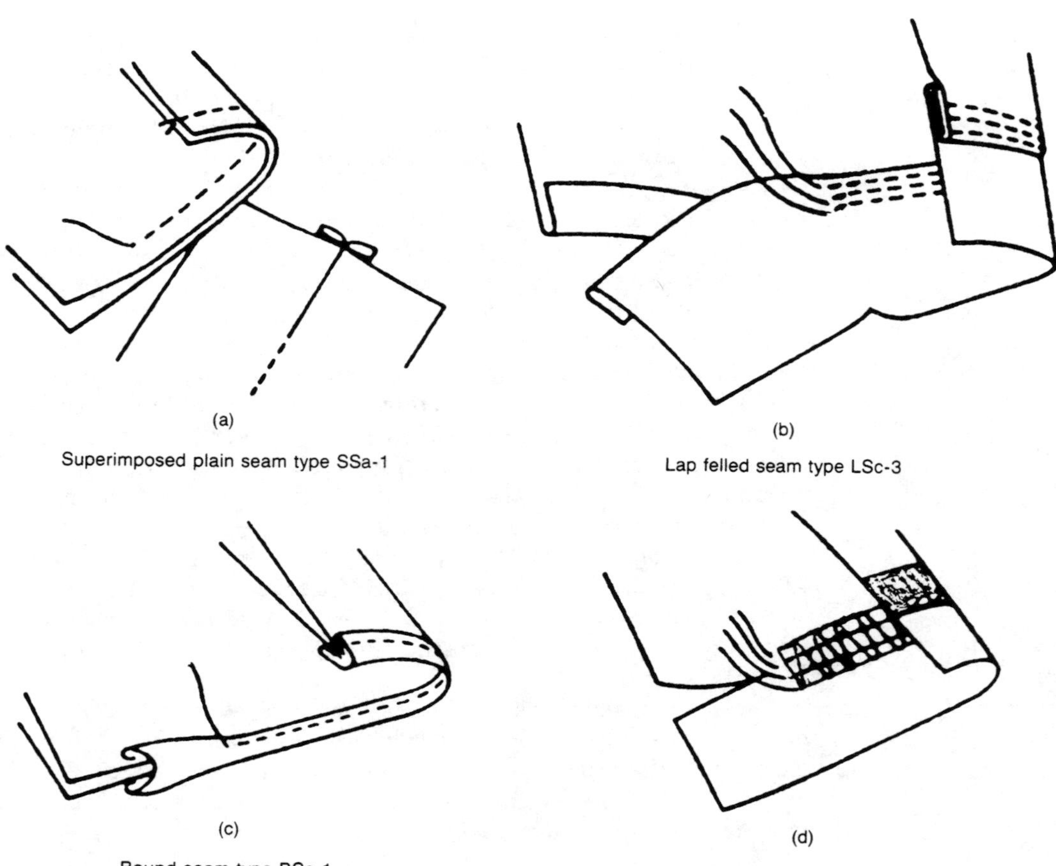

Figure 9-15 Seams are used for assembling parts in the production of sewn items. Federal Standards define and illustrate four seam classifications and their variations. (Courtesy of Coats, The Threadmakers, Glasgow, Scotland)

more rows of stitches, a *lapped seam* is formed. A popular variation, the *lap felled seam*, is typically used on blue jeans and other work and sportswear apparel where seam strength and durability are important. See Figure 9-16.

A *bound seam* is made by folding a narrow strip of material over the edge of one or more plies of material and joining the unit with one or more rows of stitches. Raw edges are neatly finished and can be hidden from view or exposed as a design feature.

The *flat seam* is formed by butting two edges of material together and sewing in such manner that stitches extend over and cover the edges of the joint. Flat seams are commonly used in underwear or foundation garments where extra fabric thickness would be uncomfortable in wear. Each seam classification has many variations and seam types are carefully selected according to design, fabric, and end-use requirements.

Federal Standards also classify stitches. Commonly used stitches are the chain stitch, lock stitch, multithread stitch, overedge stitch, and flat seam stitch. See Figure 9-17. The basic *chain stitch* is formed by passing one needle and thread through the material creating a series of loops that go through one another on the

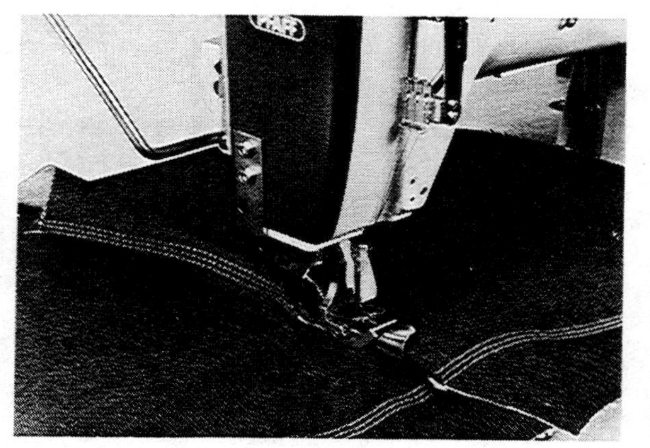

Figure 9-16 A popular variation of the lapped seam, the lap felled seam, is typically used on blue jeans and other work and sportswear apparel where seam strength and durability are important. (Courtesy of Pfaff-Pegasus of USA, Inc.)

undersurface. This stitch follows the same principle as crocheting. The single thread chain stitch is used for temporary, conventional, and ornamental stitching. The chain stitch is not as secure as other stitches. Pulling the thread from the last stitch can result in runback, the unraveling of the entire stitching unit.

The *lock stitch* operates on the same principle as a home sewing machine. The stitches are formed with two threads. The needle thread is introduced from the top side of the fabric and interlaces with a thread supplied from a bobbin on the underside creating a lock. The simple lock stitch looks the same on both sides of the fabric and produces tight, strong seams. When used to sew a superimposed seam, the lock stitch leaves an unfinished raw edge, which can create problems in fabrics that ravel easily. Consequently, the lock stitch is frequently used in apparel production operations where raw edges are finished in advance or are enclosed during assembly. Common uses include putting in zippers, stitching on facings, attaching cuffs, collars, waistbands, and applying decorative topstitching. The disadvantage of the lock stitch is that sewing operations must be stopped periodically to rewind the bobbin thread.

For applications requiring stretch, the lock stitch has several important variations. The *one-step zigzag stitch* is similar to the basic lock stitch except that successive single stitches are directed alternately to the right and then left, forming a symmetrical zigzag pattern. This zigzag stitch is used for seaming, attaching bindings, and securing collars and cuffs to knitted apparel. In the *two- or three-step zigzag stitch*, series of stitches alternate direction creating a symmetrical multiple stitch zigzag pattern. This stitch is frequently used for attaching elastic in garment areas where stretch is desirable.

The *multithread chain stitch* differs from the chain stitch in that it is formed with two or more threads. In the basic stitch, loops from a single needle thread are passed through the top side of the material and interconnect with two loops of a single underthread. This two thread chain stitch is stronger than a similar lock stitch and reduces seam pucker. Additionally, since threads are fed from long length packages, productivity is not interrupted by bobbin rewinding. Variations of the basic stitch are formed using several needle threads with one underthread. To increase seam stretch, these stitches may also be sewn in a zigzag pattern. Disadvantages of multithread chain stitches include susceptibility to runback and increased bulk under the seam.

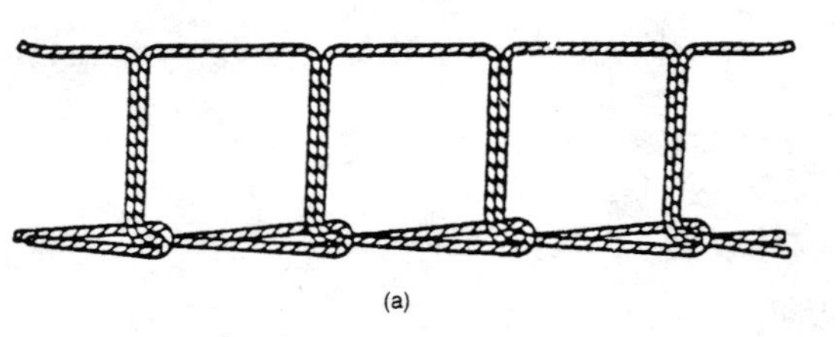

(a)

Single thread chain stitch type 101

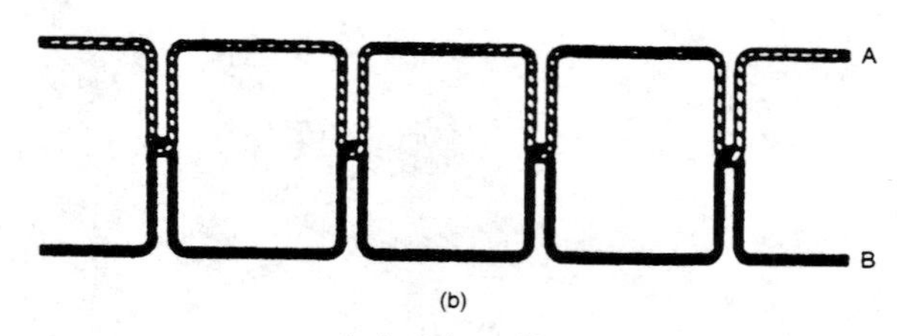

(b)

Lock stitch type 301

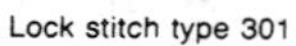

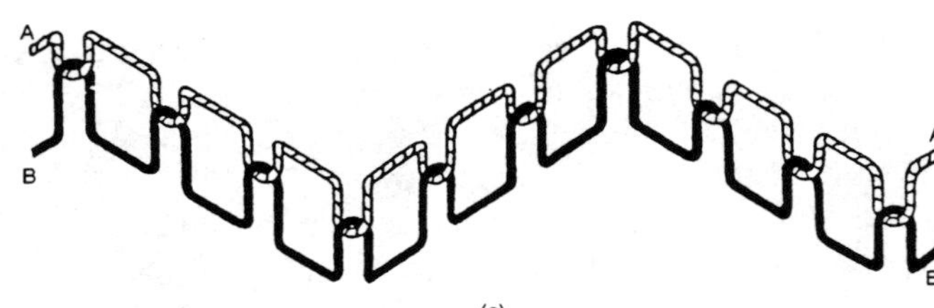

(c)

Three-step zigzag type 315

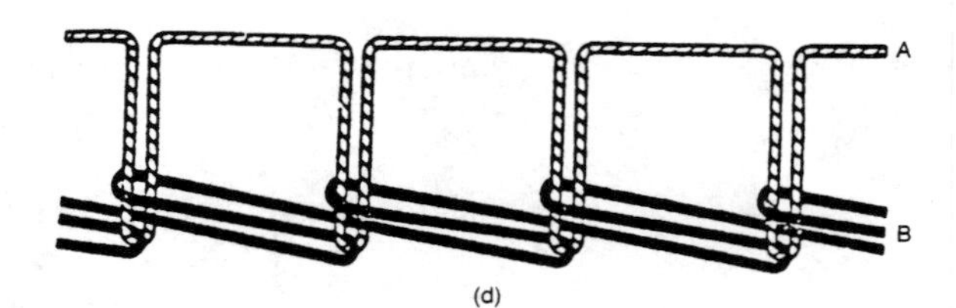

(d)

Multithread chain stitch type 401

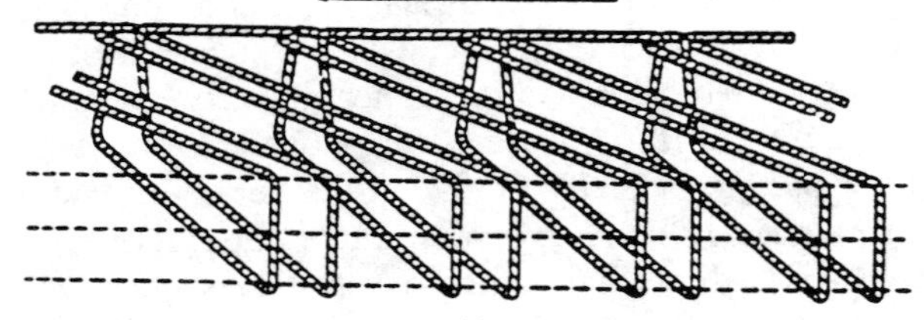

(e)

Overedge stitch type 501

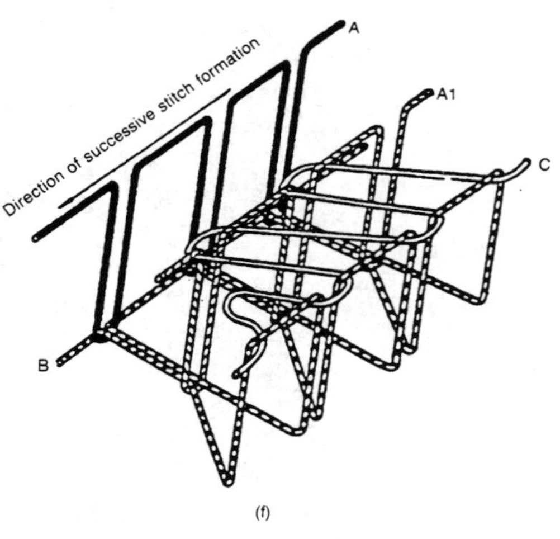

(f)

Mock safety stitch type 512

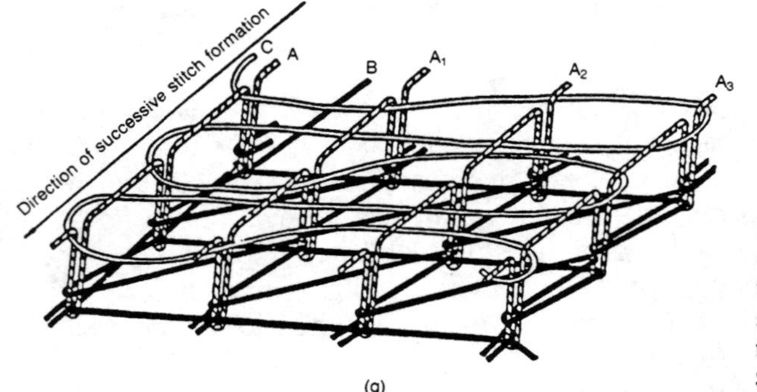

(g)

Flat seam stitch type 607

Figure 9-17 Stitches are introduced into material to join two or more pieces of material together, to finish a raw edge, or to decorate material. Federal standards are established for stitch classifications and variations within the classes. (Courtesy of Federal Standard No. 751a—Stitches, Seams, and Stitchings)

A simple *overedge stitch* is formed when one needle thread passes through and around the edge of the material interlooping with itself at the point of needle penetration. The overedge stitch finishes the cut edge of a single ply of fabric or sews the seam and finishes the edge of two plies in one operation. Since the overedge stitch has good stretch and recovery, it is frequently used in assembling knitwear. Overedge stitch variations incorporate a combination of one or more needle and underthreads with at least one thread passing around the edge of the material being sewn. When two needle threads are used, the second needle thread penetrates the material farther from the cut edge, resulting in two rows of stitching. This stitch is referred to as the *mock safety stitch* and gives greater seam security than the single needle overedge stitch. Seams formed in this manner also have good extensibility.

A simple *flat seam stitch* is formed with three threads: two needle threads and one underthread. Loops of the two needle threads pass through the material at different points and interloop with the underthread. A loop of the second needle thread extends across the top of the material to the point of the first needle's penetration and is entered by a loop of the first thread as it goes into the fabric to make the next stitch. More complex variations of the flat seam stitch utilize two to four needle threads, one underthread, and one or two cover threads. The cover threads extend across the surface of the material and are held in place by the needle threads. The underthreads interloop with all needle threads beneath the material. These stitches are used on flat seams and produce comfortable seams with good elasticity and wearability.

Power sewing machines are available to produce all of the stitch types and can also be set up to simultaneously sew two or more rows of different stitches. Manufacturers utilize a variety of sophisticated sewing machines especially designed to do very basic seaming as well as very specialized tasks such as sewing on buttons, making buttonholes, and setting in sleeves. See Figure 9-18. These operations require varying degrees of operator skill.

As modern technology becomes more and more sophisticated, repetitive tasks such as loading, positioning, and aligning parts for certain operations such as sewing centerplaits on shirts and setting pockets are done automatically. Persons operating

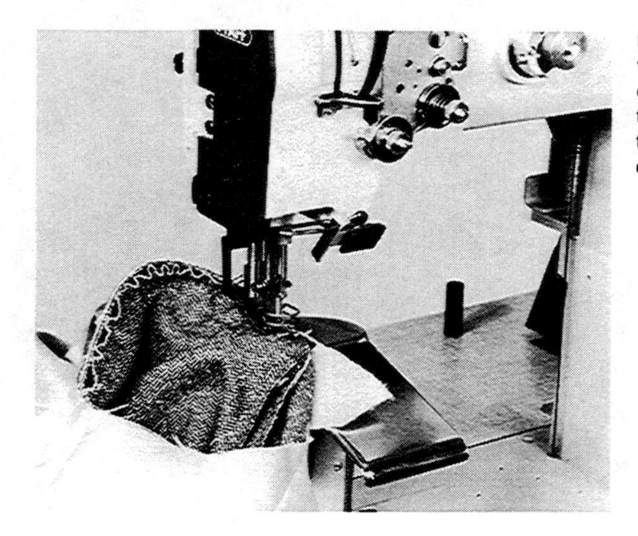

Figure 9-18 Manufacturers utilize a variety of sophisticated sewing machines designed to do specialized tasks such as sleeve setting operations on jackets and coats. (Courtesy of Pfaff-Pegasus of USA, Inc.)

this equipment require a minimum of skill and training and can frequently tend two or more machines at a time. Workstations of this type, which combine several operations into one task, increase productivity and consistency while decreasing throughput time and labor costs.

The apparel industry has been cautious in applying robotic automation to soft goods manufacturing, since handling and assembly of various fabric types has been more challenging than tasks performed in the hard goods industry. Recently, however, robotics have become a practical and cost-effective option for apparel manufacturing. Typically, the robots move and position parts and hold them through the sewing operation much as an operator would. See Figure 9-19. Robots can be programmed to perform one task and later reprogrammed to perform another without major readjustment of equipment for each new application. Proper integraton of robotics into the production process can result in important benefits to the manufacturer, such as increased speed and accuracy. Currently, robotics are used in the production of jeans pockets, washcloths, carpet samples, and rug binding. With the many equipment options available for converting materials into finished articles, the modern manufacturer has the responsiblity of determining the appropriate combination for his or her product and company's profitability (see Chapter 9 References, *Lower*).

Heat sealing. Heat sealing is a method used to join thermoplastic, limp sheetings such as polyvinyl chloride, polyethylene, polyurethane, or polyester materials. Sealing is accomplished by a hot wedge which softens the thermoplastic fabric and welds it together. See Figure 9-20. Hot wedge heat sealing can be used to make superimposed seams, lapped seams, and hems. This fast, economical and low-pollution operation

Figure 9-19 Typically, robots move and position parts and hold them through the sewing operation much as an operator would. (Courtesy of Singer Sewing Company)

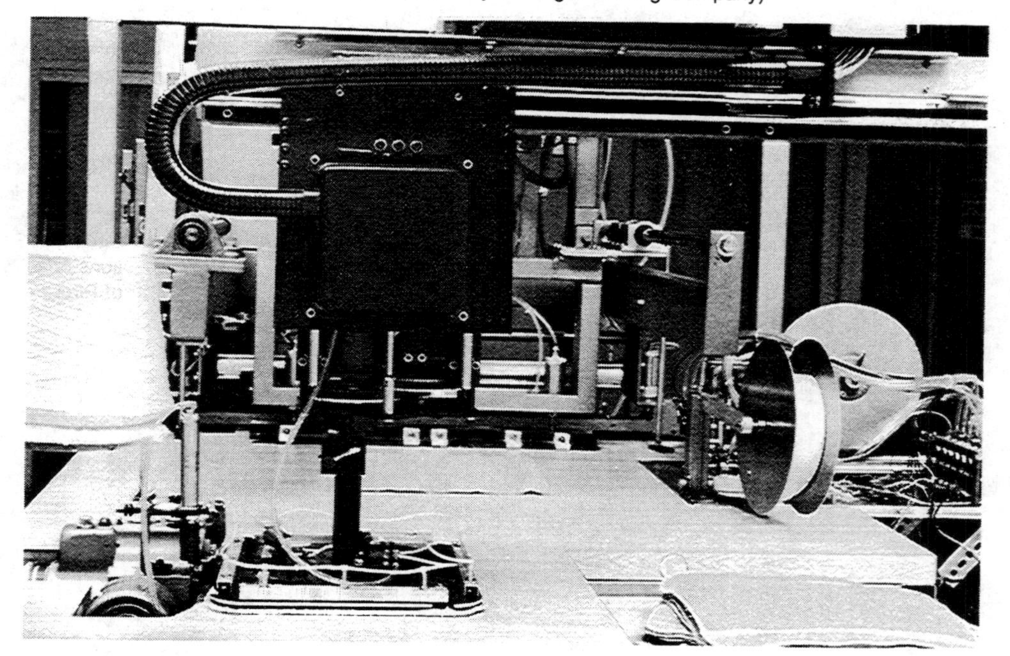

Figure 9-20 Vinyl garment parts are welded together to create moisture-proof baby pants. (Courtesy of (c) 1987 Gerber Products Company)

gives optimum seam strength and is used in the production of waterproof clothing, work and sportswear, tents, and awnings, etc. In some applications, hot air is used to weld a thermoplastic cover tape over stitched seams. See Figure 9-21. The hot air melts a substance on the tape causing it to adhere to the thermoplastic fabric. This seal inhibits the passage of moisture through openings resulting from needle penetration in stitching the seams.

Labeling. A variety of labels are found in ready to wear apparel. Labels are sewn into garments during production and identify fiber content, care requirements, size, manufacturer or designer, and employee union if produced in a union shop. Hang tags may also be attached to further identify and promote the designer or a brand. In a constant attempt to cut material and labor costs, many manufacturers are implementing an in-plant label printing system. These in-plant systems permit printing information on both sides of the label tape, thus reducing tape costs, reducing labor costs of setting additional labels, and eliminating problems associated with ordering, stocking, and inventorying purchased labels.

Figure 9-21 Hot air is used to weld a cover tape over stitched seams to inhibit the passage of moisture through openings resulting from needle penetration in stitching seams in thermoplastic sheetings. (Courtesy of Pfaff-Pegasus of USA, Inc.)

Some labels are inserted into seams during assembly, others are set independent of seams in a separate operation. Apparel manufacturers may incorporate manual, automatic, or semiautomatic systems in setting labels into garments when they are not incorporated into a seam. In manual label setting, a skilled operator utilizes a conventional single needle lock stitch machine. The operator is required to sew short distances, maintain acceptable label margins, and in some cases pivot at label corners. This series of steps makes it difficult to maintain high speed and productivity when manually setting a label.

Automatic label setting utilizes either a single needle lock stitch or zigzag machine and requires little operator skill. The operator positions the label and starts the machine. Seam and stitch length, sewing speed, pivoting, and removal of the sewn item is automatically controlled. Productivity is greatly increased. The equipment is generally designed to sew a predetermined style and size of label; therefore, consistency in label size is important with this labeling system.

Semiautomatic systems accommodate any size label, allowing greater flexibility in production; however, slightly more operator material handling is required than in completely automatic systems. Control over stitch and seam length, sewing and pivot speed, and seam margin width remains completely automatic while the equipment production rate is matched to the material handler's ability. Operator skill level is considerably less than required for manual label setting and productivity is substantially increased.

POSTPRODUCTION

Finishing, pressing, and quality inspection are important postproduction operations that add to the quality of a garment. Finishing may include sewing buttons, snaps, or linings in position, adding belts or buckles, and clipping thread ends. Hand finishing operations are rare in modern manufacturing except in high price merchandise. Moderate and lower price garments are finished by machine.

Pressing. Adequate pressing improves the overall appearance of a garment and gives it better hanger-appeal in the marketplace. High price items are generally pressed during assembly, less expensive garments are pressed after sewing is complete, while lower price garments may not be pressed at all.

Costly items such as tailored suits and coats are generally pressed during assembly to open seams and to set the shape of certain garment parts to conform to body contours. The garment is placed on a properly shaped form, or buck, and an iron or compatibly shaped pressing head is positioned on the fabric. In order to set the desired shape, steam is used to transfer heat into the the fabric, thereby relaxing the fibers and molding them to the shape created by the buck and the pressing device. Once heated and shaped, the fibers must be cooled quickly or the fabric will return to its original prepressed shape.

New pressing technology and the need to conserve time and money have made the pressing of less costly items an end-of-the-line operation in most factories. Generally, these garments do not require the shaping essential to tailored jackets and coats and are pressed only after all assembly operations are completed. End of the line pressing removes puckers and wrinkles, sets creases, and improves the general appearance of a garment after assembly.

The apparel manufacturer is provided a variety of options in pressing equipment, depending on fabric and style needs. A *vacuum pressing table* and steam iron may be used for pressing seams and small parts or for finish pressing dresses, skirts, shirts, and ties, etc. These tables are flat pressing surfaces and are available in a variety of shapes and sizes. They may be fitted with different padded forms, or bucks, depending on specialized pressing requirements.

Some pressing systems, such as the tunnel finisher and steam cabinet, are enclosed units. Garments such as lingerie, T-shirts, children's wear, and outerwear may be finished in a *tunnel finisher*. This automatic finishing system conveys hung garments through sealed chambers where steam and heated air combine to relax the fibers and remove wrinkles to produce a crisp finish. *Steam cabinets* are generally smaller than tunnel finishers and handle fewer garments at a time. The unit presses and finishes men's, women's, and children's wear and may be equipped with a dummy to support the garment during steaming or converted for use with hanging goods.

Often a combination of pressing systems is required to obtain the necessary degree of flexibility required by manufacturers who produce a broad range of apparel styles in different fabrics. As in other areas of apparel production, automation has resulted in pressing equipment that allows for precise and consistent pressing quality with less dependence on individual operator skill.

Quality control. Quality control is essential to successful manufacturing. Mass production of thousands of garments a day requires close supervision to prevent dissatisfaction and return of the finished goods. Depending on the size of the company, a separate department may be devoted to the various aspects of quality assurance.

Regardless of how quality control is handled, it is important that the quality level and the standards be well defined and that workable tolerances be determined. Every employee involved in manufacturing the product must be well informed of and committed to company quality expectations.

Prior to packing and shipping, finished goods are inspected either by 100 percent final inspection or by random sampling. See Figure 9-22. *One hundred percent*

Figure 9-22 Prior to packing and shipping, finished goods are inspected either by 100 percent final inspection or by random sampling. (Courtesy of (c) 1987 Gerber Products Company)

inspection means that upon completion, every item is checked against measurement specifications, inspected for uncut threads, and examined for satisfactory construction techniques. Inspection by *random sampling* saves time and assumes that production problems will be caught through spot checks of one or more items out of a specified number of garments. Discovery of production errors are remedied promptly and defective work is returned to the original worker for repair. Piece workers are not paid for repair work since plant productivity is reduced by the time required to make repairs.

Some companies implement an in-process quality control program which requires an observer or supervisor to insure quality work on the production floor. The quality control observer roves the sewing floor continuously, checking performance of production workers, locating unsatisfactory workmanship, and taking corrective action. The observer must be honest and have a thorough understanding of construction, machine performance, and quality standards.

Goods returned to the manufacturer by a retailer are analyzed to determine the reasons for the return. Each garment has a label so that errors can be traced and corrections made. Common causes for returns include construction errors, fabric flaws, fabric shading or streaks, and late delivery. An advantage found with the Unit Production System of garment assembly is early detection of the common causes of finished goods returns. Since garment pieces hang on carriers and are not confined in bundles during assembly, construction and fabric defects and flaws are generally discovered prior to completion. Persons responsible for quality control are anxious to eliminate problems, since a company's reputation can be damaged if problems occur too often.

Packaging and shipping. When all phases of production are completed, the garment is ready to be packaged. Some garments are placed on hangers and covered with plastic bags; others are folded either by hand or automatic equipment and inserted into protective packages. See Figure 9-23. Garments produced by a contractor, or outside shop, must be returned to the manufacturer. Upon arrival at the factory's

Figure 9-23 When all phases of production are completed, the garment is folded and inserted into protective packaging. (Courtesy of (c) 1987 Gerber Products Company)

storage area, completed garments are sorted by size, style, and color and placed in stock.

As orders are received, the specified numbers, sizes, colors, and styles of garments are pulled from stock, placed in shipping boxes, and marked for transport. A packing slip is included listing the items ordered. The boxes are then sealed, labeled, and conveyed to a loading area for shipment to the retailer.

QUESTIONS

1. What is a staple fiber? A filament fiber?
2. Name and explain the three basic weaves.
3. Briefly explain the purpose of the following finishing processes: singeing, mercerization, compressive shrinkage, and water resistance.
4. What are the three major processes involved in apparel manufacturing? What are the production tasks that must be accomplished within each process?
5. How does draping a pattern differ from drafting a pattern?
6. Explain the importance of the sample garment in apparel production.
7. Briefly define the following terms: first pattern, notch, master block or sloper, production pattern, grading.
8. Why is perfection important in a production pattern?
9. What expenses are figured in a detailed final cost analysis of a garment?
10. In pattern grading, how does a grading machine differ from a digitizer?
11. What is a marker? What is a "tight" marker?
12. Give three reasons why fabric must be inspected before being put into production.
13. Explain the following methods of cutting fabric for production: single-layer, die, straight knife, and computerized. Why might one cutting system be selected over another?
14. What is an inside shop? An outside shop?
15. Name and explain the three assembly systems available to the apparel manufacturer.
16. What is the difference between stitching and heat sealing joining methods?
17. Prepare a simple sketch of the four basic seam classifications: superimposed, lapped, bound, and flat.
18. Be able to identify the chain stitch, lock stitch, three-step zigzag, multithread chain stitch, overedge stitch, and flat seam stitch.
19. Explain why some garments are pressed during assembly and other garments are pressed after assembly is completed.
20. What is 100 percent final inspection? Random sampling?
21. Prepare a short report telling how the apparel industry has benefitted from increased utilization of computers.

part 3 MATERIAL PROCESSES: CUTTING

10 SHEARING AND SAWING

Shearing and sawing are common methods for cutting a broad range of materials, such as metal, wood, and plastic. Some applications involve the use of manual tools, such as tin snips and hand saws, when installing heating ducts or building houses. The presentation in this chapter is aimed at an overview of industrial shearing and sawing as they relate to product manufacture. See Figure 10-1.

SHEARING

Shearing is a material separation process which produces no chips so that material can be cut into several small pieces with no resultant waste or scrap. While a number of different types of shearing tools are used, all work on the principle that the workpiece is fractured by two overlapping blades, one acting on each side of the piece being cut. See Figure 10-2. Shears cut metal in the same way that ordinary scissors cut paper or cloth. Three classes of shears are employed, and each can be powered manually, mechanically, hydraulically, or pneumatically.

Guillotine shears. *Guillotine* shears have vertically mounted blades, with the bottom blade fixed in position and the top one mounted on a movable ram. The workpiece is positioned between the two blades, and the ram is driven downward. When the top blade first contacts the workpiece, the metal is plastically deformed

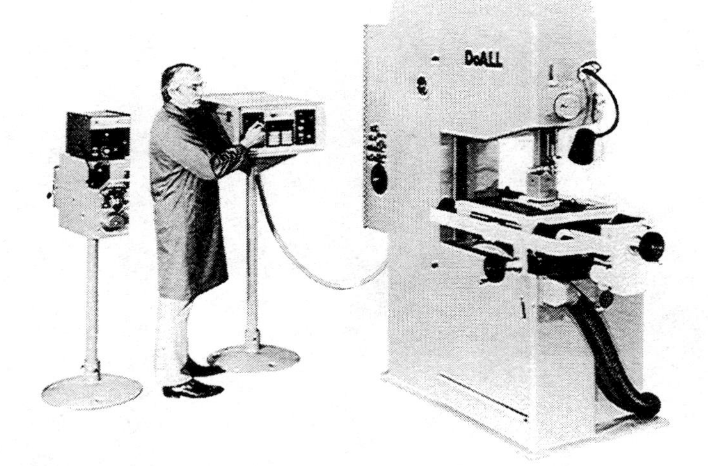

Figure 10-1 A numerically-controlled contour-cutting wire band saw cuts intricate shapes in difficult-to-machine material. (Courtesy of Do All Company, Des Plaines, IL)

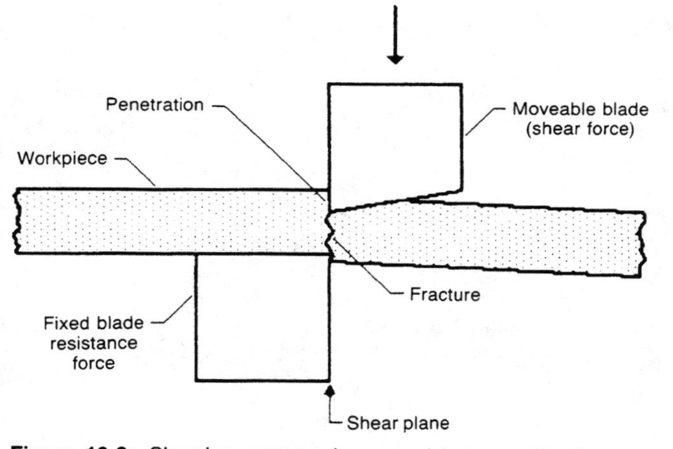

Figure 10-2 Shearing occurs when equal but opposite forces act upon a workpiece. The forces can be a punch and die, or two opposing blades. The cut occurs at the shear plane.

as it is squeezed between the two blades. The blades continue to penetrate the metal until they overcome and fracture the piece. See Figure 10-3.

The blades of guillotine shears can be shaped to match the cross section of the piece being sheared. For example, each of the two blades used for round stock contains a semicircular cutout. See Figure 10-4. The blades surround the workpiece as they cut it to reduce distortion of the piece at the shear plane. For cutting sheet and plate, the top blade is often mounted at an angle to the bottom blade so that the blade cuts progressively across the piece. Less power is required because the

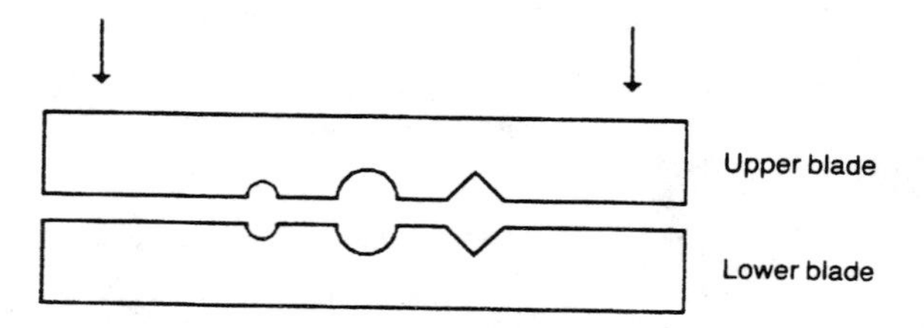

Figure 10-3 These guillotine shears are used for cutting notches in sheet metal workpieces. (Courtesy of Niagara Machine & Tool Works)

Figure 10-4 These guillotine shears are used to part or shear metal rods. Note that the blades have openings in the shapes of the workpieces which produce cleaner cuts with no distortion. Similar shears are used for cutting flat sheets and plates. (Courtesy of Metal Muncher)

material is being cut at only one point at a time. Such shears can separate metal stock up to 38 mm (1.5 inches) thick and 9 m (30 feet) wide. See Figure 10-5.

Flying shears are a special guillotine shear mounted on a movable base used to cut pieces of roll stock on a continuous production line. See Figure 10-6. The shears are regulated to move at the same speed as the stock during the cutting action. After the cut has been made, the equipment automatically returns to the starting position to repeat the process.

Alligator shears. *Alligator* shears take their name from a resemblance to the jaws of an alligator, and are comprised of a fixed lower blade and a moveable upper blade. Power is transmitted to the upper shearing arm, which moves in an

Figure 10-5 This squaring shear can be used for the straight-blade cutting of metal, wood, and plastic sheets. The capacity of this machine is 3 meters (10 feet) by 6 millimeters (0.25 inch) mild steel. (Courtesy of Niagara Machine & Tool Works)

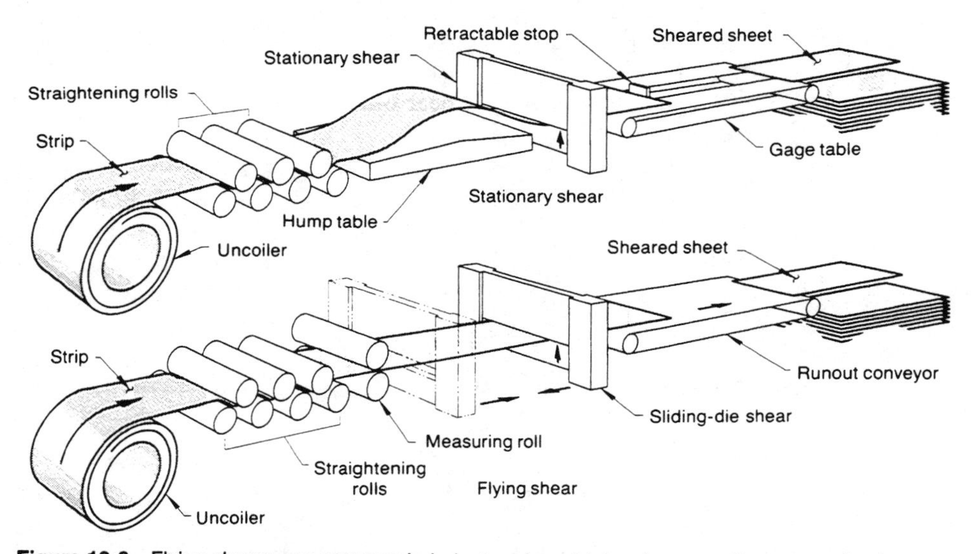

Figure 10-6 Flying shears are common in industry where high volume continuous production is essential. The stationary shear is also shown. (Courtesy of *Metals Handbook*, Vol. 6, 8th ed., ASM INTERNATIONAL (formerly American Society for Metals), 1971. Reprinted with permission.)

arc around a fulcrum pin and the leverage provides the shearing force. See Figure 10-7.

This is the most versatile of all power-driven shears. It can part plate, sheet and strip, although its greatest use is for shearing bars and bar sections and for preparing scrap where accuracy is not a factor. Alligator shears are available in a range of sizes including those that can shear plate up to 1.25 inches (32 mm) thick by 30 inches (760 mm) long.

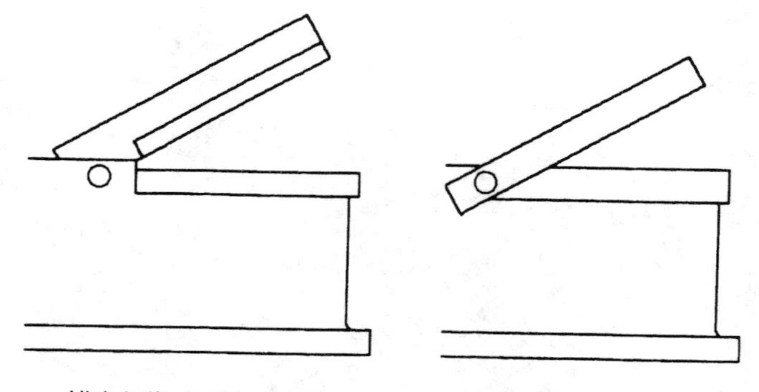

Figure 10-7 As shown here, the alligator shears work like a large pair of very powerful scissors. Two common types are shown.

The two types of shearing arms used on these shears are low-knife and high-knife. In general, the low-knife shear is preferred for cutting bars and bar sections. The high-knife shear is preferred for cutting flat stock and for use in scrap yards. A high-knife shear, by making successive cuts, can shear flat stock that is wider than the length of the shear blades.

Rotary shears. *Rotary* shears cut by the action of a workpiece passing between two revolving, tapered, round, overlapping blades, to permit the curved and circular cuts shown in Figure 10-8. These narrow blades can also be used for straight-line separating operations in sheet materials difficult to cut by any other means. See Figure 10-9. A related rotary shearing operation called slitting is described later in this chapter.

Figure 10-8 This circle-cutting shear employs rotary blades. (Courtesy of Niagara Machine & Tool Works)

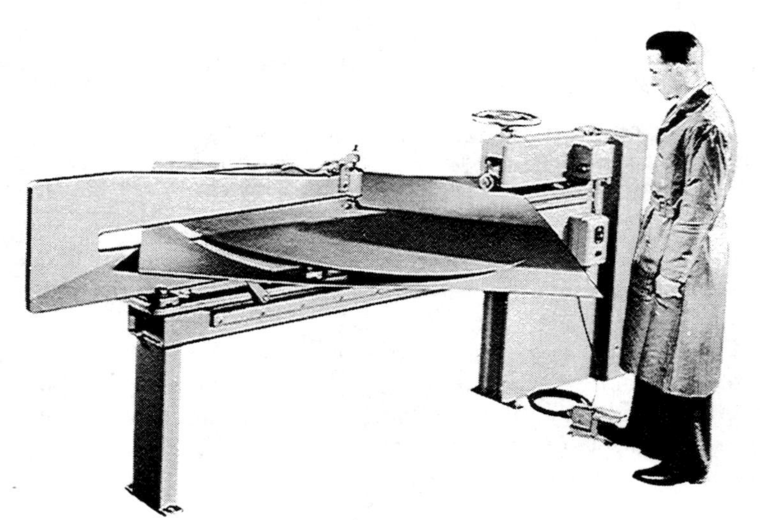

Figure 10-9 Tapered rotary blades are here employed to slit steel truss plates. As you can see, the plate spine projections rule out the use of straight bladed shears. (Courtesy of Wilder Machinery Company)

TYPES OF SHEARING

Shears can be used to cut wood, metal, cloth, paper, plastic, food, and other material. In the following paragraphs, there are descriptions of the more common shearing operations, each requiring different tooling and each designed for a specific cutting task. These methods include blanking, punching, cutting off, perforating, notching, lancing, slitting, nibbling, trimming, shaving, parting, and slotting. A chart of these appears in Figure 10-10, and it should be noted that while many of them can be applied to a variety of materials, all are applicable to metal sheet.

When a piece of sheet metal is sheared from a roll or sheet and that piece becomes the part being fabricated, the piece is called a *blank* or a *workpiece*. In some cases, the blank is the finished product and requires no further treatment, a case in point being a metal washer. Generally, however, the blank is intended as a starting workpiece to be further processed by drawing or a similar forming operation, or by cutting and fastening. Blanks are obviously important preprocess elements, and many of the shearing operations are directed at their preparation. Blanks can be simple rectangles or circles, or they can be complex shapes.

Rectangular-shaped blanks can be simply and inexpensively cut from a coil or strip by using the flying shear shown in Figure 10-5. This would be the preferred method of producing such blanks because it is fast and accurate, and no expensive dies are needed.

Blanking. *Blanking* is a shearing operation used to produce the workpieces (blanks). Blanks are usually cut from a larger piece of material on a press using a punch having the shape of the blank, and a die which mates with the punch. To be classified as blanking, the cutting action must occur about an enclosed shape.

Workpiece blanks are also produced by *cutting off*, in which shearing occurs along a line extending across the width of the stock. As shown in Figure 10-10, blanks are generally nested for greater material economy, and require only simple and inexpensive dies.

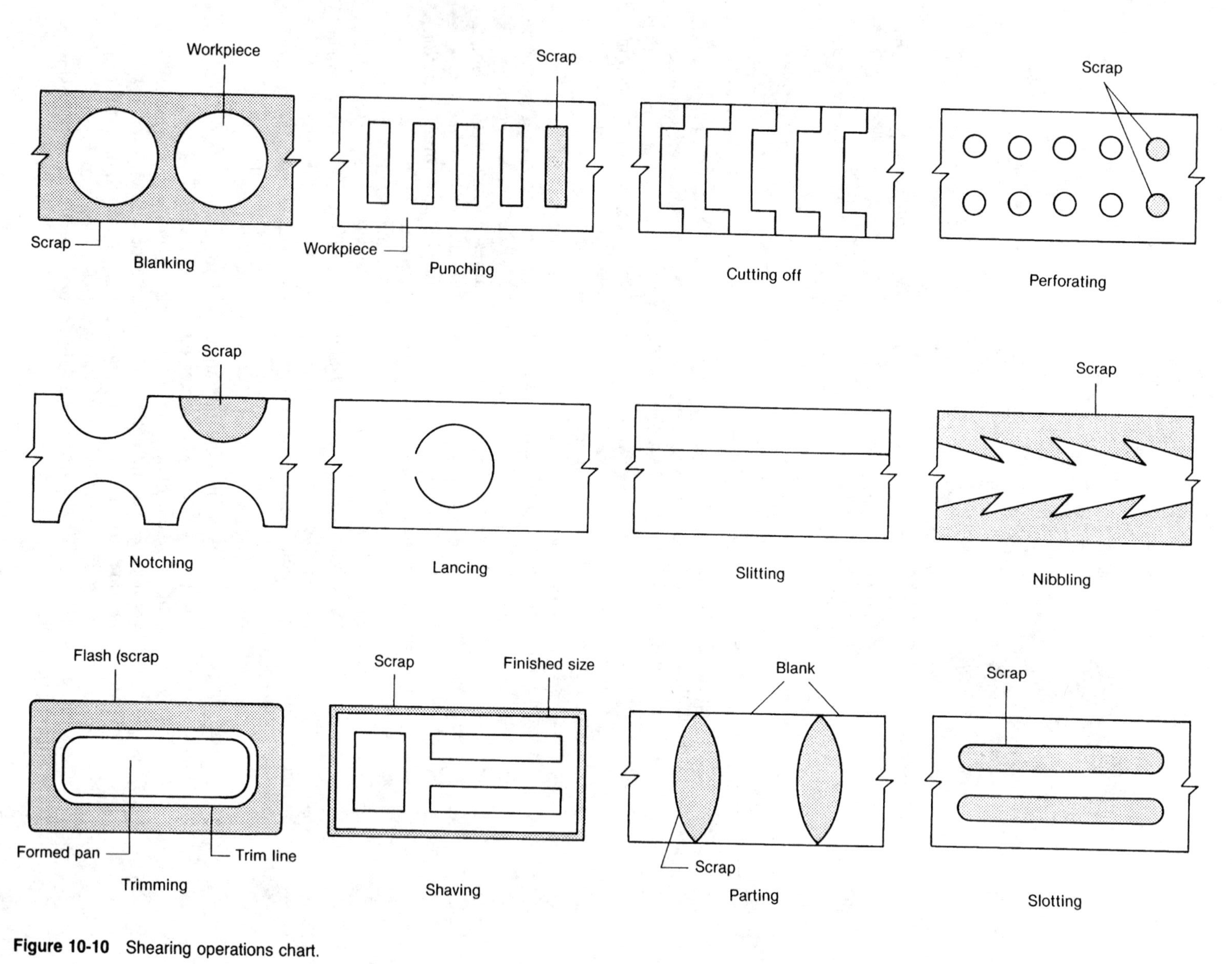

Figure 10-10 Shearing operations chart.

Parting. *Parting* is a method of producing blanks by cutting away a piece of material lying between them. The blanks so produced do not have mating adjacent surfaces, and so some scrap occurs with the process.

Punching. When holes of various shapes and sizes are cut into a workpiece, the process is called *punching*. Punching is sometimes referred to as piercing. They are not the same, however. Pierced holes are clean and smooth. It is very similar to blanking, except that in a punching operation the material around the punched hole is the workpiece, and the slug removed is the scrap. See Figure 10-11. The material removed in punching is usually smaller than obtained in blanking. As in most shearing operations, punching is done on flat workpieces, prior to forming into a finished product. The punching of elongated holes is called *slotting*.

Perforating. *Perforating* is a method of producing holes in flat sheet material. Entire sheets are perforated for use as screens, sieves, or decorative grills. In addition, selected areas are perforated to permit fastening of material with bolts or rivets. Most perforating operations generate several holes in the workpiece with one stroke of the press. (When only one hole is made with each stroke, the operation is normally called punching instead of perforating.) The ram holds a punch for each hole to be made, and the bottom die has a matching hole for each punch. To reduce the load on the equipment, the multiple punches are of different lengths so that all of them do not pass through the material at the same time.

Notching. Shearing metal slugs from the edges of workpiece is called *notching*. This process is often used to produce shapes which may be difficult using ordinary blanking methods. Notches also serve to relieve stresses in compound bend and thereby eliminate wrinkles. *Seminotching* is the process of producing cutouts from the central position of the workpiece.

Figure 10-11 The punching operation. (Courtesy of The Nolan Company)

Figure 10-12 Louvers are being formed in this metal sheet by a lancing-pressing operation. (Courtesy of The Nolan Company)

Lancing. *Lancing* operations produce an incomplete shearing cut in workpieces. The example illustrated is a small knockout for a hole in an electrical outlet box or circuit panel. Because no material is removed, no scrap is produced. Other examples include fastening tabs, tool box handles, or louvers. See Figure 10-12.

Nibbling. *Nibbling* is an operation used to produce irregular blanks, especially if only a small number of such workpieces is needed. Nibbling consists of punching a series of overlapping, small holes around the outline of the blank. The edges of nibbled blanks may be rough and usually must be smoothed by grinding or trimming. This process is often substituted for blanking when the tooling costs cannot be justified. Nibbling also is used to remove material from the interior of a workpiece. See Figure 10-13. Nibbling is commonly used in the automatic manufacturing of families of parts, where a range of patterns can be programmed into a controller. This results in a flexible fabrication system for sheet metal products such as electronic component cases, and shells or housings.

Figure 10-13 This nibbling operation is being used to remove metal from the interior of a blank which will be bent to form a sheet metal part. (Courtesy of The Nolan Company)

Figure 10-14 This slitting operation employs solid rotary shears. (Courtesy of Wilder Machinery Company)

Slitting. *Slitting* consists of cutting original coiled or sheet stock lengthwise by passing the material between circular blanks. See Figure 10-14. This provides a fast, efficient method for producing sheet stock for subsequent operations.

Trimming. *Trimming* is the process of separating excess metal from a formed part. This excess metal appears as a deformed or uneven flash surrounding a pressed metal part, where this waste area was used to grip the workpiece while it was being formed. Flash appearing on thermoformed plastic parts is also removed by trimming.

Shaving. *Shaving* is a method of improving the edge quality of a blanked part or punched hole. Shearing operations seldom leave a perfectly straight edge due to the nature of shearing. The shaving operation employs dies with extremely close tolerances so that the cut edge is straight and full. Shaving also is used to obtain accurate part dimensions.

This broad range of shearing operations is generally combined with forming operations in machine presswork. See Figure 10-15.

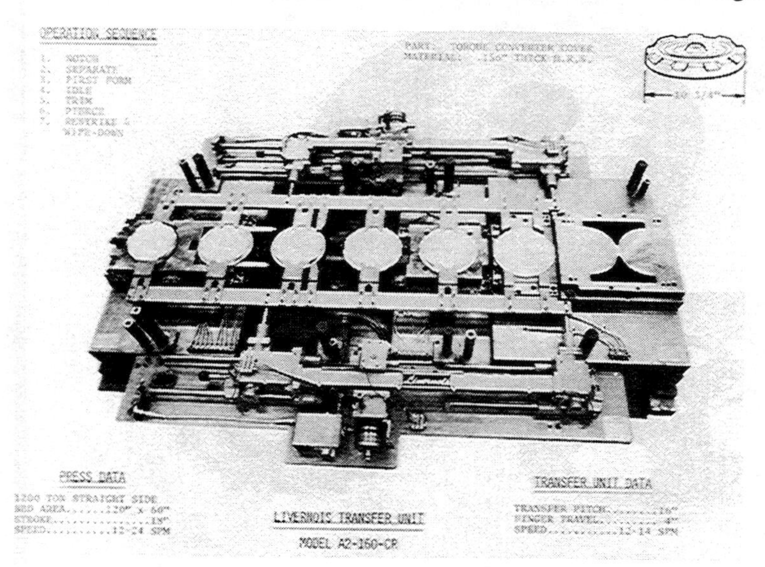

Figure 10-15 Note the several shearing operations which are combined with forming in this progressive die. The product being formed is an automotive torque converter cover. (Courtesy of Livernois Automation Company)

FINEBLANKING

Reference was made earlier to the importance of manufacturing a final part in as few separate process steps as possible to conserve time and materials. This is known as manufacturing to "net" shape, or "near-net" shape, which implies a one-process operation to final precision dimension and geometry. Fineblanking is such a net shape process which utilizes high-precision tools to clean-shear workpieces with only one press-stroke. Deburring generally is the only remaining operation needed to produce the final workpiece. The precision tooling makes possible the 100 percent clean-sheared edge, in contrast to conventional blanking where only one-third of the edge is clean-sheared, and the remainder is the rough surface of the blanking fracture. A typical part is shown in Figure 10-16. A study of this assembly reveals a precision part which is ideally suited to fineblanking operations. In fine blanking (also known as fine-edge blanking, smooth-edge blanking, or fine-flow blanking) a V-shape ring is pressed against the stock to lock it tightly against the die, and to force the work metal to flow toward the punch, so that the part can leave the strip without fracture or die break. Die clearance is extremely small, and punch speed much slower than in conventional blanking. While computer numerical control is not new in controlling the operation of most metalworking machine tools, the application of CNC to fineblanking presses is a significant technological advancement. These machines are available in a range of capacities up to 2,500 tons. See Figure 10-17. With such equipment, fineblanking technology today encompasses a greater variety of part configurations. It has progressed beyond traditional flat parts to include fineblanked components with bends, coinings, semipierced projections, and other forming operations. See Figure 10-18. Both compound and progressive dies are

Figure 10-16 This ratchet assembly consists of cam-shaped fineblanked laminations which are staked together with the fineblanked ratchet plate. Subsequently brazed, stripped, and heat treated, the unit is ready for final assembly as part of a crimping tool. (Courtesy of American Feintool)

Figure 10-17 This CNC fineblanking press is capable of multiple precision forming and shearing operations. (Courtesy of American Feintool)

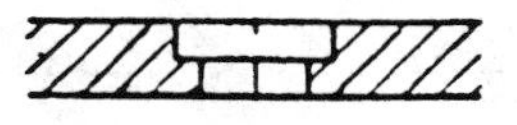

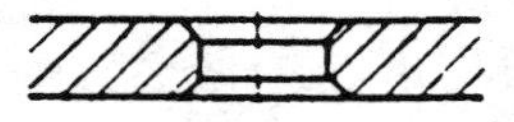

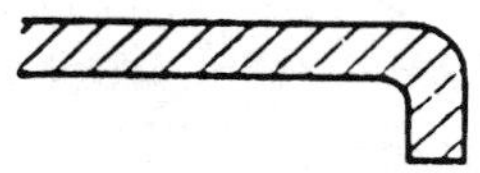

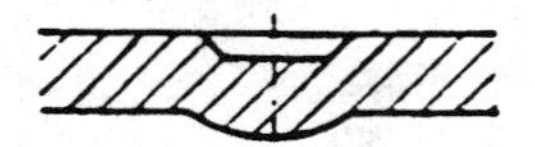

Figure 10-18 Operations capable with fineblanking dies. (Courtesy of American Feintool)

Figure 10-19 A typical fineblanking progressive die. (Courtesy of American Feintool)

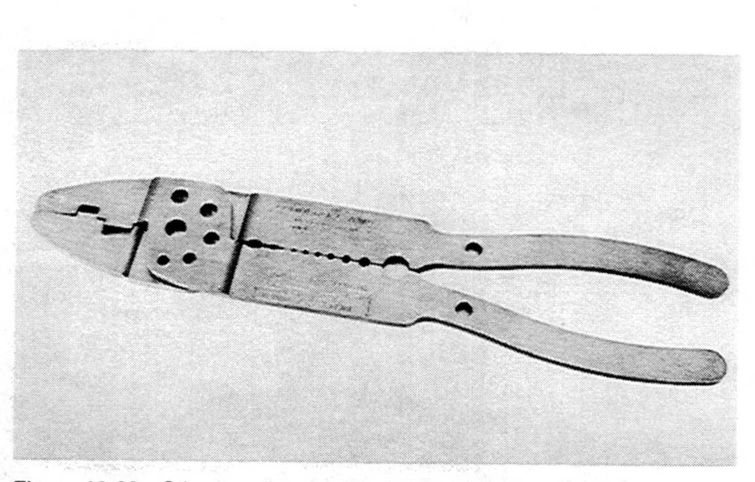

Figure 10-20 Crimping pliers for the electrical industry are completely fineblanked to heat treating and grinding. The final operation is the assembly of the two halves by riveting. (Courtesy of American Feintool)

used to produce complex shapes requiring close tolerance, dimensional integrity, and smoothly sheared faces. New techniques in tool design as well as other factors have increased the number and range of fineblanking applications. See Figure 10-19. The many attributes of this precision process lead to net shape product manufacture. See Figure 10-20.

SAWING

Sawing is a chip-generating process used to part or separate metal, plastic, or wood materials. This action is accomplished by a series of single-point, equally-spaced cutters passing through a workpiece. See Figure 10-21. The saw points or teeth are arranged either in a straight line or around the circumference of a disc, and thereby make possible the reciprocating, band, and circular saw methods.

Reciprocating sawing. *Reciprocating or hack sawing* is done with a straight, rigid blade having cutting teeth along one edge, and ranging in size from 254 to 457 mm (10 to 18 inches). The saw blade is pushed under pressure across the surface of the workpiece, and each tooth removes a small chip of metal. While small pieces are frequently cut with hand hack saws, most production hack sawing is done with powerdriven saws. The power hack saw has a reciprocating ram that moves the blade back and forth through the workpiece.

The teeth on hack saw and band saw blades are arranged so that the teeth will cut a groove in a workpiece which is wider than the blade thickness. This feature is called "set," and the term refers to the distance the teeth project from the sides of the blade. Set is necessary to prevent binding between the blade and the workpiece. The term "kerf" is used to describe the slot or groove resulting from sawing. The types of set are shown in Figure 10-22.

Figure 10-21 This horizontal slabbing cut is typical of the range of cuts possible with modern sawing equipment. (Courtesy of Hammond Machinery, Inc.)

The alternate set teeth are staggered one to the left and one to the right along the length of the blade. On the raker set blade, every third tooth remains straight and the other two are set alternately. On the wave (undulated) set blade, short sections of teeth are bent in opposite directions. The selection of set depends on the kind of material being cut.

The pitch of a saw blade refers to the number of teeth per inch of blade (or the number of millimeters of space between teeth). The blades also have a number of different tooth forms, used for cutting different materials. The higher the pitch number, the more teeth per inch of blade, and the finer the quality of cut. Harder metals are generally cut with fine-tooth blades.

Band sawing. *Band sawing* is done with a continuous blade made in the form of a thin, high speed or carbon steel band with cutting teeth along one edge. The band is looped around wheels above and below the table that holds the workpiece. One of the wheels is powerdriven to move the blade, and the workpiece is cut as it is pushed, either manually or by a power-feed mechanism, into the edge of the moving blade. See Figure 10-23. The band saw, unlike circular and rigid hack saws, can make curved or irregular cuts. The thin, narrow blade of this saw can follow almost any cutting path in a process called *contour band sawing*. Heavy contour cutting requires huge machines with special attachments to aid in moving

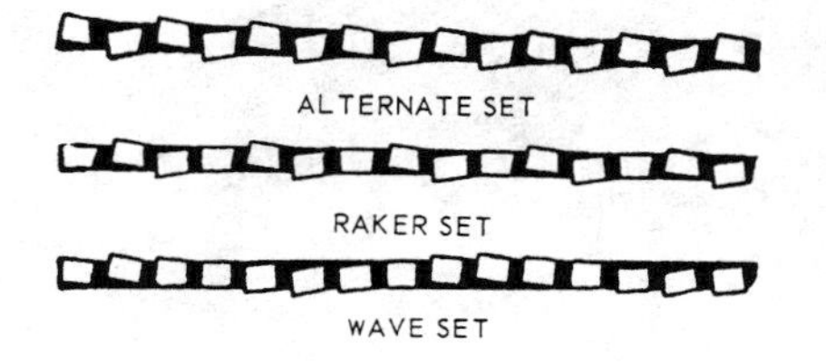

Figure 10-22 These set patterns are typical of hack saw and band saw blades. Circular saw blades generally have an alternate set arrangement.

Figure 10-23 This workpiece is being power-fed into this tilting-head band saw. (Courtesy of KTS Industries, Inc.)

the workpiece or the cutter head. See Figure 10-24. The sizes of workpieces which can be contour-sawed is limited by the distance between the supporting yoke and the blade (called the saw throat). Most saws are equipped with circulating cutting fluid systems to cool the workpiece and extend blade life. Many band saws can be used in both the vertical and horizontal positions.

Circular sawing. *Circular sawing* is done with a rotating, disc-shaped blade, and is employed to separate wood, metal, fiber, and plastic workpieces. The workpiece is secured to a table and either moved against the edge of the rotating blade or is

Figure 10-24 Worktable setup for the contour band sawing of heavy workpieces. (Courtesy of Do ALL Company, Des Plaines, IL)

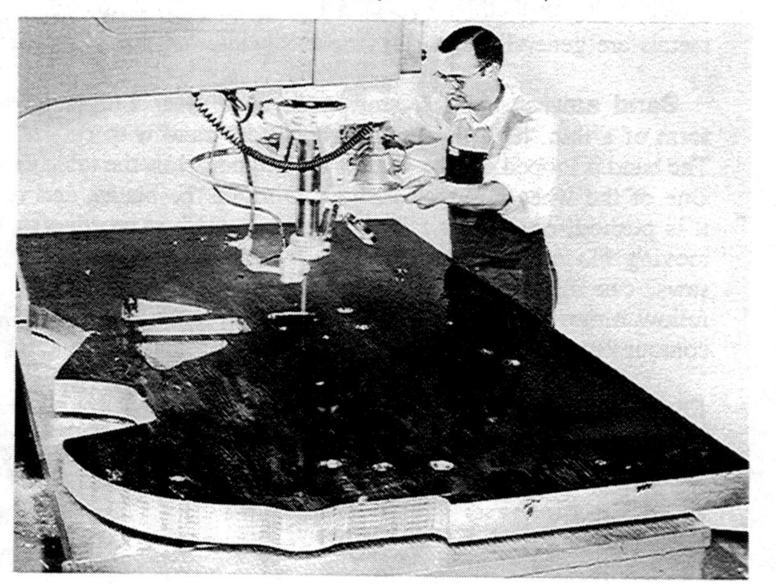

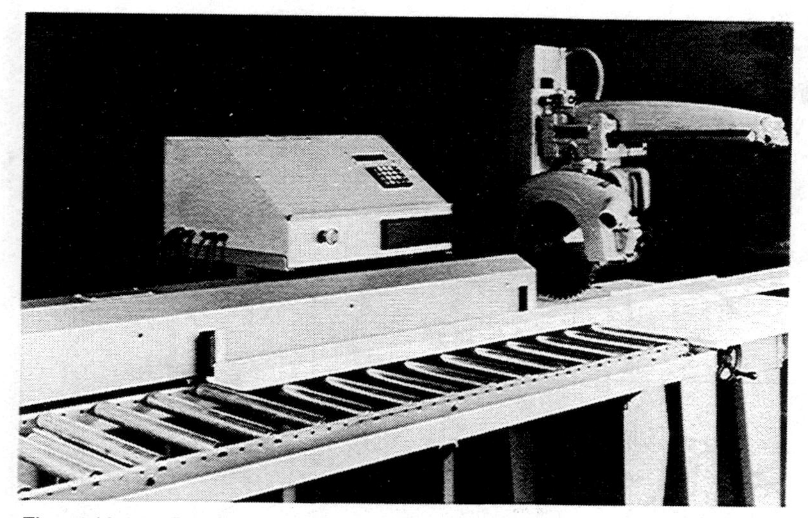

Figure 10-25 This programmable cut-off saw is used to produce wooden furniture parts of identical lengths. Note the automatic stop bar. (Courtesy of Evans Rotork, Inc.)

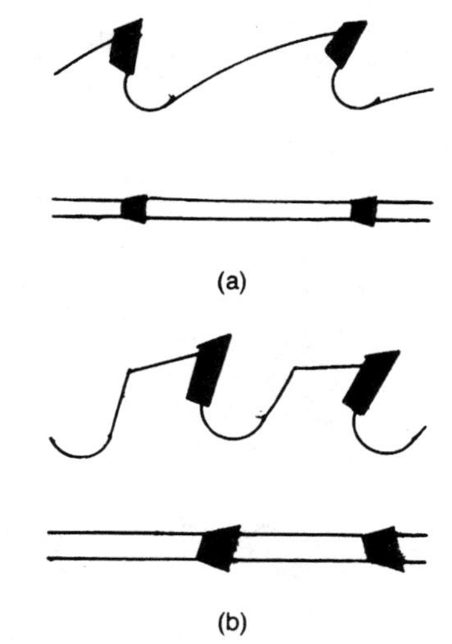

Figure 10-26 The characteristic rip (a) and crosscut (b) teeth found on wood cutting blades.

held stationary as the blade is fed into the piece. See Figure 10-25. Woodcutting blades are designed for rip-sawing (with the wood grain) or a crosscut (perpendicular to the grain). Rip teeth plow a kerf through the wood, but crosscutting teeth shear the fibers to complete the cut. See Figure 10-26.

Two types of circular saws are used for metal cutting. The cold saw uses a metal disc having cutting teeth around its edge. See Figure 10-27. Cold sawing produces a burr-free mill finish, eliminating secondary operations on tubing, channels, angles, and solid stock of most ferrous as well as nonferrous metals. The rigidity

Figure 10-27 This circular cold saw is used for slow-speed, heavy-duty metal cutting. Most such saws use a liquid coolant for cleaner cuts and to lessen blade wear and damage. (Courtesy of Kasto-Racine)

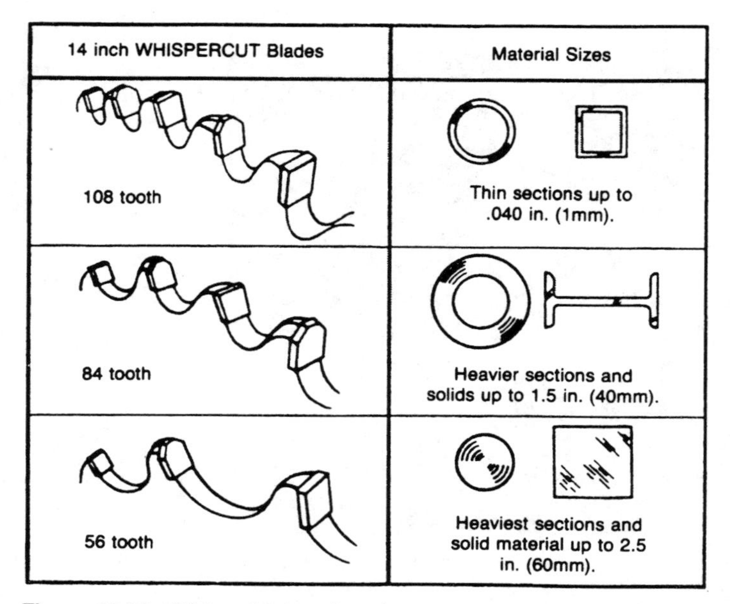

Figure 10-28 Pitch selection chart for cold saws. Note that blades with greater pitch values are used to cut thin material sections. (Courtesy of Startrite, Inc.)

of the blade produces cuts of extreme accuracy and close tolerances. The cutting operation is safe, clean, and quiet because of the slow blade revolution.

Cold saw teeth are shaped to cut much as a milling cutter would, with a scraping and shearing action of flat teeth. Some blades employ carbide-typed teeth, and most are used with liquid coolants to provide for lubrication and neater cuts. See Figure 10-28.

The abrasive saw nonmetallic blade is made from a disc of abrasive material, and is similar to a thin grinding wheel. Abrasive saws operate at much higher rotation speeds than cold saws and usually produce a smoother cut surface. See Figure 10-29.

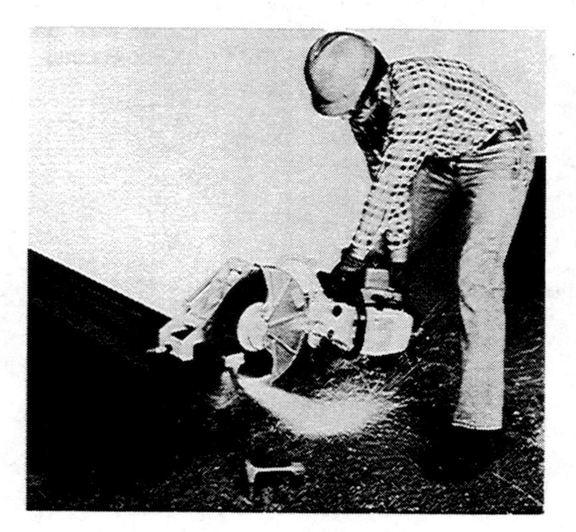

Figure 10-29 This abrasive saw is being used in steel rail manufacture. Such saws can be manually operated or automatic. Note the safety precautions taken by the operator. The machine guard is in place, and he is wearing protective shoes, gloves, helmet, goggles, and earplugs. (Courtesy of Modern Track Machinery, Inc.)

Figure 10-30 This hacksaw blade has tungsten carbide particles bonded permanently to its edge. (Courtesy of RemGrit Corporation)

Other abrasive-type saws are edge coated with tungsten carbide or diamond particles. See Figure 10-30. These blades can cut glass, ceramics, and other tough-to-machine materials. The straight line cutting action of a coated edge band saw gives exceptionally fast cutting rates as well as accuracy, parallelism, flatness, and finish of cut unobtainable through other methods. These blades provide industry with a very effective means of cutting a wide range of problem materials. See Figure 10-31. Self-sharpening, their cutting action reaches greatest efficiency at high speeds. Roller guides are recommended for use with bandsawing blades.

Friction sawing. *Friction sawing* is done on either a band saw or a circular saw. Special metal *friction* saws are designed to cut thin sections of mild steel such as sheets, tubes, and light structural shapes up to 3 mm (0.118 inch) thick. Friction sawing requires a fine tooth blade, operated at speeds of 6,000 to 15,000 SFM. This is 50 to 100 times faster than conventional sawing. Since friction blades

Figure 10-31 The grit edge circular saw is here used to cut a tough plastic-fiberglass sheet. It should not be used for metals or ceramics. (Courtesy of RemGrit Corporation)

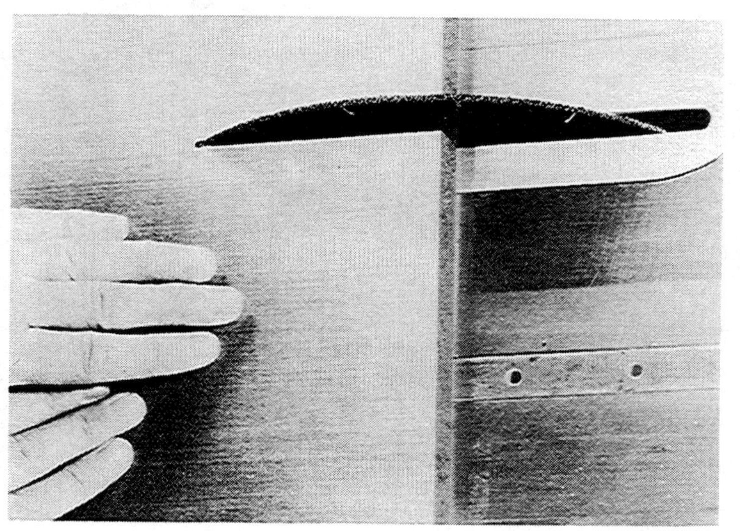

actually melt through the stock removing globules of metal rather than chips, some burring of the material may result. However, friction saws are in certain applications superior to abrasive wheels where finish and cutting life are important. The blade seldom overheats because only a small part of the band engages the workpiece, and this heat is readily dissipated. Cast iron and nonferrous metals are not suitable for friction sawing.

FLEXIBLE MANUFACTURING SYSTEMS

The availability of programmable controls adds a new dimension to fabrication operations in product manufacturing. In some cases, the work is limited to cutting process, while in others a fabrication center may involve a total range of processes from workpiece to finished product. Such systems are described in detail in Chapter 24.

A typical example of the former case is shown in Figure 10-32. This high speed, volume production vertical band saw has a mechanical bar feed that can be quickly and easily programmed for automatic operation. Programming cut-off lengths and number of cuts to be made is done on a key pad built into the operator's console. See Figure 10-33. Dimensional instructions in inches or millimeters are entered on the controller key pad to ensure cut-off length accuracy regardless of who is programming the saw. Several programs may be stored in the controller memory, and such programs can be called up by touching appropriate keys on the pad. Programmable control adds a new dimension of efficiency, accuracy, and economy to band sawing.

The following example is a program to produce 10 workpieces 219 mm (8.625 inches) long with a 45 degree angle (to the right) at each end. See Figure 10-34.

Figure 10-32 This automatic cut-off saw can be programmed to cut a variety of workpiece sizes. (Courtesy of Armstrong-Blum Mfg. Co.)

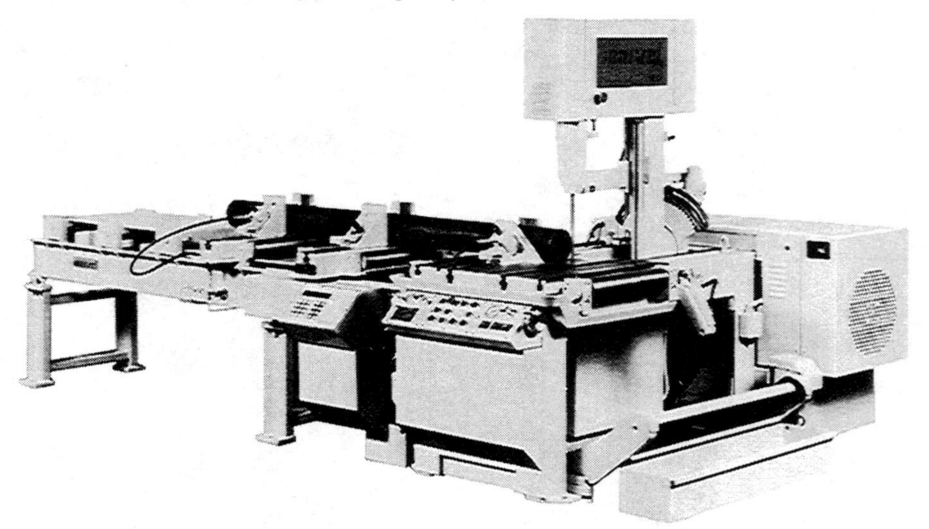

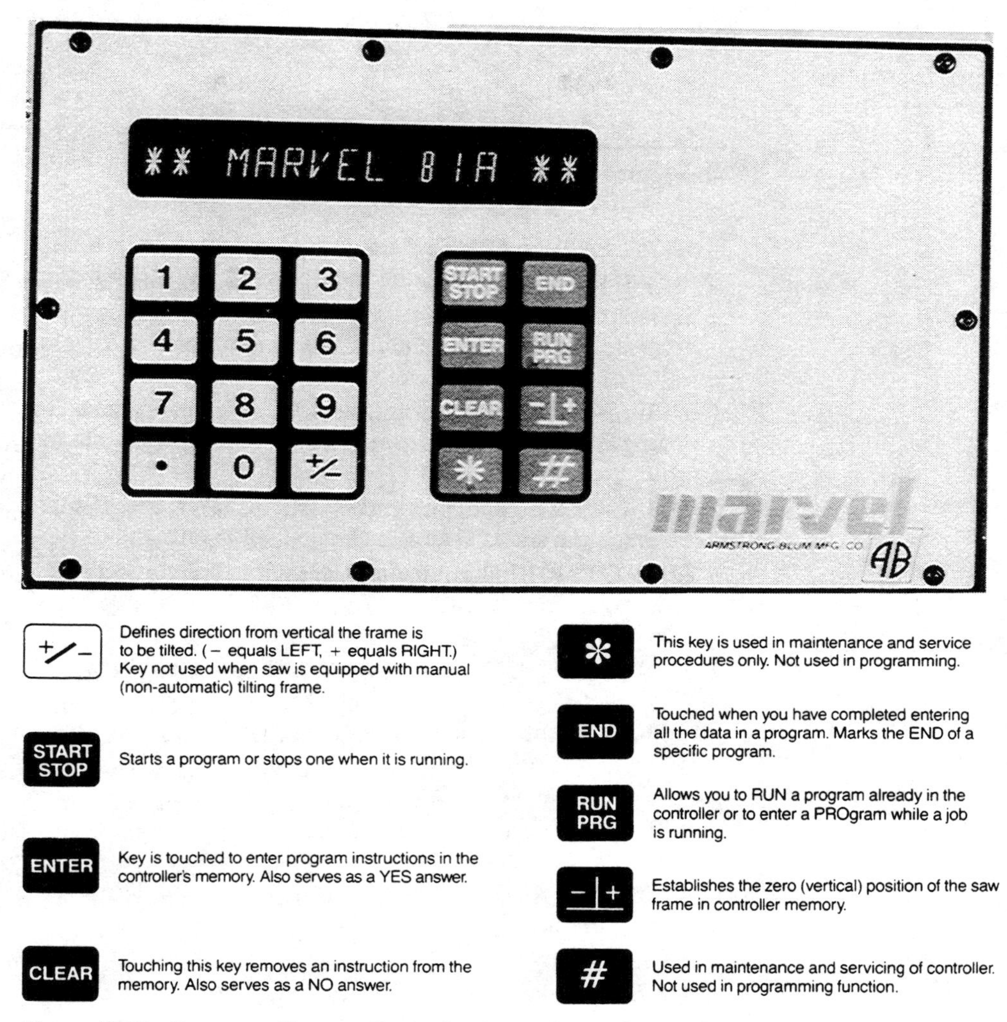

Figure 10-33 Programmable controller for the automatic cut-off saw. (Courtesy of Armstrong-Blum Mfg. Co.)

The number 3 has been assigned to the program job. The programming sequence proceeds by pressing the "START" key, and "SET TILT ZERO?" will be displayed. The operator should now set tilt to zero, and then press −/+. The display will then show:

1. PROGRAM JOB? The unit is asking if the operator wishes to program a job. Touch the 3 key and put the program number in the memory. Display changes to . . .
2. PROGRAM JOB? 3 to indicate that the job being programmed has been assigned the number 3. Touch the ENTER key and the display changes to . . .
3. 3–1 LEN? The unit is now asking how long the piece should be. Touch the 8.625 keys and the display changes to . . .

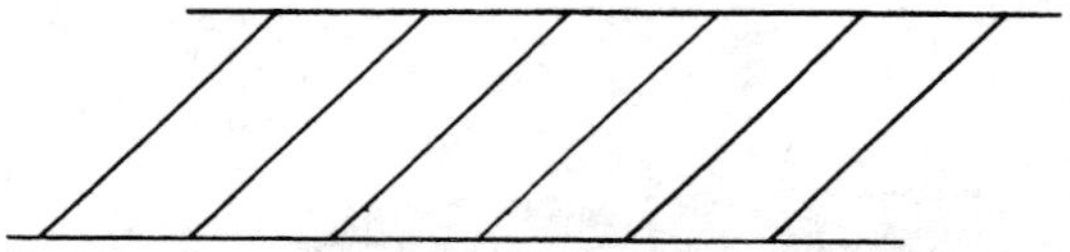

Figure 10-34 Cutting diagram for the programming job.

4. 3–1 LEN? 8.625 This confirms the length of 8.625. Touch the ENTER key to put the length into the memory and the display changes to . . .
5. 3–1 ANG? The unit is asking what angle the saw's frame should be moved to before making the cut. Touch the 4 and 5 keys and the display changes to . . .
6. 3–1 ANG? 45 Unit confirms pieces cut in Program 3 are to have a 45 degree angle at each end. Touching ENTER puts angle into memory. Display changes to . . .
7. 3–1 QTY? How many cuts are to be made? For a quantity of 10 pieces, touch the 1 and 0 keys and the display changes to . . .
8. 31 QTY? 10 Unit confirms quantity. Touching ENTER puts it into memory. Display shows 3-2-LEN? Touch END as there is no Step 2. Display now shows . . .
9. JOB 3 RPT? Unit asks how many times Job 3 is to be repeated. A number key other than 0 must be touched to answer. The program calls for 10 pieces, so touch number 1 key. Then touch ENTER and display changes to . . .
10. PROGRAM JOB? The unit is asking if you wish to program another job at this time. Since we are not going to program another job, touch the RUN/PRG key to instruct the computer you wish to run a job. Display will change to . . .
11. RUN JOB? At this point, Job 3 just programmed can be run by touching key 3. Display now shows . . .
12. RUN JOB? Confirming the command that you wish Job 3 to be run. Touch the ENTER key and the display shows . . .
13. 3–1 QTY? 10 Confirming that you wish to produce 10 pieces on Job 3, Step 1, touch the ENTER key and the display shows . . .
14. JOB 3 RPT? 1 Confirming the command that you wish Job 3 to be repeated only once, touch the ENTER key and the display shows.
15. PRESS START? Touch the START key to begin the automatic production of the 10 pieces you require. NOTE: The display will count down 10, 9, 8, etc. to 1 as each cut is made. The saw will stop after completing the last cut and wait for further instruction ordered by the controller.

That such a system is reprogrammable means that it is a *flexible system*, and it can be used in the production of a broad range of different band sawn workpieces.

A range of similar fabrication systems are employed in the manufacture of sheet metal parts. The automatic operations include material storage and transfer, cutting, deburring, inspection, and bending or forming. Examples of these and other flexible systems are found in Chapter 25.

QUESTIONS

1. Explain the difference between the cutting actions which take place in shearing and sawing.
2. Prepare a simple sketch which illustrates the action of shearing.
3. List the three primary types of shears, and make a simple sketch of each.
4. Explain the use of flying shears.
5. List the twelve types of shearing operations, and identify a typical use for each.
6. What is the difference between punching, piercing, and perforating?
7. Explain the difference between conventional blanking and fineblanking.
8. List three tools used to saw metals.
9. What is meant by pitch, set, and kerf as they relate to sawing?
10. What is the difference between the ripsawing and crosscut sawing of wood?
11. Explain the difference between cold sawing, abrasive sawing, and friction sawing.
12. Prepare a library research report on the uses of CNC equipment in industrial sawing and shearing. Visit a local industry which employs sawing or shearing operations and discuss this issue with the plant manager.

11 TURNING AND MILLING

The term commonly applied to several metal removal processes is *machining*, and includes turning, milling, shaping, planing, and drilling. All are general *bulk removal* processes, as contrasted to *separation* processes such as sawing and shearing. Four of these inclusive processes are treated in this chapter. Drilling and its related operations is the subject of the following chapter.

TURNING

Turning is the most common machining operation, and it is defined as the shaping of a rotating workpiece by forcing a single-point cutting tool against its surface to produce a cylindrical shape.

Turning is done on a lathe. There are many kinds of lathes, but all work on the same principle. The engine lathe, as shown in Figure 11-1, is the simplest and most common type. The workpiece is suspended by its ends between the headstock and tailstock of the machine. The piece is supported at the tailstock by a cone-shaped extension called a center, which fits into a matching hole in the end of the piece. The headstock has a similar center to support the other end of the workpiece. The headstock of the lathe contains a power-driven spindle to rotate the workpiece. The cutting tool is mounted on a carriage that can move back and forth between the headstock and tailstock. The tool is capable of precise movements in two directions: along the length of the workpiece or into the face of the piece. This type of work is called *turning between centers*. See Figure 11-2.

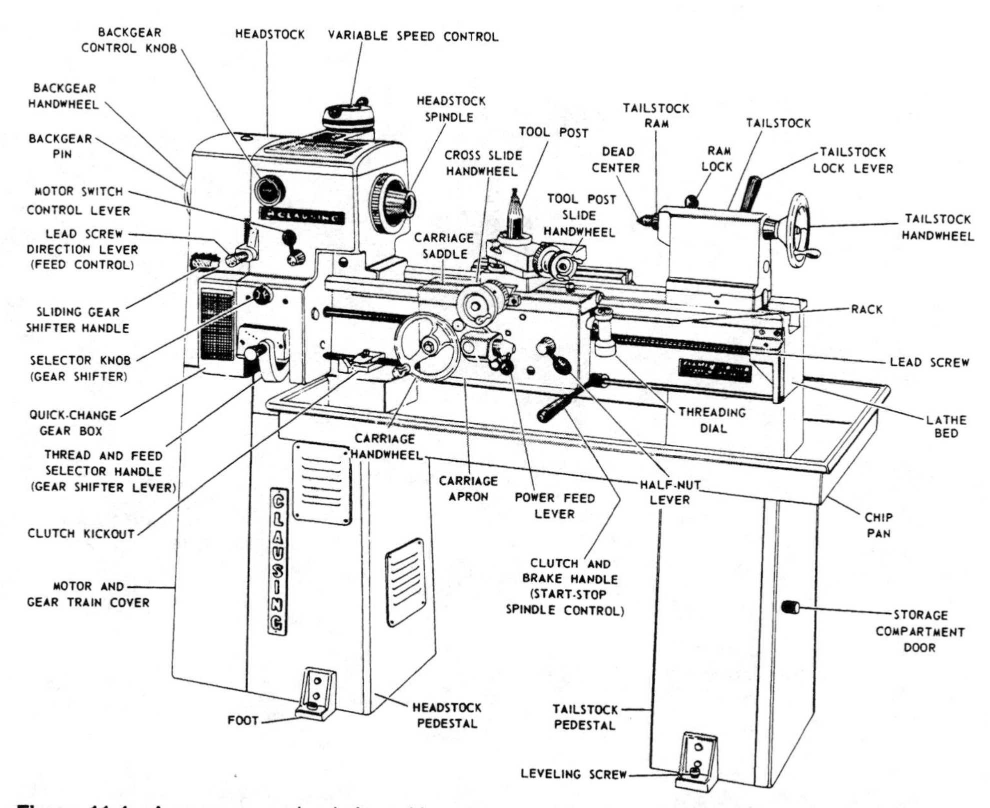

Figure 11-1 A common engine lathe, with major parts identified. There are many variations and sizes of this basic machine. (Courtesy of Clausing Industrial, Inc.)

Chuck turning is a lathe operation whereby the workpiece is secured in a chuck with adjustable jaws. The chuck is fastened to the headstock spindle, and no tailstock support is required. See Figure 11-3. Such an arrangement is common for short workpieces, or those which must be drilled or reamed.

Feed and speed. The movement of the carriage controls the cutting action of the tool. Typically, the tool is first moved into the workpiece until it is removing the desired amount of metal. The thickness of the metal being cut away is called the *depth of cut*. When the proper depth of cut is established, the carriage is started in motion along the lathe so that the tool cuts longitudinally along the workpiece. This movement of the tool is called *feed* and is expressed in millimeters or inches per revolution, or the distance the tool moves along the workpiece during each revolution of the workpiece. Feed and depth of cut are two of the three cutting variables. The third is *cutting speed,* expressed in surface feet per minute (sfpm), or the speed with which the uncut surface of the workpiece is moving past the tool. This speed factor has a metric equivalent. Cutting speed is calculated from the rotational speed of the lathe spindle and the diameter of the workpiece. See Figure 11-4.

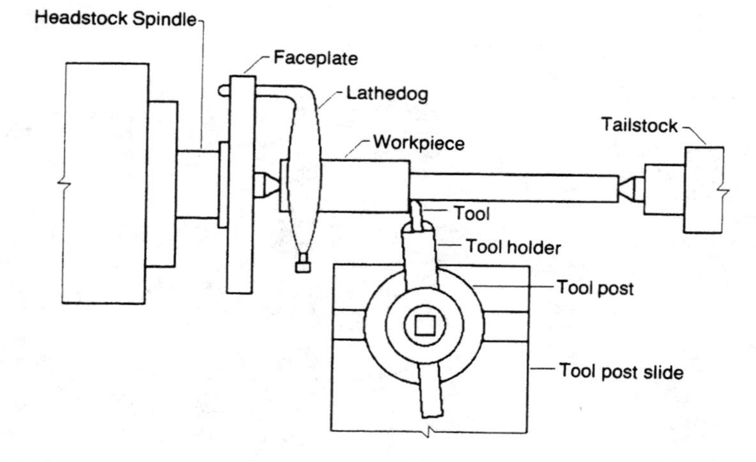

Figure 11-2 Turning between centers is a fundamental lathe machining operation. Note that the lathedog fits into a faceplate to secure the workpiece, and causes it to rotate as headstock power is applied.

Figure 11-3 A typical chuck turning operation. The workpiece is secured in the chuck, and no tailstock support is required.

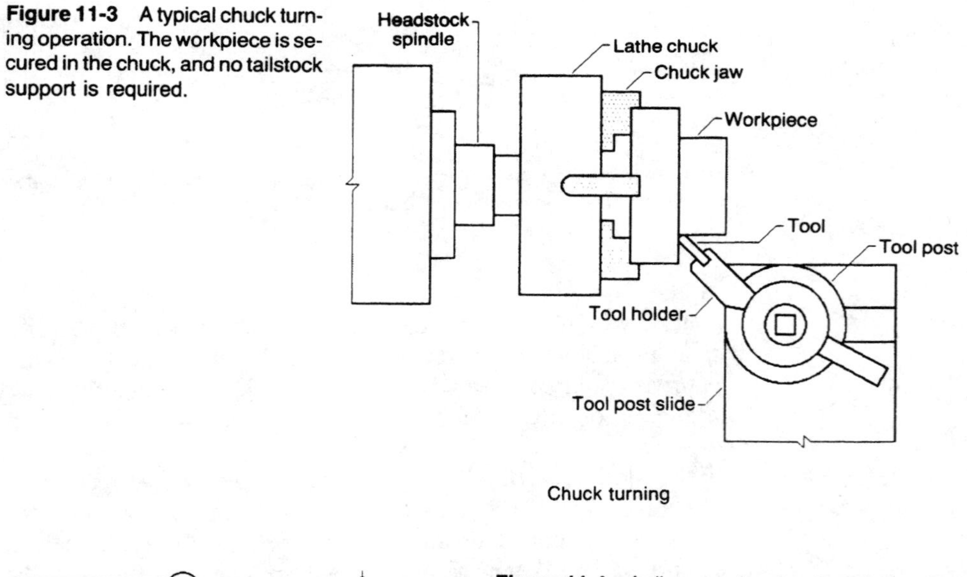

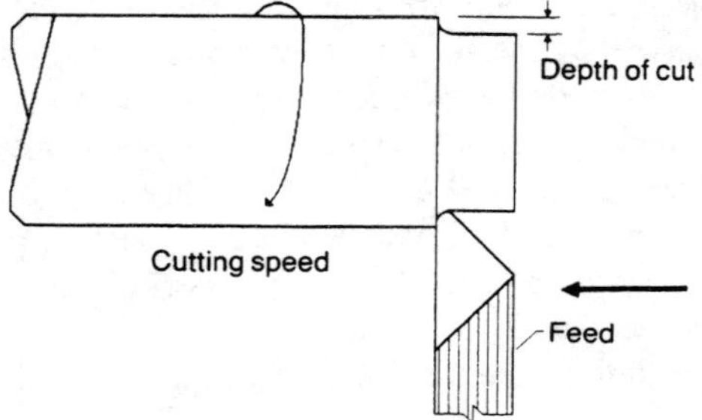

Figure 11-4 A diagram showing the relationship of depth of cut, feed, and speed.

LATHE TOOLS

Most turning operations are performed with single-point tools which shape the workpiece by lateral movement or feed along the workpiece. Such tools are shaped either by grinding the appropriate cutting edges on the end of a small bar of tool steel or by attaching a carbide or ceramic insert to the end of a steel bar. For effective cutting, the shape of the tool must suit the turning operation and the material being machined. Typical single-point tools and their uses are shown in Figure 11-5.

Other tools are commonly used for special lathe operations. *Knurling* is the process of pressing a cross-hatch pattern on a workpiece, using a knurling tool which has a set of sharpened, scribed rollers attached to it. See Figure 11-6. This modifies the surface finish on a workpiece to provide a more positive grip on a tool handle or a control lever. *Reaming* is an accurate hole enlarging or sizing operation, and can be done manually or automatically. See Figure 11-7.

TYPES OF LATHES

Specialized turning operations are performed on variations of the engine lathe, or on machine tools with unique configurations and functions.

The *hollow-spindle* lathe has a hole through the spindle of the headstock to accommodate long workpieces that would not fit between the centers of the lathe. The piece is inserted through the hollow spindle from the rear of the headstock. Only the area to be machined is in the position over the bed of the lathe, and the remainder extends from the back of the headstock. The workpiece is gripped and driven by a collet or jaws that surround the workpiece at the headstock. The collet is a special type of nonadjustable/holding device which requires a different spring collet or sleeve for each workpiece diameter. See Figure 11-8.

The *basic* engine lathe, from the previous discussion, has the capability of cutting between centers or in chuck applications, as shown in Figure 11-9. The cutting is done by one single-point tool, which can be moved in any direction. The tool can be moved manually or with power in the motion pattern shown. The machine is also available in a range of sizes, with automatic controls, and with special power requirements. See Figure 11-10. Other lathes have different cutting capabilities, as described following. This basic lathe is often used in tool room operations, rather than in production work.

Indexed tool lathes. These lathes comprise the largest and most diverse class of production turning machines. The distinguishing features are that the cutting occurs with a series of tools which can be indexed or turned, one after the other into the desired cutting position. See Figure 11-11. The machine has a turret in the tailstock position which can rotate and then move longitudinally to present a succession of tools to the end of the workpiece. Drills, reamers, countersinks, taps, and boring bars are typical turret attachments. Some machines may also have one or two cross slides to carry tools to the workpiece, and these slides may be in the form of a fixed tool mount or an indexing ram turret. Indexed tool lathes generally have no tailstocks, as the workpieces are chuck-mounted, a factor which limits the length of the workpieces. This class of machine can be hand-fed, power-fed, or cycled automatically.

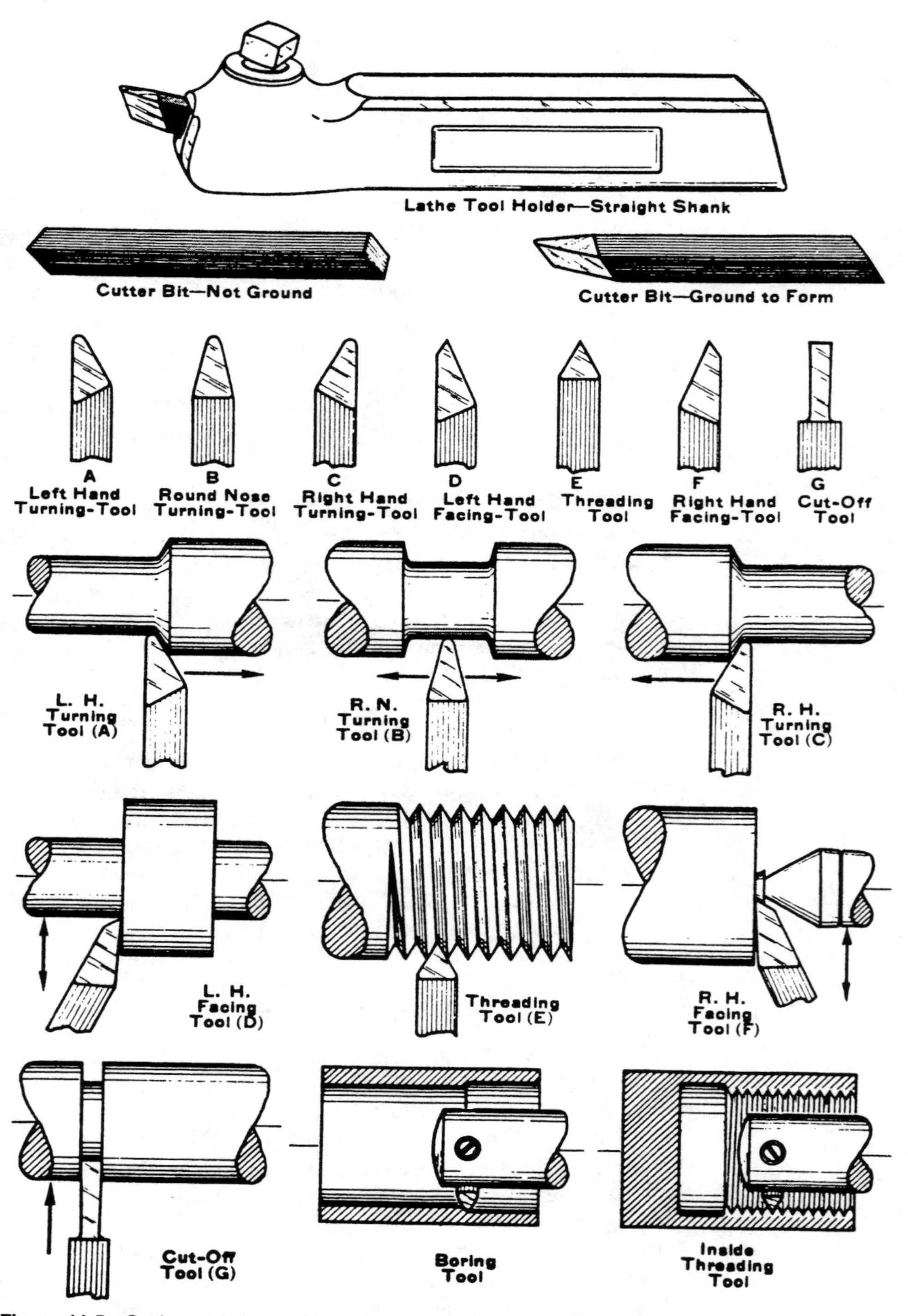

Figure 11-5 Cutting tool shapes and their applications in lathe turning. A tool holder is also shown. Note that threads may also be cut on a lathe. Boring and internal threading are typical chuck turning operations. (Courtesy of South Bend Lathe, Inc.)

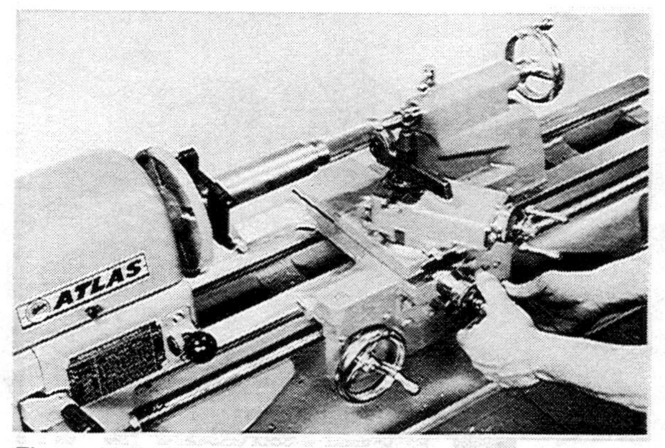

Figure 11-6 A knurling operation, here done manually, but more often done on automatic equipment. Shown is a between centers operation. It also can be performed on chuck-mounted pieces. (Courtesy of Clausing Industrial, Inc.)

Figure 11-8 A collet chuck is tightened by a special wheel adjuster at its opposite end. A quick, sure grip is provided with this workpiece holder. (Courtesy of Clausing Industrial, Inc.)

Figure 11-10 A modern CNC lathe. (Courtesy of Clausing Industrial, Inc.)

Figure 11-11 The cutting action of the basic indexed tool lathe. Such machines can utilize one or all of the rear tool, front tool, and turret configuration, and may be fed manually or automatically.

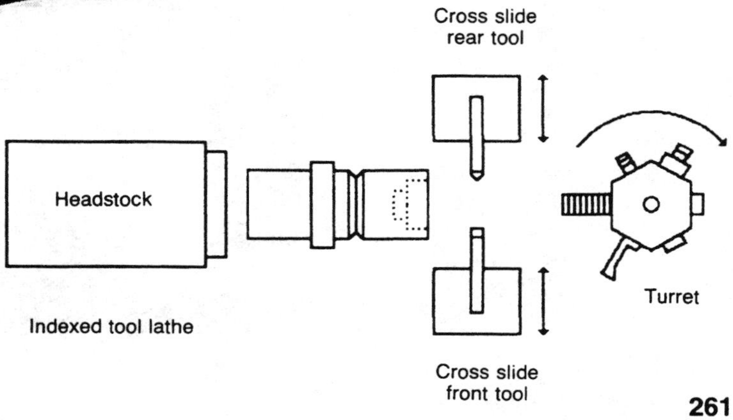

Cross slide
rear tool

Headstock

Turret

Indexed tool lathe

Cross slide
front tool

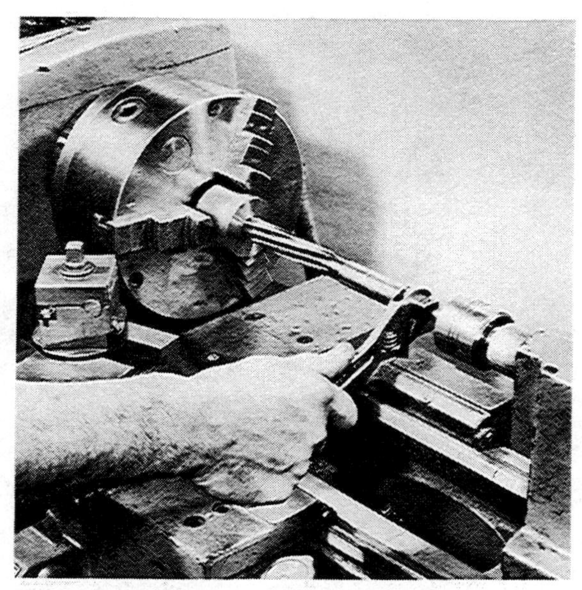

Figure 11-7 Reaming a drilled hole. Note the chuck-mounted workpiece. (Courtesy of Clausing Industrial, Inc.)

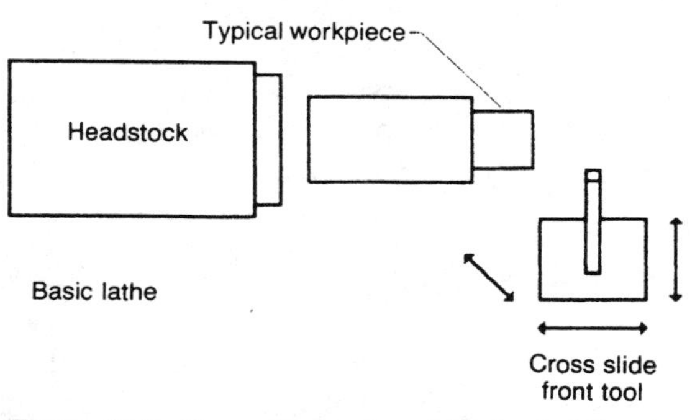

Figure 11-9 The cutting action of the basic lathe, top view.

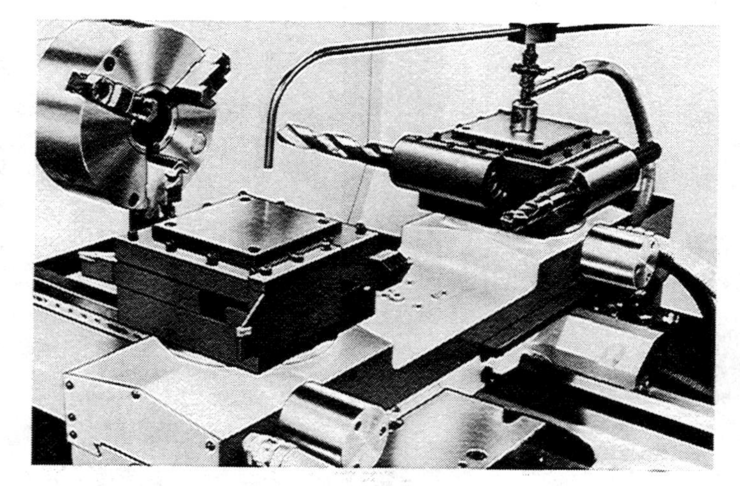

Figure 11-12 This lathe features indexable front and rear cross slides to feed a succession of tools to the workpiece. (Courtesy of Clausing Industrial, Inc.)

Hand-fed machines are used for machining small workpieces at high rates of production. The operation is such that each tool feed is stopped when the proper cutting has taken place. The operator merely retracts a ram lever to index a tool, then moves the lever to achieve the cut. Once the mechanical stops have been set, there is little chance for dimensional error. This is called a ram-type machine because the turrets are mounted on moveable rams. See Figure 11-12.

Power-fed machines commonly have either a ram turret arrangement, or a saddle-mounted turret which moves longitudinally along the lathe ways. The saddle turret is also capable of moving across the face of the workpiece. These machines include a power feed arrangement for the turrets, and generally have a larger workpiece capacity than the hand-fed models.

Automatic machines are capable of moving through an entire sequence of cutting operations without the intervention of a human operator. The completion of each cutting operation signals the machine, by means of preset mechanical stops, that another cutting phase is about to commence. There are numerous types and configurations of these machines, each with special advantages. One type features a turret which is mounted on the headstock, parallel to the spindle but offset slightly. This machine has the advantage of being able to orient a bank of cutting tools parallel to the longitudinal surface of the workpiece. See Figure 11-13. Others have a vertical

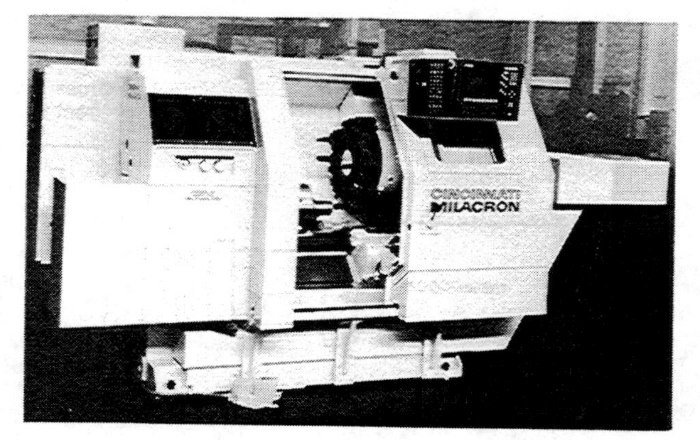

Figure 11-13 An automatic indexing lathe turning center. (Courtesy of Cincinnati Milacron)

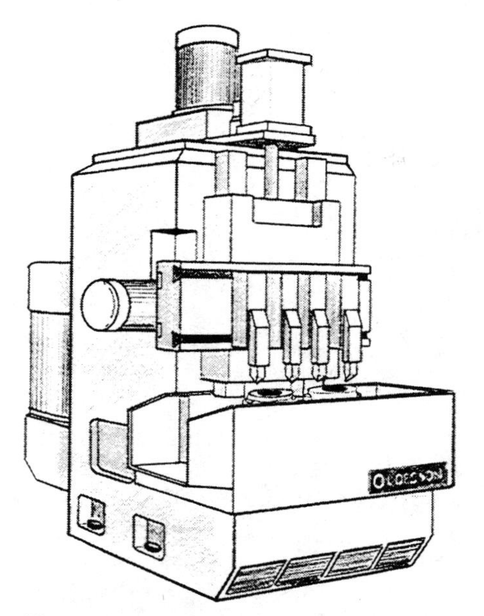

Figure 11-14 A multiple spindle, multiple tool vertical lathe. (Courtesy of The Olofsson Corporation)

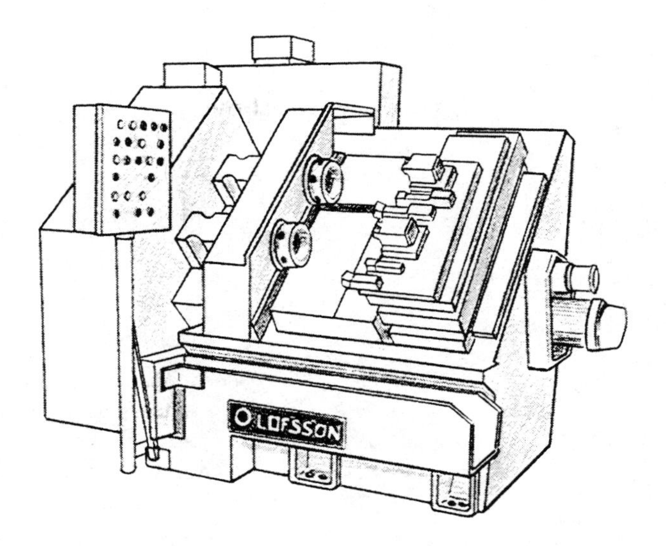

Figure 11-15 A multiple spindle, multiple tool slant bed automatic lathe. (Courtesy of The Olofsson Corporation)

configuration and cut similar to a boring mill or a drill press. See Figure 11-14. Another type features multiple spindles and tools on a slant bed, and is therefore capable of machining several workpieces for high productivity. See Figure 11-15.

The special feature of the *massed-tool lathe* is that the cutting is performed by several tools positioned in one or more tool blocks. The tool blocks are fed independent of one another, and can move along separate tool paths and feed into the workpiece to make multiple, simultaneous cuts. See Figure 11-16. Any number of combinations of tools and numbers of tool blocks are available on this versatile machine. It is a volume production automatic machine, and generally is not available in hand-fed or power-fed models. A tailstock can be used for long workpieces. Models designed to hold workpieces in chucks are called ''chuckers.''

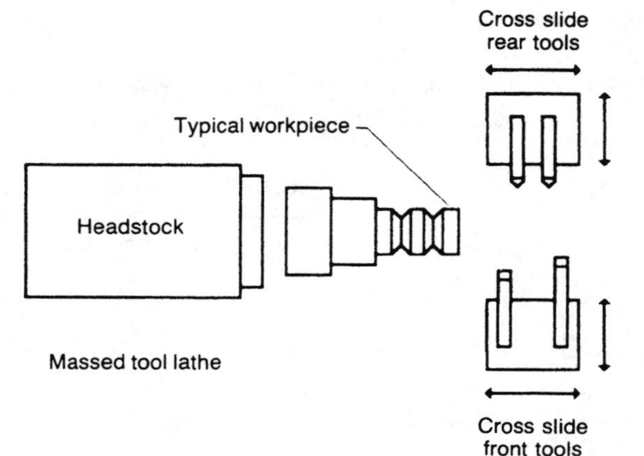

Figure 11-16 The cutting action of a massed tool lathe.

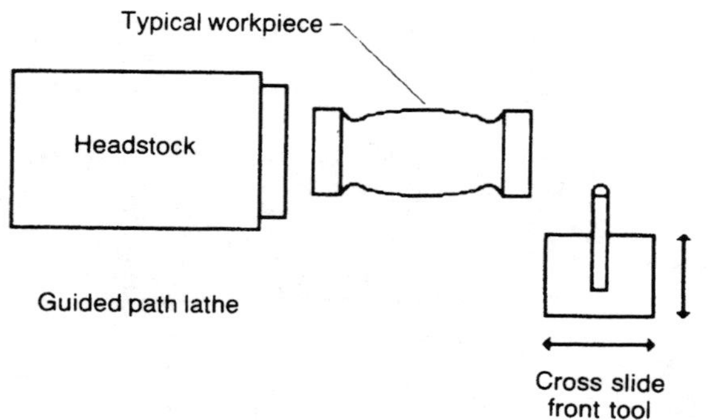

Figure 11-17 The cutting action of a guided-path, or tracer lathe.

The *guided path*, *tracer*, or *duplicating lathe* is designed to reproduce the contour of a templet or master part on a workpiece. The lathe is equipped with a tracer finger or stylus that travels along the pattern to be duplicated. The movement of the tracer is transmitted, usually by hydraulic linkage, to the controls of the lathe to guide the single-point cutting tool along a path that precisely matches the pattern. See Figure 11-17. Pieces with contours or complex shapes can be so machined at very high production rates.

The computer numerical-control lathe, like the tracer lathe, is used for repetitive machining of duplicate parts. The CNC lathe, however, requires no pattern or template. The lathe is controlled electronically by a punched tape or a computer program which contains complete instructions for making the part.

Sophisticated, automatic turning machines commonly are called *turning centers* or *machining centers*, and frequently are part of the computer integrated manufacturing systems described in Chapter 24.

MILLING

This versatile process is one of the most widely used machining methods, capable of removing bulk material to create flat or contoured surfaces on a range of workpiece sizes. Milling is done by feeding a workpiece into a fixed, multi-toothed, rotating cutter. See Figure 11-18. Milling is an operation opposite to turning. In turning, the workpiece rotates as a single-point tool is fed into it. In milling, the tool is stationary and the workpiece moves.

There are at least 25 different types of milling machines, but the most common configuration is the *horizontal* mill shown in Figure 11-19. Most others are variations

Figure 11-18 The action of a horizontal milling machine cutter. (Courtesy of Inco Alloys International, Inc.)

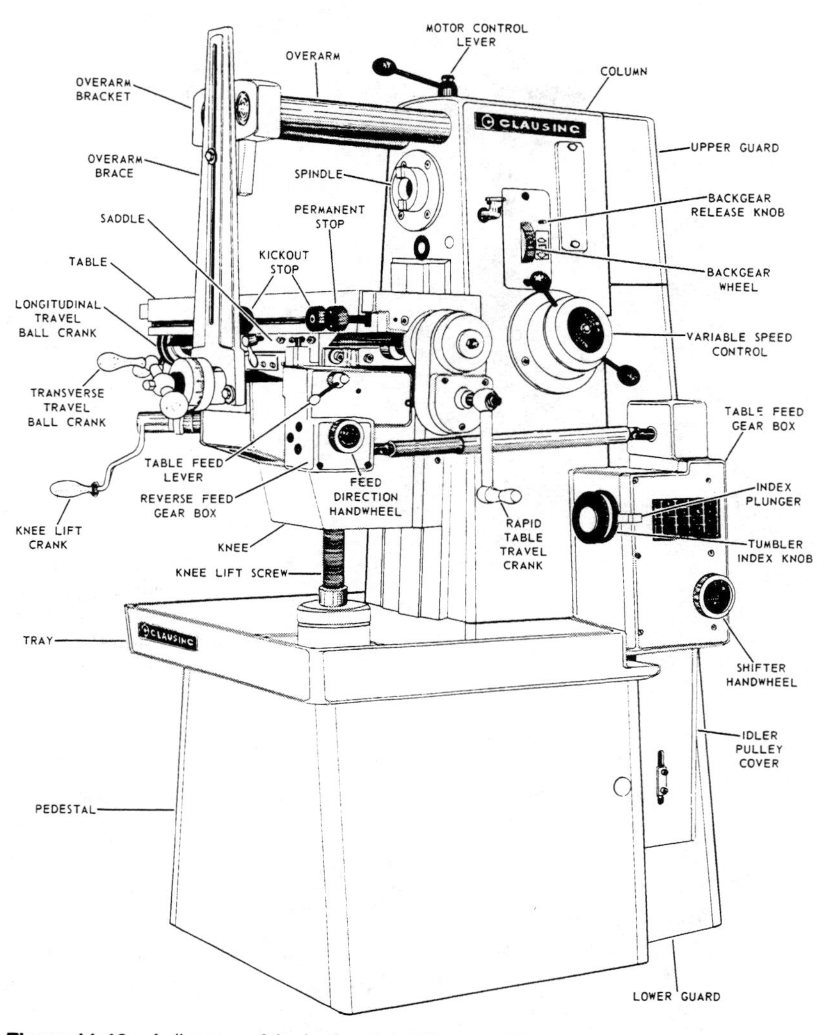

Figure 11-19 A diagram of the horizontal milling machine, with major parts identified. (Courtesy of Clausing Industrial, Inc.)

of this basic model. A second type, the vertical mill, is described later in this chapter. The basic parts of a milling machine are an arbor shaft to hold the cutting tool and a table to hold the workpiece. Cutting force is supplied by rotation of the arbor shaft, and cutting feed is supplied by movement of the table past the rotating tool. The arbor shaft is secured to the spindle, and is horizontal or parallel to the table. See Figure 11-20.

The two most common table support configurations are the column-and-knee and the fixed-bed. As shown in Figure 11-19, the table of the *column-and-knee* machine is mounted on an adjustable knee extending from a column. This permits the table to move right and left in the X-axis, which is the primary milling feed motion. The table is mounted on saddle, which provides the X-axis motion, and the saddle straddles the knee, which provides the Z-axis motion. See Figure 11-21.

Figure 11-20 A photograph of the mill shown in Figure 11-19. Note that the arbor is in place in the spindle, and is supported by the overarm brace. The rotary milling cutters are attached to the arbor. (Courtesy of Clausing Industrial, Inc.)

The table on a *fixed-bed* machine remains at a constant height; a movable spindle carrier provides vertical adjustment. Because of the absence of a moveable knee, the rigid bed can hold very heavy workpieces and accommodate higher powered driving mechanisms.

The *vertical* mill features a spindle which lies in a plane perpendicular to the table. See Figure 11-22. It too is available in knee and fixed bed models, with similar attributes. Some types have a *ram*, which is a unique spindle construction which permits movement in the Z-axis. The head may also be tilted to meet special cutting requirements. This machine generally uses end mills and shaped cutters for slotting, undercutting, and dovetailing.

Mill cutting tools. Milling operations are classified as *peripheral* milling, *end* milling, or *face* (or side) milling, depending on the shape of the cutter and the subsequent machined surface. See Figure 11-23. Furthermore, the cutting action is called either *conventional* milling or *climb* milling, depending on the direction that the workpiece moves in relation to the rotation of the cutter. A comparison of the two cutting actions appears in Figure 11-24. In climb milling (also called down milling), the cutting edges first contact the metal at the top surface of the workpiece. In conventional milling (or up milling), the cutting edges first contact the workpiece

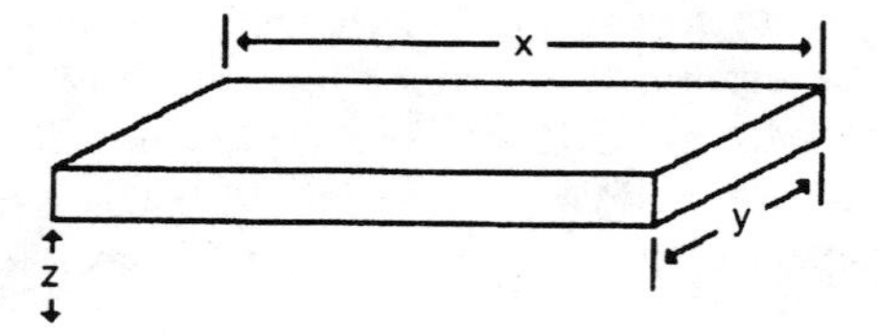

Figure 11-21 Axes of table motion. There is no Z axis table motion in a fixed-bed machine.

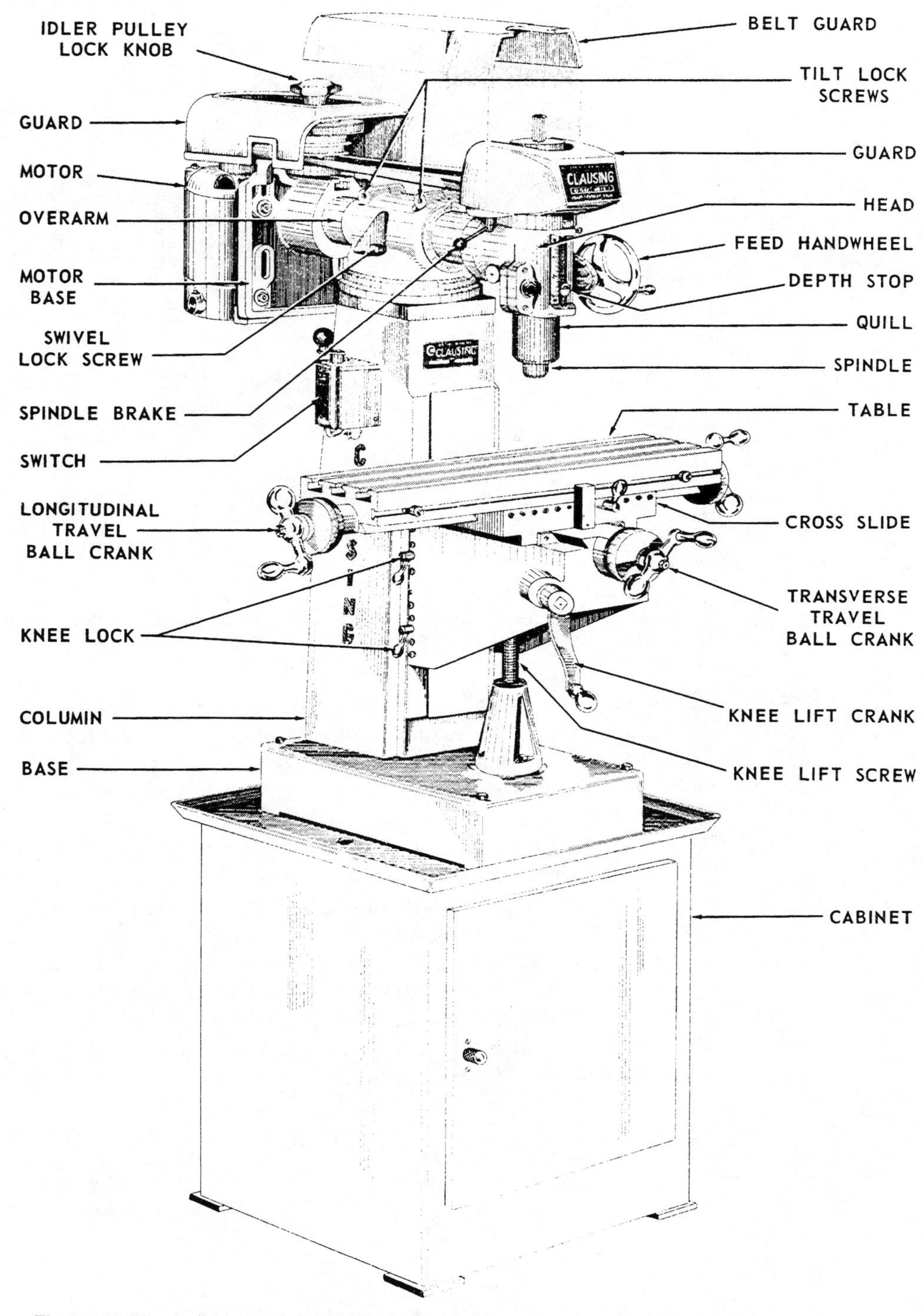

Figure 11-22 A diagram of the vertical milling machine. Compare it with the horizontal mill to observe their differences. (Courtesy of Clausing Industrial, Inc.)

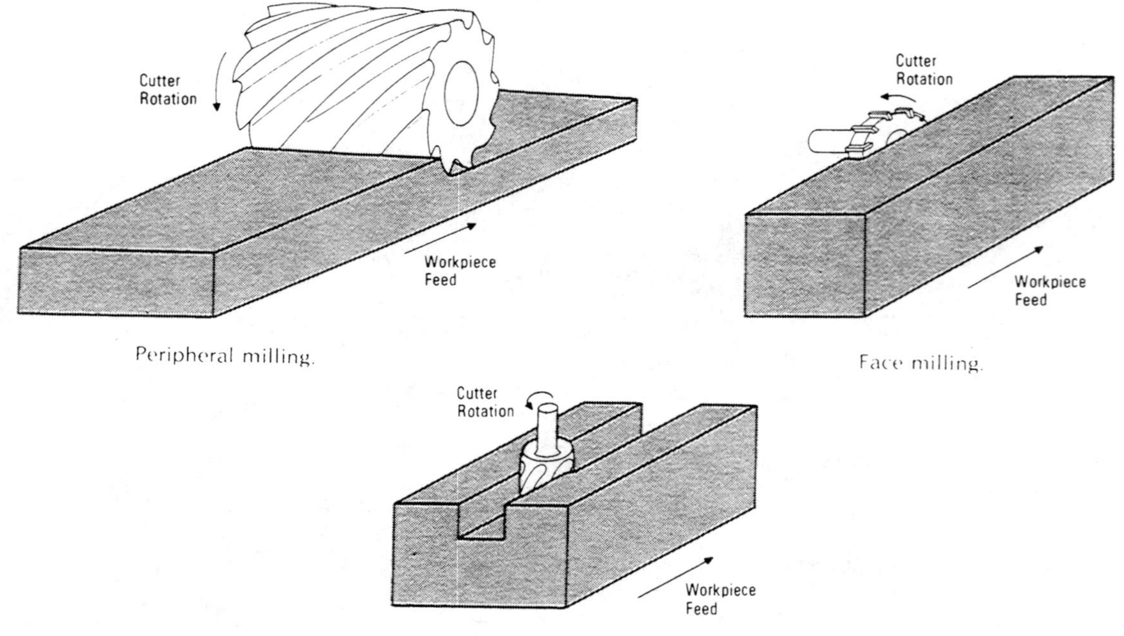

Figure 11-23 Comparison of peripheral, face, and end milling. (Courtesy of Inco Alloys International, Inc.)

at the bottom of the cut. One advantage of down milling is that the cutter forces the workpiece into the holding device so that simpler fixtures may be used, and so that a parallel cut can be maintained. This method also works better with carbide tools, or those with low cutting angles. The advantage of up milling is that scale and dirt on the workpiece are not carried into the cut.

Some typical horizontal milling cutters are shown in Figure 11-25. The cutting tool for peripheral milling is either cylindrical or disc-shaped, with cutting edges around the periphery of the disc or around the curved surface of the cylinder. The cutting edges on cylindrical tools wind around the tool in a helical pattern to reduce the cutting load on each edge and to produce a smoother milled surface. Peripheral milling with a cylindrical cutter is often called slab milling. Plain cylindrical tools are used to make wide, flat cuts. Contoured tools are used for irregular cuts. End mills are used to mill narrow cuts such as slots and keyways. Peripheral milling is usually done on machines with horizontal shafts.

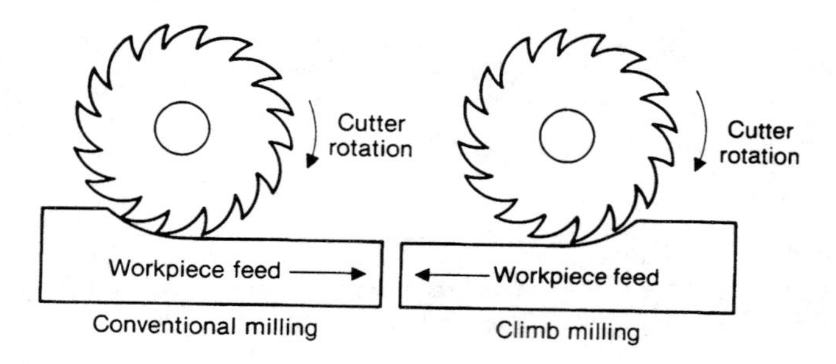

Figure 11-24 A comparison between conventional, or up milling, and climb, or down milling. (Courtesy of Inco Alloys International, Inc.)

Figure 11-25 Typical milling cutters: (a) staggered tooth cutter; (b) side milling cutter; (c) plain milling cutter; (d) shell mill; (e) single angle (RH) milling cutter; (f) double angle milling cutter; (g) convex milling cutter; (h) concave milling cutter; (i) corner rounded milling cutter.

Face milling is done with a disc-shaped tool having cutting edges on one of the flat faces of the disc, and can produce only flat surfaces. Either horizontal- or vertical-spindle machines can be used, but the spindle must be perpendicular to the surface being milled.

End milling is done with a cylindrical cutter having straight or spiral cutting edges on the cylindrical surface, much as a drill. See Figure 11-26. Cutting action can be the same as either peripheral milling or face milling, or a combination of both. End milling is usually done on a vertical-spindle machine. There are numerous different end mill shape and size configurations.

Milling fixtures. Workpieces must be fastened securely to the work table in order that precision milling may take place. Tables have T-slots in them to accommodate nuts and bolts, and special locking devices to secure fixtures in place. A common holding device is the table vise, as in Figure 11-27. This can be located anywhere

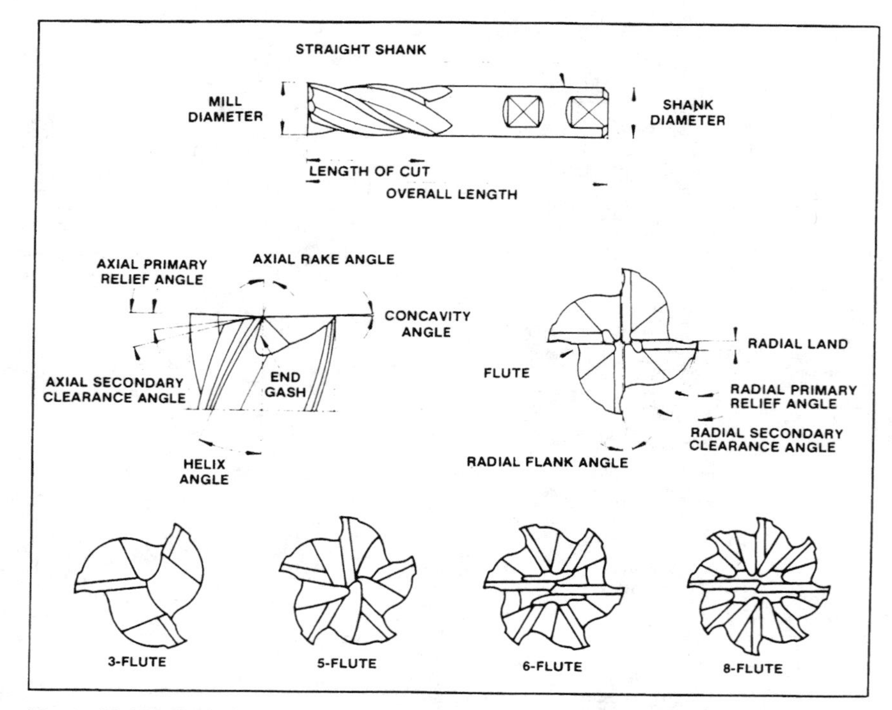

Figure 11-26 Typical end mill and terminology. Note the various flute configurations. (Courtesy of International TIC Sales Corporation)

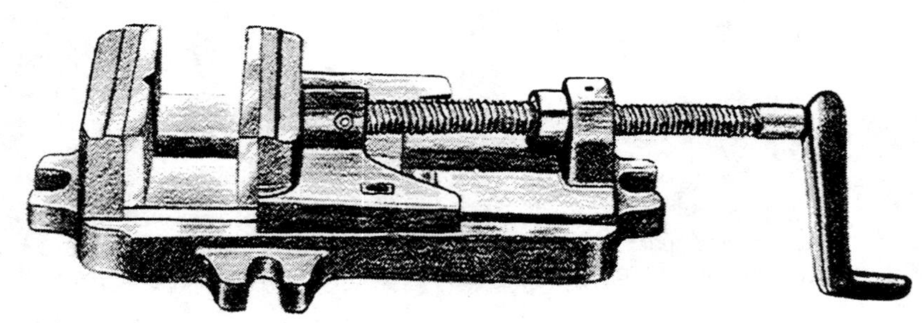

Figure 11-27 A typical milling table vise.

on a table and is held by square headed bolts which slide in the T-slots. Its use is limited to small workpieces. A variety of step clamps, step blocks, toe clamps, and high rise clamps are used to secure workpieces of any shape and size. See Figure 11-28.

MILLING MACHINE APPLICATIONS

The demands of modern manufacturing are such that new generations of sophisticated machines are constantly emerging to produce complex precision product shapes at high production rates. The computer, linked to advanced technology machines, makes it possible to produce extremely complex shapes in huge workpieces. Some striking

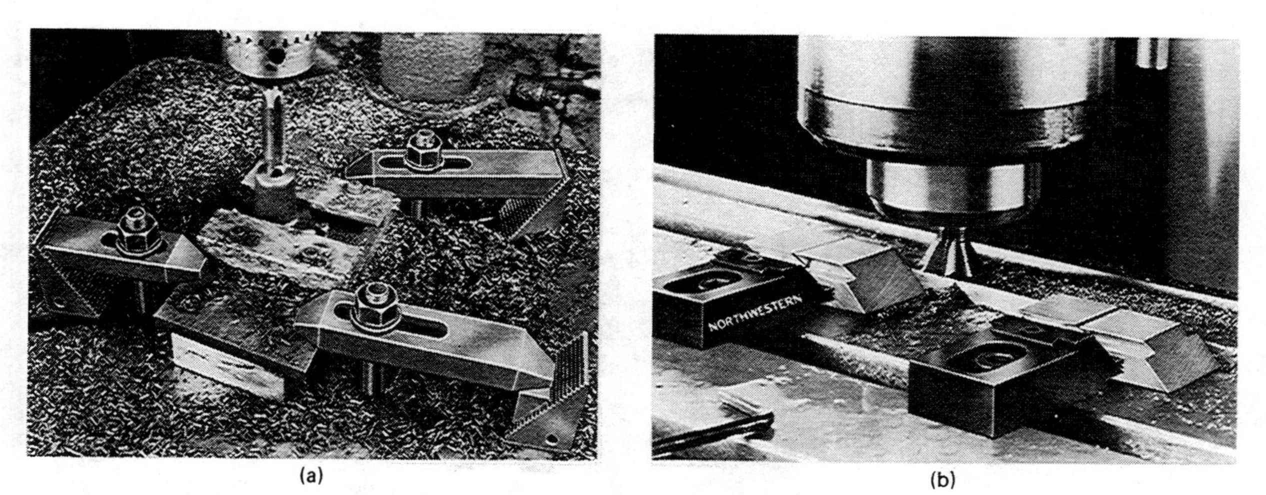

(a) (b)

Figure 11-28 Step blocks are used to secure this workpiece for a tapping operation. The steps permit a parallel orientation of the block to the workpiece (a). The toe clamp (b) locks this workpiece firmly to the table for this edge milling operation. Note the dovetail cutter in a vertical mill. (Courtesy of Northwestern Tools, Inc.)

examples are shown in Figures 11-29 and 11-30. High speed cutters are linked to special spindle heads, and CNC units direct the cutting. Some machines hold an array of tools in a vertical turret head, others in a horizontal system. All of these specialized machines depend upon computer controls to provide automatic tool selection and workpiece orientation. See Figures 11-31 and 11-32.

Planing and shaping. Planing and shaping are related machining processes whereby workpiece material removal is achieved by linear, reciprocating motion with single-point cutting tools. Consequently, these processes are used to produce

Figure 11-29 A ship's propeller blade is being machined with a special CNC mill. (Courtesy of Forest Machine Tool, a subsidiary of Brisard Machines Outils)

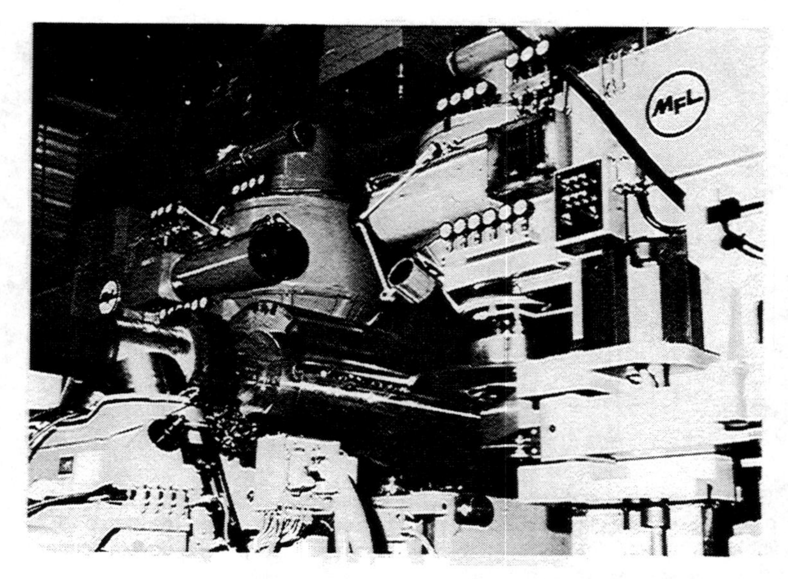

Figure 11-30 Milling slots in a rotor for heavy-duty electrical equipment. (Courtesy of Forest Machine Tool, a subsidiary of Brisard Machines Outils)

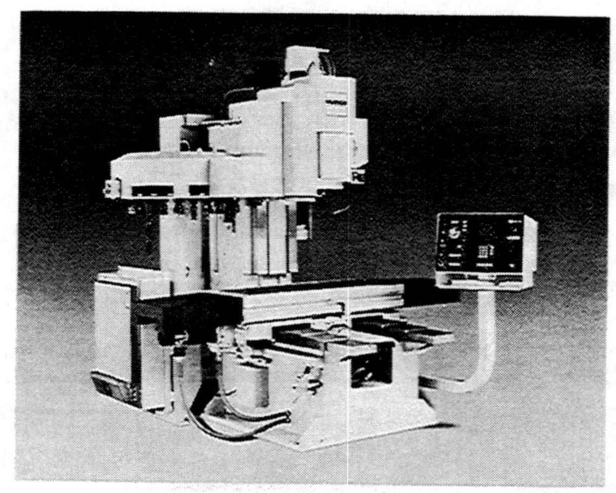

Figure 11-31 This three-axis CNC machining center carries a 24-station automatic tool changer for accurate, heavy-duty work. (Courtesy of Hurco Manufacturing Co.)

Figure 11-32 A turret head machining center. (Courtesy of MAZAK)

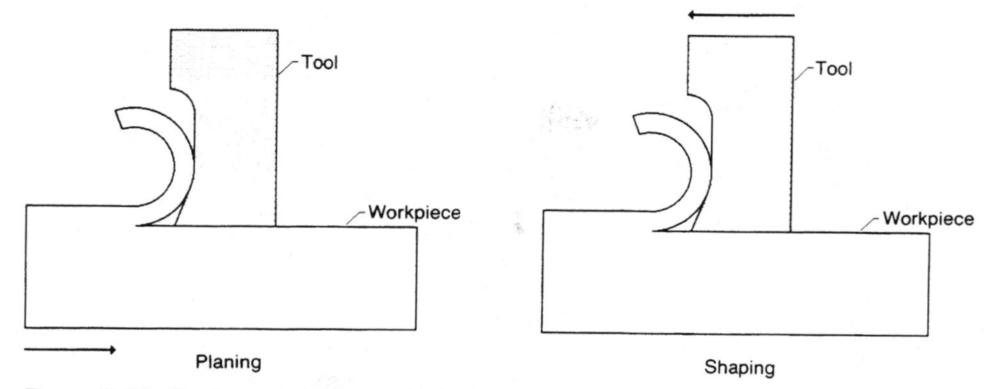

Figure 11-33 Planing and shaping cutting action diagrams.

flat, slightly contoured, or irregular workpiece surfaces. The two methods differ, however, in cutting motions. See Figure 11-33. Planing involves a stationary tool and a moving workpiece; in shaping the opposite occurs—a moving tool moves through a fixed workpiece. Most planers are very large machines for heavy-duty operations.

The *shaper* is generally smaller than the typical planer and somewhat more versatile in operation. The shaper has a reciprocating, horizontal (or sometimes vertical) ram mounted over the table of the machine, and the ram usually carries only one cutting tool. Shapers have table lengths and cutting strokes ranging from about 150 mm (6 inches) up to one meter (3 feet). They are commonly used in tool-room work. Except for gear-shaping (described later in this chapter), shaping is seldom used in production work.

Broaching is a type of machining process whereby a workpiece is cut by pulling or pushing a long tool having a series of cutting edges across the surface of the piece. See Figure 11-34. The tool, called a broach, is similar to a large file whose teeth are progressively higher from one end to the other. Each tooth removes a

Figure 11-34 A broaching operation. Note the square hole produced by pushing a square broach through a round hole. (Courtesy of The duMont Corporation)

chip of metal as it passes across the workpiece, and the surface is machined with one stroke of the tool. The size and shape of the finished cut correspond to the last tooth on the broach.

The cutting action of broaching is similar to planing or shaping, in that the tool makes straight cuts across the workpiece. The contour of the machined surface, however, is determined by the shape of the broach in some instances, and by manipulating the workpiece in others.

Broaching is normally used to shape only localized areas of a workpiece. The process is usually impractical for machining of broad surfaces. Some common applications of broaching are machining notches, slots, and keyways on external surfaces or the internal surfaces of holes, enlarging drilled holes, and machining contours or gear teeth on interior surfaces. Broaching is usually classified as internal broaching if the workpiece completely surrounds the tool during cutting, or external broaching if the tool remains on the outside surface of the workpiece. Internal broaching is often the most suitable method of machining interior surfaces, because the long tool can extend completely through a hole in the workpiece.

Broaches with contoured teeth can be used to produce almost any shape that can be made by a straight cut. Most broaches, however, are either round or flat. See Figure 11-35.

Broaching machines consist of a table or fixture to hold the workpiece and a powerful ram to push or pull the broach. The machines are classified as either vertical or horizontal, depending on the position of the broach. Most broaching machines grip the broach at both ends to keep the tool rigid. One end can be released to insert the broach through a hole for internal broaching. The machine can either push or pull the broach during the cutting stroke. Vertical broaches can cut on either the upward or downward stroke, but normally cut downward. Push broaching is usually done on a vertical machine using a short broach that is attached to the ram at only one end. Horizontal broaching machines usually have longer strokes than vertical machines and can machine heavier workpieces. Some horizontal machines use a stationary broach and move the workpiece against it. An example is the chain broach, or continuous broaching machine, in which several workpieces are mounted on an endless-chain conveyor and pulled past a stationary broach.

Figure 11-35 Round or flat broaches. (Courtesy of The duMont Corporation)

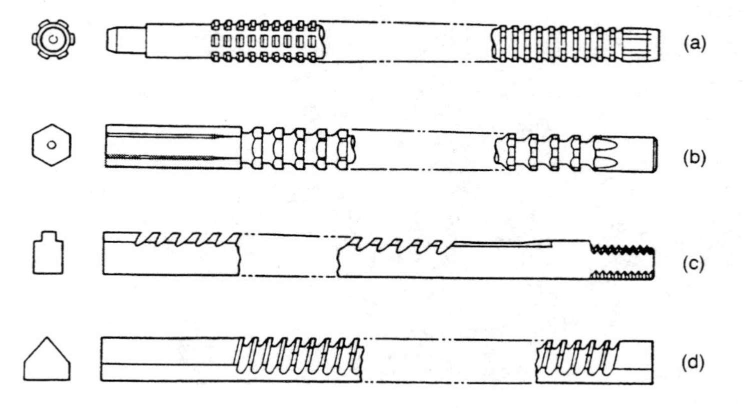

Figure 11-36 Gears. (Courtesy of The Falk Corporation)

Gear machining. Gears are machine elements which offer the most practical and dependable means of transmitting uniform angular motion. Their use in vehicles and machinery requires the precision manufacturing of gears of great size range and variety. See Figure 11-36. The most common types are the spur, the helical, the herringbone, the bevel, as well as a series of internal gears and racks. *Spur* gears, the most common type, are used to transmit motion between parallel shafts or between a shaft and a rack. *Helical* gears transmit motion between crossed or parallel shafts, or a shaft and a rack. *Herringbone* gears are double helical gears for use in heavier duty applications. *Bevel* gears transmit power between nonparallel shafts. A *rack* is a flat gear whose teeth lie along in a straight line on a plane. *Internal* gears, as the term implies, have teeth which point inward toward the center of the gear. See Figure 11-37.

Figure 11-37 Kinds of gears.

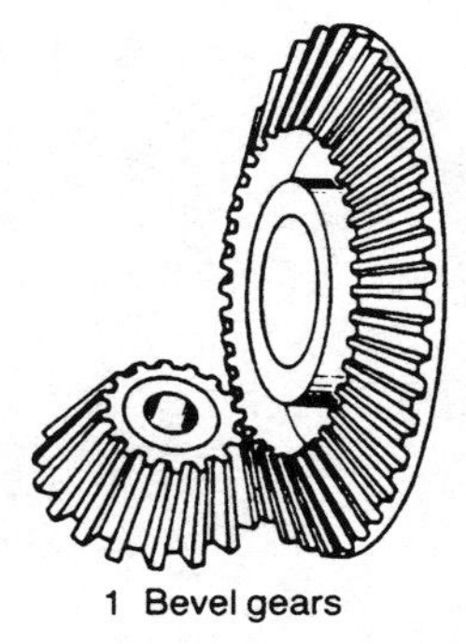

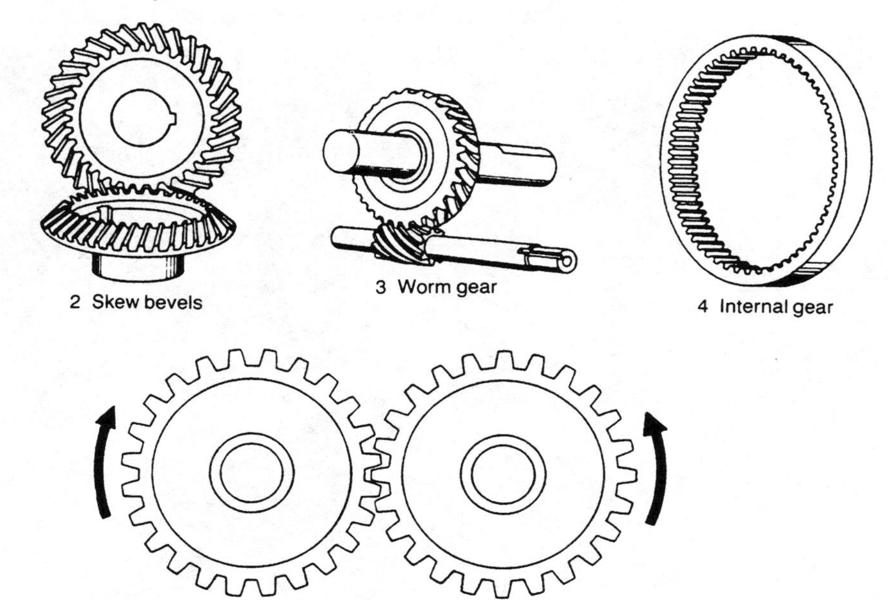

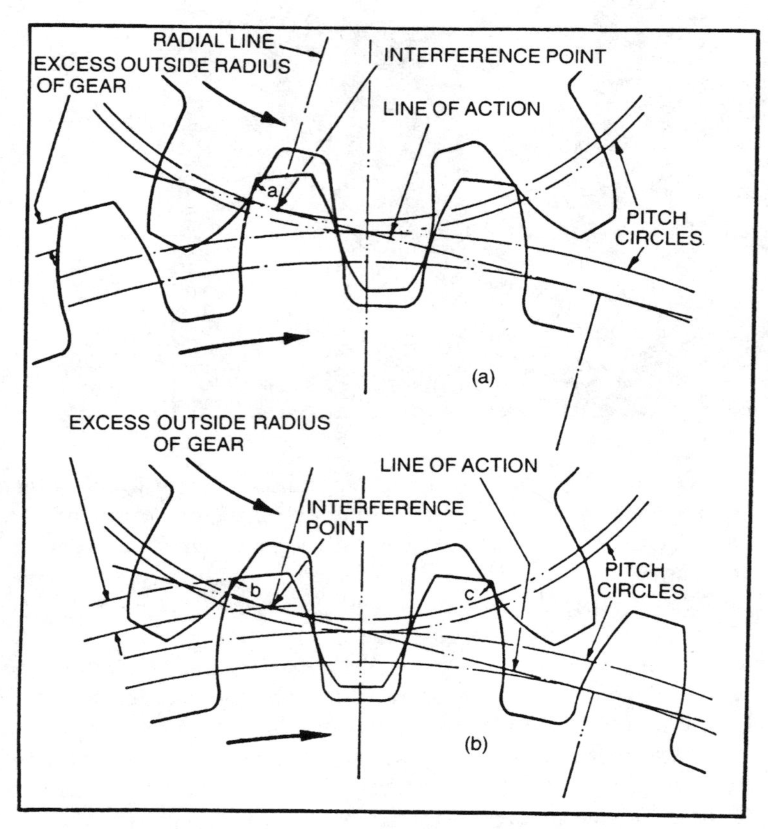

Figure 11-38 Gear teeth. (Courtesy of Fellows Corporation)

Figure 11-39 Hobbing. (Courtesy of The Falk Corporation)

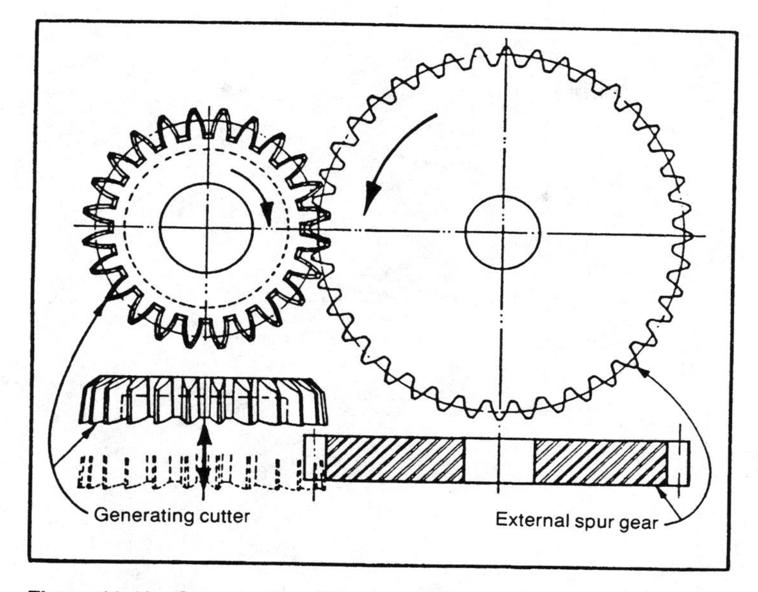

Figure 11-40 Gear shaping. (Courtesy of Fellows Corporation)

Gear teeth are complex forms which present a series of equally complex problems in order that the teeth may be cut to precise specifications. See Figure 11-38. The two principal means of high-volume precision gear production are hobbing and shaping.

Hobbing is a practical method for generating teeth on most gears except bevel and internal types. Hobbing is a process whereby the cutter (hob) and the workpiece (gear blank) revolve in constant, relative pattern as the hob is fed across the tooth surface of the gear blank. See Figure 11-39. Hobbed teeth are generated rapidly and accurately by a series of progressive cuts.

Gear *shaping* is the generation of gear teeth by the familiar reciprocating shaper action, but occurring at high speeds, and with very short strokes. This process requires a cutter which matches the tooth configuration of the final workpiece. See Figure 11-40. A typical gear shaping equipment diagram appears in Figure 11-41, and the machine in Figure 11-42. Automated gear cutting systems are shown in Figures 11-43 and 11-44.

Threading and tapping are operations by which screw threads are machined on cylindrical surfaces. Threading produces threads on external surfaces, and tapping is used to cut threads on internal surfaces. *External* threads are machined by thread cutting on a lathe, and thread chasing with a die. Both methods are widely used, and either can produce a variety of thread shapes. *Lathe threading* is done using a single-point cutting tool shaped to match the thread to be produced. See Figure 11-45. Feed controls are required to move the tool along the rotating workpiece at the proper rate to make a uniform helical cut. Several cuts along the same path may be required to finish the thread.

Die threading, or thread chasing, is done with a hollow die having internal cutters. The die is similar to a large nut with the internal threads replaced by thread-shaped cutting edges. See Figure 11-46. The end of the workpiece is placed against

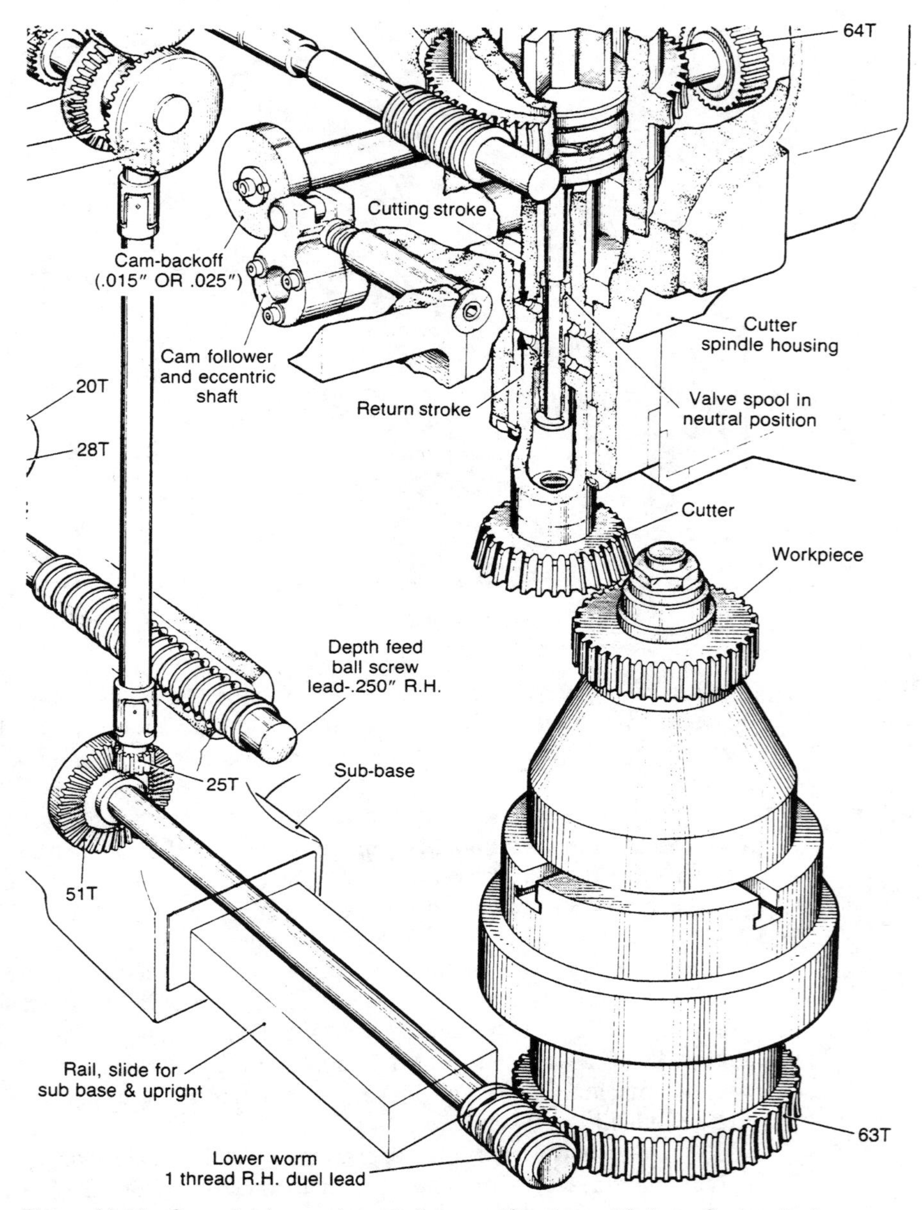

Figure 11-41 Gear shaping equipment diagram. (Courtesy of Fellows Corporation)

the opening in the die, and either the die or the workpiece is rotated as the other remains stationary. The process is the same as screwing a nut onto a bolt, except that the die cuts threads as it or the workpiece turns. Die threading is called *chasing*, because the die has a series of cutting edges, and each edge follows, or chases, the preceding edge along the same path to progressively cut the thread deeper.

Internal threads can be machined on lathes using special single-point tools, but tapping is the most common method. *Tapping* is done with a cylindrical fluted tool

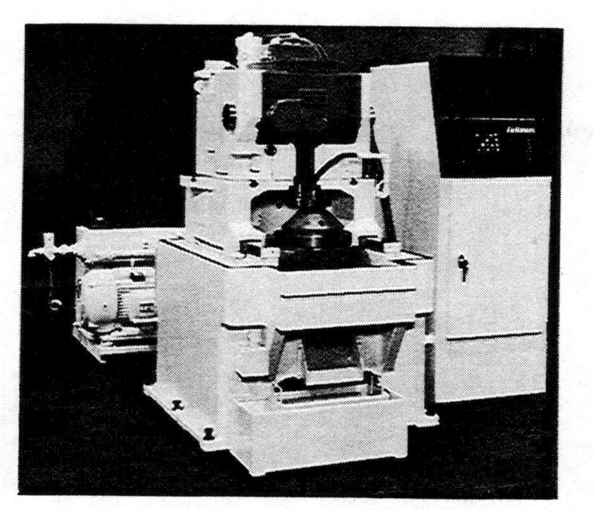

Figure 11-42 A CNC gear shaping machine. (Courtesy of Fellows Corporation)

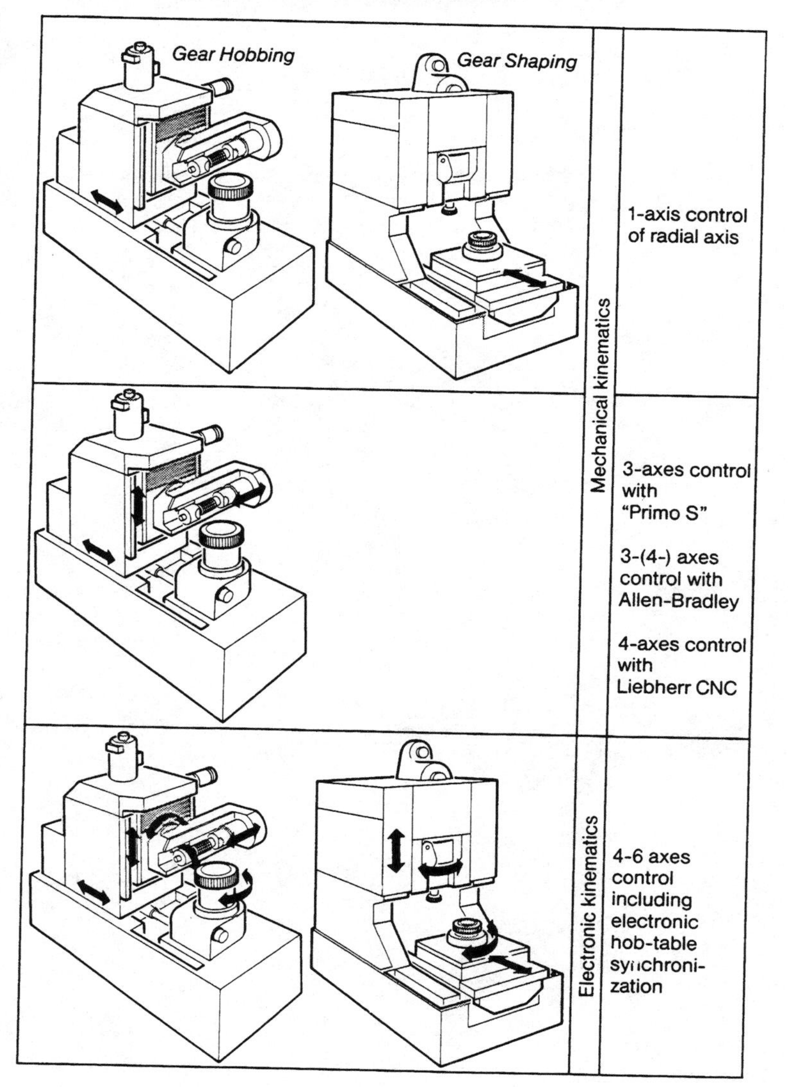

Figure 11-43 Chart of gear cutting equipment. (Courtesy of Liebherr Machine Tool)

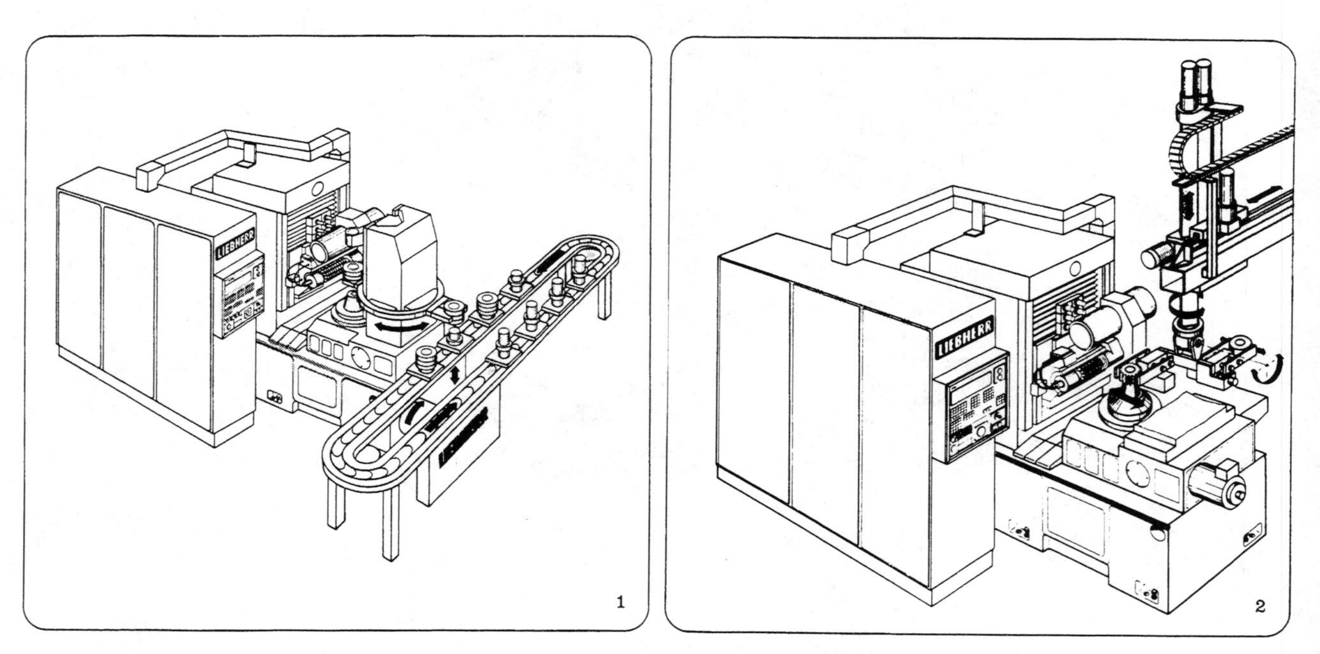

Figure 11-44 Automated gear manufacturing systems. Shown is an adjustable ring loader linked to a flexible part storage conveyor (1), and an automatic part and tool handling system utilizing a gantry robot (2). (Courtesy of Liebherr Machine Tool)

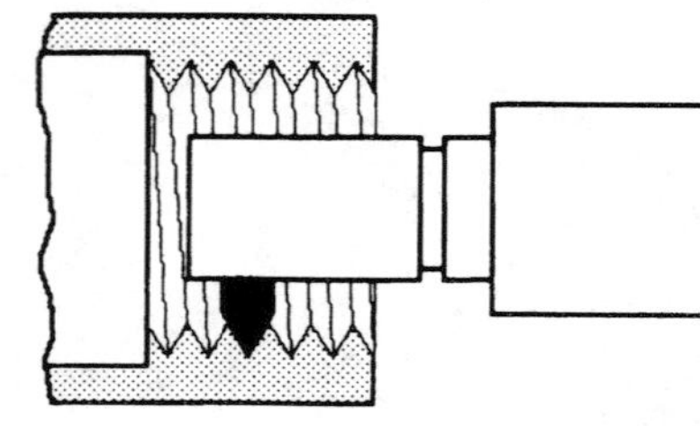

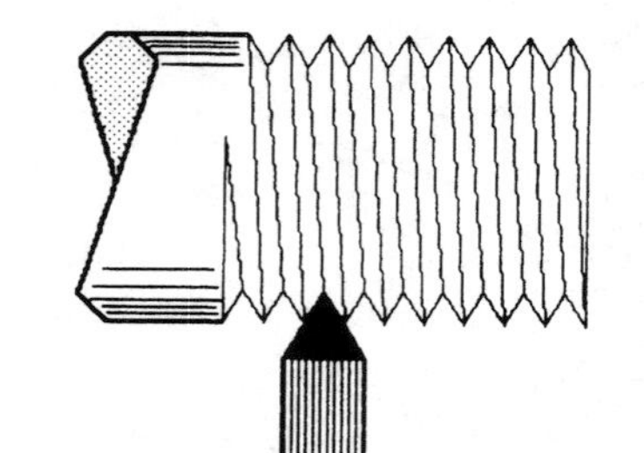

Figure 11-45 A single-point tool is used to cut both external and internal threads on a lathe.

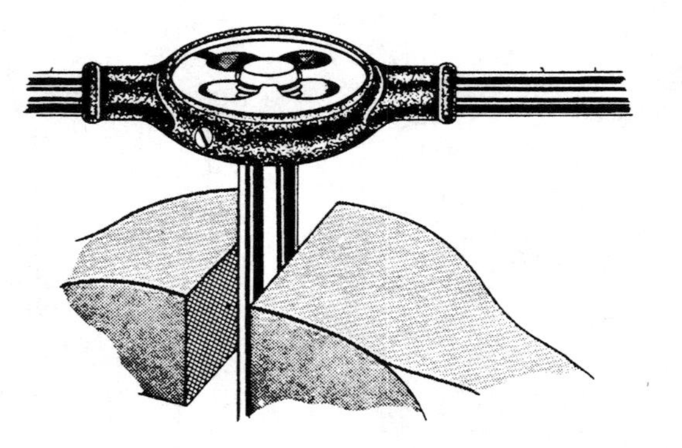

Figure 11-46 Die cutting external threads.

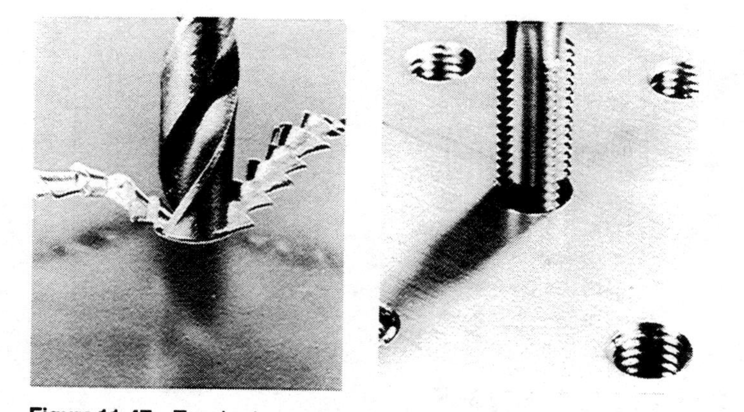

Figure 11-47 Tapping internal threads. The tap must be aligned perfectly with the drilled hole. (Courtesy of Tool Components, Inc.)

called a tap that has cutting edges shaped like threads around its surface. See Figure 11-47. Either the tap or the workpiece is rotated as the tap is fed into the hole or into the end of a tubular part. Machines used for tapping include drill presses, lathes, CNC mills, and machines similar to drill presses made especially for tapping. Threads also may be rolled and ground with appropriate equipment.

QUESTIONS

1. Describe the major difference between milling and shearing as basic cutting operations.
2. Define the terms turning, milling, shaping, and planing, and prepare simple sketches of each.
3. List and explain the two classes of turning operations.
4. Explain the following terms: depth of cut, cutting feed, and cutting speed. Prepare sketches, if necessary, to enhance your explanation. Are these terms equally applicable to milling theory?
5. Sketch a basic lathe and identify its main parts.
6. Describe the theory of indexed tool lathes.
7. Sketch a massed-tool lathe and describe its attributes.
8. Explain the primary difference between the *column and knee* milling machine and the *fixed bed* machine.
9. Explain the difference in the configuration and use of the *horizontal* and the *vertical* milling machines.
10. Prepare sketches of the following:
    a. climb milling
    b. conventional milling
    c. end milling
    d. peripheral milling
    e. face milling
11. Explain the theory of the broaching process, and list some typical applications.
12. Prepare a library research report on gear theory.
13. List four major types of gears and explain their applications.
14. Explain the difference between gear *hobbing* and gear *shaping*.

12 DRILLING AND RELATED PROCESSES

A considerable number of products require the generation of holes or openings in their surfaces. Holes can be punched, drilled, pierced, abraded, or thermally cut in almost any material. The operations to be considered in this chapter are those related to drilling. Drilling is a traditional machining process, and inasmuch as it is a material removal method, it falls into the category of the bulk or mass reduction of workpieces. See Figure 12-1. Drilling is perhaps the easiest and most common method to produce such holes. See Figure 12-2.

Drilling is the process of machining a round hole in a workpiece. The most common tool used to produce such holes is the twist drill, which is a spiral, fluted cylinder with cutting lips on one end. See Figure 12-3. The cutting force is provided by rotating the drill against a stationary workpiece, or vice versa. Chips cut at the bottom of the hole are carried away by the flutes as the drill is fed into the workpiece to progressively increase the hole depth. The flutes also serve as a passage for cutting fluids.

DRILLING TOOLS

A number of different types of drills are employed in industrial production. The selection of a specific drill type is dependent upon such factors as the kind of material being drilled, the diameter, shape, and depth of hole, whether the hole is being originated or enlarged, and the type of machinery being used. The most common drill styles are shown in Figure 12-4.

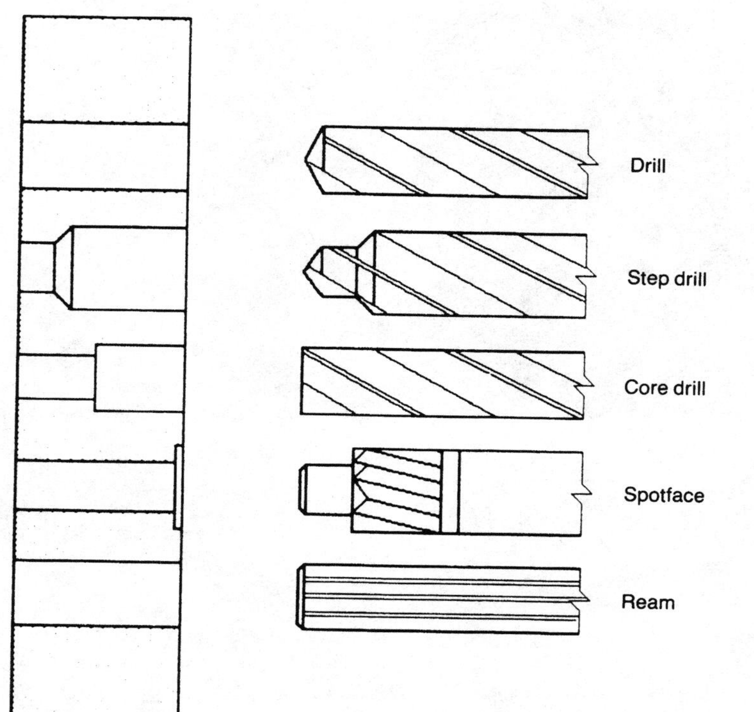

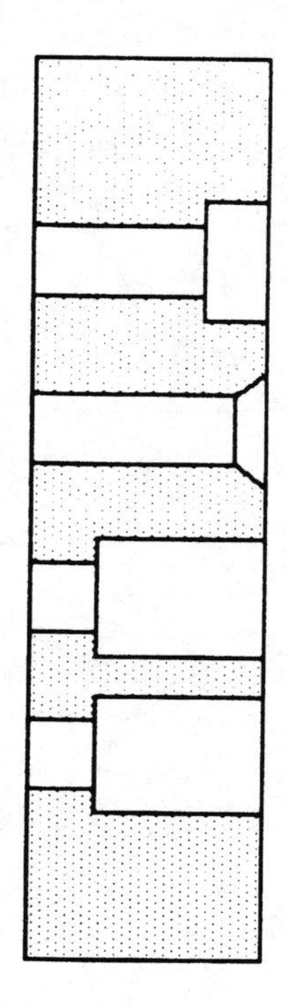

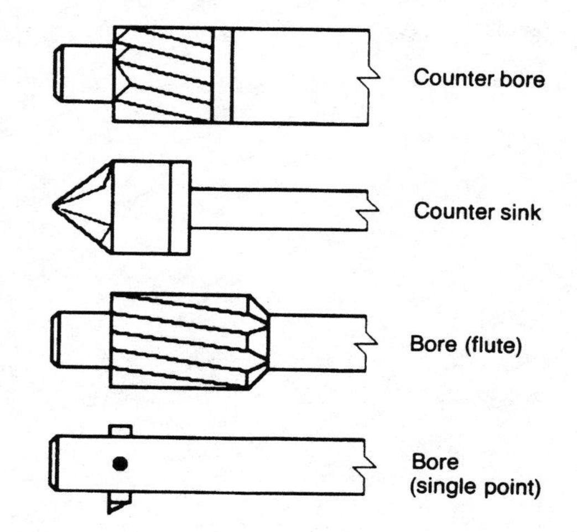

Figure 12-1 Drilling and its related processes.

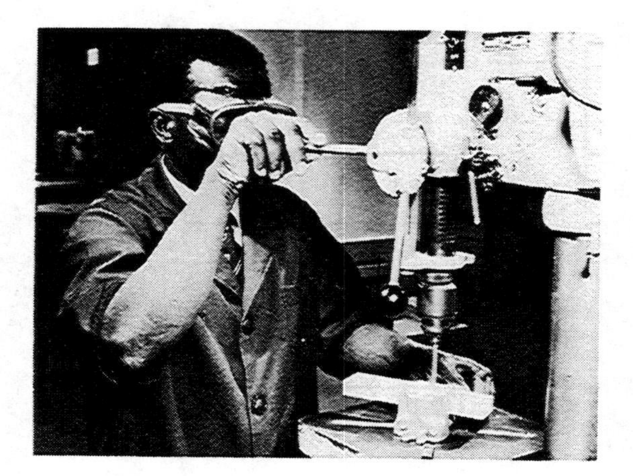

Figure 12-2 Drilling, a machining operation, is the most common method used to generate round holes in workpieces.

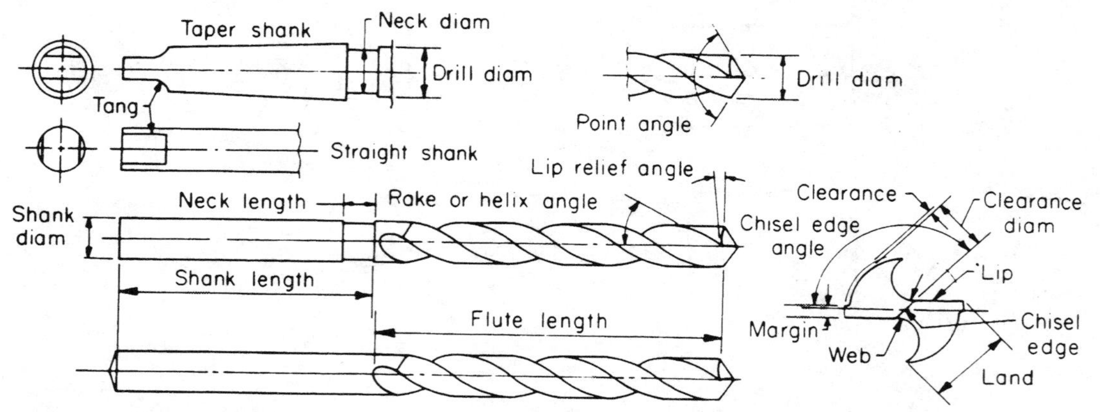

Figure 12-3 Elements of typical twist drills. (Courtesy of The Aluminum Association)

Figure 12-4 Common types of drills. Straight shanks appear on these tools, but tapered shanks are also common. (Courtesy of *Metals Handbook*, Vol. 6, 8th ed., ASM INTERNATIONAL [formerly American Society for Metals] 1971. Reprinted with permission.)

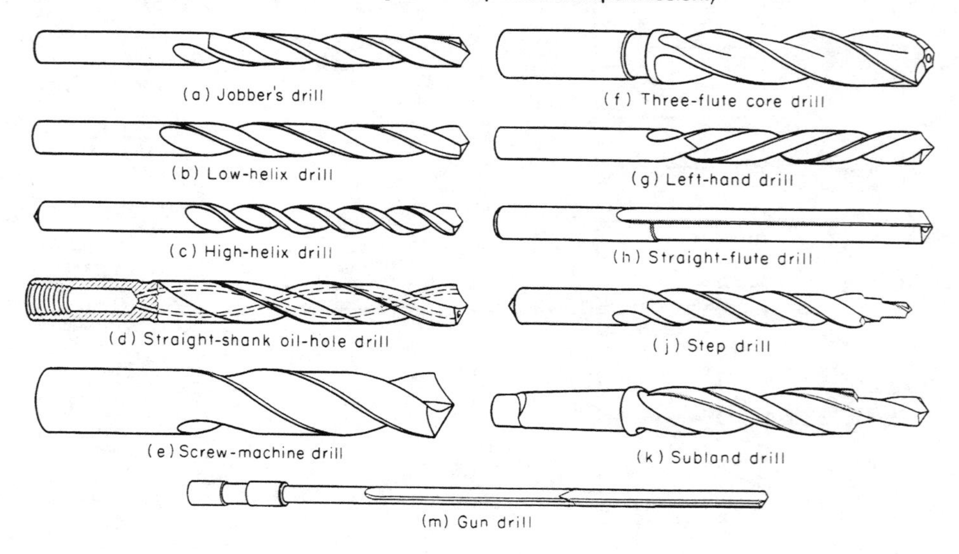

The conventional drills most commonly used in drilling tasks are called *jobbers* drills. These have two flutes, a straight shank, and their relatively short length-to-diameter ratio aids in maintaining rigidity during drilling. *Low-helix* drills, also called slow-spiral drills, have thin webs for easy penetration of soft materials. These tools are more rigid than standard drills, and their geometry causes chips to break into small pieces. This factor especially suits them to applications involving high-volume chip generation. Conversely, *high-helix* drills are fast spiral drills with wide flutes, well-suited to efficient chip removal from deep holes.

Oil-hole drills have one or more continuous body holes to permit the flow of cutting fluid under high pressure. They are particularly useful in inverted drilling operations where it is difficult to direct fluid to the cutting area. *Screw-machine* drills, or stub drills, are short, stubby, and extremely rigid. Their applications include the production of holes in tough materials with multiple-operation machines. *Core* drills have three to six flutes, and are employed to enlarge cored holes in castings, or in forgings, or previously drilled holes. *Left-hand* drills are the same as jobbers drills, but are designed to cut in an opposite or left-hand direction on a special, multiple-operations machine.

Straight flute drills are used to drill soft materials and sheet metals. Such tools do not grab or pull into the workpiece while the drilling proceeds. *Step* drills are special cutting tools with two or more diameters produced to order by grinding. These diameter variations, or steps, generally have rectangular cutting edges. These drills are used in operations requiring multidiameter holes, such as combination drilling and countersinking tools. *Subland* drills, like step drills, are combination tools which perform two or more operations within one pass. They differ from step drills in that subland tools are manufactured with the desired geometry. *Gun* drills generally are used in horizontal drilling machines to produce deep holes with a smooth finish and a minimum drift (runout) from the centerline. Typically, the drill is stationary and the workpiece rotates. *Trepanning* drills are special gun drills with one or more cutters attached to the end of a tubular shaft. The cutters do not extend across the bottom of the holes; they remove only a ring of chips and leave a solid core. As cutting progresses, the chips and the solid core pass through the hollow cutter shaft.

Drill point geometries. The actual cutting of a hole takes place at the drill point and not at the edges. While these points may look alike among an array of drills, their geometries or shapes differ according to the demands of the drilling operations and the kind of material being drilled. A conventional drill point was shown in Figure 12-3. The chisel edge acts like a blunt negative cutter, pushing the material instead of cutting it. The length of this edge is a factor of the web thickness at the drill point. Wide-web tools are rigid and therefore drill deeper, straighter, and more accurately. However, greater web thickness requires more thrust to penetrate the workpiece.

Drilling performance is enhanced by selecting the proper drill point geometries, as shown in Figure 12-5.

A. *Conventional*—industry standard; most widely used on many materials; straight chisel edge causes point to wander on the workpiece, so that a center punch is necessary; precision cuts generally require secondary reaming operations; sharp cor-

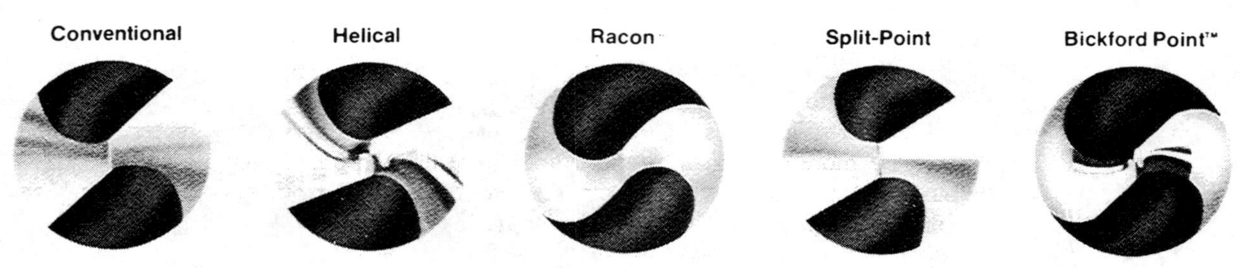

Figure 12-5 Drill point geometries. (Courtesy of Winslow ® Drill Point Grinder, Davis Tool Company, Division of Giddings & Lewis)

ners of the point break down more rapidly than other geometries; tends to produce burr when breaking through; best suited to applications where high precision and high production is not required; generally not cost-effective for machining centers and automated systems.

B. *Helical*—an S-shaped chisel and crown makes this drill self-centering; cuts very close to drill diameter, eliminating secondary operations; requires less thrust resulting in extended tool life; applicable to automated systems; produces break through burr, which must be removed.

C. *Racon*—a radiused conventional point; tool life 8 to 10 times that of conventional points; widely used in industries where drill life is a prime requirement, such as automotive; curved lips produce less heat; no burr on breakthrough; high feed rate; not self-centering; best used with guide bushing; less wear on drilling machines.

D. *Split*—a common variation of the conventional point; also called a crankshaft drill; primarily used in deep-drilling operations; self-centering; cutting edge acts as chip breaker to facilitate coolant flow; can be used on curved surfaces; must be ground on a special point-splitting machine.

E. *Brickford*—self-centering; burr-free breakthrough; high feed rate; long tool life; applicable to automated systems; combines attributes of Helical and Racon points; tendency to grab workpiece during breakthrough; not generally used in hand-held drilling tool operations.

Modern drilling operations, and the drills used to perform them, are dependent upon accurate high-quality drill grinding machines. These machines automatically grind the required point geometries. See Figure 12-6.

MISCELLANEOUS DRILLING TOOLS

Other tools are available for special drilling operations. See Figure 12-7. *Spade drills* are flat, pointed tools having two cutting edges attached to the end of a shaft or tool-holder. Spade drills have strong cutting edges and are often used for large-diameter holes, from 38 to 380 mm (1.5 to 15 inches). Similar *power bits* are used to drill holes in wooden parts. Large holes are produced in sheet materials, especially metal, with the *hole saws* and *circle, or fly, cutters*. These tools do not tear or dig into the workpiece, as would an ordinary twist drill. *Auger, lock set, expansive*, and *brad point bits* are used for producing various sized, smooth, accurate

Figure 12-6 Drill grinding machine. (Courtesy of Winslow ® Drill Point Grinder, Davis Tool Company, Division of Giddings & Lewis)

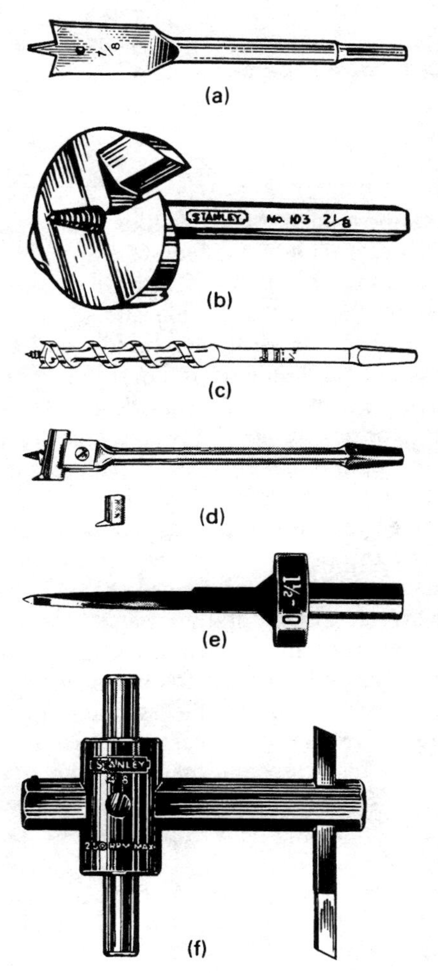

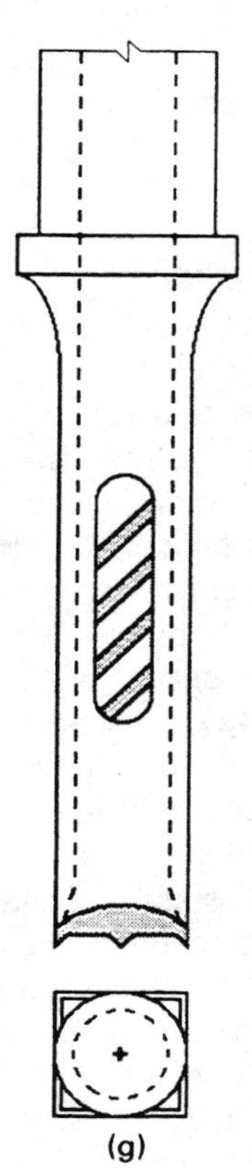

Figure 12-7 Drilling tools. (a) Spade drill. (b) Power bit. (c) Auger bit. (d) Expansive bit. (e) Screw point. (f) Circle cutter. (g) Mortice hollow chisel for drilling square holes in wood. (Courtesy of Stanley Tools)

holes in wood. Auger and expansive bits are essentially hand tools; the others are applicable to production and automatic systems.

DRILL MATERIALS AND SIZES

Production quality drills are made of high speed steel; carbon steel drills are cheaper, have shorter work lives. Some drills are tipped with TiN (titanium nitride) coatings or carbide tips to improve drilling capabilities and to hold their cutting edges longer.

Drills are manufactured in sizes denoting the body diameter of the tool and, therefore, of the size hole it will produce. These drills are available in fractional inch, numbered, lettered, and millimeter sizes. The common range of fractional sizes occur in 1/64 inch increments up to one inch. Larger fractional drills are made in sizes up to 3.5 inches. Numbered and lettered sizes range from 0.0135 to 0.413 inches in diameter, and fall between the fractional increments so that there is only a few thousandths of an inch difference between one drill size and the next in the range up to 0.5 inches. Metric drills of corresponding sizes are available. Wood bits come in sizes from 1/8 inch up to three inches, generally by 1/16 inch steps. Again, corresponding metric sizes are available.

DRILLING MACHINES

The standard drill press is the most widely used drilling machine. See Figure 12-8. The elements include a horizontal table to support the workpiece and a vertical, power-driven spindle directly above the table. The drill is secured to the spindle by a jawed chuck or tapered sleeve. Both the table and spindle are mounted on a column. To provide cutting feed for the drill, the spindle can be lowered through its housing. Vertical movement of the spindle is controlled either manually by a lever or automatically by driven gears. Manual-feed drill presses are called *sensitive* drill presses, because the operator determines the proper feed for the drill by feeling the cutting action through the control level. These machines generally are used with drills under 16 mm (5/8 inch) in diameter. See Figure 12-9.

Drill press sizes or capacities are determined by the largest diameter disk that can be drilled at the center of the press. Other specifications include the largest drill that can be accommodated. Drilling machines are available in many forms and sizes. Manual or power-driven portable tools are used for assembly operations and maintenance work. Machines used in production are described following.

Upright or vertical drill presses are designed for heavy work. The worktable is power-driven, as in the feed mechanism, and the spindles may be driven by V-belts or gears. Large drills up to 75 mm (3 inches) in diameter are as a rule used on these presses, and the drills have tapered shanks which fit into Morse taper holes in the spindle ends. See Figure 12-10. *Gang* drill presses are in effect a series of vertical presses in a row mounted on a single base, to speed the production of drilled parts.

Radial drill presses are designed for very large workpieces that cannot be moved easily for multiple drilling operations, or are too heavy for other types of presses.

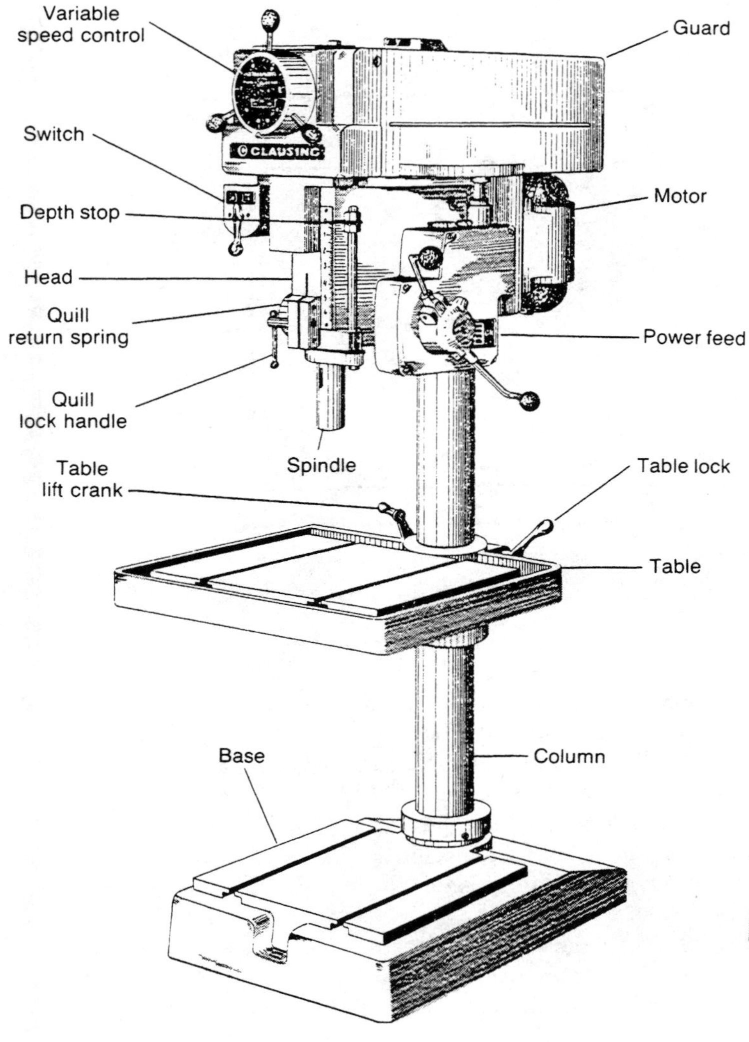

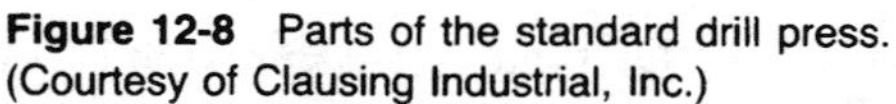

Figure 12-8 Parts of the standard drill press. (Courtesy of Clausing Industrial, Inc.)

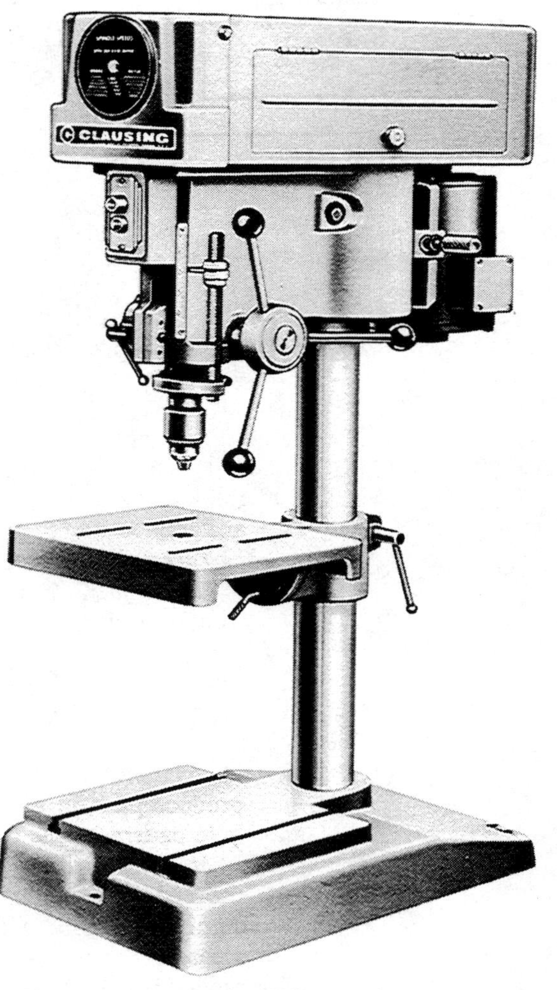

Figure 12-9 A standard drill press is commonly used for low volume production. (Courtesy of Clausing Industrial, Inc.)

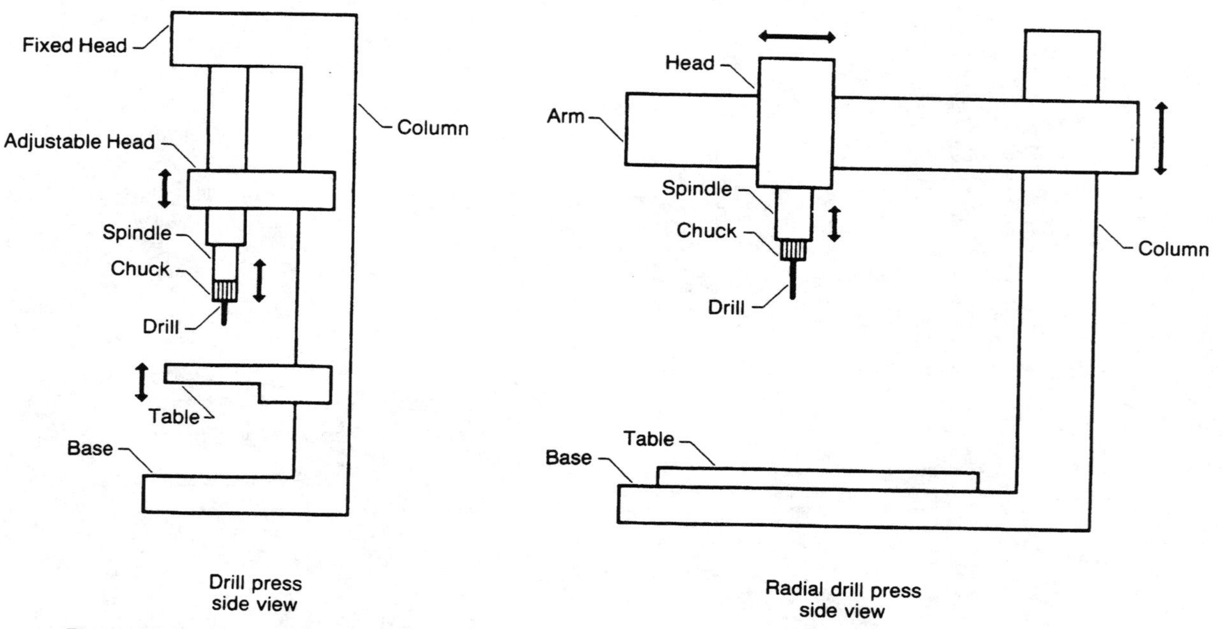

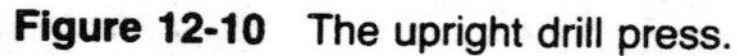
Figure 12-10 The upright drill press.

Figure 12-11 The radial drill press.

See Figure 12-11. The machine has vertical column, supported by a sturdy base, which carries a horizontal arm or beam. The spindle is held by a moveable drilling head. The head can move along the arm, and the arm can swivel on the column to permit positioning the drill over any part of the table. *Plain* radial machines drill only in the vertical plane. The head can be swivelled on the arm to produce holes at angles in a vertical plane on *semiuniversal* machines. *Universal* machines can drill holes at any angle by means of additional adjusting mechanisms on the head and arm.

Turret drill presses function like gang drills but require less space. See Figure 12-12. A number of drilling tools are stored in a rotary turret for quick and easy indexing into cutting position. Turret machines are especially adaptable to numerical control systems.

Multiple-spindle drill presses have several spindles arranged over one table. See Figure 12-13. Holes of various diameters, or different configurations such as a countersink or counterbore, can be drilled by moving the workpiece from one spindle to another. Multiple-spindle machines operate several drills at the same time to produce parts in great quantities. They are used for one-step drilling on a specific hole pattern in one or several workpieces.

Horizontal drilling machines are similar to lathes, and are used to drill holes in the ends of round rods, to produce hollow shafts, or to bore holes in wooden furniture parts. Some machines rotate the workpiece against a stationary drill, an arrangement that severely limits the position of the hole to the center of the piece. See Figure 12-14.

Reaming. *Reaming* is a process used to provide a smooth finish on the inside surface of a drilled hole, and to make the size of the hole more accurate. A rotary

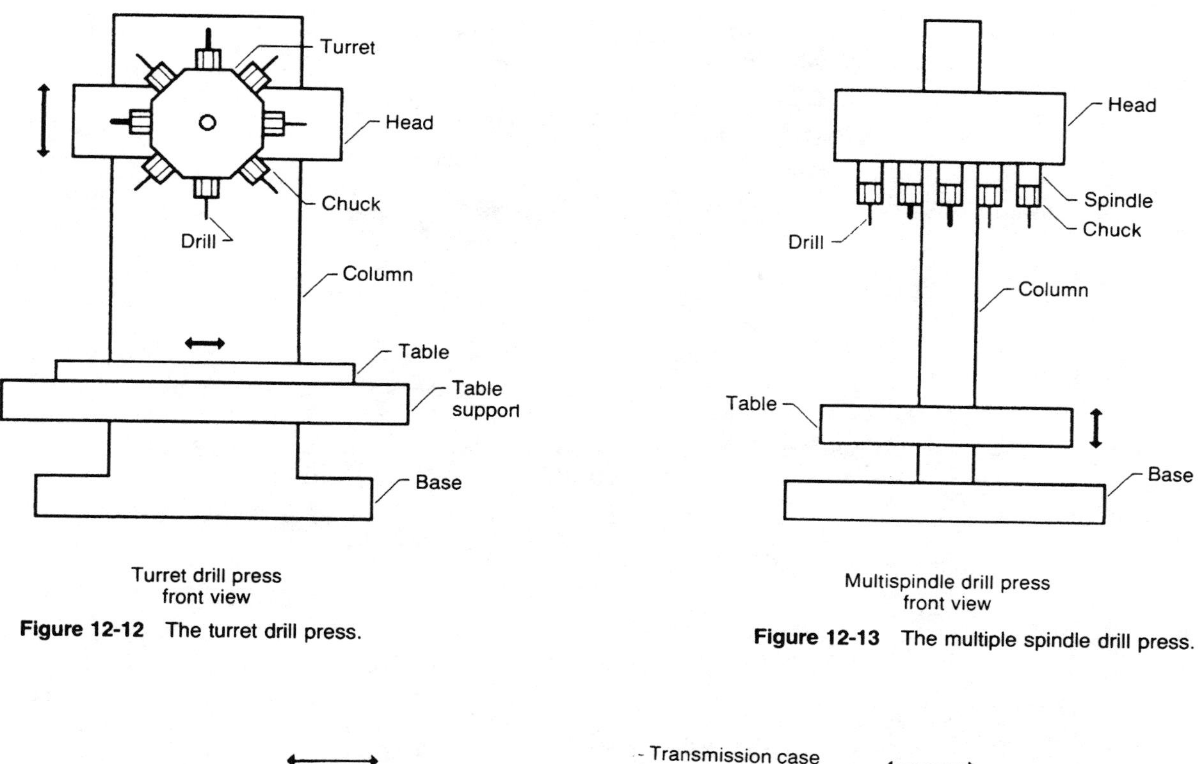

Figure 12-12 The turret drill press.

Figure 12-13 The multiple spindle drill press.

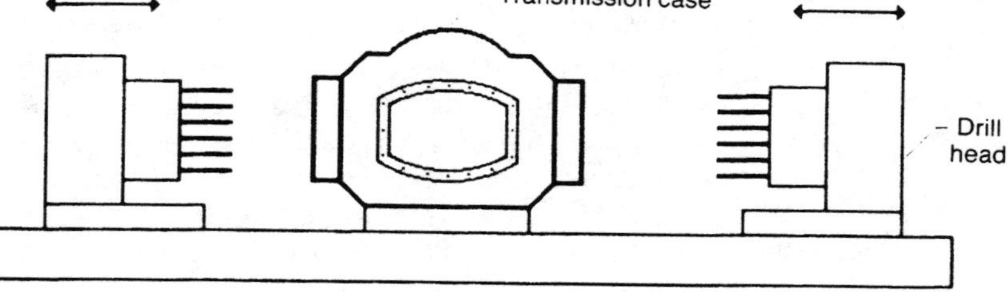

Figure 12-14 This multiple spindle horizontal drilling machine produces holes in both sides of this tractor transmission case.

cutting tool called a reamer is employed to produce a light cut on the wall of the hole. Reaming is similar to boring, but is usually used for smaller holes, and unlike boring, it cannot be used for heavy cuts or for truing. Reaming can be done by hand, or with any machine tool that can be used for drilling, as shown later in this chapter. Either the reamer or the workpiece is rotated as the tool is fed into the hole. Most reamers are cylindrical, with multiple cutting edges around their surfaces. The cutting edges are provided by flutes and can be straight or helical. See Figure 12-15. Two other common reamers are the flat reamer and the gun reamer. The flat reamer is a flat tool having cutting edges on two opposite sides. Gun reamers are cylindrical with a single flute and cutting edge. Adjustable reamers are also available.

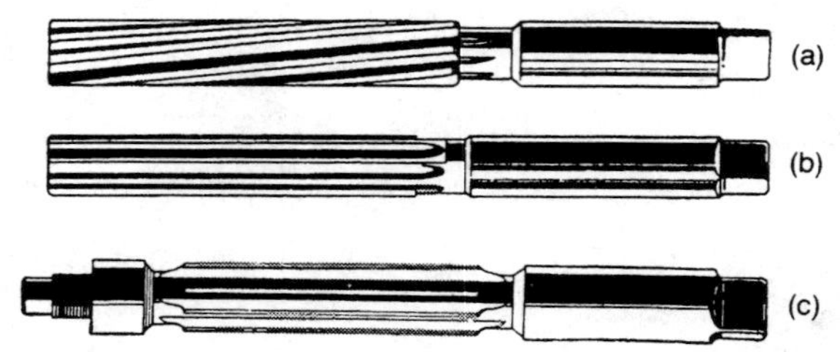

Figure 12-15 Common reamers. (a) Spiral flute. (b) Straight flute. (c) Expansion.

Boring. Boring is a technique for truing, sizing, or shaping the inside surface of an existing hole in a workpiece. The operation is performed by rotating a single-point cutting tool against the wall of the hole, and at the same time feeding the tool through the hole, or by rotating the work against a stationary tool. Cylindrical shaped boring tools are also used. Boring is used to enlarge the hole to smooth the wall, or to make the size of the hole more accurate. Holes too large to be made by drilling alone can be started by drilling a smaller hole and then completed by boring. Boring is frequently used as a finishing operation in as-cast and hot-pierced holes. Boring can also be used to cut grooves or other shapes around the wall of the hole.

Boring tools. While much boring is done with single-point tools, boring tools with multiple cutting edges are widely used. See Figure 12-16. These special rotary tools are used in such common operations as counterboring, countersinking, and spotfacing. Refer to Figure 12-1. *Counterboring* and *countersinking* are operations which enlarge or shape the openings of holes to provide space for a bolt head or nut. Both operations are similar to drilling and are performed on machines used for drilling. Counterboring produces a cylindrical enlargement around the opening of the hole, leaving a bottom shoulder perpendicular to the hole. The cutting tool, called a counterbore, has cutting edges on the face of one end and a smooth cylindrical extension in the center of the face. The extension is called a pilot. It is slightly smaller than the hole in the workpiece and serves as a guide to keep the cutting edges centered on the hole.

Figure 12-16 This array of special boring tool shapes is used in both vertical and horizontal machine applications. (Courtesy of Eclipse Industrial Products, Inc.)

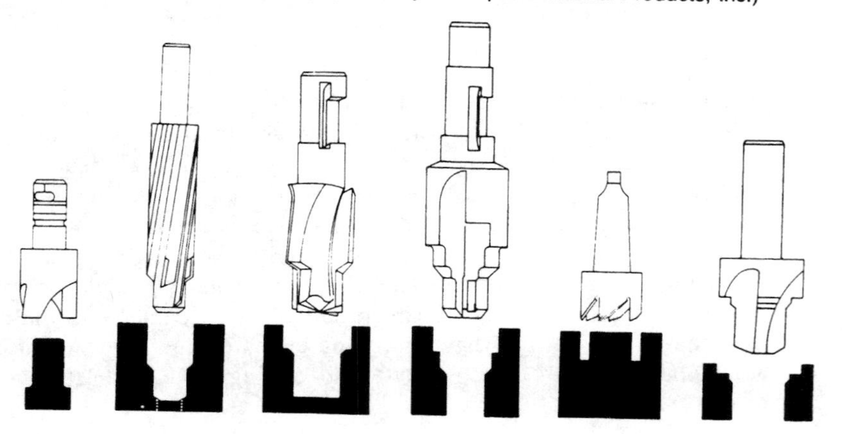

Counterboring tools as well as special tools are also used for spotfacing. Spotfacing is done by making a large-diameter shallow cut around a hole in a rough or curved surface. The operation provides a flat seat for a bolt, nut, or washer.

Countersinking produces a tapered enlargement around the opening of the hole. A cone-shaped cutting tool called a countersink is used. Many countersinks do not have pilots, since the edges of the hole support the cutting edges of the tool. These cuts are made to accept flat-headed fasteners. Numerous other specially shaped tools are designed for unique applications.

Boring machines. Boring is usually done either on a lathe or on a special machine tool called a boring mill. Lathes are used to bore holes in the ends of or through cylindrical parts where the workpiece is held by one end in the chuck of the lathe and rotated as the cutting tool is fed into the hole. See Figure 12-17.

Boring mills are of two types: vertical and horizontal. The *vertical* boring mill, also called a jig bore, is essentially an engine lathe turned on end. See Figure 12-18. The workpiece is clamped to a revolving table, with the hole to be bored centered under a vertical bar to which the fixed cutting tool is attached. Cutting force is supplied by the rotation of the workpiece. The vertical boring bar is automatically moved downward to feed the tool into the hole. *Horizontal* boring mills have a rotating, horizontal boring bar that operates a stationary workpiece, as shown in Figure 12-19. The cutting tool is fed through the hole by the horizontal movement of either the boring bar or the table which holds the workpiece. This very versatile machine can accommodate up to 90 different tool heads and a range of workpiece sizes and configurations. Programmable controllers may be used to direct the part orientation and the selection of appropriate tools. See Figure 12-20.

Figure 12-17 A typical lathe-boring operation.

Figure 12-18 The vertical boring machine.

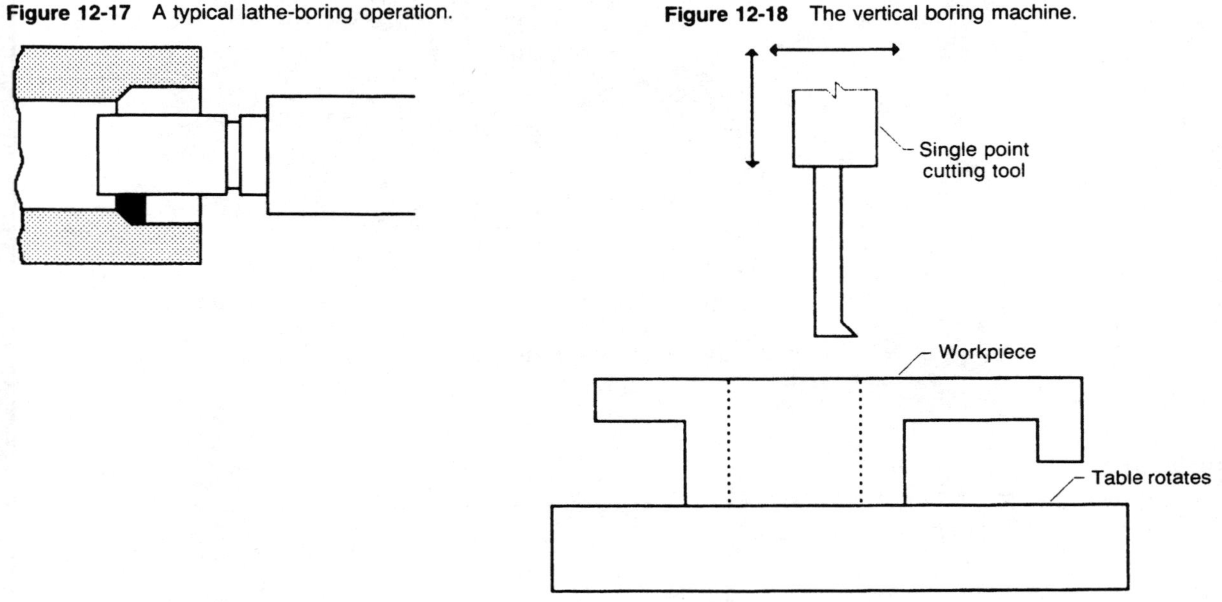

Figure 12-19 This horizontal boring machine can be used with boring attachments, or with multiple drills, as shown here. (Courtesy of Cincinnati Milacron)

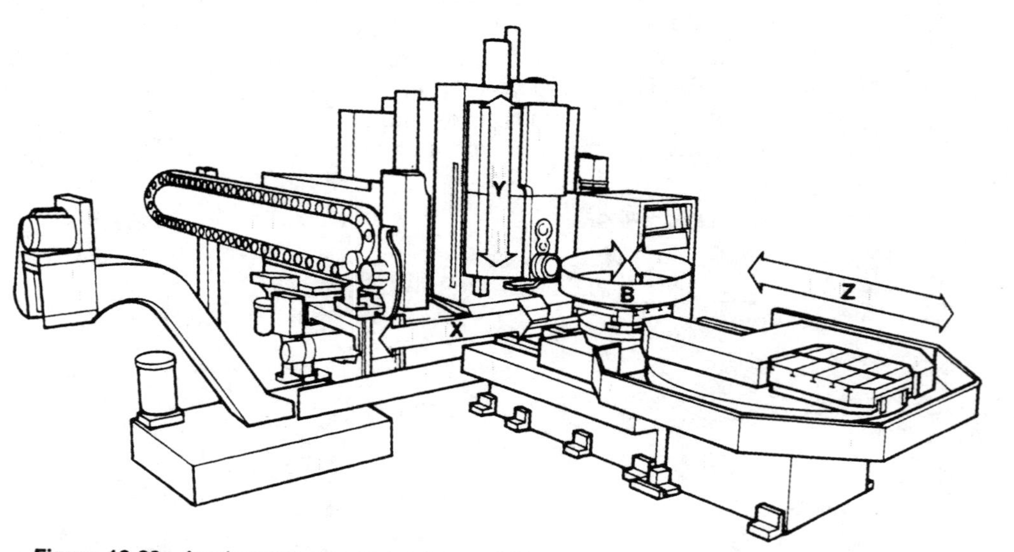

Figure 12-20 A schematic of a horizontal machining center, showing the work axes and a single pallet 45-tool automatic tool-changer. (Courtesy of Cincinnati Milacron)

QUESTIONS

1. Define or describe the following processes, identifying similarities, differences, and applications.
   a. Counterboring, countersinking, and spotfacing
   b. Drilling, core drilling, and step drilling
   c. Boring and reaming

2. Sketch the standard twist drill and identify the important parts.
3. Describe and compare the uses of these drills:
   a. Low helix and high helix
   b. Jobbers and step
   c. Straight flute and gun
4. Select two drill point geometry styles, describe them, and compare their uses.
5. Describe the unique characteristics of drilling tools designed expressly for use on wooden workpieces.
6. Describe the rationale for fractional, numbered, and lettered drills.
7. Explain the attributes of turret and multiple-spindle drilling machines in automated production applications.
8. Define boring, and list the kinds of machines used in boring operations.
9. Sketch some examples of boring tools.
10. Prepare a library research report on current developments in automatic drilling and boring equipment.

13 ABRASIVE MACHINING

Abrasive machining is a category of machining operations used to remove material by the action of rapidly-moving abrasive grains forced against a workpiece. See Figure 13-1. It is a mass- or bulk-decreasing series of processes. The abrasive is usually a rotating wheel or belt made up of small, rough grains of an extremely hard substance, such as aluminum oxide or silicon carbide. The abrasive particles in a grinding wheel are bonded together with a softer substance such as resin, vitrified ceramic, or rubber. Each of the particles on the surface of the wheel acts as a miniature cutting tool to remove a tiny chip of material. As grinding progresses, the particles on the outside of the wheel become dull and are worn away to expose fresh cutting edges.

The various grinding methods essentially duplicate the cutting action of other machining operations. Grinding produces the same shapes as turning, planing, milling, and other common metal-cutting methods, but removes the metal with a grinding wheel instead of a cutting tool. The principal advantages of grinding are much smoother finished surfaces, and more accurate shapes and dimensions on the completed part. Grinding, however, is much slower than other machining methods. Many items are produced by machining the workpiece almost to completion by standard cutting methods and then finishing the part by grinding. The most important advantage of grinding is that it can shape extremely hard materials that would be difficult to machine by other methods. Abrasive cutting methods are also employed in mechanical finishing operations. See Chapter 21. The primary difference is that abrasive machining is a precision process, whereas mechanical finishing generally is not.

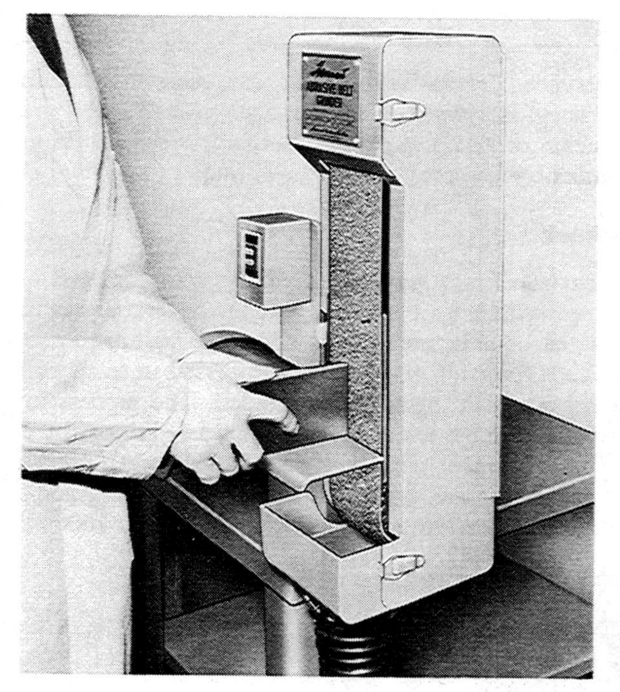

Figure 13-1 The abrasive belt grinder is one type of abrasive machining equipment. (Courtesy of Hammond Machinery, Inc.)

TYPES OF ABRASIVES

Silicon carbide (SiC) is a by-product of the reaction between high purity silica sand and coke in a resistance-type furnace at temperatures in the range of 2,040°C (3,700°F). Measuring 9+ on the Mohs' scale, silicon carbide is one of the hardest abrasives known to industry, exceeded only by diamond and boron carbide. Silicon carbide is unique in its ability to cut or grind ceramics, silicon, quartz, and other extremely tough materials. It is generally used to grind cast iron and non-ferrous metals. The inherent friability of the sharp crystalline structure assures the continuous exposure of new cutting edges to produce fast clean cutting action.

Aluminum oxide (Al_2O_3) is an inorganic compound produced by the fusion of alumina ore in an electric arc furnace at temperatures in excess of 2,000°C (3,632°F). Registering 9 on the Mohs' scale, fused aluminum oxide ranks just below silicon carbide in hardness. These characteristics produce an abrasive that is very hard, tough, and slow to dull. The difficult-to-fracture crystalline structure of an aluminum oxide makes it the most common abrasive for a variety of uses. It has the capability of producing fine, scratch-free finishes on a broad range of steel alloys.

Cubic boron nitride (CBN) is another of the manufactured abrasives, and was developed as an outgrowth of synthetic diamond research. With a Mohs' value of 9+, these abrasives approach the diamond in hardness, and are used to grind hardened carbon and alloy steels, stainless steel, and tool and die steels. The three manufactured abrasives—SiC, Al_2O_3, and CBN are preferred for industrial grinding operations.

Corundum, *emery*, and *flint* are naturally occurring abrasives, and find limited use in industrial abrading processes.

Three basic types of abrasive products are used in industrial abrasive machining operations. They are *grinding wheels* and other bonded grain structures such as bars and blocks; *coated abrasives*, occurring as sheets, drums, and belts; and *free abrasive grains* used in pressure blasting, honing, and flow abrading.

Grinding Wheels

A grinding wheel is a multiedged cutting tool comprised of numerous abrasive grains held in position by a bonding agent. The primary types of abrasives used in such wheels are the aluminum oxide, silicon carbide, and cubic boron nitride previously described. *Abrasive grain size* is expressed in terms of the mesh (screen size) by which abrasive grain particles are graded. The process involves sifting crushed abrasive material through screens of various sizes. The ultimate grit size is determined by the number of meshes per linear inch in the screen used to separate the various grains. Coarse grains are expressed in lower numbers and finer grains in higher numbers. Sizes of abrasive grit range from 4 through 1000. (See Figure 13-2.)

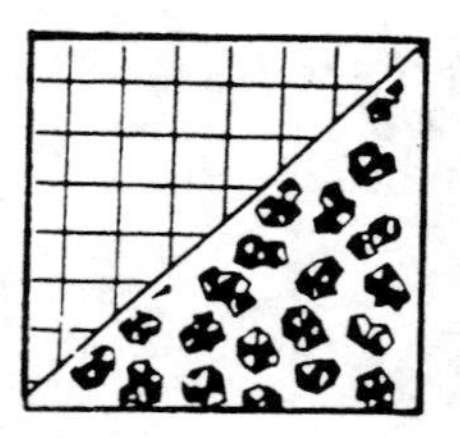

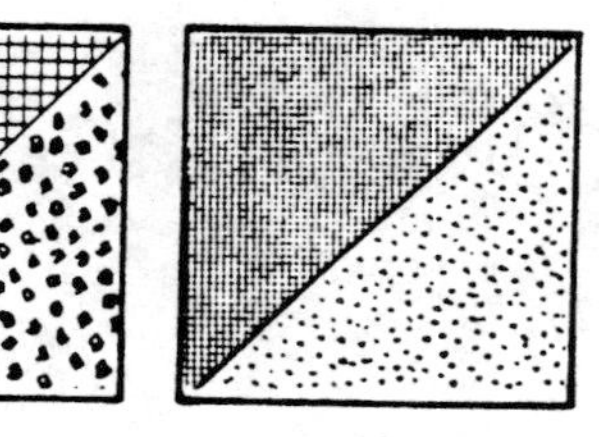

Figure 13-2 Typical screens through which grain sizes 10, 24, and 60 have been sifted. (Courtesy of Grinding Wheel Institute)

Figure 13-3 Typical identification markings for grinding wheels of the A1203 and SiC types. Similar marks are used for CBN and diamond wheels. Manufacturers also use special identification symbols with the elements on this chart.

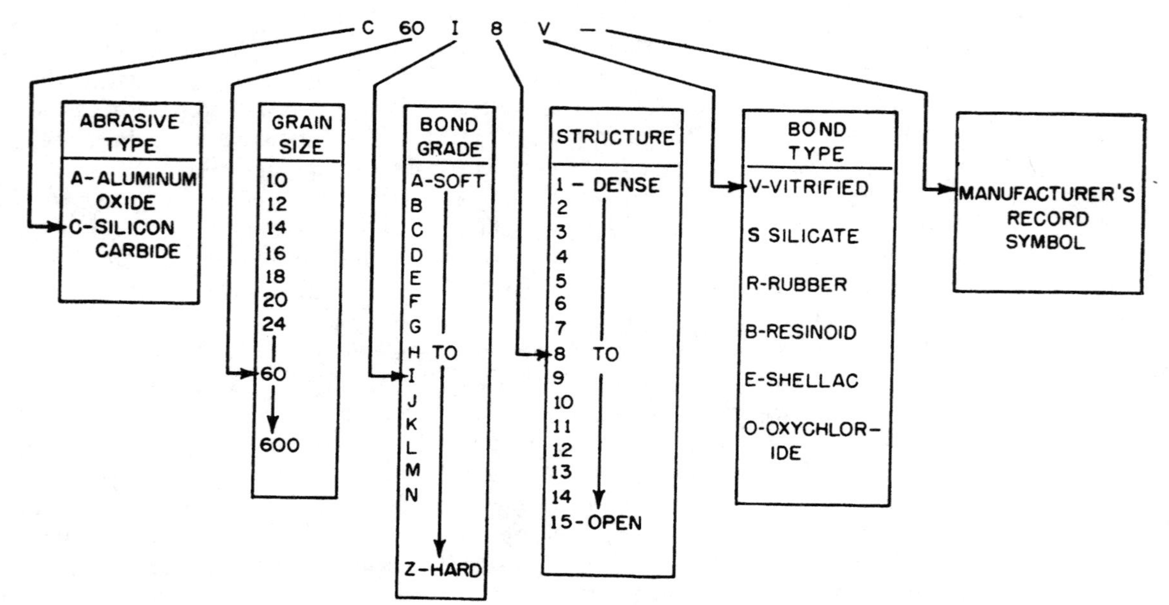

Bond type and other important grinding wheel nomenclature are shown in Figure 13-3. This data is used by manufacturers to identify the wheel markings.

The *grade* of a grinding wheel is a measure of the ability of the bond to retain the abrasive grains in the wheel. As the letter grades in the wheel mark progress from *A* to *Z*, more bond is used to retain the abrasive grain, thereby providing more holding power and resistance to grinding forces. These forces are either compressive, shear, or impact, which tend to tear the abrasive particles from the face of the grinding wheel. In any analysis of grade relationships in a given structure, the abrasive grains always remain a stable component. Void sizes and bond post sizes vary. In softer grades, for example, grinding wheels have less bonding composition—thus, weaker bond posts and larger voids. In harder grades, there is a larger volume of bond—thus, stronger bond posts and smaller voids. (See Figure 13-4).

The *structure* of a grinding wheel refers to the open space between the abrasive grains. (See Figure 13-5.) The closest grain spacing is denoted by low structure numbers, and such a wheel is considered to be dense. Higher structure numbers indicate open grain spacing.

In analyzing grinding wheel structure for any given grain size, it is important to understand that abrasive grain size remains constant. However, the volume of abrasive grain is greater in dense structure wheels and the volume is less in open structure wheels. Additionally, the number of voids (pores) and volume of bond content is

Weak holding power

Medium holding power

Strong holding power

Dense spacing

Medium spacing

Open spacing

Figure 13-4 An illustration of the meaning of the term, "grade." (Courtesy of Grinding Wheel Institute) (Left)

Figure 13-5 An illustration of the meaning of the term, "structure." (Courtesy of Grinding Wheel Institute) (Right)

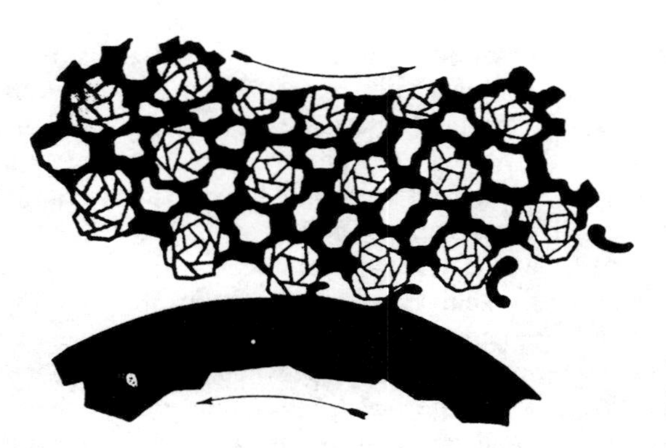

Figure 13-6 An illustration of how voids or pores in wheel structure act to provide chip clearance. (Courtesy of Grinding Wheel Institute)

reduced in dense structure wheels and increased in open structure wheels. This feature affects the promotion of cutting chip clearance, thereby avoiding the loading or glazing of the wheel. See Figure 13-6.

Bond type refers to the kinds of materials used to bind abrasive particles together. Such materials are mixed with grains pressed to shape and then solidified by firing or curing to form a porous matrix holding the abrasive grain in a definite space orientation. Ideally, the bond is of such strength as to present the abrasive grit to the workpiece until the grit approaches a dulled or worn condition. Then, it should break down to relieve the worn grit and present a new, sharp particle to the wheel's grinding surface. This process is, in effect, a "self-dressing" operation. The bond regions which surround the grains are referred to as bond "posts." See Figure 13-7.

Vitrified, designated by the letter "V," are the glassy, clay and porcelaneous bonds used in the manufacture of more than half of all grinding wheels made today. Their strength and rigidity provides fast, economical stock removal and makes them especially well-suited to precision and other general-purpose grinding operations.

Silicate bonded wheels are designated by the letter "S," and are used mainly for large, slow rpm machines where a cooler cutting action is desired. Silicate bonded wheels are said to be softer than vitrified wheels, as they release their grains more readily.

Rubber wheels, carrying the symbol "R," employ both synthetic and natural materials in their wheel manufacture. In softer types, the resiliency of this bond produces excellent polishing wheels. The harder types combine a degree of resiliency and water resistance for safe, cool-cutting cut-off wheels used where burr and burn must be held to a minimum. Other applications include grinding wheels and regulating wheels for centerless grinding.

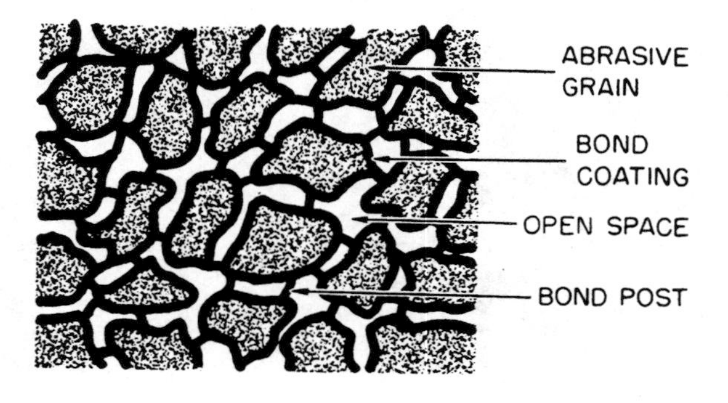

Figure 13-7 An illustration of the bond posts which secure the grain particles.

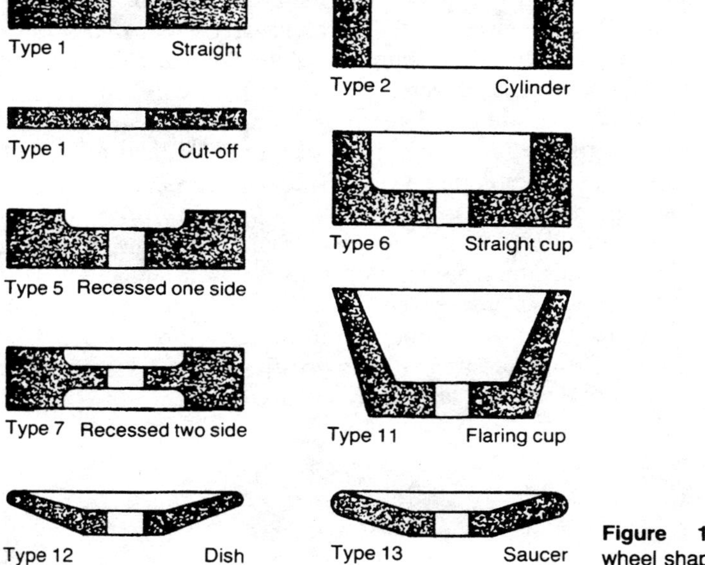

Figure 13-8 Common grinding wheel shapes.

Rubber wheels, designated "B," are made from a thermosetting phenolic resin bonding material. Resin bonded wheels, for reason of their greater toughness, are operated at higher speeds. Since this bonding material is capable of withstanding severe and abusive grinding stresses, resinoid wheels are currently the choice for high-speed, rough grinding, and cut-off operations. They also are available in fine grits to produce super finishes. For certain applications, resinoid wheels are reinforced for added strength.

Shellac, designated "E," provides a low heat resistant bond resulting in versatile grinding action, with particular qualities of free-cutting action and good finish. It has limited use, mainly in roll grinding wheels.

Oxychlorides, designated by the letter "O," are used primarily in slow grinding operations.

Wheel shapes. Wheels are manufactured in a range of shapes and sizes. The most common shapes are shown in Figure 13-8. Mounted wheels or points are used for deburring, die finishing, and hole grinding. See Figure 13-9. Other grinding tool shapes are sticks, stones, blocks, and segments.

Figure 13-9 Mounted wheels of a range of shapes and sizes are available. (Courtesy of Bay State Abrasives)

Other special grinding tools are manufactured for a range of uses. Those shown in Figure 13-10 are electroplated diamond and CBN coated products for high precision, long lasting applications. Removable thin diamond disks are perforated to accelerate cutting action and to prevent heat build-up. See Figure 13-11. This product has a variety of tool grinding and sharpening applications.

Flexible grinding disks are made of tough, resilient fiber plates impregnated with abrasive grains. See Figure 13-12. These are especially useful for smoothing welds, and on cast parts.

Coated abrasives is a term used to designate any sheet, strip, sleeve, drum, belt, or disk backing material with abrasive grains adhered to it. Backings generally are cloth, paper, or plastic, and the adhesive a glue or resin. A range of grit sizes and abrasive types are widely used in industry. Some examples of coated abrasives are shown in Figure 13-13. *Free abrasive grains* of many types and grits are also used, and are described later in this chapter.

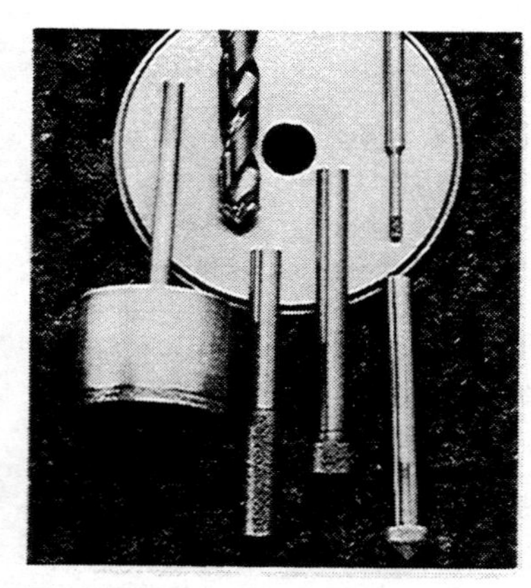

Figure 13-10 These CBN electroplate coated cutting tools are for use in precision applications. (Courtesy of Superabrasives Laboratory and Technology Corporation)

Figure 13-11 The diamond grinding disc is designed for quick, cool cutting. A carbide insert tool bit is being sharpened. (Courtesy of TBW Industries, Inc.)

Figure 13-12 Flexible grinding commonly used to smooth welded joints. (Courtesy of J. Walter Company, Limited)

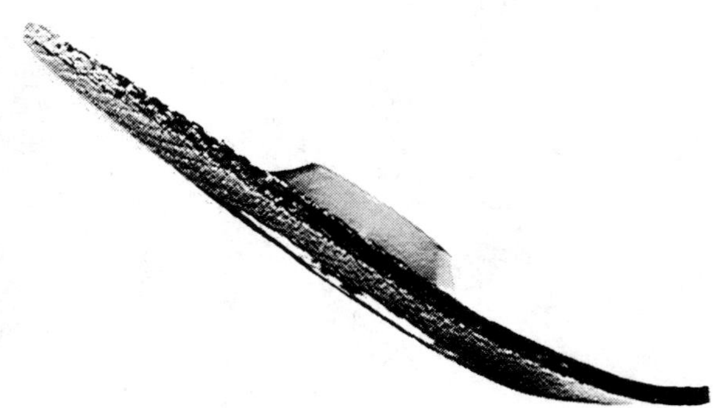

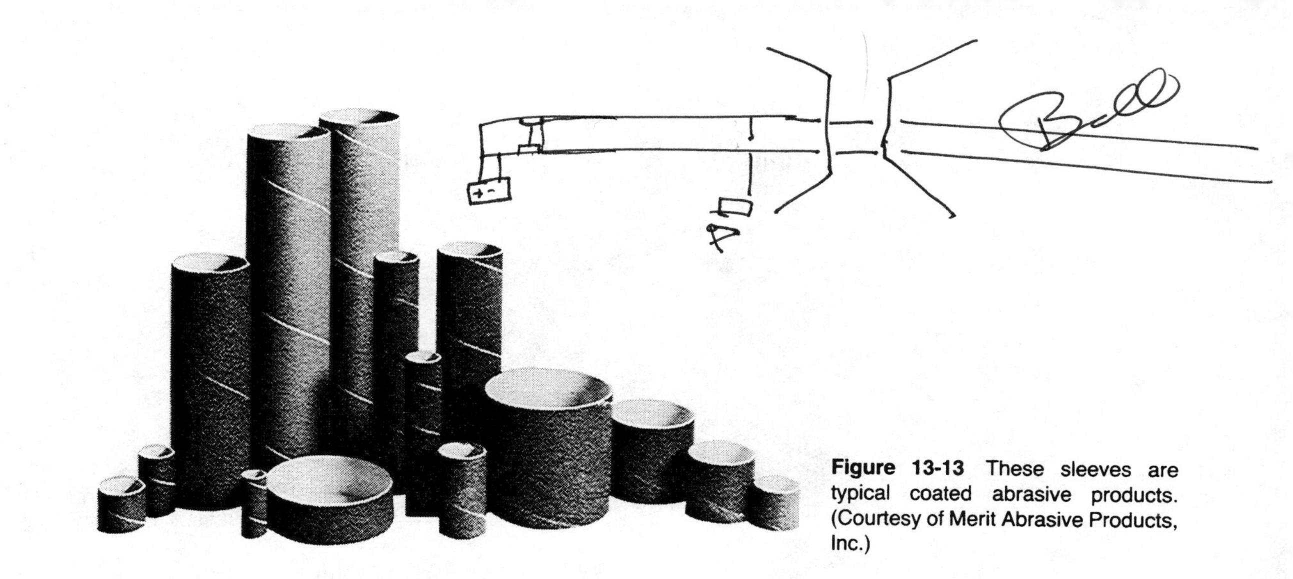

Figure 13-13 These sleeves are typical coated abrasive products. (Courtesy of Merit Abrasive Products, Inc.)

GRINDING OPERATIONS

There are literally hundreds of ways in which grinding is used in industry. Most involve the use of special equipment to treat the surfaces of precision workpieces. Grinding also has uses in the field, as indicated in Figure 13-14. Some of the more important of the industrial applications are described here.

Flat surface grinding. It can be argued that all grinding is surface grinding. The term, however, has through usage come to designate the grinding of flat surfaces (and in some cases, profiles on flat surfaces) as opposed to cylindrical ones. Surface grinding, therefore, implies the production of metal workpieces with flat, true, and precisely sized surfaces. Furthermore, the generation of such surfaces is obtained through the use of abrasive action upon that surface. See Figure 13-15.

Figure 13-14 This grinding equipment is being used to smooth the surface of a welded rail joint. (Courtesy of Nordberg Track Machinery)

Figure 13-15 The basic flat surface grinder. (Courtesy Do ALL Company, Des Plaines, IL)

Machines. There is a variety of equipment available to grind flat surfaces. A comparative schematic illustration of the four basic machine configurations appears in Figure 13-16.

The *horizontal spindle, reciprocating table* surface grinder is the most common type of machine in use today. In operation, the wheel-spindle position is fixed, and the workpiece, held in a magnetic chuck secured to the table, is reciprocated back and forth under the wheel. Abrasive machining occurs each time the workpiece

Figure 13-16 The above illustrations show how spindles are mounted and tables move on the four basic, different types of precision surface grinding machines: (a) horizontal spindle and reciprocating table, (b) horizontal spindle and rotary table, (c) vertical spindle and reciprocating table, (d) vertical spindle and rotary table. (Courtesy of Do ALL Company, Des Plaines, IL)

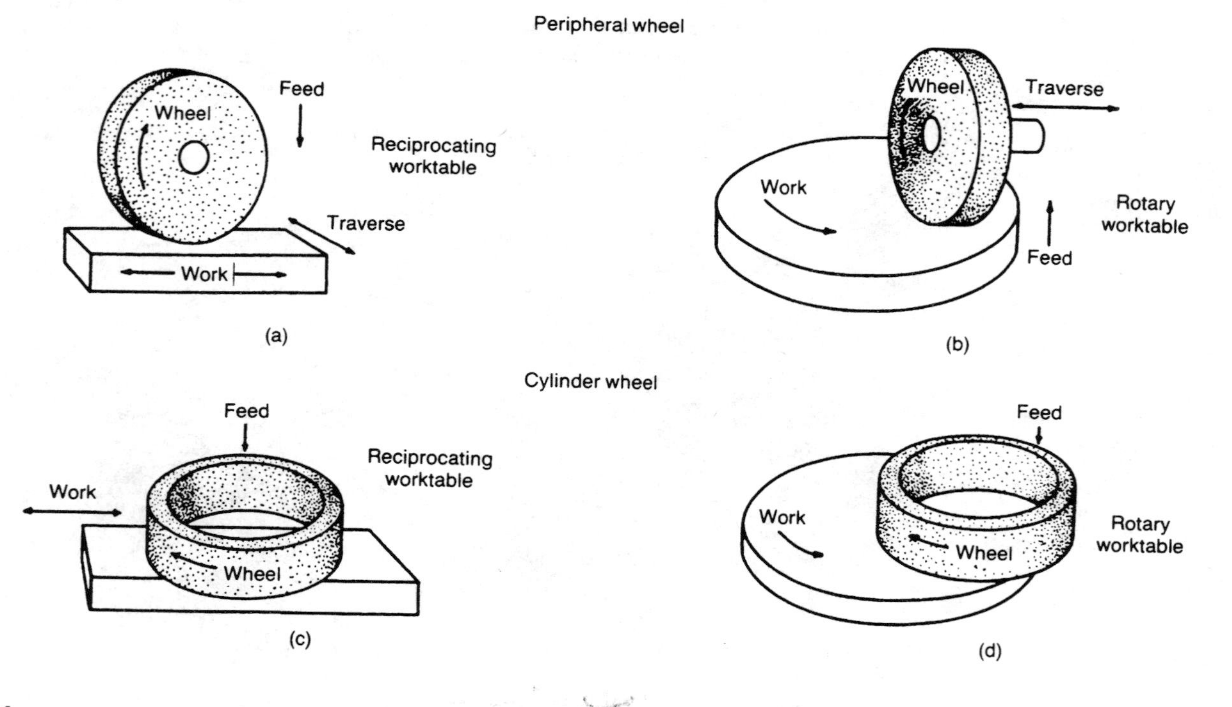

Bell

moves to engage the periphery (extreme bottom area) of the grinding wheel. With each pass, the workpiece also cross-traverses to effect a series of overlapping cuts. The wheel generally, and sometimes the table, is moved up and down in a vertical traverse to feed into the workpiece. The versatility of this machine is such that profiles may be cut on flat surfaces by using a special tilting headstock to orient the wheel to the desired contour. This is still flat surface grinding, because the workpiece is flat and the grinding plane is always parallel to the table surface. Another measure of the versatility of this machine is its range of available sizes. Smaller machines have a longitudinal travel limit of about 400 mm (16 inches). Larger models are capable of travels of 75 mm (300 inches) or more. Wheels vary in size from 175 mm (7 inches) in diameter and 12 mm (0.5 inch) wide, to 500 mm (20 inches) by 200 mm (8 inches).

The *horizontal spindle, rotary table* machine features a table which carries small workpieces in a circular path beneath the grinding wheel, and in opposition to the wheel's direction of travel. This grinder is also used for precision work applications where extreme flatness is desired in the removal of material in larger, single workpieces. An example would be the hollow grinding of circular saw blades, where the outer area of the blade is thinner than the central portion near the mounting hole. This feature relieves the severe contact with the wooden workpiece to prevent saw binding.

The *vertical spindle, reciprocating table* type of grinder is typically a larger production machine with a larger rectangular table which moves the workpiece to the wheel with a horizontal-traversing action. The basic difference between this vertical spindle machine and the horizontal type is that the vertically mounted grinding wheel cuts with the wheel face instead of the periphery. The result is that two flat surfaces contact one another—the flat wheel face and the flat workpiece face. Consequently, the rate of stock removal is high, sometimes as much as 12 mm (0.5 inch) in one pass. Many of these machines are designed to permit the tilting of the spindle a few degrees to increase the wheel pressure on the workpiece contact area. This speeds stock removal. The spindle returns to its "true" position for final finishing.

The *vertical spindle, rotary table*, also called the Blanchard grinder, creates a surface pattern which is a series of concentric circles within circles on a workpiece which is centered on the rotary table. A series of intersecting arcs are formed on a small workpiece mounted off center around the table. This tool is also capable of a high stock removal rate, but is more accurate because of the rotary table motion. For more efficient, high volume applications, these machines are equipped with multiple spindles, as many as five mounted around the periphery of the rotating table. The result of this arrangement is that the first spindle (and grinder) to engage the workpiece performs rough grinding, and the remaining spindles do gradually finer work.

Profile grinding. As stated earlier, flat surfaces may be ground to produce a shaped profile. This requires that the wheel be crush-formed by pressing a peripheral grinding wheel against a shaped, hardened steel roller, which imparts its shape to the wheel. Another method is to shape the wheel with a diamond dressing tool.

Profile, or form grinding, provides one of the most efficient methods of producing hardened steel dies and punches for cutting and forming metal parts. One method

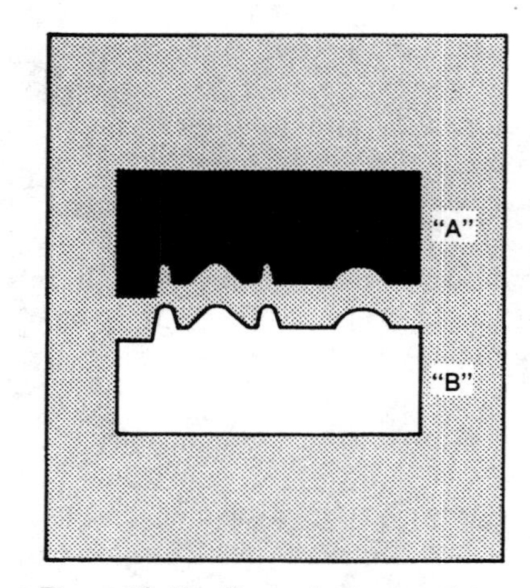

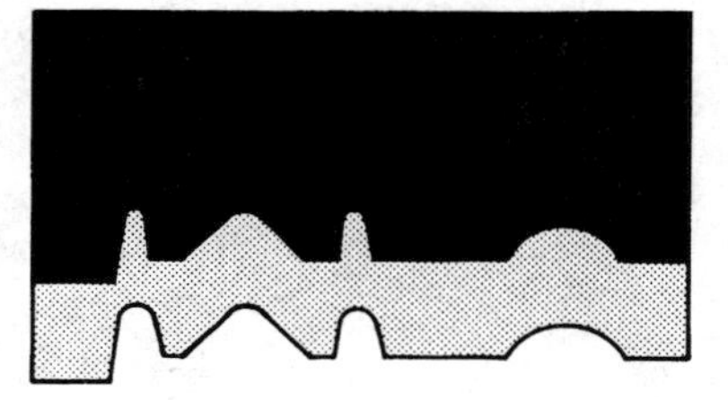

Figure 13-17 Form grinding templates are made in matching sets. (Courtesy of Engis Corporation, Diamond Tool Division)

of preparing such tooling is accomplished by a process called *diaforming*, which is an accurate, rapid, and inexpensive method of shaping grinding wheel profiles using enlarged templates. The shape is transferred pantographically from a template to the grinding wheel by means of diamond tool. The operator moves a tracer over the profile of an enlarged template. The tracer path is transmitted to the truing diamond, which accurately cuts the grinding wheel to the desired shape and size. This process example shown and described here involves the preparation of the tooling needed to manufacture the common razor blade.

1. The work begins with the preparation of mating templates. See Figure 13-17. Template "A" for the punch is made by cutting 6 mm ($\frac{1}{4}$ inch) mild steel 10 times larger, and to the shape required for the finished part. Template "B" for the die

Figure 13-18 Drawing of razor blades. (Courtesy of Engis Corporation, Diamond Tool Division)

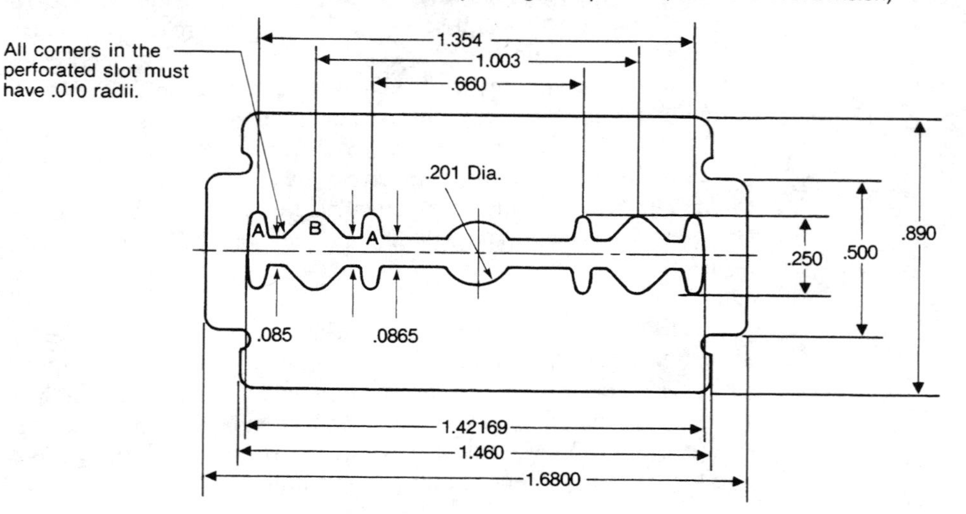

Figure 13-19 A template in place on the pantograph. (Courtesy of Engis Corporation, Diamond Tool Division)

Bill

must match Template "A" exactly for proper punch and die fit. The blade drawing is shown in Figure 13-18. Note that the templates represent one-half of the configuration.

2. Template "A" is mounted to the carrier slide of the diaform, which is permanently mounted over the spindle of the surface grinder. See Figure 13-19. The form is transferred pantographically from the template to the wheel.

3. The diamond roughing tool is mounted and centered on the diamond spindle by using the 0.002 mm (0.0001 inch) dial indicator. The diamond tool is now properly positioned, thereby completing the preparation for diaforming. See Figure 13-20. By making successive passes of the tracer along the template edge, the punch form is gradually cut into the grinding wheel by the diamond roughing tool.

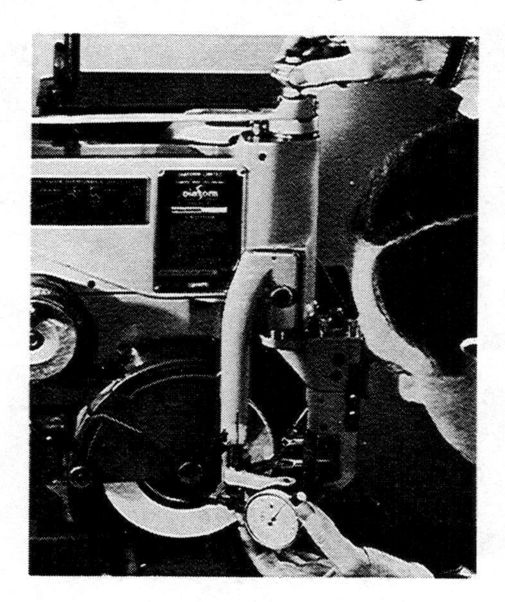

Figure 13-20 Positioning the diamond roughing tool with a dial indicator. (Courtesy of Engis Corporation, Diamond Tool Division)

Figure 13-21 Cutting the punch form into the wheel with the diamond tool. (Courtesy of Engis Corporation, Diamond Tool Division)

Bill

4. After the roughing and semifinishing has been completed, the final diaforming is performed using the finishing diamond tool. See Figure 13-21. The diamond tool trues the wheel to a precision form, and the wheel is then ready to rough grind the punch form.

5. The diaform tracer arm is set in its rest position, and the hardened punch blank is placed on the magnetic chuck of the grinder. By plunge grinding one quarter of the punch form at a time, the rough grinding of the punch is completed to within a few thousandths of the finish size. The wheel is retrued, and the punch is finish-ground to size. Now using Template "B," and adjusting the diaform as previously described, the die profile is cut in the wheel.

6. The two die halves are secured in a chuck, and are ground in tandem in the same manner as the punch form. See Figure 13-22. The tool parts are dimensionally accurate to 0.0127 mm (0.0005 inch), and have the required clearance between the punch and die for 0.1016 mm (0.004 inch) blanking stock. The completed tooling appears in Figure 13-23.

Figure 13-22 Grinding the punch blank. (Courtesy of Engis Corporation, Diamond Tool Division)

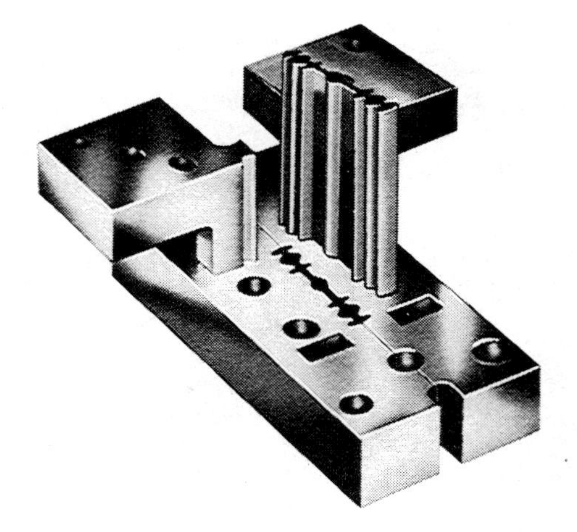

Figure 13-23 The completed die set matches precisely. (Courtesy of Engis Corporation, Diamond Tool Division)

Abrasive belt grinders. Flat surfaces can be generated on most materials with efficiency and precision by using coated abrasive belts, disks, and drums. Theoretically, the grinding wheels used on vertical and horizontal spindle machines could be replaced with wheelheads comprised of coated abrasive components, and indeed sometimes are. In practice, coated abrasives are used in polishing operations on workpieces to produce a smooth, true finish which could not practically be done with solid grinding wheels. The sanding of wooden furniture components is a good example. Examples of these finishing machine systems are shown in Chapter 8. It should be noted that the term *grinding* applies to both the traditional solid grinding wheels, and coated abrasive media.

Lapping machines give unsurpassed accuracy in controlling parallelism, roundness, flatness, and surface finish quality. The machine is especially suited to applications where parts are too small to hold on a magnetic chuck, too thin to grind without warping them, or for use with nonmagnetic materials which would normally be required for complicated fixturing. See Figure 13-24. The lapping of flat surfaces

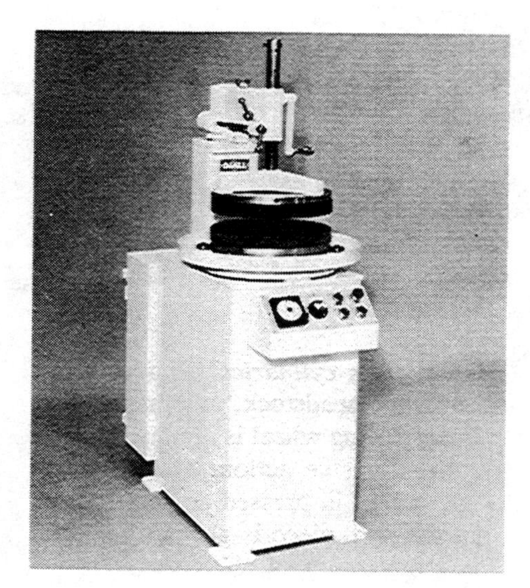

Figure 13-24 A vertical lapping machine is used for the production of accurate surfaces on workpieces. (Courtesy of Do ALL Company, Des Plaines, IL)

Figure 13-25 A plate holder used to control these gears for lapping. Two lap machines are used to finish both the top and bottom surfaces. (Courtesy of Do ALL Company, Des Plaines, IL)

is performed on vertical spindle machines having a horizontal *lap*. A lap is a metal plate, usually cast iron, which carries abrasive grains in a vehicle, usually oil or grease. The process involves the rolling of the suspended abrasives between the workpiece and the lap. Although lapping is a material removal process, it is not economical for that purpose, inasmuch as the material to be ground away is in the order of 0.03 mm (0.001 inch). Lapping can also be performed on cylindrical surfaces, or on flat parts where both surfaces must be smoothed. Machines for this purpose have two laps, a *lower* one to support and slowly rotate (at about 80 rpm) the workpieces against a *stationary upper* lap, which supplies pressure control. A plate type holder is used to contain the workpieces as they move between the two laps. See Figure 13-25.

Free-abrasive machining is related to lapping, except that the back-up plate (lap) is generally made of a hardened steel alloy. Consequently, there is no imbedding of the abrasives in the lap, and coarser abrasives are used. All the cutting edges of the grains are available for stock removal. Also, the workpiece does not contact the plate, but rides instead on a rolling layer of abrasive slurry.

Cylindrical grinding. Flat surface grinding, described earlier, has to do with the action of abrasives on essentially flat surfaces. The information presented here deals with the grinding of the external or internal surfaces of cylinders, or other cylindrical shapes such as shafts and drums. These processes generally occur by rotating workpieces against a grinding wheel or an abrasive drum, and both the workpiece and the wheel rotate to generate a cylindrical shape. There are several kinds of external cylindrical grinding, and there are appropriate machines to accomplish these operations. See Figure 13-26.

External grinding. The *plain cylindrical* grinder supports the workpiece between centers, and turns with the headstock, as a metal shaft might be turned on a lathe. See Figure 13-27. The grinding wheel is mounted on a cross slide and traverses the workpiece to achieve the abrasive action. *Plunge* grinding may also be done on this machine, where the wheel is pressed or plunged to full cutting depth and no traversing occurs. Here the workpiece is generally shorter than the width of the wheel. *Peel* grinding is a combination of plunge and traverse grinding, where the

Figure 13-26 The workpiece is held between centers on this plain cylindrical grinder, much like lathe-turning between centers. (Courtesy of Cincinnati Milacron)

plunged wheel traverses the length of the workpiece to "peel" a layer away. The *step* grinding, the wheel is plunged into the workpiece, withdrawn, traversed, and then plunged again in a series of steps to complete the grinding. Single or multiple grinding diameters can be so achieved. Workrests must be used to support long workpieces, which may bow during grinding and thereby distort the accuracy of the ground shaft. Smaller workpieces may also be chuck-mounted.

The *centerless* grinder has a grinding wheel, a regulating wheel, and a workrest blade mounted between them. The workpiece is driven by the regulating (or feed) wheel, which is typically a rubber bonded abrasive wheel, and is slower moving than the grinding wheel. See Figure 13-28. The grinding wheel does the cutting, and the workpiece has a lower rpm and is therefore subject to abrasion and moves in an opposite direction to the grinder. There are three variations of centerless

Figure 13-27 The basic cylindrical grinding machine. (Courtesy of Grinding Wheel Institute)

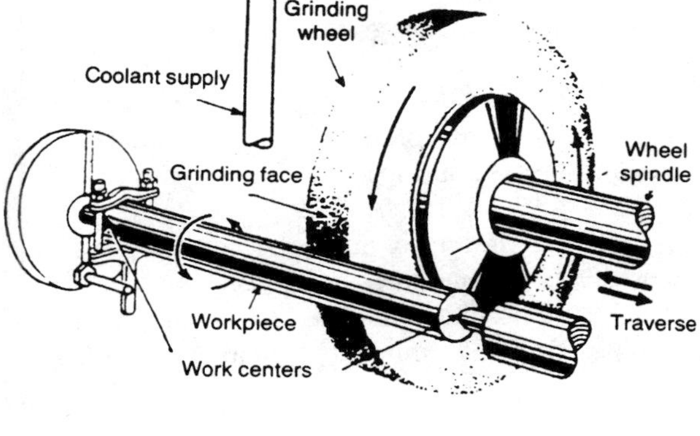

Figure 13-28 Basic centerless grinding diagram. (Courtesy of Grinding Wheel Institute)

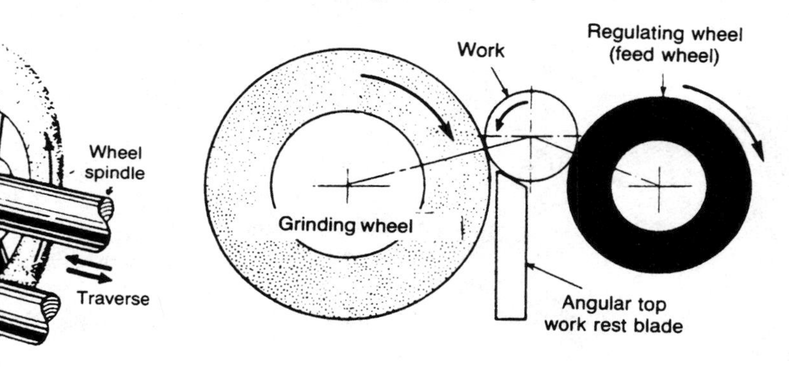

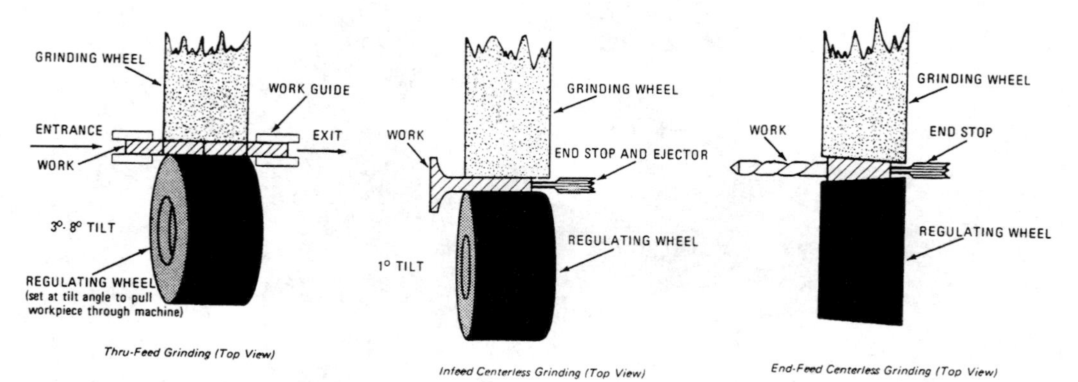

Figure 13-29 Three types of centerless grinding. (Courtesy of Grinding Wheel Institute)

grinding: throughfeed, end-feed, and in-feed. See Figure 13-29. *Throughfeed* grinding is achieved by tilting the rotational axis of the regulating wheel a few degrees from the parallel orientation of the grinder. This gives the workpiece longitudinal as well as rotary movement, thereby passing the workpiece completely through the space between the two wheels. This method is used on cylindrical pieces with straight, parallel sides, where complete finishing is desired. This, of course, could not be achieved efficiently with between centers mounting. It also lends itself to rapid, automatic part feeding.

End-feed grinding is used to produce tapered workpieces, where the grinding wheel, and occasionally the feed wheel, are ground to an appropriate taper. The part-piece is fed into the workspace, where it advances and is ground until it reaches its end stop. *In-feed* grinding is employed for ground parts having multiple diameters or special shapes. The technique is used to produce ball bearings, or parts with shoulders or heads. This is a plunge type of grinding where the workpiece is forced against the formed grinder by the regulating wheel.

Abrasive belt centerless grinders operate on the same principle as the centerless solid wheel grinder, but they are not the same, and do have some limitations. They do not produce close workpiece tolerances, because of the resilience in both the finishing medium and the regulating wheel or belt. Also, although the regulating head may be secured in a fixed position for some applications, it is usually permitted to float against an adjustable spring pressure. Such floating allows the head to follow eccentric (out-of-round) workpieces, but at the same time minimizes the correction of eccentricity. Therefore, these finishes are best suited to roughing prior to centerless grinding, and to fine surface improvement after centerless grinding. See Figure 13-30.

The regulating belt is used for heavy work and provides a more positive workpiece engagement. The regulating head can be set at 0° for plunge grinding, and from 0°–15° for throughfeed operations. Finishing brushes are also used.

Internal grinding. Implicit in this type of grinding is the sizing and truing of holes with an abrasive wheel, as the workpiece either rotates on its axis, or is held stationary while the wheel makes a planetary motion. The axial cross-section of the hole may be parallel, contoured, or tapered. Furthermore, they may be through

READ TO CLASS

Figure 13-30 Abrasive belt centerless finishing. Note the regulating belt (not a wheel), the large abrasive belt pulley which controls the cutting, and the workrest shoe. (Courtesy of Hammond Machinery, Inc.)

holes, blind holes, or holes with several diameters. There are three common types of internal grinding, and each will be treated here. See Figure 13-31.

The *chucking* type of internal grinder employs a variety of techniques to hold the workpiece, such as round collet chucks, sliding jaw collets, jaw chucks, face plates, finger chucks, and magnetic chucks. These devices are similar to those used in lathe turning operations. The grinder is secured in a wheelhead mounted on a slide, so that as the revolving grinder traverses the hole, it is simultaneously fed on the slides to engage the holes and bring it down to size. The chuck-mounted workpiece rotates in a fixed position, typical of this grinding method.

A common problem with chuck-held grinding is that excessive containment pressure can squeeze a workpiece out of shape, as can occur with a three-jaw chuck. It is important to select the proper holding mechanism.

The *roll-support centerless* internal grinder produces accurate holes in a workpiece rotated on its own outside diameter (OD), as it is held between three rolls. The fixed *regulating* roll provides the driving motion and bears the force of a workpiece squeezed against it by a *pressure* roll, and held in relative position by a *support* roll. A revolving grinder is reciprocated in the hole as it is fed against hole's inner surface. The particular advantage of this method is that the inside diameter (ID) of the hole depends on the OD of the workpiece and not by the accuracy of the chuck-holding method. The result is that perfect concentricity between ID and OD can be attained. Very precise bearing races and bushings can be produced in this way, as the orientation of the rolls provides a rigid support for the workpiece. This centerless method overcomes distortion problems occurring with three-jaw chucking.

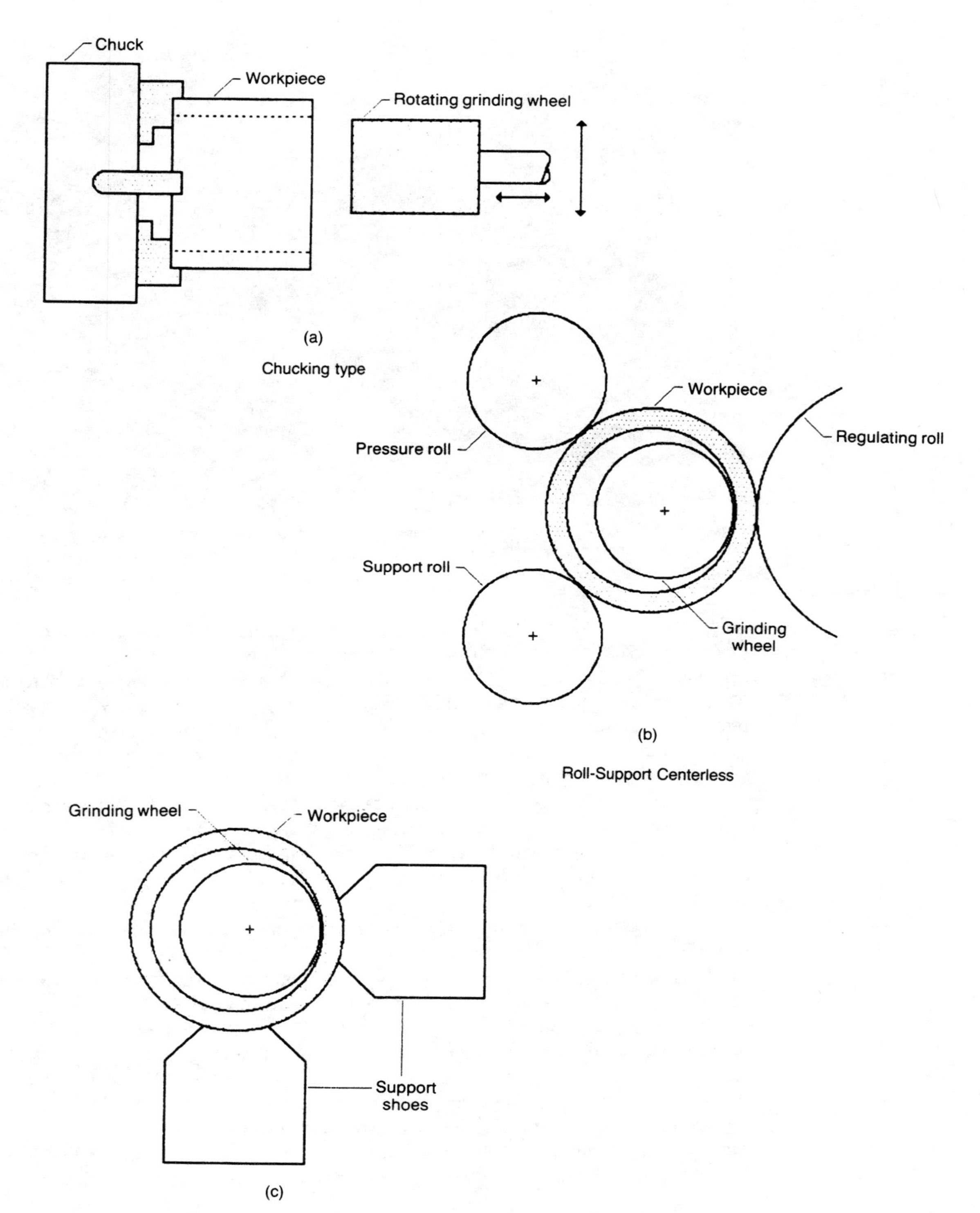

Figure 13-31 The three basic types of internal grinding.

Shoe-support internal grinding is a variation on the roll-support type. The workpiece is held against a magnetic faceplate by two fixed, hardened shoes which provide maximum support and reduce the waviness or chatter of the ground surface. The magnetic faceplate turns the workpiece at a low rotational speed in one direction, in opposition to a faster-turning grinder.

Honing. Honing is a low-velocity, precisely controlled abrading process used to accurately finish a hole. Honing is done by passing an abrasive tool, called a honing stone (or honing stick), over the surface of the workpiece. The honing stone is composed of fine abrasive particles and is shaped to fit the part to be honed, usually the inside surfaces of holes or tubular parts. For those applications, one or more honing stones are mounted on an expandable holder attached to the end of a shaft. The tool is inserted into the hole and is expanded by springs or cones to press the stones against the interior wall. The cutting action is of two types. Either the tool or the workpiece is rotated, and simultaneously the tool is reciprocated in the hole, or the workpiece is reciprocated over the tool. The operation can be performed on drill presses, lathes, or special honing machines. This rotating and reciprocating motion produces a series of cross-hatched minute scratches, which retain lubricants in parts such as automotive engine cylinders. Other applications include the honing of gun barrels, bearings, shafts, and ring gauges.

QUESTIONS

1. Define the term abrasive machining.
2. List the types of natural and manufactured abrasives, and describe the special attributes and uses of each.
3. Define the following grinding wheel terms:
   a. Grain size
   b. Grade
   c. Structure
   d. Bond type
4. Describe a grinding wheel which has the following markings: A 36L 5V.
5. Sketch five grinding wheel shapes.
6. Define the term "coated abrasives."
7. Prepare sketches of the four basic flat surface grinding machines.
8. Describe the process of profile grinding.
9. Describe the uses of the lapping and honing processes.
10. Prepare sketches of some basic external cylindrical grinding processes.
11. Prepare sketches of the three basic internal grinding processes.
12. Prepare a library research report on current advances in abrasive machining.

14 SPECIAL MACHINING METHODS

Despite the substantial number of common mechanical techniques for material cutting, an array of nontraditional methods has emerged to meet the demands of special materials and complex workpieces. Some are basically electrical systems, others thermal or chemical, and yet others combinations of these. Whatever their unique characteristics, all are employed in material removal and separation functions, or related processes, and are described here. All are to be considered mass or bulk decreasing techniques.

ELECTRICAL DISCHARGE MACHINING

Electrical Discharge Machining (EDM) involves the removal of metal from a workpiece by the erosive action of a controlled electrical spark to produce holes, slots, and cavities. See Figure 14-1. There is no direct contact between the electrode and the workpiece, and no physical force is exerted. The rate at which metal is removed is influenced by the electrical conductivity of the workpiece and not by its material hardness. One terminal of the power supply is connected to the workpiece and the other to the electrode that controls the cutting. The shape of the hole will be identical to the shape of the electrode. The workpiece and the electrode are separated by a dielectric fluid—usually deionized water or a special hydrocarbon—which acts as an electrical insulator until the spark occurs. The dielectric also cools the area after the sparking, and flushes away metal particles before the next spark sequence.

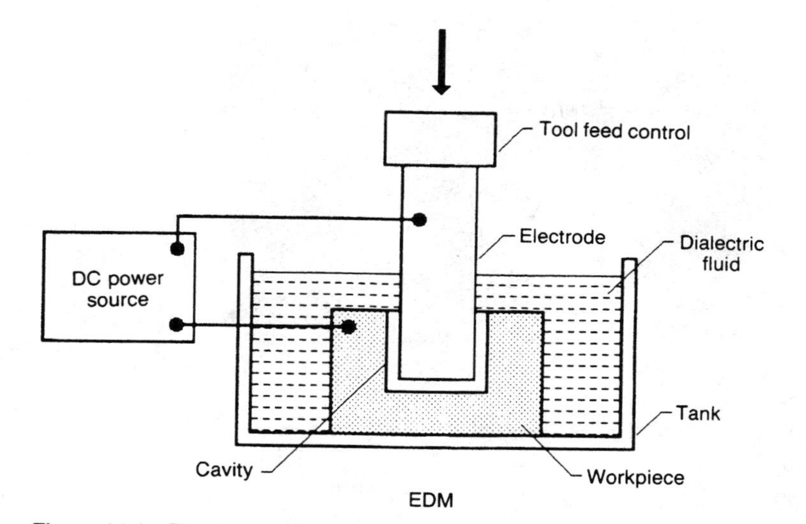

Figure 14-1 Electrical discharge machining system, EDM.

A DC voltage is applied between the electrode and the workpiece. Initially, no electric current flows because the two pieces are insulated by the dielectric. An electric field, however, does build up across the gap. If the gap is made narrower, a point will occur where the voltage can ionize the dielectric and an electric spark will then jump between the electrode and the workpiece.

When the spark or discharge first occurs, a large amount of energy is released, vaporizing material from the surface of the workpiece. As current continues to flow, the intense heat melts additional material. Whenever the voltage drops to zero, the current ceases, and the spark is quenched.

As soon as the spark is quenched, the vapor bubble begins to collapse. The dielectric cools the area, thereby solidifying the material that was melted. This material is carried away by the movement of the dielectric fluid, leaving a small cavity at the point where the discharge occurred. As one discharge follows another, a series of connected cavities are cut to complete the erosion process. It is important to mention that it is not simply the flow of an electric current that removes material and does the cutting; it is turning that current ON and OFF that actually vaporizes and melts the metal. In this fashion, die cavities are shaped. The efficiency of an EDM system is measured by its material removal rate in cubic millimeters or inches per minute. This process lends itself especially well to numerical control for the production of complex tooling patterns. See Figure 14-2.

An important variation of conventional EDM is the *wire EDM* process, variously known as electrical discharge wire cutting (EDWC), or travelling wire EDM. See Figure 14-3. Here the electrode is a moving wire of brass, tungsten, or copper of 0.08 to 0.3 mm (0.003 to 0.01 inch) diameter. The cutting action approximates that of a bandsaw, but where sparks instead of teeth perform the cutting, to produce such pieces as an intricate progressive die. See Figure 14-4. A machine for this process appears in Figure 14-5. The slot or kerf produced is somewhat larger than the wire diameter, due to the spark erosion process. Generally, wire EDM is used to produce through holes and solid shapes, and conventional EDM is used to generate cavities.

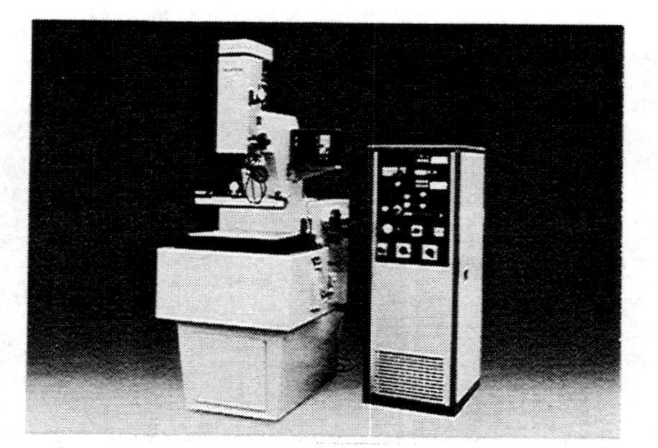

Figure 14-2 A precision spark erosion, or EDM, machine. (Courtesy of Hurco Manufacturing Company, Inc.)

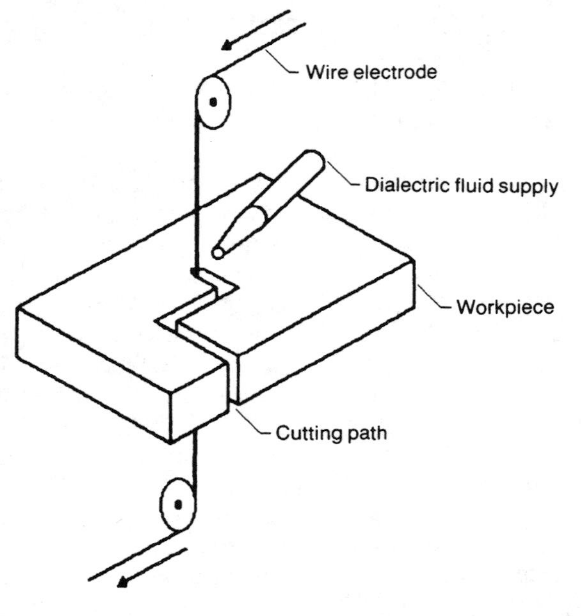

Figure 14-3 Electrical discharge wire cutting system, EDWC.

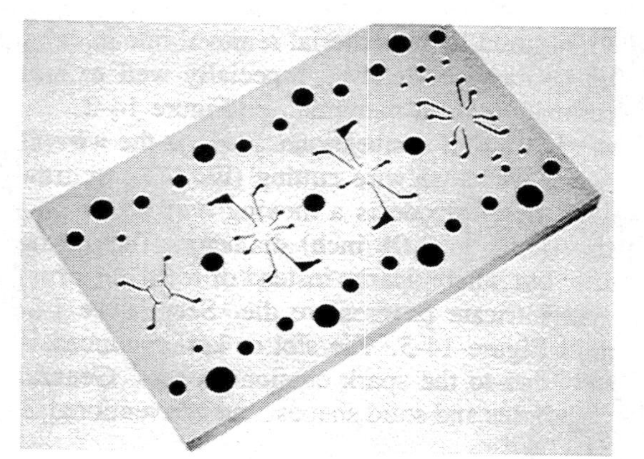

Figure 14-4 Intricate progressive dies such as this are machined by using the EDWC process. (Courtesy of Agietron Corporation)

Figure 14-5 An EDWC machine. (Courtesy of Agietron Corporation)

Electrical discharge grinding. *Electrical discharge grinding* (EDG) is similar to EDM, except that a rotating graphite wheel is used as a tool. The wheel does not touch the workpiece; it delivers a shower of sparks across a gap to the workpiece surface.

ELECTROCHEMICAL MACHINING

Electrochemical machining (ECM) is the removal of metal from a workpiece by dissolving the metal in a chemical solution with the aid of an electric current. Metal removal is controlled by a tool, but the tool does not contact the workpiece. The metal removed is in the form of atom-sized particles instead of chips. See Figure 14-6.

This process is achieved by passing an electrical current between a tool and a workpiece while both are immersed in an electrolyte. The area of the tool that faces the workpiece is shaped in the mirror image of the contour to be produced on the workpiece. As the electric current travels from the workpiece to the tool, particles of metal are carried away from the surface of the workpiece and into the electrolyte. As metal is removed, the tool is fed into the workpiece. A bulge on the tool produces a depression on the workpiece; a depression on the tool produces a bulge on the workpiece.

The process works on the same principle as electroplating, and the metal removed from the workpiece would normally be plated on the tool. To prevent the metal from being deposited on the tool, the electrolyte is continuously circulated through the space between the tool and workpiece. The metal therefore is carried away by the electrolyte before it reaches the tool.

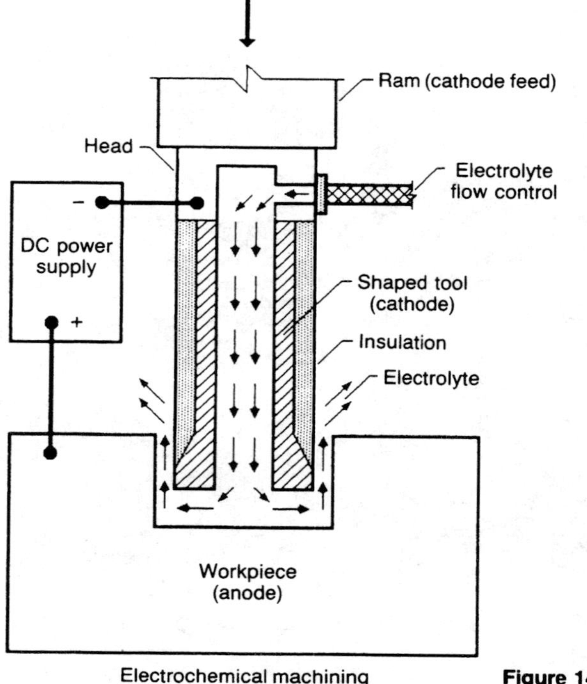

Figure 14-6 Electrochemical machining, ECM.

ELECTROCHEMICAL GRINDING

Electrochemical grinding (ECG), also called electrolytic grinding, is a process of removing metal through the chemical action that occurs when an electric current is passed through an electrolyte. The reverse of electroplating, ECG deplates material from a workpiece and deposits it in the electrolyte. A high-current, low-voltage electrical process, ECG facilitates the manufacture of parts that were not practical by other abrasive methods. As pertains to all electrochemical processes, ECG requires the immersion of two electrodes in an electrically conductive solution or electrolyte. In this process, the workpiece is the positive electrode or anode, and the electrically conductive abrasive wheel is the negative electrode, or cathode. Both electrodes are connected by a direct current power source. See Figure 14-7.

The electrolyte generally is a solution of chemical salts and other additives, and water. The solution is directed to the workpiece and the wheel through a nozzle. As the current flows between the work and the wheel, a chemical action causes the surface metal to change to a soft oxide that is removed by the abrasive grains in the cathode wheel. As the oxide film is brushed away, more surface metal is exposed to the chemical oxidation. Though often confused with electrical discharge machining (EDM), ECG is not a discharge process.

While ECG is faster, cleaner, and less distortive than conventional grinding methods, it is generally economical only for those workpieces which are too difficult for ordinary grinding. An example is the regrinding of carbide insert bits used in metal machining. See Figure 14-8. Additional applications include the deburring of welding torch tips, and finishing and sizing electrical relay parts.

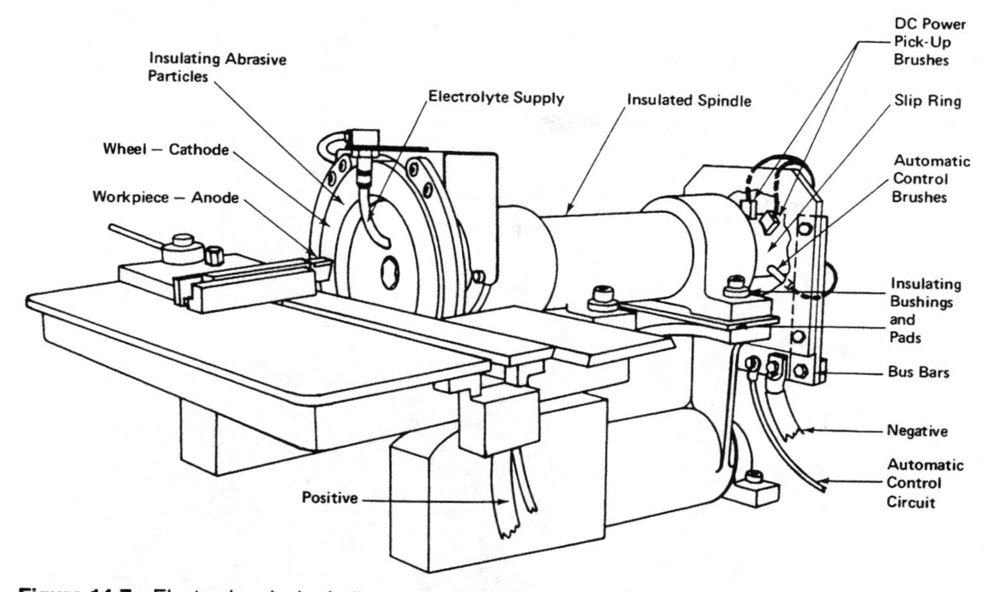

Figure 14-7 Electrochemical grinding system, ECG. (Courtesy of Hammond Machinery)

Figure 14-8 Carbide insert bits used in metal machining are ground efficiently with the ECG process. Note the inserts locked in the tool head for grinding. (Courtesy of Hammond Machinery)

CHEMICAL MACHINING

Chemical machining (CHM) is the removal of metal by the chemical attack of corrosive liquids such as acids. No tool or electricity is used. The workpiece is submerged in the chemical, and metal is uniformly dissolved from all exposed surfaces. Areas of the workpiece that are not machined are covered with a material, such as plastic, that is not affected by the chemical. See Figure 14-9.

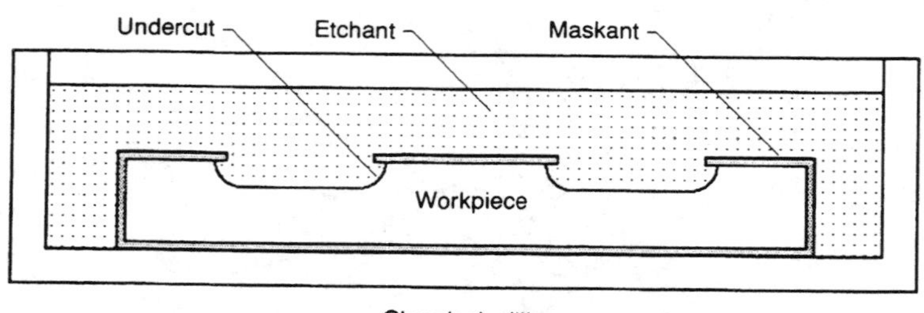

Figure 14-9 Chemical machining system, CHM. This is also called chemical milling.

Figure 14-10 An example of parts produced by the photochemical milling process. (Courtesy of Microphoto, Inc.)

Thin material can be completely penetrated by CHM. Complex shapes such as electrical circuit boards are separated from thin sheets of material by masking all of the material except the outline of the shape. That procedure is usually called *chemical blanking*. When metal is removed from broad areas of a workpiece, the process is often referred to as chemical milling.

Most fine, precision CHM is a variation of the conventional photoengraving process, and therefore is more properly termed photochemical milling. The process involves the preparation of a scale drawing from 2 to 100 times the size of the finished part. The drawing then is reduced photographically to the exact size of the final part. A metal workpiece is cleaned thoroughly and coated with a photosensitive resist. An image of the reduced photographic negative is produced on the sensitized metal workpiece, or plate, and then developed in a solution that dissolves all the coating except that in the photoexposed region. The plate is then etched to dissolve all unprotected metal, to produce the final part. Finally, the resist is removed from the part. See Figure 14-10.

ULTRASONIC MACHINING

Ultrasonic machining (USM) is the process of removing material by the action of abrasive particles vibrating in a water slurry, which circulates through a narrow gap between a workpiece and a tool, which oscillates at about 20,000 cycles per

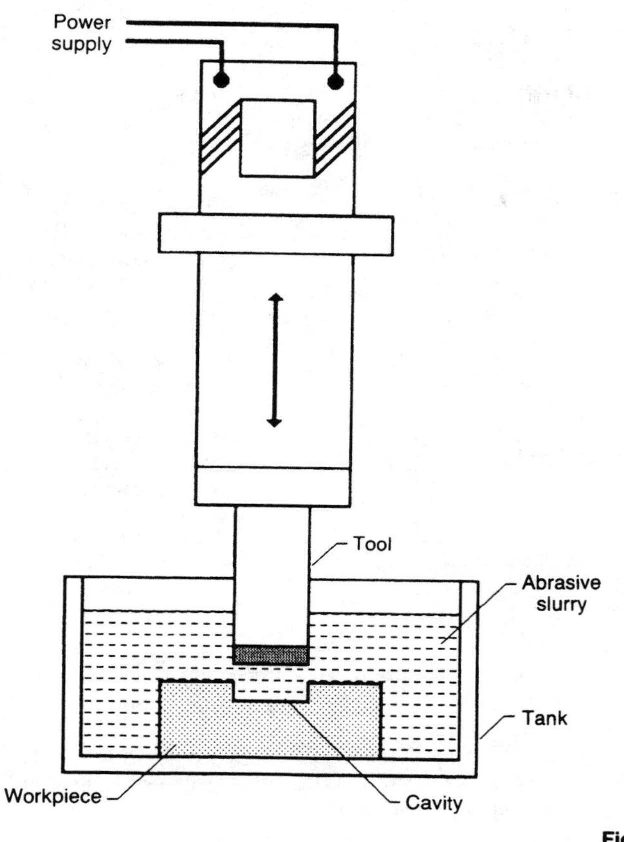

Figure 14-11 Ultrasonic machining system, USM.

second. The tool shape is reproduced in the workpiece at an accuracy of 0.25 mm (+/− 0.001 inch). See Figure 14-11. This process is used chiefly on hard, brittle materials that do not conduct electricity, such as glass and ceramics. However, it also can be used on hardened alloy steels. It is particularly applicable to the production of shallow, irregular cavities, and for machining fragile products such as silicon chip wafers.

THERMAL DRILLING

Thermal drilling is a process whereby holes are produced using the heat energy generated from the friction of a rotating tool forced against the surface of a workpiece. See Figure 14-12. Tungsten carbide tools are used, and the ratio of the tool diameter to material thickness can range from 3:1 to 6:1. There are several advantages of this drilling method.

1. Thermal drilling is chipless, eliminating clean-up and reducing contamination.
2. The cone and body of the drilling tool are ground to close tolerances and

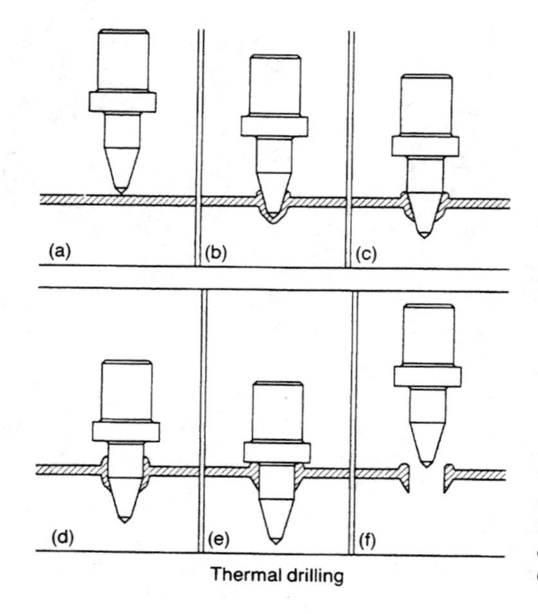

Figure 14-12 In the thermal drilling process, the rotating bit is pressed against the workpiece, thereby generating a high localized heat. The drill advance is slowed while the workpiece temperature increases (a). As the contact area widens, the rapid temperature rise melts the workpiece and permits the drill point to advance (b). The softened material flows up the drill until the point breaks through (c). The remainder of the displaced metal is extruded into a precision bushing below the surface of the workpiece (d). When the desired depth is reached, the top of the bushing is flattened by the collar of the drill tool (e). The tool is withdrawn when the drilling has been achieved (f). Courtesy of Foremost Machinery Corp.)

therefore retain their dimensions, and never need resharpening. Also, the conical shape reduces wander and provides exceptional accuracy of the resulting bushing.

3. Typical life expectancy is 10,000 to 20,000 holes. The factors which affect tool life are temperature, speed, axial force, material, lubrication, and the quality of the drilling equipment. Conventional equipment is used in thermal drilling.

4. Thermal drills may be used on all ferrous and nonferrous metals including stainless steel, copper, bronze, brass, aluminum, and mild steel. Thermal drilling of soft materials such as aluminum and copper requires a lubricant to prevent material build-up on the tool surface.

5. The time required to drill a hole with a thermal drill is comparable to that of a conventional drill. In fact, thermal drilling must be completed quickly, to prevent high temperatures which may damage the tool.

The protrusion produced in a workpiece undersurface is called a *bushing*. The bushing can be used for a load bearing application, or threaded for use with a high-strength fastener. Thermal drilling provides designers with new ways to reduce costs and improve fabrication methods in products made of thin walled tube or sheet metal.

WATER JET CUTTING

Water jet cutting involves the use of a high-velocity stream of water for cutting, and has some distinct advantages over conventional methods. Waterjet systems employ a finely focused jet of water pressurized to around 375 MPa (55,000 psi) to separate nonmetallics such as plastic, boxboard, fibrous cloth, leather, fiberglass, and food

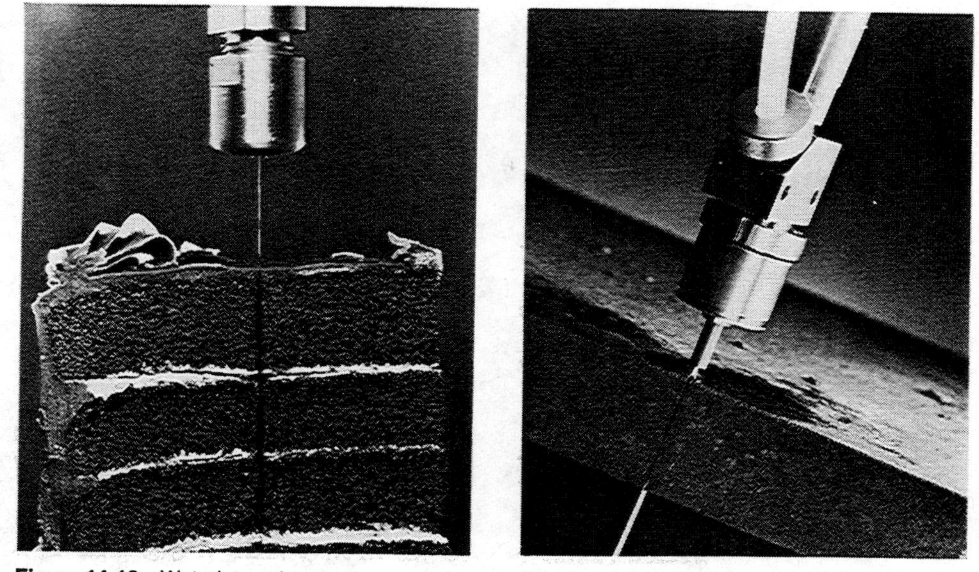

Figure 14-13 Waterjet cutting can be used on many nonmetallics, including foods. (Courtesy of Flow Systems, Inc.)

Figure 14-14 Cutting steel with an abrasive waterjet. (Courtesy of Flow Systems, Inc.)

products. See Figure 14-13. The high-pressure jet travels at a velocity of 900 m/s (3,000 fps) to cut with a compressive shearing action. The attributes of such a system are that it is dustless, clean, non-heat-generating, omnidirectional, and noncontacting. A tiny sapphire orifice in the cutting nozzle directs a noncrushing stream to the workpiece to produce a narrow kerf of around 0.15 mm (0.006 inch).

The introduction of an abrasive powder such as garnet makes possible the cutting of steel plates up to 75 mm (3 inches) thick, as well as dense plastic, graphite, and glass. See Figure 14-14.

THERMAL CUTTING

Thermal cutting, or flame cutting, is the melting and removing of a narrow section of a metal workpiece to separate it into two parts. Although it is used for separating instead of joining, thermal cutting is closely related to those welding processes in which the workpiece surfaces are melted. See Chapter 20. Both welding and cutting are performed in much the same manner, but the major difference between the two is that, in cutting, the molten metal is removed from the workpiece, whereas in welding it remains on the workpiece and solidifies into a weldment.

Nearly all of the methods used to melt metal for welding can also be used for cutting. Even some of the consumable-electrode methods, such as the shielded-metal-arc and gas-metal-arc processes, are used for cutting, but the procedures must be modified to blow the molten metal from the workpiece.

Oxygen cutting with the same torches used for gas welding (an oxyacetylene torch, for example) is a common method for metal cutting. The process is called

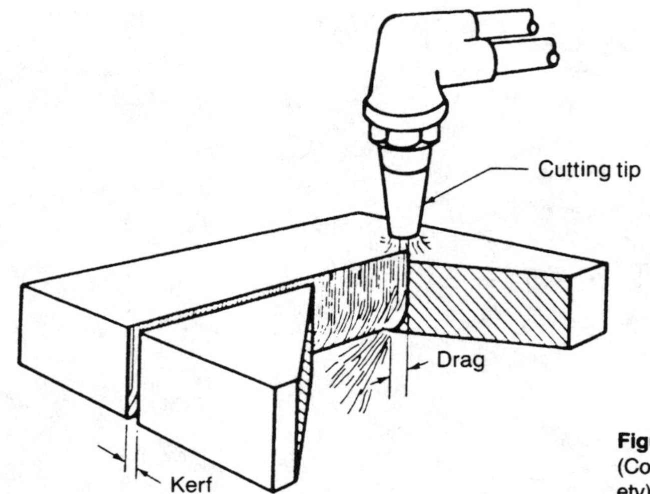

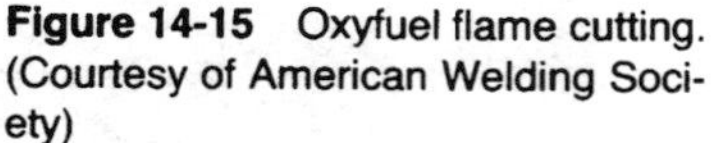

Figure 14-15 Oxyfuel flame cutting. (Courtesy of American Welding Society)

oxygen cutting because the oxyfuel flame is used only to begin the cut. After the metal is heated to the desired temperature, the fuel gas is turned off, and a steam of pure oxygen is directed against the metal. The metal along the cutting path then ignites and burns away with the aid of the oxygen. See Figure 14-15. Conventional oxygen cutting cannot be used on nickel alloys, because not enough heat is generated by their reaction with oxygen to sustain the cutting action. Special tooling can be used to cut special shapes. See Figure 14-16. In general, oxyfuel methods are used to cut metal plate over 25 mm (1 inch) thick. For workpieces up to that size, shearing is the preferred cutting method.

A modification of oxygen cutting, called metal *powder cutting* (POC), is used for nickel alloys. A fine powder of iron oxide and aluminum is added to the oxygen stream, where the powder burns and creates enough heat for effective cutting. The process is not often used for accurate cutting or for cutting finished products. Burned metal from the powder and the workpiece accumulates on and around the cut edges and must be removed after separation. Powder cutting is illustrated in Figure 14-17.

Figure 14-16 This special fixture permits the oxyfuel cutting of circles. Other special tooling is available for a variety of shapes. (Courtesy of Weld Tooling Corp.)

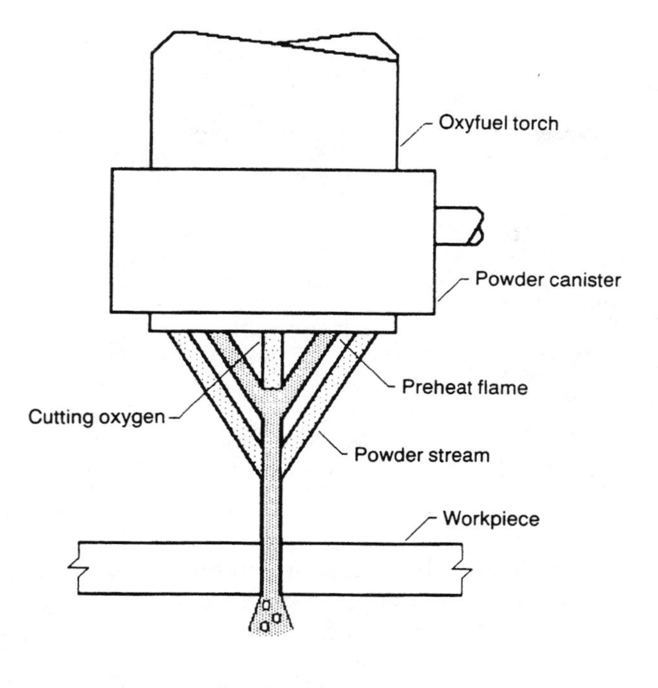

Powder cutting

Figure 14-17 Metal powder cutting.

Other common methods for thermal cutting use an electric arc as the heat source. The three processes having the widest use are *plasma arc cutting* (PAC), *gas tungsten arc cutting* (GTAC), and *air carbon arc cutting* (AAC). Plasma arc and gas tungsten arc cutting are both done with the same type of equipment used for welding. The cutting equipment, however, is usually much larger, especially for PAC. Increased current and gas pressure are used so that the metal melted by the arc is blown off the workpiece before it solidifies. Thicker material can be cut with the plasma arc process. Cutting speeds are also faster with plasma arc than with gas tungsten arc cutting. See Figure 14-18.

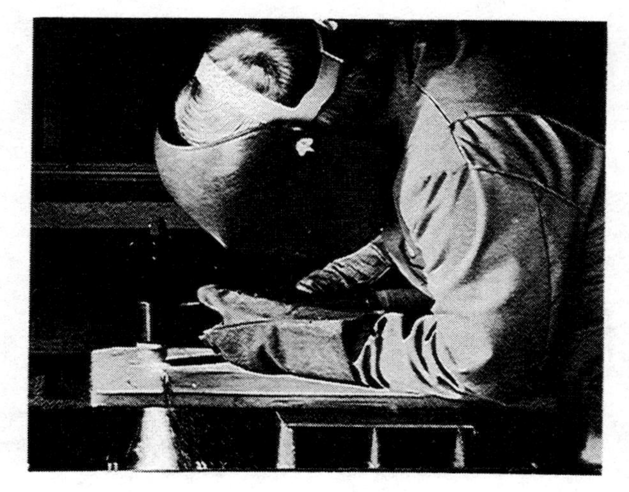

Figure 14-18 Plasma arc cutting. (Courtesy of Thermal Dynamics Corporation, West Lebanon, NH)

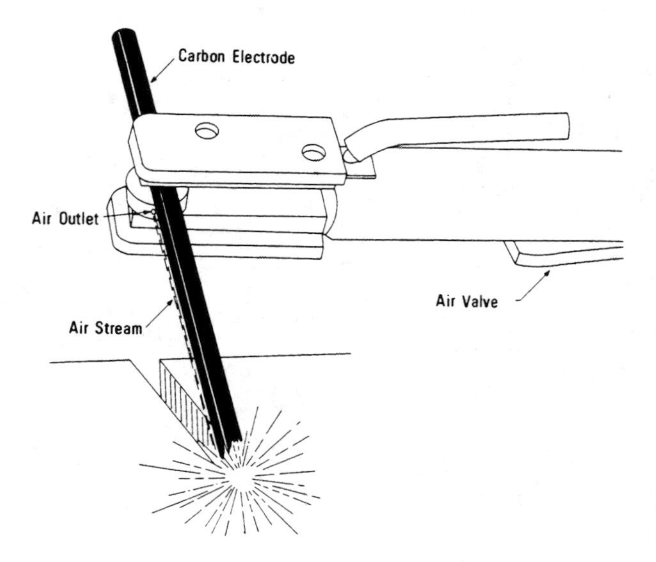

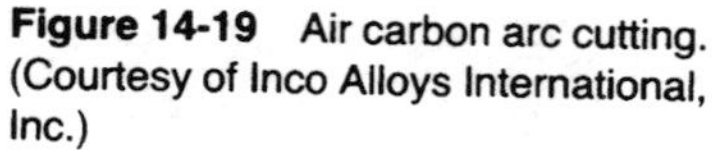

Figure 14-19 Air carbon arc cutting. (Courtesy of Inco Alloys International, Inc.)

Air carbon arc cutting, like other arc cutting methods, consists of melting a path through a workpiece by heating the metal with an electric arc discharged between an electrode and the workpiece. The carbon electrode is mounted in a holder that also contains an air jet. In operation, the metal at the cut is melted by the arc and is immediately blown away by the high pressure air stream. See Figure 14-19. While manual methods are often used for thermal cutting, most volume production is done by computer controlled systems. See Figure 14-20.

LASER SYSTEMS

The laser is a special form of high intensity energy which has many applications in communications, medicine, and manufacturing. Laser light travelling through a finger-thick cable containing 144 optical fibers can transmit 50,000 conversations. The same telecommunication level would require a 415 mm (4.5 inches) thick cable

Figure 14-20 Automatic thermal cutting systems are generally employed in industrial manufacturing. (Courtesy of MG Industries)

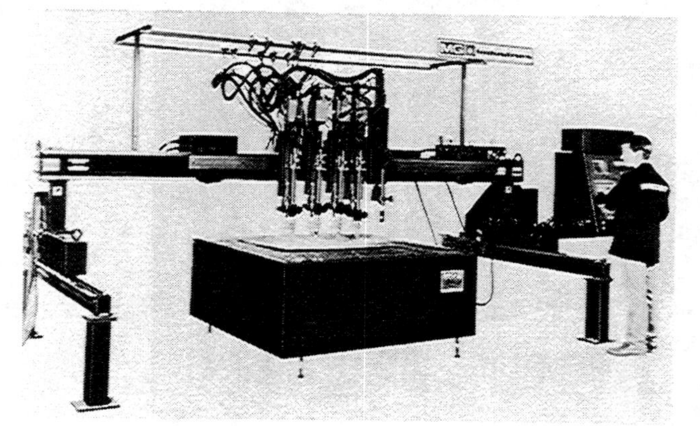

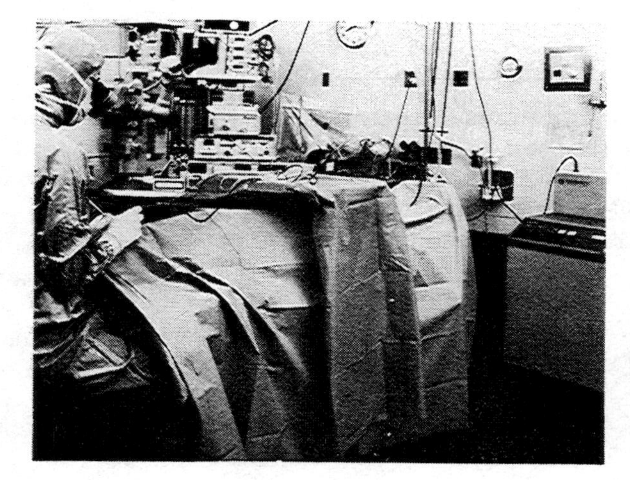

Figure 14-21 Laser technology is employed in numerous medical applications, such as this ophthalmic photocoagulator for eye surgery. (Courtesy of Coherent Medical Division)

of 8,100 conventional copper wires. The bed of a large precision milling machine can be aligned by measuring the variations in a laser beam flashed parallel to the machine ways. A laser beam focused by the lens of the human eye upon the retina can spot weld a detached retina into place. See Figure 14-21. The plastic insulation covering tiny copper wires can be accurately burned away to facilitate area specific electronic soldering. Lasers find industrial applications in welding, cutting, drilling, cladding, heat-treating, and measuring. Laser systems have a number of advantages over conventional processing methods. See Figure 14-22.

Comparison of Laser and Conventional Processes

Process	Laser's Advantages
CUTTING	
Oxy-Acetylene	Narrow kerf, minimal heat-affected zone, cuts sharp profiles, minimal part distortion
Plasma Arc	Narrow kerf, minimal heat-affected zone, cuts sharp profiles, minimal part distortion
Band Saw	Narrow kerf, no burrs, minimal distortion, sharp profiles, faster rates
Nibbling	Smooth edges, no burrs, narrow kerf, no part distortion
Punching	No dies necessary, complex profiles processed, adaptable to short runs
WELDING	
Gas Metal Arc	No filler metal required, low distortion, faster welding rates, single-pass two-sided welding
Submerged Arc	No flux and filler metal required, low distortion, faster welding rates
Resistance	No work contact required, reach inaccessible locations, faster rates
Electronbeam	Operates in air, on-line processing, welds magnetic materials
SURFACE HARDENING	
Flame Hardening	Precise control of case depth, no quenchant needed, minimal part distortion, faster process rates, hardens most iron and steel alloys, selective area hardening
Induction Heating	No coils necessary, working distance not critical, no quenchant necessary, low distortion, reach inaccessible locations, process most steels
Electronbeam	Operates in air, working distance not critical, reaches inaccessible locations
Carburizing Carbonitriding Nitriding	On-line processing, no protective environment necessary, can process cast iron

Figure 14-22 The laser has certain advantages over conventional processing methods. (Courtesy of Control Laser Corporation)

LASER THEORY

Laser light differs from ordinary light in several ways. One is that the light waves of a laser beam are all in phase, whereas those of the typical light beam are not. Another difference is that the light from a laser scatters far less with distance than does ordinary light. For example, laser light beamed from the Earth to the Moon theoretically would spread out to a spot on the Moon 3 kilometers (1.8 miles) in diameter. A common light beam theoretically would have expanded to a circle diameter of 40,000 kilometers (240,000 miles) at the same distance. A laser beam is also monochromatic, meaning that its waves all have the same frequency, or, if visible, color. The reason for the qualification "if visible" is that there are lasers that emit beams in invisible infrared or ultraviolet frequencies. Ordinary light, as well as infrared and ultraviolet light, is usually a mixture of frequencies.

Laser technology is the result of many researchers applying scientific theory to gain practical results. Such scientists as Max Planck, Albert Einstein, and Niels Bohr laid the theoretical groundwork for the laser early in this century. Those associated with applying this theory to develop the laser are Charles H. Townes (who built the first maser—*M*icrowave *A*mplification by *S*timulated *E*mission of *R*adiation), Arthur L. Schawlow (who with Townes developed a plan for a working laser), Theodore H. Maiman (who demonstrated the first crystal laser), and Ali Javan (who demonstrated the first gas laser).

The word *laser* is an acronym derived from the descriptive phrase, *L*ight *A*mplification by *S*timulated *E*mission of *R*adiation. Laser theory is based upon the ability of light energy photons to stimulate the emission of additional photons, each possessing the same wavelength and direction of travel.

Atoms and molecules have various energy levels. They are capable of changing from one level to another by absorbing or releasing electromagnetic radiation. Under normal conditions, most atoms or molecules maintain their lowest energy level or *ground state*. However, when they are excited by electrical energy (called *pumping*), they are driven to a higher state, and as they return to their ground state, they emit energy in the form of protons. This phenomenon is called *population inversion*.

In the laser chamber, emitted photons are trapped between two highly polished parallel mirrors, forcing them to move back and forth within the confines of the chamber. See Figure 14-23. When a photon passes close to another excited particle, the second particle will also be stimulated to emit a photon that is identical in wavelength, phase, and spatial coherence to the first. This amplification process continues, increasing the number of active photons, leading to the generation of a laser beam.

Figure 14-23 A typical gas laser unit. (Courtesy of Coherent Medical Division)

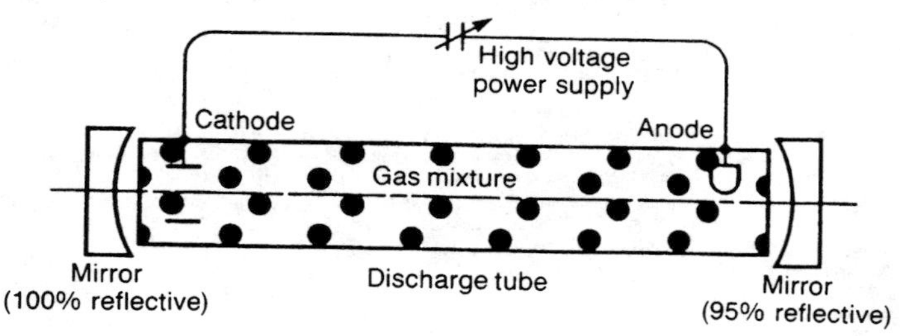

All lasers are comprised of three basic elements: a lasing medium, which provides atoms molecules; an energy source to excite the medium; and an optical resonator to provide feedback of the amplified light. The light wave builds up in the resonant cavity, or chamber, as it makes hundreds of reflective trips between the mirrors. On each of these trips, a fraction of the energy is emitted from one of the mirrors, which is only partially reflective. This fraction becomes the laser beam. An illustration of this action is shown and described in Figure 14-24.

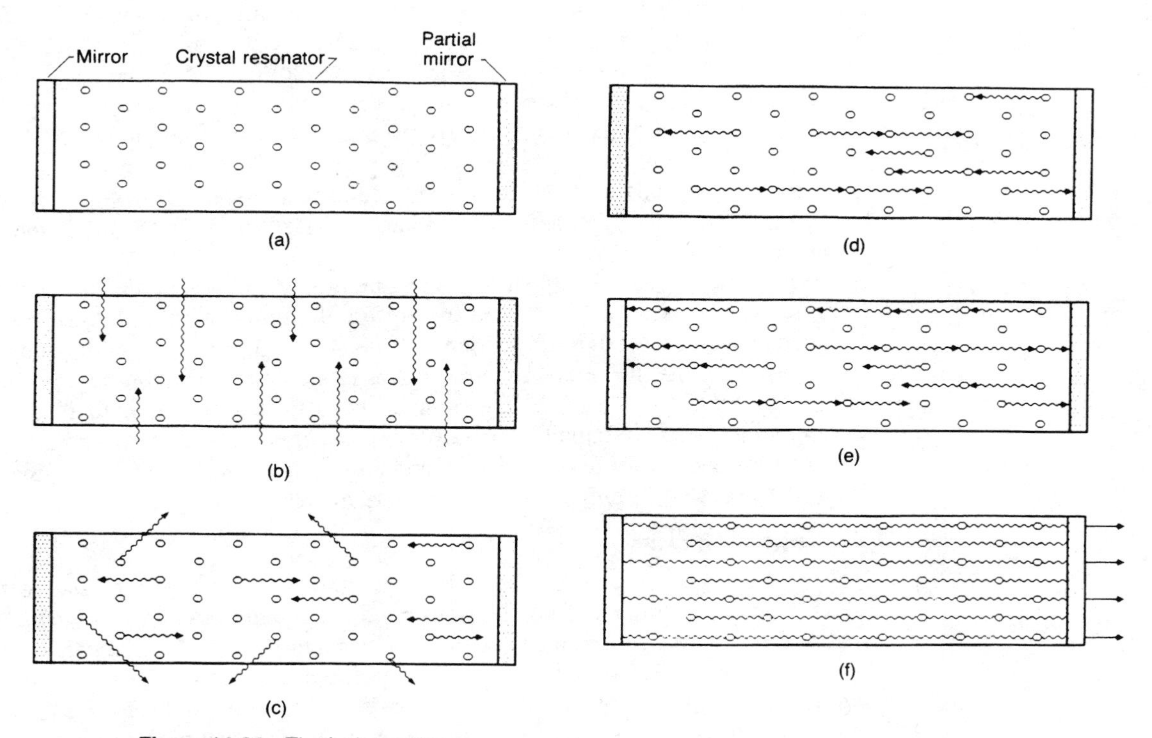

Figure 14-24 The lasing action.

a. A laser beam is generated when the atoms in a lasing medium (in this example a ruby crystal), are driven to a higher energy level. The atoms shown here are in their normal ground (lowest) energy level because no excitation energy is present.
b. Here the atoms are excited, or pumped, by saturating the crystal with energy from an external source such as a flash lamp. The atoms are raised to a higher energy level, and photons are emitted spontaneously. This is called population inversion.
c. Those photons emitted in directions not in line with the mirrored surfaces will pass through the sides of the crystal without contributing to the laser action.
d. When a photon travelling parallel to the axis of the crystal medium collides with another excited atom, amplification occurs, because that atom will be stimulated to emit a photon identical to the first, and moving in the same direction.
e. These photons produce amplified, single-frequency, coherent light waves in the crystal as they reflect back and forth between the mirrors, and trigger the emission of more identical photons.
f. Some of these photons pass as amplified light through the partially mirrored surface to emerge as an intense laser beam of parallel rays. Others are reflected back into the laser cavity to generate more photons. This phenomenon will continue as long as population inversion is maintained in the crystal medium.

TYPES OF LASERS

Lasers can be classified according to the medium used in their design and construction, and the most common are the gas, solid state, liquid, and semiconductor lasers. Of these, the gas and solid state types are most used in industrial processing. Liquid dye lasers are used in research and in medical applications, while the semiconductors find use in communication systems.

Gas lasers typically are comprised of an optically transparent tube filled with a gas, or a gas mixture, as the lasing medium. The pumping mechanism is an electronic discharge system. The gas laser most commonly used in industry is the CO_2, which uses a mixture of helium, nitrogen, and carbon dioxide. This laser can be operated in either the pulsed or the continuous wave (CW) mode. See following. This is a powerful industrial laser and is widely used as a welding and cutting tool.

The *helium-neon* (HeNe) laser is a low power gas unit used in bar code reading, such as in supermarket checkout counters, and to transmit newspaper photographs. The *argon* gas laser is another low power unit used in research, medicine, and for commercial light displays.

Solid-state lasers are crystalline in structure. The typical lasing medium is a commercially produced crystal rod of yttrium-aluminum-garnet (YAG), doped with a material such as neodymium which will lase. It replaces the original ruby crystal. The correct name for this medium is Nd:YAG, but it is commonly referred to simply as YAG. These lasers are pumped by a flashlamp mounted in a reflecting cavity, which also contains the crystal rod. See Figure 14-25. It has many industrial applications because of high efficiency and pulse rates, and its simple cooling system. It can be operated in either the CW or pulse modes.

Operational Modes

Lasers produce heat either as a short, intense burst light energy called pulsing, or as a constant stream of steady power, called continuous wave. *Pulsing* implies that the excitation mechanism, such as a flashlamp, stimulates the laser medium for very brief, high power peaks. This permits the laser to emit short, intense bursts of energy, or pulses, sufficient to generate enough heat to overcome the absorption threshold of most materials. In this fashion, welding or cutting can take place. Furthermore, this short pulse duration assures minimum heat input, to control the heat affected zone (HAZ).

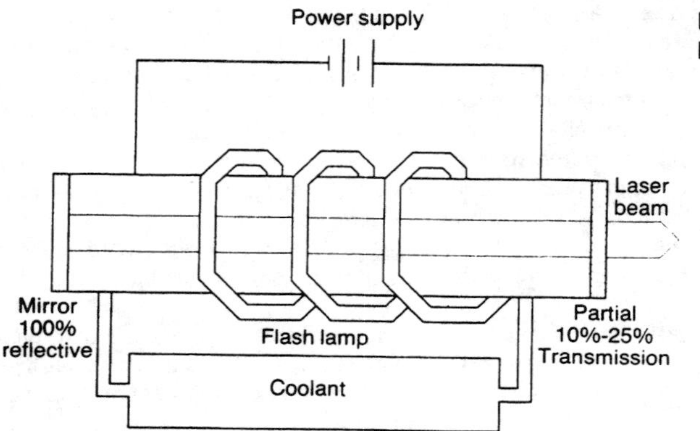

Figure 14-25 A typical solid state laser unit.

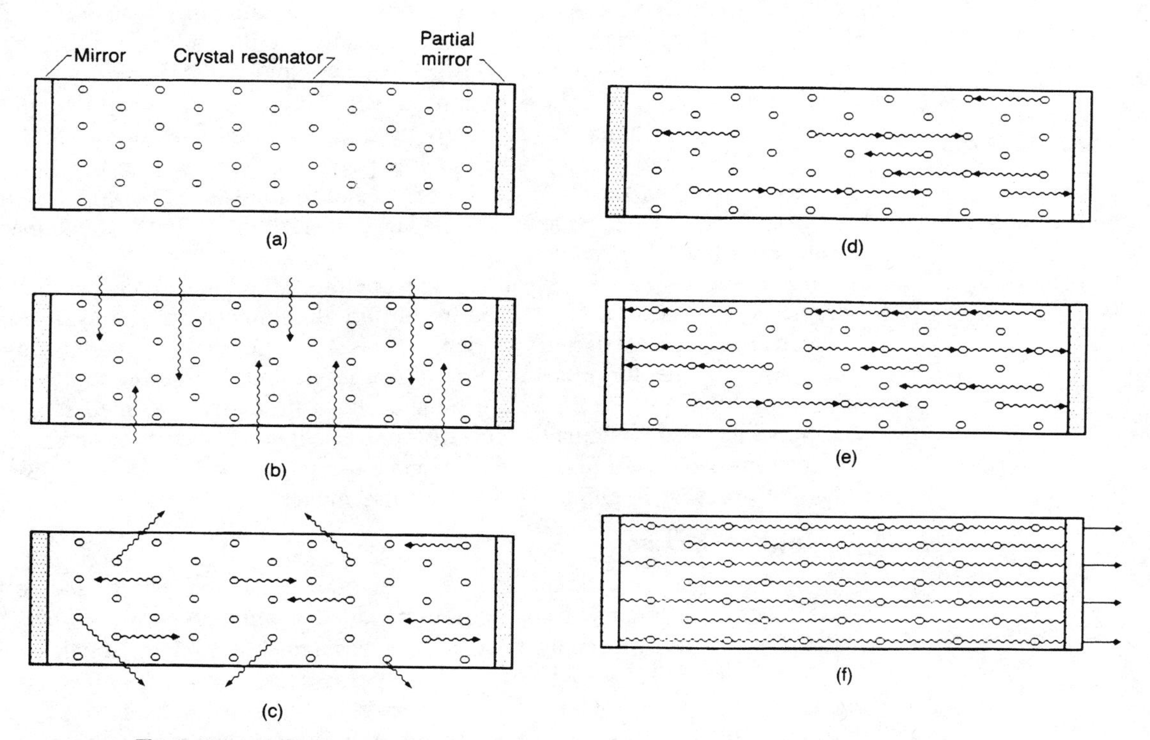

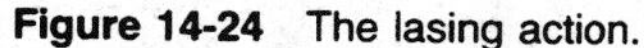

However, pulsed lasers have limited power, a factor which restricts their depth of penetration. This precludes their use in efficient seam welding, because slower weld overlaps are necessary.

Continuous wave (CW) lasers maintain their levels of energy output over the entire duration of the process cycle. They are, therefore, better suited to the constancy required for seam welding. The key to effective, high-speed production is the capability to maintain a constant weld puddle or keyhole. At present, CO2 lasers only are capable of sustaining enough power to generate and maintain a keyhole and permit it to be moved along the weld joint area.

The two types of pulsing in general use are the enhanced and the gated. With *enhanced* pulses, higher current pulses are generated to excite a CO2 medium resulting in higher peak power. This enhancement permits low-to-medium CW lasers to achieve enough energy to vaporize most materials. In *gated* operation, the current is switched on and off to produce high energy pulses. This is called *Q-switching*. A Q-switch is a device which permits a medium to be pumped until it holds the maximum energy level. These switches generally are used with solid state lasers. Q-switches are available as mechanical rotating prisms, polarized optical units, or acoustical optical pressure units. These functions are identical to control the sequence of light beam travel.

As a summary to the preceding discussion, CO_2 and YAG lasers are the most common in use in industry. Each system has attributes and limitations which dictate their selection for specific applications. See Figures 14-26 and 14-27. There follows a treatment of specific industrial laser beam machining (LBM) processes.

Laser cutting is done with either YAG or CO_2 equipment, utilizing both the CW or pulsed modes. In the pulse mode, material removal occurs when a high energy light pulse of very short duration—around 1 millisecond—impinges on the material surface. The absorption of this energy causes the material to heat rapidly, melt, and vaporize, thereby generating a small hole about 0.3 mm (0.01 inch) in diameter. The hole size is a factor of the power level, beam focus, and the type of material. The laser beam is moved laterally a fraction of a millimeter, where the process repeats to produce a second hole or bite. A laser cut is in reality a series of overlapping laser pulses as shown in Figure 14-28. The kerf thereby produced

Figure 14-26 Chart of laser types and uses. (Courtesy Raytheon Laser Products)

Industrial Lasers—A Profile of Available Equipment		
Laser Type	**Power**	**Principal Uses**
CO_2	Up to 100 W 100–500 W 1000–5000 W 10,000 W and up	Ceramic scribing, cutting, drilling Welding, some cutting applications Heat treating—heavy section welding, cutting Potentially useful for large scale heat treating, cladding, alloying
Pulsed YAG	Up to 50 W 100–400 W	Spot welding, small hole drilling and soldering Seam welding, high speed spot welding, cutting, and larger hole drilling
CW YAG (q-switched)	Up to 100 W	Ceramic scribing, resistor trimming, diamond sawing, noncontact marking
CW YAG	200–800 W	Welding and heat treating

Laser Cut Quality

Material	Quality of Cut CO_2	YAG
Alumina	Excellent	Fair to good
Aluminum	Fair	Good to excellent
Brass	N/A	Fair to good
Carbon steel	Excellent	Excellent
Composites	Variable	Fair to excellent
Copper	N/A	Fair to excellent
Glass	Poor to good	N/A
Paper/cardboard	Good-excellent	N/A
Plastic (thermoforming)	Good to excellent	N/A
Plastic (thermosetting)	Poor to good	N/A
Stainless steel	Fair to good	Excellent
Titanium	Fair to good	Good
Wood	Excellent	N/A
Zircalloy	Excellent	Excellent

Poor—considerable dross—ragged cut
Fair—moderate dross—little roughness to cut
Good—little dross—cut relatively smooth
Excellent—essentially no dross—cut smooth
N/A—not applicable

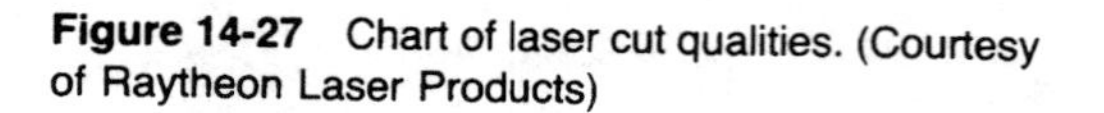

Figure 14-27 Chart of laser cut qualities. (Courtesy of Raytheon Laser Products)

is the same size as the hole diameter, and can be smooth or jagged, depending on the bite size, which is regulated by the lateral laser motion. The *bite size* is the amount of additional material attacked with each laser pulse. The *kerf* is the outer width of the cut. The *Q-rate* (see Q-switching) is the number of laser pulses issued per second.

Laser cutting (or trimming) is especially applicable to electronic components. If hole size is small, as would be desired in high resolution circuits, bite size must be lowered to achieve a quality cut. This results in a lower cutting speed if the same Q-rate is maintained. Sufficient overlap of holes must be maintained to assure a "clean" kerf; that is, all the material within the kerf has been removed. Almost

Figure 14-28 Laser cutting using a series of overlapping hole cuts. This method is typical of the method used to trim electronic circuit components such as resistors. Note that a ragged kerf results if the bite size is too large. (Courtesy of Chicago Laser Systems, Inc.)

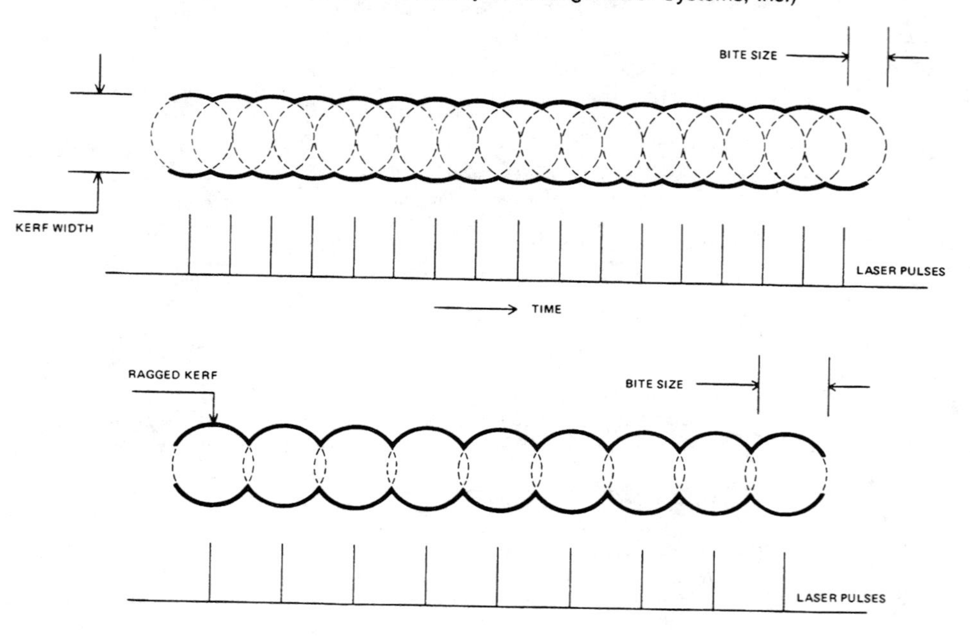

all trimming requires hole overlap. A notable exception is crystal trimming. Here it is desirable to remove material in a hole pattern to maintain a conducting path in the metallization layer. Excessive overlap, as would occur if bite size is too small for the hole size, results in ''overpowering'' the cut. This may assure a smooth clean kerf, but generates excessive heating, which can cause drifting of the resistance value, or micro cracking of the uncut material after the trim is complete.

The CW laser operates in a similar fashion, except that the beam is constant and moves in a continuous path to complete the cut, as discussed earlier.

As stated, most laser cutting operations in industry are done with either CO_2 or the YAG laser. The choice of laser is dependent upon such factors as type of material to be cut, width of cut (kerf) restrictions, and ability to cut with the aid of various gases. Laser cutting takes place in one of three ways. For those materials with low thermal conductivity, the cutting occurs by vaporization. Typical here are the cuttings of wood, paper, or cloth using a CO_2 laser. A second class of materials are those which are nonreactive and have high thermal conductivity. The material removed in this case is the form of a boiling liquid blown away by a high-velocity gas stream. A good example of this is cutting stainless steel with a VAG laser.

The third method is similar to the second, but with the addition of an oxidizing gas stream. This stream assists the laser beam and greatly enhances the speed of cutting. Typical here is the cutting of carbon steel with a CO_2 laser. Assist gases also protect the laser lens from spatter.

Most materials can be cut by laser. Omnidirectional, the laser has no contact with material to be cut and requires no starting hole to produce smooth cuts. See Figures 14-29 and 14-30. Almost every intricate pattern, shape, or contour desired may be cut without the need of special fixtures or dies. Cutting rates are extremely rapid, and cuts in excess of 25 mm (1 inch) can be made. Very narrow kerf widths, minimum heat-affected zones, and the absence of chips or slivers are features of the products of laser cutting. See Figure 14-31.

Figure 14-29 This computer controlled laser metal cutting operation is assisted by high pressure gases. (Courtesy of Westinghouse Electric Corporation, Marine Division)

Figure 14-30 The precision and omnidirectional cutting capabilities of the laser are exhibited by these circular holes being cut on a metal cylinder. (Courtesy of Westinghouse Electric Corporation, Marine Division)

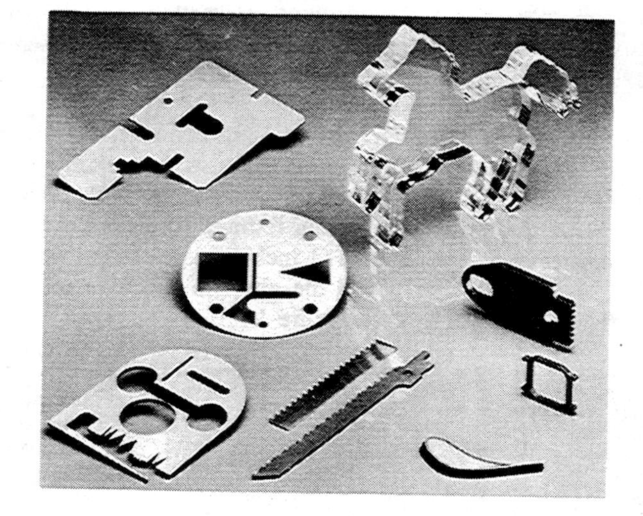

Figure 14-31 Some examples of laser cut materials. (Courtesy of Leybold-Heraeus Vacuum Systems, Inc.)

The laser is especially adaptable to precision work. Such sharp-edged cutting can be done on workpieces up to 6 mm (0.25 inch) thick, with kerf widths or about 0.1 mm (0.0039 inch). There is virtually no thermal distortion in these cuts. Complex forms can be noncontact cut in workpieces of any shape—sheet, tube, or close hollow objects. See Figure 14-32.

Laser drilling is somewhat of a misnomer, for the term ''drill'' generally implies a cylindrical contact tool which produces a hole. A better term is laser hole cutting or generating, for it is achieved in the same manner as cutting, except that the technique is modified depending on the hole size and, as shown in Figure 14-32, hole shape.

A small hole is ''burned'' by the laser in a single pulse in thin material. A thicker workpiece may require a sequence of controlled laser pulses to gradually increase the hole depth. The taper of holes created by lasers is negligible, and the

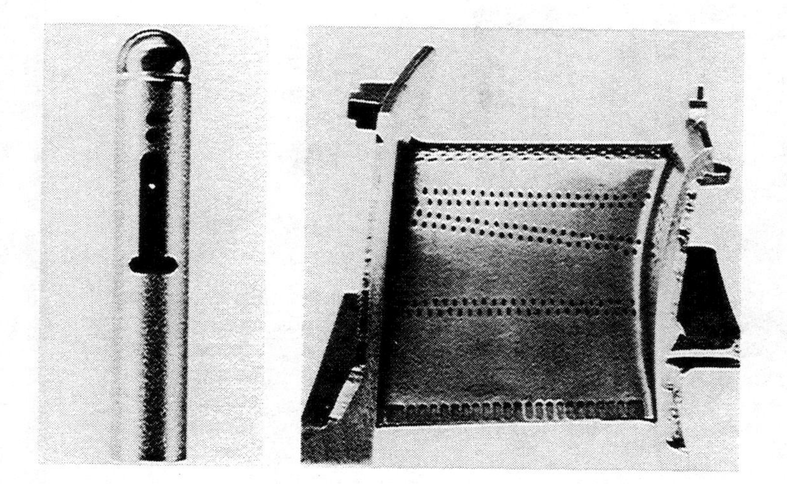

Figure 14-32 (Left) Complex geometric holes were cut into the thin, hollow tube which is the endpiece of an endoscope. The steel tube diameter is 0.14 mm (0.06 inch), and the cutting time was 3 seconds. (Courtesy of LASAG Corporation)

Figure 14-33 (Right) The inclined laser cut holes on this aircraft engine turbine guide vane are between 0.3 mm (0.0012 inch) and 0.8 mm (0.003 inch) in diameter. (Courtesy of LASAG Corporation)

ratio of depth to hole diameter is around 15:1. In hole cutting, nearly all vaporized material is expelled from the hole, especially with gas jet assistance. What is not expelled remains as a smooth coating on the walls of the holes, and this coating is called *recast*. An example of precision hole cutting appears in Figure 14-33. Holes above 0.8 m (0.003 inch) in diameter are cut out, utilizing the sequenced overlapping holes described earlier. Here the workpiece is placed on a CNC coordinate table, and the table is directed under a stationary beam to achieve the cut.

LBM also is being applied to lathe turning operations. See Figure 14-34. According to this system, a beam of laser lights is split into two beams, which are then directed along different axes of a rotating cylindrical workpiece. One beam is directed against the end face, while the second beam plays against the side of the workpiece. When the kerfs created by the two cutting beams intersect, a solid ring of material is removed. The side beam is then stepped back by a predetermined increment, and the process continues along the length of the workpiece.

Figure 14-34 A comparison of conventional and laser turning. In laser turning, the axial beam cuts a trepanning kerf on the front surface of the workpiece, while the radial beam simultaneously cuts a transverse kerf on the circumferential surface. When the kerfs intersect, a solid ring of material is separated from the workpiece. (Courtesy of *Manufacturing Engineering* Magazine)

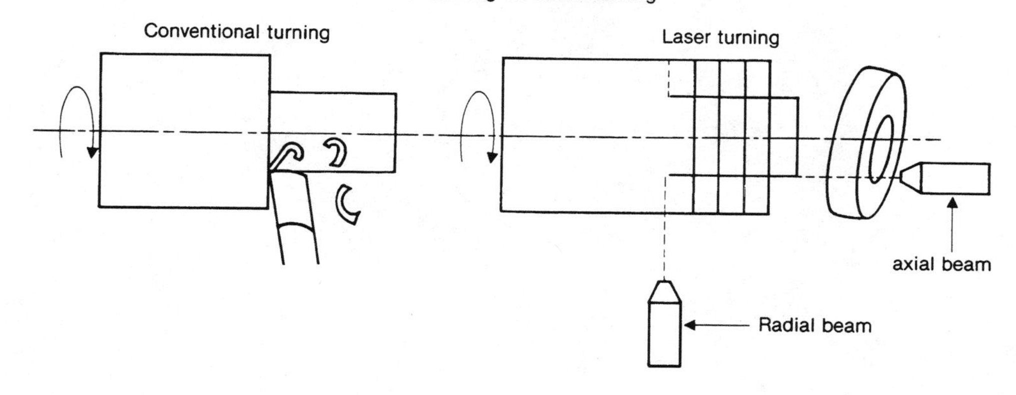

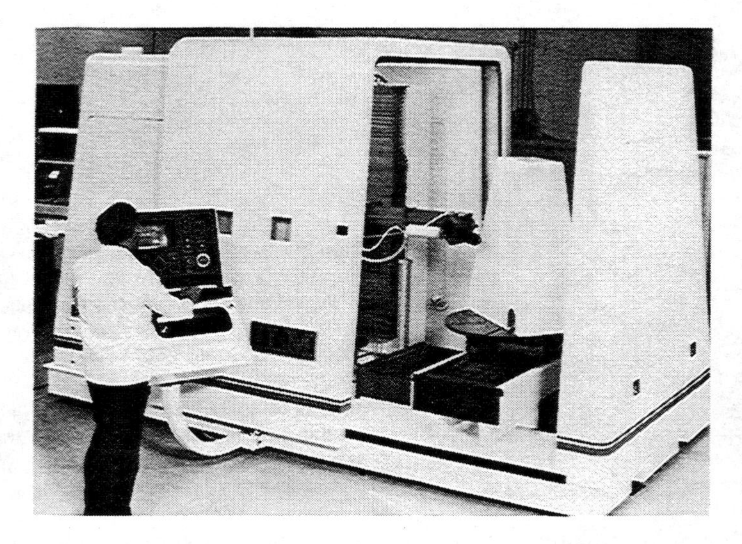

Figure 14-35 This five-axis drilling, cutting, and welding laser is being programmed via a pedestal-mounted CNC controller. The sliding hood provides complete accessibility of the work chamber. (Courtesy International Laser Machines Corporation, A Cincinnati Milacron Company)

Figure 14-36 A typical laser marking application. (Courtesy of Control Laser Corporation)

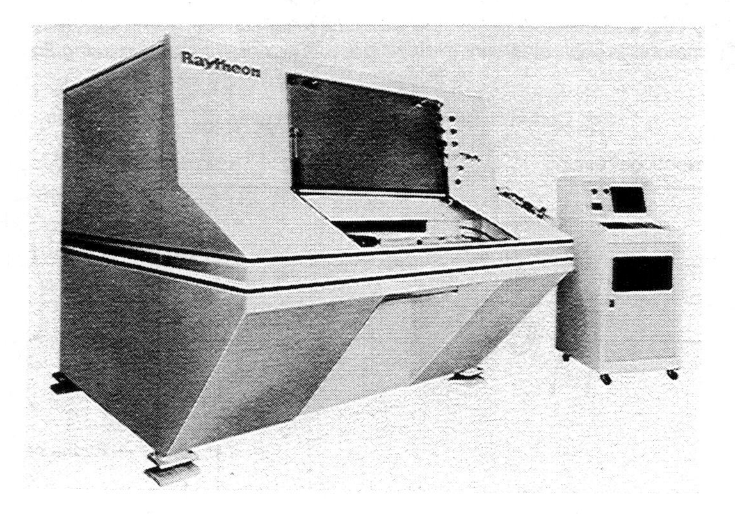

Figure 14-37 Laser marking or engraving equipment. (Courtesy of Raytheon Laser Products)

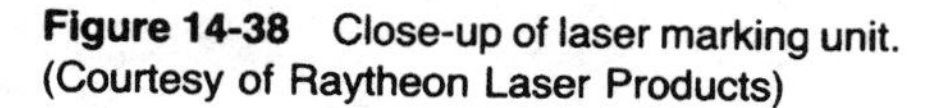

Figure 14-38 Close-up of laser marking unit. (Courtesy of Raytheon Laser Products)

As indicated earlier, sophisticated equipment is available to process materials speedily, accurately, efficiently, and safely. See Figure 14-35. The precision of the laser, augmented by the precision controls of CNC equipment, facilitates the production of extremely high quality production rates.

Scribing and marking are lasing processes similar to cutting. For example, a pulsed CO_2 laser is used to produce shallow, overlapping holes or marks on a ceramic sheet to create a perforated *scribe* line, along which the sheet can be snapped to fracture and separate the pieces. Sometimes the holes are close together and pierce the sheet, but do not overlap, to produce a similar scribe line. This is a good laser application, because ceramic sheets are very brittle and do not lend themselves to separation with ordinary cutting tools.

Laser *marking* or engraving is the production of controlled surface patterns on a workpiece, such as company logos, parts numbers, bar codes, or serial numbers. See Figure 14-36. Metal, glass, and paper cans can be so marked using CO_2, krypton, and liquid die lasers. See Figures 14-37 and 14-38.

QUESTIONS

1. Describe the differences between EDM and EDWC.
2. Describe the differences between ECM and CHM.
3. Describe the differences between EDG and ECG.
4. List the steps in the photochemical machining process.
5. Prepare a sketch of the USM process, and state the applications of the method of machining.
6. How does thermal drilling differ from conventional drilling? From thermal cutting?
7. List some uses for water jet cutting.
8. Differentiate between the common types of thermal cutting.
9. Define the term ''laser.''
10. Define these laser terms: pumping; ground state; population inversion; pulsing; CW; and Q-switching.
11. List some applications of laser beam machining, and explain the attributes of each.
12. Prepare a library research report on current laser technology.

part 4 MATERIAL PROCESSES: FORMING

15 BENDING AND DRAWING

Two of the most common and versatile methods of cold forming sheet metal products are bending and drawing. Both systems are mass deformation processes and involve some type of rigid or flexible forming elements to shape such products as heating ducts, food containers, and numerous others. See Figure 15-1. Such operations typically are performed on cold metal workpieces. Cold forming is defined as the deformation (plastic flow) of a metal at a temperature lower than its recrystallization point. Plastic flow at such temperatures strains and thereby alters the internal structure of the metal, instead of simply rearranging it as in hot forming. This cold work strengthening process is known as *strain-hardening*.

Cold forming has several advantages over hot forming. It is generally more economical, because there is no need for heating furnaces and fuels. The surfaces of cold-formed parts are cleaner and brighter than the oxidized surfaces of hot-formed parts. Because cold forming can be done to closer tolerances, parts can be formed to the actual size required. Hot-formed parts shrink as they cool, and so must be formed oversize. Another important advantage is the strengthening of the metal that results from strain-hardening. Tremendous increases in strength can be gained by cold work.

An important disadvantage of cold forming is that for the same amount of deformation, more energy is required than is necessary for hot forming. The larger power requirements make it impossible to achieve the extreme deformation that is common in hot extrusion and forging. Also, strain-hardening becomes a disadvantage if the work is hardened so much before it is formed to the final shape that annealing is

Figure 15-1 This motorcycle is made from a number of cold-formed parts, such as deep drawn fuel tanks and mud guards, bent exhaust tubes, and roll-formed wheels. Hot-formed parts include forged engine piston rods and wheel axles. (Courtesy of Harley Davidson, Inc.)

required. Despite these disadvantages, cold forming processes are a common and valuable means of giving shape to metal articles. The range of bending and drawing operations, also called *presswork*, is the subject of this chapter.

BENDING

Very simply, bending is a process by which a material is uniformly strained along a straight axis, and it is applicable to sheet, plate, bar, rod, wire, tube, and structural shapes. It is a common forming operation used to produce hundreds of shapes. See Figure 15-2. In contrast with more complex forming operations, bending requires

Figure 15-2 Bends such as these can be produced in metal, plastic, ceramic, and other materials. All such bends are made in long or short straight workpieces.

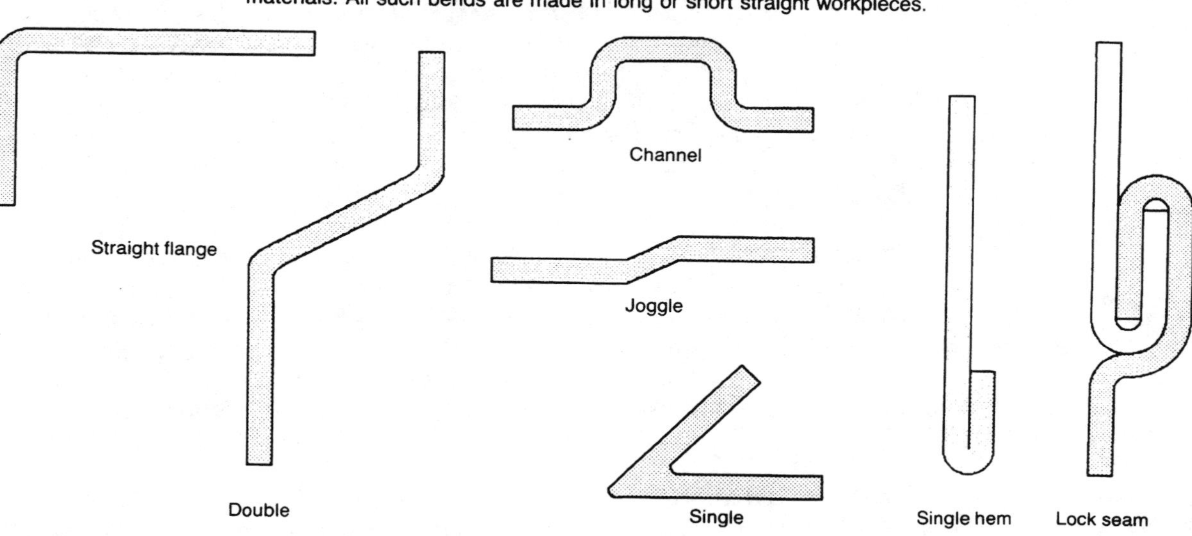

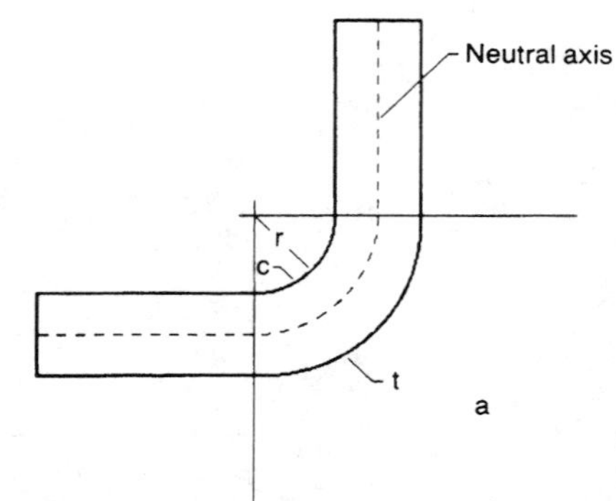

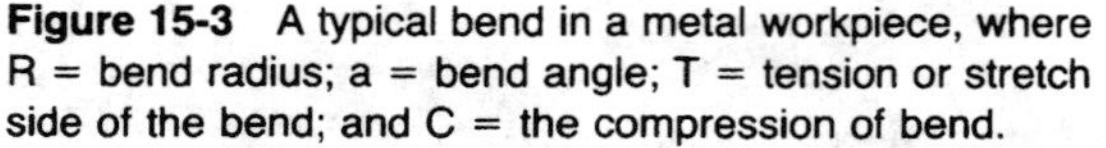

Figure 15-3 A typical bend in a metal workpiece, where R = bend radius; a = bend angle; T = tension or stretch side of the bend; and C = the compression of bend.

little plastic flow and does not appreciably alter the structure of the workpiece. The forming usually occurs along a straight line with little or no change in the thickness of the material.

The condition illustrated in Figure 15-3 is typical of bent metal workpieces where the metal is stressed beyond its elastic limit in tension on the outside of the bend and in compression on the inside surface. Certain precautions must be taken to avoid cracking the metal when bending it, such as by bending with the grain direction of a coiled sheet. Softer metals usually can be strained to a 180 degree bend, with a bend radius approximately equal to the workpiece thickness. Bend angles must be smaller and radii larger on thicker, harder metals. Forming difficulties also can be avoided by attempting no bends closer to the edge of the sheet than $1\frac{1}{2}$ times the thickness of the metal plus the bend radius. Tabular data in handbooks should be consulted to determine working values for specific bend situations. Bends are characteristically rounded at the bend point. Excessively sharp bends will fracture the tension side workpiece.

Metals are elastic when cold, and consequently have a greater tendency to return to their original shapes after forming. For example, a bend made to 90 degrees can be expected to spring back to a smaller angle of 88 or 89 degrees. This condition is called *springback*, and is greatest with heavier workpieces, smaller bend radii, and larger bend angles. Where springback is a factor, it is compensated for by *overbending* the material beyond the limits specified in the final workpiece shape. In press-brake bending, springback compensation also may occur with modified die designs. Several of the more common bending operations and machines are described here.

Brake bending. Metal sheets can be formed into heating ducts and large rectangular containers on *bending brakes*, called bar folders, folding brakes, or box and pan brakes. See Figure 15-4. The capacities of these machines range from simple hand-operated models to huge power equipment capable of forming straight bends in steel plate. One limitation of such machines is the force needed to lift the bending wing and workpiece to form large, heavy objects. This bending action is illustrated in Figure 15-5.

Press-brake bending. The most widely used method to form lengths of sheet metal into both simple and complex shapes is by *press-braking*. Precisely mated dies of hardened tool steel are made in suitable lengths to produce shapes in one

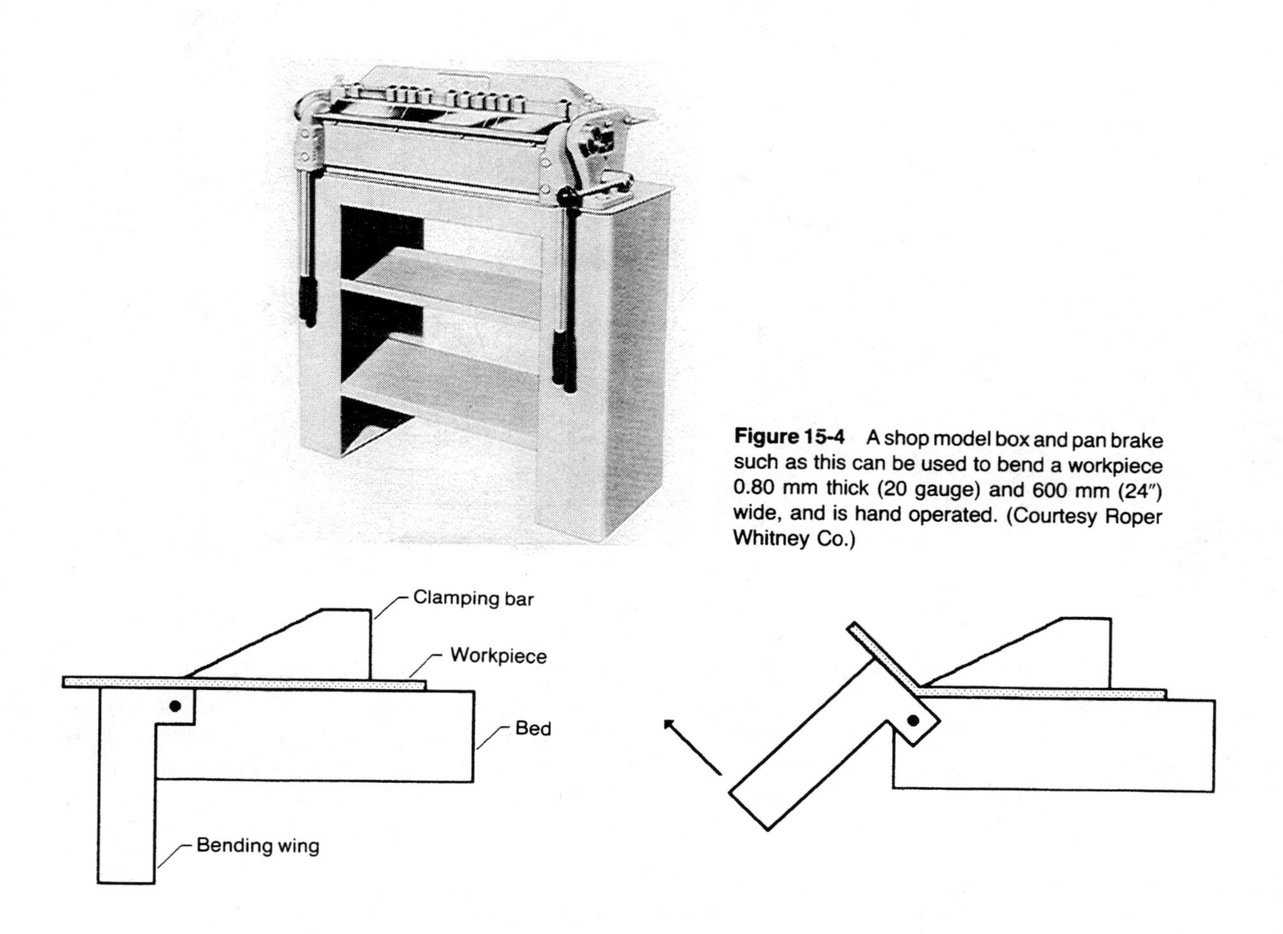

Figure 15-4 A shop model box and pan brake such as this can be used to bend a workpiece 0.80 mm thick (20 gauge) and 600 mm (24″) wide, and is hand operated. (Courtesy Roper Whitney Co.)

Figure 15-5 Bending brake action. The workpiece is held within this clamping bar, A, and bent by lifting the bending wing, B.

or more passes on the press. Bends made on these presses generally employ air-bending, bottoming, or three point techniques. See Figure 15-6. In *air bending*, the punch has an acute angle between 30 and 60 degrees to provide sufficient play so that springback compensation can be made by press adjustments alone. Angles formed by this method depend upon exact stroke depth settings to obtain the required punch penetration into the lower die. The accuracy of the bend is affected by the precision of the press, and variations in the tensile strength and gauge of the workpiece. The term "air bending" is used because the workpiece spans the gap between the nose of the punch and the edges of the mating die.

Bottom bending demands a greater press capacity to overbend the workpiece, and requires different die sets for each angle size to be formed. Additionally, the final product can be adversely affected by die variations and press frame deflection. In bottoming, the workpiece is in contact with the entire functional surfaces of both punch and die. The result is that very accurate angular tolerances can be maintained. Bottoming methods generally need three to five times greater pressing force than does air bending.

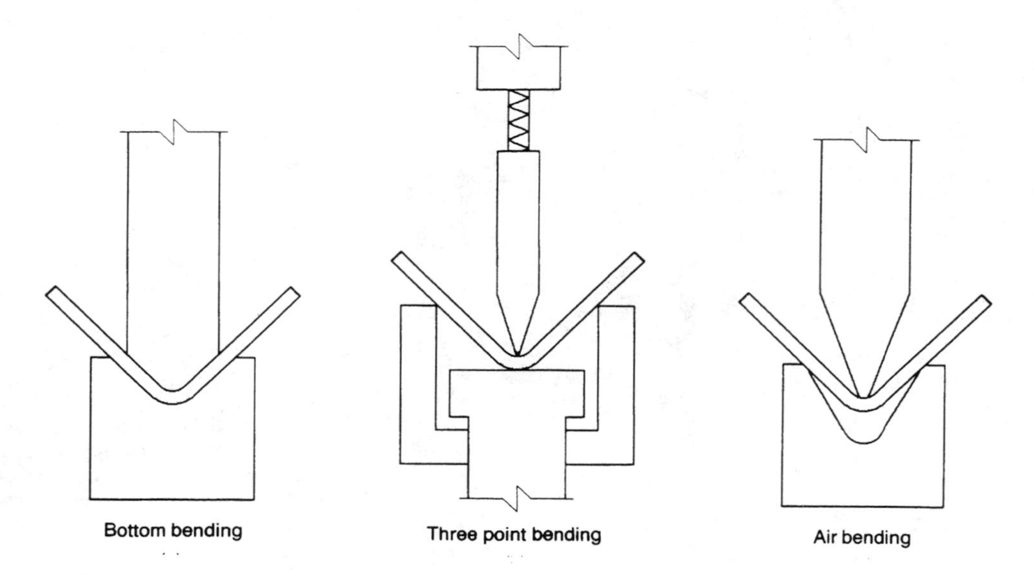

Figure 15-6 Classes of rigid press brake dies.

Three point bending incorporates three support points in the bottom tool. The bend angle is determined by adjusting the depth of the groove in the bottom tool, so that the full length accuracy of the bend is assured by the geometry of the bottom tool. A continuous action oil cushion in the ram provides constant pressure distribution on the sectioned punch to eliminate any errors which may result from press inaccuracy, frame deflections, and material variations. See Figure 15-7.

Figure 15-7 A 100 ton capacity, programmable hydraulic press brake such as this is used to produce bends in metal sheets and plates. Both rigid and flexible dies can be used. (Courtesy of Niagara Machine & Tool Works)

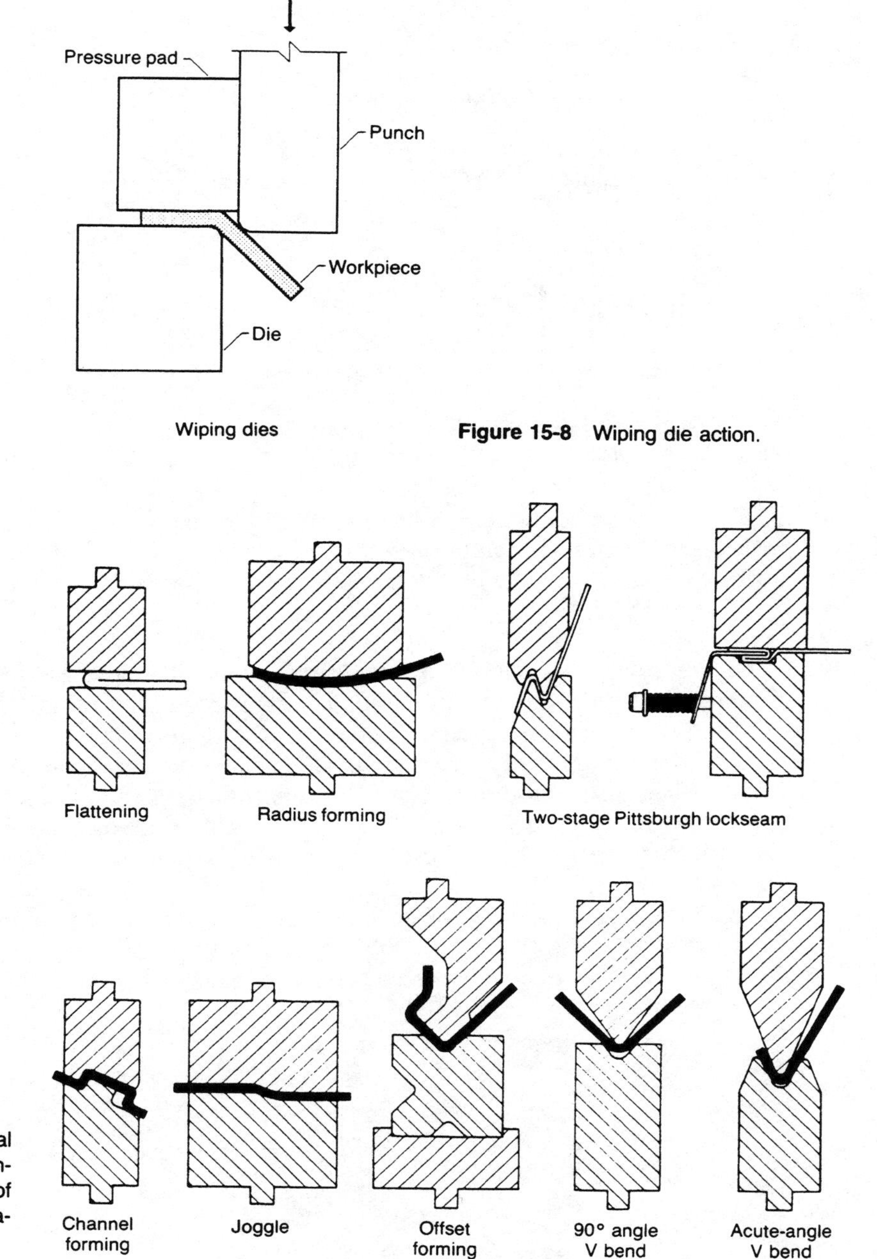

Figure 15-8 Wiping die action.

Figure 15-9 Typical rigid punch and die configurations. (Courtesy of The Aluminum Association)

Wiping dies feature a pressure pad which locks the workpiece in place. A punch then descends to contact the workpiece and "wipe" or press it down and around a die. See Figure 15-8.

Both rigid and flexible dies are used in press brake work. Surfaces of steel dies must be kept smooth and clean to avoid undesirable marks in the workpieces. Typical rigid punch and die sets are shown in Figure 15-9. Flexible dies employ a rubber

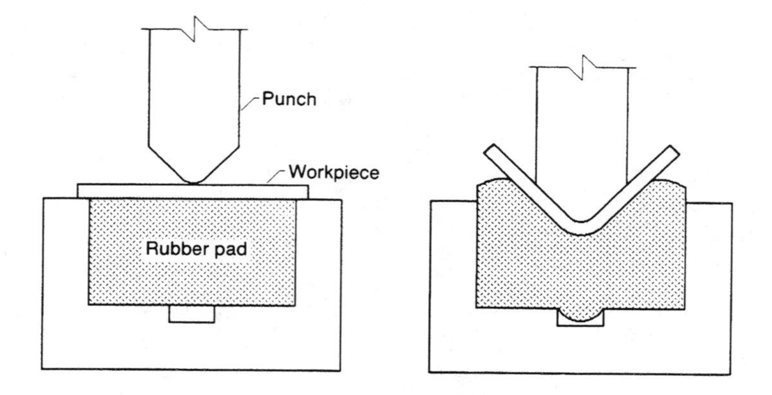

Figure 15-10 The action of a flexible press brake die. (Courtesy of The Aluminum Association)

pad in place of a bottom die, a feature which facilitates a range of punch shapes to make metal bends and eliminates the need for a mating die. See Figure 15-10. Special tooling set-ups are used to produce multiple bends in rectangular workpieces. See Figure 15-11.

Roll bending. Circular, curved, and cylindrical shapes are produced by roll bending methods. Roll bending is not to be confused with rolling, ring-rolling, or roll forming, which are processes described elsewhere in this book. Basically, roll bending employs three or four rollers to effect shape changes in material. See Figure 15-12. Wheel-shaped rolls are used to bend narrow bar sections, and sheet materials are formed with cylindrical rolls. The employment of specially shaped wheel rolls makes it possible to form almost any bar such as rod, tube, channel, and angle. See Figure 15-13.

There are four basic types of bending roll configurations: the pyramid; the three roll single pinch; the three roll double pinch; and the four roll double pinch. See Figure 15-14. The *pyramid* machine has two fixed lower rolls and one movable

Figure 15-11 Special tooling is here employed to make eight bends in one operation to form this rectangular pin. (Courtesy LAGAN Press, an ASEA Group Company)

Figure 15-12 Hydraulic four roll bending machine shown forming a heavy steel plate cylinder. (Courtesy of COMEQ, Inc., Baltimore, MD)

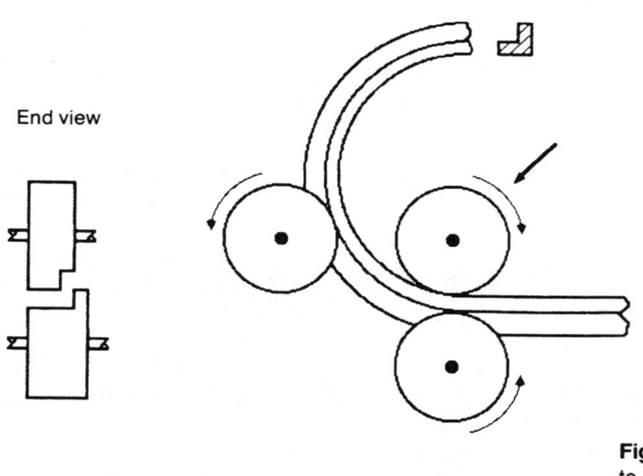

Figure 15-13 Wheel rolls are used to form circular bends in bar shapes, such as angles.

top roll. Its main limitation is the lack of prebending, or pinching, capabilities. Prebending is the capability to reduce the flat end to a minimum on the leading and trailing end of the section being rolled. This is accomplished by pinching, or clamping the material between the top roll and one lower roll, and using the other lower roll to bend the clamped portion of the material to the correct radius. This is especially important on larger machines, where the material loss due to long flat ends could be significant.

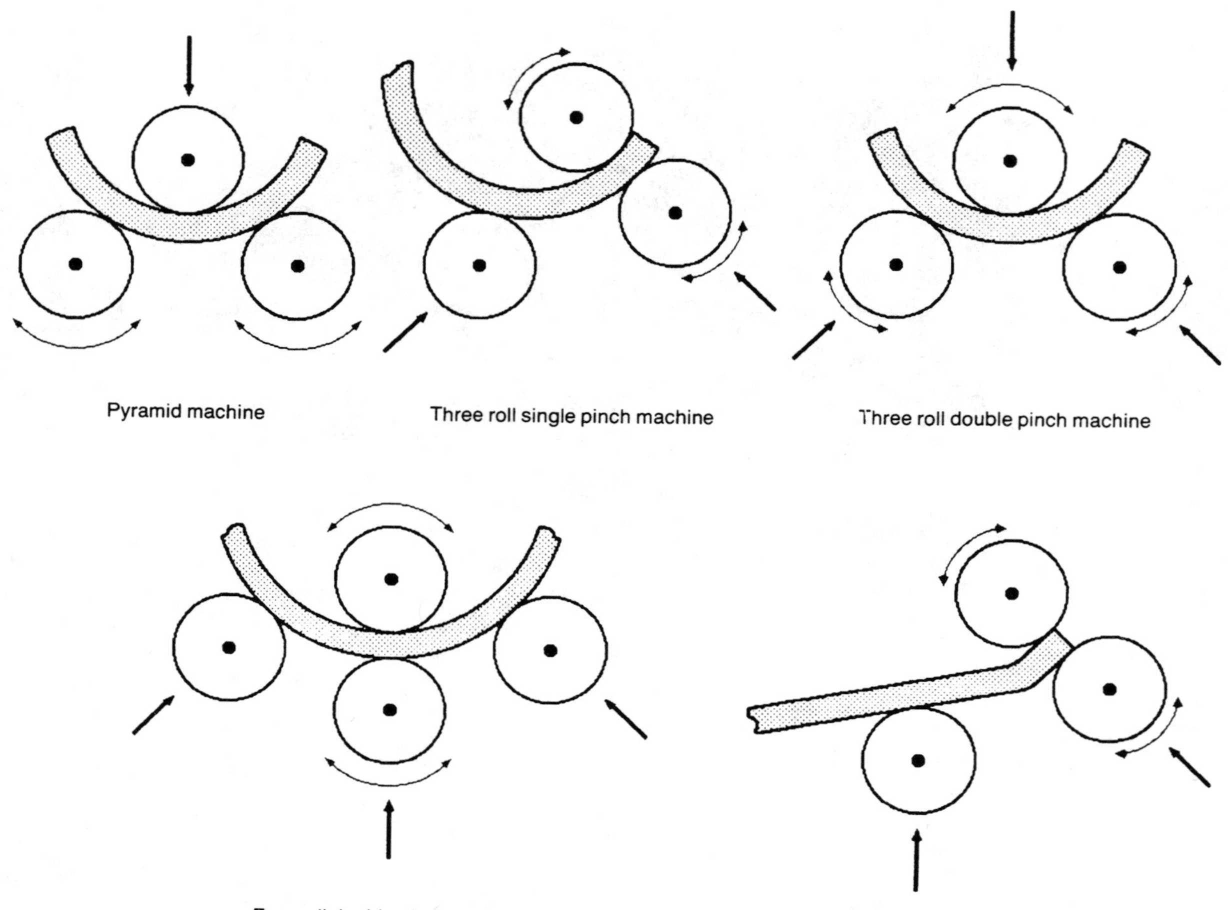

Figure 15-14 Roll bending machine arrangements. The prebending feature also is shown.

The *three roll single pinch* machine has a fixed top roll, a movable bottom roll to clamp the material against the top roll, and an adjustable side roll, to control the bending diameter. Its major advantage over the pyramid style is its ability to prebend the material. However, only one end can be prebent with each insertion of the material into the machine. This need to remove the workpiece and turn it end for end to prebend the second end adds considerably to material handling time and labor costs.

The *four roll double pinch* machine can prebend both the leading and trailing ends of the workpiece without having to remove it from the machine. This machine has the same basic roll geometry as the single pinch machine, and most importantly, adds another movable side roll. It is the best machine to mate with a NC-CNC unit, because once the material is secured between the top and bottom rolls, a fixed reference point is established. The disadvantages of the four roll machine are its comparatively complex construction and the resulting comparatively high cost,

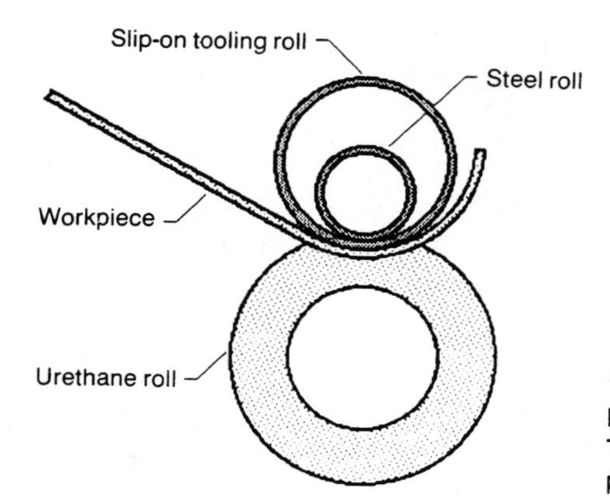

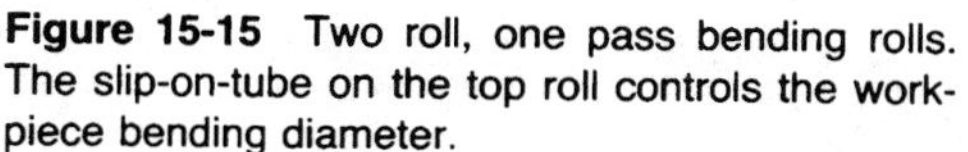

Figure 15-15 Two roll, one pass bending rolls. The slip-on-tube on the top roll controls the workpiece bending diameter.

and that, because of its geometry, it is limited in the size of the tooling that can be fitted on the roll shafts.

The *three roll double pinch* machine has become the standard of the industry during the last 20 years. This style machine, incorporating a fixed top roll and two independently adjustable lower rolls, is the most universal and versatile. It can be used for symmetrical bending by arranging the rolls in the pyramid configuration, and for asymmetrical bending by arranging the rolls as they are on a three roll single pinch type. Since it is a double pinch machine, both ends of the workpiece can be prebent with one insertion into the machine by alternating the clamping and bending functions between the bottom rolls. This machine is not well-suited to NC-CNC applications because of the absence of a fixed reference point when bending.

A unique roll bending and curving machine for sheet metal parts has only *two rolls*. See Figure 15-15. The upper steel roll controls a slip-on-tube of the desired finished workpiece size. This roll acts as a punch to force the workpiece into a tough, flexible urethane-covered lower roll. The urethane deflects under high pressure, and this deflection forces the workpiece to conform to the radius of the upper roll tooling. The upper roll can also be used independently of the slip-on-tube. This two-roll action results in superior control. The size of the formed part is determined by the diameter of the upper tool plus the natural springback of the material. Consequently, part duplication is easily attained when the mechanical properties of the material are held within normal ranges. The system advantages include the elimination of prebending operations, the production of mar- and scratch-free parts, and the ability to form perforated or cut-out materials without flattening or deformation. Additionally, only one pass is needed to secure accurate, continuous curves in workpieces.

Roll bending tubing. Curved bends in tubing are difficult to make, because the hollow tube collapses at the bend, and therefore some means of supporting the tube is necessary to form short-radius bends without flattening. Equipment for making short-radius bends in tubing is of two general types. The *stationary-die machine* consists of a fixed bending die and a movable wiper block, and the tube is wrapped around the die by the wiper block. Stationary-die bending is also called compression bending. The *rotating-die machine* has a moveable bending die and stationary wiper block. The tube is clamped to the die and bent as the die rotates. See Figure 15-16.

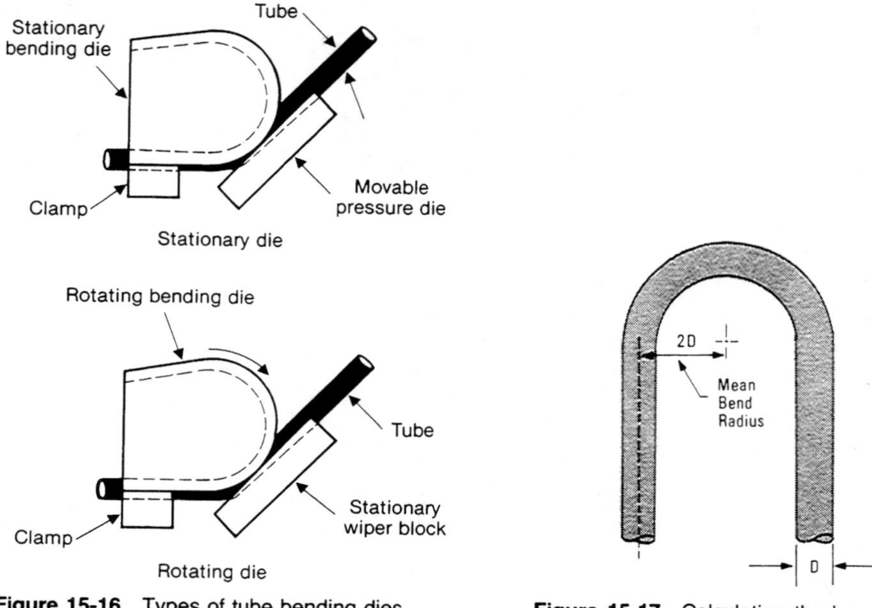

Figure 15-16 Types of tube bending dies. (Courtesy of Inco Alloys International, Inc.)

Figure 15-17 Calculating the bend radius for tubing. (Courtesy of Inco Alloys International, Inc.)

The sharpness of bends in tubing is expressed as a measure of the *mean bend radius*. This radius is the distance to the center of the tube from the center of the circle represented by the bend. See Figure 15-17. The radius is usually expressed in multiples of the outside diameter of the tube being bent. For example, a 4D bend in a 50 mm (2 inch) diameter tube has a mean radius of 200 mm (8 inches).

ROLL FORMING

This method differs from roll bending, and is used to form sheet or strip stock into straight lengths of various cross sections, such as angles, channels, and trim parts. Nearly any shape can be roll formed that can be made by parallel bends in flat stock. The material is shaped as it passes through a series of paired convex and concave contoured rolls, called *roll sets*. The roll forming machine can be compared to a series of small rolling mills arranged in a straight line. See Figure 15-18.

Stock for roll forming is usually a coil, strip, or short sheets. One end of the strip is placed between the first roll set which grips the strip, forms it to the profile in the roll surfaces, and passes it along to the next set of rolls. Each succeeding set of rolls more closely approaches the final shape, so that the strip is progressively formed into the desired cross section. Although simple shapes such as angles can be formed by one or two sets of rolls, most of the shapes made by roll forming require from 5 to 15 sets. See Figure 15-19.

One of the most important uses of roll forming is in the production of welded tubing. Welding equipment is combined with the roll forming machine for automatic production. Strip stock is roll formed into tubular shape, and the edges are joined by welding in one continuous operation.

Figure 15-18 A roll forming machine employs a series of roll sets to progressively form a workpiece of uniform cross section. A vertical outer wall panel for prefabricated metal buildings is shown entering (upper photo) and emerging (lower photo) from the machine. (Courtesy of The Lockformer Company)

Figure 15-19 Typical roll-formed part sequence. (Courtesy of The Aluminum Association)

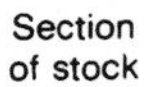

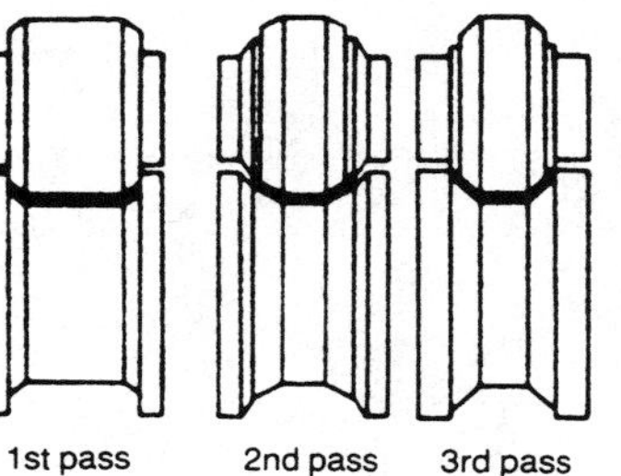

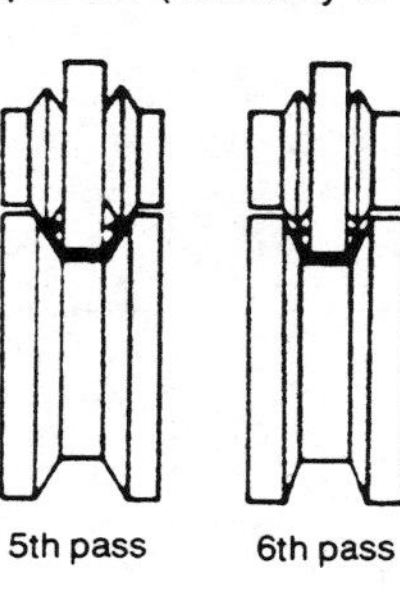

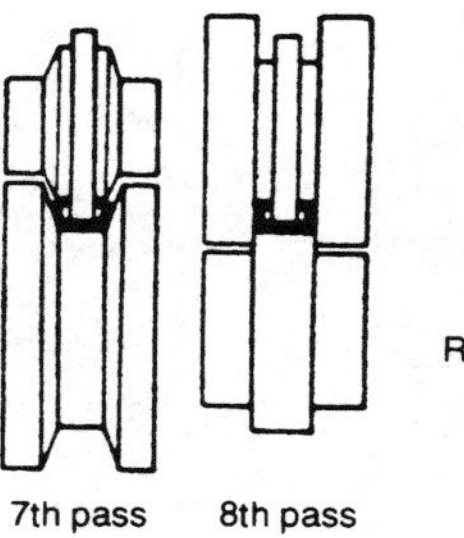

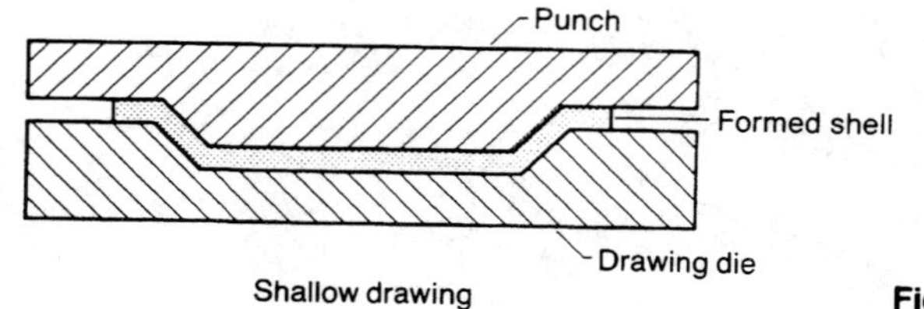

Figure 15-20 Shallow drawing diagram.

Drawing

This process is used to form a flat metal sheet into a thin-walled seamless hollow shape. When cup-shaped articles of considerable depth are formed in this manner, the process is referred to as *deep drawing*. The forming of shallow shapes is often called pressing or stamping, but the distinction is not clear. For purposes of consistency, the terms shallow and deep drawing will be used to describe these processes. Compare with Figures 17-8 and 17-9.

Shallow drawing. Shallow drawn products include such items as automobile trim, hub-caps, trays, and pans. Because these operations result in only a moderate movement of material, there is seldom a problem with undue material stress and work hardening. See Figure 15-20. Shallow as well as deep drawing is accomplished by using either mechanical or hydraulic presses.

Practically all drawing operations begin with a workpiece called a *blank*. Blank preparation is described in Chapter 11. The surface area for a blank should be as close to that of the finished article as possible, allowing only sufficient extra material to facilitate the holding of the blank during drawing and for final trimming. Excessively large blanks cost more and require increased working pressures, which may result in fractures. Ideally, there should be no appreciable reduction in thickness between the blank and the finished part. Any necessary gauge reduction should be done by ironing, treated later in this chapter.

Deep drawing. The deep drawing process is illustrated in Figure 15-21. The bottom die on the drawing press has a recess or opening called a *drawing die*, and the upper die is called a *punch*. This bottom die is also called the draw ring or die block. When force is applied, the punch fits into the opening in the bottom die. The blank is placed over the hole in the drawing die, and then the descending

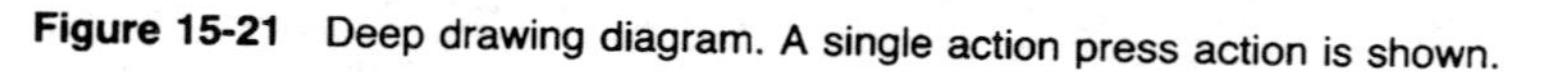

Figure 15-21 Deep drawing diagram. A single action press action is shown.

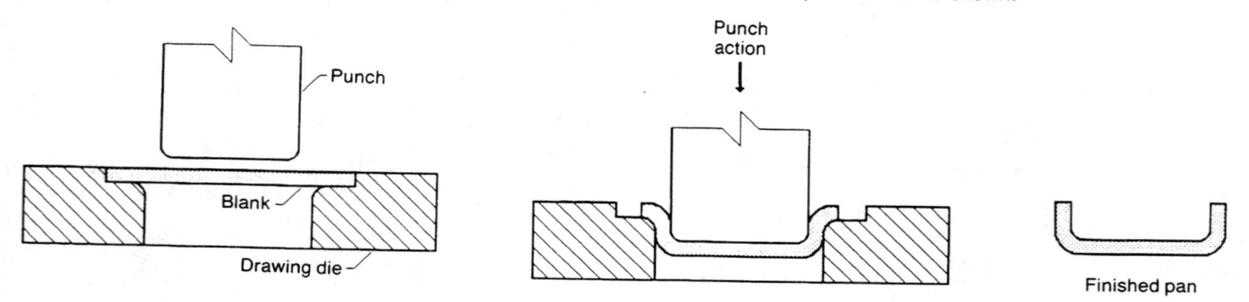

punch presses the blank into and through it. Parts of almost any shape can be produced this way.

A common deep-drawing operation is that used to make cylindrical vessels. In such cases, the opening in the drawing die is round, and the punch is a solid cylinder. During forming, the circular blank is pressed through the drawing die and wrapped around the punch. The drawn article then has the shape of the punch.

Because the blank is necessarily larger than the hole through which it is pressed, the edges of the blank have a tendency to wrinkle. While thick blanks may be strong enough to resist wrinkling, thin blanks usually require support, dependent upon the depth of the draw. The support is provided by disk-shaped clamps, called hold-downs. Examples of deep-drawn parts are shown in Figure 15-22.

Types of equipment. Single-, double-, and triple-action presses are used in drawing, and they are available in a range of types, sizes, and capacities. See Figure 15-23. Factors which determine press selection include gauge, alloy, temper, and kind of material; shape and depth of the part to be drawn; and quantity of parts required.

Single-action presses have one drawing action and no hold-down pressure. Refer to Figure 15-21. Shallow-drawn articles generally are formed on such presses because of the moderate drawing depth and because of the high speeds at which they can operate. Large single action presses can operate at about 60 strokes per minute, while smaller models can exceed 400 strokes per minute.

Double-action presses employ two working motions. The hold-down descends first to secure the blank against the bottom die. The blank is formed by the second

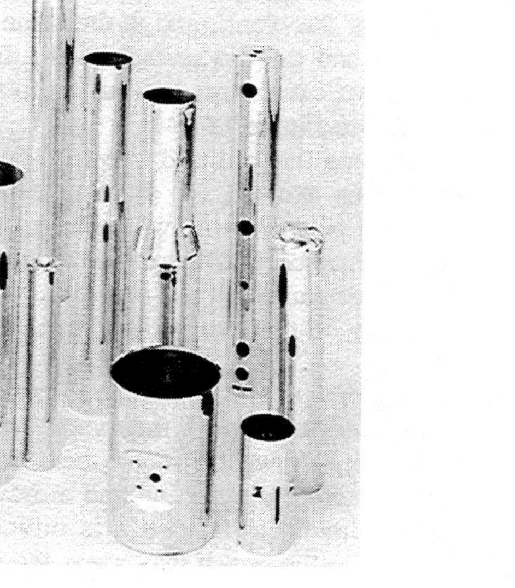

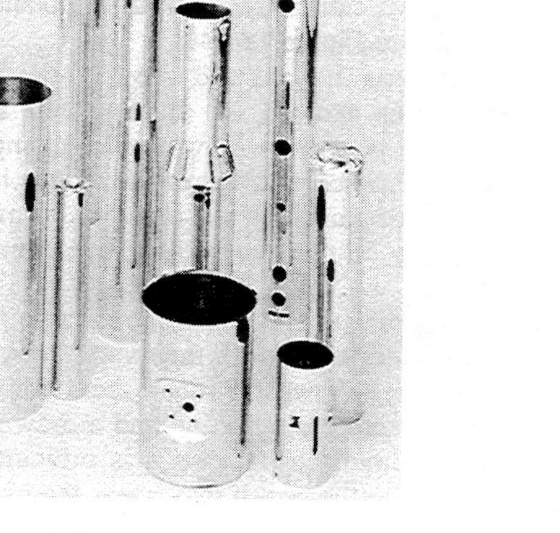

Figure 15-22 These are excellent examples of high-quality deep drawn, chrome plates parts, some drawn to a 20:1 length to diameter ratio. (Courtesy of Shape Form, Inc., Plain City, OH)

Figure 15-23 A typical 200 ton, four-post guided ram hydraulic forming press. (Courtesy of Greenerd Press and Machine Company, Inc.)

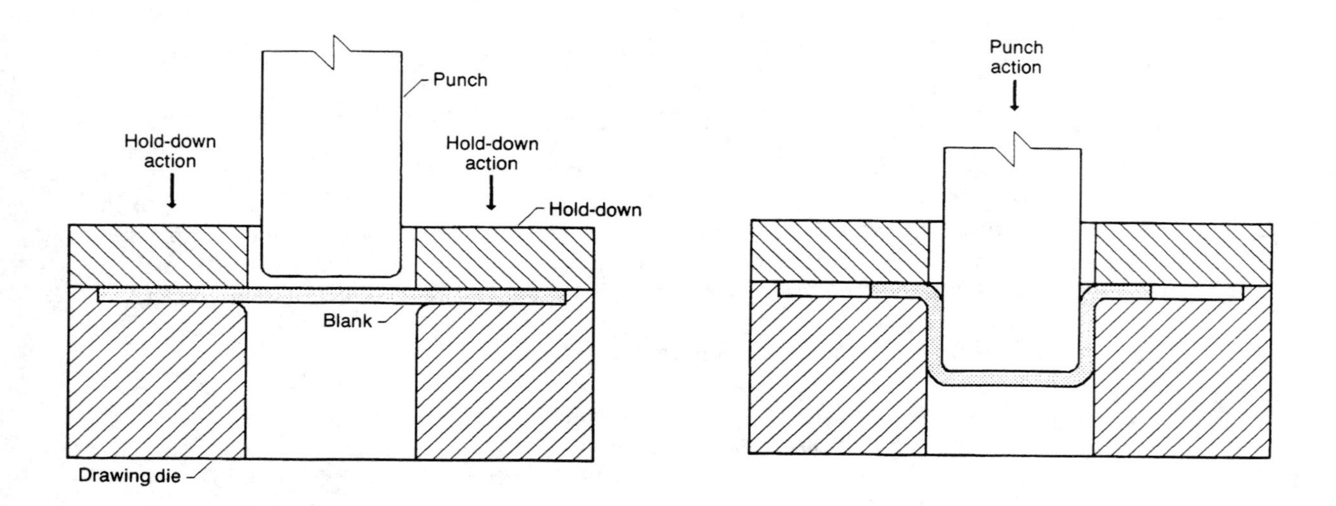

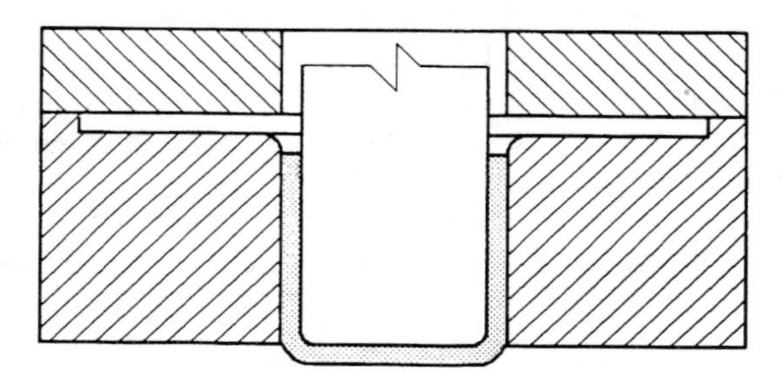

Figure 15-24 Double action press diagram, showing hold-down operation.

motion as the punch draws the metal through the die. The hold-down exerts just enough pressure to prevent wrinkling of the blank, and at the same time allow the blank to slide between the hold-down and the top surface of the drawing dies. In some automated systems, the hold-down can be used to cut the blank from a coil of stock that is fed into the machine. See Figure 15-24.

Triple-action presses have two forming motions and one hold-down action. The primary forming motion pushes the cup through the drawing die. In the second forming motion, a die moves up from below the drawing die to impart a special shape to the bottom of the cup, to provide extra strength to the bottom, or to provide a foot for added stability of a container. When a second forming motion is not required, the third press action may be used to eject the cup from the die. See Figure 15-25.

Additional information on metal forming equipment is found later in this chapter. While most deep-drawn objects are cylindrical cups or their variations, other shapes, such as rectangular boxes, are also deep-drawn. A square box, for example, can be formed by using a punch with a square cross section and a drawing die having a square hole. Sharp corners must be avoided because it is difficult for metal to flow into these areas, and sharp edges on the punch may shear the blank.

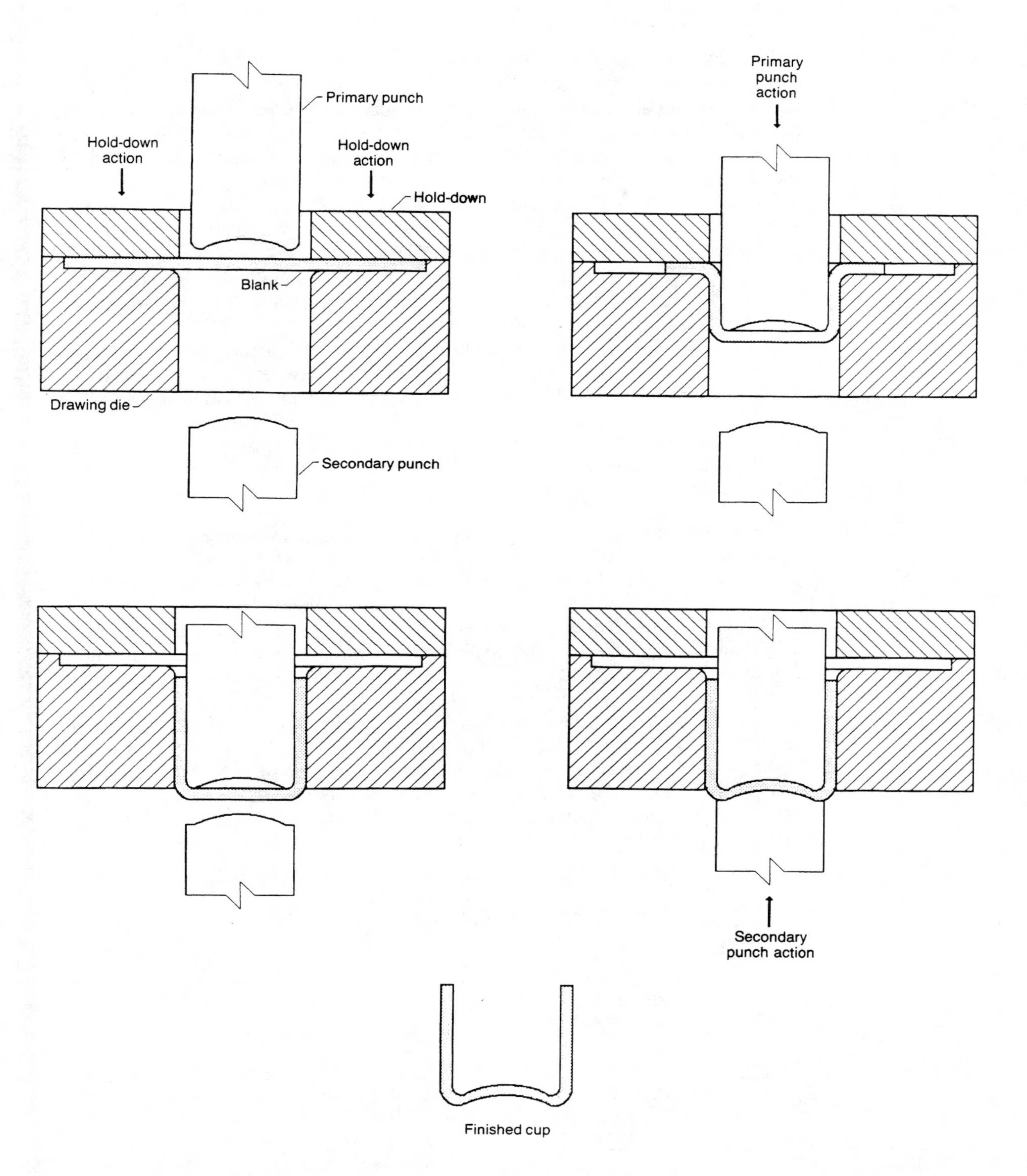

Figure 15-25 Triple action press diagram, showing hold-down and secondary punch operations.

Redrawing. Many parts are deep-drawn to the final size and shape in one draw, while others require additional operations to achieve the final form. In multiple-draw operations, the blank is first cupped to form a shell that is larger in diameter but shorter than the final article. Redrawing reduces the diameter and increases the height of the shell. Smaller punches and drawing dies are used for each redraw until the desired size is reached. Redrawing can be done in two ways. In *standard redrawing*, the cup is drawn through a smaller die in the same direction as the first draw. In *reverse redrawing*, the cup is turned over and forced through the die by punch pressure on the bottom of the cup. See Figure 15-26. Reverse redrawing reduces wrinkling and also facilitates the formation of difficult contours on the bottom of the cup. Reverse redrawing obviously is a more severe forming operation and often results in work hardening.

Redrawing is also classified as either ironing or sinking, depending on the effect of the redraw on the cup wall. In *ironing*, the space between the punch and die is less than the thickness of the workpiece, and consequently the wall of the cup is stretched and thinned. This technique is used to produce vessels whose walls are thinner than their bottoms. In *sinking*, the punch and die clearance is greater than the thickness of the blank, so that the diameter of the cup is decreased with little or no change in wall thickness. See Figure 15-27.

Flexible die forming. As described earlier, many drawn shapes are produced by pressing workpieces between two rigid dies, each of which approaches the contour of the finished article. However, under high pressure, materials such as rubber act as an hydraulic medium to exert equal pressure in all directions. See Figure 15-28. In some drawing operations (as well as press braking operations discussed earlier), rubber can act as an effective concave die to form a workpiece around an appropriately contoured punch. The rubber transmits the pressure because it resists deformation, an attribute which serves to control local elongation in the workpiece being formed. The use of such flexible dies has a number of advantages, such as low tool costs, the elimination of die-matching, reduced tool wear, and the absence of die marks on the finished product. Also, identical parts of different material gauges can be made without costly tool changes. Flexible die production runs are generally lower than in other drawing operations. ''Short'' production runs refer to quantities of several hundred. ''Long'' production runs usually mean 1,000 or more pieces.

Figure 15-26 Redrawing operations diagram.

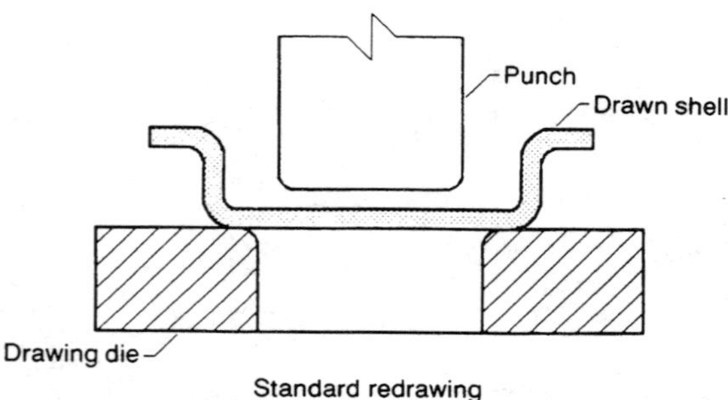

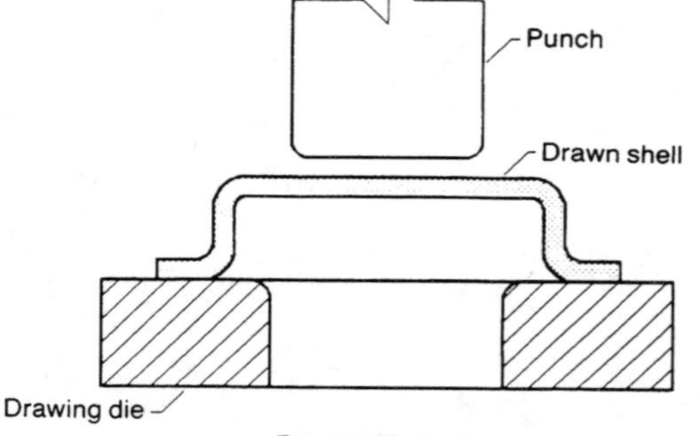

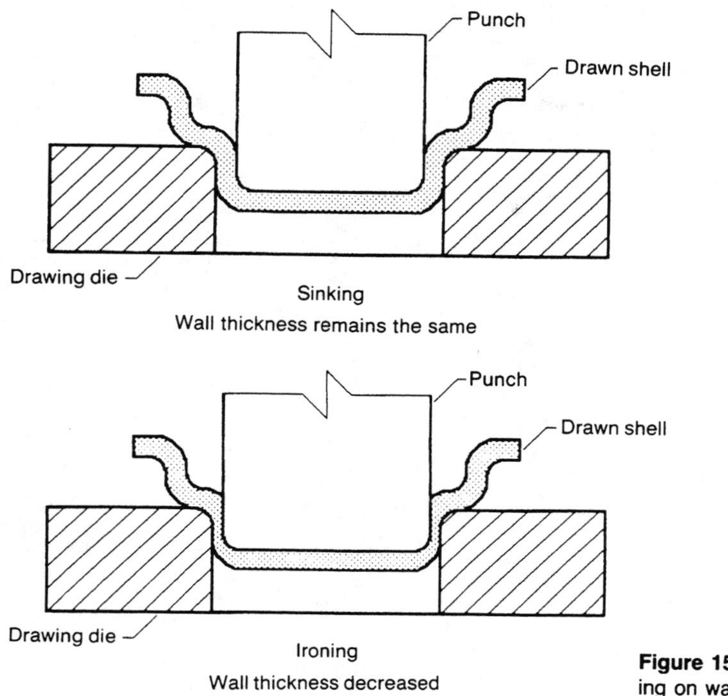

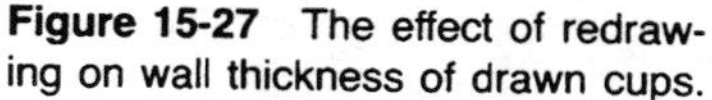

Figure 15-27 The effect of redrawing on wall thickness of drawn cups.

The range of available flexible-die processes can be classified under two broad categories: (1) shallow-drawn techniques, which utilize the pressure exerted against the rubber pad to both hold the workpiece and form it, such as occurs with the Guerin and the Verson-Wheelon methods; and (2) deeper-draw techniques, which employ independent workpiece holding mechanisms, such as found in the Marform and Hydroform processes.

The *Guerin* process employs a rubber pad contained by a metal shell as the concave die, and which is mounted on the ram of a hydraulic press. The workpiece

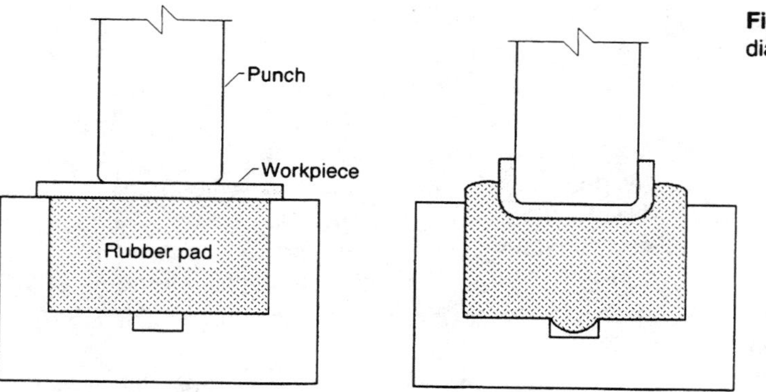

Figure 15-28 Flexible die forming diagram.

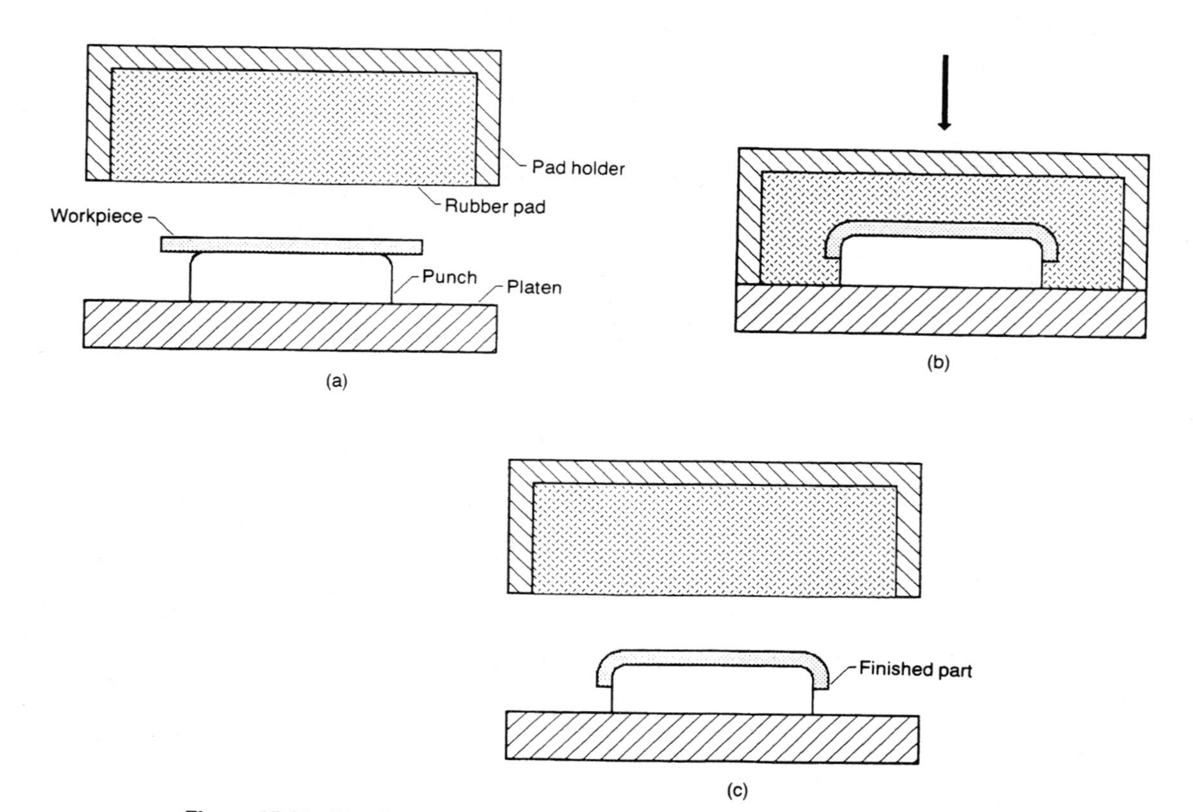

Figure 15-29 The Guerin process. (a) Operation ready. (b) Die closes to form part. (c) Die opens.

is placed over the punch, which is affixed to the press platen. See Figure 15-29. When the press is activated, the ram presses the rubber pad against the blank, and down around the punch to form the part. The maximum pressures used in the Guerin process are in the order of 7–14 MPa (1,000–2,000 psi). Products made by this method can range from a few square inches to a few square feet in size, but the flanges or rib heights are restricted to about 30 mm ($1\frac{1}{4}$ inches). The method can also be adapted to blanking and piercing operations. The Guerin process was the first of the rubber die techniques, and was introduced in 1935 by the Douglas Aircraft Company.

The *Verson-Wheelon* process utilizes a special press where force is applied to a rubber pad by a flexible fluid sack instead of a hydraulic ram. The press is operated by mounting the punch and workpiece on a feed table, which is then slid under the rubber pad in the pressing chamber containing a neoprene fluid cell. The cell is inflated with hydraulic fluid when the operator opens a valve, thereby forcing the pad and workpiece around the punch to form the part. Maximum working pressures are 35–53 MPa (5,000 to 7,500 psi). A combination of higher pressures and softer rubber pads makes possible the forming of larger articles and higher flanges. See Figure 15-30.

The *Marform* process, a variation of the Guerin method, has a distinctive blank-holding mechanism that automatically controls the operating pressure. The rubber

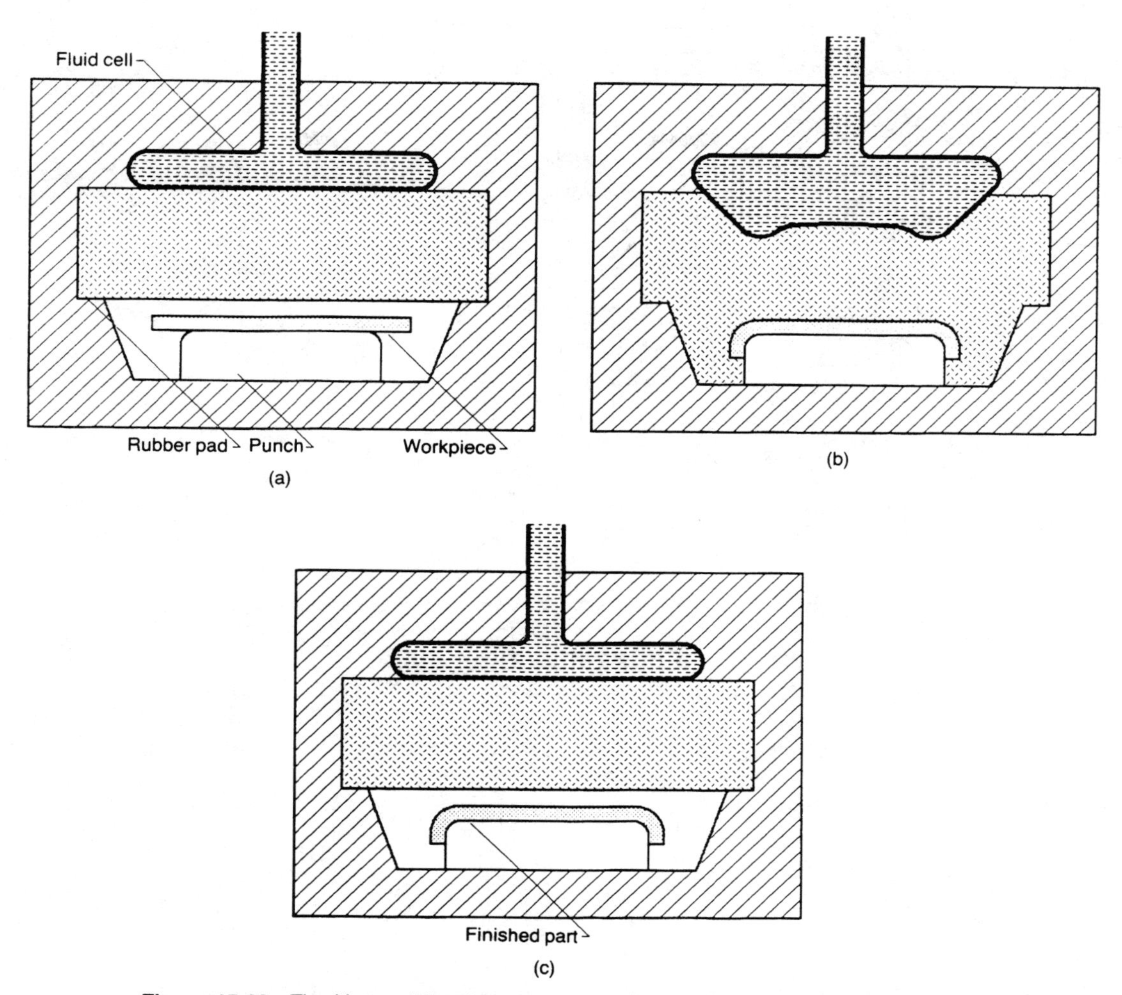

Figure 15-30 The Verson-Wheelon process. (a) Operation ready. (b) Fluid sack inflates to form part. (c) Fluid sack deflates.

pad is fixed to the press ram. A hydraulic cushion unit supports a punch and a blankholder, which has a mating hole in its center to fit around the punch. This holder is fixed to support rods extending through the press bed to a hydraulic piston below. The workpiece is set on the blank holder, and as the press closes, the upper ram forces the rubber pad and workpiece blank around the punch. Simultaneously, the piston compresses the hydraulic fluid in the cylinder to apply the required blank-holder pressure, monitored by a pressure control valve. See Figure 15-31. Draws with depths 3 times the shell diameter are possible with the Marform process, utilizing pressures of about 42 MPa (6,000 psi).

The *Hydroform* process requires a hydraulic press specially constructed to deep-draw shells with rubber dies. See Figure 15-32. This machine is used to form workpieces up to 810 mm (32 inches) in diameter and shells up to 300 mm (12 inches) deep, at pressure between 70 and 100 MPa (10,000–12,000 psi). The drawing

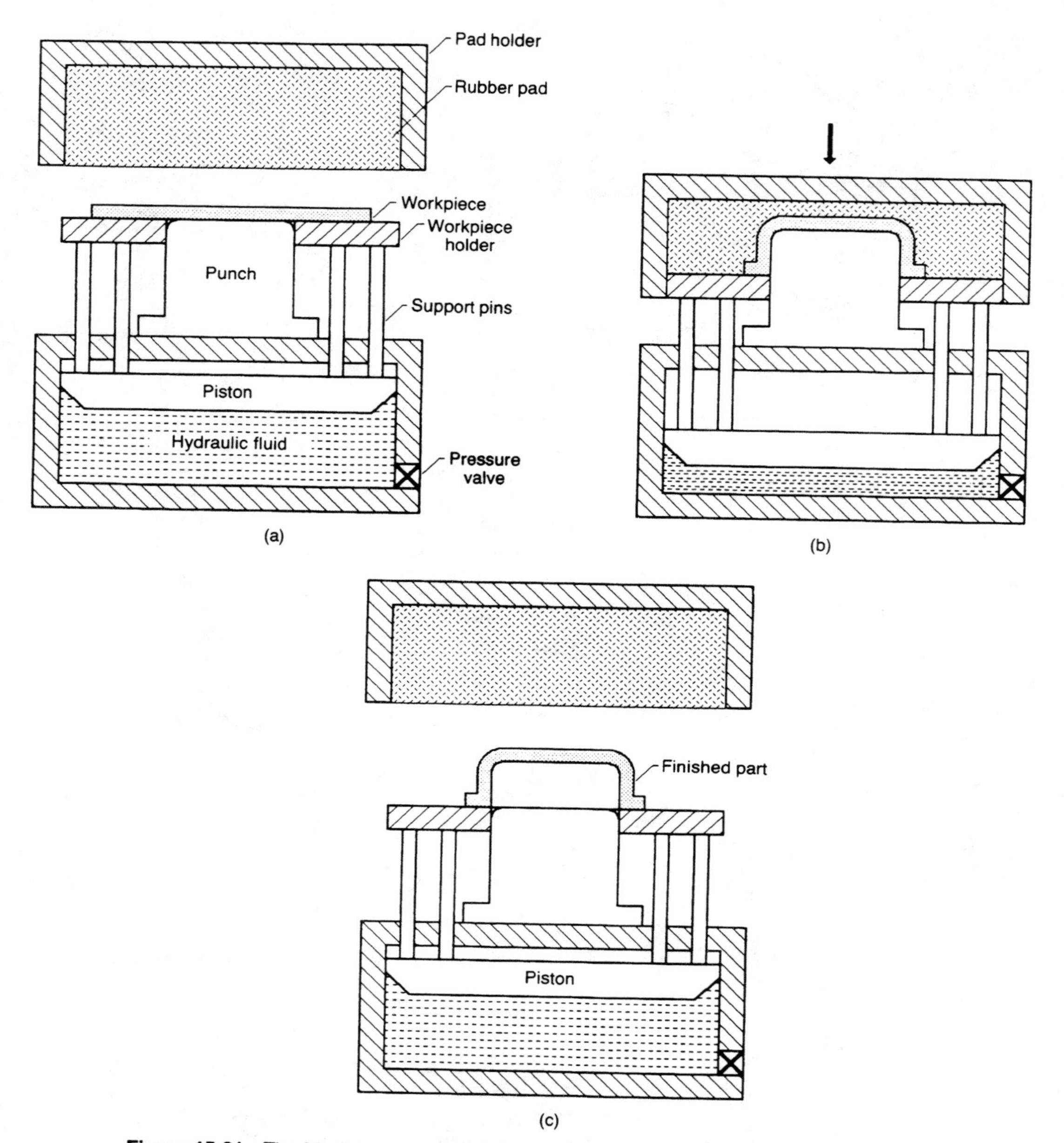

Figure 15-31 The Marform process. (a) Operation ready. (b) Die closes to form part. (c) Die opens.

cycle begins by placing the workpiece on the blankholder and dropping the forming chamber to close the press. Blankholder pressure is applied in the upper cylinder to set the rubber diaphragm firmly against the workpiece and punch. Next, the pressurized lower cylinder raises the punch, thus forcing the workpiece and diaphragm upward into the forming chamber and thereby drawing the part. Similar to the Marform process, Hydroform provides a constant control of the workpiece deformation, an attribute which prevents distortion in the finished part.

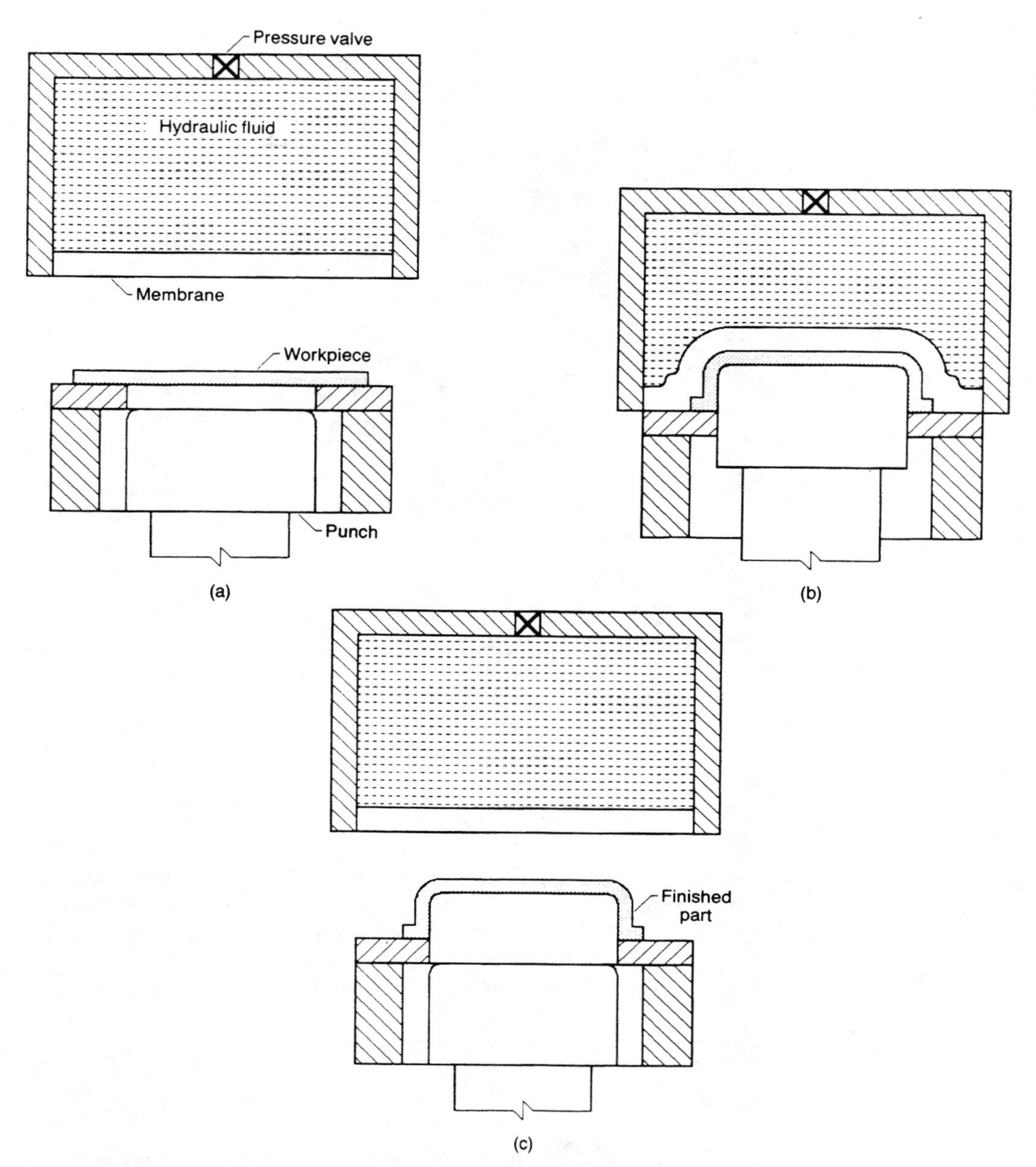

Figure 15-32 The Hydroform process. (a) Operation ready. (b) Die closes. (C) Die opens.

Presswork applications. The aforementioned types of drawing dies are used in a variety of ways to produce formed parts. *Conventional dies* employ either a common rigid punch and die setup, or one of the flexible die methods. They are also called single operation or simple dies (though some are hardly simple in configuration) and are used to produce one part in one operation. In the example shown in Figure 15-33, a rigid punch is used to press a workpiece into a flexible, hydromechanical die to produce a deep-drawn cylinder.

Figure 15-33 A conventional die system is used to produce these drawn cylinders. Note the variety of punches used in combination with a flexible die. (Courtesy of LAGAN Press, an ASEA Group Company)

Progressive dies are used in applications where a sheet coil or strip is fed into a machine which has a series of connected die stations, any of which performs a separate operation on the coil workpiece. The movement is from station to station, automatically, and at each station the piece is blanked, notched, pierced, formed, and cut off, progressively, as dictated by the part requirement. See Figure 15-34. Note in the illustration that the workpiece remains attached to the feed coil until the forming is completed. This system is expensive to construct, but is cost-justified by an exceptionally high production rate.

Transfer die methods are employed on workpieces which, because of their sizes or complexity, are not adaptable to progressive die methods. See Figure 15-35. In the example shown, a blank is transferred from die station to die station, where it receives a cutting or forming treatment. See Figure 15-36. Simple assembly operations such as press-fitting or staking can also be performed on such systems. See Figure 15-37. Other automatic transfer systems are also used. See Figure 15-38. The requirements of the workpiece define the requirements of the transfer system.

Figure 15-34 A typical product of a progressive die system. Note that the workpieces remain attached to the metal flat stock until the completion of the forming process. (Courtesy of the Minster Machine Company)

Figure 15-35 Transfer die forming sequence for an automotive part. (Courtesy of LAGAN Press, an ASEA Group Company)

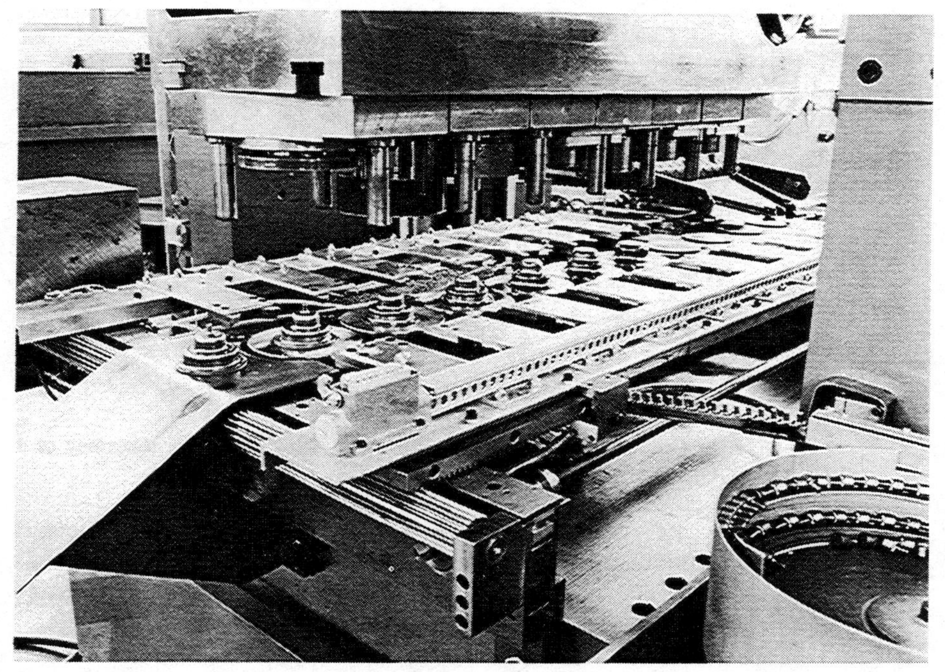

Figure 15-36 Transfer die stations for the automotive part in Figure 15-35. (Courtesy of LAGAN Press, an ASEA Group Company)

Figure 15-37 Pick and place robotic arms are used in this transfer die system. (Courtesy of Livernois Automation Company)

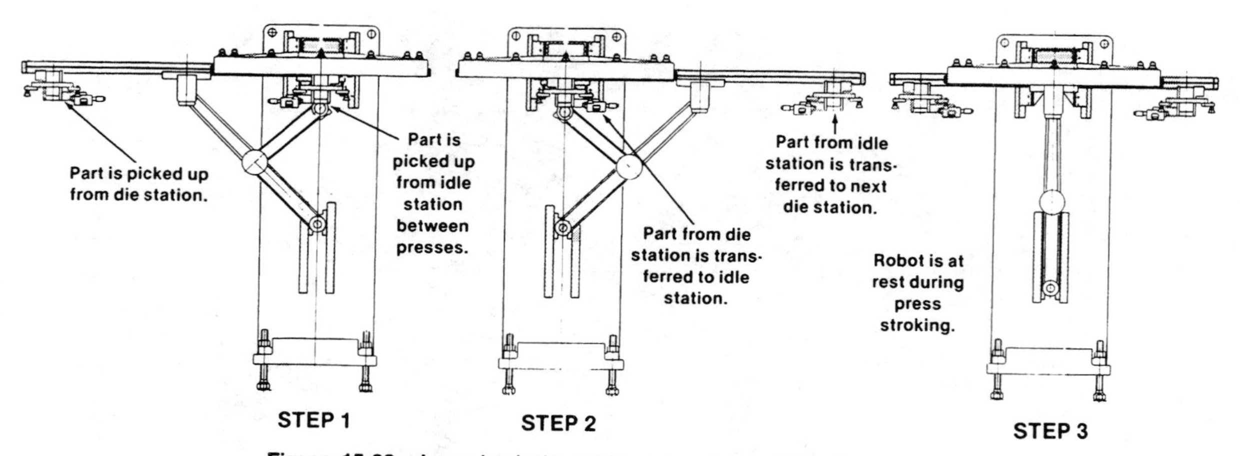

Figure 15-38 A mechanical workpiece transfer system. (Courtesy of the Minster Machine Company)

All die systems utilize presses, some with computer controls. See Figure 15-39. Additionally, special coil feed systems must be employed to facilitate the automatic system. See Figure 15-40. Improved coil handling systems can permit better feed performance and save valuable floor space.

OTHER METHODS

There are a number of specialized applications and variations of flexible-die forming. *Hydrodynamic*, or flexible punch, forming is employed to produce cup or dish shapes without the use of pads or diaphragms. All of the forming pressure is delivered to the workpiece by a liquid, usually water. See Figure 15-41. A water-tight cavity is formed when two hollow chamber halves are brought together. The inside surface of one of the cavity halves is made in the shape of the article to be formed. The

Figure 15-39 A mechanical press for automatic cutting and forming operations. (Courtesy of the Minster Machine Company)

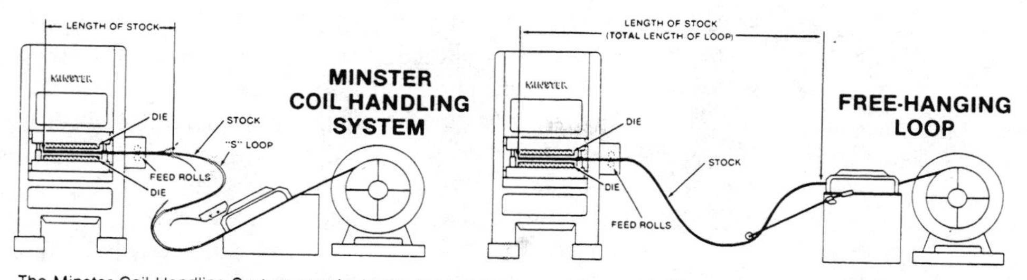

Figure 15-40 The respective attributes of the S-loop and the free-hanging loop coil feed systems. (Courtesy of the Minster Machine Company)

Figure 15-41 The Hydrodynamic process. (a) Operation ready. (b) Fluid enters die cavity to form part. (c) Fluid leaves die cavity.

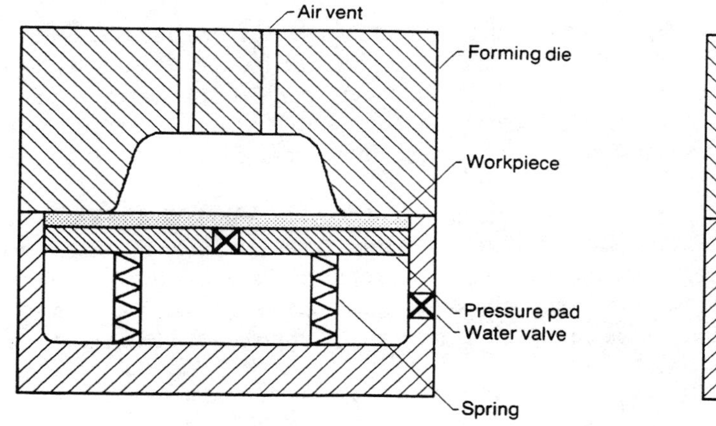

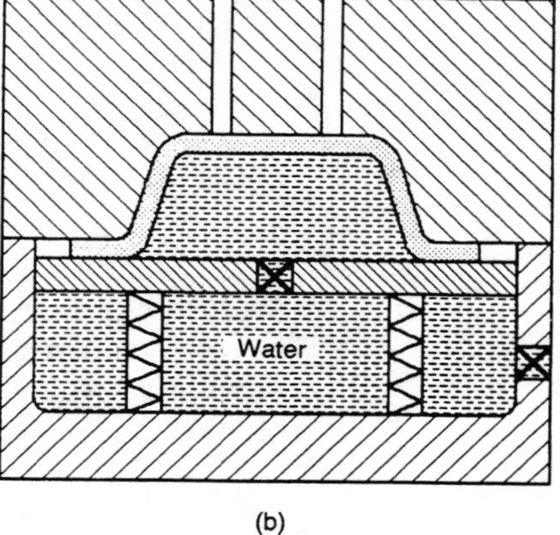

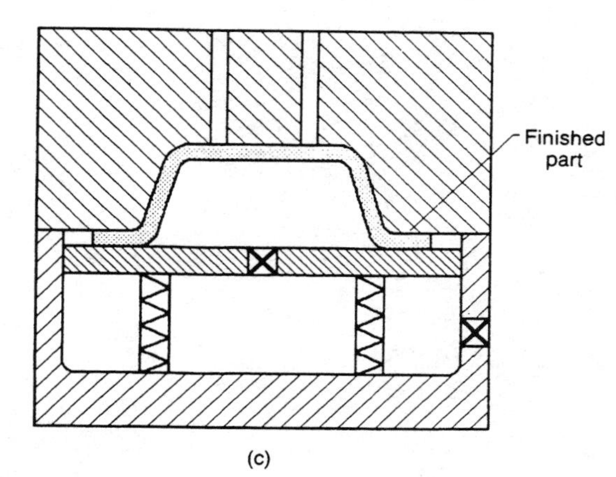

workpiece is clamped over the shaped cavity by a hold-down plate, and water under high pressure is pumped into the unshaped side of the chamber. The water pressure acts as a fluid punch to force the workpiece into the shaped cavity.

Other applications are used to shape the sides of tubular shapes, such as a process called *bulging*. Bulging is used to form such products as corrugated tubes and bellows, where the workpiece is a tubular shell produced by deep drawing, or by bending and welding a flat sheet into a hollow cylinder, or by cutting a piece from tube stock. In one bulging method, a two-piece die is clamped around the tubular shell. The inside surfaces of the die contain the contours to be formed in the tube wall. The tube is then expanded into the die by hydraulic pressure from within. The same result can be achieved by applying force to a rubber block inserted into the tube. Another bulging method is used to produce large bellows. Retaining rings are clamped at intervals along the tube, and the tube is expanded by hydraulic pressure. The pressure bulges the tube wall at the areas between the retaining rings. See Figure 15-42.

Forming achieved by high velocity force over a short time period can give exceptional results. This process category is called *high energy rate forming* (HERF), and includes several types.

Explosive forming is generally used to make large parts from sheet or plate stock, a process similar to flexible-die forming, but where the force is supplied by detonation of an explosive material such as blasting powder. Its greatest advantage is the ability to deliver large amounts of power at low cost, and at speeds up to 100 meters per second. Conventional forming takes place in the range of 0.5 to 10 meters per second.

A common explosive forming method is shown in Figure 15-43. The workpiece is placed over a hollow die in a tank of water. A vacuum line is connected to the die cavity to remove the air from the space between the workpiece and the die. An explosive charge is suspended in the water a predetermined distance from the workpiece into the die, and is detonated to form the workpiece. Sometimes an electrical

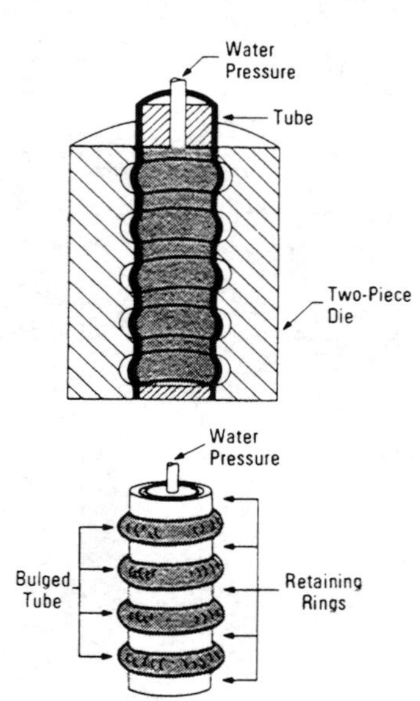

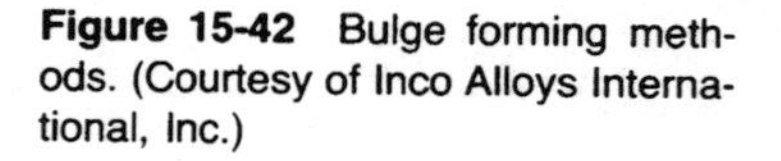

Figure 15-42 Bulge forming methods. (Courtesy of Inco Alloys International, Inc.)

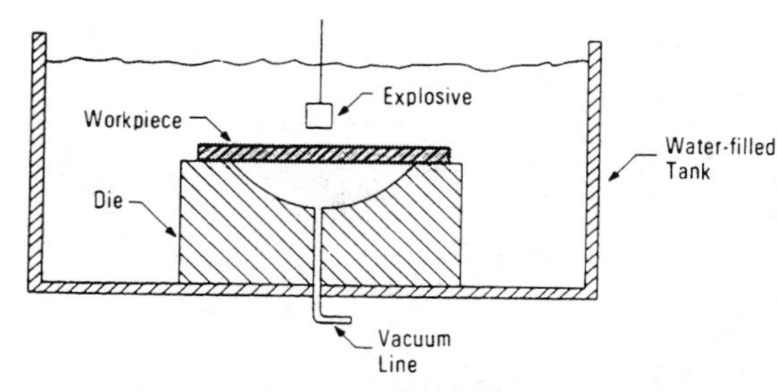

Figure 15-43 Explosive forming in a fluid. (Courtesy of Inco Alloys International, Inc.)

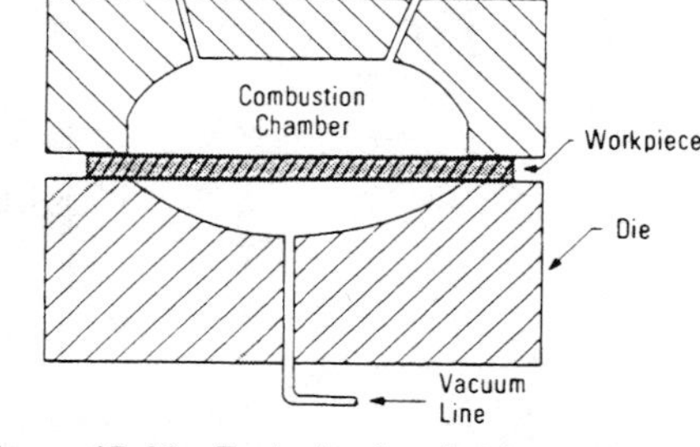

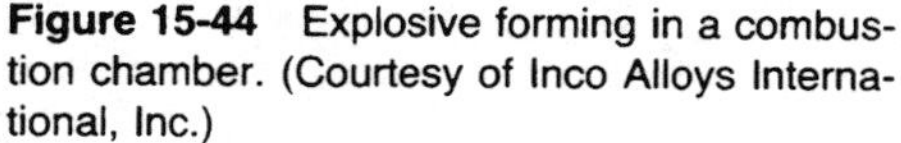

Figure 15-44 Explosive forming in a combustion chamber. (Courtesy of Inco Alloys International, Inc.)

spark is used instead of an explosive, where stored high-voltage energy is released between two electrodes suspended in the water. The resulting shock waves act in the same way as an explosion to form a part. This process is called *electrospark or electrohydraulic forming*.

Another technique employs combustible gases as the energy source. See Figure 15-44. The workpiece is placed over an evacuated hollow die. A gas-tight combustion chamber is mounted on the other side of the workpiece and directly over the die cavity. The combustible gas is ignited within the combustion chamber. No liquid is used to transmit the force; the explosive energy acts directly on the workpiece.

Electromagnetic forming is accomplished with banks of capacitors used to build a large electrical energy potential, which, when passed through coils, produces a high intensity, pulse magnetic field. Eddy currents are induced in the conductive metal workpiece, which is secured over a concave die made from a dielectric material. The coil is placed next to or inside a workpiece, and the strong repulsion between the workpiece and the magnetic field forces the metal into the die cavity. See Figure 15-45. A primary use for this process is to join parts by locking them together by expanding or compressing metal into grooves.

Spinning. Metal spinning involves the clamping a circular sheet workpiece against a contoured chuck or mandrel mounted in a lathe. As the chuck rotates at high speed, pressure applied with a forming tool progressively forces the workpiece

Figure 15-45 Electromagnetic forming. (a) Operation ready. (b) Part formed.

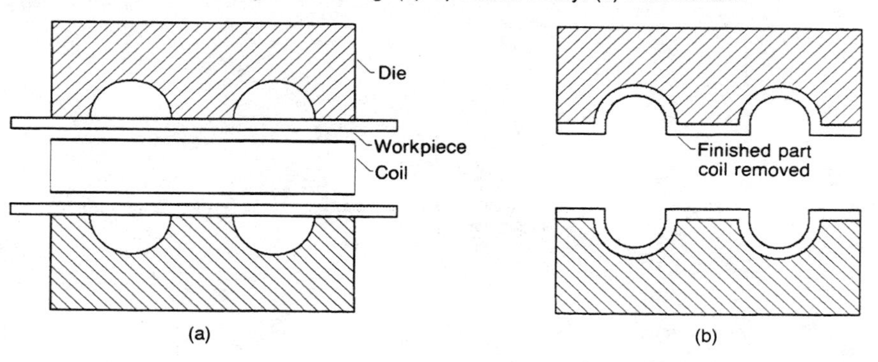

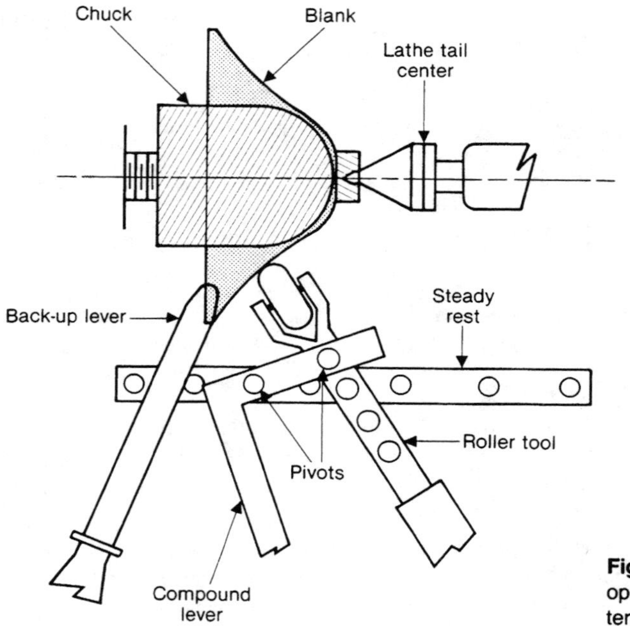

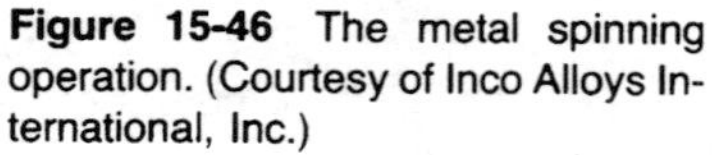

Figure 15-46 The metal spinning operation. (Courtesy of Inco Alloys International, Inc.)

against the chuck to displace the metal, and causes it to conform to the contour of the chuck. See Figure 15-46.

Spinning operations are of two types, as shown in Figure 15-47. In *standard* spinning, the blank is formed around the chuck primarily by bending, and is not reduced in thickness. In *shear* spinning (also called shear forming, floturning, or spin forging), the blank thickness is reduced during forming. Shear spinning is used to produce cone-shaped articles with flat bottoms, where the bottoms are thicker than the side walls. Another use is to decrease the wall thickness of tubular shapes, as in tube forming.

Thin blanks of soft material can be spun manually, where an operator presses the roller tool against the blank by using the tool as a lever. Thick blanks or hard materials require more force than can be applied by hand. Power or mechanical

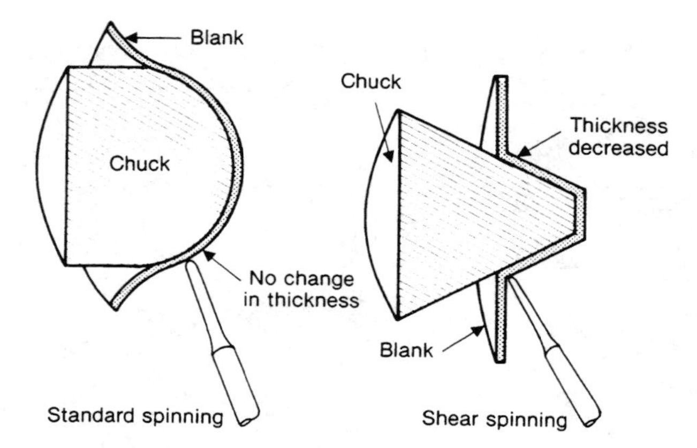

Figure 15-47 The difference between standard and shear spinning. (Courtesy of Inco Alloys International, Inc.)

spinning is performed in the same manner as manual spinning, except that the tool is moved over the blank by motor-driven gears or by a hydraulic system. Shear spinning usually must be done with a power-driven tool.

Many shapes can be made either by spinning or by deep drawing. For small runs, spinning is more economical because the expensive dies needed for deep drawing are eliminated. Spinning chucks are often made of inexpensive materials such as wood or cast iron. Some formed parts are made by first deep drawing the blank to nearly the final shape, and then completing the shape by a light spinning operation. In that way, more complex contours can be produced on the final shape than could be achieved by deep drawing alone. Also, large parts that are impractical for forming presses can be made by spinning.

Stretch forming. The operating principle of this process is that when metal is stretched beyond its yield point, it enters a plastic range and will retain a formed shape with little or no springback.

In stretch forming, the workpiece is gripped at two opposite ends by powerful jaws, hydraulically powered to move in opposite directions and pull the workpiece until the metal is stretched a small amount. The workpiece is held in this stretched condition, and the forming is completed by wrapping the workpiece around a forming punch. Stretch forming is illustrated in Figure 15-48.

Stretch-draw forming is a combination of stretch forming and deep drawing, a process often used to produce large sheet metal panels having varying contours. The equipment consists of a conventional hydraulic press with gripper jaws mounted on each side, which stretch the workpiece over the bottom die. The ram is then lowered to press the top die or punch against the workpiece. See Figure 15-49.

The advantages of stretch forming are reduced wrinkling and buckling of the workpiece and relatively low forming forces. Wrinkling is reduced because the

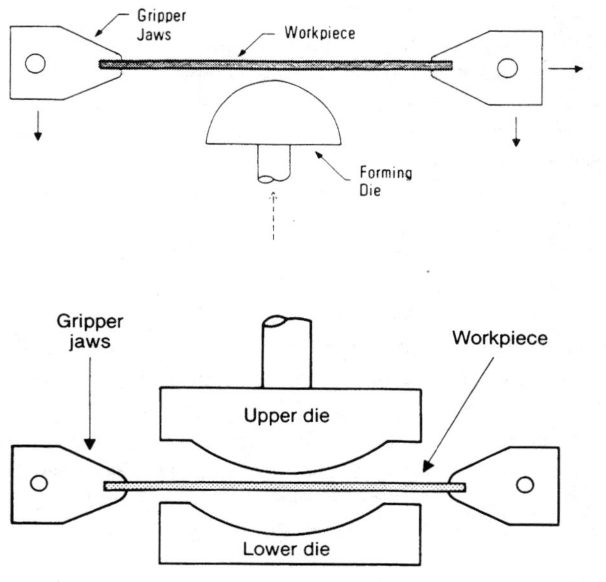

Figure 15-48 Stretch forming. The gripper jaws first stretch the workpiece slightly. The material is then wrapped around the forming die by moving the workpiece against the die or by moving the die against the workpiece. (Courtesy of Inco Alloys International, Inc.)

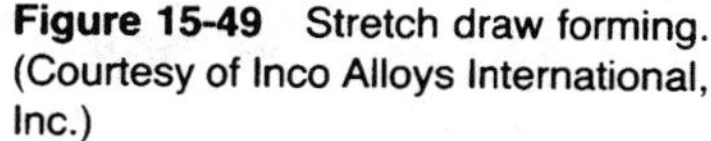

Figure 15-49 Stretch draw forming. (Courtesy of Inco Alloys International, Inc.)

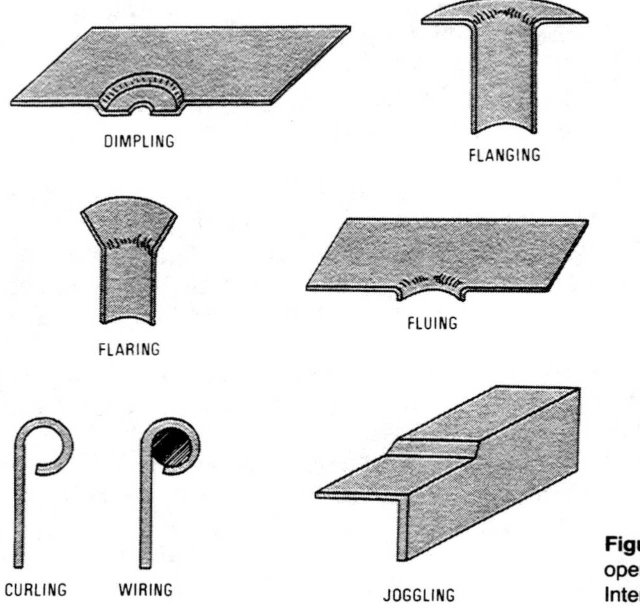

Figure 15-50 Miscellaneous form operations. (Courtesy of Inco Alloys International, Inc.)

stretch condition of the material tends to cancel the wrinkling action. Forming pressure is reduced because the stretching operation holds the material on the verge of plastic flow, so less pressure is needed for the actual forming operation.

Miscellaneous operations. Forming also is done either as part of another forming process, or as a separate operation on an article made by a different method. Some examples of these operations are shown in Figure 15-50.

Dimpling is forming a depression around a hole in sheet stock. The hole is punched in a separate operation, and the dimple is formed around the hole on a press using dimpling dies. The upper die is a punch that mates with a cup-shaped bottom die. Dimpling is often used to form countersinks on parts to be joined with flathead rivets or bolts, where a smooth surface is desired. The head of the fastener fits in the dimple, resulting in a flush fit.

Flanging consists of bending over the edge of a shape to make a lip (flange) at a sharp angle to the body of the part. *Flaring* spreads out the end of a tubular part to form a cone-shaped end. *Fluing* is forming a flange around a hole by pressing a cylindrical die through the hole, thereby enlarging the hole and bending out the edges.

Curling and *wiring* are done to avoid sharp edges on sheet-metal articles or to strengthen the edges, as obtained with hemming. Both operations are performed by rolling over the edges to produce a tubular rim. In curling, the rim is left hollow. In wiring, the edge is rolled over a wire to add strength to the rim.

Joggling is producing an offset in a structural part such as an angle or channel. The joggle permits two pieces to fit together with a flush joint. The offset in one piece is made equal to the thickness of the piece to be joined to it.

PRESS EQUIPMENT

The machinery used in the pressworking of metal is, of necessity, of a variety of sizes, designs, and operations to match the tremendous range of products produced with them. Classifications generally are made according to the frame design, the manner in which power is applied, and the function. The common types of press frames are shown in Figure 15-51. The press *bed* provides a rigid support for the forming die, and rests on the lower part of the frame. The *frame* is the structure which contains and orients all of the functional parts of the press. The drive mechanism controls the action of the movable *ram*, which is positioned on the upper frame and delivers the force necessary to form a workpiece on the die.

The several frame configurations are so constructed to provide the necessary rigidity, and to permit access to the die space. The C-type frame is typical of presses with capacities up to 1.8 MN (200 tons), and includes the knee, the horn, the gap, and the inclinable styles.

The *knee* press has an adjustable bed or table (knee) which can be adjusted to accommodate various sizes of workpieces and dies. The *horn* or horning press features a round projection which acts as a bed support for round or curved workpieces.

Figure 15-51 Common types of press frame geometries.

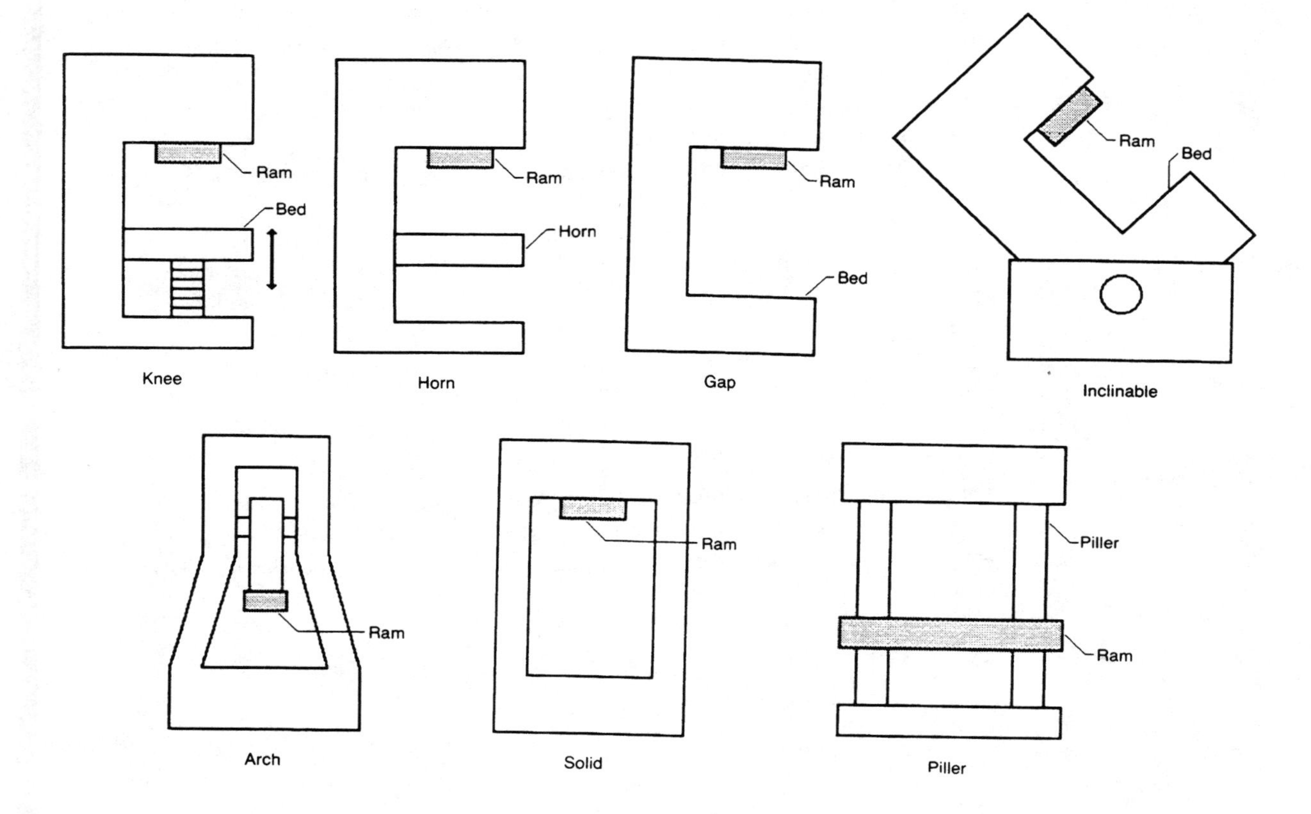

The *gap* design has an open throat and open back frame for use with long workpieces. This is the most rigid of the C-type presses. The *inclinable* press provides access for odd-shaped workpieces, and for convenient additions of feed chutes and gravity removal of finished parts.

Multiple column press frames have less die accessibility, but are more rigid, and are of the arch, solid, and pillar styles. The *arch* press is a low capacity rating machine, whose small top and large bed permits the use of light, small moving parts for light-gauge sheet metal forming. The *solid* press is similar to the arch, but for heavier work. *Pillar* or open frame pressure generally is hydraulically powered, and features a movable platen or ram.

The press drives generally are mechanical or hydraulic, to actuate devices such as cams, eccentrics, screws, or cranks, similar to those used on forging presses.

QUESTIONS

1. Define cold forming, and explain its advantages and disadvantages regarding hot forming processes.
2. Define bending, and sketch a diagram to describe the imposed stresses.
3. What is springback, and how does one compensate for it?
4. Explain the essential differences between brake bending and press-brake bending.
5. Prepare simple sketches of the air, bottom, and three-point methods of brake bending.
6. Prepare simple sketches of the four basic bending roll configurations, and explain the advantages of each.
7. Compute the mean bend radius of a 6D bend in a 25 mm diameter tube.
8. What is the primary purpose of roll forming?
9. Explain the difference between shallow and deep drawing, and prepare simple sketches of each.
10. Describe the operational differences between single-, double-, and triple-action presses.
11. Prepare sketches to illustrate the differences between ironing and sinking operations.
12. What is the primary advantage of the use of flexible dies in drawing operations?
13. List four flexible die systems and explain their operational differences.
14. Explain the differences and typical uses of progressive and transfer die systems.
15. Prepare simple sketches of three HERF systems.
16. Explain the main difference between standard and shear spinning.
17. What are the attributes of sketch forming?
18. Prepare simple sketches to illustrate a flare, a dimple, a flange, and a flute.
19. Prepare simple sketches of the seven basic styles of press machinery.
20. Write a library research report of current advances in press equipment or techniques.

16 CASTING PROCESSES

Very simply, *casting* is the process of pouring a liquid or a viscous material into a cavity where it sets or hardens to form a part. See Figure 16-1. It also is called *founding*, and a *foundry* is an assembly of all the apparatus needed to produce metal castings. *Molding* is another term used to describe this process, but actually a *mold* is the prepared cavity into which the casting material is poured. Metals, plastics, rubbers, and cementitious materials all may be cast. Casting is a bulk conserving process, because the final mass of the object essentially equals the mass of the molten material.

Casting is an ancient process, beginning perhaps when a ring of copper-bearing stones surrounding a primitive campfire leached out the copper, which ran off into a depression in the soil. Early humans eventually used this technology to produce copper arrowheads and tools. Contemporary uses of casting processes are numerous, and comprise the subject of this chapter. They include molding preparation and castings design, sand casting, shell molding, plaster mold casting, investment casting, centrifugal casting, permanent mold casting, die casting, slush casting, and powder metallurgy. The discussion generally is centered around the casting of metals. The treatment of plastic molding methods appears in Chapter 7.

Figure 16-1 These high quality wood planes have cast bottoms, machined to the desired smoothness. (Courtesy of Stanley Tools.)

PATTERNS

A *pattern* is a replica of the part to be cast, and from this pattern a mold is made. There are a number of design rules to follow when making patterns, and these rules are described in the following discussion of the kinds of patterns in general use.

Loose patterns are models of the pieces to be cast, and may be of the flat or split pattern types. Flat patterns are simple and are the easiest to use. See Figure 16-2. Split patterns are made in two pieces, a cope half and a drag half, so designed to facilitate moldmaking. These are employed for curved parts where a flat pattern would be impossible to withdraw from; for example, a rammed sand mold. See Figure 16-3.

Fixed or *mounted* patterns are best exemplified by the *match-plate*. Here the pattern is permanently fixed to a plate, which makes mold-making much simpler. See Figure 16-4. An advantage of this type of pattern is that the gating system to allow for the flow movement of the molten metal to the mold cavity can be added to the plate. A typical set-up appears in Figure 16-5.

Patterns are made of wood, plaster, or plastic, with wood being one of the more common materials. Foamed polystyrene plastic and wax patterns also are used in evaporative casting processes. Here the patterns are imbedded in the sand or plaster molding material, and the patterns vaporize as the molten metal fills the mold. A separate pattern obviously must be used for each mold, as it is "lost" in the casting process.

Core prints are attached to the pattern to support the dry sand cores that will form any required interior casting voids. The pattern is also made slightly oversize

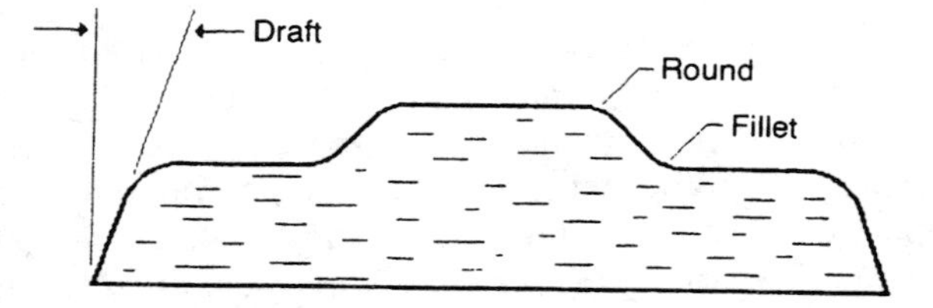

Figure 16-2 A flat pattern is a simple replica of the cast piece. The terms draft, round, and fillet are illustrated in this drawing.

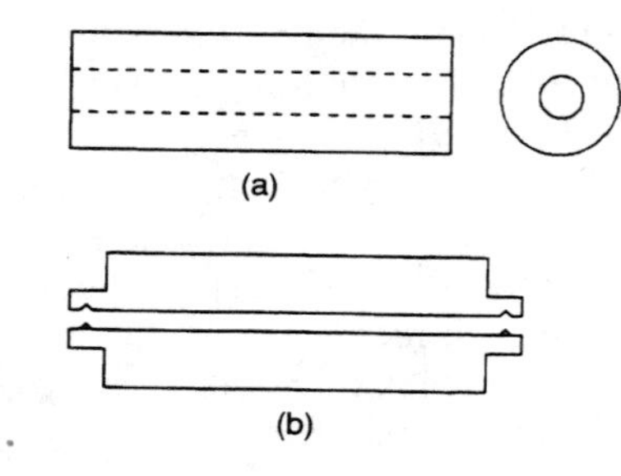

Figure 16-3 A split pattern is so-designed to permit easy removal from a sand mold. The cast part is shown at "a," the split pattern at "b." Note the core prints.

Figure 16-4 A match-plate pattern provides for quick and sure alignment of the cope and drag pattern sections.

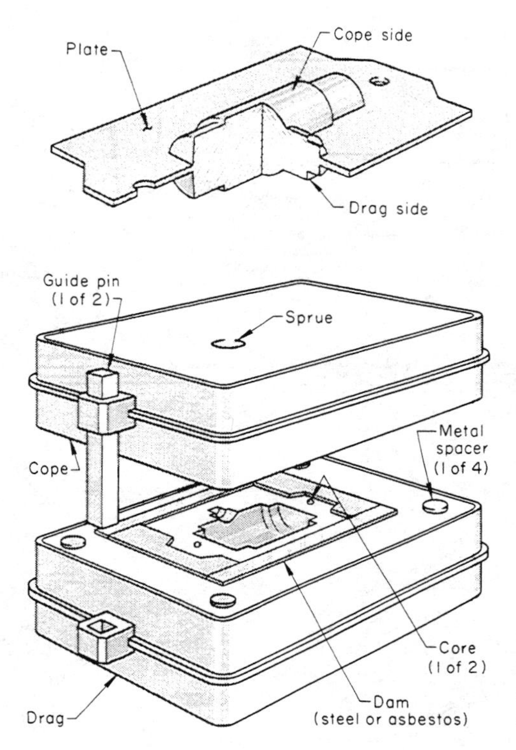

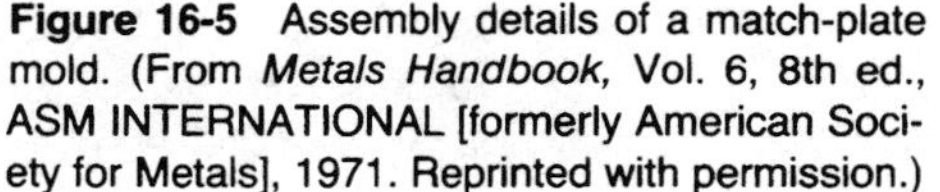
Figure 16-5 Assembly details of a match-plate mold. (From *Metals Handbook,* Vol. 6, 8th ed., ASM INTERNATIONAL [formerly American Society for Metals], 1971. Reprinted with permission.)

to compensate for the natural shrinkage or contraction of the metal as it cools. All of the size modifications made on a pattern are usually standard for a given type of metal or alloy. However, there are many ways to construct a mold, and the pattern will largely determine which way is the most economical.

The *mold* is actually a negative print of the pattern in molding sand or other material. The pattern must be removed from the sand, leaving an undamaged cavity that can be filled with molten metal. To withdraw the pattern from the mold, a parting line and draft must be provided on the pattern. Prior to the construction of the pattern, the need for draft and parting lines must be determined. In effect, the plan for the molding procedure is set up before the pattern is even started.

FOUNDRY PRACTICE

Founding involves four basic procedures: molding, coremaking, melting and pouring molten metals, and cleaning and finishing.

The *mold* is the heart of the process, because it represents the center of all activities, and other phases of foundry practice are grouped around it. The purpose of a mold is to form a cavity with accurate dimensions that will hold and support molten metal until it becomes solid. A mold is constructed of sand, plaster, or metal, with provisions made for the opening of the mold and the withdrawal of the pattern. An entry port or sprue is provided for introducing the molten metal into the mold cavity left when the pattern is removed. See Figure 16-6.

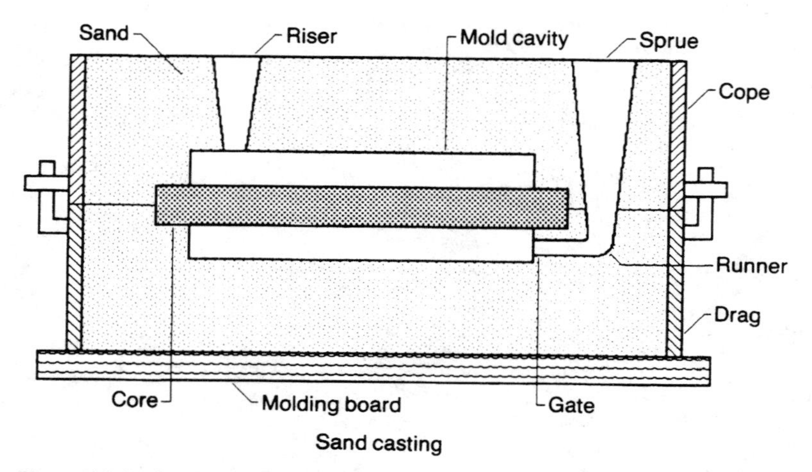

Figure 16-6 A common sand casting arrangement with important features identified.

Coremaking is closely related to the mold, because the core actually becomes a part of the mold prior to pouring the molten metal. The purpose of a core is to form a cavity within the casting or to act as a support when the surface of the mold cavity is irregular or difficult to form from molding sand. The core must be strong, and therefore core sand mixtures require a special treatment. The sand from which the core is made is prepared from materials that provide the core with the ability to occupy these cavities in the mold without collapsing. The prepared sand mixture is rammed or blown into *core boxes* that will give the core the desired shape. See Figure 16-7. The core box is opened and the core is placed in a metal or refractory drier and baked in an oven to fully develop its setting properties. The core is then placed in the mold cavity on a core-print, where it serves to form part of the casting.

Melting and *pouring* metals is a science that requires the utmost accuracy and skill if sound, usable castings are to be produced. Melting is done in oil, gas, coke, or electric furnaces. Heat is prepared from appropriate metals and placed in

Figure 16-7 Production of a blown sand core in a two-piece core box. (From *Metals Handbook,* Vol. 6, 8th ed., ASM INTERNATIONAL [formerly American Society for Metals], 1971. Reprinted with permission.)

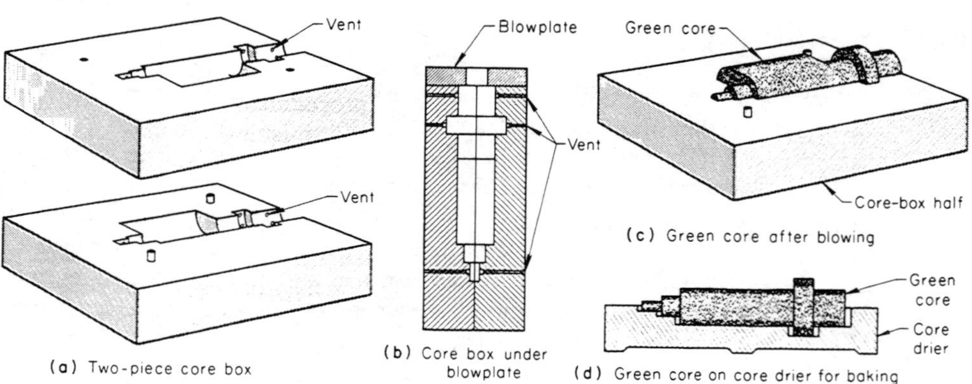

the furnace, where it is brought to required pouring temperature. When the metal or alloy is ready for pouring, the metal is tapped from the furnace and poured into the mold cavity through the opening in the mold. Automatic pour systems are commonly used. As the metal becomes solid, the casting will assume the shape of the original pattern from which the mold was made. After cooling, the mold is opened, and the casting is removed, ready for cleaning and touch-up as necessary.

CASTING DESIGN

Molten metal solidifies in the mold as a gradually thickening envelope or skin. The rate of solidification is a factor of the section mass, mold surface area, and the solidification rate of the metal. Other things being equal, thinner sections will solidify before the thicker ones.

During solidification, a pronounced contraction takes place. This means that additional metal, more than that required for the initial filling of the mold cavity, must be fed to the solidifying mass to assure internal soundness. This additional metal, stored in the risers, must remain liquid until the casting has solidified. It is important that the casting sections be proportioned and positioned so that the sections most distant from the risers solidify first. Subsequent solidification then progresses toward the risered section, where the hottest metal is located.

With solidification, the cast metal taken on an increasingly rigid form, accompanied by contraction. The pattern shrinkage allowance must compensate for this shrinkage, or the casting will undersize. This contraction in the cast metal is opposed by some parts of the casting because of any shape irregularities. This may result in severe contraction stresses, requiring that castings be heat-treated. The different cooling rates of thin and heavy sections result in the cooling and severe contraction of the thin sections prior to the complete solidification of the heavy sections. This results in stressing the partially solidified, and still very weak, heavy section.

Besides solidification, the crystal structure of metals requires consideration. Most metals set by the formation and growth of crystals, where the relative size is determined largely by the time required for solidifying and cooling in the mold. As this time is greater for heavy sections, the crystalline structure of a heavy section is correspondingly coarser than that of the lighter sections. In steel, coarse crystallization means lower physical properties. With the nonferrous metals, a separation of the lower melting point constituents may occur.

In designing a casting, specific design rules must be followed, leading to a simplification which reduces the patternmaking and molding costs. If a complicated design is unavoidable, the possibility of making the part in sections which later can be bolted, riveted, or welded together should be considered.

Design rules. There are a number of general rules to follow in designing a casting. One rule is to avoid sharp angles by using gradual contours, as the shape of the casting section affects the metal structure. Rounded corners, or rounds, are advantageous in the formation of the crystal grains.

A design must also provide for the shrinkage which occurs in metals as they change from a liquid to a solid state. The parts should increase in thickness progressively to points where risers will provide the metal needed to offset the metal shrinkage. A riser cannot feed a section of uniform thickness for a distance greater than approximately 4-$\frac{1}{2}$ times the section thickness. Therefore, it is important that casting sections having considerable length, but not accessible for risering, be tapered rather than uniform in cross section. Further, the larger portion of the section should be near the riser.

Castings should be designed so that large differences in cross sections should be as uniform as possible. At the same time, the length of uniform sections must not exceed the ability of the riser to feed the section. Heavier sections should be tapered into the lighter one gradually, never abruptly.

A proper casting thickness must be maintained. The minimum cross-section thickness through which a molten metal will normally flow is indicated in the following guide:

Aluminum . . . . . . 3 mm (0.125 inch)
Brass and bronze . . . 2.5 mm (0.094 inch)
Cast iron . . . . . . 3 mm (0.125 inch)
Steel . . . . 5 mm (0.187 inch)

The sections should be no thicker than necessary, but should be sufficient to permit the proper flow of the metal in the mold.

When designing adjoining sections, sharp corners should be curved to avoid heat and stress concentration. Intersection members of equal cross-sectional thickness do not create a molding problem if the joint location can be directly fed by a riser. All too frequently, though, it is impossible to feed these members directly. One way to avoid an area of excessive mass and at the same time obtain a more uniform section thickness is to stagger the intersecting members. Stagger the cross members or ribs, and eliminate sharp corners at adjoining sections. Do not bring more than three sections together, because shrinkage and porosity troubles occur most frequently at member junctions. If the gradual blending of the sections is not possible, use fillets at the junctions.

Use fillets at all sharp angles, found at inside corners, to make the corners more moldable and to eliminate a plane of weakness resulting from a peculiar type of grain growth which occurs at sharp internal angles. The arrangement of the crystal growth is such that the lines of strength are perpendicular to the face of the casting. The size of the fillet depends on factors such as the kind of metal, the thickness of the wall section, and the shape and size of the casting. Large fillets produce nonuniform metal thicknesses and a subsequent irregular cooling, resulting in a weak casting. A good rule is to make the radius of the fillet one-half to one-third the size of the thickness of the sections joined.

Projections such as bosses or pads should not be included in the casting design unless absolutely necessary. They increase the metal's thickness and create hot spots which may lead to improper solidification and to coarse grain structure. If bosses and pads are used, they should be blended into the casting by tapering or flattening the fillets.

If possible, casting designs should be such that the surfaces to be machined are cast in the drag section of the mold. If such surfaces must be cast in the cope section, an extra allowance for the finish must be included on the pattern.

In addition to bosses and pads, there is the problem of designing ribs. The primary use of ribs is to reinforce the casting without increasing overall wall thickness. Properly designed ribs also reduce the tendency of large flat areas to distort. Some typical design configurations appear in Figure 16-8.

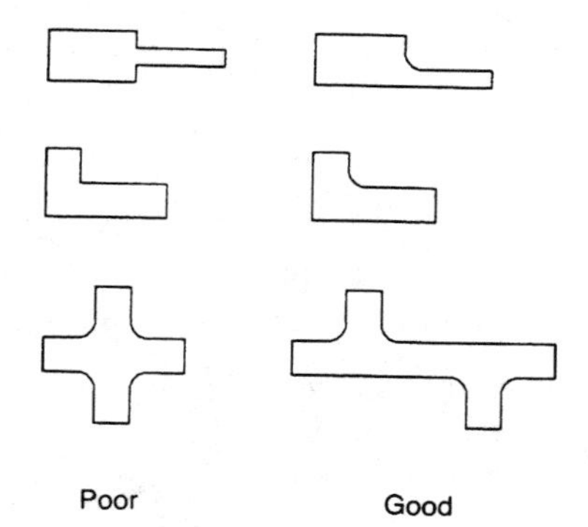

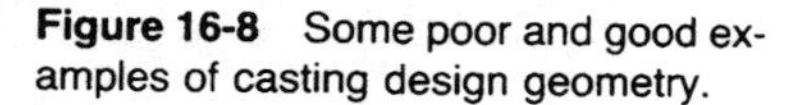

Figure 16-8 Some poor and good examples of casting design geometry.

Sand Casting

Of all the processes used in the production of castings, the most common is the *sand casting* method. Here the sand used contains sufficient refractory clay to bond strongly without destroying venting quality when it is rammed to the required degree of hardness according to casting size and geometry. Based upon the sand conditioning prior to use, molds may be green sand, dry sand, or skin-dried.

Green sand molds may be of natural bonded sand or synthetic sand, and can be poured as soon as they are rammed. Molds made from natural sand contain a sufficient amount of clay bond, either present when the sand was taken from its deposit site or added before shipment, to make the sand suitable for use. Adding moisture and tempering is the only treatment necessary before use. Molds made from synthetic sand are made by mixing correct proportions of an unbonded sand and additives such as clay binders, cereals, or rosin, and tempering or allowing the mixture to stand before use.

A *dry sand* mold is slowly baked in an oven before it is used, and is generally used for heavy castings. Dry sand molding has both advantages and disadvantages. The rigidity of the mold resists metal contraction during the solidification of the casting, and the resistance is often great enough to cause the casting to crack. The hard surface of a dry sand mold enables it to withstand the eroding action of the force of the flowing metal and support the weight of large volumes of metal. The baking of the mold removes moisture, lessening the possibility of the formation of mold gases and rapid chilling of the metal.

A *skin-dried* mold is one that has been surface heated with a torch. It is used when the requirements call for a mold having the surface characteristics of a dry sand mold, combined with the collapsibility of a green sand mold. Skin-dried molds may also be used when an oven is not available for baking a dry sand mold. When using a skin-dried mold, the melt (liquid metal) must be ready to pour as soon as the mold is completed. The effect of skin-drying will be lost if the mold is allowed to stand, since moisture from the backing sand will penetrate to the mold cavity surface.

Molds may be classified according to size, such as *bench molds*, *floor molds*, or *pit molds*. Bench molds are those small and light enough for easy handling. A mold that is too large for one person to handle is usually constructed on the foundry floor. Pit molds are used when the size of the casting requires a mold constructed in a large pit in the foundry floor.

The procedure for producing a simple, flat-pattern sand casting is as follows (see Figure 16-9):

1. A flat pattern is placed on a molding board, surrounded by the inverted drag half of a flank. Parting powder is dusted over the pattern to facilitate its later removal.

2. Sand is riddled (screened) to cover the pattern with a layer of fine, lump-free sand. This *facing* sand will improve the surface texture of the cast part.

3. The drag is now filled with unscreened *backing* sand, which is rammed or pressed to compact the sand tightly around the pattern. A stiff wire is used to press venting holes into the sand, being careful not to touch the pattern.

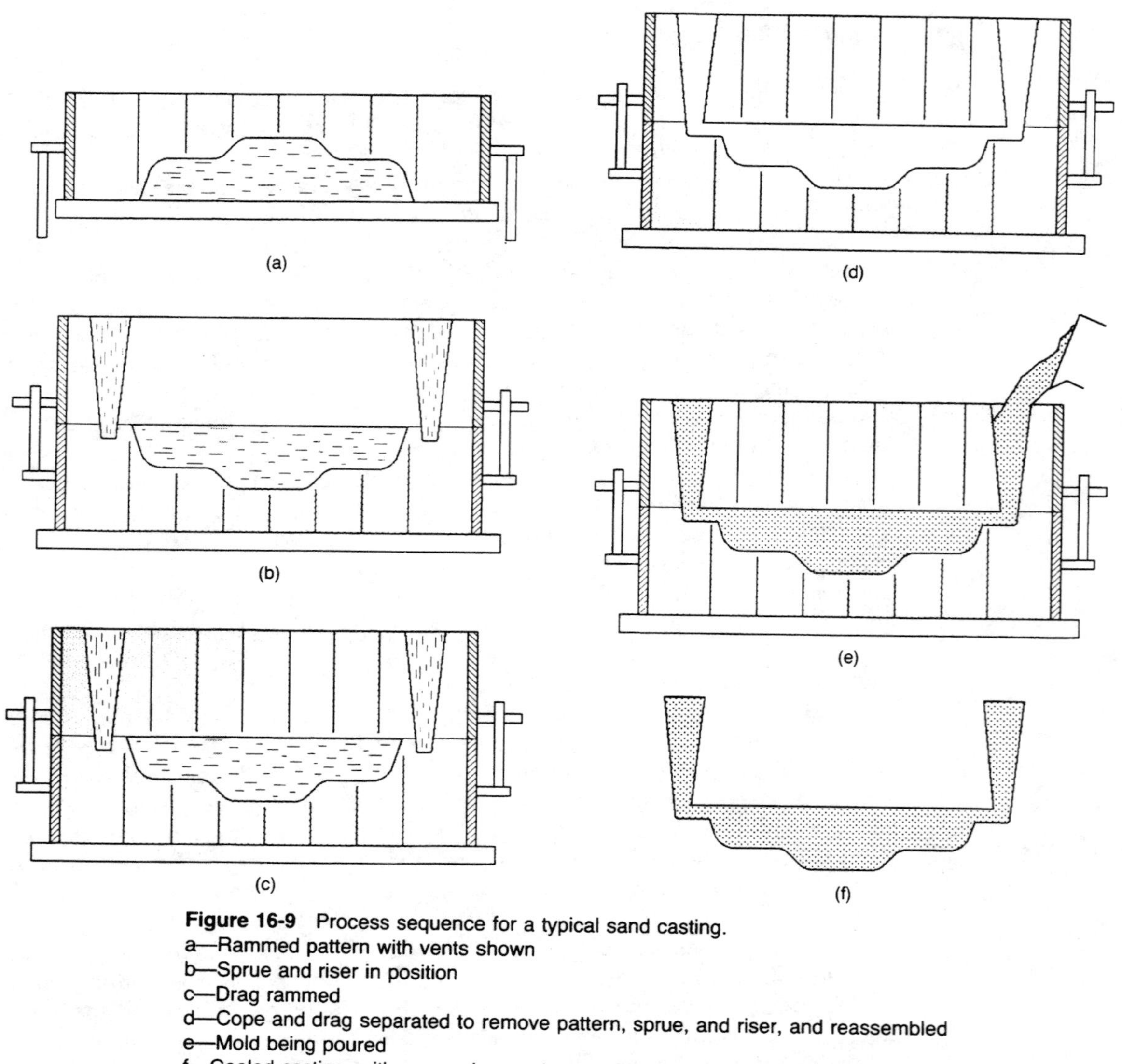

Figure 16-9 Process sequence for a typical sand casting.
a—Rammed pattern with vents shown
b—Sprue and riser in position
c—Drag rammed
d—Cope and drag separated to remove pattern, sprue, and riser, and reassembled
e—Mold being poured
f—Cooled casting, with sprue, riser and runner attached, after removal from flask

4. Next, a second or bottom molding board is placed on top of the rammed drag, and the cope is positioned on the drag guided by the flask pins. More powder is dusted over the sand mass and exposed pattern base to provide a mold parting line. This will ease the later separation of drag and cope.

5. Sprue and riser dowel pins are placed and appropriately located by pressing them into the sand so that they stand upright. The cope is now filled with backing sand and rammed.

6. The cope is carefully lifted from the drag, and the pattern is removed to expose the molding cavity.

7. The sprue and riser pins are removed from the cope, and the necessary gates or runners are cut into the sand. These elements provide the ways through which the molten metal travels.

8. The cope is replaced on the drag to create a mold ready for pouring.

9. Once the metal has been poured and allowed to cool, the cope and drag are separated to expose the cast piece.

10. The sprues, risers, and runners are cut away from the casting, which is now ready for post-processing.

Shell Molding

Shell molding is a modified sand casting process whereby a mold is formed from a mixture of sand and resin, which is placed against a heated metal pattern to bake it to a rigid shell. A pair of these matching shell halves, when glued or clamped together, form the mold cavity into which molten metal is poured to create a cast product. A match-plate pattern system is used to form the sturdy sand shell molds. See Figure 16-10. The advantage of this technique is that the resultant castings have greater dimensional accuracy and are of a smoother surface texture. The procedure for shell molding is described following:

1. A precision metal match pattern is *prepared* to design specifications. One of the drawbacks of shell molding is the high cost of the machined patterns.

2. The metal pattern plate is *heated* to around 260°C (500°F) and sprayed with a silicone release material to facilitate the removal of the shell.

3. The heated pattern is *invested* with the resin-sand mixture by clamping the pattern to a sand container or dump-box, and inverting it. This causes the sand to drop onto the heated pattern and adhere to it. Shell thickness, usually 3 to 9 mm

Figure 16-10 A schematic of the shell molding process.

(0.125 to 0.375 inch), is a factor of 30-second investment time in the dump-box. The unit is returned to its upright position, causing the excess sand to fall back into the dump. Sand also may be sprayed onto the pattern to form the shell, a process which provides a more uniform coat and forces sand into pattern crevices. By either method, the result is a soft but firm shell attached to the plate.

4. The firm shell is next *cured* by moving the pattern plate to an oven for about 30 seconds at a temperature of about 370°C (700°F).

5. The cured shells are removed from the plate and *assembled* by bonding the two shells together with an adhesive.

6. Molten metal is next *poured* into the mold cavity, placed in a bed of sand, or shot to provide a firm pouring arrangement.

7. Upon cooling, the casting is *removed* from the shell, and cleaned and processed to its finished form.

Shell molded casting methods are applicable to parts of sizes up to 180 kg (400 pounds), and of almost any type made by sanding casting. See Figure 16-11.

Plaster Mold Casting

A specialized process called *plaster mold casting* employs disposable, one-time-use plaster molds to produce nonferrous metal castings. The mold material is a slurry composition of gypsum plaster (plaster of paris), a chopped fiber strengthening material, additives to control setting time and workability, and water. Patterns typically are made from polished aluminum or brass, sealed wood, plastic, or plaster. They are similar in design and construction to those used in sand casting, and can be of loose-piece or matched plate variations. Similarly, the flasks and molding boards are much like those employed in sand casting.

The process advantages are the production of parts with unusually smooth surfaces, and of intricate design detail and dimensional accuracy. Because such castings have a slow cooling rate, thinner-walled pieces can be cast with a minimum of warpage. A major limitation of the project is that it is suitable only for use with nonferrous metals, generally aluminum. Because plaster molds are not reusable, their costs are usually higher than metal molds, which may be used repeatedly. A brief overview of the process sequence follows.

1. The pattern is *positioned* on the bottom board in the flask, and sprayed with a release film. Appropriate gates, sprues, and risers are properly oriented to the pattern.

2. The plaster slurry is *poured* directly onto the pattern to fill the flask. The flask is agitated to assure slurry entry into all pattern areas to remove bubbles.

3. After the plaster has set for about 10 minutes, the pattern is *removed* from the mold, with great care to avoid cracking. Both parts of the mold, the cope and the drag, are prepared in this way.

4. The plaster molds are *baked* in a drying oven to remove any moisture. The drying time varies according to mold size and other casting design factors. These

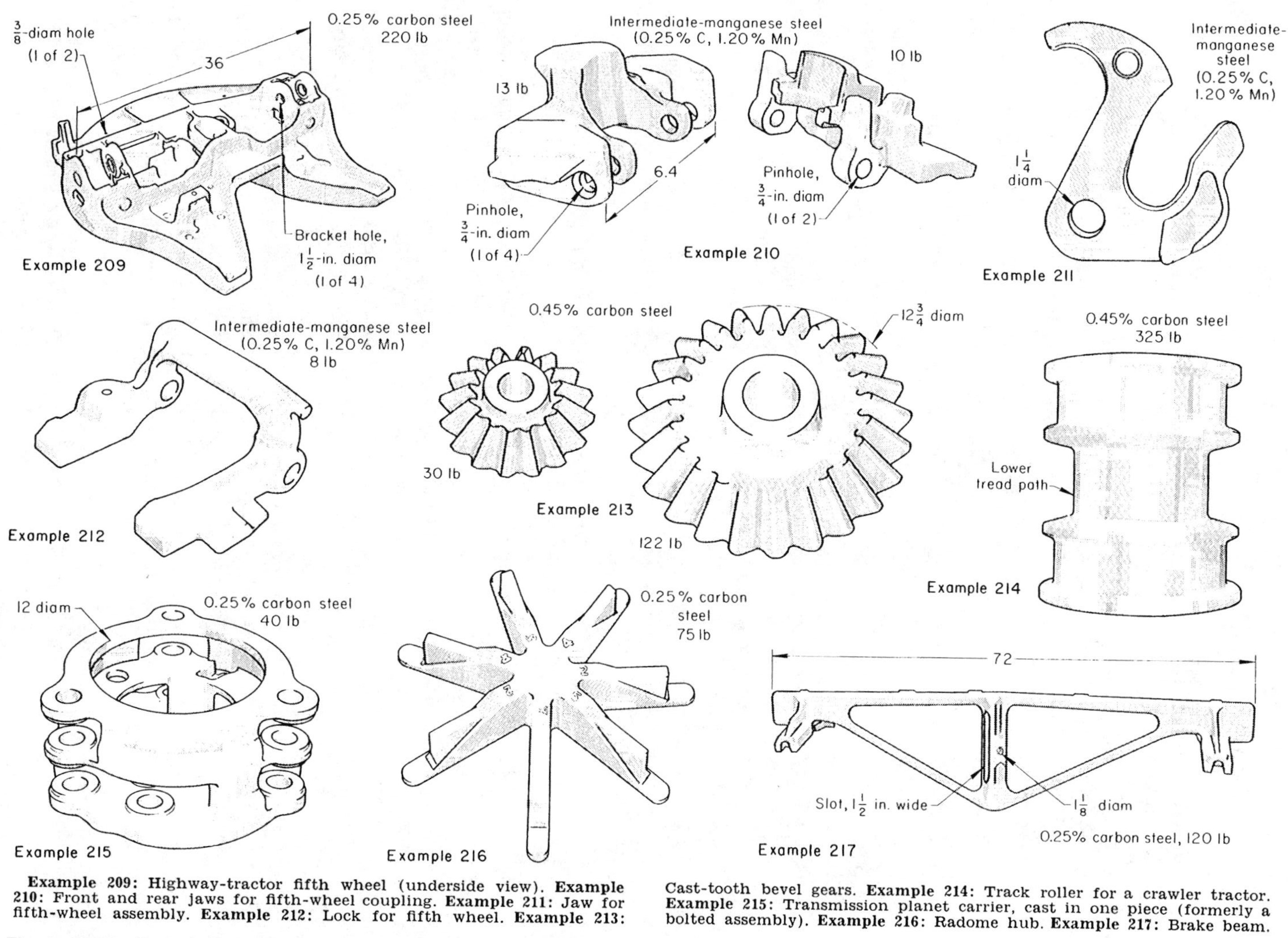

Example 209: Highway-tractor fifth wheel (underside view). **Example 210:** Front and rear jaws for fifth-wheel coupling. **Example 211:** Jaw for fifth-wheel assembly. **Example 212:** Lock for fifth wheel. **Example 213:** Cast-tooth bevel gears. **Example 214:** Track roller for a crawler tractor. **Example 215:** Transmission planet carrier, cast in one piece (formerly a bolted assembly). **Example 216:** Radome hub. **Example 217:** Brake beam.

Figure 16-11 Typical shell-molded parts. (From *Metals Handbook,* Vol. 6, 8th ed., ASM INTERNATIONAL [formerly American Society for Metals], 1971. Reprinted with permission.)

ranges may be from about one to 72 hours, at temperatures from 170°C to 870°C (350°F to 1,600°F).

5. After baking, the molding flask is *assembled*, and any necessary cores are added. Molten metal is then *poured* into the assembled mold. The completed casting is removed and cleaned as required.

Product applications include those with complex, irregularly-shaped surfaces, difficult to machine, for which an as-cast smooth surface is especially desirable. Typical products include aircraft and automotive parts, plumbing components, hand tools, locks, propellers, and tire molds.

Investment Casting

Perhaps the most familiar example of this process is the gold crown made by a dentist to cap a tooth. The dentist carefully shapes a precise wax pattern of the crown or filling, invests it in a metal container of plaster, heats the flask or container to melt the wax, and forces molten gold under pressure into the mold cavity formerly occupied by the wax pattern. This is *investment casting* (or lost wax casting), an ancient technique used for bronze sculptures which is applied to industrial products. The process is used to manufacture precision parts of almost any configuration. Indeed, it often is referred to as the ''precision casting'' technique. See Figure 16-12. The process sequence follows.

1. Wax, and often plastic, is *injected* into an aluminum die to produce a disposable wax pattern which is an exact replica of the part to be cast. For every part, a wax pattern must be made.

2. The individual patterns are *clustered* around a central sprue and gating system to produce a ''tree'' of assembled patterns.

3. The pattern tree is placed in a metal flask and *invested* by pouring a quick-setting ceramic slurry into the flask until full. The flask is agitated during pouring to insure an integral cover of the pattern and to remove any air bubbles which may affect the casting quality. Solid patterns (and therefore parts) are invested in this manner. Larger parts with shell configurations are invested by repeatedly dipping them in agitating vats of slurry until the required mold thickness is achieved, generally around 9.5 mm (0.375 inch). A robot or other automated equipment generally is used for this operation.

4. The molds next are *dewaxed* by flash firing at temperatures around 760°C (1400°F), or by autoclaving with pressure and steam. This process causes the wax patterns to melt and drain out of the mold to create the casting cavities.

5. Upon cooling, the mold material is broken or abrasive-blasted away and the cast parts *removed*. Gates and sprues are cut away, and the parts are ready for secondary operations.

Investment castings are widely used in products for the aerospace, automotive, jewelry, and appliance industries. Their sizes range from a few grams to over 450 kg (1,000 pounds). Almost all metals can be investment cast, and this process permits the mass manufacture of complex shapes that would be difficult, if not impossible, to cast by any other method.

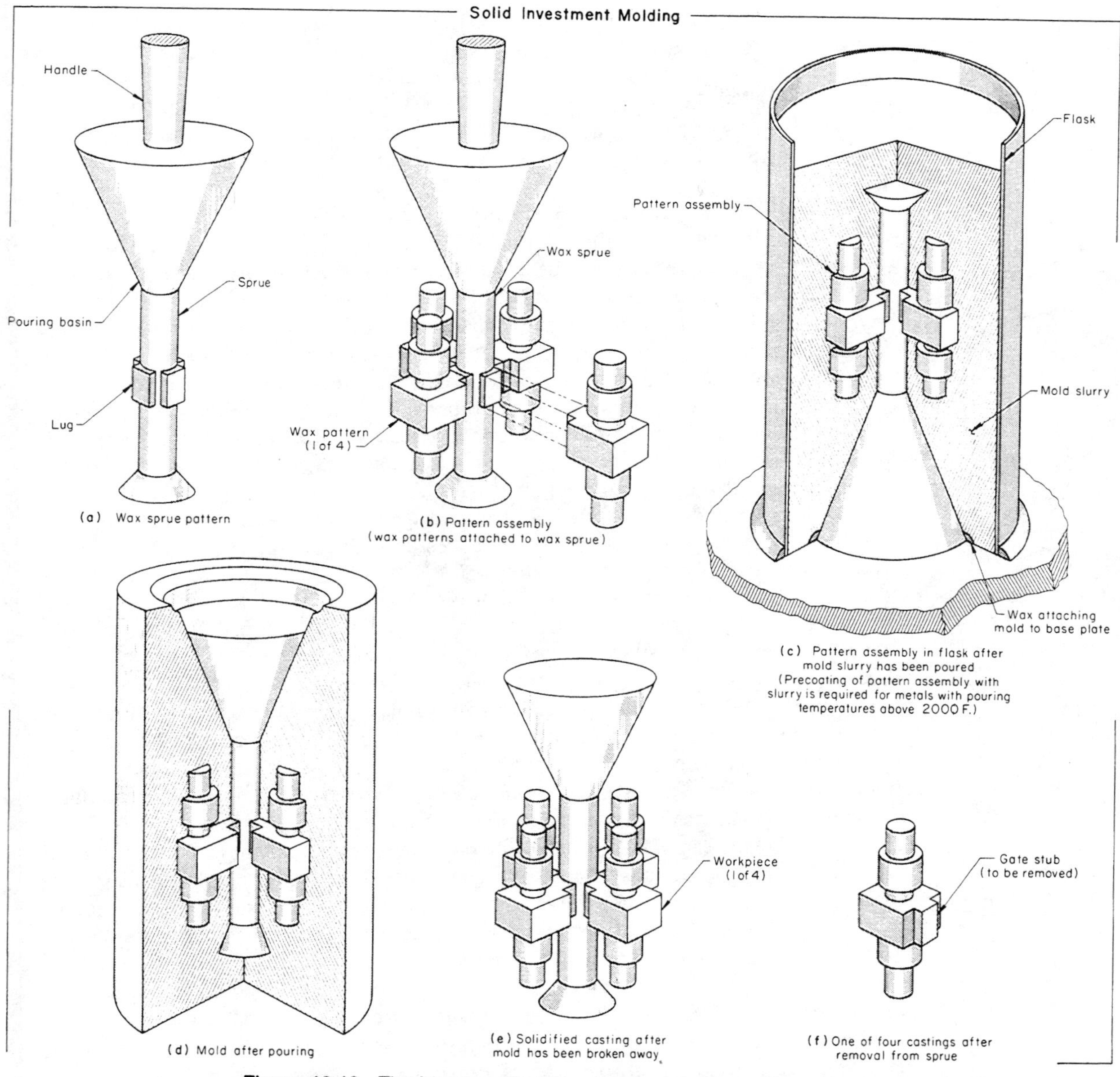

Figure 16-12 The investment casting process sequence. (From *Metals Handbook,* Vol. 6, 8th ed., ASM INTERNATIONAL [formerly American Society for Metals], 1971. Reprinted with permission.)

Centrifugal Casting

A unique casting method involves the introduction of molten metal into a rotating mold, where the metal is picked up and held firmly by centrifugal force against the inside surface of the mold cavity until solidification occurs. This process is called *centrifugal casting*.

Solidification of the casting starts when the metal is first poured. Heat immediately flows from the hot metal into the wall of the cold mold. As a result, solidification begins at the outside surface of the casting and progresses across the wall to the inner surface. The cooler the outer mold surface, the more steep will be the temperature gradient from its outer to inner surface and the more rapid will be the rate of heat withdrawal from the casting. A cooling water spray or jacket is often used to enhance this.

During the casting process, centrifugal force establishes a pressure gradient radially across the wall thickness of the casting. Solidification shrinkage pores are essentially eliminated by the pressure of the molten metal against the solidifying metal as solidification advances inwardly across the wall section of the casting. Pressure is lowest at the inner surface. Since particles of varying densities are subject to different pressures at a given rotational velocity, there is an outward movement of the higher density material, which displaces the lower density material toward the inner surface. In this manner, small slag particles and other light nonmetallic inclusions gravitate toward the inner surface. This process of solidification shrinkage occurs under conditions far more ideal than ever attained in static castings, producing a casting with a high degree of integrity. See Figure 16-13.

Figure 16-13 The setting action of a centrifugal casting. The molten metal is shown as a darkened area solidifying against the walls of a cold mold. The inner surface remains rough and grainy upon implementation of the process.

There are three discrete methods of centrifugal casting, and in each the process is essentially as just described. The differences lie with the types of molding systems employed, and the kinds of parts produced.

True centrifugal casting involves the use of hollow cylindrical tube molds which rotate about their horizontal axes. A steel flask provides the main containment vessel, and it is lined with a refractory mix rammed around a pattern, and coated with a thin refractory wash after pattern removal. See Figure 16-14. The mold is baked, the molten metal is poured into the rotating mold, and solidification occurs. This method is used to produce cable drums, pipes, paper machinery rolls, fabric finishing rolls, and hydraulic cylinders. The shapes may be round, hexagonal, tapered, or fluted. The rough inner surfaces of such castings are often removed by machining.

Semicentrifugal casting is used in the manufacture of wheels with spokes and hubs, sprockets, gears, pulleys, and other similar complex shapes. Tubular shaped flasks packed with refractory sand are used, as are special copes and drags. Several molds may be stacked one atop the other and connected with gates for better economy. Cores are frequently part of such a mold arrangement. Molten metal is introduced at the top of the mold in semicentrifugal casting.

Centrifuge casting is a method whereby molds for small, irregularly shaped parts are arranged in a balanced, wheel-like manner around a central sprue and gating system. As molten metal enters the top of the spinning mold, it follows the sprue and gates and is forced into several mold cavities. The mold position generally is vertical, though horizontal systems are in use.

Permanent Mold

Permanent mold casting employs a metal mold of two or more parts, which provides a reusable cavity system for producing identical parts. This method is especially suited to the volume manufacture of small, simple pieces of uniform wall thickness, which do not require intricate coring.

The simplest metal molds are comprised of two hinged sections which can be clamped together to provide a reliable mold cavity. See Figure 16-15. Rough molds

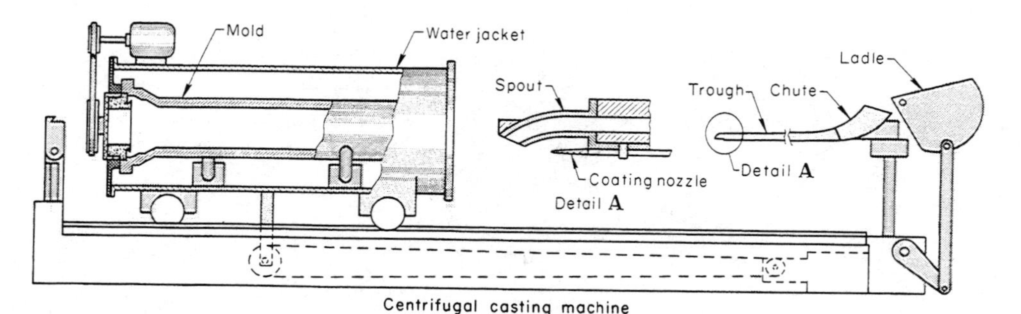

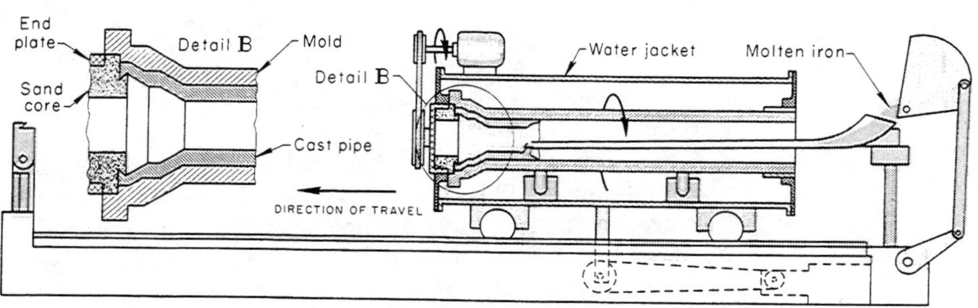

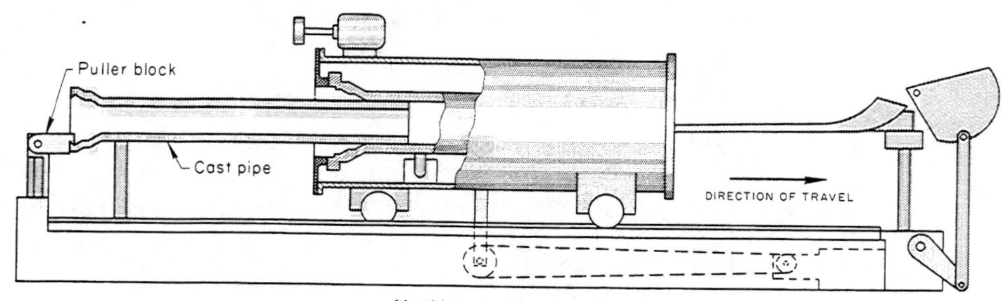

Figure 16-14 Equipment for the centrifugal casting of iron pipe in a rotating, water-cooled mold. (From *Metals Handbook,* Vol. 6, 8th ed., ASM INTERNATIONAL [formerly American Society for Metals], 1971. Reprinted with permission.)

Figure 16-15 This common book-type, manually operated permanent mold casting machine is used principally for molds with shallow cavities. (From *Metals Handbook,* Vol. 6, 8th ed., ASM INTERNATIONAL [formerly American Society for Metals], 1971. Reprinted with permission.)

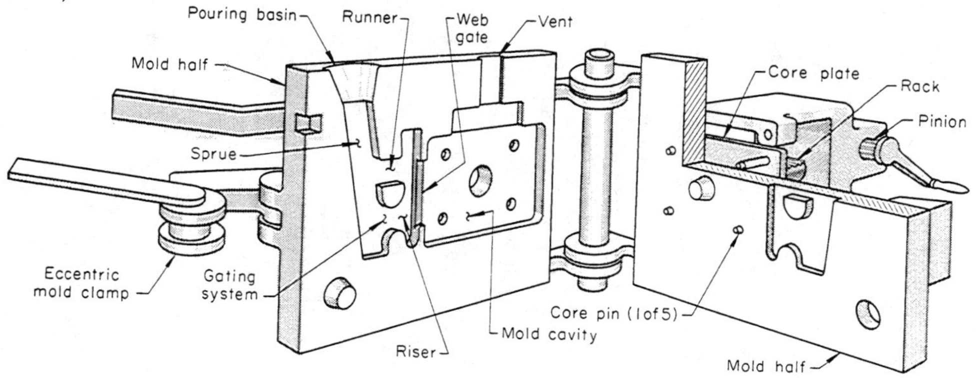

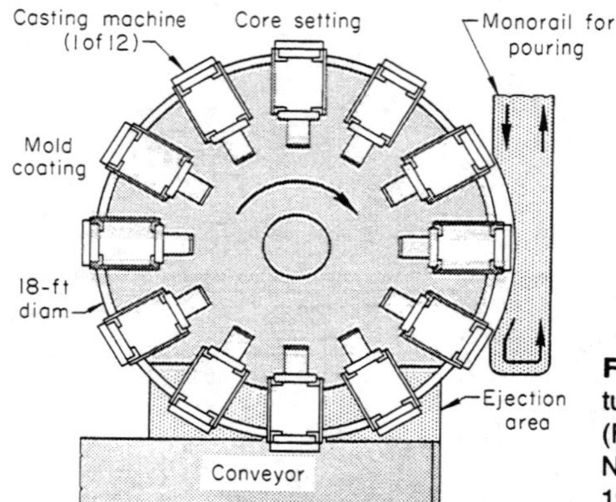

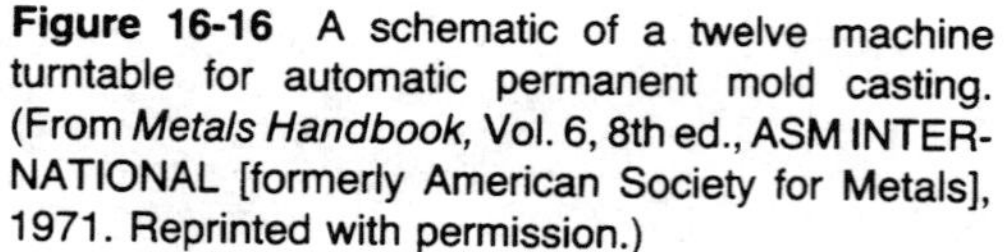
Figure 16-16 A schematic of a twelve machine turntable for automatic permanent mold casting. (From *Metals Handbook,* Vol. 6, 8th ed., ASM INTERNATIONAL [formerly American Society for Metals], 1971. Reprinted with permission.)

frequently are produced by casting and then machined to final size and geometry. Bronze and iron are typical mold materials. The casting process is fast and convenient, and is readily applicable to automatic systems, including the placement of cores. Further, the process provides uniform cast parts with excellent surface finish qualities and close dimensional tolerance. Metals used in these castings include aluminum, zinc, and copper alloys. Part sizes range up to 135 kg (300 pounds). This system is readily adaptable to automated casting systems. See Figure 16-16.

Die Casting

Die casting is the process of forcing molten metal rapidly, under high pressure, into a metal die cavity and then allowing it to solidify. Upon solidification, the die is opened and the casting is ejected. The process is fast and permits complex shapes to be cast ready or very nearly ready for use. Numerous identical castings can be produced economically from one set of dies.

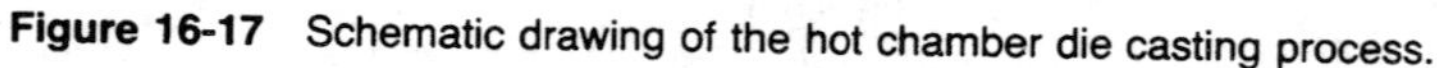
Figure 16-17 Schematic drawing of the hot chamber die casting process.

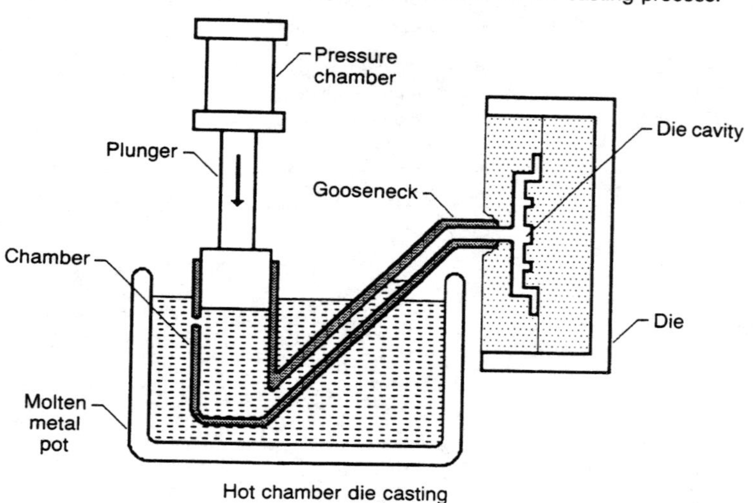

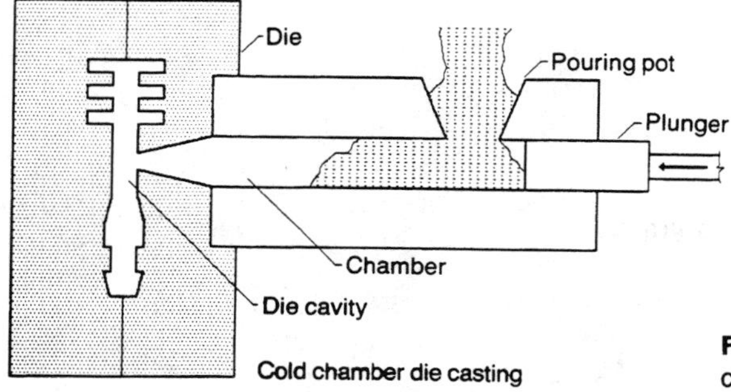

Figure 16-18 Schematic drawing of the cold chamber die casting process.

There are two basic die casting methods. The *hot chamber process* is used for die casting at lower melt temperature metals such as zinc and lead. The *cold chamber process* is used for metals that melt at higher temperatures, such as aluminum, magnesium, and brass.

The basic components of the *hot chamber* system are shown in Figure 16-17. The plunger and cylinder are submerged in the molten metal in the holding furnace. The casting sequence is as follows: when a shot is made, a control valve opens and causes the cylinder to force the plunger down. Molten metal flows through the nozzle, through the runners and gates, and into the die cavity. The gases that were in the system, and some of the molten metal, flow through the cavity and out through vents. After the cavity is filled, the metal is allowed to set, the casting is ejected, and the cycle is repeated. Since the gooseneck and plunger are immersed in the molten metal, the system refills automatically when the plunger is withdrawn.

The casting method for *cold chamber* casting is shown in Figure 16-18. Molten metal is ladled into the cold chamber, and the plunger advances to force the metal into the die. Except for the manner in which molten metal is fed into the shot system and injected into the die, the casting sequences for the two processes are similar.

Die casting dies. Die casting dies consist of two sections, the cover half and the ejector half, which meet at a parting line. The cover half of the die is secured to the front or stationary platen of the machine. The sprue for filling the die cavity is in this half, and it is aligned with the nozzle of the machine. The ejector half of the die is attached to the movable platen of the machine. It contains the ejection mechanism and, in most cases, the runners.

The die cavity which forms the part being cast is machined into both halves of the die block or into inserts that are installed in the die blocks. The die is designed so that the casting remains in the ejector half when the die opens. The casting is then pushed out of the cavity with ejector pins that come through holes in the die and are actuated by an ejector plate. Guide pins extending from one die half enter holes in the other die half as the die closes, to insure alignment between the two halves.

Dies that produce castings with complex shapes may contain stationary or movable cores. The movable cores are positioned by cam pins or hydraulic cylinders and are locked in place when the die is closed.

Since die casting machines operate at high rates, heat must be removed from the die at a high rate by circulating water or other coolant through channels drilled in the die blocks. Heat sometimes is added to the die by electric or gas heaters, for warm-up or for making thin sections which transmit insufficient heat to maintain the die at the proper operating temperature.

Die casting system requirements. In order to produce high quality castings, the die-casting system must have the following characteristics:

1. The shot system must be large enough to deliver sufficient molten metal to fill all of the die cavity when a shot is made.

2. The complete system must be designed so that the cavity is filled uniformly and rapidly, to insure that the cavity is filled with zinc which has not started to solidify when flow stops.

3. The system must be designed to eliminate or at least minimize the entrapment of air in the die cavity when flow stops.

Die cast products include small zinc alloy parts, transmission cases, pump components, rack and pinion automotive steering housings, escalator treads, and computer chassis.

Slush Casting

Slush castings are made by utilizing sand or metal molds to produce thin walled parts primarily for ornamental or statuary work. In this process, the metal is poured into a split bronze mold and allowed to cool long enough to form a shell. As soon as the desired thickness of the shell is obtained, the mold is inverted and the remaining liquid metal is poured out, leaving a hollow center in the casting. This process is used primarily for castings of the lead and zinc alloys. See Figure 16-19.

Powder Metallurgy

Powder metallurgy, or P/M, is a unique method for manufacturing complex, accurate, and reliable ferrous and nonferrous components. See Figure 16-20. P/M parts are made by mixing metal alloy powders and compacting the mixture in a die. The resultant shapes are then sintered or heated in a controlled-atmosphere furnace to metallurgically bond the particles. P/M is basically a casting process, inasmuch as the powdered shapes are the result of die or mold containment. Furthermore, it is a bulk conserving and very cost-effective process, because typically more than 97 percent of the starting raw material is used in the finished part. Production rates range from a few hundred to several thousand parts per hour.

P/M parts have additional unique attributes. Ferrous and nonferrous pieces can be oil impregnated to function as self-lubricating bearings, resin impregnated to seal interconnecting porosity, infiltrated with a lower melting point metal for greater strength and shock resistance, and heat treated or plated as required.

Most parts weigh less than 2.25 kg (5 pounds), although parts weighing as much as 16 kg (35 pounds) can be fabricated with conventional equipment. Many of the early P/M components, such as bushings and bearings, were very simple shapes as contrasted to the complex contours that are produced economically today. In many cases, functions that normally would require intricate multiple process and assembly

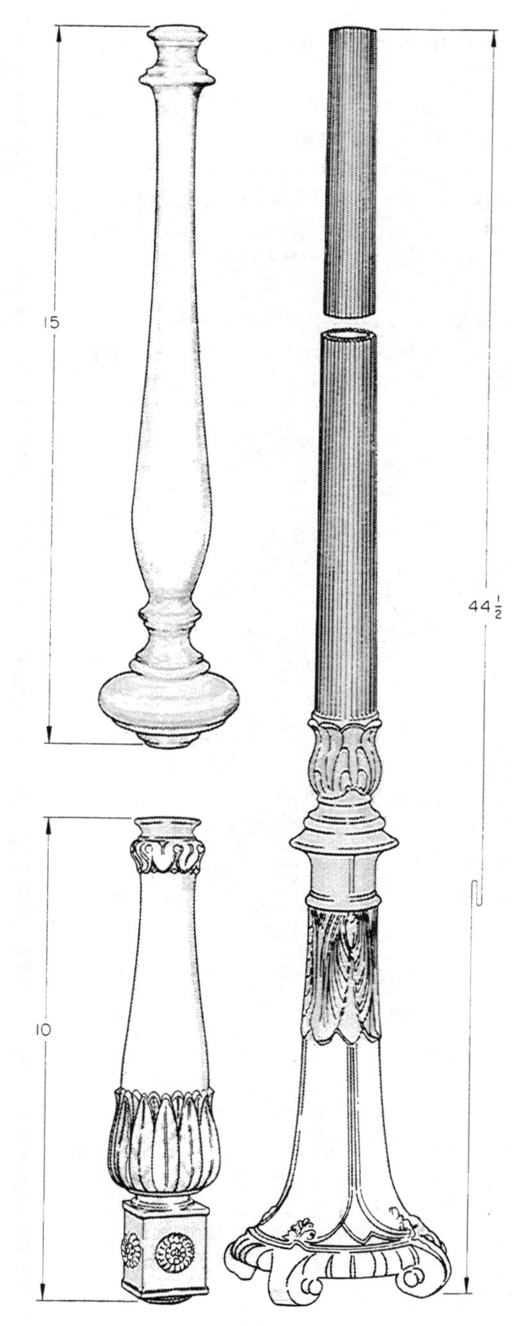

Figure 16-19 Lamp bases cast by the slush casting process. The floor-lamp column at right is about the maximum size commercially producible by slush casting. (From *Metals Handbook,* Vol. 6, 8th ed., ASM INTERNATIONAL [formerly American Society for Metals], 1971. Reprinted with permission.)

Figure 16-20 Many of the mechanical components for this mortise lock are P/M parts. (a) Turn knob hub, nickel, steel, and copper. (b) Hub lever, nickel steel. (c) Hub, nickle steel. (d) Latch bolt, chrome-plated brass. (e) Auxiliary bolt, oil-impregnated stainless steel. (f) Deadbolt, chrome-plated brass. (Courtesy of Best Lock Corporation)

steps are consolidated in a single P/M part that minimizes production procedures and reduces cost.

Since the P/M process is not shape-sensitive and normally does not require draft, parts like cams, gears, sprockets, and levers are produced efficiently. In many cases, part-pairs such as a cam and gear, or a spur gear and a pinion gear, can be joined together by a secondary assembly method. However, these additional assembly steps very often can be eliminated with designs that combine the separate shapes into a one-piece P/M part. In other instances, two P/M parts may be assembled after compacting, then bonded into a one-piece part during sintering.

The versatility of P/M is evidenced through its applications in a range of industries, including automotive, business machines, aerospace, electrical and electronic equipment, small and major appliances, agricultural and garden equipment, and hand and power tools.

The advantages of the P/M process include:

- eliminates or minimizes machining and scrap losses,
- maintains close dimensional tolerances,
- achieves a wide variety of alloy systems,
- produces good surface finishes,
- provides materials which may be heat-treated for increased strength or enchanced wear resistance,
- provides controlled porosity for self-lubrication or filtration,
- facilitates manufacture of complex or unique shapes which would be impractical or impossible with other metalworking processes, and
- meets moderate-to-high volume part production requirements.

Basic P/M steps. The three basic steps for producing conventional density parts by the powder metallurgy processes are mixing, compacting and sintering. See Figure 16-21.

1. **Mixing**—Metal powders are first mixed with lubricants or other materials to produce a homogeneous blend of ingredients.

2. **Compacting**—A controlled amount of mixed powder is automatically gravity-fed into a precision die and compacted, usually at room temperature at pressures as low as 138 MPa (20,000 psi) or as high as 827 MPa (120,000 psi) depending on the density requirements of the part. Normally, compacting pressures in the range of 414–690 MPa (60,000–100,000 psi) are used. Higher pressures limit tool life, especially with thin or fragile tool components.

Squeezing the loose powder results in a "green compact," which has the size and shape of the finished part when ejected from the die, and sufficient strength for in-process handling and transport to a sintering furnace. The most widely-used compacting techniques involve rigid dies and special mechanical or hydraulic presses.

Tool sets are made of either hardened steel and/or carbides, and consist of a die body or mold, an upper and lower punch and, in some cases, one or more core rods. A typical set of tools for producing, for example, a flat spur gear with a double "D" shaped hole, is shown in Figure 16-22.

The pressing cycle for producing this part is shown in Figure 16-23. The die receives a charge of mixed powder, delivered to the cavity by a feeder shoe. The

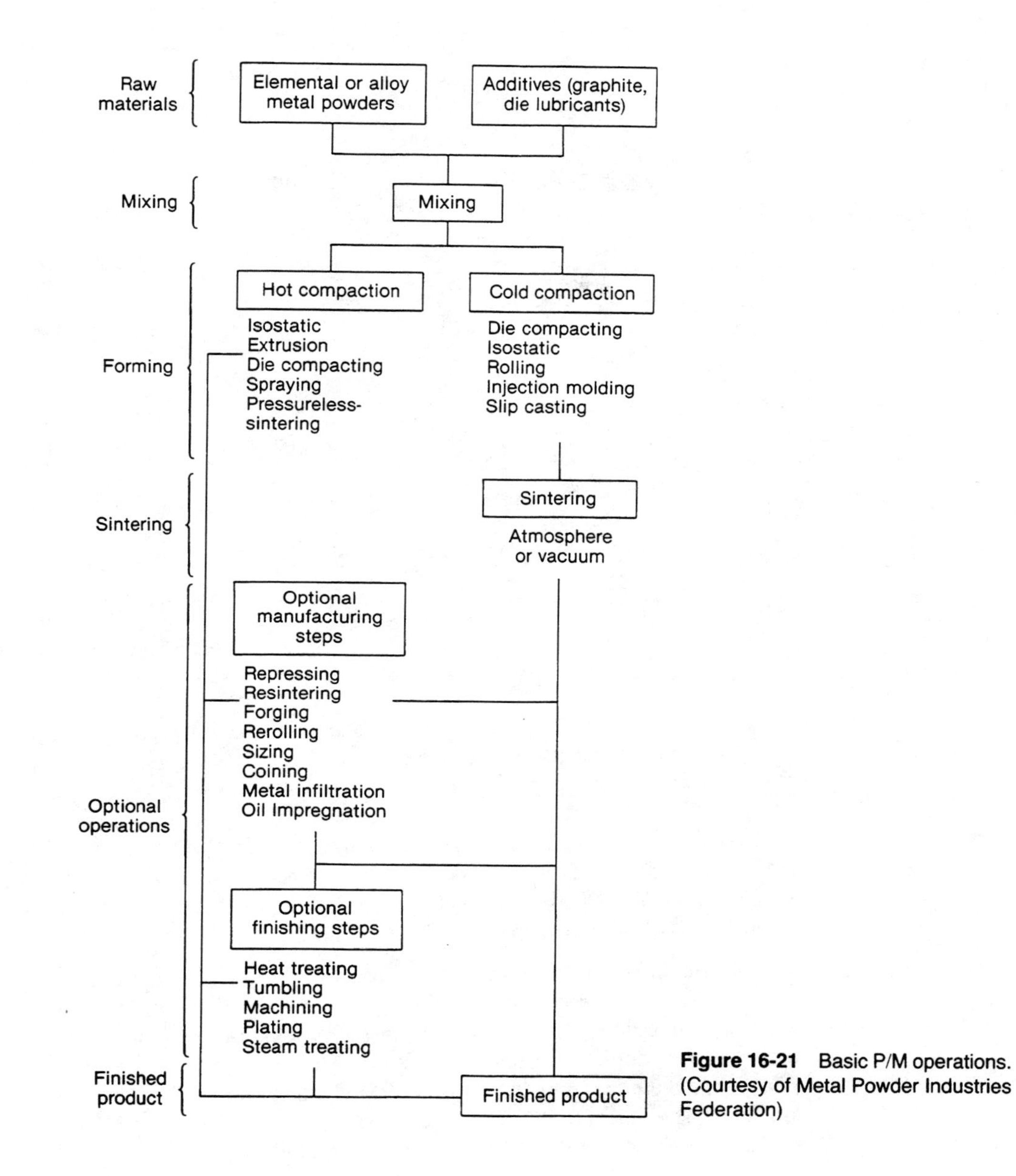

Figure 16-21 Basic P/M operations. (Courtesy of Metal Powder Industries Federation)

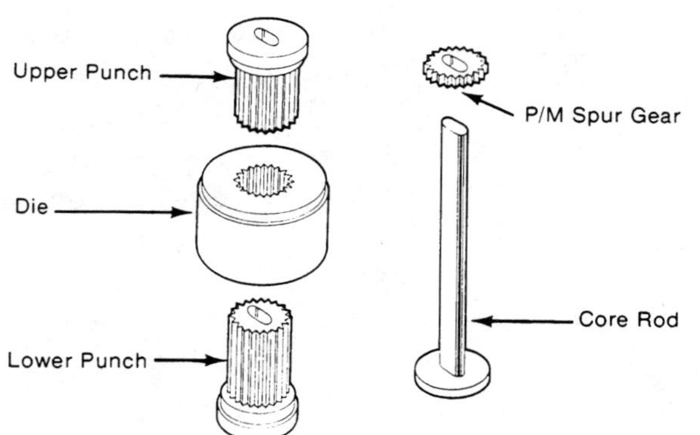

Figure 16-22 A typical tool set for a flat spur gear. (Courtesy of Metal Powder Industries Federation)

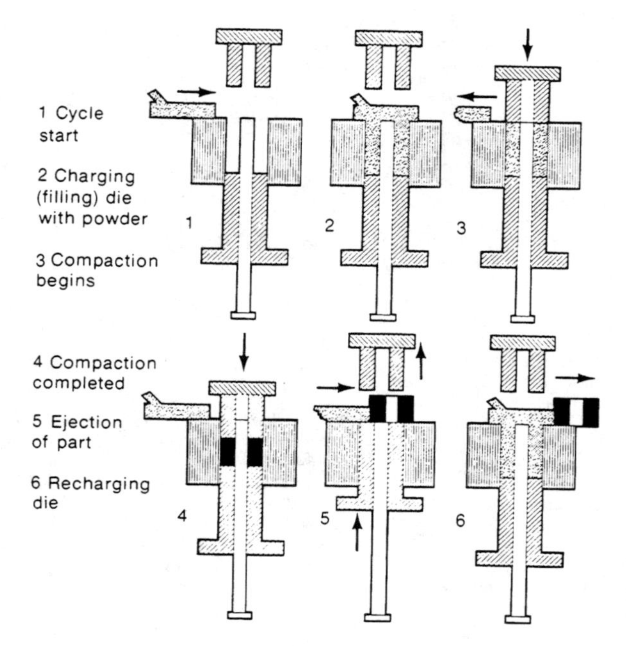

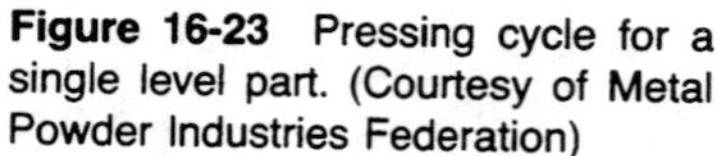
Figure 16-23 Pressing cycle for a single level part. (Courtesy of Metal Powder Industries Federation)

upper and lower punches compress the powder. The upper punch is withdrawn, the lower punch ejects the pressed compact, and the shoe slides the piece away from the die cavity. The shoe continues forward and refills the cavity with another charge of powder, and the cycle automatically repeats.

This compacting cycle is typical. However, when more than one pressing level is needed, as for example in making flanged shapes or cluster gears, multiple punches with separate actions may be needed. Holes in the direction of pressing can be molded by using additional core rods.

3. **Sintering**—In the typical sintering step, the green compact, placed on a wide-mesh endless belt, slowly moves through a controlled-atmosphere furnace. The parts are heated to below the melting point of the base metal, held at the sintering temperature, and then cooled. Basically a solid-state process, sintering produces metallurgical bonds between the powder particles. P/M parts generally are ready for use after sintering. However, to provide special properties, the parts can be repressed, impregnated, machined, tumbled, plated or heat treated.

Secondary operations. Several operations may be necessary to complete the P/M part.

1. **Coining and Sizing**—Coining a P/M part reduces its volume, increases density, and improves strength, hardness, and dimensional accuracy. Sizing, by contrast, primarily ensures dimensional accuracy without increasing density or strength.

2. **Impregnation**—The controlled porosity of P/M parts permits their impregnation with oil or a resin. Oil-impregnated P/M bearings have been used in automobiles since the late 1920s. Conventional P/M bearings can absorb from 10 to 30 percent by volume of, generally, additive-free, nonautomotive engine oils. Impregnation is

achieved by soaking the parts in heated oil, or by vacuum techniques. When friction heats the part, the oil expands and flows to the bearing surface. On cooling, the oil returns into the metal's pores by capillary action.

3. **Infiltration**—Mechanical properties can be increased by infiltrating P/M parts with another metal. Preforms of the infiltrant material, with a lower melting point than the porous P/M parts, are positioned on or under the pressed green compact prior to sintering. Upon melting, the infiltrant is absorbed into the part's pores by capillary action, producing a composite metal structure.

4. **Heat Treating**—Like wrought products, P/M materials can be annealed, and quench- and surface-hardened. A neutral, dry atmosphere should be maintained in the heat-treating furnace for porous parts.

P/M parts are normally produced to finished dimensions and geometry. However, machining operations such as turning, milling, drilling, and grinding may be necessary to produce special shapes, holes, threads, and other modifications. Similarly, P/M parts may be finished by deburring, burnishing, coloring, and plating.

P/M materials. Metal powders are precisely engineered materials available in many types and grades to meet required performance specifications. The resultant parts possess mechanical properties which compare favorably with parts produced by other manufacturing techniques. The powders are produced by electrolysis, atomization, or reduction methods, to include materials such as tin, iron, nickel, copper, aluminum, and ceramics. These powders are mixed in correct proportions, and when compacted and sintered, emerge as metallurgically bonded alloys.

Improved performance methods. P/M technology has advanced to include a generation of parts with improved densities, based upon special compacting processes such as hot forging and isostatic pressing.

Hot forging is employed to produce components with wrought properties, utilizing steel powders with appropriate carbon levels, and heat treatment when necessary. Such parts find wide use in the automotive industry. See Figure 16-24. The basic

Figure 16-24 Typical hot-forged P/M parts. Left: steel chain drive sprocket. Right: race and cam for an automatic transmission. (Courtesy of Metal Powder Industries Federation)

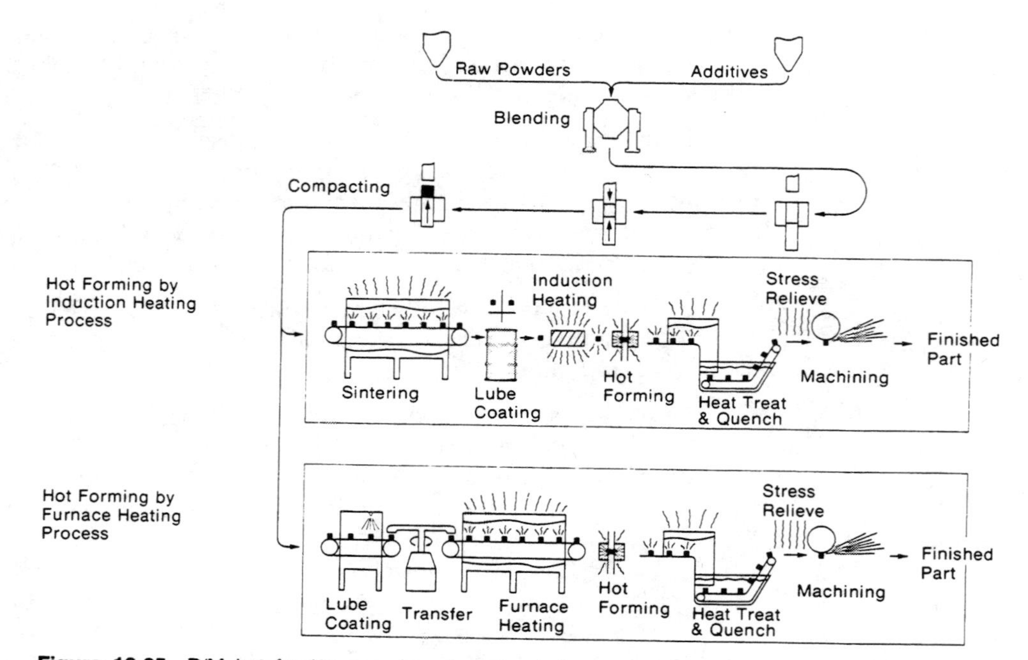

Figure 16-25 P/M hot forging processes employ induction and furnace heating methods. (Courtesy of Metal Powder Industries Federation)

process commonly employed to produce these high density P/M parts is to make a green compact called a *preform*, and to heat and then forge (restrike) the preform to the required density. Two common processes are shown in Figure 16-25.

Isostatic pressing is performed by surrounding the powder mass with a flexible membrane containing a fluid such as oil, gas, or water. A pressure applied to the membrane provides the necessary compactive force. See Figure 16-26. Some of the attributes of isostatic pressing are:

1. capability of producing complex shapes with materials which are difficult to compact,
2. minimal use of expensive materials,

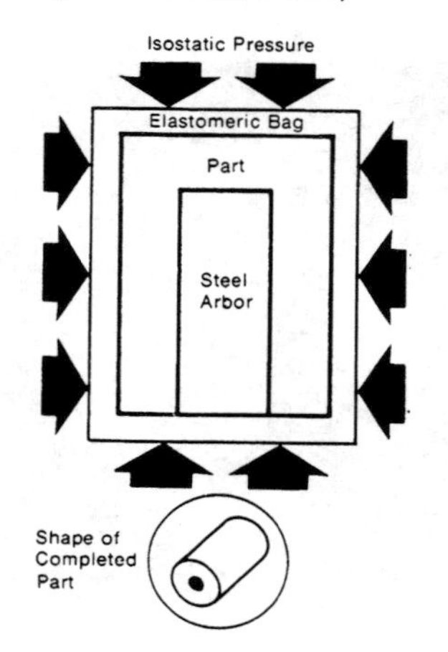

Figure 16-26 P/M isostatic compacting. (Courtesy of Metal Powder Industries Federation)

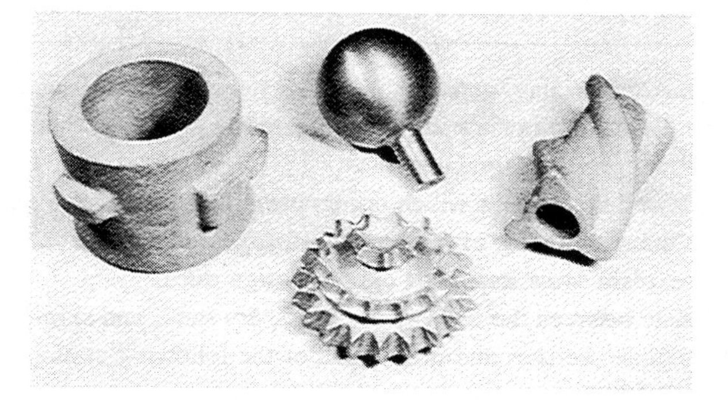

Figure 16-27 Shapes produced by isostatic compacting. (Courtesy of Metal Powder Industries Federation)

3. highly effective compacting, generally without the addition of special binding materials, and
4. uniform densities and geometries.

Some typical parts appear in Figure 16-27.

OTHER CASTING PROCESSES

Numerous other cast products are made from cementitous materials. Railroad ties and structural beams are produced from concrete cast in molds containing prestressed steel rods. These components are extremely strong, load bearing, and flexible, due to the steel rod matrix. Toilets, sinks, and other plumbing fixtures are made from liquid clays or slips poured into plaster molds. Ceramic plates and cups are similarly made. The plaster absorbs much of the moisture to accelerate the setting of the ceramic product. These products are removed from the split molds, dried, and fired in special furnaces or kilns. Glazing and further firing operations follow to produce a resultant strong artifact impervious to moisture and dirt. Some plumbing fixtures are also made of cast iron. A typical slip casting operation used to produce ceramic holloware is shown in Figure 16-28.

Figure 16-28 Slip, or liquid clay is being poured into these plaster molds to produce ceramic holloware such as vases and pots. (Courtesy of Syracuse China Corporation)

QUESTIONS

1. Define the term casting, and relate it to the terms founding and molding.
2. Describe the function of a pattern and relate it to the terms core, core print, and mold.
3. Describe the process of making cores.
4. Briefly describe the action which occurs during metal solidification.
5. Describe the relationship of the terms shrinkage and draft to patternmaking.
6. List and explain some important casting design rules.
7. Differentiate between the terms green sand, dry sand, and skin-dried.
8. Prepare simple sketches and description of the following casting processes:
   a. Shell molding
   b. Plaster mold casting
   c. Investment casting
   d. Centrifugal casting
   e. Permanent mold casting
   f. Slush casting
9. Prepare simple sketches of the hot chamber and cold chamber die casting processes.
10. Define P/M, and list the basic steps in the process.
11. Describe the function of the following P/M secondary operations:
    a. Coining
    b. Sizing
    c. Impregnation
    d. Infiltration
12. Select one of the several casting methods described in this chapter, and prepare a library research report on the current status of the method in industry.

17 FORGING AND RELATED PROCESSES

The word "forging" conjures up images of great size, strength, and toughness. While it is true that parts of considerable dimension can be formed by this process, most forged products are of modest size, yet possessing this attribute of strength. See Figure 17-1. Forging is a bulk-deformation process whereby the mass of the final workpiece is equal, or very nearly equal, to the mass of the raw material from which it is formed. It is a superior shaping process which results in an extremely tough product. See Figure 17-2.

Generally, forging is considered as a hot forming technique, and large workpieces must be worked hot for practical reasons. However, it also is performed as a cold working process. Similarly, extruding, rolling, and wire drawing, among others, are deformation and material-conserving processes and will be treated here.

In the early chapters of this book, there was an examination of material properties, the nature of metals, and how they behave when subjected to temperatures. As a very practical matter, a cold steel bar 13 mm (0.5 inch) in diameter can be beaten with a large hammer on an anvil to flatten it. It is easier to do this if the metal is at a red heat. The techniques used to work hot and cold metals are described following.

FORGING

Forging is the most common method for producing hot formed shapes. It is the process of heating a metal workpiece and shaping it by impact or squeeze pressure. The forging methods used today are basically the same as those used by preindustrial

Figure 17-1 Forged aluminum truck wheels are tough and lightweight to increase the cargo payload. (Courtesy of ALCOA)

blacksmiths, where the metal is brought to a high temperature and then shaped by the force applied by one surface (the hammer) as the metal rests on another surface (the anvil). This shaping is done by powerful forging presses and hammers, and can be generally classified as either an open die or closed die operation.

Open die forging is a method of shaping a workpiece between flat dies that do not completely confine the metal. See Figure 17-3. In this type of forging the workpiece does not take the shape of the dies, but is formed by manipulating the workpiece. For example, a round bar can be forged into a square bar by alternately turning and hammering along its length between the open, flat dies. While many open die forgings are struck between flat dies, these dies are often vee or half-round shaped. See Figure 17-4. Open dies are far less costly than closed dies, and therefore more economical for forming limited numbers of parts, such as very large shafts and gear blanks. Open die forging is also known as flat die, hand, or smith forging.

In *closed die forging*, also called impression die forging, the workpiece is shaped between dies that completely enclose the metal and control its flow. The upper and

Figure 17-2 A comparison of manufacturing methods. Note how forging imparts directional strength to a workpiece, different from casting and machining. (Courtesy of Forging Industry Association)

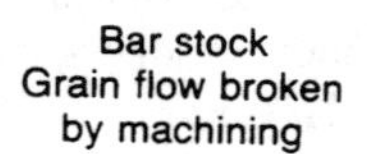

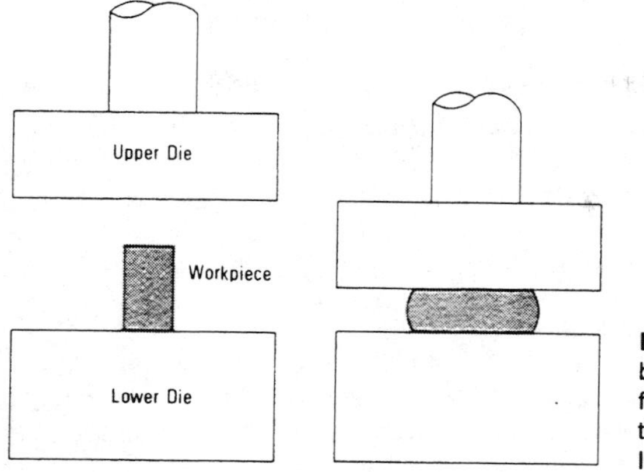

Figure 17-3 A pancake forging produced by pressing a cylindrical workpiece between flat, open dies. This is an "upsetting" operation. (Courtesy of Inco Alloys International, Inc.)

lower dies have cavities which, when the two come together, create a hollow space having the shape of the forged piece, a feature which precludes workpiece manipulation during forging. The work is set between the dies, and pressure is applied until the metal entirely fills the die cavity.

In the typical closed die operation, the original workpiece contains more metal than is needed for the finished article. In order to accommodate this excess, the dies have grooves around the perimeter of the impression. After the metal fills the cavities of the dies, the excess is squeezed into the grooves as a thin plate. The groove is called the *gutter*, excess is called *flash*, and the flash is later trimmed off. In some impression die operations, the flash gutter is omitted and the flash occurs as a spacing between the upper and lower dies. The line around the outside edge of the workpiece where the flash forms is called the *parting line*, and it is the separation line between the top and bottom dies.

Typical open and closed die operations are shown in Figure 17-5. Many forged parts are made by first producing a rough shape by open die forging and then completing the part with a closed die operation. While the terms "closed die" and

Figure 17-4 Open die shapes. (Courtesy of Forging Industry Association)

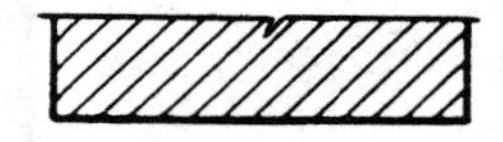

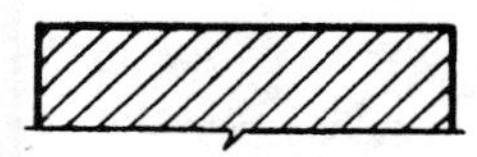

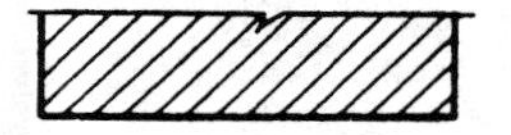

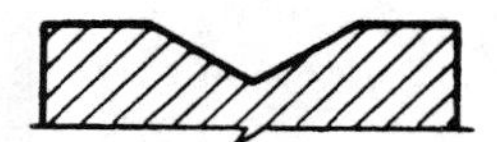

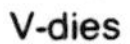

Figure 17-5 A comparison of open-die and closed forging operations. Most closed die operations are similar to "a" where flash is squeezed out between the dies as the metal fills the die pockets. Flashless forging, as in "b," produces no flash, and all of the workpiece is used to make the finished shape. (Courtesy of Inco Alloys International, Inc.)

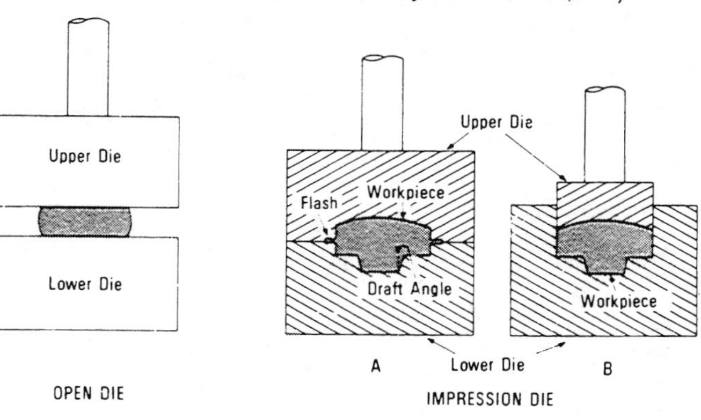

"impression die" are generally used interchangeably, a true closed die operation appears in Figure 17-5(b).

Dies used to forge parts that have deep holes or bosses with straight sides are made with a slight taper, called *draft*, on the sides of those areas. Draft is necessary for easy removal of the forged piece, and is expressed in degrees of draft angle. The draft concept is also used in metal casting and plastic molding.

Forging Operations

A variety of operations are employed to cause heated metal to flow into closed die cavities, or to move in a less restricted fashion in open dies. Described here are some of the common operations which result in finished forgings.

Flattening (or pancaking) between flat dies is the simplest type of forging and a common open die operation. This is forging "on end," where the height of the workpiece is reduced as the metal is flattened between the dies. This operation is sometimes called "upsetting." See Figure 17-3. This is essentially a workpiece or blank-producing or preforming operation.

Edging employs cup-shaped dies that trap the metal at the ends of the workpiece and force it toward the center of the cavity. See Figure 17-6. This inward metal flow fills the die pocket and increases the cross section of the workpiece. Edging is often called gathering because it "gathers" the metal to form a bulge.

Fullering is the opposite of edging. It reduces the thickness of the workpiece between the two ends, because the metal is displaced away from the center. As shown in Figure 17-7, fullering increases the length of the workpiece and produces a characteristic dumbbell shape.

Drawing is similar to fullering, except that the workpiece cross section is decreased at one or both ends, or continuously, rather than at the center. Drawing can be done between two flat dies or in slots cut in the dies. See Figure 17-8.

Wire drawing is a cold-forming method whereby wire is formed by pulling a rod through an opening in a die. See Figure 17-9. The die hole is smaller than the size of the workpiece, and as the workpiece is pulled through the opening it is stretched so that both directional and shape changes take place. One end of the

Figure 17-6 Edging. (Courtesy of Inco Alloys International, Inc.)

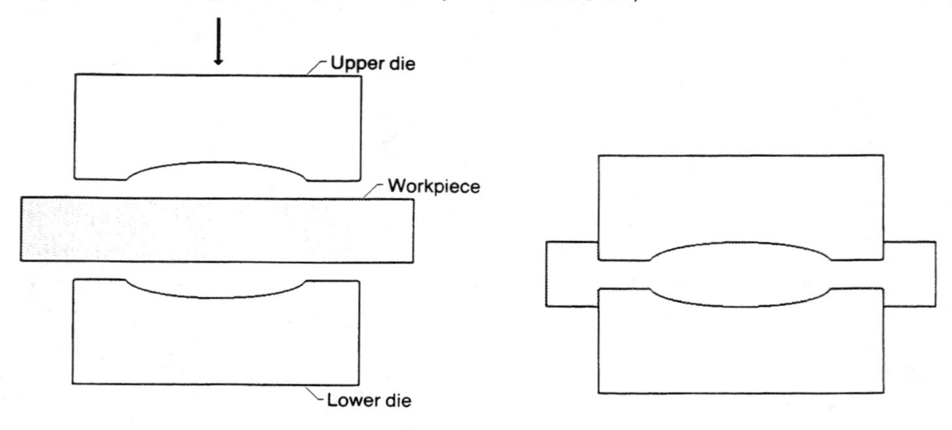

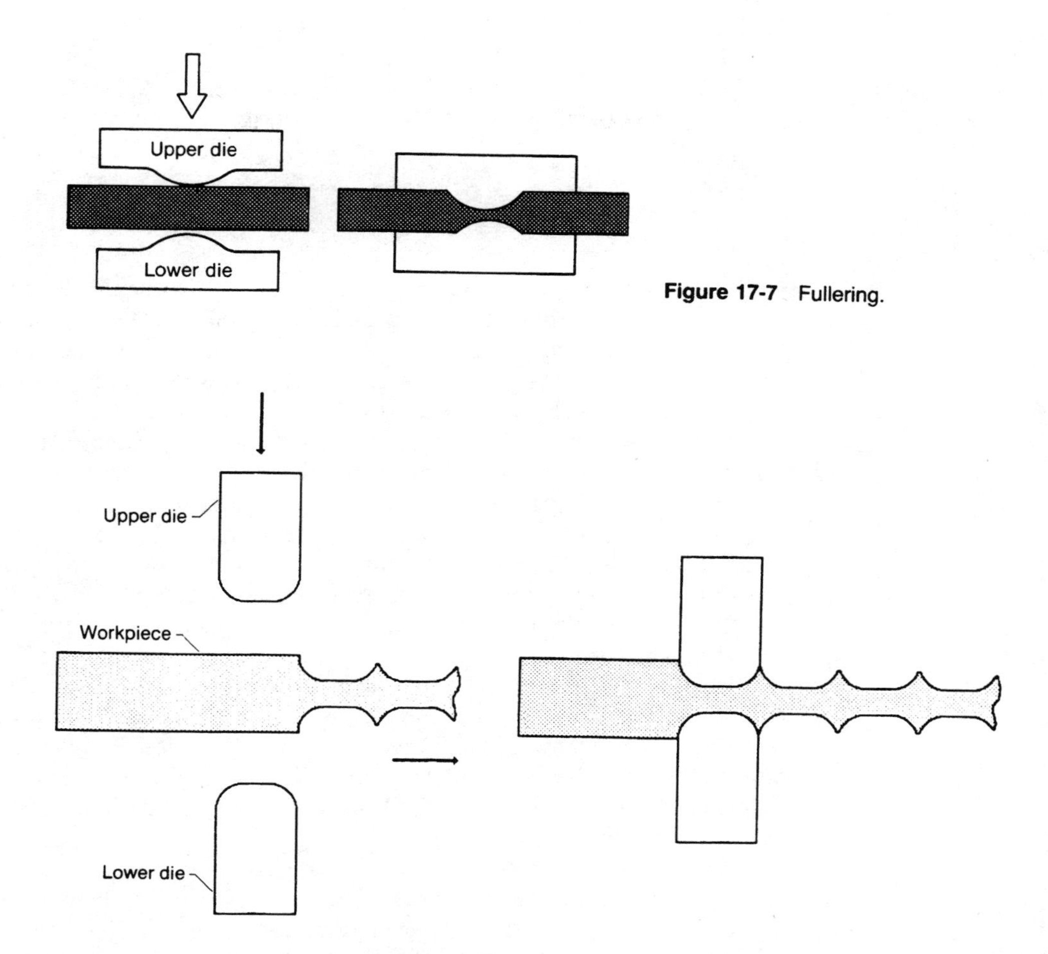

Figure 17-7 Fullering.

Figure 17-8 Drawing stretches and changes the cross-sectional shape of a workpiece. This is a typical forging process.

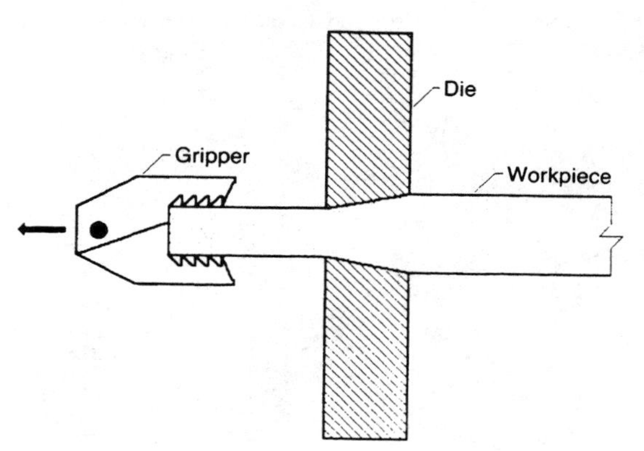

Figure 17-9 Wire drawing is generally a cold-forming operation.

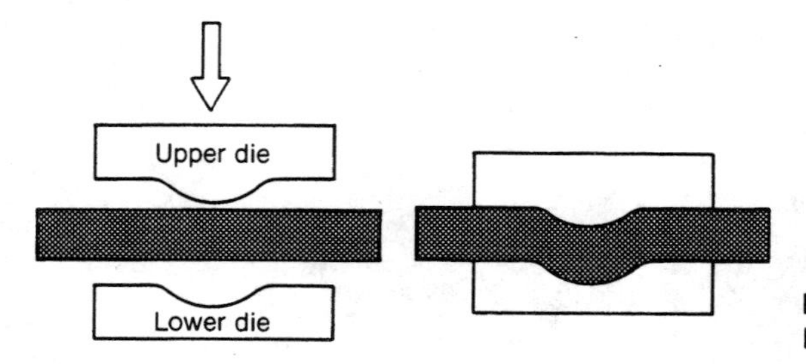

Figure 17-10 Bending. (Courtesy of Inco Alloys International, Inc.)

workpiece is tapered for easy insertion into the die to begin drawing. Tough jaws are used to secure the pointed end extending from the die and pull the workpiece through. Square, rectangular, round, hexagonal, and similar shapes, as well as tubing and wire, can be drawn by using appropriate dies. The machine used in this forming method is called a *draw bench*.

Bending takes place in dies shaped to directionally change a workpiece rather than to alter its thickness or cross section. Bending is often used for parts that are asymmetrical or that have large offsets (such as a crankshaft), where the metal is "moved" as required to conform to the general shape of the finished article. Forging in finishing dies is employed to impart a final shape to such articles. See Figure 17-10. Bending usually follows other preliminary operations such as edging, fullering, and drawing. Blocking and finishing are the final steps in forging, and are usually impression die operations.

Blocking is employed to produce an approximation of the finished shape, complete with corners, holes, and fillets. Blocking facilitates metal flow into all parts of the finishing dies, thereby reducing their rate of wear. The blocking operation can sometimes be omitted, but is usually required if the final piece has complex contours and shapes. *Finishing* produces the final shape, using dies having the exact configuration of the end product.

All of the forging operations needed to produce an article are usually performed on a small machine. The die blocks are made large enough to contain several impression dies for drawing, fullering, an edging, a blocking, and finishing. The workpiece is transferred from one die cavity to another between strokes of the ram. An example of the sequence of forging operations appears in Figure 17-11. After final forging in finishing dies, one or more additional postforging operations usually must follow before the part is ready for use.

Figure 17-11 The forging operation sequence used to produce an automotive connecting rod. (a) Tag forging. (b) Fullering. (c) Flattening. (d) Rolling. (e) Blocking. (f) Die forging. (g) Die trimming.

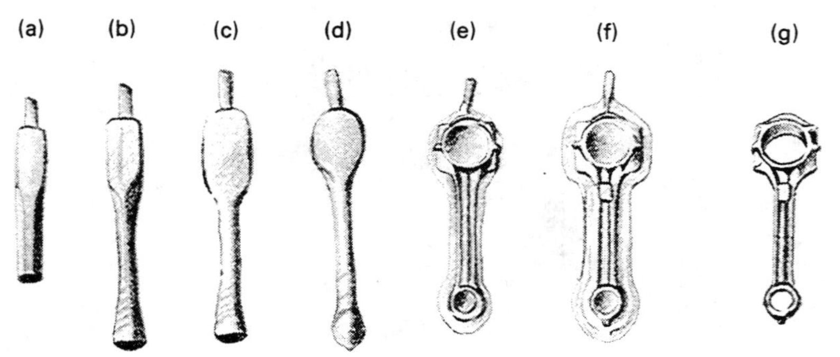

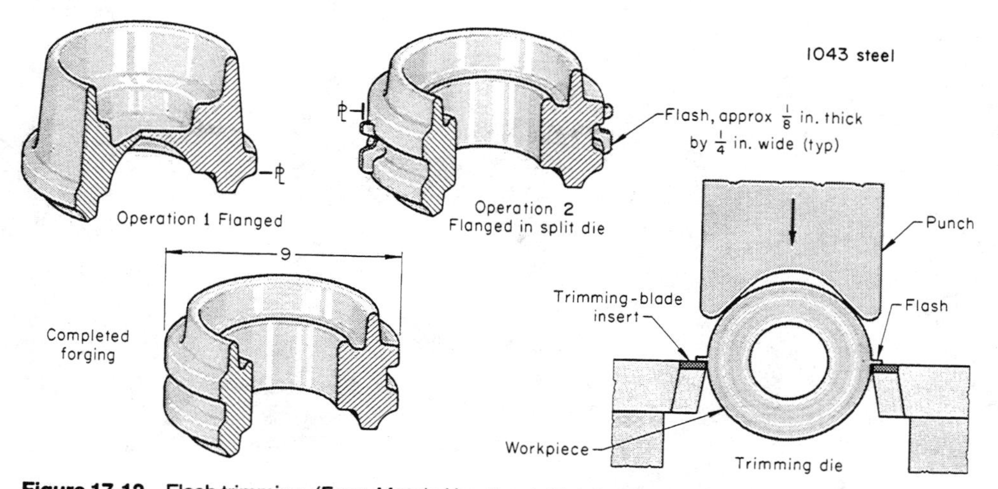

Figure 17-12 Flash trimming. (From *Metals Handbook,* Vol. 6, 8th ed., ASM INTERNATIONAL [formerly American Society for Metals], 1971. Reprinted with permission.)

Postforging Operations

Trimming is a shearing operation, usually is the first of these operations, and generally is performed on small mechanical presses featuring trim dies to remove flash. The bottom die, called the blade, has a sharp shearing edge of the exact contour as the finished piece. The top die, called the punch, is also shaped to fit the forging, but is slightly smaller so it will fit in the hole. The workpiece is placed between the dies so that the flash rests on the blade surface. The punch forces the workpiece through the hole, shearing off the flash. See Figure 17-12.

Planishing is often done to remove the rough burr left by trimming, to smooth the surface of the part, and to obtain closer part tolerance. The operation consists of striking the forging with light, rapid blows using polished dies shaped to fit the workpiece. The workpiece is either cold or heated to a lower temperature. Planishing can also be done by hot or cold rolling.

Coining is a sizing and smoothing operation which receives its name from the method used to imprint images on coins. The workpiece is pressed between dies to obtain closer tolerances or to align various areas of the workpiece. The greatest use of coining is to size areas such as bosses that protrude from a workpiece. Coining may be done while forgings are hot or cold. See Chapter 15. If draft is undesirable on the finished article, the taper can be removed by planishing or coining.

RELATED FORMING OPERATIONS

A number of bulk forming techniques similar to forging are utilized to shape metal workpieces while they are hot. While these are not, in the strictest sense, forging operations, the forming often is facilitated by heating the workpieces to appropriate temperatures. Other processes work cold parts.

Rolling is accomplished by squeezing a workpiece between two or more rotating cylinders. Metal sheet and plate is produced this way. See Figure 17-13. In this process, the friction force between the workpiece and the rollers draws the work

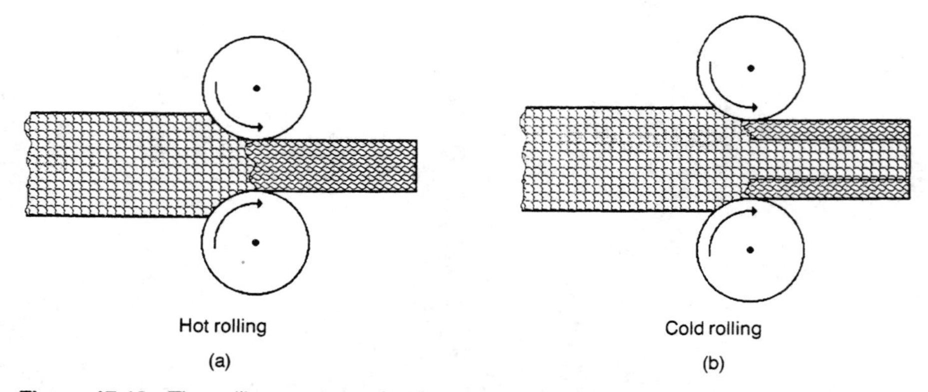

Figure 17-13 The rolling operation involves the squeezing of a metal workpiece between two rolls to produce a new shape. Note the different grain structure produced by hot and cold rolling.

between the rollers, thereby reducing its thickness. *Hot rolling* is typically a roughing operation used to reduce very large, heated ingots to shapes and sizes suitable as raw material for further processing.

Cold rolling is normally used to produce thin sheets and strips. Cold rolling requires considerable force and, therefore, smaller rolls are required to concentrate the pressure over a smaller area. Smaller rolls also enable thinner material to be rolled. Thin material does not absorb all of the rolling force so that, in effect, one roll acts directly against the other.

The machine used in hot and cold rolling is called a rolling mill or a roll, and of these there are several styles. The most common are the "two-high." Others are the three-, four-, and six-high types which can be used for greater efficiency, for handling larger sheets, and for harder materials. See Figure 17-14. The machines can reciprocate the workpieces between the rolls, and simultaneously adjust the rolls for greater pressure to achieve the proper material thickness.

With appropriate grooves in the rolls, rolling mills can produce a variety of shapes. Some of the more common are round rods, flat bars, angles, half-rounds, channels, T-bars, and I-bars. Almost any shape having a continuous cross section can be made on a rolling mill.

Ring rolling is a method of shaping seamless rings to be used for ring-gear blanks, rims, and heavy wheels, among other similar parts. A hot, doughnut-shaped blank is worked between two rolls to reduce its thickness (the distance from the outer edge to the hole) and to increase the circumference or overall diameter. See Figure 17-15. The outside roll is usually larger in diameter and power-driven. The

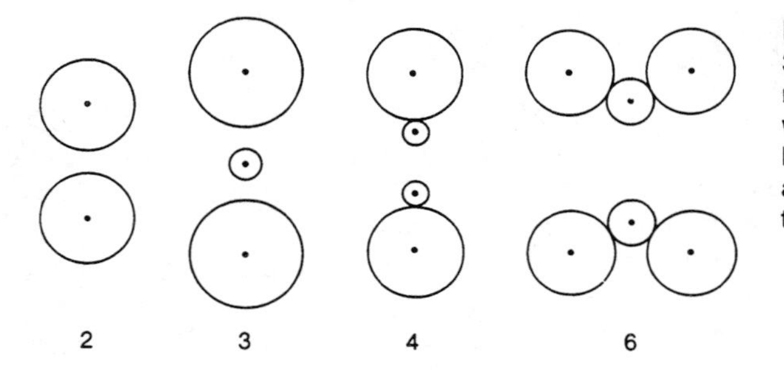

Figure 17-14 A schematic of the 2, 3, 4, and 6 high rolling mill arrangements. The 3-high setup allows the workpiece to move in one direction between the bottom and middle rolls, and in the reverse direction between the top and middle rolls for efficiency.

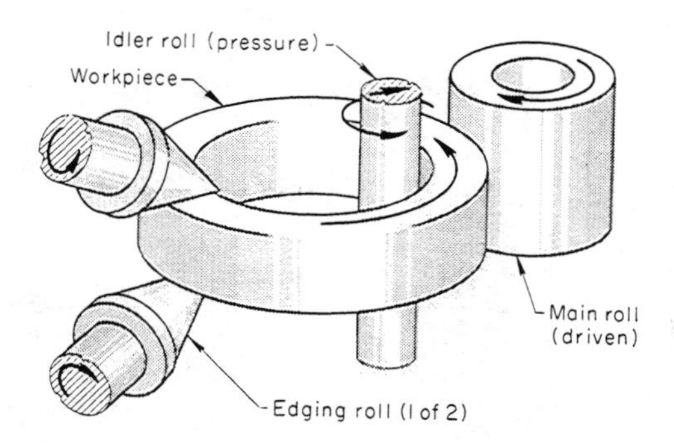

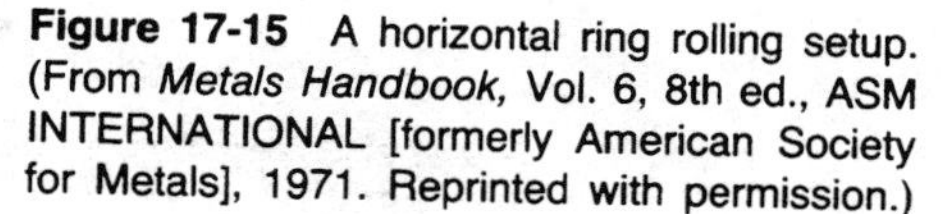
Figure 17-15 A horizontal ring rolling setup. (From *Metals Handbook,* Vol. 6, 8th ed., ASM INTERNATIONAL [formerly American Society for Metals], 1971. Reprinted with permission.)

smaller, inside or idler roll is mounted on a slide so that it can move toward the outside roll and apply the pressure. The inside roll is not power-driven, but idles against the inner surface of the ring. Either one or both rolls can be grooved to produce desired shapes on the edges of the rings. A forging sequence for this operation is shown in Figure 17-14.

Extrusion is the process of compressing metal above its elastic limit and forcing it through an opening in a die, or in a punch, or through the space between a die and punch, in order to produce a workpiece of given cross section. The workpiece is placed in a hollow die called a container, and pressure is applied. The workpiece assumes the shape of the die orifice through which it flows. As a practical example, toothpaste is extruded from its tube opening by extrusion.

The operation is called *forward extrusion* if the metal is forced in the same direction as the punch travels. It is called *backward extrusion* if metal flow is opposite to the motion of the punch. Other directions of metal flow are possible, depending on the location of the hole in the container. Some typical extrusion methods are illustrated in Figure 17-17. To extrude parts that require metal flow in directions other than straight forward or backward, the container must be made in two pieces to that the workpiece can be removed. Extrusion may be done hot or cold.

Figure 17-16 Steps in the forging and rolling of a railroad freight-car wheel. (From *Metals Handbook,* Vol. 6, 8th ed., ASM INTERNATIONAL [formerly American Society for Metals], 1971. Reprinted with permission.)

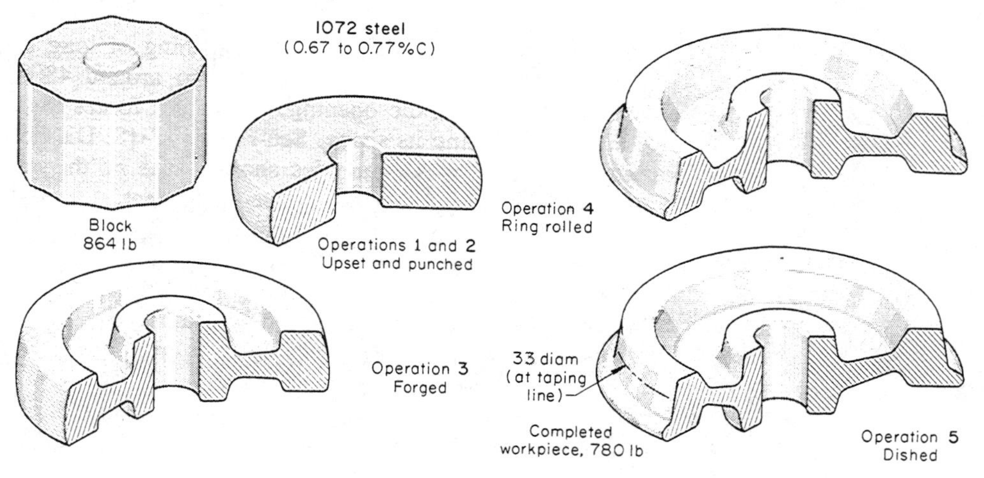

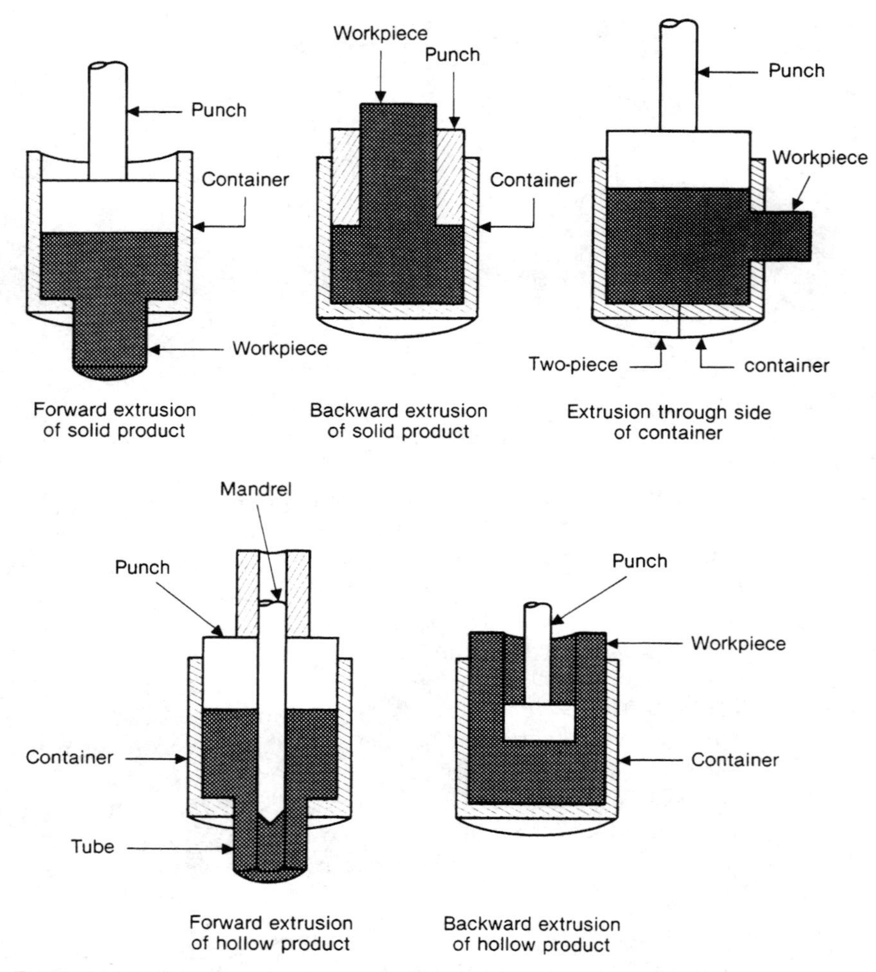

Figure 17-17 Some common extrusion methods. (Courtesy of Inco Alloys International, Inc.

A typical hot extrusion process is the forming of long aluminum members. In this process, a billet (workpiece) is heated to around 480°C (896°F) and placed into a chamber against a die opening, and a ram forces the aluminum through the orifice, thereby assuming its shape. See Figure 17-18. Die design considerations are important here, as per the examples shown. One of the attendant problems with hot extruding is the wear and tear on the equipment.

Cold extrusion, also called cold forming, cold heading, or cold forging, occurs with unheated blanks. There are three common types. *Forward extrusion* causes the simultaneous reduction in cross section and lengthening of the part along the axial dimensions of the die walls which contain the workpiece. See Figure 17-19. *Backward extrusion* forces the material from one die into and around the interior geometry of the punch. See Figure 17-20.

Cold heading is an *upsetting* operation where a workpiece or blank of rod or wire is struck on its end by a die. The blow spreads out the end of the blank to

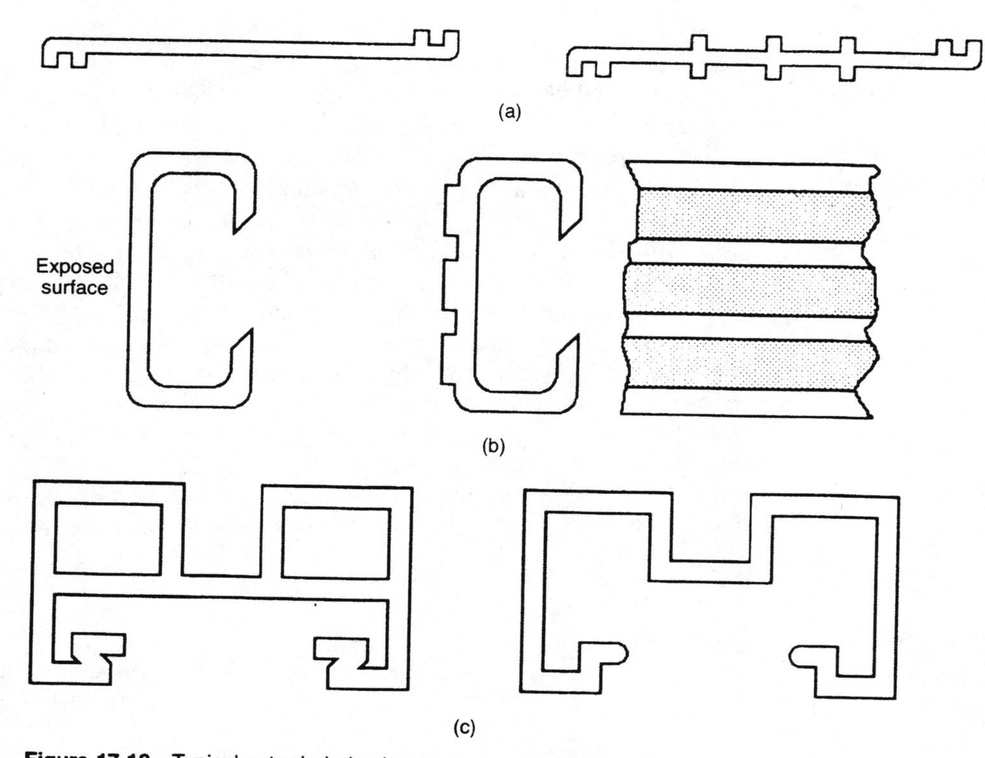

Figure 17-18 Typical extruded aluminum cross sections. In each example, the shape at the right is preferred. (a) Thin, wide sections can be hard to straighten after extruding. Ribs help to prevent twisting. (b) Appearance and strength are enhanced by adding grooves and ridges. (c) Avoid hollows and complex, sharp corners.

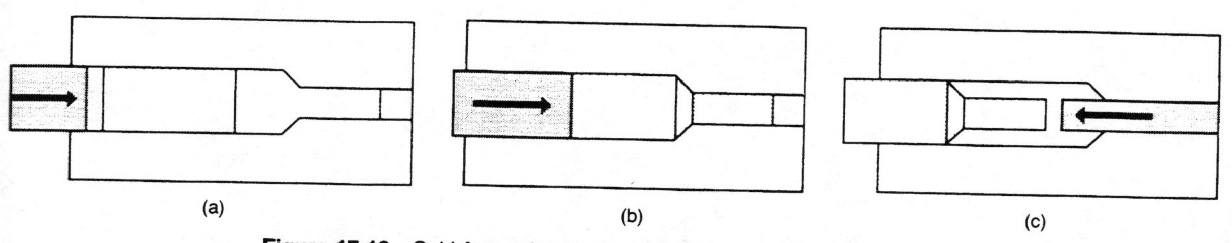

Figure 17-19 Cold forward extrusion. (a) Blank inserted in die. (b) Punch forces metal into die. (c) Finished part is ejected.

Figure 17-20 Cold backward extrusion. (a) Blank inserted in die. (b) Punch forces metal into die and around punch. (c) Finished part is ejected.

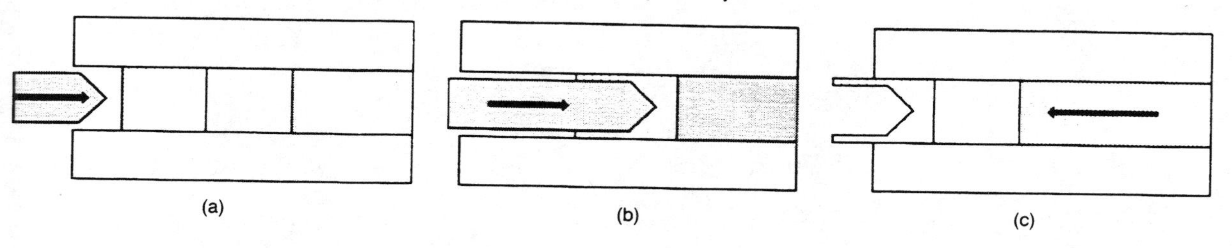

form the head of the fastener. Most cold heading is done on automatic equipment that cuts the blank from a coil of stock, forms the head and shank of the fastener, and ejects the formed part. Two dies are required to upset the end of the blank, one to hold the blank and one to deliver the forming blow. The holding die can be either solid or two-piece. The solid die is more often used, and has a cylindrical hole of the desired size to hold the blank. A knockout pin is placed in the hole from the back side of the die. The blank is inserted from the front of the die so that its end rests against the knockout pin with the proper amount of the blank protruding from the die. The protruding end is then struck with the forming die or punch. The knockout pin moves forward to push the headed part out of the die.

The two-piece die has a groove in each half, so that when the two come together the grooves form a hole through the assembled die to hold the blank. The blank is normally placed between the die halves from the back side when the two halves are apart. The die closes to grip the blank while it is being headed and separates to release the cold-headed part. See Figure 17-21.

Three types of heading machines are in common use: the single blow header, the double-blow header, and the transfer header. Such equipment often is referred to as a *forging machine*.

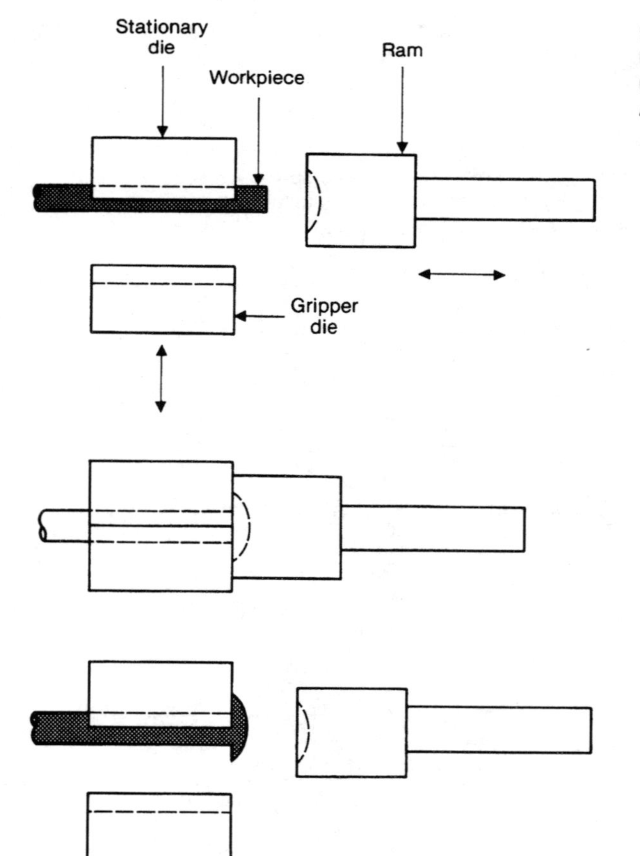

Figure 17-21 The two-piece cold heading die in use in a forging, or upsetting, machine. (Courtesy of Inco Alloys International, Inc.)

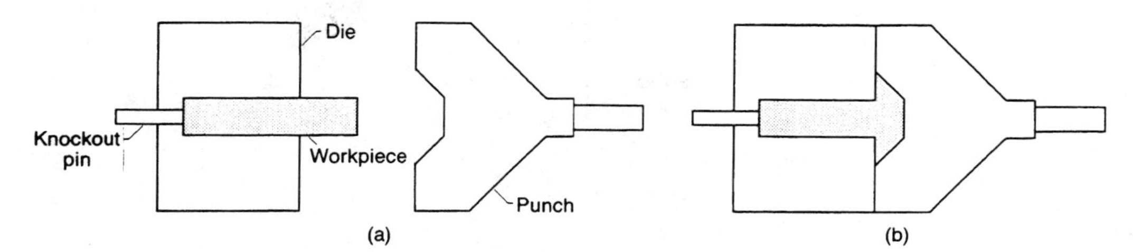

Figure 17-22 Single blow cold heading.

Single-blow heading is shown in Figure 17-22. The protruding end of the blank is formed by a single blow of the punch. The cavity is formed when the closed punch and die have the same configuration as the finished head. The cavity can be contained entirely within the punch or die or partly in each. Heads with smooth undersides (such as most rivet heads) are usually formed within the punch; those with shaped undersides but flat tops (wood screws) are usually formed in the holding die. Parts such as carriage bolts that require shaping on both sides of the head are formed partly in the punch and partly in the die.

Large or complex heads require more than one forming blow. The *double-blow header* upsets the blank in two steps. The blank is held in the same die for both blows, but is formed by two different punches. The first punch preforms (upsets) the head into a cone. A cone shape is used because it results in less bending of the blank, especially when large blank extensions are required. The second punch finishes the head. Figure 17-23 shows the steps in double-blow heading.

The transfer heading, or progressive headers, uses a series of forming stations. The blank is transferred from one die to another and receives either one or two blows in each die. Such multiple-station headers are generally used only for complex parts such as those that require extrusion in addition to severe heading.

Various combinations of these methods can be employed, and they can occur simultaneously to produce heads, threaded fasteners, collars, and multiple diameter,

Figure 17-23 Double blow cold heading.

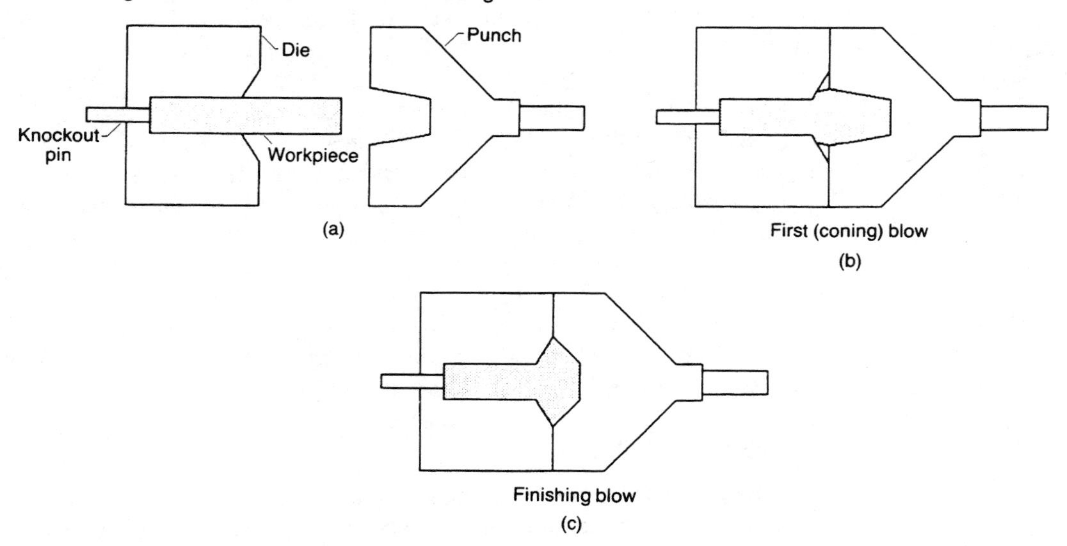

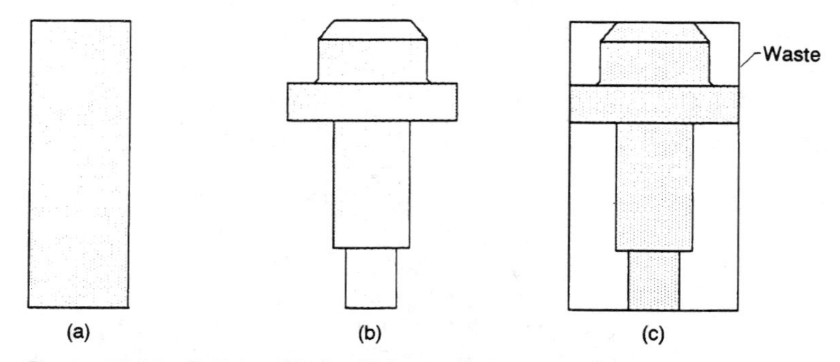

Figure 17-24 Cold forming is efficient, with no machining waste. (a) Workpiece blank. (b) Finished part. (c) Machined part waste.

asymmetrical parts. This is an efficient means of producing small parts which are tough and precise, and produce no scrap. See Figure 17-24.

Extrusion requires high pressures to force the metal through the relatively small openings involved. The amount of deformation in extrusion is greater than in any other hot-forming process. Presses are rated at from 2 MN to 220 MN (250 to 25,000 tons). A lubricant, such as graphite, which can withstand high temperatures, is often used on the dies to reduce friction.

Piercing as a method of producing hollow parts from solid stock is a forming, not a cutting, operation, because no metal is removed from the workpiece. It is closely related to reverse extrusion, but differs because of the greater movement of the punch relative to the movement of the workpiece.

The operation is performed by placing the workpiece in a hollow die and forcing a punch into the metal. The metal can flow in either of two directions, depending on whether the workpiece is the same size or smaller than the die. If the workpiece completely fills the die, the metal displaced by the punch is forced to flow around the punch and up the sides of the die. If the cavity in the die is larger than the workpiece, the metal will flow away from the punch and toward the sides of the die. Outward metal flow is normally preferred because it results in less wear of the tools. See Figure 17-25.

Swaging is a means of tapering or decreasing the cross section of the ends of round metal bars or tubes by repeated hammering from rotating dies as they close and open against the work. In this fashion such products as needles, bicycle spokes, flag and awning poles, fishing rods, and furniture legs are manufactured.

The swaging machine contains a ring with small hammers mounted around the inside edge. The workpiece is clamped securely and one end is inserted into the center of the die. As the ring spins, a cam system intermittently forces the hammers toward the center of the ring die and against the stock, thus working it through a combination of impact and pressure. The continual, very rapid blows of the whirling hammers quickly reduce the material to a smooth round shape. Swaging is often a cold forging process. See Figure 17-26.

Thread Rolling

Bolts and screws made by cold heading usually are threaded by thread rolling. Thread rolling is a forming process and no metal is cut from the piece. See Figure

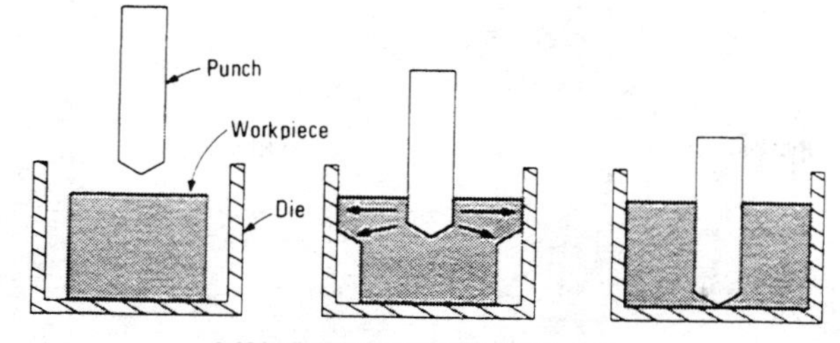

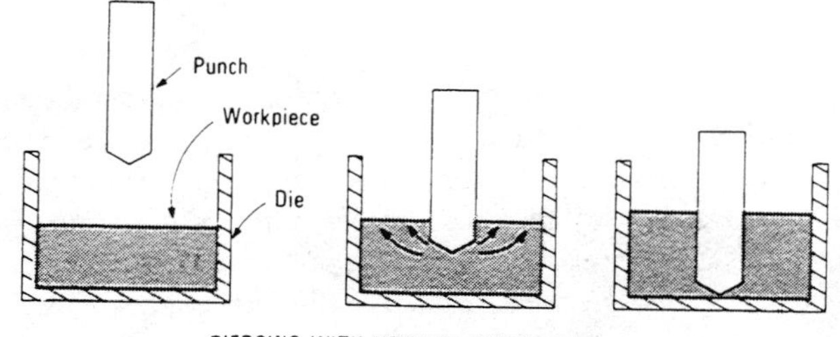

Figure 17-25 Piercing methods. (Courtesy of Inco Alloys International, Inc.)

Figure 17-26 Swaging die configurations, with sample swaged parts shown. (Courtesy of Manca, Inc.)

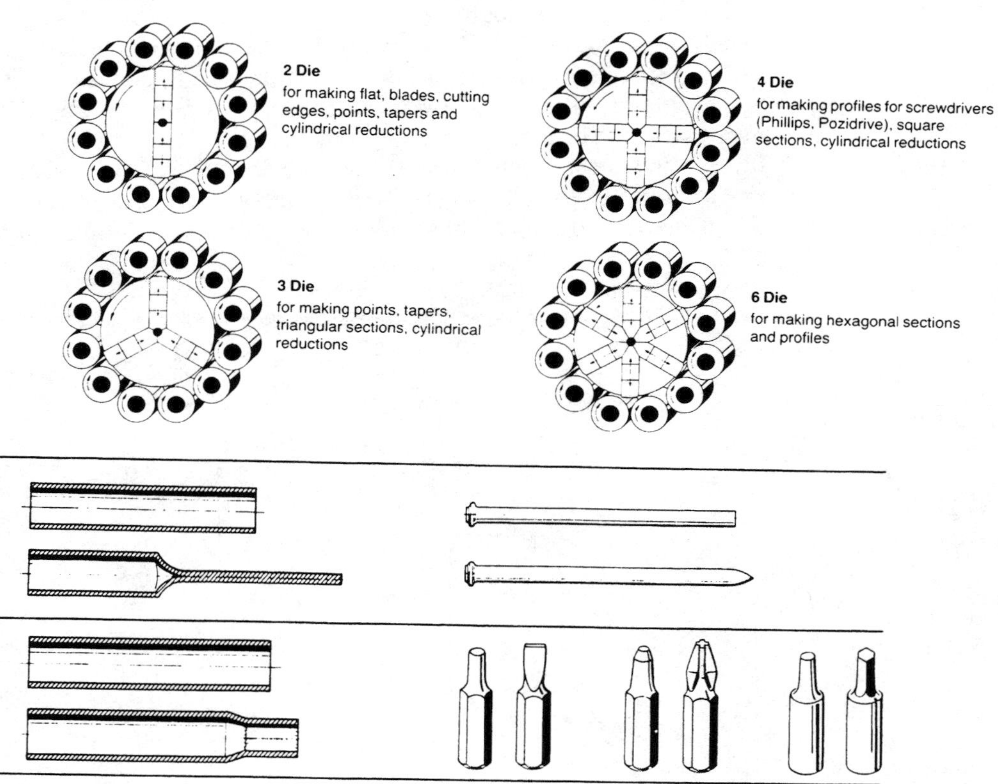

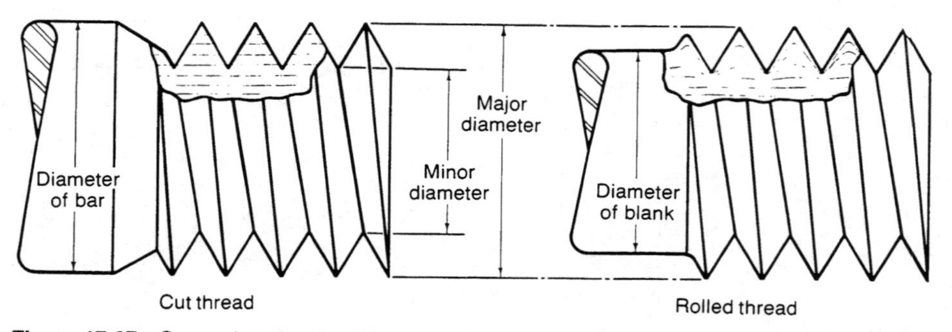

Figure 17-27 Comparison between cut and rolled threads. (Courtesy of Reed Rolled Thread Die Co.)

17-27. As seen in the illustration, cold thread rolling produces an unbroken grain structure resulting in a thread form less susceptible to fatigue than a cut thread. The rolling action also work hardens the surface of the thread to make it tougher. Rolling also is considerably faster and more efficient than cutting, resulting in higher production rates. One thread rolling method is performed by rolling the part of the fastener to be threaded between two *flat dies*. The opposing faces of the dies are cut with the pattern of the thread to be formed. One die is held stationary while the other die moves over it. The space between the two dies is less than the diameter of the unthreaded bolt or screw. As the fastener is rolled between the dies, metal is forced into the grooves of the dies to form the threads. See Figure 17-28.

Flat die rolling primarily is employed in producing small fasteners such as sheet metal, machine, and wood screws, with a variety of geometries. See Figure 17-29. To a lesser extent, flat dies are used for rolling large diameter bolts.

Threads are also rolled by using rotating, *cylindrical* dies with threads cut in their surface. Three dies are used, mounted in a threading head so that the surface of each is almost touching the surfaces of the other two (a triangular arrangement similar to a three-roll bender). The fastener is inserted at the ends of the cylinders into the space between the three dies. The rotating dies form the threads. See Figure 17-30.

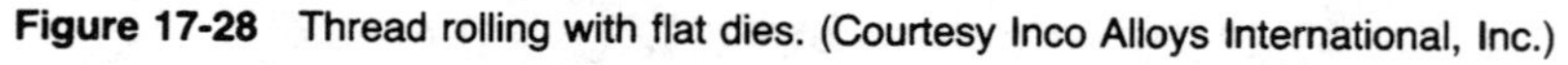

Figure 17-28 Thread rolling with flat dies. (Courtesy Inco Alloys International, Inc.)

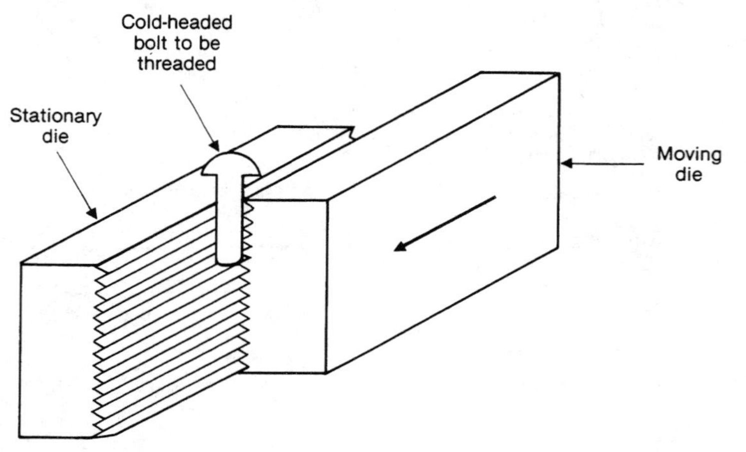

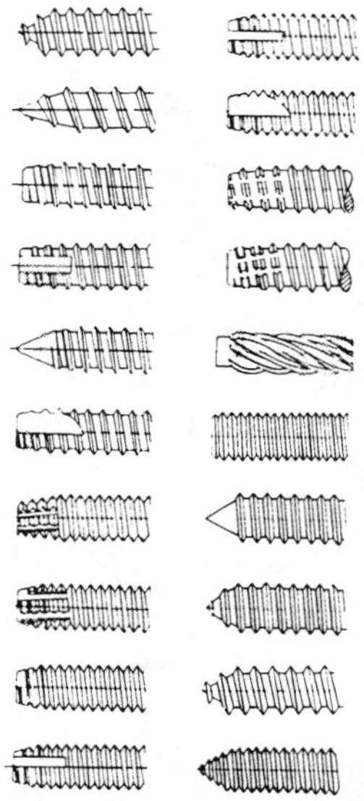

Figure 17-29 Types of small fasteners produced with flat dies. (Courtesy of Reed Rolled Thread Die Co.)

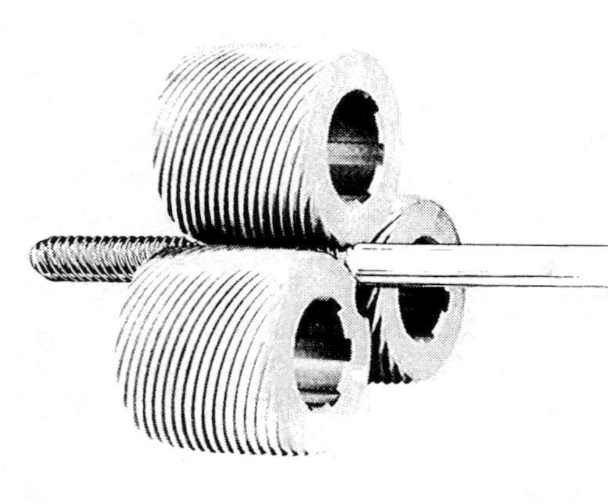

Figure 17-30 Cylindrical thread rolling die. (Courtesy of Reed Rolled Thread Die Co.)

These dies are especially valuable for rolling large diameter pieces where thread tolerance and accuracy are important. Two- and three-die machines are in common use to roll automatic fasteners, and to thread hollow spark plug shells. In this latter application, a three-die machine is employed in conjunction with a mandrel to prevent the collapse of the fragile shell while rolling. Cylindrical machines also are employed to roll threaded bar stock up to six meters (20 feet) in length.

A typical high production machine is shown in Figure 17-31. Some of this equipment can produce parts at the rate of 1,400 pieces per minute.

Figure 17-31 A typical precision thread rolling machine. (Courtesy of Reed Rolled Thread Die Co.)

Forging Equipment

Powerful machinery is used to forge metal products of varying shapes and sizes, replacing the arm and hammer of the blacksmith. The hammer and the press are the two general types of forging machines. Both are used to shape the metal between two dies, one to support the workpiece and the other to provide the force. As stated earlier, dies may be flat or they may have the contours of the finished product. The hammer delivers a high-velocity, impact pressure, while the press delivers a slower, squeeze pressure. Impact pressure is greatest when the hammer first strikes the workpiece. Squeeze pressure increases as squeezing progresses and reaches a maximum just before the pressure is released.

Forging hammers bring force to the workpiece with a heavy weight or ram that is raised and then dropped onto the workpiece. Common types are the *board hammer*,

Figure 17-32 The principal parts of a board drop hammer. (From *Metals Handbook,* Vol. 6, 8th ed., ASM INTERNATIONAL [formerly American Society for Metals], 1971. Reprinted with permission.)

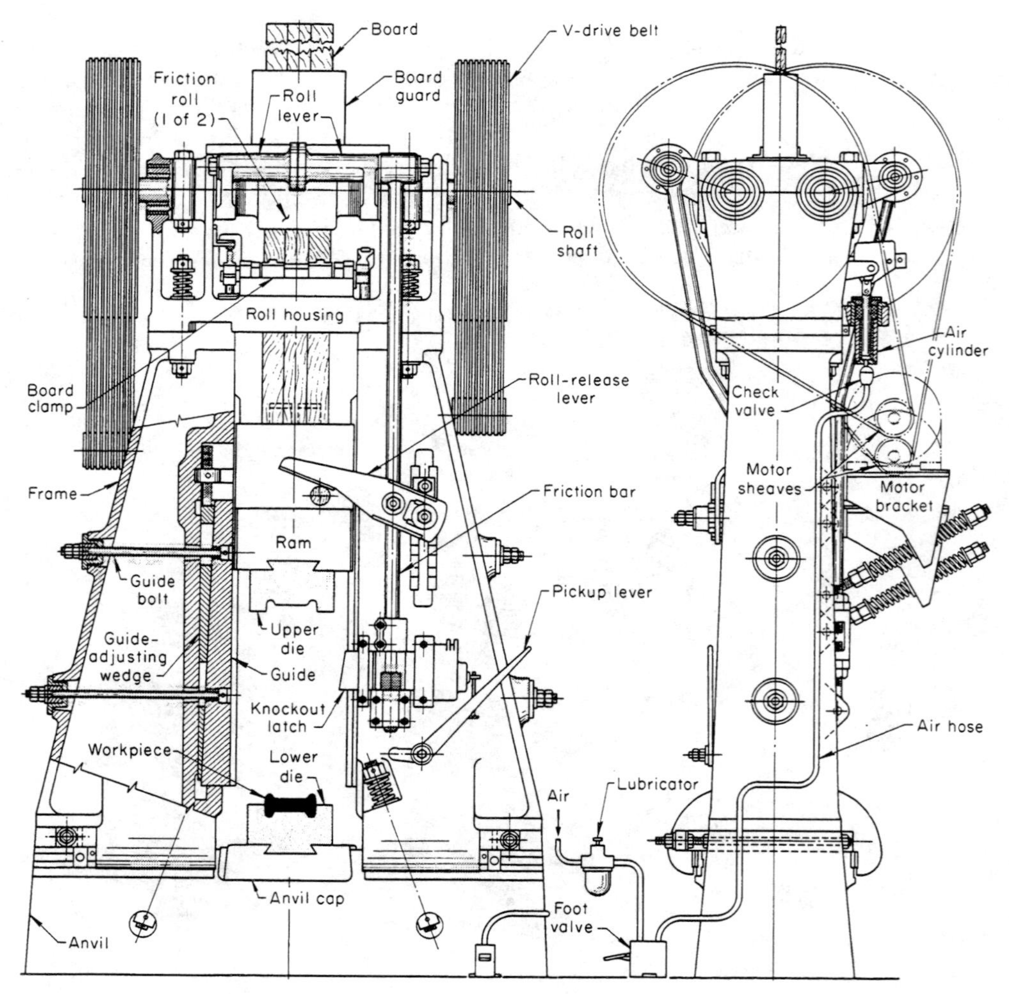

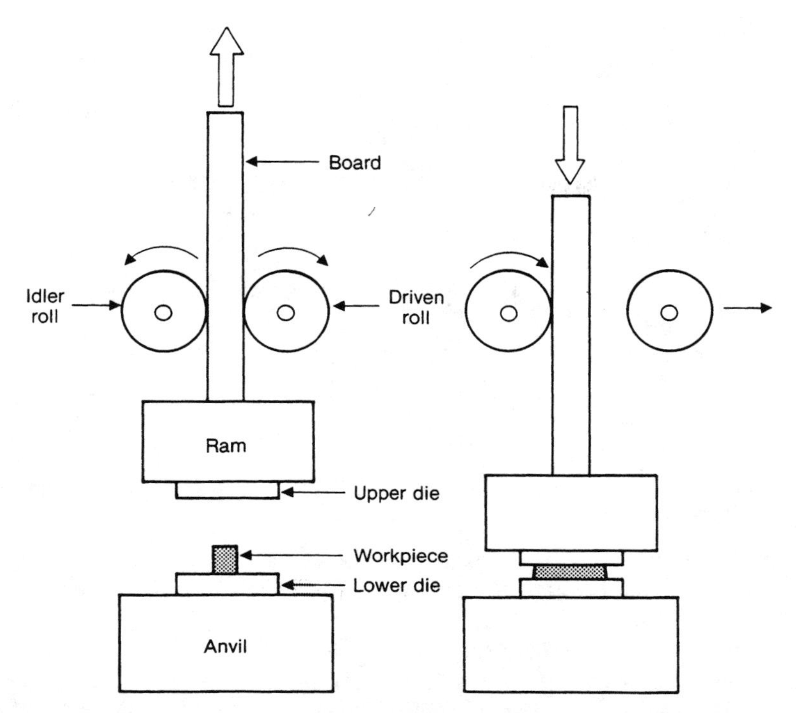

Figure 17-33 The action of a board drop hammer. The air-lift hammer operates in much the same way, except that the ram is raised by air pressure rather than friction rolls. (Courtesy of Inco Alloys International, Inc.)

the *air-lift hammer*, and the *steam hammer*. A board hammer (or board drop hammer, hence the term "drop forging") has a ram attached to the ends of hardwood boards which provide a friction surface. This can be gripped by power-driven rolls which press against the boards to raise the ram. See Figure 17-32. The hammer is held at the top of the stroke and released to deliver a blow. The resultant forging energy is a factor of the mass and velocity of the freely-falling ram and die. The hammer delivers around 70 strokes per minute, and all are of equal intensity, because the hammer is raised to the same point on each stroke. Board hammers are rated by the weight of the ram and range from 220 to 16,500 kg (100 to 7,500 pounds). They are used mainly for small forgings weighing only a few kilograms. See Figure 17-33.

The *air-lift hammers* provide a free-falling ram force similar to board drop hammers, but are somewhat larger. They differ in that the ram is raised by the action of an air cylinder attached to the top of the ram. The ram is held at the top of its stroke by a piston rod clamp, and the clamp is released by the operator to allow the hammer to drop on the workpiece. The length of the stroke can be varied so that the amount of impact can be changed with each stroke. The ram weight of air-lift hammers ranges from 1,000 to 22,000 kg (500 to 10,000 pounds).

Steam hammers are the largest and most powerful forging hammers. The hammer is similar to the air-lift hammer in construction, but differs in operation. See Figure 17-34. The ram is raised by steam on the bottom of the piston. At the top of the stroke, steam pressure is introduced to supplement the gravity force of the ram in

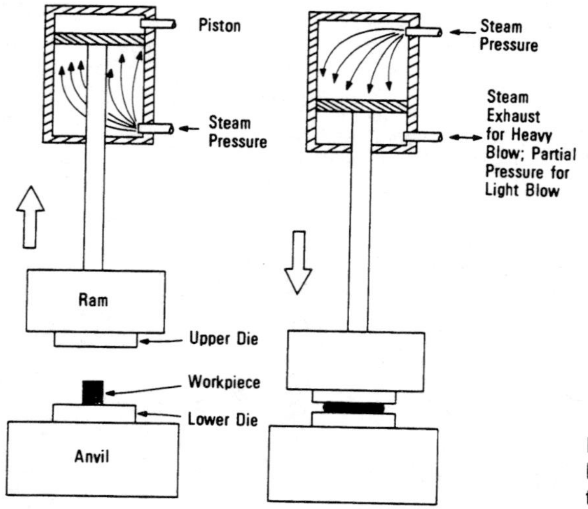

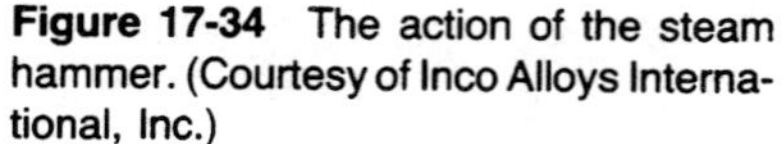

Figure 17-34 The action of the steam hammer. (Courtesy of Inco Alloys International, Inc.)

the down or work stroke. The stroke can be varied from a light tap to a heavy blow by controlling the amount of pressure released from the bottom of the piston cylinder and the amount introduced into the top. Because of the extra boost provided by the steam pressure, the working force of the hammer is about twice that obtained from the weight of the falling ram alone. The ram weight ranges from 2,200 to 110,000 kg (1,000 to 50,000 pounds).

Other Types of Hammers

Helve and trip hammers are designed to deliver quick, light blows to produce small forged parts, to preshape parts before they are forged on larger hammers, and to perform postforging operations such as planishing and coining. The hammers are rated at from 33 to 1,100 kg (15 to 500 pounds). See Figure 17-35.

Forging machines (or upsetters) are a type of horizontal trip hammer, and were originally developed to upset or spread out the ends of rods to form bolt heads and similar shapes. See Figure 17-36. The die that holds the workpiece is split so that the work can be positioned so that the proper amount protrudes on the impression

Figure 17-35 The actions of the helve hammer (a) and the trip hammer (b) produce rapid, light-hitting blows on a workpiece. (Courtesy of Inco Alloys International, Inc.)

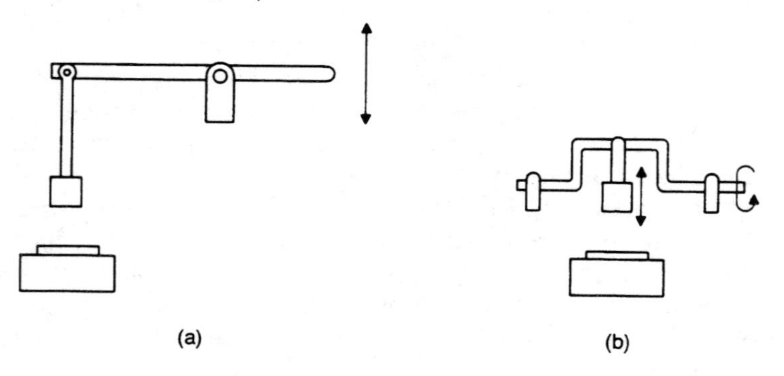

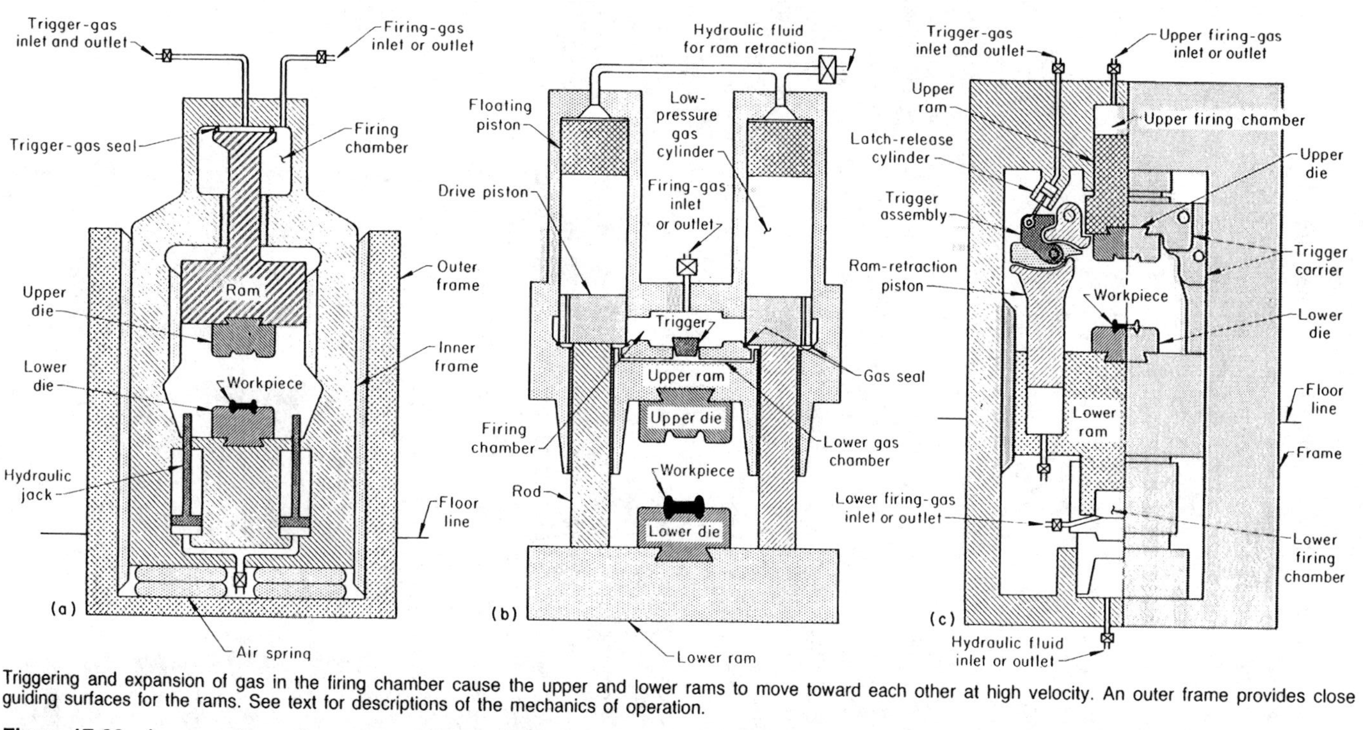

Triggering and expansion of gas in the firing chamber cause the upper and lower rams to move toward each other at high velocity. An outer frame provides close guiding surfaces for the rams. See text for descriptions of the mechanics of operation.

Figure 17-36 A ram-and-inner-frame type of HERF machine. Other styles are the controlled-energy-flow and the two-ram machines. (From *Metals Handbook,* Vol. 6, 8th ed., ASM INTERNATIONAL [formerly American Society for Metals], 1971. Reprinted with permission.)

side. The die halves then close to grip the stock, and the ram is driven against the workpiece to forge the desired shape. Other uses of this machine include preshaping parts before final forging on other equipment, and producing precision blanks for machining into gears. Forging machines are rated as the maximum diameter of stock they can upset, and range from 6 to 250 mm (0.25 to 10 inches) in capacity.

High-energy-rate forging machines are designed to deliver high-velocity blows to the workpiece, and are similar to steam hammers in that they employ piston-driven rams. The piston is actuated by highly compressed inert gas, and has a quick release mechanism to drive the ram downward at speeds up to 10 times those of ordinary hammers. A hydraulic system brings the ram back to the top. During the workstroke, the anvil moves upward to cancel the force of the blow and eliminate the need for a heavy base to absorb the impact. This counterblow principle also conserves energy.

The force of the ram can be duplicated from stroke to stroke with great accuracy. Consequently, parts can be forged to small tolerances without the upper die striking the lower die. Such die-to-die contact is usually necessary to make accurate forgings with ordinary hammers. A typical machine style is shown in Figure 17-36.

Forging rolls, however, do not produce shapes of continuous cross section. Part of the circumference of each roll is contoured to shape the workpiece. The remainder of the circumference is notched so that there is more space between the two rolls than the thickness of the workpiece. As the rolls turn, the workpiece is placed between them at the notched area. The metal if forged as the rolls continue to turn and the working surfaces of the rolls contact the workpiece. The forged part emerges from the rolls, on the same side from which it was first placed between them, at the end of one revolution, when the notches are again opposed. Roll forging is shown in Figure 17-37.

Forging presses, the second major type of forging equipment, provide a squeezing action on workpieces to impart special qualities to a product. Some alloys retain much of their toughness at forging temperatures, and the slow movement of the press allows more time for the metal to flow into the contours of dies. Also, the

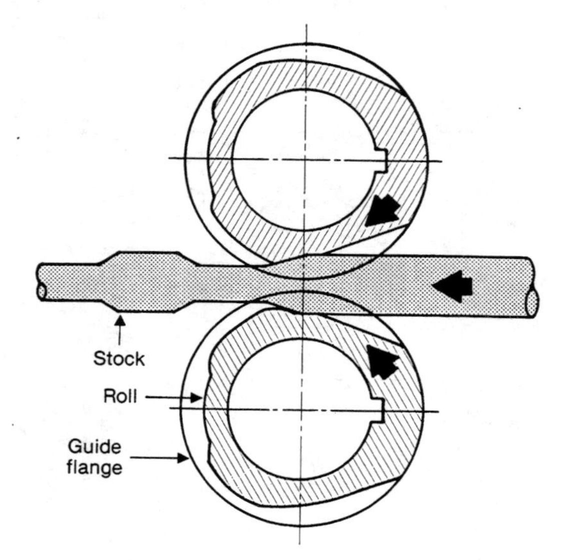

Figure 17-37 Roll forging. (Courtesy of Forging Industry Association)

press has a kneading effect on the metal, causing a more complete deformation in the workpiece. In general, larger forgings can be produced on presses than with hammers. The two types of forging presses are the mechanical and the hydraulic.

The *mechanical press* operates much like a large, slow-moving trip hammer. The ram (or slide) is moved vertically by a crankshaft, levers, or gears which are powered by an electric motor. Unlike hammers, the mechanical press blow is a squeeze rather than impact, delivered by a stroke of length and speed. Maximum pressure is delivered at the bottom of the stroke, and presses are rated by the pressure at that point. Capacities range from 90,700 to 907,000 kg (100 to 10,000 tons).

The *hydraulic press* has a large piston attached to the top of the ram with the result that hydraulic pressure on the top of the piston moves it down for the workstroke, and pressure is applied to the bottom of the piston for the return stroke. The presses are capable of fast ram speeds, and the return stroke and the first part of the workstroke (until the upper die contracts the metal) are often performed at high speeds. The presses are sometimes equipped with programmable controls to vary the forging speed and pressure. Hydraulic presses are available in many sizes, the larger ones being huge machines that deliver up to 220,000,000 kg (100,000 pounds) of pressure. See Figure 17-38.

Heating Methods for Hot Forming

The metal workpieces used in forging and other hot forming operations must be raised to an appropriate temperature to facilitate metal deformation. A number of techniques are employed to do this. Some automatic forging processes utilize induction and electrical resistance systems, but the most common heat source is the gas or oil burning furnace. The workpiece is placed inside the furnace, heated to the proper temperature, removed from the furnace, and worked immediately. The two common types are the batch and the continuous furnaces.

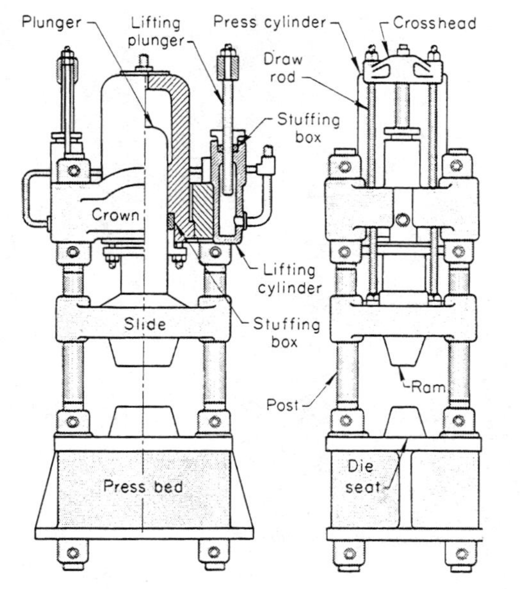

Figure 17-38 The components of a four post hydraulic press for closed die forging. (From *Metals Handbook,* Vol. 6, 8th ed., ASM INTERNATIONAL [formerly American Society for Metals], 1971. Reprinted with permission.)

A *batch furnace* consists of a rectangular heating chamber, with heat sources at the sides or back, and an opening in front through which the stock is loaded and unloaded.

A *continuous furnace* has an opening at each end of the heating chamber and a system of rails or rolls for automatically moving the stock through the chamber. Continuous furnaces are normally used only for high-volume production, where the movement of the stock is controlled so that the time required for it to travel from the entrance to the exit is sufficient to heat the metal to the correct temperature. Once in operation, the furnace discharges a continuous supply of properly heated stock.

Both batch and continuous furnaces may be either directly or indirectly heated. *Directly* heated furnaces have burners that fire into the chamber that contains the work. *Indirectly* heated furnaces have the burners mounted inside of combustion chambers so that the fire is separated from the material being heated. Indirectly heated furnaces are used when the metal being heated could be damaged by contact with the flame or the burned gases. Typical furnace designs are shown in Figure 17-39.

The *muffle* furnace is an indirectly heated furnace with two chambers, one within the other. The burners fire into the outer chamber, as in a direct-fired furnace. The work is placed inside the inner chamber, or muffle. The muffle protects the work from direct flame contact. The *semimuffle* furnace is a direct-fired furnace with an open-top container for the work. The sides of the container protect the workpiece, but conditions are otherwise the same as in a direct-fired furnace. A widely used indirectly-heated furnace is the *radiant-tube* furnace, which has tubes mounted on the walls of the heating chamber. Fuel is burned inside the tubes, and the hot tubes radiate heat to the work.

Direct-fired, batch furnaces are most often used to heat material for hot forming.

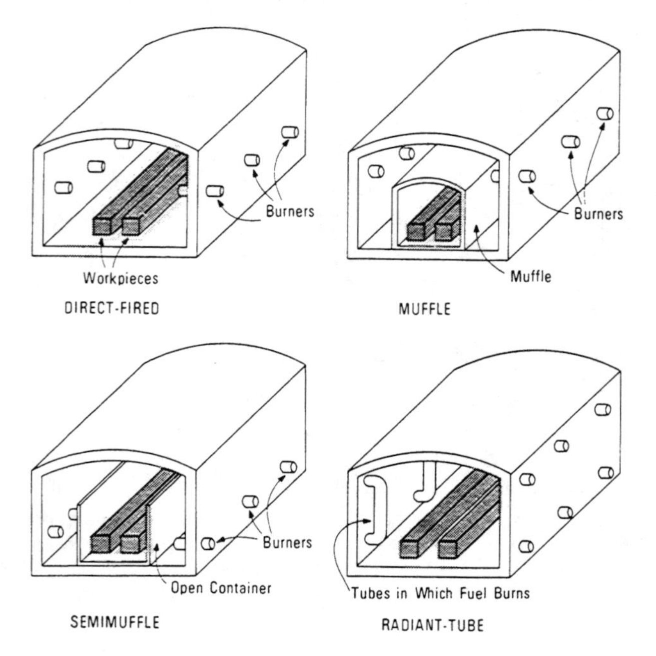

Figure 17-39 Furnace designs.

QUESTIONS

1. Describe the difference between open-die and closed-die forging.
2. What is flash? What is draft?
3. Prepare simple sketches of these forging operations: flattening; edging; fullering; and drawing.
4. Describe the purpose of some typical part-forging operations.
5. Describe the ring-rolling process and some typical uses.
6. What is the difference between forward and backward extrusion?
7. Sketch a typical cold-heading operation.
8. Describe the difference between piercing as a forging and as a shearing process.
9. What is the purpose of swaging?
10. Describe the difference between the two thread rolling methods.
11. Describe the difference between a forging hammer and a forging press.
12. Prepare a library research report on current industrial forging techniques.

part 5 MATERIAL PROCESSES: JOINING AND ASSEMBLING

18 MECHANICAL FASTENING SYSTEMS

A wide variety of mechanical devices are used in industry to provide permanent and semipermanent methods for assembling or joining workpieces. Typical examples are nails, staples, rivets, threaded fasteners, and lock joints, among others. Wood screws are used to join wooden furniture components permanently. Nuts and bolts provide a means of securing metal parts, with the added advantage of convenient disassembly. Such are the types of mechanical fasteners which are the subject of this chapter. Other special wood fasteners are described in Chapter 8. See Figure 18-1. These fastening systems are examples of mass-increasing component joining processes.

NAILS

Nails are primarily used in wood construction, and achieve their fastening power when they displace wood fibers in a part piece. The pressure exerted against the nail by these fibers, as they attempt to return to their original position, provides the considerable holding power. The lengths of the most commonly used nails are designated by the penny system. The abbreviation for the word "penny" is the letter "d." Thus, the expression "a 2d nail" means a two-penny nail. The penny sizes, corresponding lengths, and thicknesses of common nails are shown in Figure 18-2.

Nails larger than 20d are called *spikes* and are generally designated by their lengths in inches (such as 5 inches or 6½ inches); nails smaller than 2d are specified

Figure 18-1 Air-powered staplers are in general use in wood cabinet and furniture industries. (Courtesy of Senco Products, Inc.)

Figure 18-2 Chart of common nail sizes.

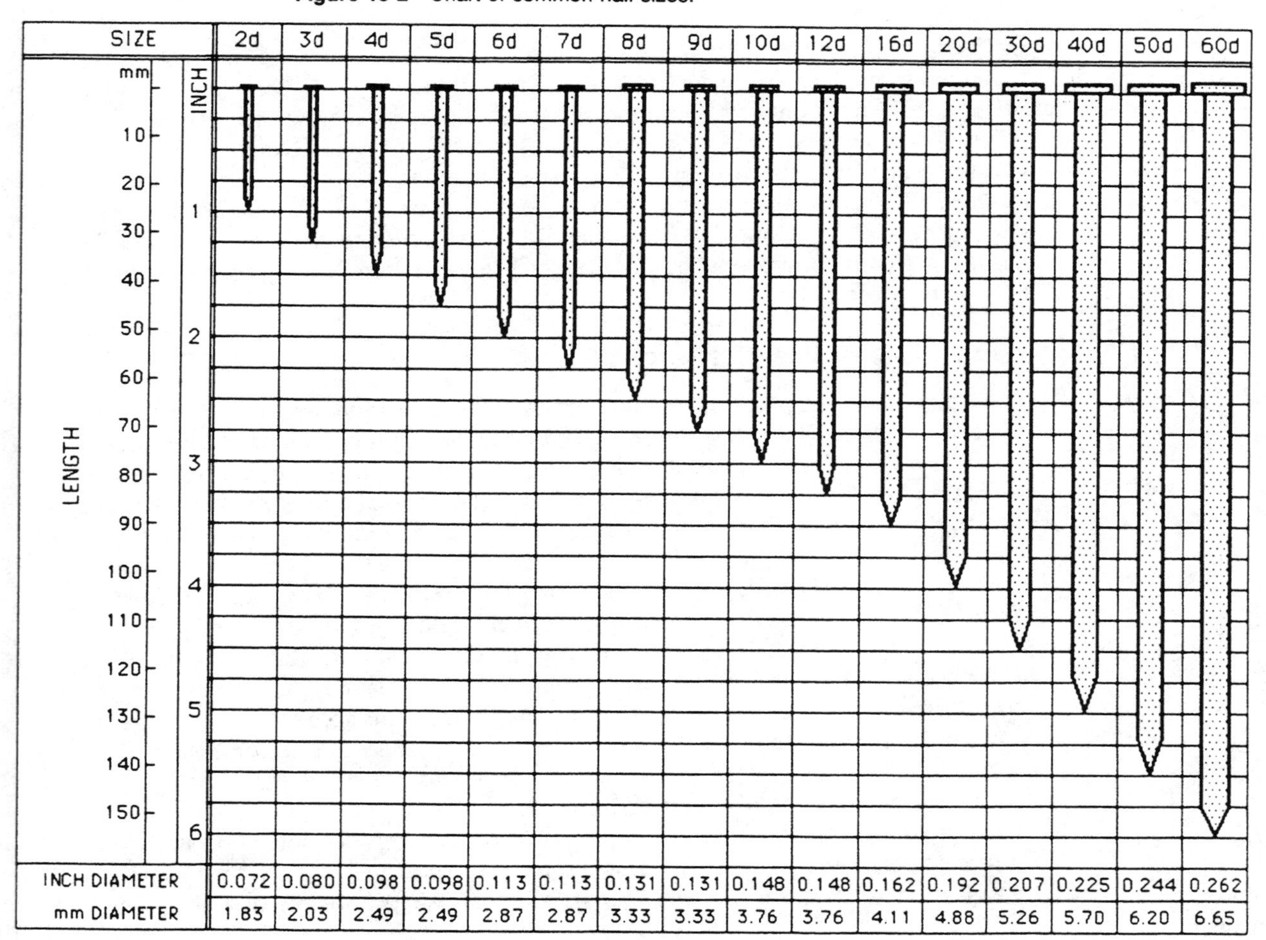

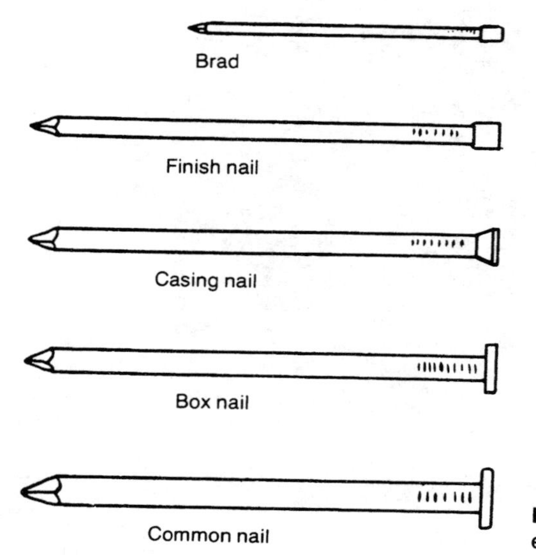

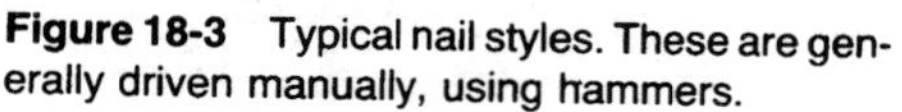

Figure 18-3 Typical nail styles. These are generally driven manually, using hammers.

in fractions of an inch instead of in the penny system. Metric sizes also are in use.

The more widely-used nail styles are shown in Figure 18-3. The *brad* and *finish* nails have deep countersink heads designed to be ''set'' below the surface of the work. These nails are used for interior and exterior trim work, and furniture case construction where the nails are ''set'' and puttied to conceal their location. The *casing* nail is used for the same purpose, but because of its flat countersink head, may be driven flush and left that way. *Box* and *common* nails have larger, flat heads for general purpose wood fastenings. These nails are usually driven by hand, using a hammer.

Automatic Nails and Nailing

Automatic nailing systems are more generally employed in manufactured or modular home construction, and in furniture assembly, where high production is a factor. See Figure 18-4. The nails used with this equipment are glued together in staggered magazines or clips for convenient handling, and to accommodate the automatic machine. See Figure 18-5. Staples are used in a fashion similar to nails, as shown in Figure 18-1.

Nails in use resist either withdrawal loads or lateral loads, or a combination of the two. The surface condition of nails often is modified during manufacture to improve the resistance to withdrawal. This modification generally takes the form of surface coating, surface roughening, or shank deformation. A common *surface coating* is so-called cement coating, which in reality is an application of a resin to increase the friction between the nail and the wood. Zinc and plastic coatings are used. The adhesives used to hold nails together for use in automatic nail cartridges also act to secure the nail in a wooden piece.

Surface roughening includes chemical etching and sand blasting to pit the shank for increased holding power. Mechanical *shank deformation* is typified by the imposition of annular or spiral grooves, barbs, or square shapes on the nail shank. See Figure 18-6.

Figure 18-4 Automatic nailing machines are used in high volume production applications. (Courtesy of Senco Products, Inc.)

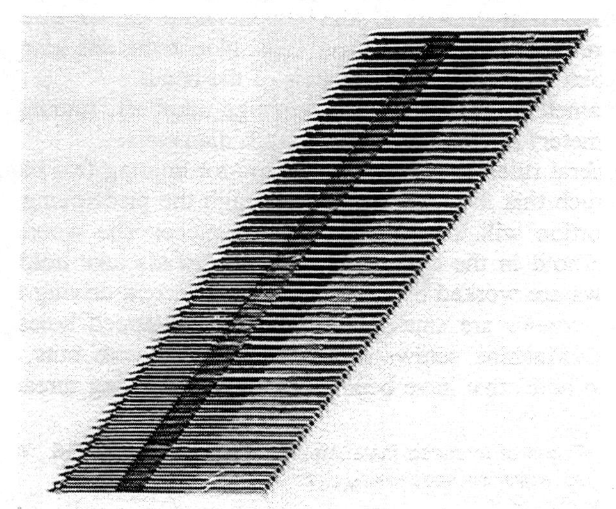

Figure 18-5 Automatic nailing systems use nail cartridges such as these. (Courtesy of Senco Products, Inc.)

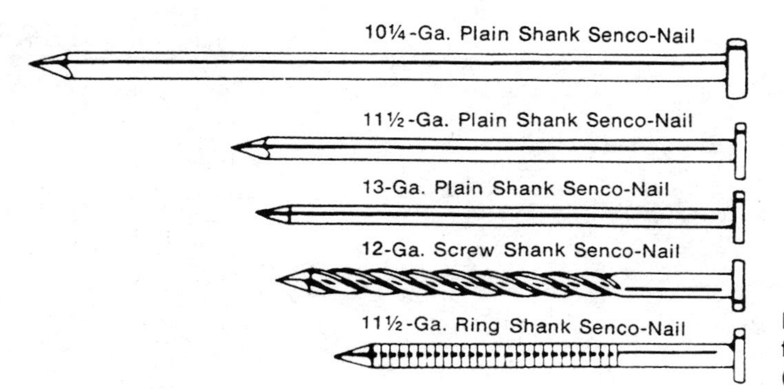

Figure 18-6 Nail shanks are modified to improve their holding punch. (Courtesy of Senco Products, Inc.)

THREADED FASTENERS

This fastener category includes all those cylindrical shapes whose surfaces have threads cut or rolled into them. See Chapter 17. Wood screws, cap screws, studs, sheet metal screws, and bolts are typical of these.

Although the words are often used interchangeably, there is a difference between a screw and a bolt. A *screw* is a fastener with external threads capable of insertion *into* holes in assembled parts, and of mating with preformed internal threads, or of forming its own thread, by torquing the fastener head. A stud is a special class of screw. A *bolt* is a fastener with external threads designed for insertion through holes in assembled parts, and secured by a tightened nut to perform its intended function. There is a great variety of types and styles of these two major threaded fastener classes. See Figure 18-7.

Wood screws have several advantages over nails. They may be easily withdrawn at any time without injury to the material. They also hold the wood more securely, can be easily tightened and, generally, are neater in appearance.

Wood screws are designated by material, type of head, and size. Most wood screws are made of steel or brass, but other metals are used as well.

The size of an ordinary wood screw is indicated by the length and body diameter (unthreaded part) of the screw. The nomenclature and the three most common types of wood screws are shown in Figure 18-8. Notice that the length is always measured from the point to the greatest diameter of the head.

Body diameters are designated by gauge numbers, running from 0 (for about a $\frac{1}{16}$ inch diameter) to 24 (for about a $\frac{3}{8}$ inch diameter).

As a general rule, the length of a screw for holding two pieces of wood together should be such that the body extends through the piece being screwed down so the threaded portion will then enter the other piece. The wood screw simply passes through the hold in the top piece, and the threads take hold in the bottom piece. Wood screws are worked by manual or power screw driving tools.

Machine screws are small fasteners used in tapped holes for the assembly of metal parts. Machine screws may also be used with nuts, but usually they are inserted into holes that have been tapped with matching threads.

Figure 18-7 Types of threaded fasteners. Nut and bolt (a). Stud, which is a special kind of screw (b). Cap screw (c).

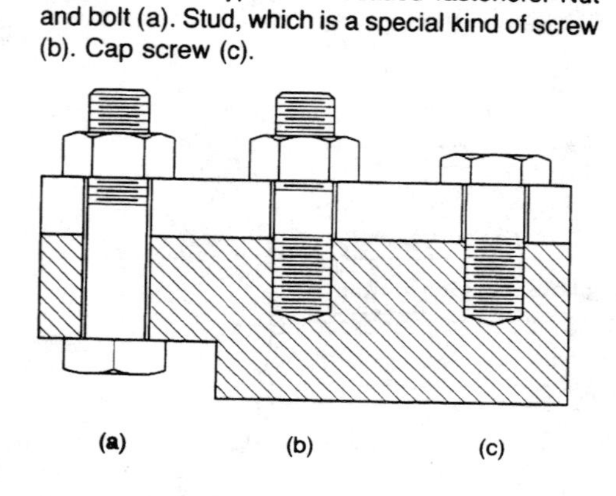

Figure 18-8 Wood screw nomenclature.

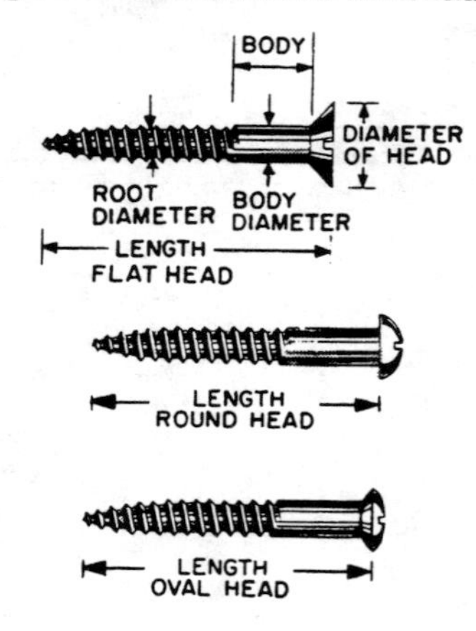

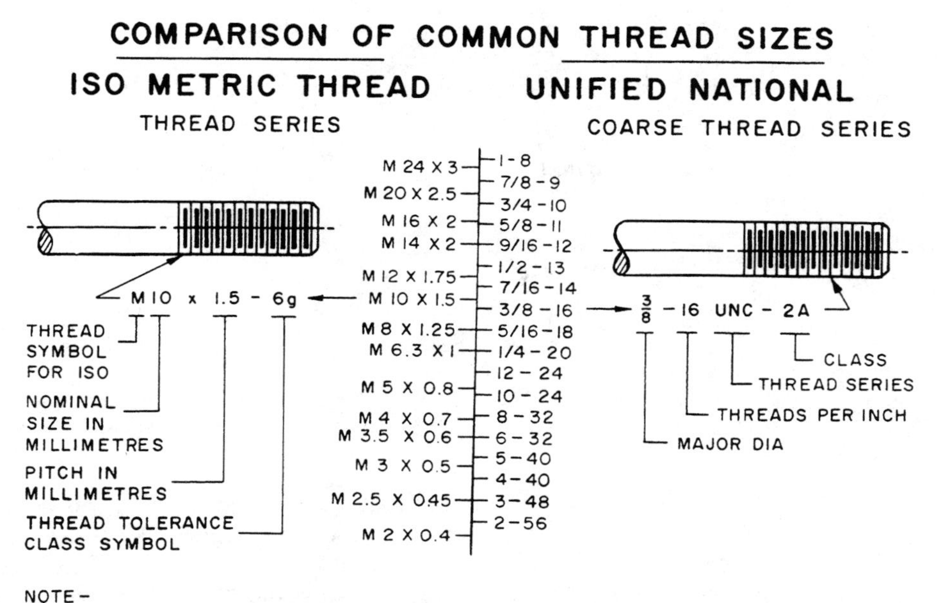

Figure 18-9 A comparison of metric and customary machine screw threads.

Machine screws are manufactured in a variety of lengths, diameters, pitches, materials, head shapes, finishes, and thread fits. A complete description of machine screws must include these factors. Comparative thread designations are shown in Figure 18-9.

The diameters of machine screws are expressed in millimeters, or in gauge numbers or fractions of an inch. Note that metric thread pitch is the distance between thread crests, while customary pitch is the number of threads per inch. Most machine screws are made of steel or brass, and may be plated to help prevent corrosion.

A variety of common and special machine screw head shapes are shown in Figure 18-10. Some of the heads require special tools for driving and removing. The slotted head screws are tightened with a standard screwdriver. The recessed styles are designed for greater, surer tightening power, and need the special tools. See Figure 18-11. These same head shapes and styles are found on wood screws, cap screws, sheet metal screws, and some bolts. Most bolt heads however, have square or hexagon shapes.

Figure 18-10 These head shapes are typical of all screws.

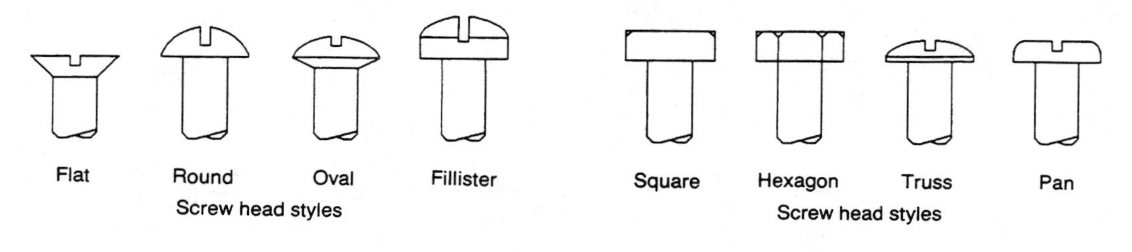

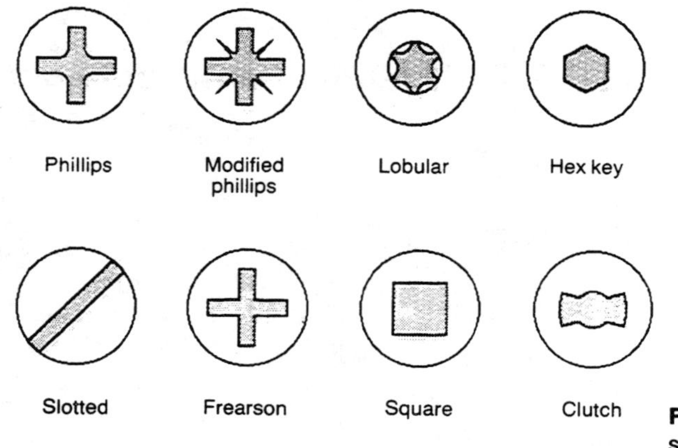

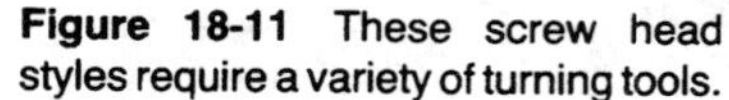
Figure 18-11 These screw head styles require a variety of turning tools.

Considerable effort has gone into the design of recessed heads to improve the torquing of the fastener and to protect the integrity of the head. While the conical shaped recessed heads, such as phillips and modified phillips, have the advantage of ease of tool use, they do display some inherent deficiencies. Due to their conical or cruciform design, and torque applied to them tends to force the turning tool upward and out of the recess. This is called "cam-out," and can only be overcome by the application of considerable downward pressure, or "end-load." This not only causes worker fatigue, but it also causes the tool to act as a reamer to strip and destroy the recessed hole. Other recessed heads such as lobular, hex, square, and clutch provide for a more positive tool contact with the recess, resulting in less fastener damage and operator fatigue. For this reason, common cruciform and slotted screw heads are restricted to fastening applications such as door hardware and part assembly, where extreme torquing of the fastener is not required. In heavier structural applications, the more positive hexagon and lobular recesses are used, or the hexagon headed styles. See Figure 18-12.

One other important element in screw thread designation is the *class of fit*, or tightness. There are three main classes of fit for metric threads. The 5H internal thread and the 4g external means a close fit for precision work. A 6H internal and a 6g external is a medium or general purpose fit. The 7H and the 8g external is a free fit for easy assembly and disassembly, even when the threads are dirty.

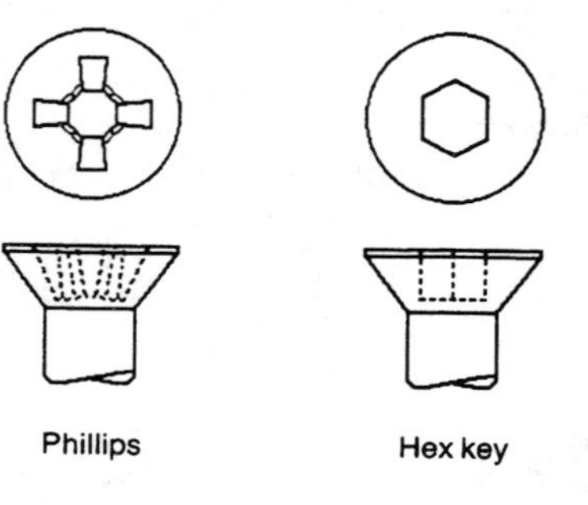

Figure 18-12 The hex key recessed head, and other similar styles, provide a more positive tool contact area than the phillips for heavy loading.

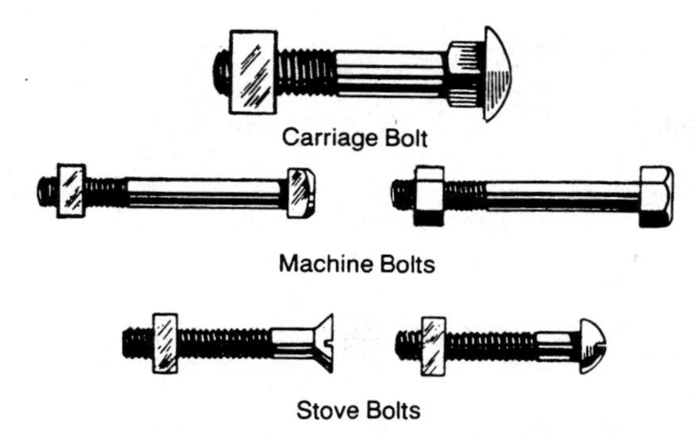

Figure 18-13 Common types of bolts.

The customary inch threaded fasteners have four classes of fit. They are: Class I, loose fit; Class II, free fit; Class III, medium fit; and Class IV, close fit.

Capscrews perform the same functions as machine screws, but come in larger sizes for heavier work. Sizes range up to 25 mm (1 inch) in diameter and 150 mm (6 inches) in length. Cap screws are usually used without nuts. They are screwed into tapped holes, and are sometimes referred to as tap bolts. Capscrews may have square, hex, flat, button, or fillister heads. Fillister heads are best for use on moving parts when such heads are sunk into counterbored holes. Hex heads are usually above the surface and used where the metal parts do not move.

A *bolt* is distinguished from a *machine screw* in that it does not thread into the material, but goes through and is held by a nut threaded onto the end of the bolt. Four common types of bolts are shown in Figure 18-13. *Stove* bolts are rather small, ranging in length from 10 mm to 100 mm ($\frac{3}{8}$ to 4 inches) and in body diameter from 3 mm to 10 mm ($\frac{1}{8}$ to $\frac{3}{8}$ inch). *Carriage* and *machine* bolts run from 20 mm to 500 mm ($\frac{3}{4}$ to 20 inches) long, and from 5 mm to 20 mm ($\frac{3}{16}$ to $\frac{3}{4}$ inch) in diameter. The carriage bolt has a square section below the head, which is imbedded in wood to prevent the bolt from turning as the nut is drawn up. The machine bolt has a hexagon or square head, which is held with a wrench to prevent it from turning. Carriage bolts are used in wood and other soft materials. Stove bolts are used for light assembly, and machine bolts for heavy work.

Setscrews are used to secure small pulleys, gears, and cams to shafts, and provide positive adjustment of machine parts. They are classified by diameter, thread, head shape, and point shape. The point shape is important, because it determines the holding qualities of the setscrew. See Figure 18-14. Setscrews hold best if they have either a *cone point* or a *dog point*, because these points fit into matching recesses in the shafts against which they bear. Headless setscrews are used with moving parts because they do not project above the surface. They are threaded all the way from point to head. Common setscrews, used on fixed parts, have square heads. They have threads all the way from the point to the shoulder of the head. See Figure 18-15.

Thumb screws are used as setscrews, adjusting screws, and clamping screws. Because of their design, they can be loosened or tightened without the use of tools.

Sheet metal screws are used to secure thin workpieces made of metal or plastic. See Figure 18-16. They have a variety of points, and may be self-drilling. See Figure 18-17. Others have washers locked to their heads. Similar fasteners for similar applications are made of plastic and have serrated or webbed edges instead of threads. They too are available in many styles for common and special applications.

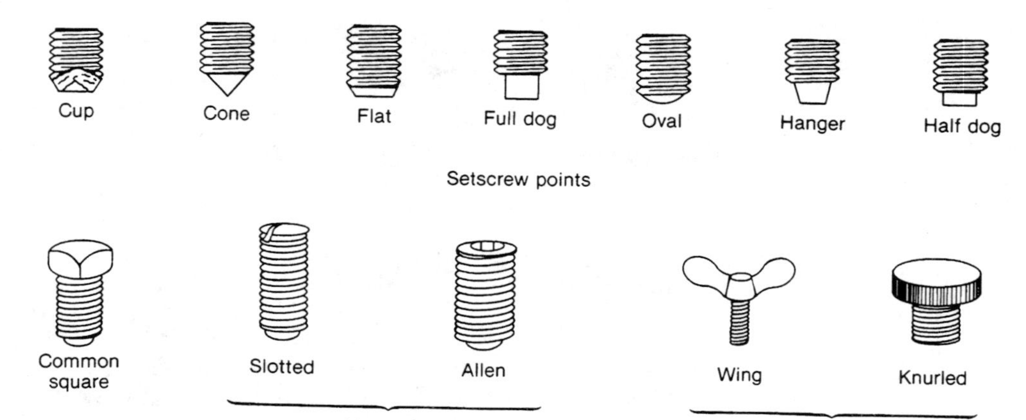

Figure 18-14 Types of setscrews.

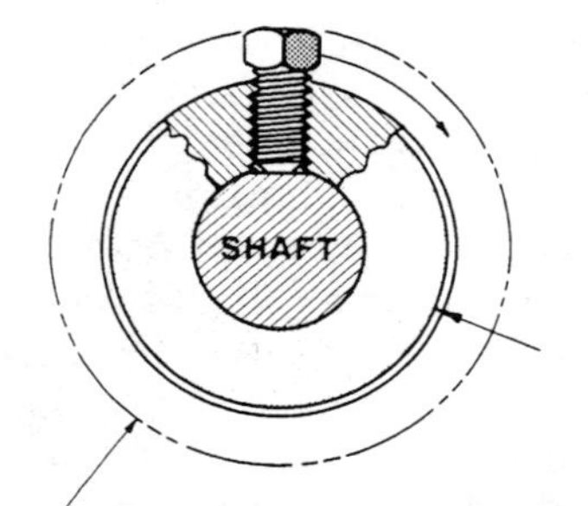

Figure 18-15 Note that for safety reasons, a headless setscrew would be preferred to the common square head type shown here.

Figure 18-16 A typical sheet metal screw application.

Figure 18-17 A common sheet metal screw, with several point styles shown.

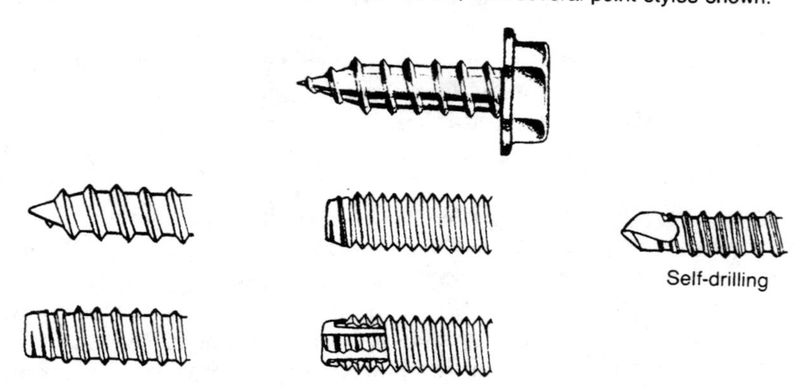

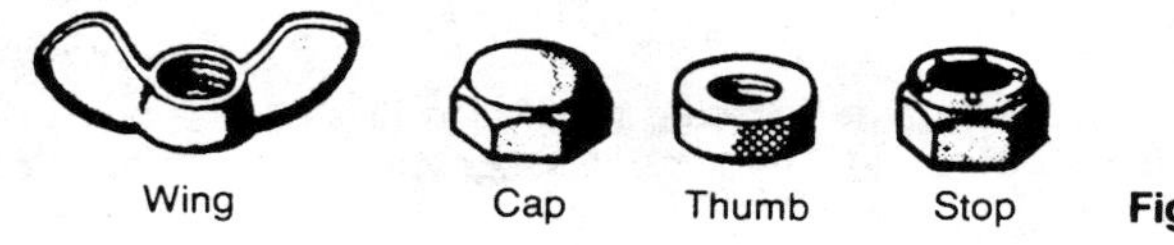

Figure 18-18 Types of nuts.

Square or hexagonal *nuts* are commonly used with screws and bolts, but they are supplemented by special nuts. See Figure 18-18. One of these is the jam nut, used above a standard hex nut to lock it in position. It is about half as thick as the standard hex nut, and has a washer face.

Castellated nuts are slotted so that a safety wire or cotter key may be pushed through the slots and into a matching hole in the bolt. This provides a positive method of preventing the nut from working loose. For example, you will see these nuts used with the bolts that hold the two halves of an engine connecting rod together.

Wing nuts are used where the desired degree of tightness can be obtained by the fingers. Cap nuts are used where appearance is an important consideration. They are usually made of chromium plated brass. Thumb nuts are knurled, so they can be turned by hand for easy assembly and disassembly.

Elastic stop nuts are used where it is imperative that the nut does not come loose. These nuts have a fiber or composition washer or insert built into them, which is compressed automatically against the screw threads to provide holding tension.

Thread inserts are special nuts used to provide durable assembly systems in wood and plastic materials, where normal threads would be likely to strip and fail under load, or fracture the workpiece. Aluminum and brass inserts are used in soft materials, and steel for harder workpieces. See Figure 18-19.

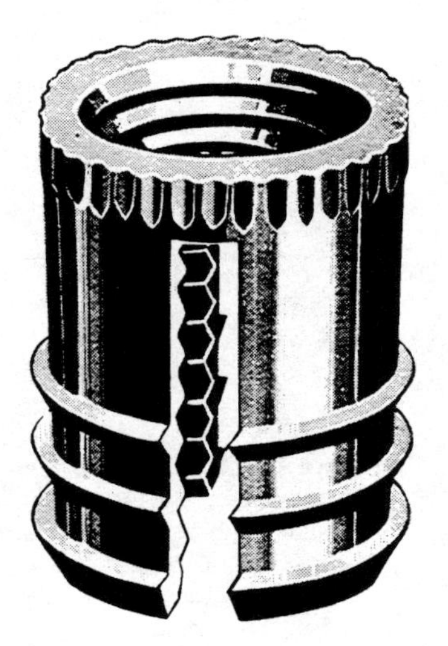

Figure 18-19 Thread insert installation diagram. These systems provide a positive locking arrangement in wood and soft plastic. Similar inserts can be fixed in threaded holes in metal workpieces. (Courtesy of Tool Components, Inc.)

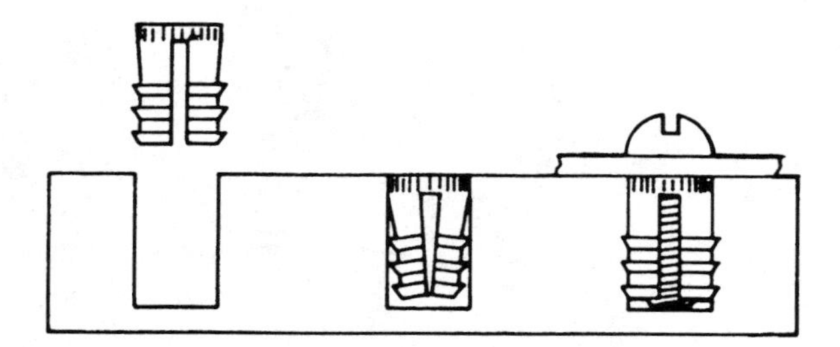

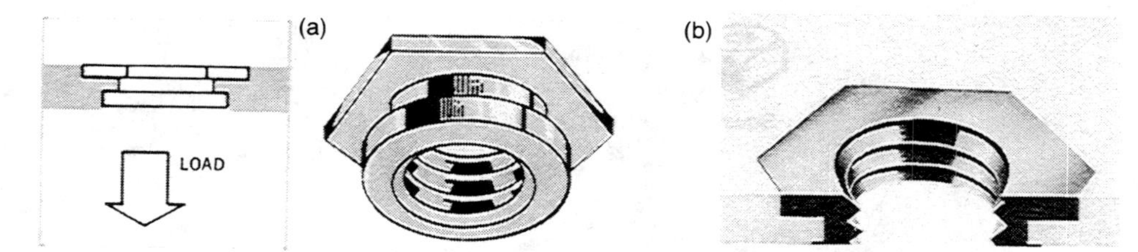

Figure 18-20 Typical captive hardware installation in sheet materials. The nut insert and load diagram are shown at (a). The firmly locked nut is shown at (b). (Courtesy of Precision Metal Products Co.)

Another type of insert is the *press nut*, used to join metal and plastic panels. There are many types and styles. The nut is forced into a predrilled hole, and the cold flow of the panel material into the nut recesses locks it firmly and permanently in place. See Figure 18-20. The integral design of these nuts speeds handling and assembly operations in fastening components, technical boards, circuit boards, and tube sockets. Their extreme durability makes them especially valuable in applications where repeated component assembly and disassembly occurs. See Figure 18-21.

Many other captive fasteners, push nuts, spring nuts, T-nuts, and self-clinching devices are used to secure threaded fasteners in various assemblies. Typical styles and applications are shown in Figure 18-22.

Metal "nails" or tacks are hardened steel points which are impact-driven into metal to secure parts. See Figures 18-23 and 18-24. As such, they replace metal

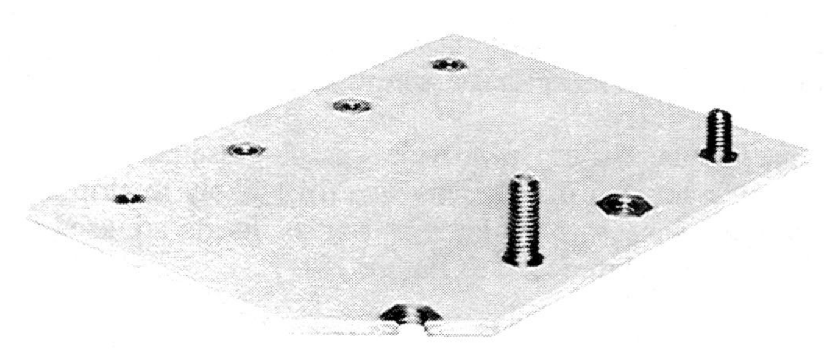

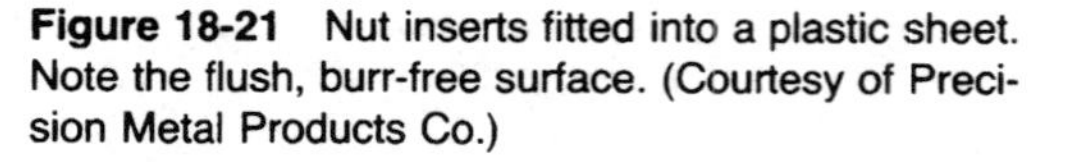

Figure 18-21 Nut inserts fitted into a plastic sheet. Note the flush, burr-free surface. (Courtesy of Precision Metal Products Co.)

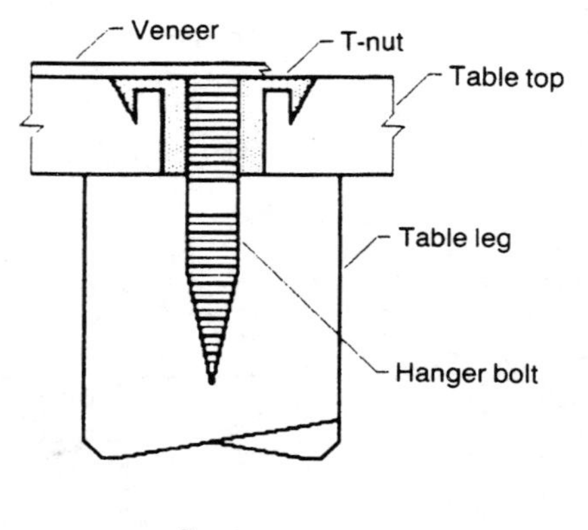

Figure 18-22 A T-nut assembly. The nut is forced into a wooden table-top. A special hanger bolt has wood screw threads on one end and machine screw threads on the other. The bolt is screwed into a wooden table leg, and the leg is then screwed into the T-nut. A veneer sheet hides the T-nut.

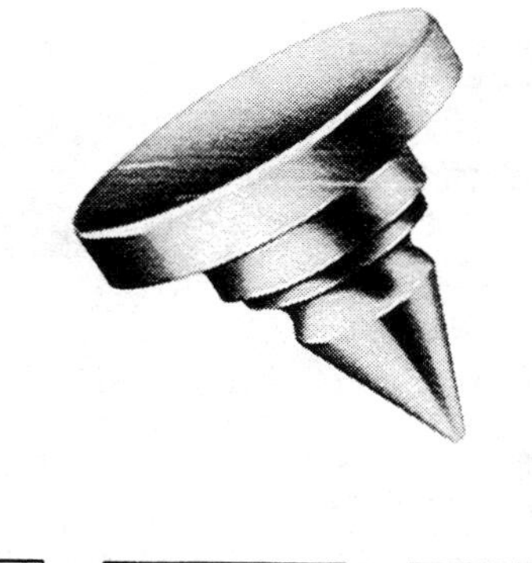

Figure 18-23 A hardened steel nail or tack can be used for sheet metal assemblies. (Courtesy of Amtak Fasteners)

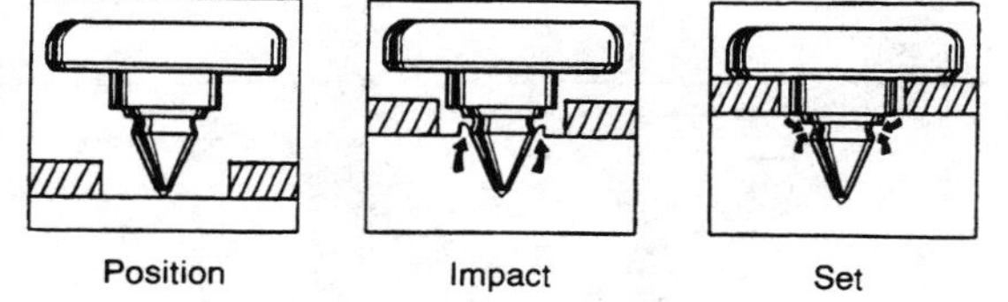

Figure 18-24 The clinching action of the Metal Tack®. (Courtesy of Amtak Fasteners)

screws in similar assemblies. Typical applications include the fastening of cable clamps, electrical grounds, name plates, bar codes, and serial togs. See Figure 18-25. A similar fastener, shown in Figure 18-26, has sharp cutting teeth which securely grip sheet metal parts under an impact tool. This pin system provides tight, tamper-proof, permanent joints in metal sheets with thicknesses ranging from one to 0.40 mm (18 to 26 gauge), and can carry a 113 kg (250 pound) load. Manual and semiautomatic equipment is used with both of these fastener types. There are similar plastic push nuts used to assemble plastic sheet components.

SwissLok is a unique fastener designed to simplify the joining of sheet metal at right angles. See Figure 18-27. It speeds perpendicular sheet metal assembly by eliminating flanging, notching, drilling, welding, or riveting. The fastener is secured by positioning it over a prepared hole in the horizontal sheet, and with an impact hammer locking the perpendicular sheet firmly in place. See Figure 18-28.

Figure 18-25 This bar code label is secured with Metal Tacks®. (Courtesy of Amtak Fasteners)

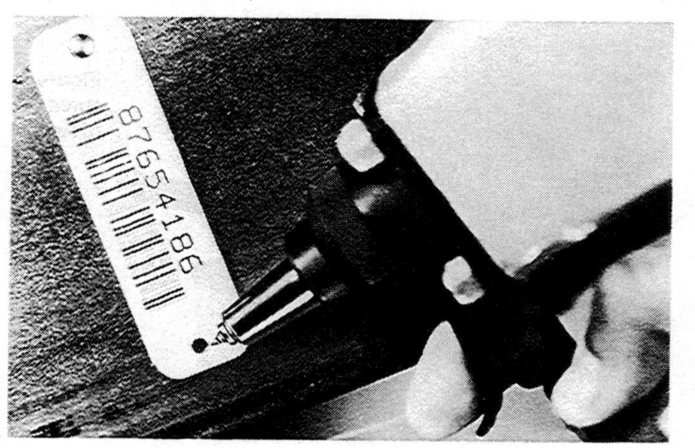

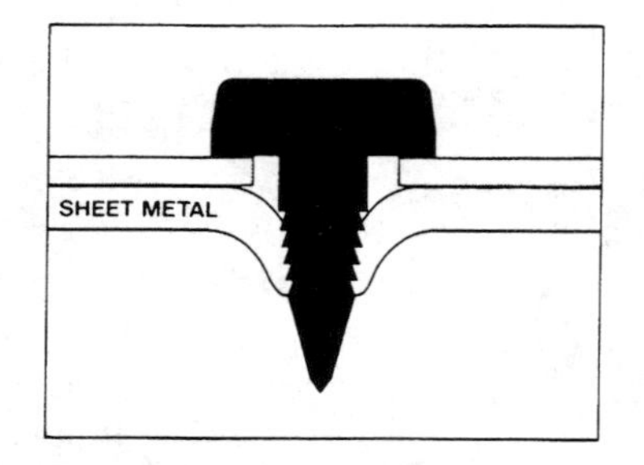

Figure 18-26 This punch pin is used to join sheet metal parts. (Courtesy of Amtak Fasteners)

Figure 18-27 This SwissLok clip is used in sheet metal assemblies. (Courtesy of Amtak Fasteners)

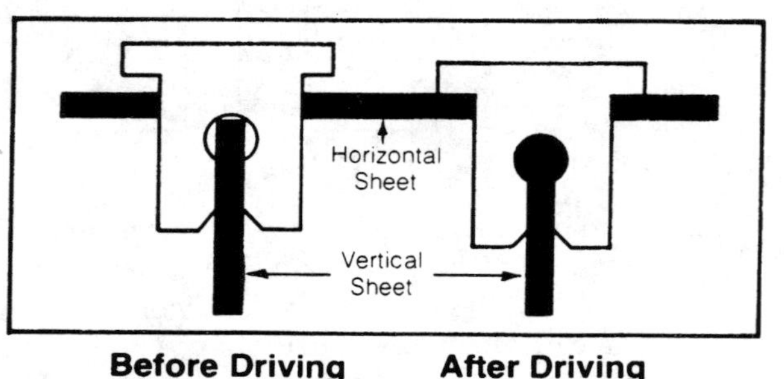

Figure 18-28 The holding action of the metal clip. (Courtesy of Amtak Fasteners)

The four fastener "legs" are inserted into the hole of the horizontal sheet and straddle the vertical sheet. A single impact from a hammer drives the fastener down, upsetting the metal of the vertical sheet to fill the circular cavity of the fastener and lock the two sheets together. This cold-forming action creates a permanent mechanical bond, capable of holding pulls up to 73 kg (160 pounds). See Figure 18-29.

A variety of *washers* are used in conjunction with threaded fasteners. See Figure 18-30. Flat washers are used to back up bolt heads and nuts to provide larger bearing surfaces, and to prevent damage to the surfaces of the workpieces. Split

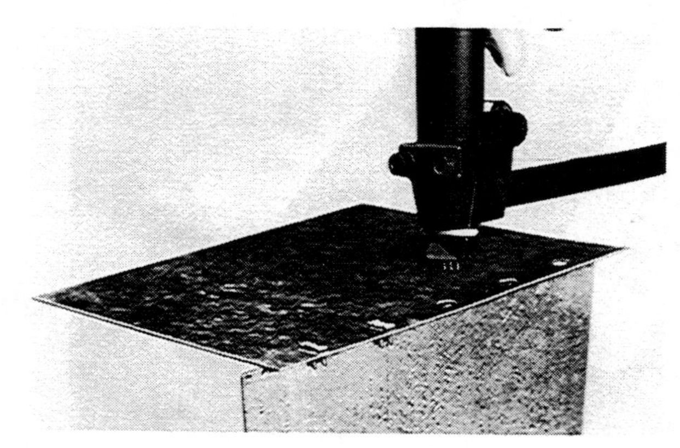

Figure 18-29 The SwissLok is secured by air impact tools. (Courtesy of Amtak Fasteners)

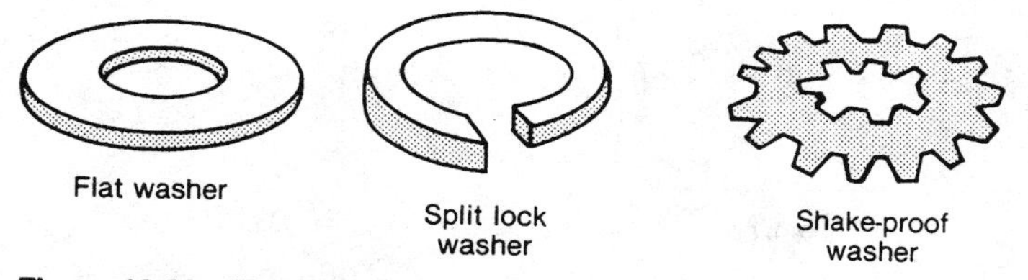

Figure 18-30 Types of washers.

lock washers are used under nuts to prevent loosening by vibration. The ends of these spring-hardened washers dig into both the nut and the work to prevent slippage. Shakeproof lock washers have teeth or lugs that grip both the work and the nut. Several patented designs, shapes, and sizes are obtainable. A special type of wedge-surface washer is used in pairs to secure fasteners. See Figure 18-31.

Metal *keys* and *pins* are employed in a variety of assembly operations. See Figure 18-32. Cotter keys are used to secure screws, nuts, bolts, and pins. They are also used as stops and holders on shafts and rods. Square keys and woodruff keys are used to prevent hand wheels, gears, cams, and pulleys from turning on a shaft. They are strong enough to carry heavy loads if they are fitted and seated properly. Taper pins are used to locate and position matching parts. They are also used to secure small pulleys and gears to shafts. Dowel pins are used to position and align the units or parts of an assembly. One end of a dowel pin is chamfered, and it is slightly larger in diameter than the size of the hole into which the pin will be driven, to provide a force fit. Cotter keys or pins are placed through holes in the ends of bolts to prevent the nuts from working loose.

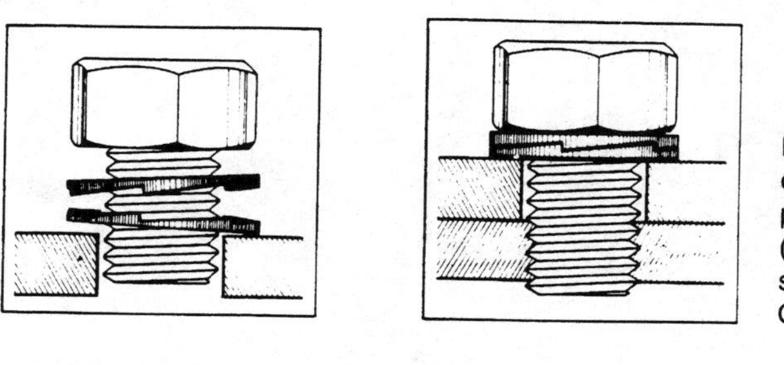

Figure 18-31 These washers have cams or wedges on their surfaces for positive, shock-proof engagement. (Courtesy of Nord-Lock Sales Division, Lake Michigan Wire and Chain Corp.)

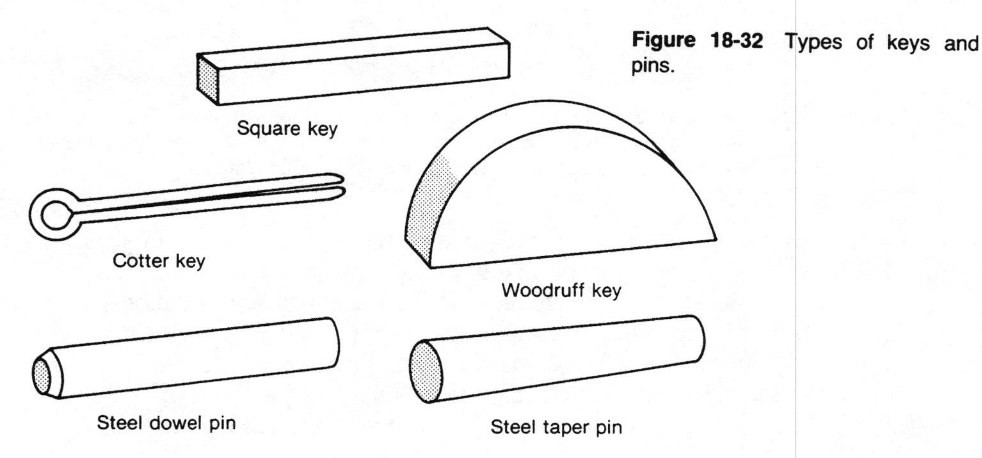

Figure 18-32 Types of keys and pins.

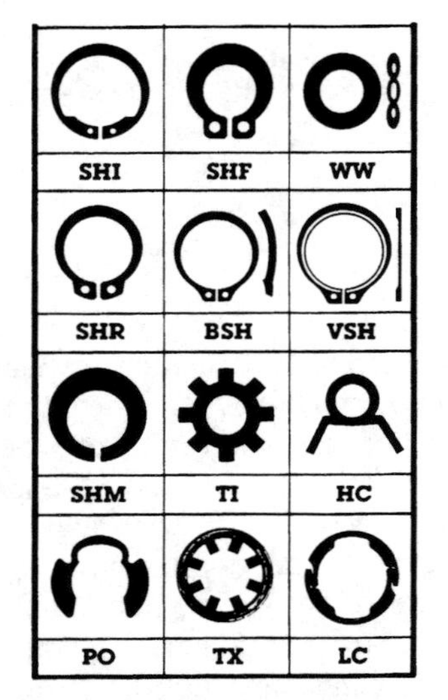

Figure 18-33 Some typical retaining ring styles. (Courtesy of Rotor Clip Company, Inc.)

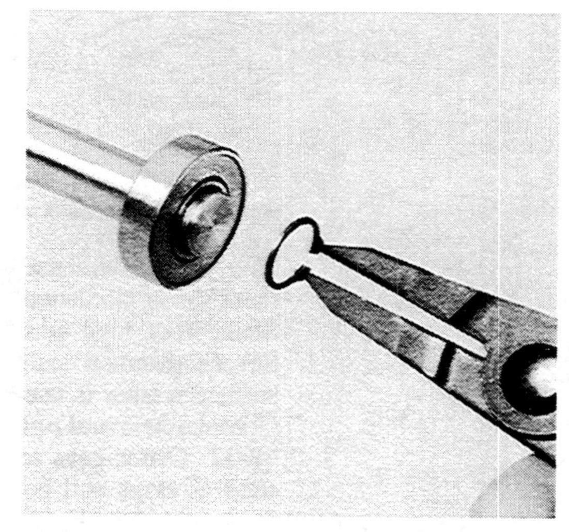

Figure 18-34 An external retaining ring application. The special pliers also is shown. (Courtesy of Rotor Clip Company, Inc.)

Figure 18-35 This internal ring is used to retain a hydraulic damper assembly. (Courtesy of Rotor Clip Company, Inc.)

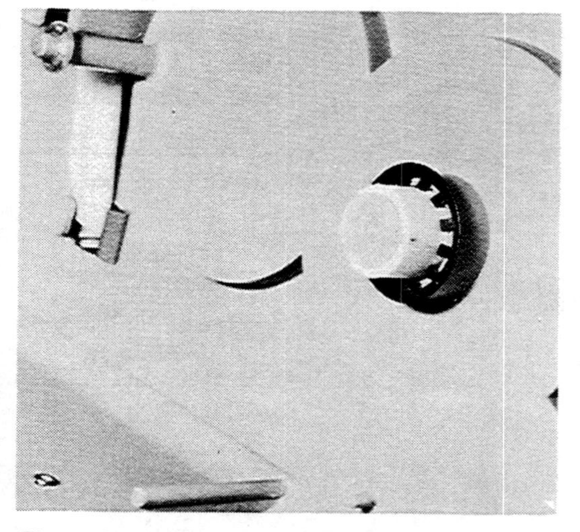

Figure 18-36 This ungrooved plastic shaft is held in place by a TX type retaining ring. (Courtesy of Rotor Clip Company, Inc.)

Retaining rings provide a semipermanent means of securing various shaft assemblies, and are of many configurations for an equal number of applications. See Figure 18-33. External and internal rings are commonly used, and require special ring-spreading pliers to install them. See Figures 18-34 and 18-35. Others can be press fit over soft plastic shafts, and are of a more permanent nature. See Figure 18-36. An advantage of most such rings is that they permit the easy assembly and disassembly of components, while providing a secure joint.

RIVETS

Basic riveting involves passing a headed pin, or rivet, through aligned holes in workpieces, and forming a second head on the protruding end to draw the parts together in a permanent mechanical joint. See Figure 18-37. They are used on sheet materials, structural shapes, and soft materials. Rivets perform similar functions as do threaded fasteners, but rivets are not practical for joints requiring disassembly. Rivets are generally cold headed, while others must be heated red in larger structural applications.

Rivets are less expensive than threaded fasteners, are easy to install, and provide a permanent, strong joint. Rivet joints also provide a degree of flexibility in structures such as bridges, where wind-imposed stresses would cause a rigid weldment to fail.

The major types of rivets are the standard and the blind rivets. Standard rivets must be driven using a bucking bar, whereas the blind rivets have a self-heading capability and may be installed where it is impossible to use a bucking bar. Wherever possible, rivets should be made of the same material as the material they join. They are classified by lengths, diameters, and their head shape and size, similar to the classes of threaded fasteners.

Some rivets for light load applications, such as in leather and bindings, have hollow ends for quick and easy setting with lightweight equipment. See Figure 18-38. *Grommets* are hollow rivets made to strengthen the rope passage and fastening points on canvas and tents, awnings, and sails.

Staking is similar to riveting, except that it involves the setting or heading of a protrusion on a shaft or axle. See Figure 18-39.

Blind rivets facilitate the joining of parts accessible from only one direction, and can be inserted and set from one side of the workpiece. A typical installation is shown in Figure 18-40. While this blind feature makes their use mandatory in

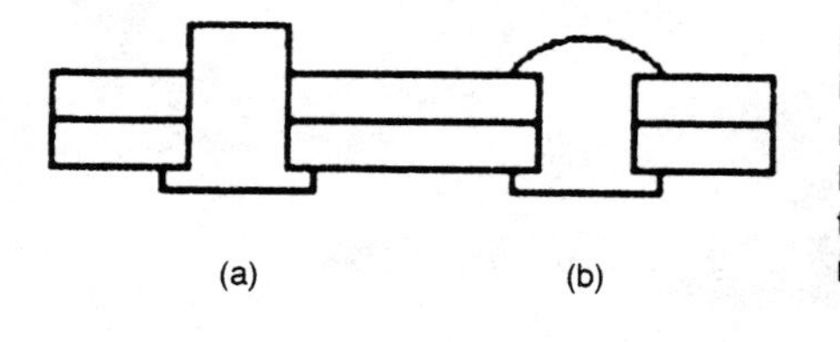

Figure 18-37 A rivet operation diagram. The rivet is inserted into holes in metal workpieces (a). The rivet is headed to form a permanent metal joint (b). Note that these common rivets require support as they are being headed. The rivet shown is a tinners rivet, which features a flat, protruding head.

Figure 18-38 The use of hollow end rivets and grommets. The raw fasteners are shown before (a) and after (b) heading.

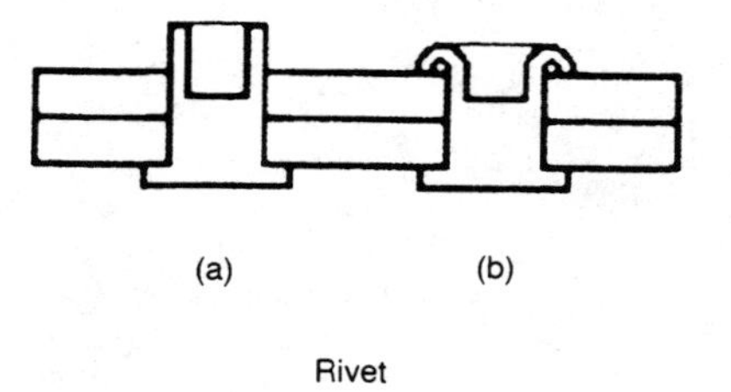

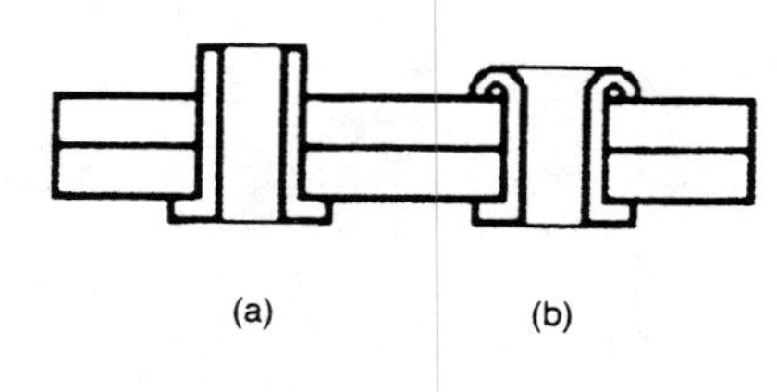

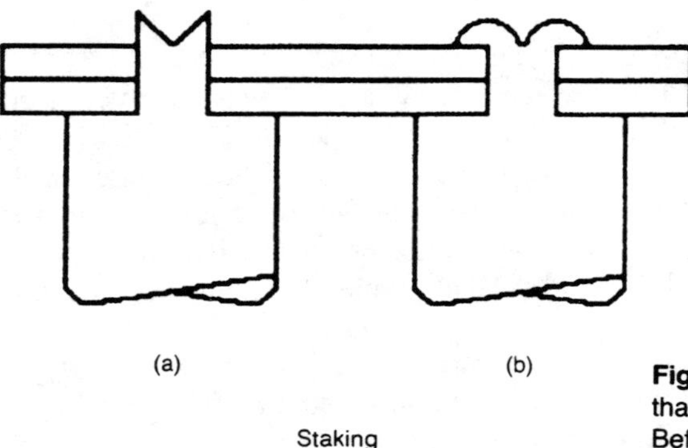

Figure 18-39 Staking similar to riveting in that it provides a permanent, headed fastening. Before (a) and after (b) views are shown.

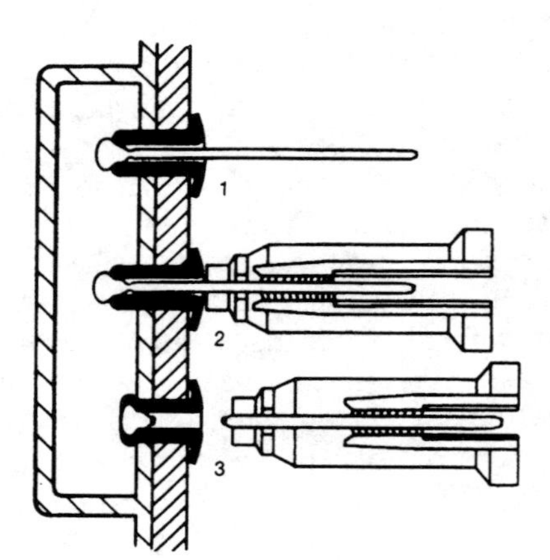

Figure 18-40 Installation diagram for blind rivets. (1) Insert rivet into workpiece holes. (2) Grip rivet mandrel with tool jaws. (3) Actuate tool to pull mandrel head rivet body, to expand and set the rivet. (Courtesy of Emhart Fastening Systems Group, POP Fasteners Division)

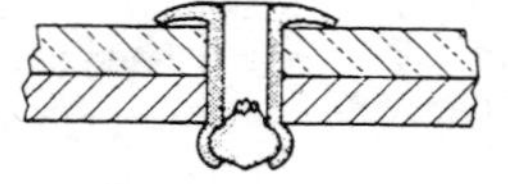

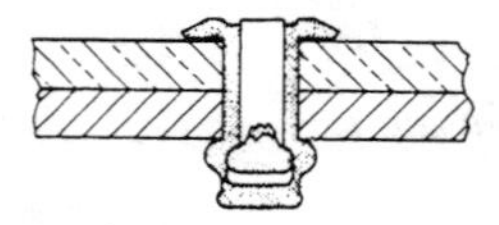

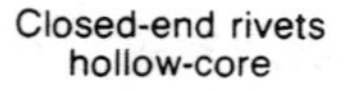

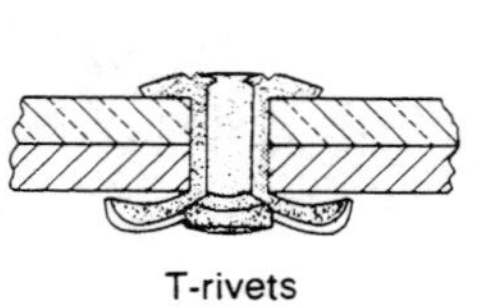

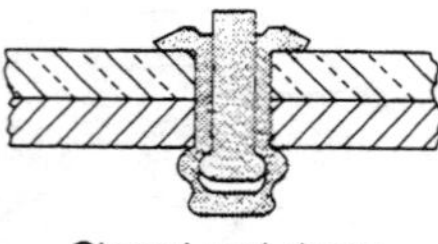

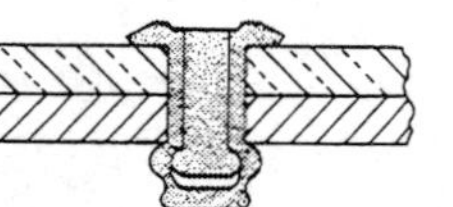

Figure 18-41 Rivet body configurations. Note that closed-end systems are self-sealing. An air hydraulic riveting tool, lightweight and well-balanced for operational ease, and sensitively designed. (Courtesy of Emhart Fastening Systems Group, POP Fasteners Division)

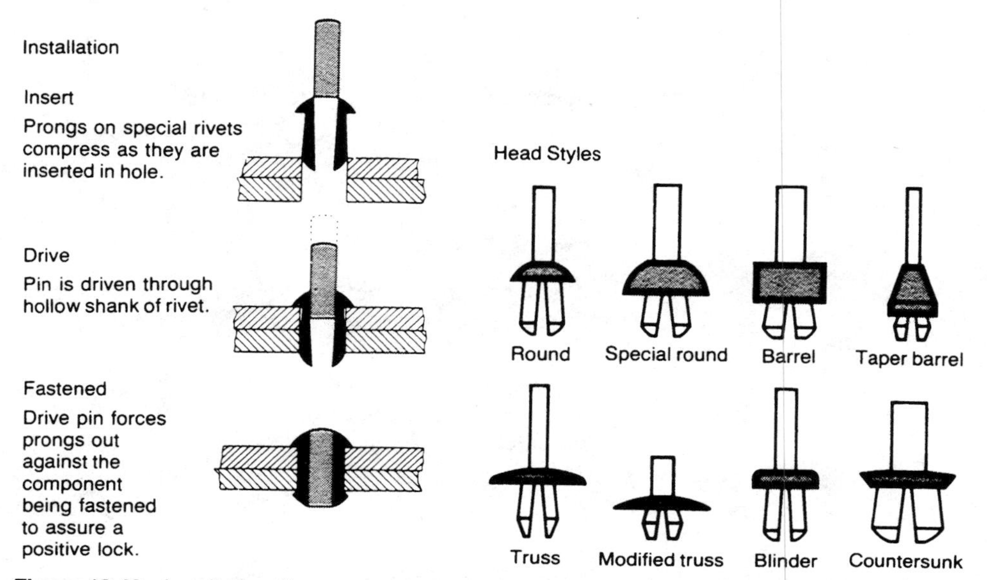

Figure 18-42 Installation diagram and head styles of plastic rivets. Note that their design lends itself to automatic assembly. (Courtesy of Illinois Tool Works)

many instances, their utility, speed, ease of installation, and range of styles make them desirable in numerous other structural applications. See Figure 18-41. Closed end models provide a vapor or liquid seal. Variations in head styles facilitate the employment of blind rivets with soft or brittle materials, and in applications requiring flush surfaces. These rivets can set with spring loaded, manually operated tools, or automatic air hydraulic equipment. An array of plastic rivets are in use for a wide range of assembly applications. See Figure 18-42.

OTHER SYSTEMS

Metal sheets are joined with a number of special *seams*, many of which have been described in other chapters in this book. Some such joints are sealed and strengthened with adhesives and solders. See Figure 18-43. A range of *tab* fastenings also is employed, some with bends and some with twists. See Figure 18-44.

Figure 18-43 Typical metal joint seam styles. These may be used with adhesives and solders.

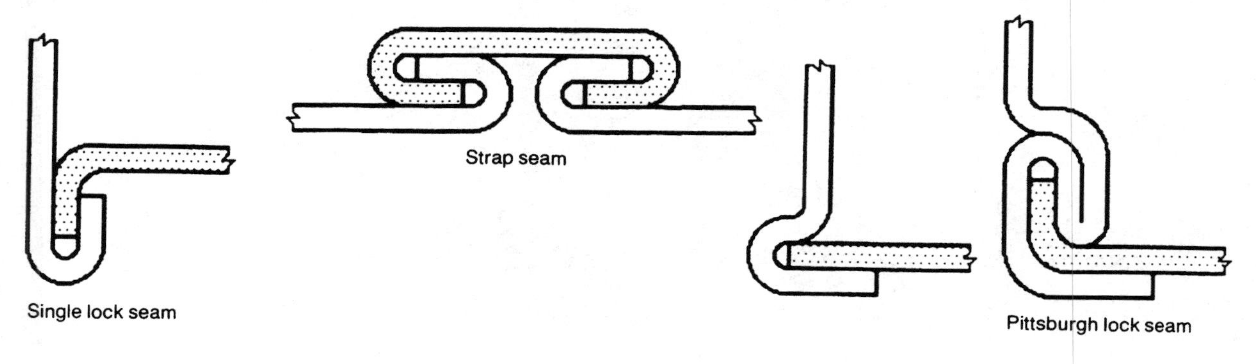

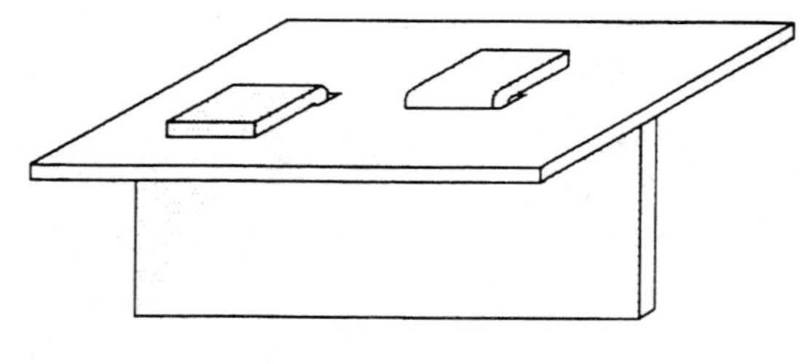

Figure 18-44 Folded tab fastenings can be used to join sheet metal parts. Their styles and configurations are numerous.

Special integral locking systems are also used in sheet metal part joinery. These are especially suited to automated assembly applications. See Figures 18-45 and 18-46.

Heat-shrink tubing is used to provide a skin-tight, secure, plastic protective jacket in assemblies that will be subjected to abrasion, heat, shock, corrosion, moisture, part movement, and other critical environments. A sleeve is slipped over the joint and, with a brief application of heat, molds itself snugly to the joint. See Figure 18-47. The system is ideal for use in such widely diverse fields as automotive, aerospace, medical, electronic, and chemical.

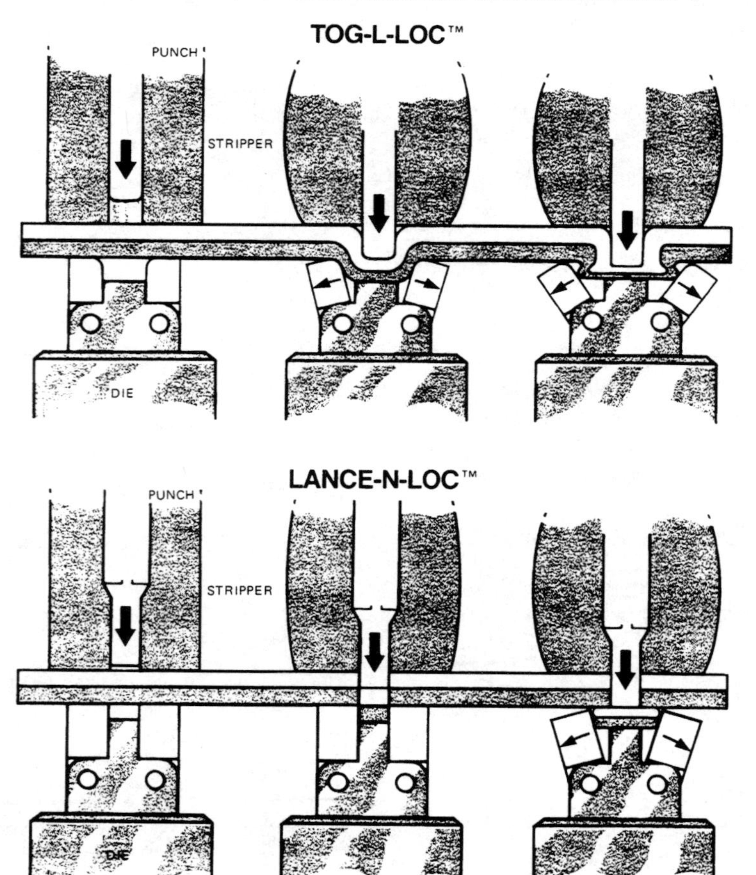

Figure 18-45 Two mechanical metal locking systems. Note that in the upper diagram, the joint is watertight because no metal is broken. (Courtesy of BTM Corporation)

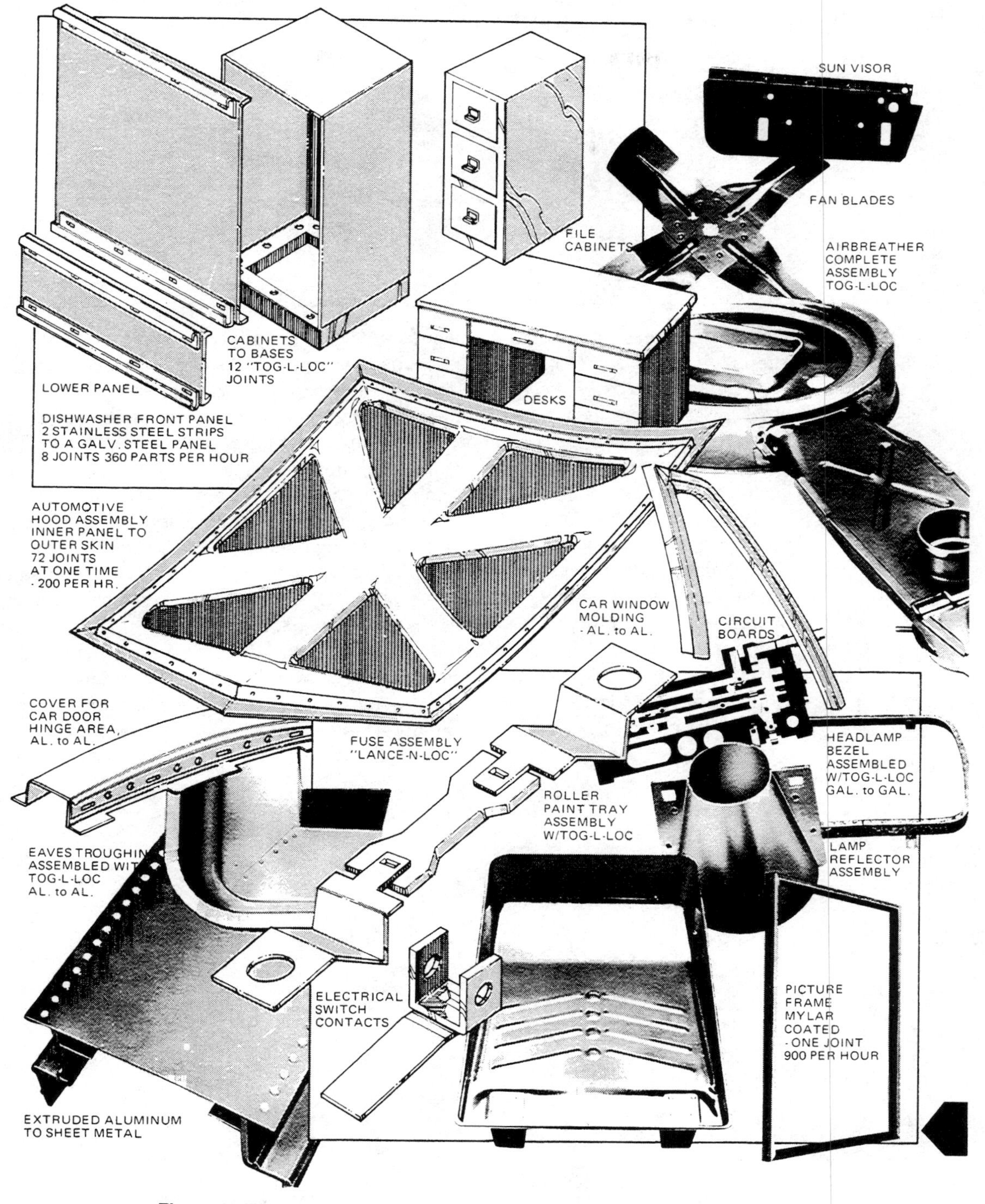

Figure 18-46 Locking system applications. (Courtesy of BTM Corporation)

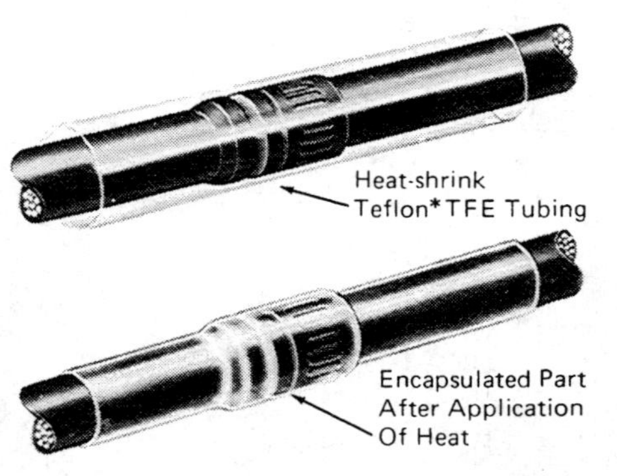

Figure 18-47 A typical electronic heat-shrink tubing application. (Courtesy of Zeus Industrial Products, Inc.)

AUTOMATIC ASSEMBLY SYSTEMS

One of the areas of concern in manufacturing is the study of ways and means of simplifying and automating the component assembly process. Assembly is time-consuming and costly, because it requires that parts be fastened securely and with great care to assure the integrity of the completed unit, and to maintain its appearance. Many problems can be solved in the product planning stage, with special care being given to the issue of design for assembly. This topic is treated further in Chapter 23. Some examples of assembly systems will be presented in this chapter.

Automatic screw driving and nut running units have been employed in industrial assemblies for many years. A typical nut fastening machine features handling systems which place components in the proper orientation (or it can be done manually), feed nuts to the assembly, and run the nuts onto the threaded fastener and torque them properly. The driving of screws is performed similarly.

Automatic riveting is also common. The common *hammer riveting* is noisy, and can subject the workpiece to undue stresses. *Radial riveting* has unique motions which produce a kneading action, with the rivet head formed from the center outward, rather than around the center. See Figure 18-48. There is less resultant pressure on

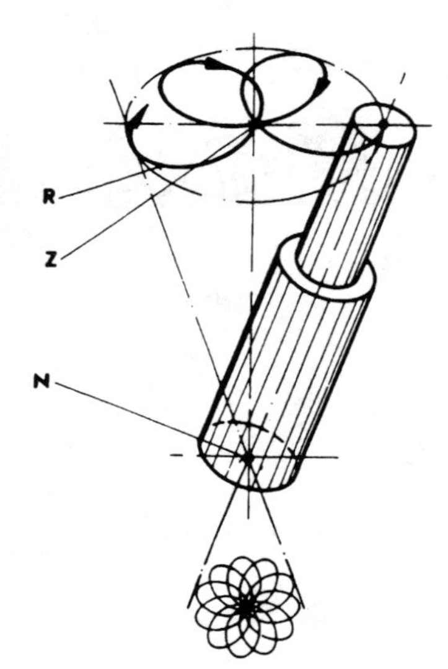

Figure 18-48 The action of the radial riveting process. (Courtesy of Bracker Corporation)

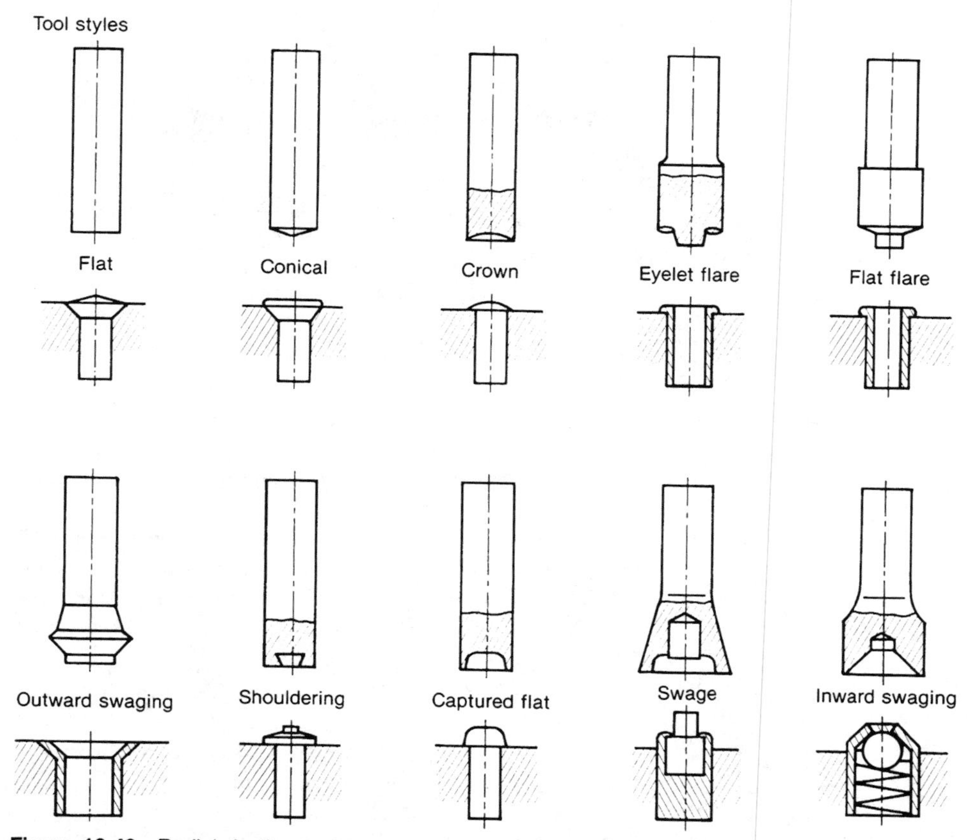

Figure 18-49 Radial riveting tool styles are available to produce a range of head shapes. (Courtesy of Bracker Corporation)

the riveting tool, with no rotation and therefore no friction against the rivet blank. Instead, the peen quietly but firmly produces rivet heads, or flares, flanges, and other heading forms as required. See Figure 18-49. A typical machine and product appears in Figure 18-50.

Advanced assembly systems are multifunctional, with capabilities for depositing solders and adhesives, inserting parts, positioning surface mounted devices, and joining components, using robots or other automated equipment. These high precision, programmable systems are used to assemble delicate parts for watches, cameras, and computers, and printers, and are easily integrated into work cells and automated manufacturing lines.

The example described here involves the use of robots in the mounting of electronic surface components, typically the jointed or flexible arm, and rectiliner or cartesian types. See Figure 18-51. Precision assembly robots can either be programmed off-line, saving valuable production time, or can be led through the assembly sequence so the robot "learns" the required steps and actions with a teach pendant. In addition, the robots may be linked to a CAD/CAM system for direct programming of the operating coordinates.

Figure 18-50 A radial riveting machine and assembled product example. (Courtesy of Bracker Corporation)

A typical surface mounting application of a conveyor rail PCB assembly line is shown in Figures 18-52 and 18-53. Two robots are used, one to deposit adhesive dots and solder paste and the other to attach the necessary electronic components.

At the left is a SSR-H253 robot, a 3-axis unit with a 250 mm reach, with the work envelope shown. The robot has two independent dispensers for adhesive and solder paste, and a fixed effector for moving the printed circuit boards (PCB) along the conveyor rail.

The robot on the right is a 4-axis SSR-H414 unit with a 410 mm reach, with a larger work envelope. This robot has two vacuum chucks for picking up and positioning components, and a fixed effector for moving the printed circuit boards.

Surrounding the robot are a vibratory bowl feeder for chip capacitors, a vibratory magazine feeder for integrated circuits, a linear pallet for miniature molded transistors, and a matrix pallet for flat-pack integrated circuits.

The use of such a large number of different types of component feeders is made possible by the robot's unusually large work envelope. Flexible arm robots can

Figure 18-51 Typical robots used in automatic assembly include the flexible arm (left) and the cartesian (right). (Courtesy of Accusembler Robotic Systems, A Division of Kanematsu-Gosho (USA) Inc.)

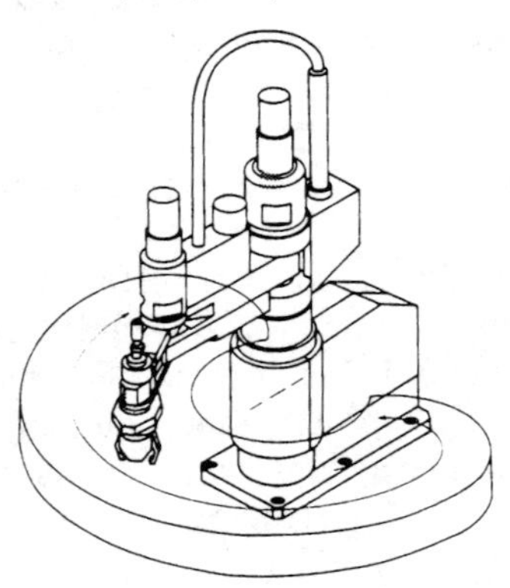

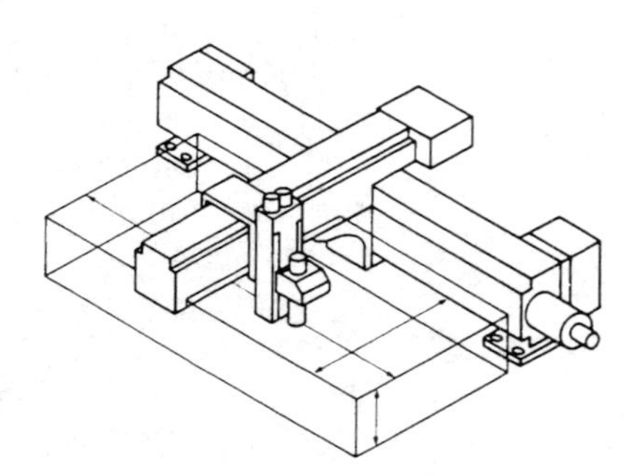

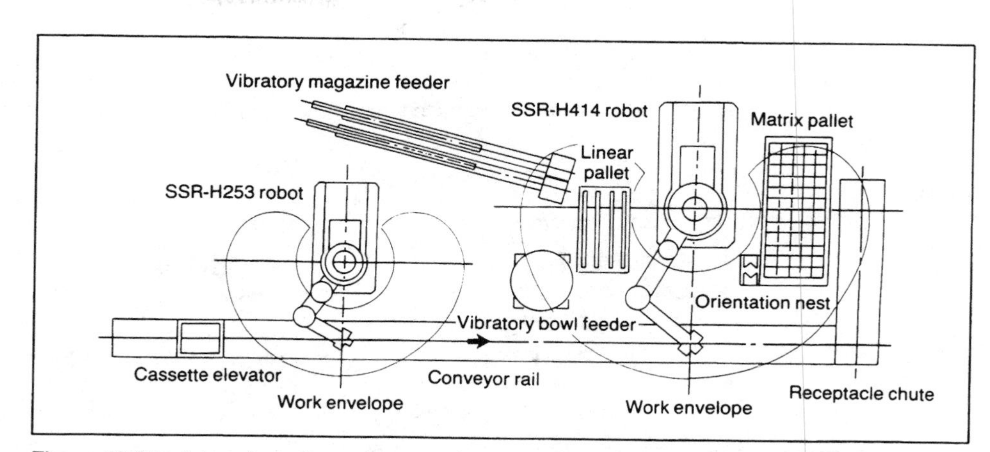

Figure 18-52 A top view of a surface mounting assembly line showing the work envelopes of the two robots, and the positioning of the parts feeders around the robot at the right. Such systems can be reprogrammed to adjust a specific assembly task. (Courtesy of Accusembler Robotic Systems, A Division of Kanematsu-Gosho (USA) Inc.)

pick up components from surrounding feeders, as compared to Cartesian robots, which are restricted to components inside the rectangular work envelope.

Both vacuum chucks may be rotated to position the components correctly before the parts are mounted. For example, the larger vacuum chuck picks flat-packs from the matrix pallet and then positions the integrated circuit (IC) in the orientation nest adjacent to the pallet. The correctly positioned IC is then placed on the printed circuit board. When component mounting is complete, the PCB is sent to either a receptacle chute or an infrared wave soldering oven.

As is the case with so many other processes, the joining or assembling (whether manual or automatic) of components is a matter of selecting among options. Two sheet metal components may be joined with adhesives, spot welds, solders, lock seams, sheet metal screws, bolts and nuts, or lock tabs. The choice of method is

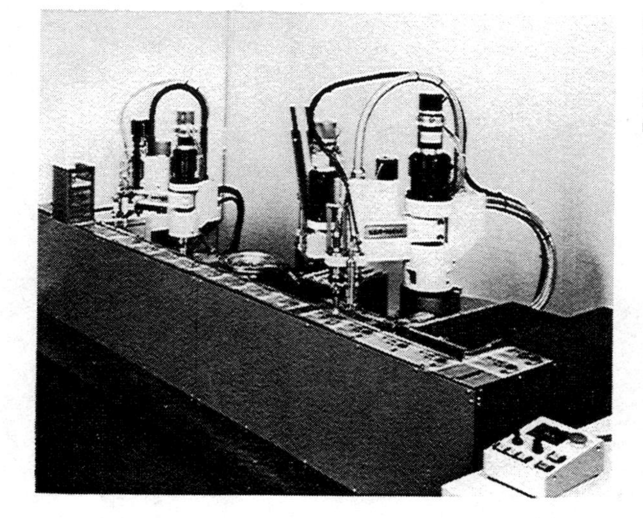

Figure 18-53 Camera view of the assembly system depicted in Figure 18-51. (Courtesy of Accusembler Robotic Systems, A Division of Kanematsu-Gosho (USA) Inc.)

dependent upon such factors as joint strength, permanence of joint, manual or automated assembly considerations, and the importance of finish. These elements must be weighed against cost, time, and numbers of product to make the final decision.

QUESTIONS

1. Sketch some typical nail styles and describe their uses.
2. Explain how nails derive their holding power, and what effects coatings and shank shapes have upon this.
3. Describe the difference between a bolt and a screw, and prepare sketches to illustrate each.
4. Describe the attributes of recessed screw heads over slotted types, and prepare sketches to illustrate each system.
5. Prepare sketches of three types of bolts.
6. Write the thread specifications for a metric and a customary threaded fastener.
7. Explain and illustrate the purpose and use of setscrews.
8. Sketch some special lock systems used to join sheet metal parts.
9. Explain the purpose of washers in conjunction with threaded fasteners.
10. Explain the function of rivets in assembled components.
11. Sketch several riveting types or systems.
12. Explain the difference between rivetting and staking.
13. Prepare a library research report on current automatic assembly systems.

19 ADHESIVE BONDING

The use of adhesion bonding in component assembly implies that two workpieces are permanently attached by a distinctly different substance. According to this definition, brazing, soldering, and adhesive joining are similar methods of assembling workpieces. *Solder* bonding denotes that a thin coat of a tin and other nontoxic metallics alloy joins two faying metal workpieces, generally sheets. *Braze* bonding employs a joint for metal parts comprised of a bronze coating or fillet. *Adhesive* bonding, as considered in its most refined sense, involves the use of a nonmetallic interfacial coat. Soldering and brazing techniques and applications are described in Chapter 20. The uses of common, and some not so common, adhesives is the subject of this chapter. See Figure 19-1. Adhesive assembly method are of the mass-increasing category.

The terms ''glue,'' ''cement,'' and ''adhesive'' are generally employed interchangeably in modern parlance, and therefore are loosely synonymous. The occasional exception is the strict use of the word ''cement'' as an inorganic mixture of lime, clay, and water used as construction material or as a mortar for brick and tile. Some writers suggest that *glues* are made from natural materials, *adhesives* are synthetic materials, and *cements* are rubber-based. However, in this book the term ''adhesive'' will be used to describe these assembly media.

While there is no precise theory of just how adhesives act to join two substrates (or workpieces, or adherends), it generally is concluded that the bond may be a result of a mechanical interlocking of surface irregularities; molecular forces such as covalent, electrostatic, or vanderWaals; a chemical reaction between the adhesive

Figure 19-1 This metal can is a product which frequently has adhesive bonds in combination with mechanical joints. The drawn can and lid are shown. (Courtesy of Reynolds Metals Company)

and the adherend; or a combination of these. Despite this theoretical void, adhesives do work satisfactorily in a range of applications.

There are many reasons for using adhesives, all of which attest to their increased utilization in manufacturing assemblies.

1. They provide for a uniform distribution of forces over an entire bonded area, thereby eliminating the stress concentrations caused by rivets, bolts, or spot welds.
2. They are effective in joining both similar and dissimilar materials, those which are thin and fragile, and a range of shapes and sizes.
3. They serve as environmental seals, heat insulators, and vibration dampers, and they prevent electrolytic corrosion.
4. They produce joints with smooth contours and no surface imperfection.
5. They cure at relatively low temperatures.
6. They are lightweight and inexpensive.

There are some disadvantages in their use, such as lower service temperatures than most metals, rigid process control and cleanliness, and the difficulty of joint inspection. In reality, adhesives are no more difficult or costly to use than most other fastening systems, and have a considerable number of attributes.

ADHESIVE CLASSIFICATIONS

The tremendously broad range of adhesive types can be made more understandable by applying certain descriptive classifications to them. They can, for example, be placed in natural and synthetic categories. *Natural* adhesives have organic origins such as animal and vegetable matter, and natural fiber gums. Bone, hide, blood, starch, and wood resins commonly are used. They are inexpensive, water soluble, easy to apply, but may have low bond strengths, and typically are nonstructural, with some exceptions. *Synthetics* apply to all adhesives other than the natural types. All structural adhesives are synthetic, and are generally chemical compounds.

Adhesives are also commonly categorized as either thermosetting or thermoplastic, depending upon their chemical classifications. *Thermosetting* adhesives are cured by chemical reactions at specified temperatures, and they cannot be heated and softened repeatedly after they have initially cured. They are available as either one-part systems, which must cure at elevated temperatures and have a short shelf life, or as two-part systems, the components of which require careful mixing before application. The dual component types cure slowly at room temperatures, and have longer shelf lives. Structural adhesives are thermosets.

Thermoplastic adhesives can be heated and softened repeatedly after setting, and are of the single component type. They set by cooling, as with hot melts, or by solvent or water evaporation, typified by wood glues.

The adhesives used in manufacturing may also be either structural or nonstructural. High strength *structural* adhesives are those which will join almost any material—metal, wood, plastic, or ceramic—over extended periods of time, and are fundamentally load-bearing. These adhesives will strengthen structures generally so that the bond becomes as strong as the materials being joined. See Figure 19-2. The list of typical applications is virtually endless, but the following are noteworthy:

1. Aircraft: wing and stabilizer panels, interior bulkheads, and fuel tanks.
2. Automotive: gas tanks, headlights, refrigeration coils, brake linings, wheels, body panels, and fuel filters.

Figure 19-2 Structural epoxy adhesives are used in a variety of joint applications. (a) Thread sealant for fire extinguisher pressure gauge threads. (b) Potting tank antenna. (c) Aerospace bearing. (d) Hydraulic pump port plate bond. (Courtesy of H. B. Fuller Company, St. Paul, MN)

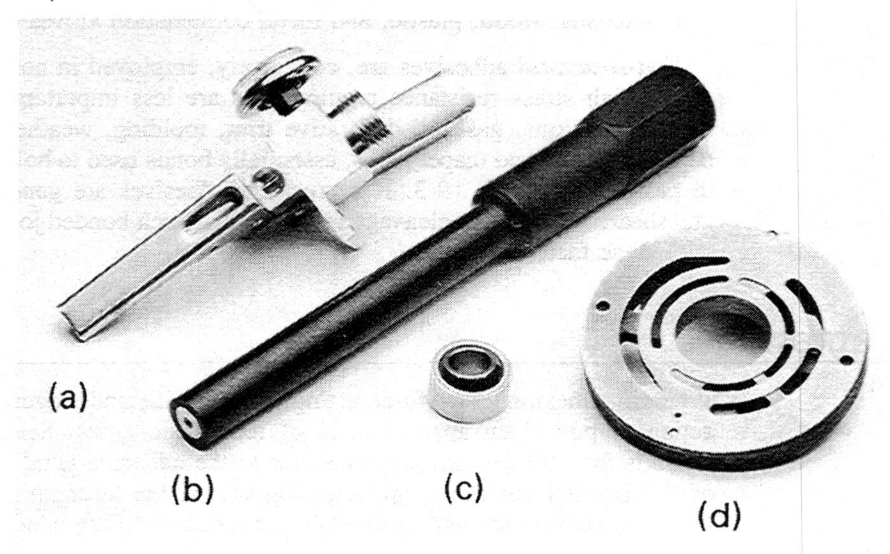

Figure 19-3 Typical nonstructural adhesive joints are exemplified by these filters. (Courtesy of H. B. Fuller Company, St. Paul, MN)

3. Electronics: speakers, gear and shaft assemblies in timing mechanisms, relays and controls, coils, and cabinets.
4. Hardware: water faucets, lawn sprinklers, pipes, and tools.
5. Building Construction: metal window frames, sandwich building panels, and roofing systems.
6. Packaging: shipping and food containers, and containers for scientific and electronic equipment.
7. Pneumatics and Hydraulics: motors, pump castings, control valves, and filter end caps.
8. Sports: golf clubs, bows, and telescopic rifle sights.
9. Utensils: wood, plastic, and metal combination knives and forks.

Non-structural adhesives are, conversely, employed in non-load bearing systems where high stress resistance requirements are less important. Typical applications include cushions, gaskets, decorative trim, molding, weather stripping, bottle and package labels, and diaper tapes, essentially bonds used to hold lightweight materials in place. See Figure 19-3. Nonstructural adhesives are generally strong in tensile and shear, and weak in cleavage and peel, and such bonded joints should be designed with these factors in mind.

JOINT STRESSES

Whereas adhesion is the force acting to hold adherends together, *stress* is the force acting to pull them apart, and is of four basic types. See Figure 19-4. *Tensile* stress is the pull exerted perpendicular to the adhesive bond line. Each square unit of the bonded surface contributes equally to the strength of the joint, with the result that tensile strength is directly proportional to the bond surface area.

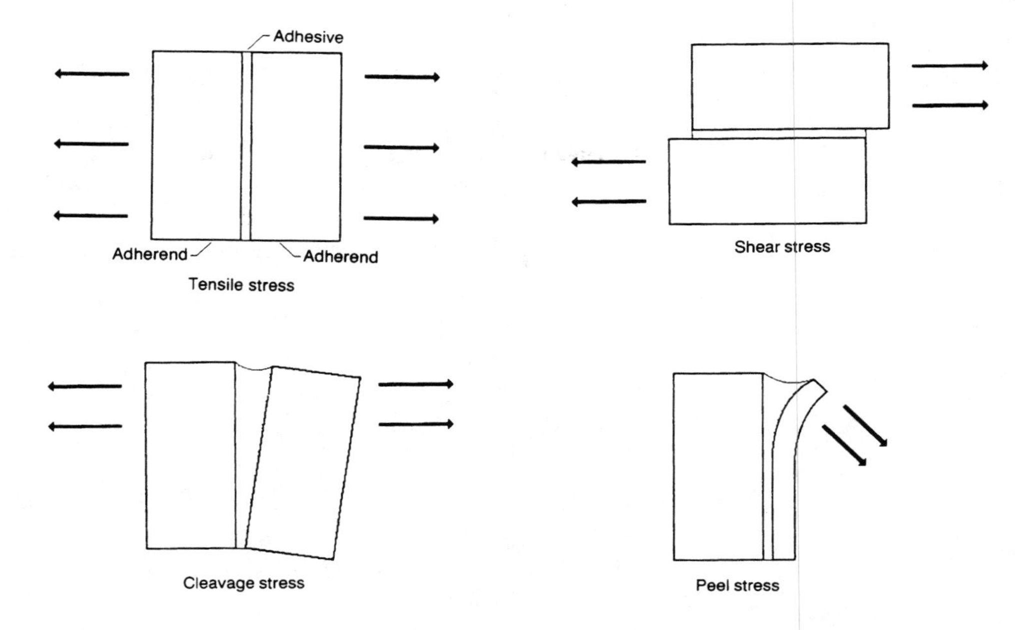

Figure 19-4 Stresses which act to separate an adhesive joint are tensile, shear, cleavage, and peel.

Shear stress is the pull exerted parallel to the bond line, so that the adherends are forced to slide over and away from each other. Here, the stress is not uniformly distributed across the joint, and consequently the bond strength does not vary directly with the bond area. *Cleavage* stress is the perpendicular pull concentrated at one edge of the joint, and exerts a prying force on the bond. The other edge of the joint is theoretically under zero stress. This type of bond failure occurs progressively from the cleaved end, and generally is a result of workpiece deformation. *Peel* stress requires that one or both adherends be flexible. Peel stress is much like cleavage stress, in that the stress is concentrated at the failing edge of the bonded joint. This action is similar to the peeling of an orange.

Temperature is another type of stress which can affect the strength of a bonded joint by melting the adhesive and causing joint failure. *Chemical* reactions can have the same detrimental results. *Compressive* stresses are less injurious, but constant squeeze pressures can force adhesive from a joint and thereby weaken it.

Adhesive performance is usually better when the primary stress is tensile or shear. However, in most engineering applications a combination of stresses is involved. Best performance occurs when the entire bond surface area carries the bulk of the stress. This factor indicates the importance of joint design.

JOINT DESIGN

The aim of joint design is to provide maximum strength for a specific bond area in a specific application. Hence, the geometry of a given joint must take into account the bond width, the overlap length, the shapes and thicknesses of the workpieces (for example, sheet, or bar, or tube), workpiece flatners and trueness, and the adhesive thickness. For example, the strength of a lap joint stressed in shear has a direct relationship to the width of the overlap, with the result that a doubled bond width will double the strength of the joint. This would indicate that, all other factors being equal, the designer would specify joint width of 50 mm (2 inches) and a 25 mm (1 inch) lap, rather than a 25 mm width and 50 mm lap, because the greater width would result in a stronger joint.

Some other important general joint design considerations are:

1. Orient the bondline so that it is stressed in the strongest direction, for example, tensile or shear rather than cleavage or peel.
2. Use the largest possible bond surface area.
3. Avoid parts with complex geometries which may make bonding difficult.
4. Keep the number of separate bonding operations to a minimum.
5. Consider stress factors in all joint designs.
6. Design joints specifically for use with adhesives in mind. Combined joints using mechanical systems also should be considered.
7. Use film adhesives or gap-filling adhesives as dictated by the surface conditions of the adherends. See Figure 19-5.
8. Good joint design is largely a matter of common sense and experience.

Examples of typical joints, with their limitations and attributes, are described here. The *butt joint* is impractical because it offers a minimum surface bond area and poor cleavage resistance. See Figure 19-6. It can be improved by adding a

Figure 19-5 Thin film adhesives work best on "perfectly" flat and smooth workpieces, or in assemblies where one workpiece is thin enough to assume the contour of the heavier member, as in (a) and (b). Gap or void filling adhesives must be used where workpiece surfaces are not true, as shown in (c).

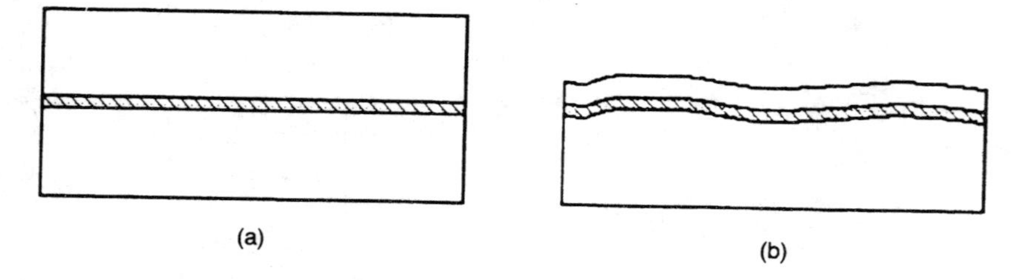

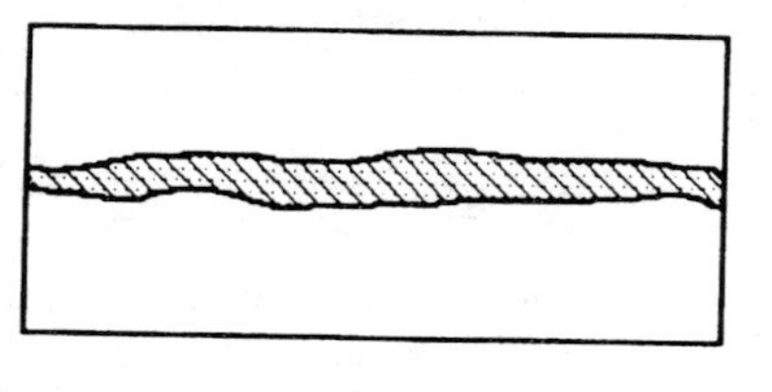

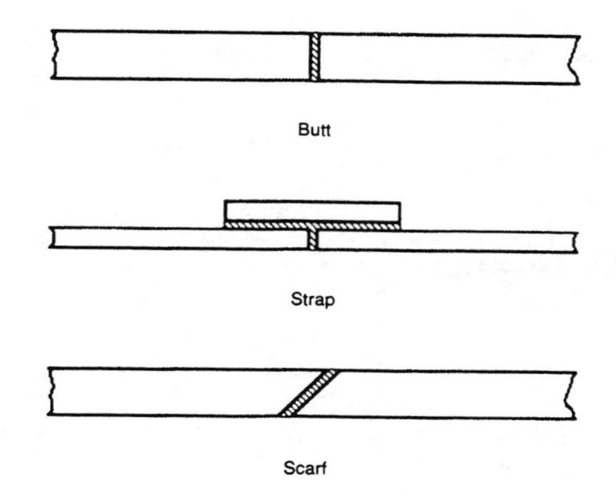

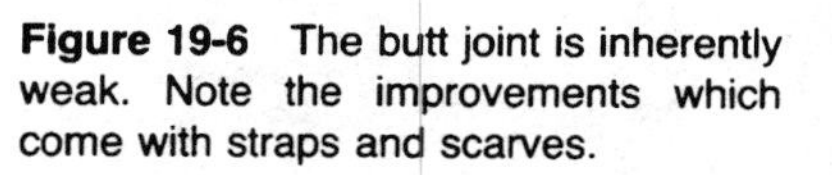

Figure 19-6 The butt joint is inherently weak. Note the improvements which come with straps and scarves.

strap, or by cutting a *scarf* to increase the bonding area. There are additional means for the strength enhancement of butt joints. All those shown in Figure 19-7 are of the *tongue and groove* variety, a feature which both increases surface bond area and resists bending. These recesses are self-aligning, a factor which facilitates assembly, they serve as adhesive reservoirs for thick mastic adhesives, and they tend to control the amount of joint adhesive applied. Similar tongue and groove assemblies are commonly used in woodwork.

Lap joints, also called shear joints, are an improvement over the butt variations and are characterized by laying one part of the surface over the other to extend the bond area. See Figure 19-8. *Simple* laps are offset, and consequently the shear forces are not in line. This factor can result in cleavage or peel stresses under load. The *tapered single* lap joint is more effective than the simple lap, because it permits the distortion of the joint edge under stress. However, it is more costly to prepare this joint and may therefore be impractical. The *joggle* lap offers a more

Figure 19-7 Tongue and groove joints are improved butt joints.

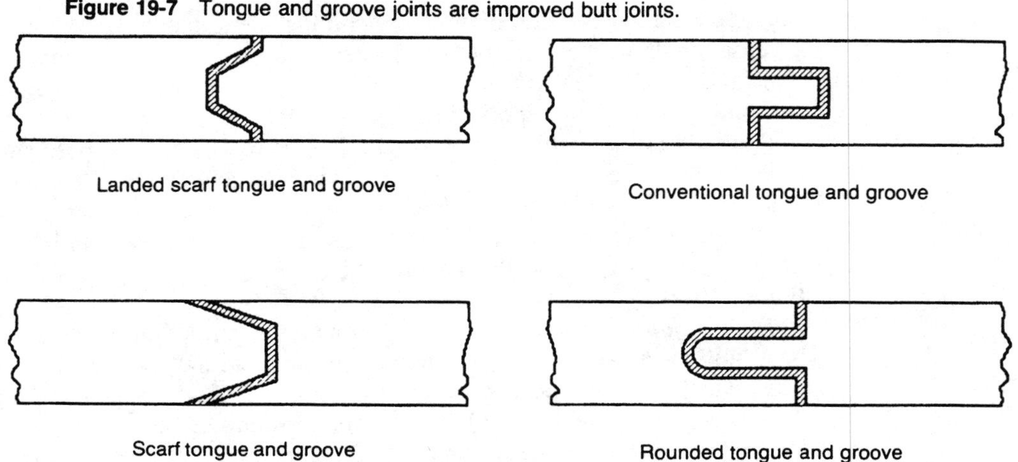

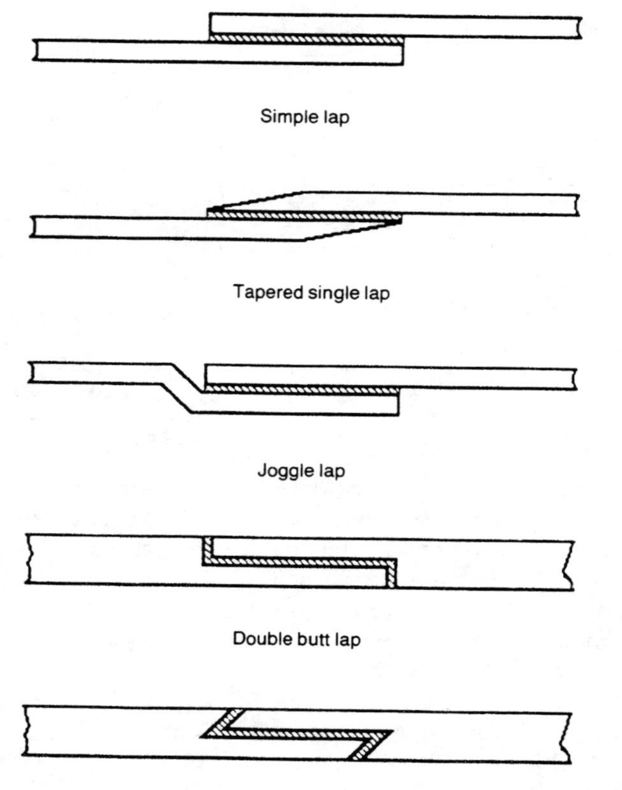

Figure 19-8 Some typical lap joints generally used in sheet bonding.

uniform stress distribution than either the simple or tapered laps, by placing the entire load-bearing surface in the same plane as the adherends. The joint can be prepared by simple sheet metal bending operations, and the curing pressure is easily applied. The *double butt* lap joint provides the same load-bearing distributions as the joggle lap, but requires complex machining operations and therefore is not practical for use on thin sheet metal workpieces. *Double scarf* lap joints have a greater resistance to bending forces than do the double butt joints, but again require costly machining and are not applicable to thin sheets.

Angle joints are employed to assemble adherends which lie perpendicular to each other, and can be used on workpieces of various gauge sizes. They provide adequate tensile and shear strength, but some are subject to cleavage and peel stresses according to the workpiece thickness. Thinner pieces will flex more, and in use may fail. Three basic systems and their approaches to cleavage and peel reduction appear in Figure 19-9.

Corner joints provide a means of constructing rectilinear containers, housings, boat hulls, vehicle bodies, building components, aircraft bulkheads, and other structures which do not lend themselves to metal and plastic sheet forming techniques. This is accomplished by employing corner laps or attachment devices, which permit joining and sealing in one operation. See Figure 19-10. Such joints increase structural

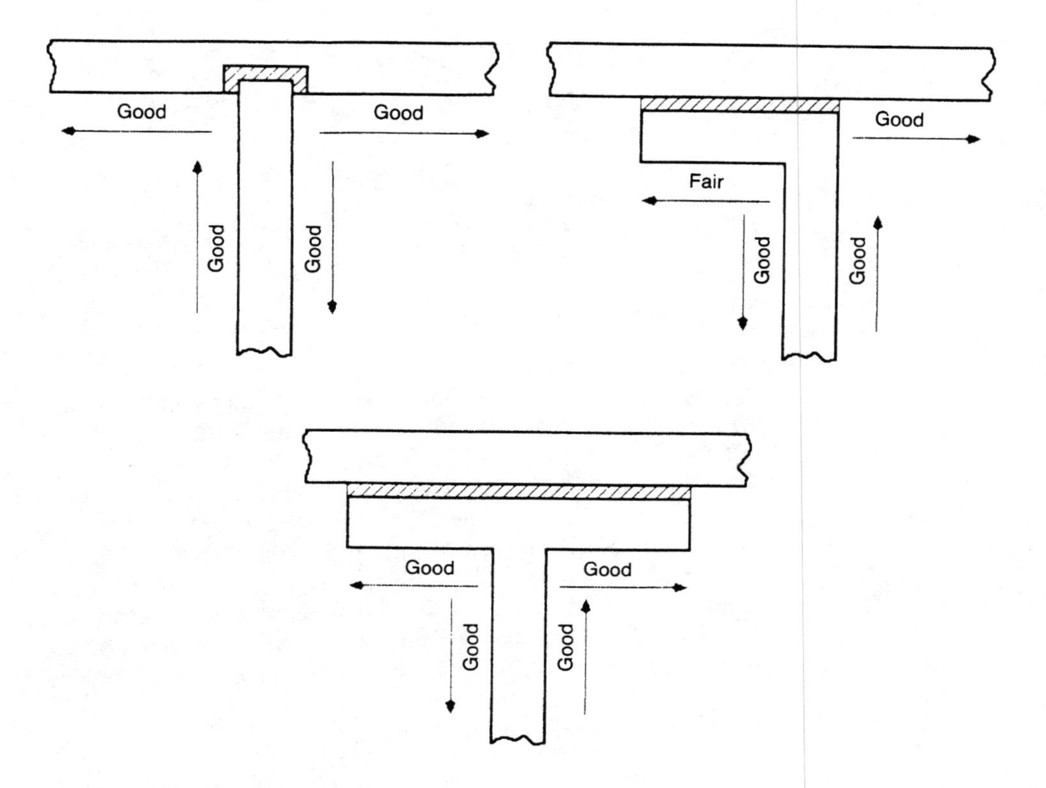

Figure 19-9 Angle joints can be used with sheets and structural shapes.

Figure 19-10 These corner joints are used in rigid assemblies, such as storm doors and windows.

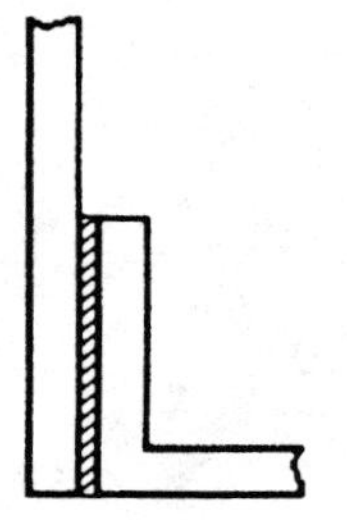

Right
angle butt

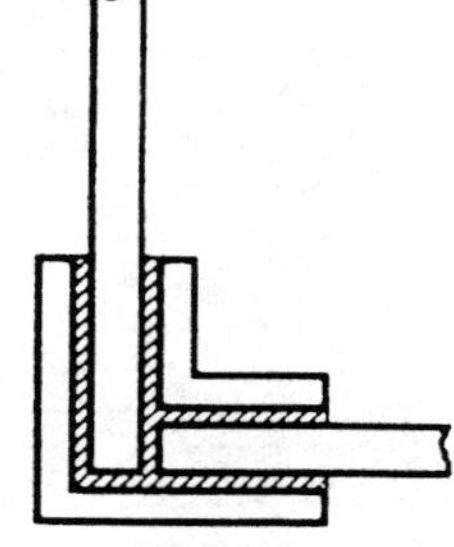

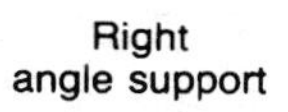

Right
angle support

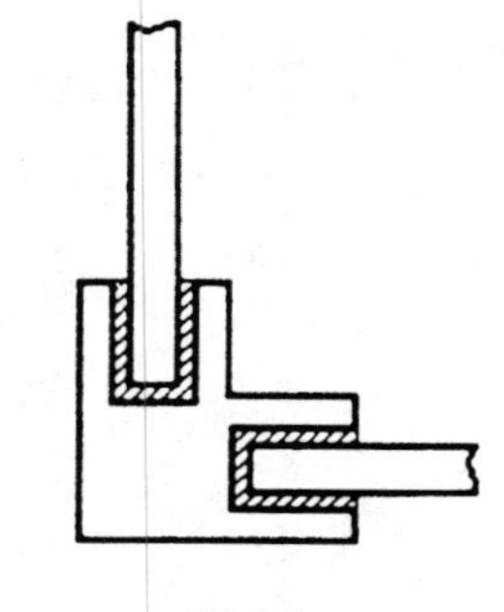

Slip joint

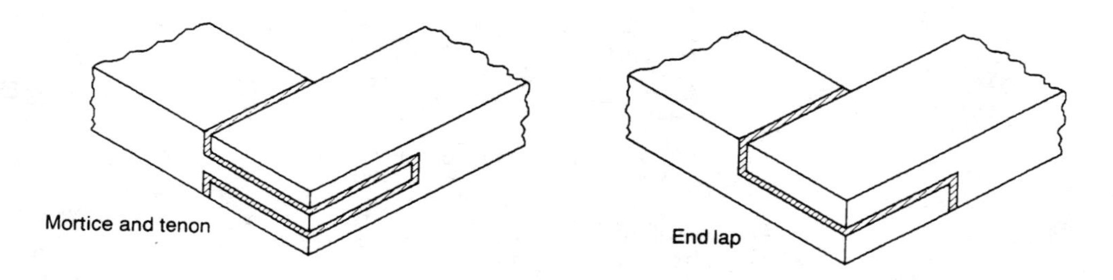

Figure 19-11 Special corner joints such as these are applicable to metal and wood component construction.

rigidity, and generally require the use of void or gap-filling adhesives. These corner systems also facilitate the convenient joining of dissimilar materials.

Other corner joints such as end lap, mortise and tenon, and spline miter, similar again to those employed in woodwork, are used with rigid members to construct storm doors and frames. See Figure 19-11.

Stiffener joints are used to minimize whipping and deflection in thin sheet metal structures. See Figure 19-12. Various styles of stiffening sections can be employed, depending upon the degree of rigidity required in the structure.

Sandwich structures are commonly used in the manufacture of wooden interior or exterior doors, wall panels, bulkheads, and other lightweight, rigid assemblies.

Figure 19-12 Stiffener joints.

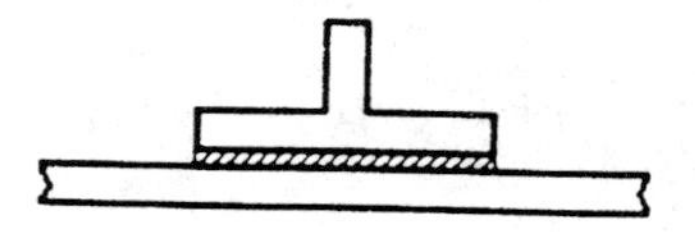

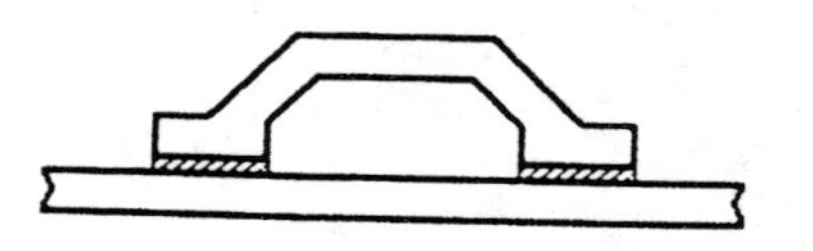

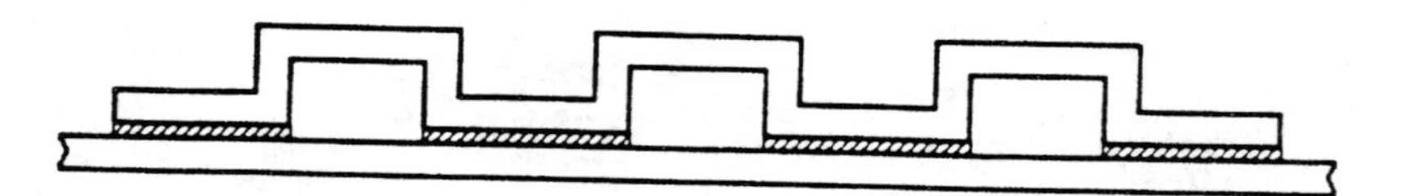

Figure 19-13 Sandwich structures can be made of hollow core or honeycomb members covered with sheet materials such as aluminum and marble. (Courtesy of H. B. Fuller Company, St. Paul, MN)

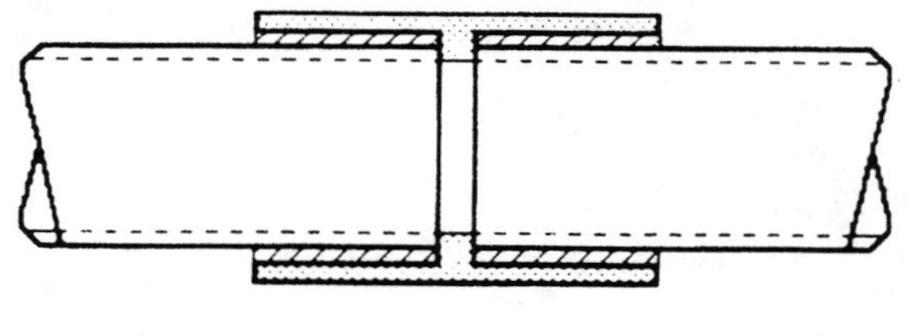

Figure 19-14 Round and rectangular tubes and shafts can be joined with special fittings and adhesives.

The core materials can be wood cross members, plastic honeycomb, or wood or metal slats, with sheet materials bonded to each surface to create the familiar "sandwich." A variety of edge treatments are employed to enhance and stiffen the structure. See Figure 19-13.

Tubing joints generally are created by using tee, elbow, and splice fittings along with adhesives to create assemblies and to secure shafts. See Figure 19-14. Any shape of tube or bar member can be so joined. Special epoxy adhesives are available for use in hot and cold potable water pressure plumbing systems, as are the common solders.

TYPES OF ADHESIVES

A considerable number of adhesives are available to provide bonding systems for the range of required industrial applications. The most common of these are described in Table 19-1. This table serves as a reference to the adhesive types, their characteristics, and their uses.

TABLE 19-1
INDUSTRIAL ADHESIVES

Type	Characteristics	Uses
1. acrylic	thermoplastic, quick-setting, tough bond at room temperature; two component; good solvent, chemical and impact resistance; short work life, odorous, ventilation required	solar panels, ceramic magnets, fiberglass and steel sandwich bonds, tennis racquets, vehicle body components, metal parts, plastics
2. anaerobic	thermoset, easy to use, slow curing, bonds at room temperature; curing occurs in absence of air, will not cure where air contacts adherends; easy to use; one component; good cohesive, low adhesive strength; not good on permeable surfaces, can flow into areas and cause unexpected bonds	close-fitting machine parts, such as shafts and pulleys, bolts, and nuts, studs, keys, bearings, bushings, core plugs, and pins; (not for use on plated metals; glass, ceramic, and plastic assemblies)
3. cyanoacrylate	thermoplastic, quick-setting, tough bond at room temperature; one component; easy to use; colorless; no clamps or fixtures needed because of rapid cure; low shock, peel strength, moisture, and temperature resistance	electronic components, gaskets, musical instruments, locking of nuts and bolts, lipstick cases, most plastics, metals, rubber, ceramics, hardwoods, and automotive body trim
4. epoxy	thermoset, one or two component, tough bond; strongest of the engineering adhesives; very versatile, wide variety of formulations to provide different pot life, cure and temperature rates, and bond rigidity; more expensive than plastisols; familiar as two-component household adhesives; low creep and shrinkage; high tensile and low peel strengths; resists moisture and high temperatures; resists most solvents and chemicals; difficult to use, requiring measuring and mixing equipment; one component epoxies cure at elevated temperatures	metal, ceramic, and rigid plastic parts, rubber
5. formaldehyde —urea —melamine —phenol —resorcinol	thermosetting, strong wood to wood bonds, *urea* is inexpensive, versatile, available in powder or liquid forms, must be combined with a catalyst to produce hard, rigid, water-resistant bond at room temperature; *melamine* is more expensive, cures with heat, bond is waterproof; *phenol* cures with heat, bond is waterproof; *resorcinol* forms waterproof bonds at room temperature; *urea-melamine* and *phenol-resorcinol* combinations can be used with radio frequency heating systems for fast, sure bonds	all types of wood joint and plywood bonding
6. hot melt	thermoplastic, quick-setting, rigid or flexible bonds; joins permeable and impermeable adherends; easy to apply; special equipment needed; familiar hot melt gun apparatus for household use; low heat, creep, and solvent resistance; good moisture resistance; solid at room temperature, becomes fluid when heated, bonds adherends upon cooling; no solvents; gap filling; brittle at low temperatures; based	bonds most materials, packaging, bookbinding, textiles, metal can joints, furniture, footwear, carpets, oil and air filters

TABLE 19–1 (cont'd.)

Type	Characteristics	Uses
	on ethylene vinyl acetates, polyolefins, polyamides, and polyesters	
7. phenolic	thermoset, high-pressure, oven-cure, strong bond; high tensile and low impact strength; high shrinkage; brittle; easy to use; long shelf life; cures by solvent evaporation; one component; resistant to elevated temperatures	acoustical padding, cushioning, brake lining, clutch pads, oil and air filter webs, abrasive grain bonding, foundry sand bonding, core bonding, honeycomb structures
8. pressure sensitive	thermoplastic, variable strength bonds, applied by low pressure; primer anchors adhesive to roll tape backing material, a release agent coated on back of webb permits unwinding of roll; peel, shear, and tack strengths vary according to demands of product, for example, permanent or removable; based on polyacrylate esters and various natural and synthetic rubbers	surgical, masking, binding, and electrical tapes, packaging and mailing labels, metal foils, clear acetate film, protective coverings for acrylic and high quality metal sheets, shelf and wall coverings, wood tape, duct tape
9. silicone	thermoset, slow-curing, flexible, bonds at room temperature; high impact and peel strength; rubber-like; one and two component; easy to apply; excellent sealer; functional at high and low temperatures; cures when exposed to atmospheric moisture; long pot life when covered; good chemical and environmental resistance; expands when immersed in oil	gaskets, sealants
10. urethane	thermoset, flexible, bond at room or oven cure temperatures; water resistant; difficult to mix and use; good gap-filling qualities, work areas should be vented; good at low temperatures	fiberglass body parts, rubber, fabrics, and other materials to metal parts
11. water base —animal —vegetable —polyvinyl acetate —natural and synthetic rubbers	thermoplastic, water-soluble and water-dispersed bonds; water based systems frequently used where at least one adherend is porous, are inexpensive, nontoxic, nonflammable, and set by the evacuation of water from the joint area; *animal* bone and hide glues are protein by nature, water soluble, and form strong bonds with low moisture resistance, common as "brown" liquid glues; others of animal origin include casein (milk) and fish glues; *vegetable* glues are water soluble, based on starches, soy beans, and resins, have poor moisture resistance, used as inexpensive paper bonds; "rubber" cement and other cellulosic cements have vegetable origins, but are not water soluble; *polyvinyl acetate* is a water emulsion, dries quickly, tough bond, low resistance to heat and moisture, common as "white" liquid glues; *rubbers* are used as contact cements and pressure resistive tapes, have good impact and moisture resistance, water dispersion, limited load bearing qualities	wood, fabric, paper, and leather workpieces, dry seal envelopes, carpet backing, tire cord bonding, and packaging

METHODS OF APPLYING ADHESIVES

Adhesive application techniques must be selected with the same care and consideration as the adhesive itself. The process selected depends upon viscosity or form of the adhesive, for example, liquid or hot melt, the size and shape of the adherends, the specific areas of application, the number of assemblies required, and the desired production rate. It should be noted that many of these manual or automatic adhesive application methods are identical to those used to apply finishes to workpieces. See Chapter 21.

Brushing is a common method for small and specific area application of liquid adhesives. It is an inexpensive and simple technique, though the resultant adhesive cost may be uneven. It is also slow and not suited to high volume production. For higher production rates, a *flow brush* is recommended. A flow brush is a hollow tool where the adhesive is pressure-fed for manual, continuous application of broad coats.

Flowing refers to the application of narrow ribbons, such as is necessary for the bonding of weatherstripping and insulation. High speed production requires a continuous flow gun, fed by a remotely operated pump which maintains a constant adhesive pressure flow. For intermittent or slow volume application, a manual caulking gun is employed. Flowing lends itself well to automated systems (see Figure 19-15), and can be used with liquid and paste adhesives. Hot melts also can be so applied.

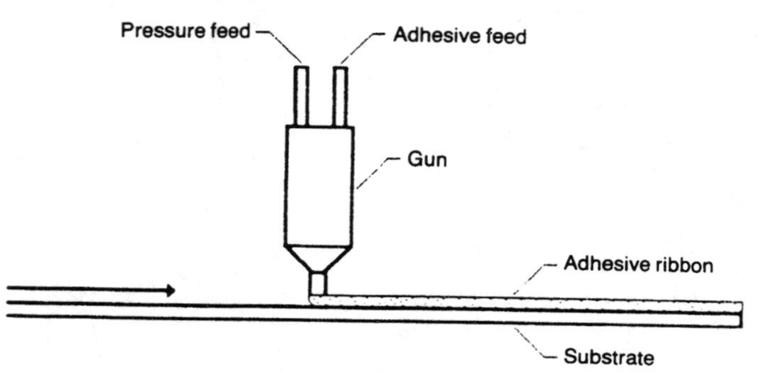

Figure 19-15 This robot is being used to flow an adhesive ribbon on the floor shell of an automobile to bond components. A schematic of the process also is shown. (Courtesy of Cincinnati Milacron)

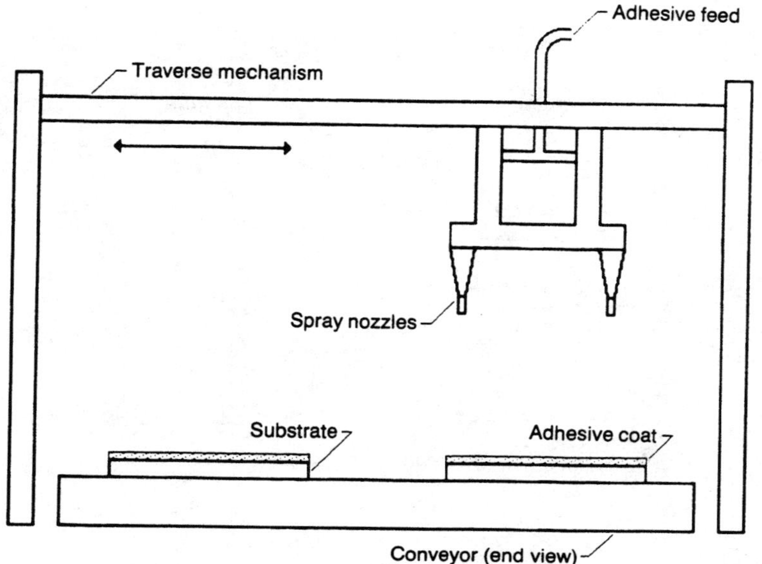

Figure 19-16 Adhesive spraying is done with handguns, or automatic equipment, as shown here.

Figure 19-17 Roll coating.

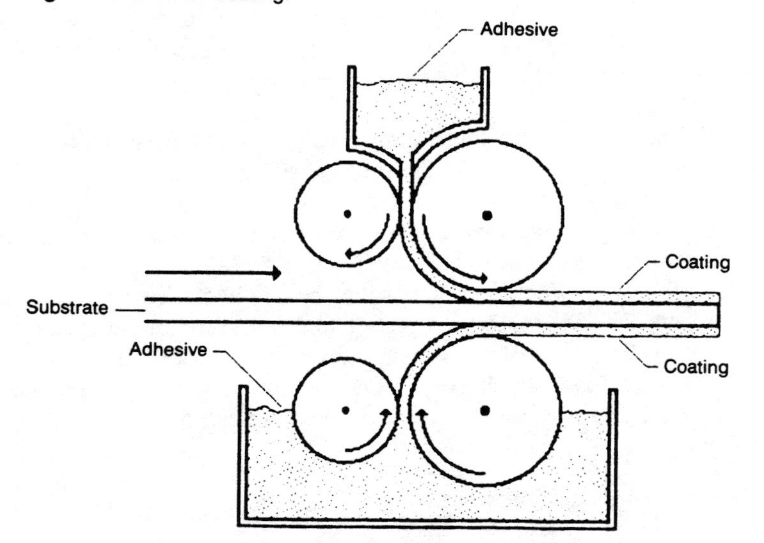

Spraying is one of the fastest and most economical methods of applying adhesives to large areas. Sprayable versions of all basic adhesive classes are available. This method is faster than flowing and provides a more uniform coat thickness. Manual air and airless spray systems are used, as well as automatic equipment. See Figure 19-16.

Roll coating is based upon the transfer of an adhesive from a storage container by a pick-up roller partially immersed in it, to an adherend web in contact with the pick-up roller. See Figure 19-17. Specially formulated adhesives are necessary

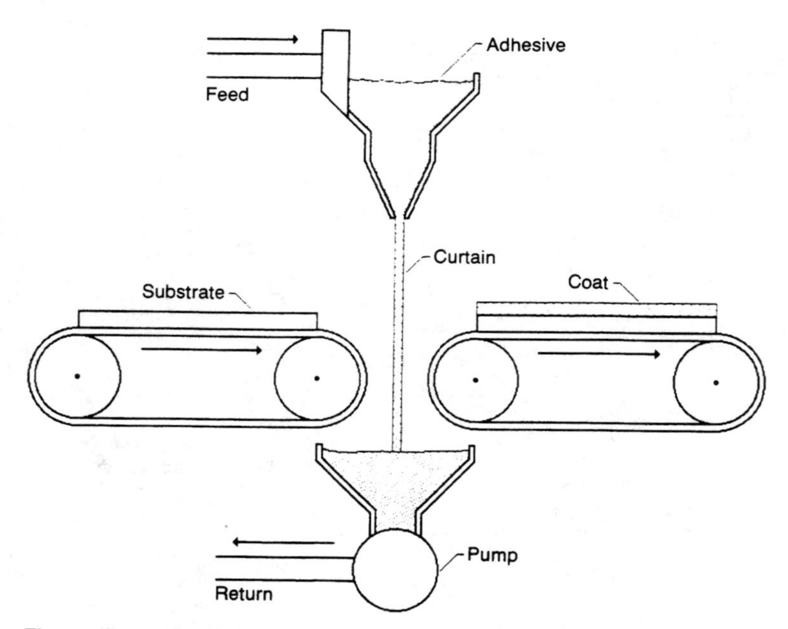

Figure 19-18 Curtain coating.

for this transfer method, which can be used to coat one or both surfaces of a substrate. This is a high volume, high speed application method for broad, flat areas. Ordinary paint rollers can be used, some pressure-fed, for low volume production jobs.

Curtain coating involves the application of an adhesive film to a workpiece travelling through a "curtain" of adhesive falling from a reservoir. See Figure 19-18. This method is faster than roll coating, but is used to coat one surface only. As with roll coating, special adhesive formulations must be used, and curtain coating is generally used on flat surfaces.

Knife-coating techniques involve the application of an excess of adhesive to a substrate. The substrate then passes under a knife spreader-scraper, which removes all material except that necessary for the required coat thickness. See Figure 19-19. This mechanically driven unit is used to apply high-viscosity adhesives at volume production rates. Similarly, mastic adhesives can be manually applied with a trowel.

Figure 19-19 Knife coating.

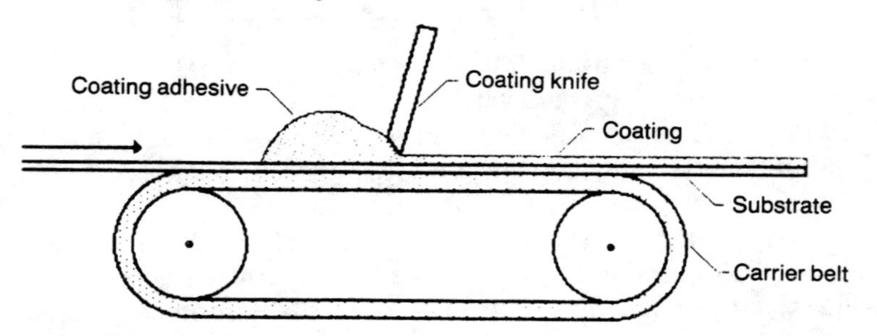

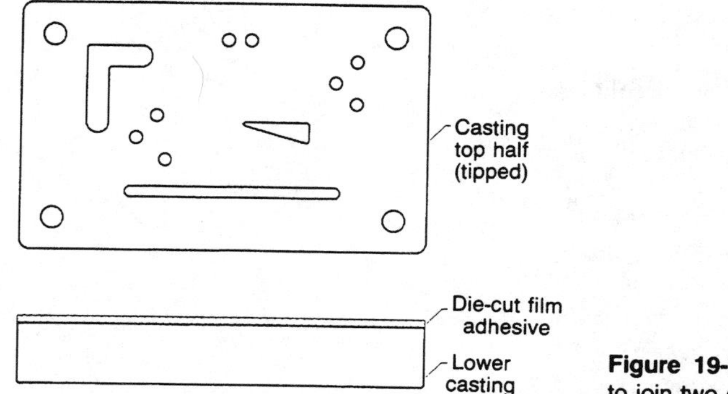

Figure 19-20 Adhesive film is being used to join two cast mating manifold sections.

Adhesive films are applied as dry sheets, can be cut to size (and therefore involve little waste), and are easy to use. Care must be exercised in their application to avoid wrinkles and voids. See Figure 19-20.

Powder adhesives can be sifted into preheated substrates pressed together as they melt, and appropriately cured. Combinations of lock joints and adhesives are in common use in vehicle body construction. See Figure 19-21. Such techniques assure the proper component orientation as the adhesive bonding operation proceeds, and add considerable strength to the joint. A variety of adhesives and lock systems can be used.

Figure 19-21 Vehicle body panel bonding system employing a sheet metal lock joint and an adhesive. A ribbon of adhesive is flowed into the outer panel joint (a). The inner panel is pressed into place (b). The outer panel joint is partially bent to secure the pieces (c). The bending process is completed (d). The assembly is placed under heater coils to cure the bond (e).

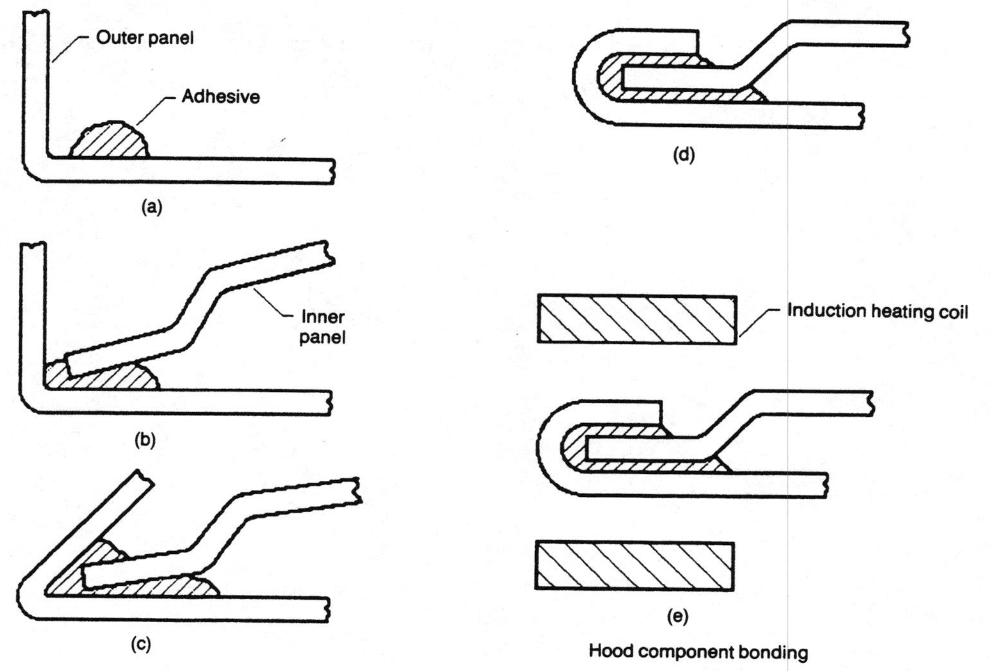

QUESTIONS

1. Define the following terms:
   a. adhesive
   b. adherend
   c. structural
   d. nonstructural
   e. thermosetting
   f. thermoplastic
2. Prepare sketches of tensile, shear, cleavage, and peel stress diagrams, and explain their meanings.
3. State some of the important advantages and disadvantages of adhesives.
4. Prepare sketches of the following joints and their applications or characteristics:
   a. butt
   b. lap
   c. angle
   d. corner
5. Consult Table 19-1. What adhesive types could be used for the following applications?
   a. wood furniture joints
   b. tennis racquets
   c. packaging and bookbinding
   d. foundry sand cores
   e. gaskets
   f. fiberglass body parts
6. List three methods of applying adhesives to substrates, and describe their important advantages and uses.
7. Prepare a library research report on an example of automation in industrial adhesive applications.
8. Prepare a library research report on current developments in industrial adhesives.

20 WELDING AND RELATED PROCESSES

Welding is a mass-increasing, materials joining method used to make permanent, continuous metallic bonds between workpieces. While materials other than metals can be welded, this chapter deals primarily with metals. Other material welding techniques are covered elsewhere in this book. The official American Welding Society (AWS) definition is that a weld is "a localized coalescence of metals or nonmetals produced either by heating the materials to the welding temperature, by the application of pressure alone, and with or without the use of filler metal." The term coalescence implies the growing together into one body of the materials being welded. Welding is therefore unique as a metal joining process, in that the two or more workpieces being joined become one homogeneous article, strong and inseparable. As such, welding is a primary method for joining metallic parts. See Figure 20-1.

CLASSES OF WELDING

As shown in Figure 20-2, there are three basic welding classes. *Fusion welding* implies the actual melting of the faying surfaces of a joint, causing them to flow together and harden into a solid piece. (The faying surfaces of a joint are the areas where the metals will be welded together.) This metallic bonding of the workpieces is made possible because in the liquid state, the aggregate of the metallic atoms can intermingle and remain close enough to form a solid upon cooling. A term commonly used to describe this is *cohesion*.

Figure 20-1 Welding is an important way of joining metal parts. Here, the submerged arc welding process is used to manufacture railroad tank cars. (Courtesy of Ransome Company)

There are three common sources of the heat used in fusion welding. These sources, incidentally, serve to identify the three basic fusion welding types. In *oxyfuel gas welding*, the heat given off by burning gases is used. In *arc welding*, the heat is supplied by an electric arc discharged across a gap between two conductors, one of which is usually the piece being welded. In *resistance welding*, heat is generated within the pieces being joined by the resistance of the metal to electric current flow. The electric arc is the most widely used heat source.

The second welding class is *solid state*. This group of processes achieves coalescence at temperatures below the melting point of the workpiece, without the use of filler metals. Pressure may or may not be used. What distinguishes this class from other forms of welding is that *no* melting takes place. Instead, coalescence occurs

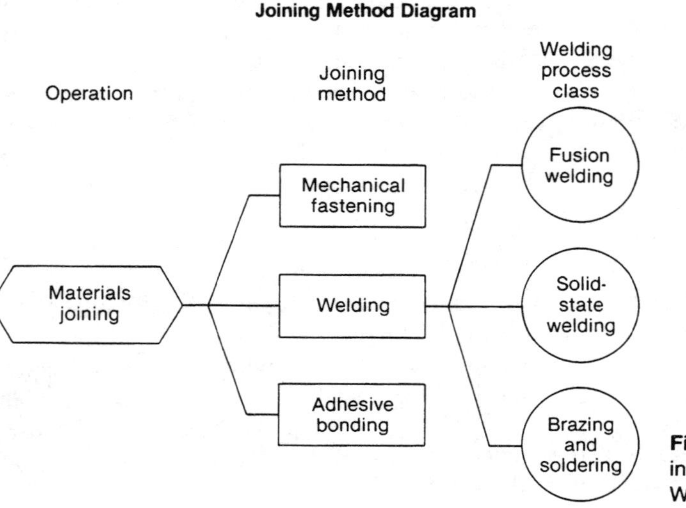

Figure 20-2 Chart of materials joining systems. (Courtesy of American Welding Society)

by the slow diffusion of atoms across and between the faying surfaces. For example, a blacksmith heats two steel workpieces to a glowing white-hot temperature, places them together on an anvil, and strikes them with a hammer. The repeated heating and hammering of the workpieces gradually causes them to merge together as one solid unit. This is forge welding, and is a solid-state process.

The third class includes *brazing* and *soldering*, which though identified as a welding process class is in fact a special form of it. Brazing and soldering are joining methods in which the joint surfaces are bonded by filler metal placed between the surfaces. The operations are carried out at temperatures which melt the filler metal but not the base metal. Brazing is done with relatively high strength filler metals which melt at temperatures above 450°C (840°F) and always below the melting point of the workpieces. Soldering is done with filler metals of lower strength which melt at temperatures below 450°C. Fusion does not occur; the workpieces do not melt, but are heated to the liquidus of the filler metal. Consequently, they are much weaker than weldments and must be used appropriately. A term commonly used here is *adhesion*.

In addition to these three classes, a fourth class of "other welding" is included in the lexicon of available processes. It will be described later in this chapter.

TERMINOLOGY

The *filler metals* mentioned earlier are often required in addition to material obtained by melting the workpiece surfaces. One purpose of filler metal is to help fill the space between the pieces being joined. Another is to improve the characteristics of the weldment after it becomes solid, inasmuch as filler metals frequently contain elements that help prevent porosity and cracks in the completed joint. A joint made without the use of filler metal is called an *autogeneous* (self-generated) weld.

To distinguish them from filler metal, the workpieces to be joined by welding are often called *base metal*, *base material*, or *parent metal*. All of the metal, both filler and base, that is melted during welding is called *weld metal*. The area containing the weld metal is often called the *fusion zone*.

The heat from welding often alters the properties of some of the unmelted base metal along each side of the welded joint, such as hardening or annealing it. That portion of the base metal is known as the *heat-affected zone* (usually abbreviated HAZ).

One of the major problems in welding is the harmful effect air can have on the hot weld metal, because the chemical reactions of the metal with atmosphere gases can weaken the finished joint. Oxide films, for example, can form on the metal and prevent the atomic contact necessary for effective metallic bonding to take place. While several methods are used to prevent or reduce such oxidation, the most prevalent include flooding the weld area with a flux or with an inert gas. Both serve to protect the weldment by excluding air from the weld metal.

WELDING JOINTS

Workpieces can be arranged to produce five primary joint types, as shown in Figure 20-3. Basic product designs generally involve such butt, tee, corner, lap, or edge welds. Additionally, some basic weld types appear in Figure 20-4. Welds in which

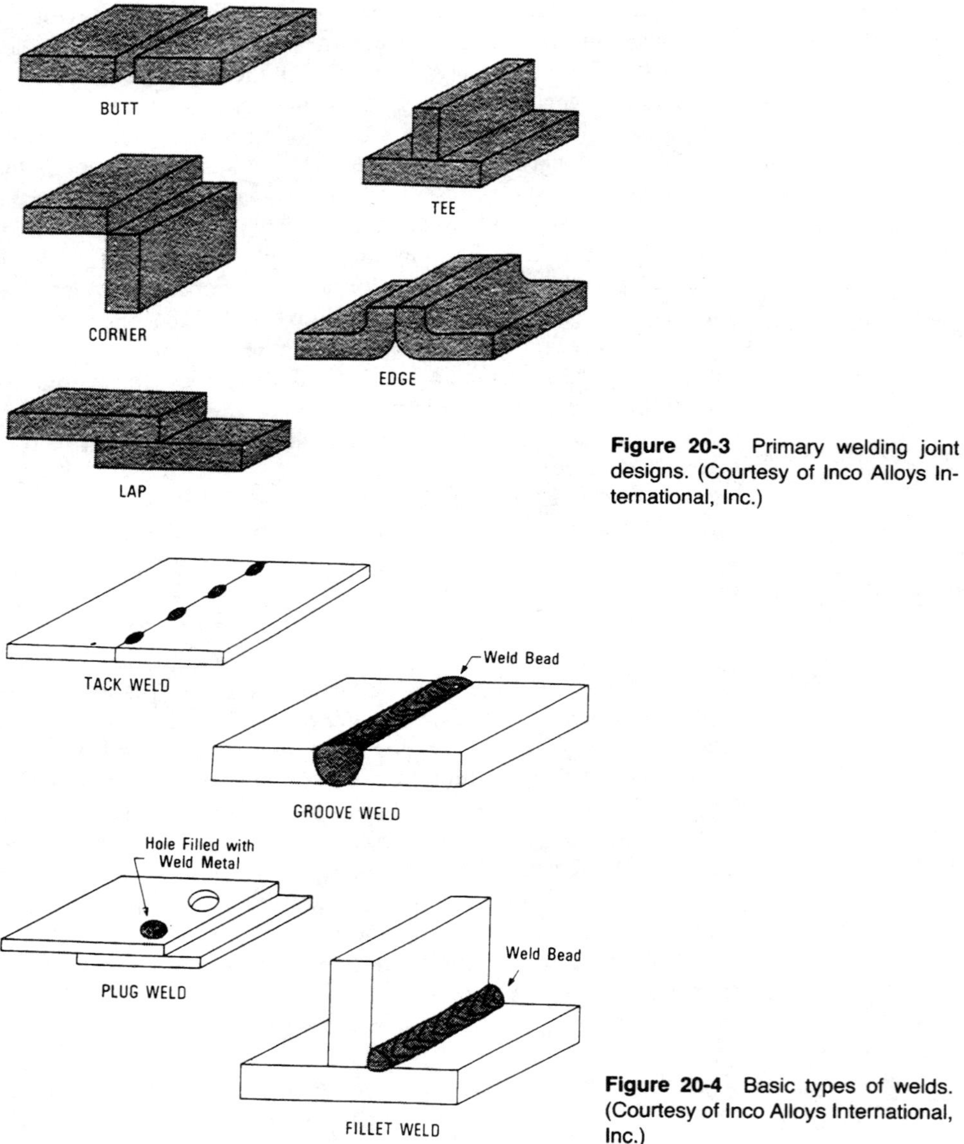

Figure 20-3 Primary welding joint designs. (Courtesy of Inco Alloys International, Inc.)

Figure 20-4 Basic types of welds. (Courtesy of Inco Alloys International, Inc.)

filler metal is deposited in a groove along the joint area are called groove welds. Butt and edge welds are so joined. Fillet welds are produced by adding filler metal to a corner formed by the placement of two workpieces at angles to one another. Fillet welds would apply to tee, corner, and lap joints.

Welding can be done in four basic positions, depending on the orientation of the joint and base metal surfaces. In the flat (also called downhand) position, the joint is in a flat, horizontal plane, with welding done from above. In the vertical position, both the joint and base metal are vertical. In the horizontal position, the joint is horizontal, but the base metal is vertical. In the overhead position, the joint is in a horizontal surface, with welding done from below. Flat welding is the optimum position, because it is usually faster and requires the least operator skill. In other positions, the flow of the molten weld metal can be adversely affected by gravity. Welding done in other than the flat position often is referred to as "out-of-position" welding.

There are some two dozen different welding processes available for use in permanent metal joint applications. See Figure 20-5. The decisions regarding their selection are based on the types of materials to be joined, their gauges and sizes, strength and appearance requirements, allowable distortion, and available equipment. Some processes lend themselves to automated assembly better than others, and this may be a consideration. See Figure 20-6. The important elements of selected, important welding processes will be found in the following paragraphs. Not all are included; some are no longer used, having been superceded by more modern and efficient

Figure 20-5 Master chart of welding and allied processes. (Courtesy of American Welding Society)

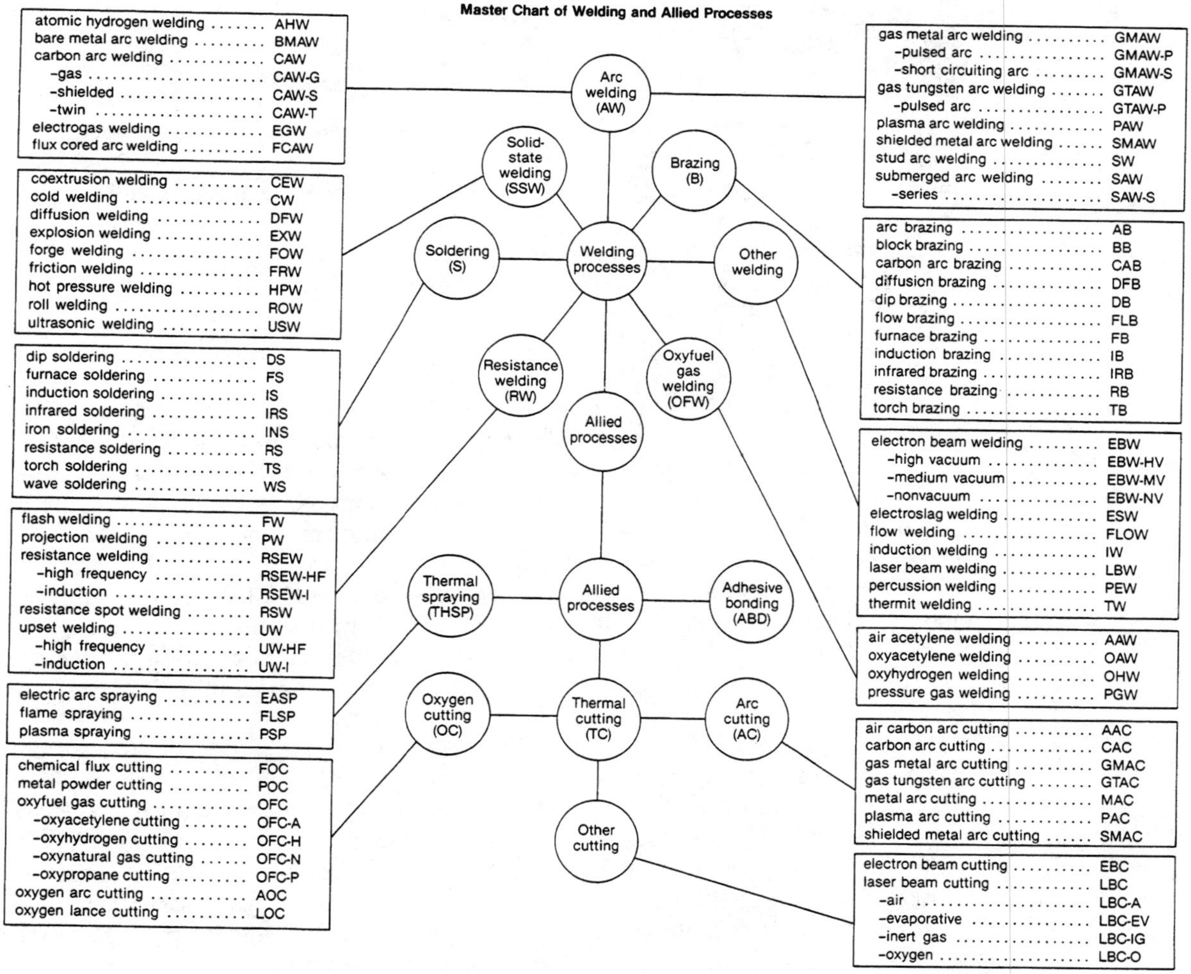

Master Chart of Welding and Allied Processes

Welding processes: Arc welding (AW); Brazing (B); Other welding; Oxyfuel gas welding (OFW); Allied processes; Resistance welding (RW); Soldering (S); Solid-state welding (SSW)

Allied processes: Adhesive bonding (ABD); Thermal cutting (TC); Thermal spraying (THSP)

Thermal cutting (TC): Arc cutting (AC); Other cutting; Oxygen cutting (OC)

atomic hydrogen welding .... AHW
bare metal arc welding .... BMAW
carbon arc welding .... CAW
-gas .... CAW-G
-shielded .... CAW-S
-twin .... CAW-T
electrogas welding .... EGW
flux cored arc welding .... FCAW

coextrusion welding .... CEW
cold welding .... CW
diffusion welding .... DFW
explosion welding .... EXW
forge welding .... FOW
friction welding .... FRW
hot pressure welding .... HPW
roll welding .... ROW
ultrasonic welding .... USW

dip soldering .... DS
furnace soldering .... FS
induction soldering .... IS
infrared soldering .... IRS
iron soldering .... INS
resistance soldering .... RS
torch soldering .... TS
wave soldering .... WS

flash welding .... FW
projection welding .... PW
resistance welding .... RSEW
-high frequency .... RSEW-HF
-induction .... RSEW-I
resistance spot welding .... RSW
upset welding .... UW
-high frequency .... UW-HF
-induction .... UW-I

electric arc spraying .... EASP
flame spraying .... FLSP
plasma spraying .... PSP

chemical flux cutting .... FOC
metal powder cutting .... POC
oxyfuel gas cutting .... OFC
-oxyacetylene cutting .... OFC-A
-oxyhydrogen cutting .... OFC-H
-oxynatural gas cutting ...... OFC-N
-oxypropane cutting .... OFC-P
oxygen arc cutting .... AOC
oxygen lance cutting .... LOC

gas metal arc welding .... GMAW
-pulsed arc .... GMAW-P
-short circuiting arc .... GMAW-S
gas tungsten arc welding .... GTAW
-pulsed arc .... GTAW-P
plasma arc welding .... PAW
shielded metal arc welding ...... SMAW
stud arc welding .... SW
submerged arc welding .... SAW
-series .... SAW-S

arc brazing .... AB
block brazing .... BB
carbon arc brazing .... CAB
diffusion brazing .... DFB
dip brazing .... DB
flow brazing .... FLB
furnace brazing .... FB
induction brazing .... IB
infrared brazing .... IRB
resistance brazing .... RB
torch brazing .... TB

electron beam welding .... EBW
-high vacuum .... EBW-HV
-medium vacuum .... EBW-MV
-nonvacuum .... EBW-NV
electroslag welding .... ESW
flow welding .... FLOW
induction welding .... IW
laser beam welding .... LBW
percussion welding .... PEW
thermit welding .... TW

air acetylene welding .... AAW
oxyacetylene welding .... OAW
oxyhydrogen welding .... OHW
pressure gas welding .... PGW

air carbon arc cutting .... AAC
carbon arc cutting .... CAC
gas metal arc cutting .... GMAC
gas tungsten arc cutting .... GTAC
metal arc cutting .... MAC
plasma arc cutting .... PAC
shielded metal arc cutting ...... SMAC

electron beam cutting .... EBC
laser beam cutting .... LBC
-air .... LBC-A
-evaporative .... LBC-EV
-inert gas .... LBC-IG
-oxygen .... LBC-O

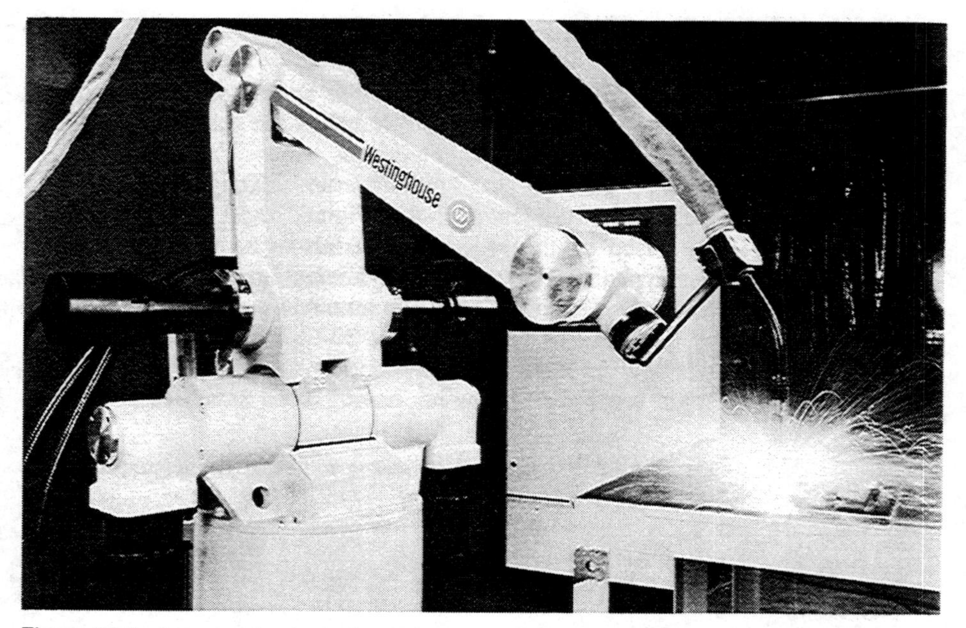

Figure 20-6 An example of robotics technology applied to welding operations. (Courtesy of Westinghouse Electric Corporation, Marine Division)

methods. The purpose here is to describe and illustrate the basic techniques and their attributes, along with major considerations relative to process selection.

It should be noted that, in reference to the chart in Figure 20-5, the thermal cutting processes discussion appears in Chapter 14.

Oxyfuel Gas Welding

Oxyacetylene welding (OAW) is a process by which coalescence is achieved by heating a workpiece with flame produced by the combination of oxygen and acetylene gases. The process may be used with or without pressure or filler materials. OAW is the most widely used oxyfuel method, whose flame is hotter than the flame of any other common gas and is the only one of sufficient temperature to gas-weld many alloys.

Equipment for oxyacetylene welding includes tanks to hold the supplies of oxygen and acetylene, a torch, hoses to carry the gases from the tanks to the torch, and valves and regulators to control the gas flow. Filler metal is normally used to make an oxyacetylene weld, and usually is a length of wire or rod 6 mm (0.25 inch) or less in diameter.

The welding process is illustrated in Figure 20-7. With the workpieces in position, the flame is directed at the joint. The welder holds the torch in one hand and the filler metal in the other. When the flame has melted a small amount of the base metal, the end of the filler metal is placed in the pool of molten metal. As the base metal and filler metal melt and flow together, the torch and filler metal are moved slowly along the joint. The puddle then cools and forms a continuous bead of solid weld metal. If more than one bead is needed for joints in thick material, filler metal is deposited on top of the previous beads until the weld is completed.

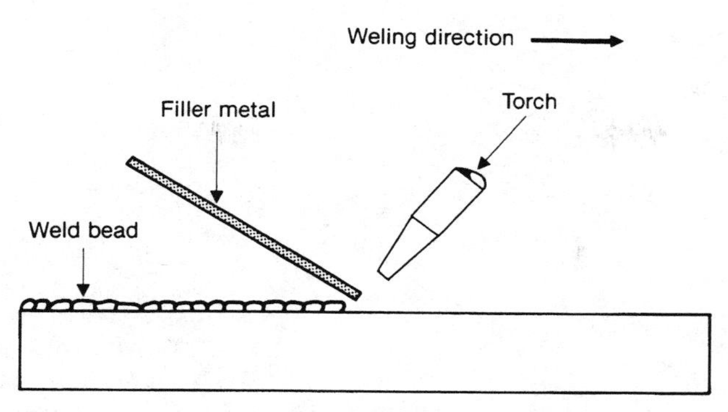

Figure 20-7 Oxyacetylene welding diagram. (Courtesy of Inco Alloys International, Inc.)

Oxyacetylene welding can be done in all of the standard positions (flat, vertical, horizontal, and overhead). Thin material is sometimes joined without using filler metal. The faying surfaces are heated with the torch until they melt and flow together.

Fluxes used with oxyacetylene welding are pastes that are brushed on the base-metal and filler metal. During welding, the flux melts and floats on top of the molten weld metal to prevent air from contacting the metal. Another important function of the flux is to clean the base metal to permit more thorough wetting of the solid metal by the molten weld metal. The flux solidifies with the weld metal and is scraped or wire-brushed off after the joint is completed. Some materials can be oxyacetylene-welded without the use of flux; the burned gases from the torch are sufficient to protect the weld metal.

Three flame adjustments are used in OAW, and each has its own practical significance. See Figure 20-8. A *neutral flame* is produced by burning approximately equal parts of oxygen and acetylene, and has a temperature of about 3,258°C (5,850°F). This flame is ideal for a wide range of cutting and welding. The *carburizing flame* contains an excess of acetylene, and reaches temperatures of about 3,090°C (5,550°F),

Figure 20-8 OAW flame patterns. (Courtesy of American Welding Society)

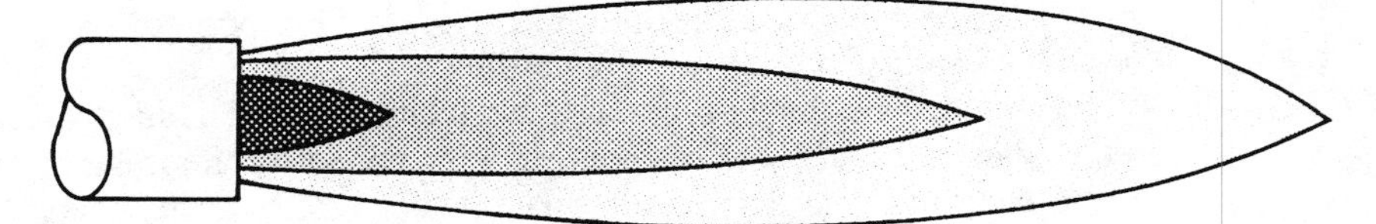

	Neutral flame	Oxidizing flame	Reducing flame
(dark swatch)	White blue cone	White cone	Intense white cone
(medium swatch)	Nearly colorless	Orange to purplish	White or colorless
(light swatch)	Bluish to orange		Orange to bluish

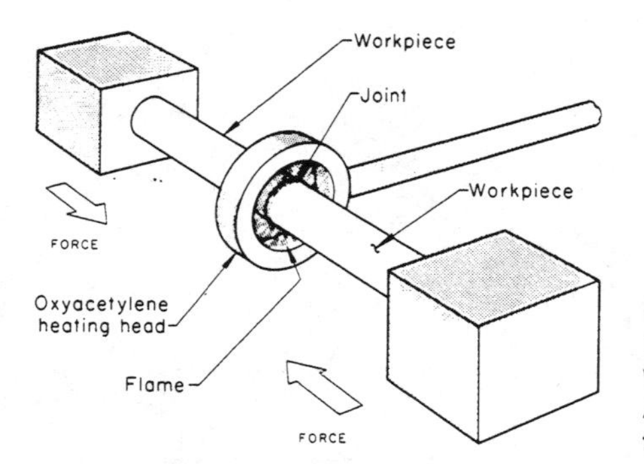

Figure 20-9 A typical arrangement for pressure gas welding. (From *Metals Handbook,* Vol. 6, 8th ed., ASM INTERNATIONAL [formerly American Society for Metals], 1971. Reprinted with permission.)

a "soft" flame suitable for silver brazing and soldering. The *oxidizing flame* burns an excess of oxygen (about 1.75:1), with a temperature approaching 3,342°C (6,000°F). Its primary use is in welding copper and certain copper alloys, and it should never be used on steel.

OAW is a common welding method, and one in which the human welder has considerable control over the weld zone temperature, filler-metal deposition rates, shaping and positioning of the weld, and flame envelope direction. These attributes make this gas welding method very versatile for field work, repairs and maintenance, and short production runs.

Pressure gas welding (PGW) produces coalescence simultaneously over an entire faying surface area using heat supplied from the combination of a fuel gas and oxygen, and by the application of pressure. No filler meterial is used. See Figure 20-9. The method is fast and effective, and is commonly applied to the joining of pipe and bar sections, machine components, and small railroad rails.

Arc Welding

Metal arc welding involves fusing the faying surfaces of workpieces, using temperatures of approximately 5,980°C (10,000°F) produced by an electric arc between an electrode and the workpiece. See Figure 20-10. Metals are readily joined by a variety of arc welding processes, all employing the arc to melt the edges of the work and the electrode or filler metal to form a pool of molten metal. A solid, continuous workpiece joint is created upon cooling.

In most arc welding processes, the weld area is shielded to protect the molten metal from oxidizing. Such shielding is provided by the melting of the flux coating on the electrode, by a granular flux blanket which forms a cover of molten slag, or by an atmosphere of shielding gas. Shielding arcs produce cleaner and tougher welding joints. The techniques and attributes of the several arc welding processes emerge in the following paragraphs.

Shielded metal arc welding (SMAW) is the most common of the arc welding processes, because of its wide range of metal joint application, and because it requires only relatively low-cost equipment. This method uses heat supplied by an electric arc generated between the workpiece and a consumable electrode. See Figure 20-11. The electrode, which also serves as the filler metal, is a metal rod coated with a flux, which melts as the electrode is consumed. Part of the flux is converted into a gas, and the remainder forms a slag deposit on the weld as it cools. Both protect the molten metal from oxidation by excluding air from the weldment.

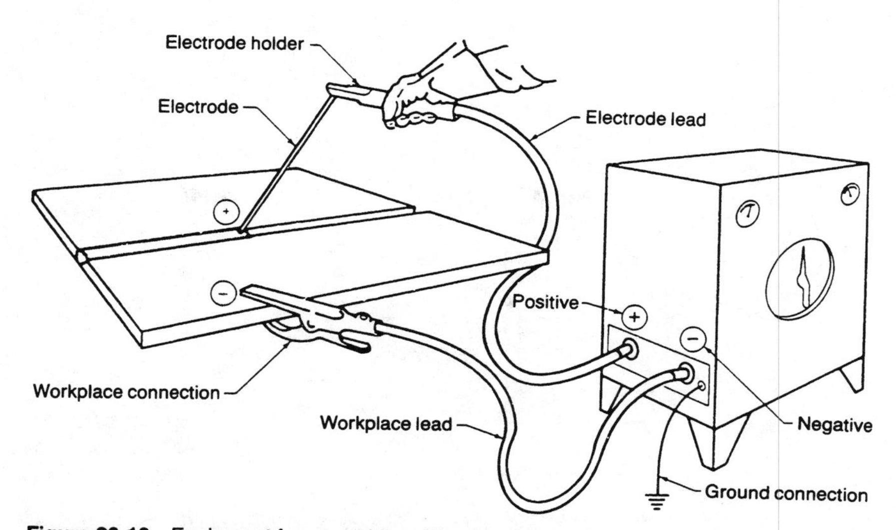

Figure 20-10 Equipment for arc welding. The direct current electrode may be either of positive or negative polarity. (Courtesy of American Welding Society)

SMAW is a fast, very versatile process. Covered electrodes ranging from plain carbon to special alloy and stainless steels are available to make this process applicable to most steels.

Flux cored arc welding (FCAW) is closely related to the SMAW process, except that the electrode is a flux-cored wire instead of a flux-coated stick. Because the wire is flexible and can be stored on spools, and because the flux core is less affected by moisture, FCAW provides a faster, higher quality automatic welding process than does SMAW.

Additional advantages of this welding method include the constant visibility of the weld puddle and arc to the human welder, and fewer restrictions on joint type and welding position.

Gas metal arc welding (GMAW) employs the heat of an electric arc between the workpiece and a consumable electrode that supplies the filler metal. See Figure 20-12. It is a type of gas shielded arc welding, and consequently no flux or coating

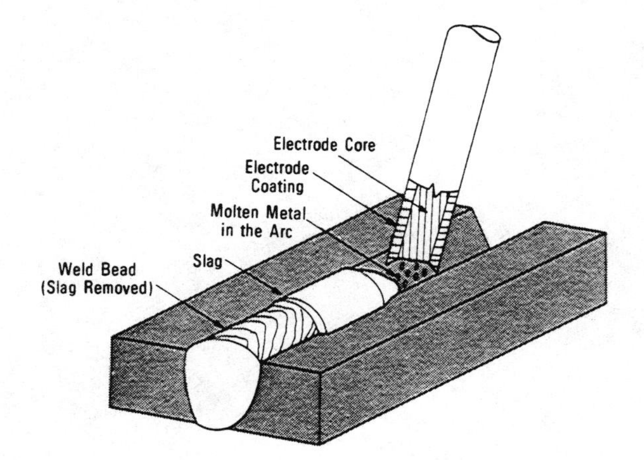

Figure 20-11 The features of the shielded metal arc weld. (Courtesy of Inco Alloys International, Inc.)

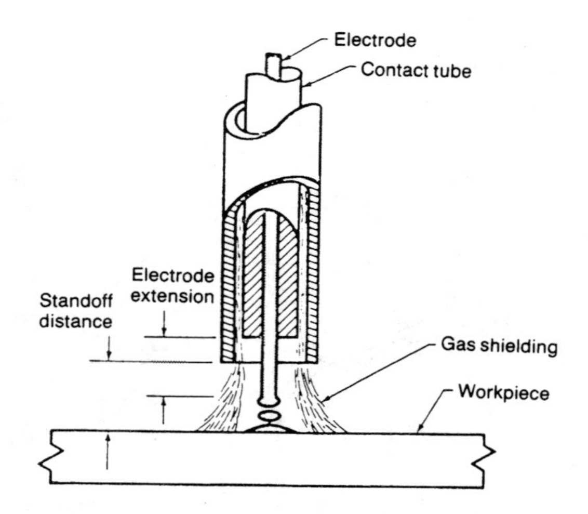

Figure 20-12 The setup for gas metal arc welding. (Courtesy of American Welding Society)

is used on the filler metal. Welding takes place in an atmosphere of gas that shields the metal from the air. The process is often called MIG (for metal inert gas) welding.

The GMAW process as applied to automatic circular weld systems is illustrated in Figure 20-13. The continuous coil of filler metal is stored on a reel and is fed through a flexible tube to the welding gun. Shielding gas, usually argon or a mixture of argon and helium, is carried from a pressurized tank to the gun nozzle by a hose. During operation, the filler metal slides against an electric contact inside the feed tube to maintain the welding circuit. The wire feed, current, and gas flow are automatically controlled. The table turns to produce the circular welded parts shown in Figure 20-14.

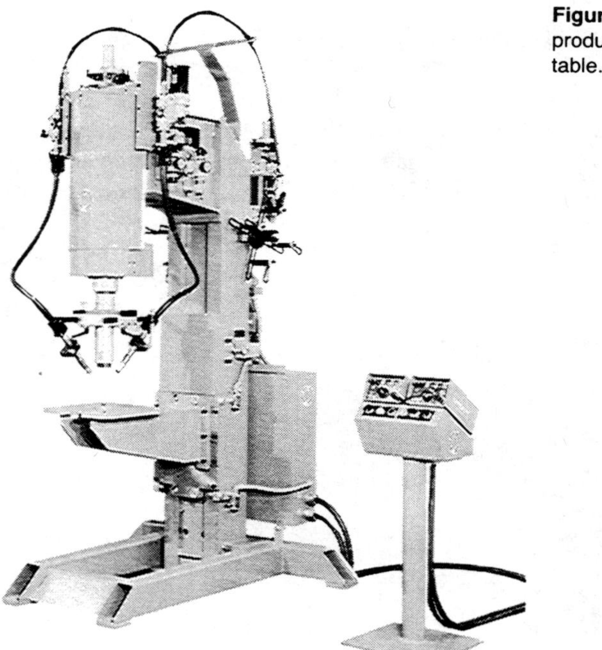

Figure 20-13 This twin torch automatic unit produces circular welds by using a rotary table. (Courtesy of Bancroft Corporation)

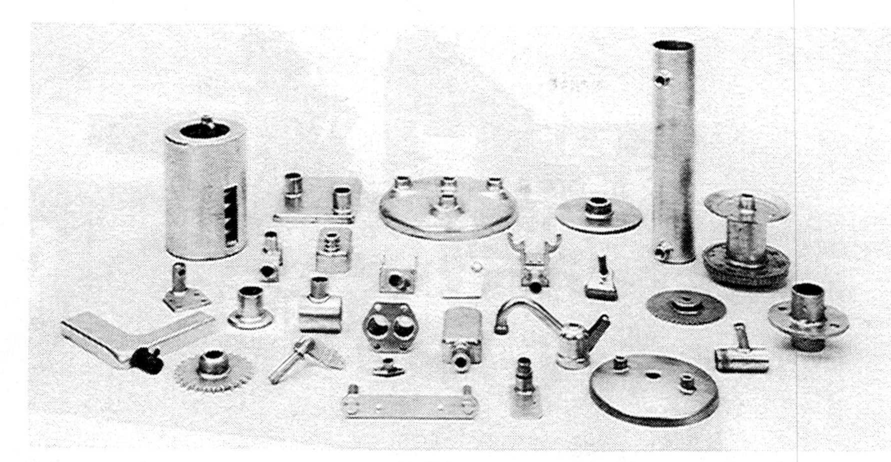

Figure 20-14 Typical circular welded parts. (Courtesy of Bancroft Corporation)

The advantages of this process are the comparatively clean welds because there is no slag to remove, and the elimination of distortion because of low heat input in the weld area.

Gas tungsten arc welding (GTAW) is performed with the heat from an electric arc discharged between the workpiece and a nonconsumable electrode made of tungsten. See Figure 20-15. This process, like GMAW, is gas-shielded arc welding, where an atmosphere of inert gas is maintained around the arc and no flux is used. The process is often called TIG (for tungsten inert gas) welding. Tungsten is used for the electrode because of its resistance to high temperatures, and, because the electrode does not melt, the arc is extremely steady and can be accurately directed into the joint. This welding can be done with or without filler metal. An important use of the process is for joining thin pieces of material without the use of filler metal.

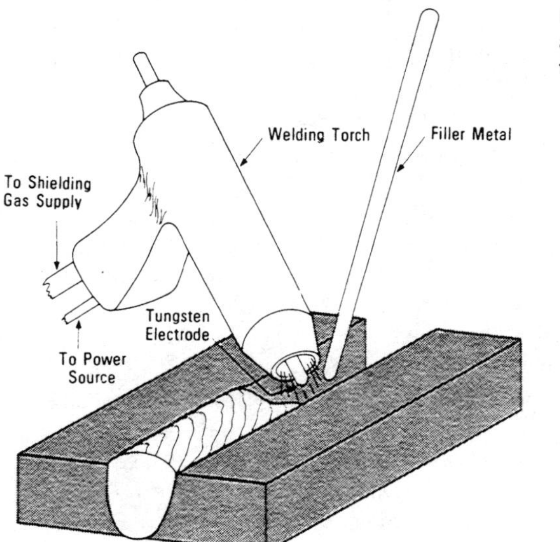

Figure 20-15 The setup for gas tungsten arc welding. (Courtesy of Inco Alloys International, Inc.)

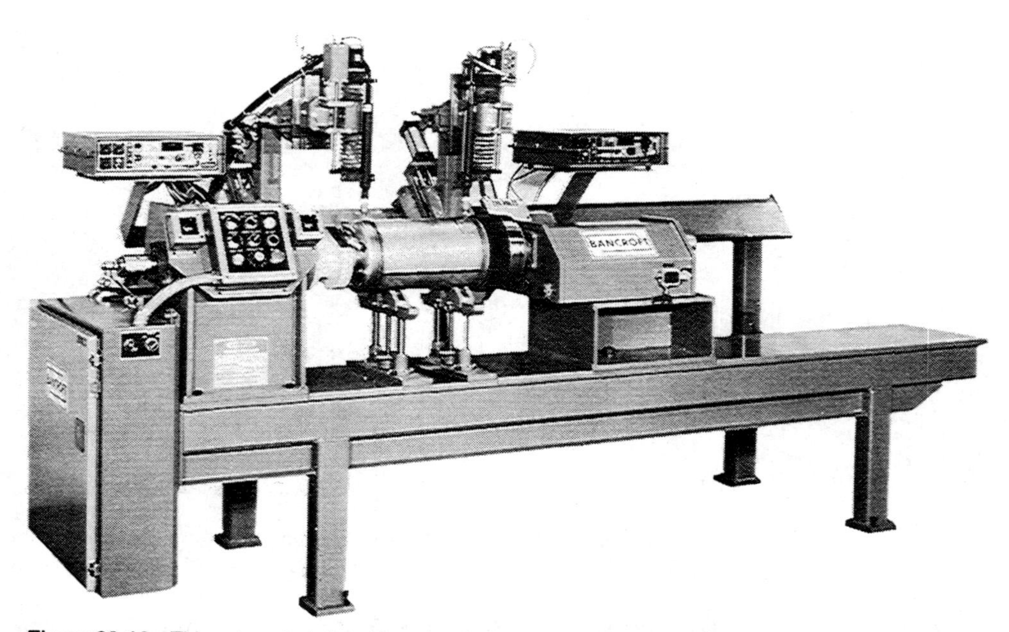

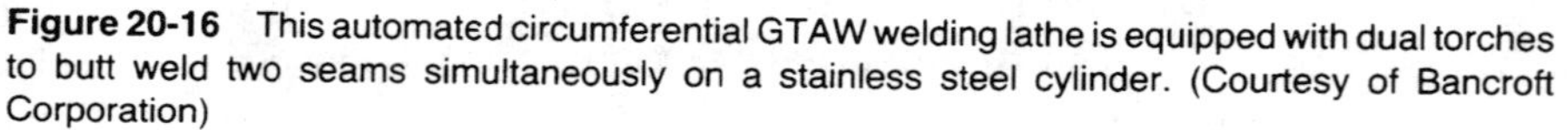
Figure 20-16 This automated circumferential GTAW welding lathe is equipped with dual torches to butt weld two seams simultaneously on a stainless steel cylinder. (Courtesy of Bancroft Corporation)

Equipment for GTAW welding is similar to that used for GMAW welding. The electrode holder is connected to a supply of shielding gas, and the electric circuit is provided by cables that connect the workpiece and electrode to the power source. An automated GTAW system is shown in Figure 20-16.

Submerged arc welding (SAW) is flux-shielded arc welding, where the heat is furnished by an electric arc occurring between the workpiece and a consumable filler metal electrode. The weld metal is shielded from the air by granular flux that surrounds the arc.

Submerged-arc welding is similar to GMAW, in that filler wire is used as the electrode, and the wire is fed automatically into the joint. The method of shielding and the nature of the arc differ greatly, however. In SAW, the faying surfaces along the joint are covered by a thick blanket of flux. Because the end of the electrode extends into the flux, the arc is not visible during welding.

The SAW process typically is a mechanized process and is illustrated in Figure 20-17. The power source is connected to the workpiece and filler metal. The filler metal slides against an electrical contact inside the tubular electrode holder. The flux is normally stored in a hopper and poured into the joint through a tube mounted in front of the electrode holder. This flux, in addition to shielding the weld metal from the air, transmits the arc to the workpiece. The flux is designed to be a nonconductor of electricity when solid, but becomes a conductor when heated to the molten state.

The SAW process is most often used to join thicker sections of carbon, or low alloy steel plate. The method provides high welding speeds, but all welding is done in a horizontal position unless special flux-holders are used.

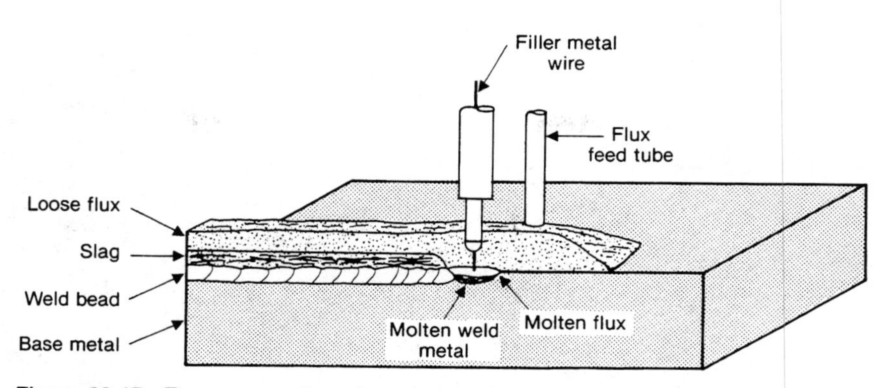

Figure 20-17 The cross section of a submerged arc welded joint. (Courtesy of Inco Alloys International, Inc.)

Plasma arc welding (PAW) is similar to gas tungsten arc welding, in that an electric arc is discharged between the workpiece and a nonconsumable tungsten electrode. The arc for plasma arc welding, however, is conducted to the workpiece by a controlled stream of plasma (electrically charged gas capable of conducting electricity).

The electrode in the torch is recessed behind a nozzle, as shown in Figure 20-18. A stream of gas, usually a mixture of argon and hydrogen, flows around the electrode and out the nozzle. During welding, both the arc and the gas are simultaneously forced to pass through the small nozzle opening. As the gas passes through the opening, it is ionized (converting to plasma) by the arc. While conducting the arc to the workpiece, the plasma stream is heated by its electrical resistance to 29,980°C (50,000°F) or higher.

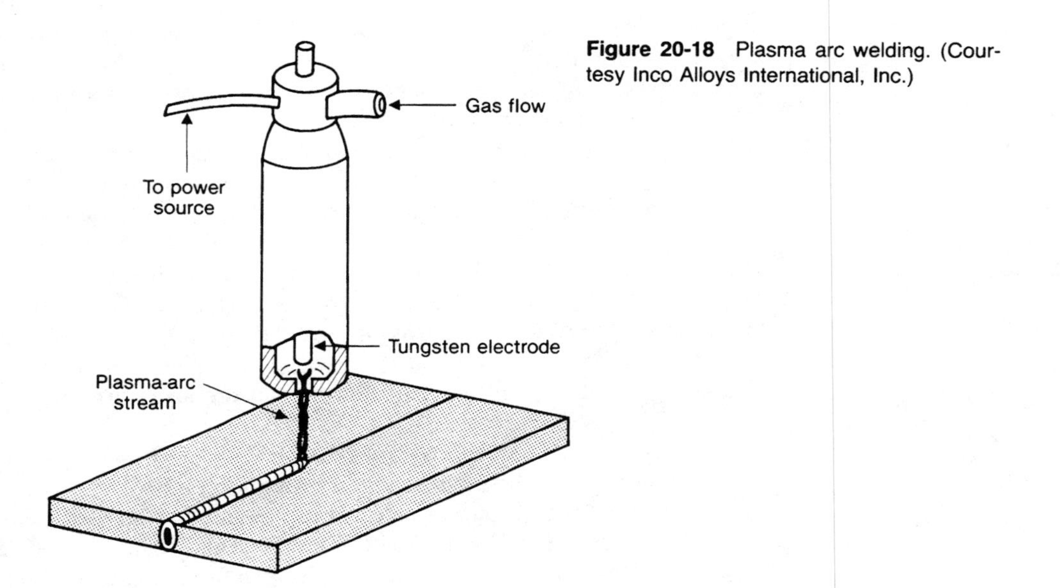

Figure 20-18 Plasma arc welding. (Courtesy Inco Alloys International, Inc.)

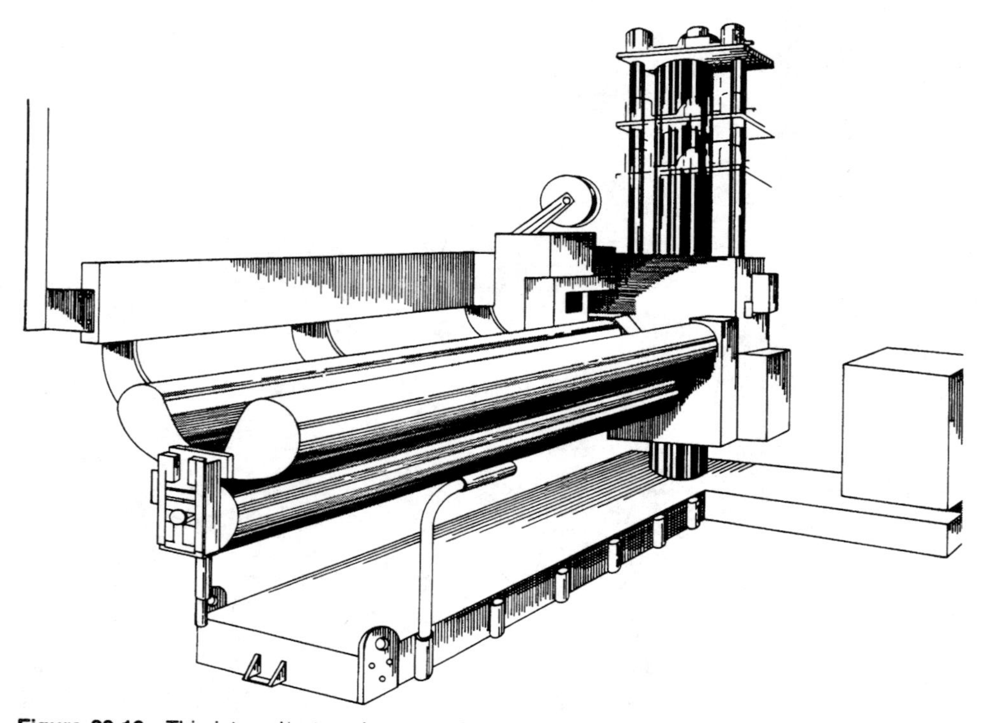

Figure 20-19 This internal/external seam welder employs GTAW, GMAW, or plasma arc to produce accurate seams in round, rectangular, and flat work. The welding head rides between the two curved pressure supports. (Courtesy of Four Corp., Green Bay, WI)

This welding process is used primarily for autogeneous welds made with automatic equipment. See Figure 20-19. The high temperature of the plasma arc permits rapid welding, and the nozzle orifice focuses the arc and plasma into a thin column that can be directed into joints with great accuracy. Thicker autogeneous welds can be made with the plasma arc process than with gas tungsten arc welding.

Stud arc welding (SW) is a specialized welding method used to attach studs such as bolts, pins, and other fasteners to a metal surface. This welding method employs heat from an electric arc produced between the end of the stud and the surface to which the stud is to be welded. The stud is held against a compressed spring in a welding gun. When the arc has melted the surfaces of both parts, the spring is released to plunge the stud into the workpiece. The weld metal is shielded either by a stream of inert gas from the welding gun or by a ceramic collar mounted around the end of the stud. Stud arc welding saves considerable time over drilling and tapping, or fillet welding techniques, and produces a strong, fused joint. See Figure 20-20.

Atomic hydrogen welding (AHW) is done with the heat released from hydrogen gas as it undergoes a chemical reaction brought about by an electric arc. A stream of hydrogen is brought through an electric arc discharged between two nonconsumable tungsten electrodes. See Figure 20-21. The arc separates the molecules of hydrogen, each of which normally contains two hydrogen atoms, into individual atoms. The separation process (called dissociation) causes the atoms to absorb heat. Welding

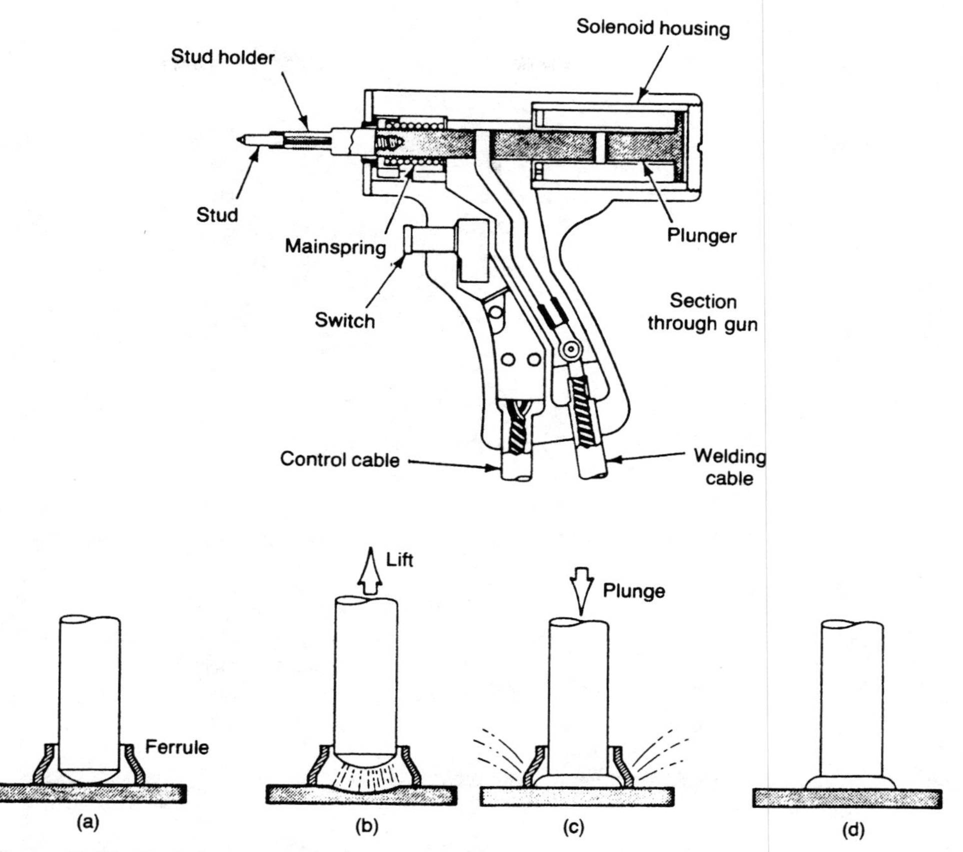

Figure 20-20 Typical gun used for stud welding. The sequence is also shown. (From *Metals Handbook,* Vol. 6, 8th ed., ASM INTERNATIONAL [formerly American Society for Metals], 1971. Reprinted with permission.)

is performed by directing the stream of hydrogen into the joint. Filler metal is added by feeding a bare rod into the molten base metal, and shielding is provided by the burning hydrogen gas. AHW is seldom used today, because of the availability of other less complicated processes.

Electrogas welding (EGW) is quite similar to electroslag welding, but operates on the principle of the gas metal arc (GMAW) process. As shown in Figure 20-22, molding plates are clamped to the joint, and filler metal is fed into the cavity

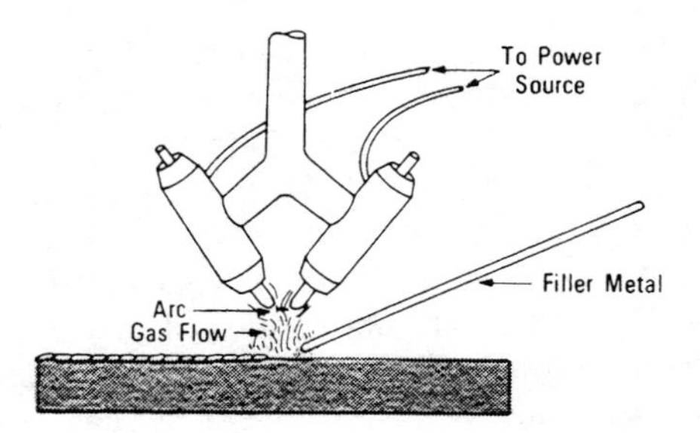

Figure 20-21 Atomic hydrogen welding. (Courtesy of Inco Alloys International, Inc.)

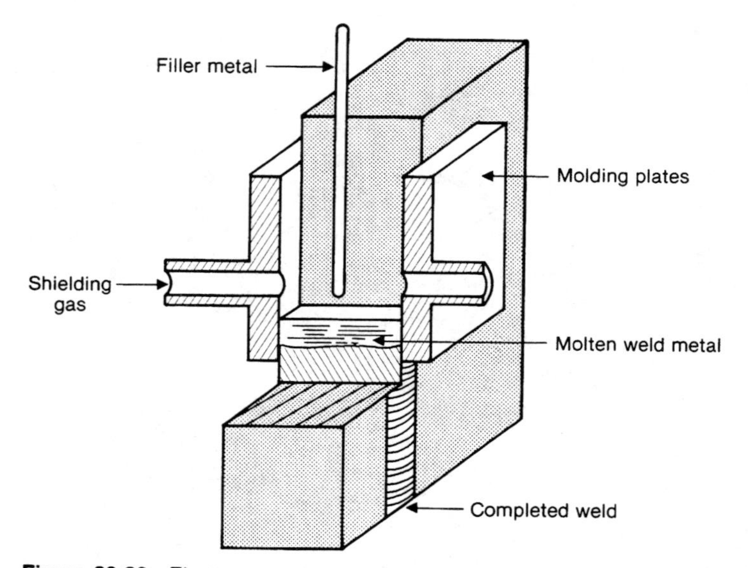

Figure 20-22 Electrogas welding. (Courtesy of Inco Alloys International, Inc.)

from above. Welding heat is supplied by an electric arc discharged between the consumable filler metal electrode and the molten pool of weld metal. An atmosphere of argon or helium shielding gas is maintained over the weld metal. EGW is a rapid automatic welding process used to join vertical metal workpieces, generally up to 25 mm (1 inch) in thickness.

Resistance Welding

This welding method employs a combination of heat and pressure to join metal parts. The heat, instead of being applied by an arc or a flame, is generated within workpieces by their resistance to the flow of electricity. The work is heated until molten while being held under pressure, and these molten surfaces flow together and solidify to form the welded joint. No filler metal is used. The tight contact maintained between the pieces shuts out air and eliminates the need for fluxes and shielding gases.

Resistance welding methods are economical and efficient, and readily adaptable to mass production. One of the common uses of resistance welding is to weld lap joints in relatively thin sheet metal. Three variations of the process—spot, seam, and projection welding—are used for such joints. Two other methods—flash and upset—are used mainly for butt welds.

Resistance spot welding (RSW) is a process whereby lapped metal sheets are joined by local fusion caused by electric current flowing between opposing electrodes. See Figure 20-23. As illustrated, the electrodes hold the workpiece in position, and transmit pressure and current to the work. During welding, the current flows from one electrode, through the workpieces, to the other electrode. The greatest electrical resistance, and therefore the greatest amount of heat, is at the contacting, or faying, surfaces of the work. This heat continues until the metal at that point is melted, and the electrode pressure holds the molten pieces together. The current is turned off, the molten metal solidifies to form a weld nugget, and electrodes are

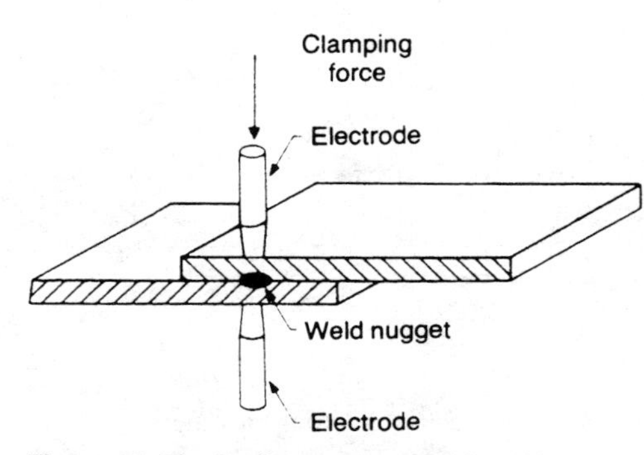

Figure 20-23 Resistance spot welding. (Courtesy of Inco Alloys International, Inc.)

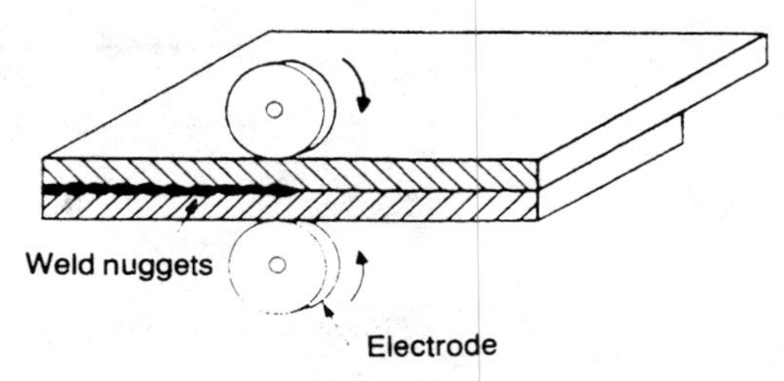

Figure 20-24 Resistance seam welding. (Courtesy of Inco Alloys International, Inc.)

retracted to release the pressure. The electrodes are water cooled to draw heat from workpieces, thereby minimizing distortion.

Spot welding typically is used to join sheet metal parts up to 3 mm (0.125 inch) thick, where a sealed joint is not required. This type of welding causes a slight depression or dimple in the workpieces, which is difficult to conceal, and is therefore generally used where appearance is not a factor. Automotive body structures are usually spot welded.

Resistance seam welding (RSEW) produces a series of overlapping spot welds on workpieces. See Figure 20-24. The work is gripped between two narrow copper wheels that serve as electrodes. The current passes through the wheels to the workpieces, and the flow is interrupted to form a distinct line of weld nuggets. Wheel rotational speed, pressure, and current flow are carefully regulated to produce the desired weld pattern.

Seam welds generally overlap to create a water or airtight joint. Related techniques include the continuous butt welding of pipe and tubing seams, and the high rolling pressure mash welding of lapped seams. Refrigeration evaporators and condensers are among the many sheet metal products joined by seam welding.

Resistance projection welding (RPW) employs special spot welds in which the placement of weld nuggets is controlled by projections or embossments on one or both of the workpieces. The parts to be welded are held under pressure between electrodes, but contact each other only at the projections. The metal of the areas of contact is melted to form the weld nuggets. See Figure 20-25. Projection welding

Figure 20-25 Resistance projection welding. (Courtesy of Inco Alloys International, Inc.)

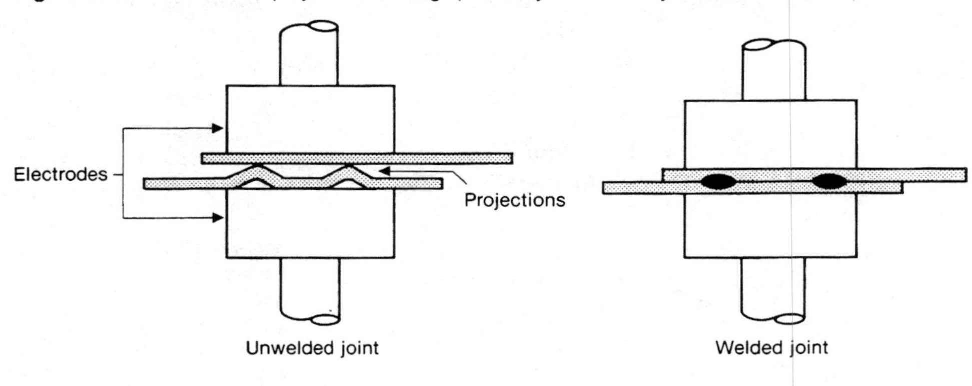

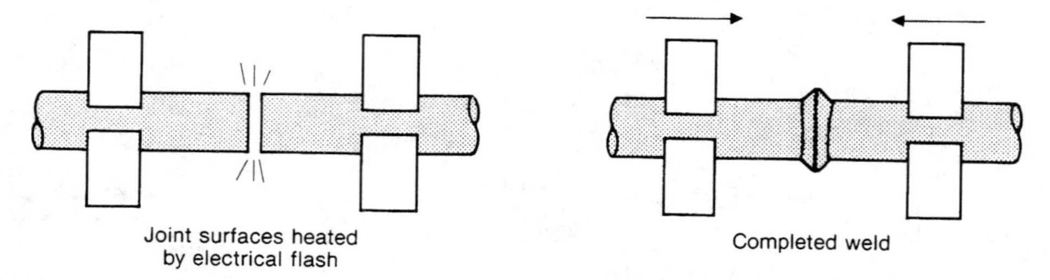

Figure 20-26 Flash welding. (Courtesy of Inco Alloys International, Inc.)

enables two or more spot welds to be made at the same time with one pair of electrodes. Also, projection welding is especially applicable to welding zinc-coated steels, where larger, flat electrodes, less current, and lower pressure extends electrode life.

In *flash welding* (FW), the two parts to be joined are secured in current-carrying jaws, as shown in Figure 20-26. The workpieces are held in light contact or are slightly separated while electric current is passed through them. The arcing at the joint melts the surfaces of the parts, which are then pressed tightly together, and the electric current is turned off. The joint is held under pressure until the weld metal solidifies.

In *upset welding* (UW), parts are gripped in the jaws of a welding machine in the same manner as for flash welding. For upset welding, however, the parts are held under heavy pressure while the current is being passed through them. The electrical resistance at the contacting surfaces heats the metal until the surfaces flow together; no arcing occurs. Molten metal is not squeezed out of the joint, but the upsetting force spreads the material in the joint area and creates a bulge.

Solid State Welding

Friction welding (FRW) is effected by rubbing together two joint surfaces until they are melted by friction and flow together. The process is easily applied to the welding of round workpieces such as axle shafts, or to joint rod ends to flat surfaces. As shown in Figure 20-27, one workpiece is held stationary while the other is rotated against it in a machine resembling a lathe. Within a few seconds there is sufficient heat generated to achieve weld temperatures, and a marked resistance to continued rotation develops. By stopping the rotation suddenly and increasing the pressure, the weld is produced.

The pressure upsets the joint ends of the workpieces and molten metal is squeezed out, resulting in a rough joint which must be polished. The mechanical properties of such joints are excellent, and the process is readily applied to automation.

Explosion welding (EXW) employs the impact resulting from the detonation of an explosive to drive two pieces together and create a bond between them. An

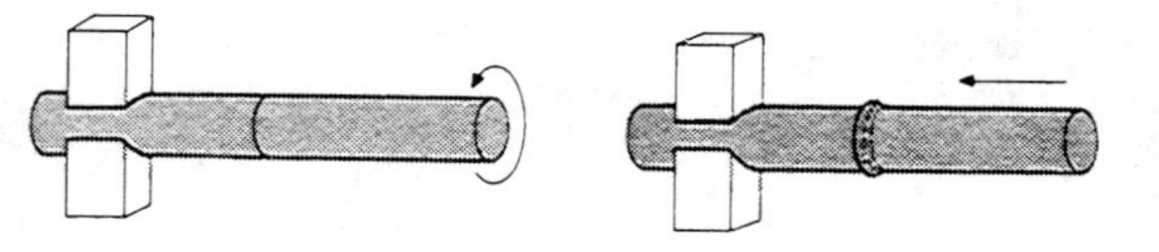

Figure 20-27 Friction welding. (Courtesy of Inco Alloys International, Inc.)

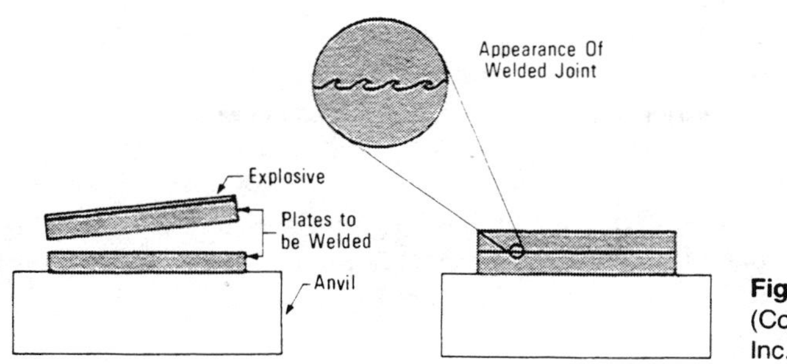

Figure 20-28 Explosion welding. (Courtesy of Inco Alloys International, Inc.)

important use of the process if for the production of large clad plates, in which the flat surfaces of two plates are bonded together to make a single piece. Special alloys are often used as cladding materials to give corrosion resistance to structures made of less resistant metals.

And explosion welding operation is shown in Figure 20-28. The two pieces are arranged with a space between them and with one slightly inclined to the other. One piece is covered with an explosive material, and the other is supported by a rigid base or anvil. When the explosive is detonated, the top piece is driven toward the bottom one. Because of the angle between their surfaces, the two pieces contact progressively from one edge to the other. The force of the impact causes plastic flow of the workpieces together in a series of interlocking ripples.

Cold welding (CW) produces a bond of continuous grain structure in two metal workpieces by the use of pressure alone. The theory is that when pressure is applied, the faying metal surfaces are sealed off. Additional pressure causes the joint surfaces to flow together by cold plastic deformation, a condition similar to the ''seizing'' of an engine due to improper lubrication.

Lap joints are welded by squeeze or impact pressure delivered by two dies, one on each side of the work. For example, two roller dies can be employed to weld aluminum sheets, and simultaneously turn the edge. Butt welds are produced when two workpieces are held in powerful grippers and forced together under heavy pressure. The plastic flow of the interface forms the weld. The joint surfaces must be smooth and very clean in order for quality cold welding to take place.

Ultrasonic welding (USW) is done by subjecting the pieces to be joined to high-frequency vibrations while the joint is held under pressure. The combination of clamping force and vibration of one surface against the other results in a metallic bond without the use of heat. See Figure 20-29.

The greatest use of USW is to weld lap joints in thin material. Welding is carried out in a manner similar to spot or seam resistance welding. The workpieces are clamped between an anvil and a punch or wheel that transmits the vibrations to the

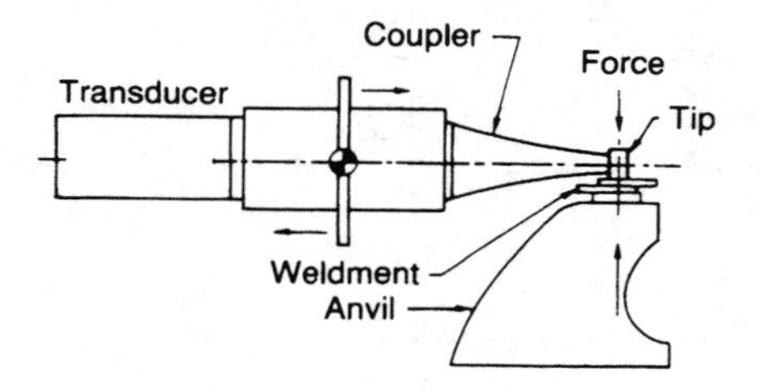

Figure 20-29 Ultrasonic welding. (From *Metals Handbook,* Vol. 6, 8th ed., ASM INTERNATIONAL [formerly American Society for Metals], 1971. Reprinted with permission.)

top piece. The punch or wheel resembles the electrodes used for resistance welding, and is usually called a sonotrode. The sonotrode is vibrated by a device called a transducer, which converts electricity to mechanical vibrations.

Diffusion welding (DW) is a special solid-state welding process whereby heat and pressure are applied to workpieces for sufficient time to cause a diffusion, or transmigration of atoms, of abutting surfaces and a subsequent weld. Because no part of the joint ever reaches the molten state, this is referred to as solid-state weldment. The primary application for this process is the butt or end-to-end joining of rods and tubes.

Other Welding Processes

Electron beam welding (EBW) is a process whereby coalescence is obtained from a concentrated beam of high velocity electrons directed to the weld area. In a common system example, the free electrons are propelled from a heated tungsten filament (cathode) across a vacuum to an anode which draws them into a beam and accelerates their speed. The electrons leave the anode arranged into a dense beam travelling about half the speed of light, and are further focused as a fine beam against the workpiece by the magnetic focusing coils. See Figure 20-30. The welding sequence is shown in Figure 20-31.

The electron beam must be generated in a vacuum because the electrons are easily deflected by air and gas molecules, which tend to scatter the beam. EBW is done by three methods, depending on the type of vaccum system used, and according to the workpiece requirements. In *high vacuum* welding, both the electron gun (the cathode, anode, and focusing coils) and the workpiece are inside the airtight chamber from which nearly all air is evacuated before welding begins. See Figure 20-32. The method has the advantages of minimum disruption of the electron beam and complete protection of the weld metal from contamination by air. In this system, the beam range is approximately 900 mm (36 inches), with a welding penetration

Figure 20-30 An electron beam welding system. (Courtesy of Inco Alloys International, Inc.)

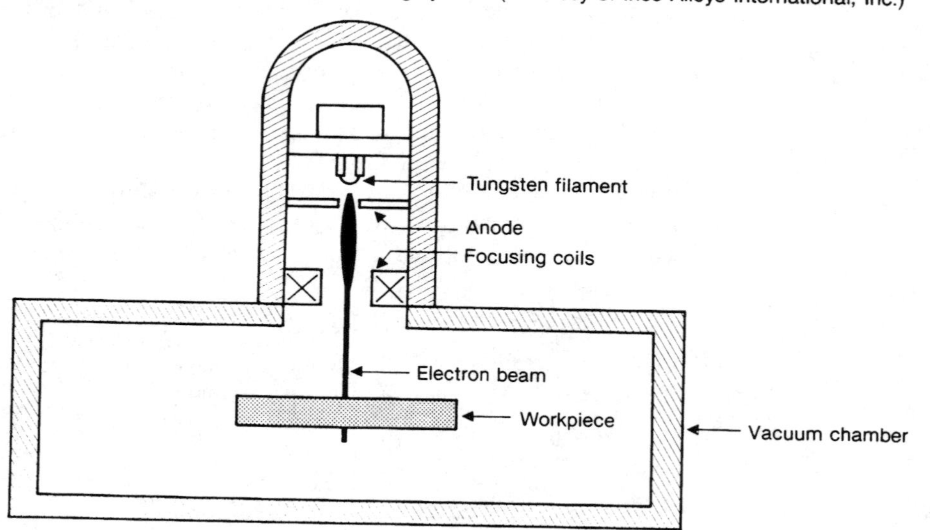

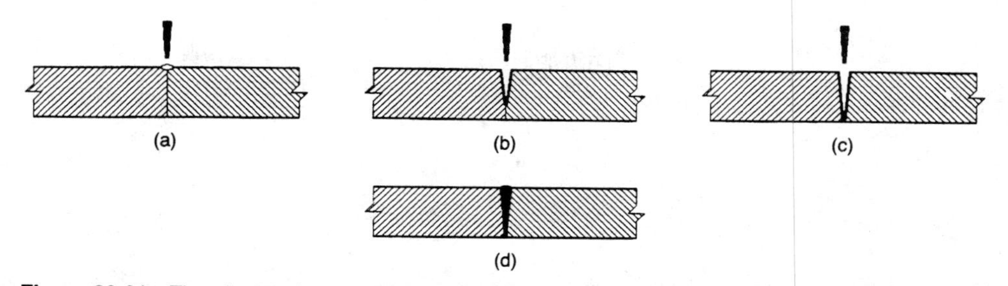

Figure 20-31 The electron beam welding sequence. An electron beam is instantly converted from kinetic to heat energy as it contacts the surface of a workpiece, with the following results: (a) the surface spot begins to melt; (b) the melt is pushed aside by high vapor pressure to facilitate greater beam penetration; (c) the channel is worked deeper into the material to form a capillary vapor tube surrounded by a fused wall; (d) the fused material flows and closes the tube with a nearly parallel zone of solidification.

of about 150 mm (6 inches). The disadvantages of the method are the time required to pump air from the chamber and the limitation on workpiece size imposed by the chamber this method is restricted to.

In *partial vacuum* welding, a nearly complete vacuum is maintained around the electron gun, but only a partial vacuum is produced in the workpiece chamber. See Figure 20-33. The electron beam passes into the chamber through a small hole. Less time is required to prepare the chamber for welding, but the usable length of the beam and penetration is decreased about 30 percent. The partial vacuum provides adequate protection for the weld metal. In *nonvacuum welding*, the electron beam is generated in a high vacuum and passed through a series of chambers having

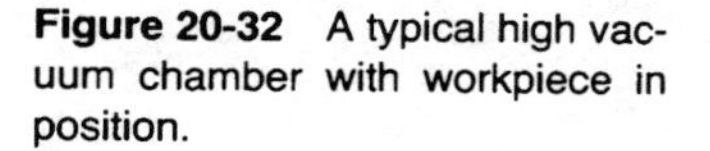

Figure 20-32 A typical high vacuum chamber with workpiece in position.

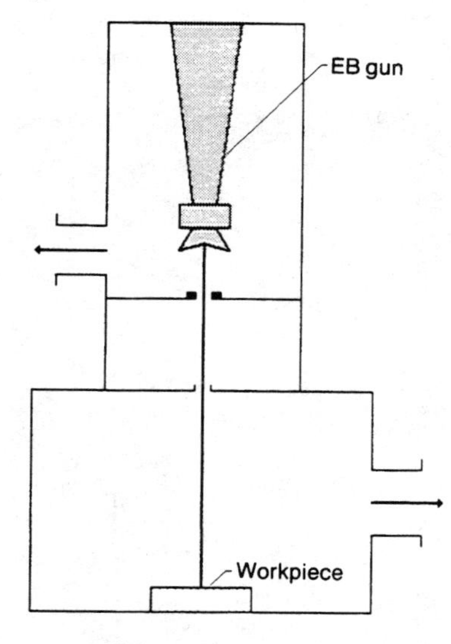

Figure 20-33 The partial vacuum method.

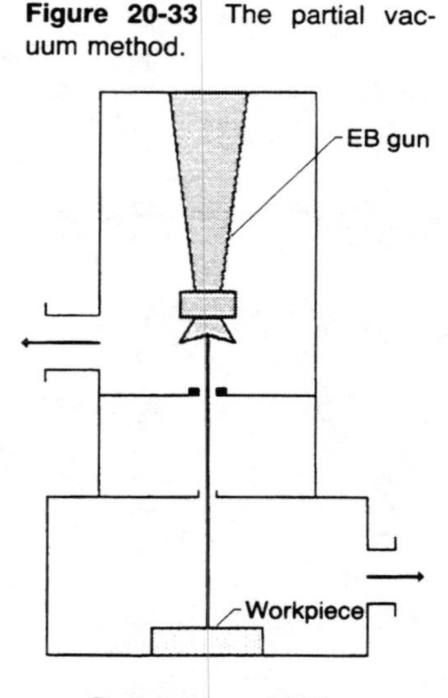

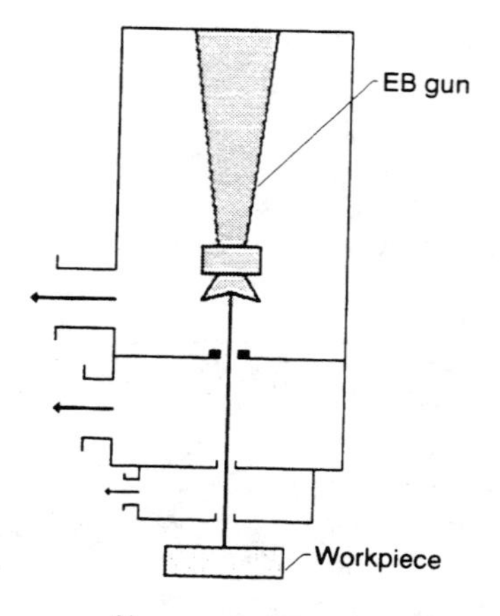

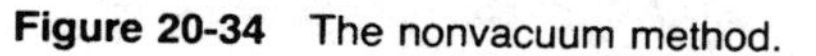
Figure 20-34 The nonvacuum method.

progressively softer vacuums. After leaving the last chamber, the beam travels through the air to the workpiece. See Figure 20-34. The method eliminates the workpiece chamber, but reduces the intensity of the beam and penetration to about 75 percent. The workpiece must be placed close to the beam nozzle, and thick pieces cannot be welded. Also, a shielding gas must be used to protect the weld metal.

Although EBW equipment is relatively expensive, it does offer the advantage of speed, automatic control, and very narrow fusion zones. See Figure 20-35. For this reason, this welding system has found use in aircraft and automotive industries where weld quality and production rates generally offset costs. Electron beam welding does generate small amounts of radiation, and therefore the operator must be protected against any possible danger.

Figure 20-35 An important attribute of EBW is very narrow weld zone, (a), as contrasted with typical fusion weld zone, (b). (Courtesy of Leybold-Heraeus Vacuum Systems, Inc.)

(a)

(b)

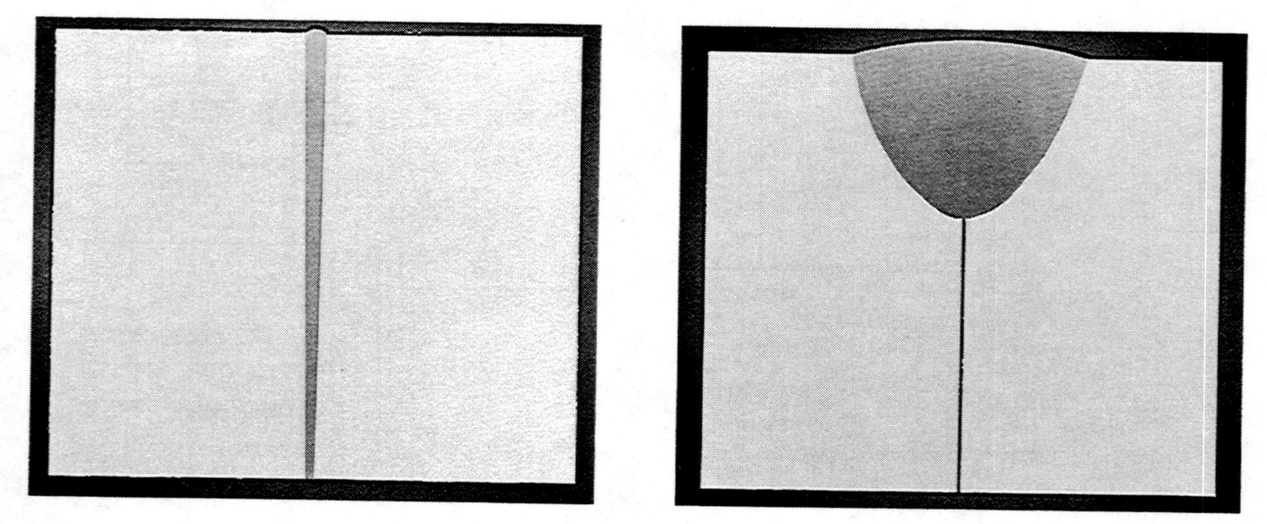

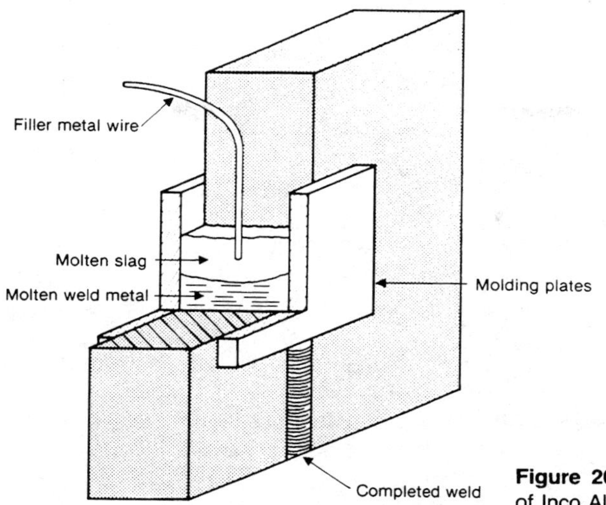

Figure 20-36 Electroslag welding. (Courtesy of Inco Alloys International, Inc.)

Electroslag welding (ESW) is a special type of submerged arc welding used to join sections of steel plate from 25 to 350 mm (1 to 14 inches) in thickness. In this process, water-cooled copper molding plates or shoes are fitted against the workpieces to form a mold to contain the molten weld metal. See Figure 20-36. A filler wire is fed into the weld area to create a molten puddle, which is held in place by the copper plates. The plates move up the joint as the weld metal solidifies, and flux is continuously fed into the weld area where it forms a molten slag float to shield the weldment from oxidation. The welding heat is generated by the resistance of the slag to the flow of electric current from the filler wire to the workpieces. This molten slag layer, therefore, provides the heat necessary for the electroslag process. ESW is typically used in shipbuilding, the construction of storage tanks, or in the building and repair of large industrial machinery. See Figure 20-37.

Figure 20-37 Automatic ESW of storage tank sections. Both horizontal and vertical welds are feasible with this system. (Courtesy of Ransome Company)

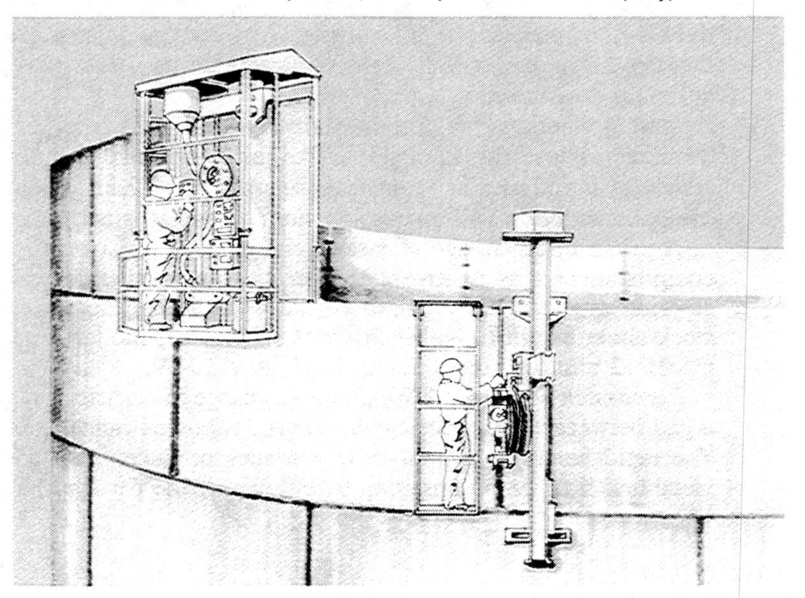

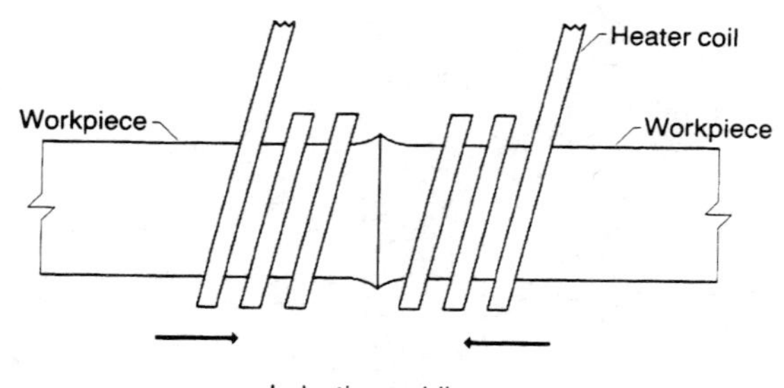

Figure 20-38 Induction welding.

Induction welding (IW) is a process which produces welding temperatures with the heat generated from the resistance of a workpiece to high frequency alternating current flowing through a coil which surrounds the workpiece. The effect of the induction work coil is to concentrate the welding heat in the specific weld area. A workpiece to be joined by this process is placed within coils formed to approximate the shape of the workpiece. See Figure 20-38. Pressure may or may not be used, and filler metal is added to the joint by capillary action.

Laser beam welding (LBW) is a welding process that produces coalescence of workpieces by directing a narrow, coherent, concentrated beam of amplified light to the weld area. (For basic information on laser technology, see Chapter 14.) The laser resembles the electron beam, in that it too is a well-collimated beam of intense energy; its main difference is that the light energy of the laser beam is converted to the thermal energy when it strikes a nontransparent substance, such as metal. This attribute makes it possible to weld tiny components, such as filaments inside a clear vacuum tube. LBW also has excellent controllability, because the light beam can be intensified or reduced speedily, and is easily interrupted.

An important attribute of laser welding is that the sharply focused beam and the speed of the operation minimizes thermal damage to the workpiece, because the heat-affected zones (HAZ) are greatly reduced. As a result, little thermal distortion occurs, and sensitive components, such as electronic ones, are rarely damaged. In normal operation, a small weld puddle is melted and sets up in milliseconds, and the HAZ is restricted to 0.25 mm (0.010 inches) or less. Because of such a short duration of the molten state, any chemical reaction which may occur will not adversely affect the weld quality.

During welding, the high-intensity, precisely focused laser beam melts the two or more workpieces that comprise the joint. An inert gas, such as helium or argon, generally is introduced to protect the weld pool from oxidation. Laser welding is classified as *autogeneous*, because no filler metal is added to the joint.

A prime disadvantage of laser welding systems is the size and sensitivity of the equipment, and its relatively high cost. Careful attention must be given to the fit of joint area, and to the sizes of the workpieces; the system is most effective in stock about 1 mm (0.040 inch) thick. However, modern equipment is used to join pieces 12 mm (0.5 inch) thick. See Figure 20-39.

Percussion welding (PEW) utilizes a charge of electricity rapidly discharged across a gap between the surfaces to be joined, followed immediately by an impact blow. The rapid heating and joining of surfaces produces only a small amount of weld metal and little or no upsetting or bulging of the parts at the joint. This welding is

Figure 20-39 Laser beam apparatus used to butt weld stainless steel cylinders. (Courtesy of Westinghouse Electric Corporation, Marine Division)

limited to butt weld areas of less than 13 mm (0.5 inch) thick, and typically is used to join bars or tubing.

Thermit welding (TW) is a process by which welding temperatures are produced by pouring a superheated liquid metal, or thermit, around the parts to be welded. The thermit is the result of a chemical reaction between a metal oxide (usually iron) and aluminum. Basically, the process involves fitting a metal casting flask or mold around the workpiece joint area. See Figure 20-40. A hard wax pattern defines the weld area, and the flask is rammed full of molding compound, surrounding the wax pattern and wooden sprue pattern. The mold is heated to melt the wax, and form the weld cavity, and the sprue is removed to create a pouring hole.

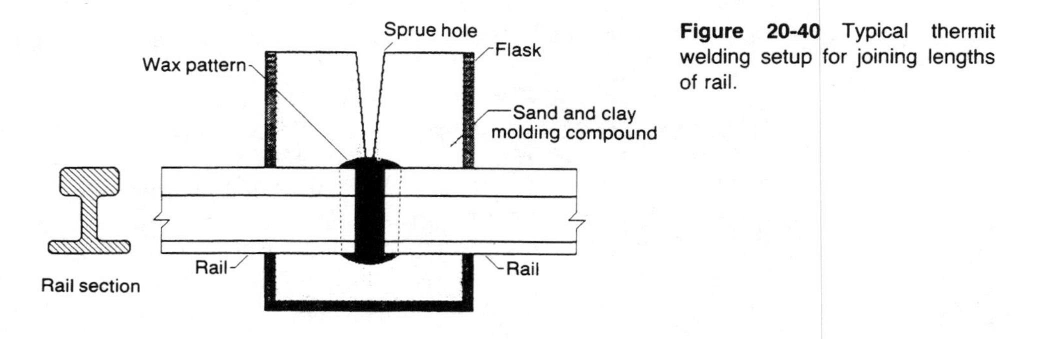

Figure 20-40 Typical thermit welding setup for joining lengths of rail.

The thermit is poured into the sprue and down to the weld area, where the superhot metal surrounds the workpieces, coalescence is achieved, and a homogeneous weld results upon cooling. The joint is then ground and dressed to the degree of smoothness required.

An important application of thermit welding is the field welding of standard 15 meters (50 feet) lengths of steel railroad rails into long, continuous sections. The method is fast and efficient, and results in lowered maintenance costs, and smoother train rides at high speeds.

BRAZING AND SOLDERING

It was stated earlier that while brazing and soldering are joining processes related to welding, they are not true welding methods, because fusion of the workpiece and filler metal does not occur. Brazing and soldering may be best described as processes whereby workpieces are joined by a metallic bond using filler metals below the melting point (liquidus) of the workpieces. They both are permanent joining systems, but because of the nature of the processes, the resultant joints are not as strong as true weldments. Refer to Figure 20-5 for a chart of brazing and soldering processes.

Brazing (B) is performed by filling a small space between two surfaces with molten filler metal, typically a bronze rod. Health requirements currently do not permit brazing alloys which contain cadmium or other hazardous materials. The completed joint consists of two pieces of base metal connected by a thin layer of filler metal. Although the process is similar to welding methods in which filler metal is added to a joint, there are two important differences. One is that no base metal is melted during brazing, which results in little mixing of filler metal and base metal. (Some mixing does occur, because the molten filler metal dissolves a small amount of the solid base metal.) The second difference is that brazing filler metal is distributed along the joint surfaces by its own wetting action (capillary action), whereas welding filler metal is placed into the joint where needed.

Brazing produces smooth, neat joints without rough and protruding weld beads. It is especially suitable for joints in small parts, joints in articles that must have good visual appearance, and joints that have inaccessible surfaces. Lap joints and slipover connections in tubular parts can be completely filled with metal by brazing, because the filler metal is drawn into the crevice by capillary action and heat, or is placed in the joint before the parts are assembled. The lap joint is the most widely used joint type for brazed connection. Butt and scarf joints are also used. See Figure 20-41.

Brazing operations are identified by the type of filler metal used and by the method used to heat the metal. Terms such as copper brazing, silver brazing, and nickel brazing indicate that the filler metal contains the named element. Heating methods for brazing include torch, furnace, resistance, and induction heating. Various forms of filler metal are used for brazing, including rods, strips, powder, and pastes or powders mixed with semiliquid materials. The filler metal may be applied manually during heating of the joint, or it may be placed in the joint before the metal is heated. Preplaced filler metal is often shaped to conform to the joint surfaces. These are called *preforms*. Brazing requires the use of a flux, unless it is performed in a

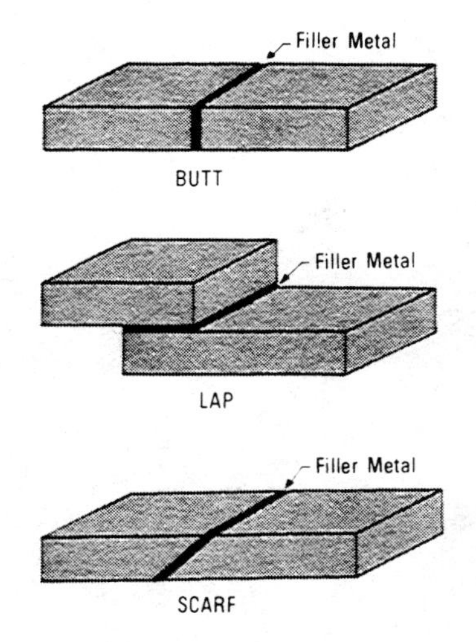

Figure 20-41 The principal joint designs used for brazing. The lap joint is usually strongest because it provides the greatest area of contact. The scarf joint is a butt joint modified to give more surface area. (Courtesy of Inco Alloys International, Inc.)

vacuum or in an atmosphere such as hydrogen that will prevent oxidation of the parts. Fluxes are usually applied as pastes that can be brushed on the workpieces. Small pieces are sometimes dipped in the flux. See Figure 20-42.

Torch brazing (TB) is done with a torch which burns acetylene or other gases. The flux-coated joint surfaces are positioned slightly apart and heated with the torch. The end of the rod or wire of filler metal is placed against one edge of the joint, and the hot base metal melts the filler metal. The molten filler metal is drawn into the joint and distributed evenly over the surfaces by capillary action. Preforms can be placed between the joint surfaces, and the joint heated with the torch. As it melts, the filler metal spreads over the surfaces.

Torch brazing also lends itself to automatic control systems. In one example, aluminum "tee" unions are braze-joined as components for medical equipment. See Figure 20-43. Each "tee" assembly consists of two short lengths of aluminum tubing with a 3 mm (0.1250 inch) wall thickness. A concave die cut is made in the end of one piece to form the saddle-shaped joint. One operator loads and unloads the parts, aided by special fixtures on a rotary brazing machine. As the fixtured parts index clockwise, precise deposits of paste brazing alloy are applied to each joint. This paste is comprised of flux, powdered base metal, and binders. Next,

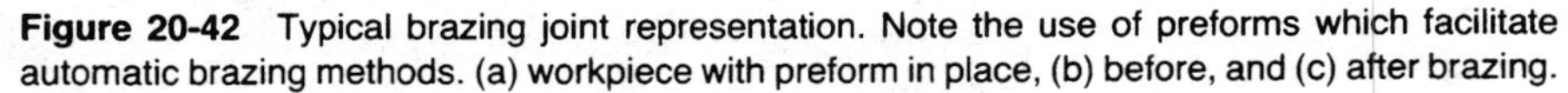

Figure 20-42 Typical brazing joint representation. Note the use of preforms which facilitate automatic brazing methods. (a) workpiece with preform in place, (b) before, and (c) after brazing.

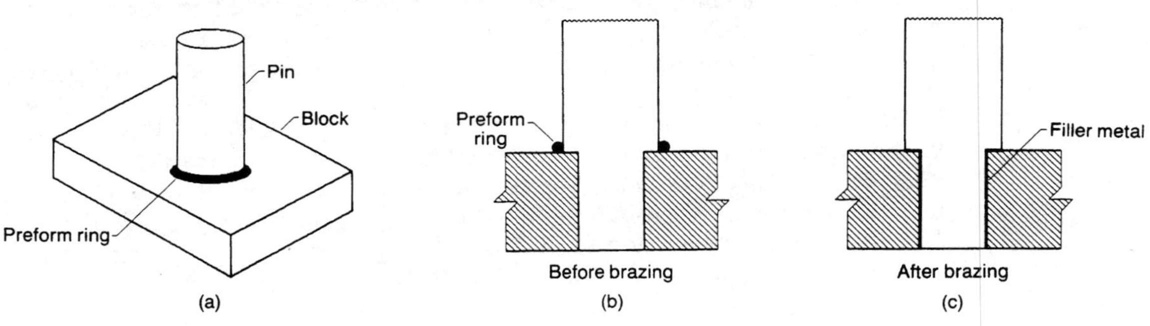

Figure 20-43 Automatic brazing system workpieces.

the parts index through 6 natural gas-fired heating stations, where the aluminum brazing alloy is progressively heated to its 580°C (1,080°F) melting point. This progressive heating technique cuts the time each part spends at a given station, enabling the system to keep moving at 500 cycles per hour. After heating, the parts enter an ambient cooling station, allowing the molten brazing alloy to solidify. Two water spray stations then thoroughly cool the parts and the fixtures, for safe handling by the operator. See Figure 20-44.

Furnace brazing (FB) can be used to complete several joints at the same time. Most production brazing is done in furnaces, and by using preforms. The workpieces are arranged in the desired position with filler metal placed in the joint, and the entire assembly is put into a furnace and heated to the brazing temperature. For an assembly requiring a series of brazing operations, filler metals with progressively lower melting points are used to prevent remelting of previously brazed joints during each heating cycle.

Electrical resistance and induction heating are also used to employ heat for brazing. In *resistance brazing* (RB), the parts are heated by their resistance to the flow of an electric current. In *induction brazing* (IB), the parts are placed in or near an induction coil, which generates heat within the workpieces.

Soldering (S) is essentially brazing with low-strength, low-melting point filler metals. The filler metals, called solders, usually were alloys of lead and tin. Solders now in use preclude those alloys containing lead, for health and environmental reasons. Modern solders are alloys of tin, silver, antimony, and copper. In soldering, as in brazing, a small space between the surfaces to be joined is filled with molten filler metal, and no melting of the base metal occurs. The bond formed by soldering, however, is a weaker bond created as the molten solder wets the base metal.

Soldering is normally used only to seal a joint or to provide good electrical contact between connected parts. Joint strength is usually achieved by the use of fasteners, spot welds, or designs that interlock the parts. Most soldering is done manually with either a torch or soldering iron, which has a metal tip heated by electrical resistance or a flame. The base metal is heated with the torch or soldering iron, and the solder is fed into the joint. Paste or liquid flux is applied to the joint surfaces before the metal is heated. The solder is available as wire, strip, bar, or preforms. Flux-cored solder, tubular wire filled with flux, is sometimes used. Other heating methods for soldering include furnace, resistance, and induction heating,

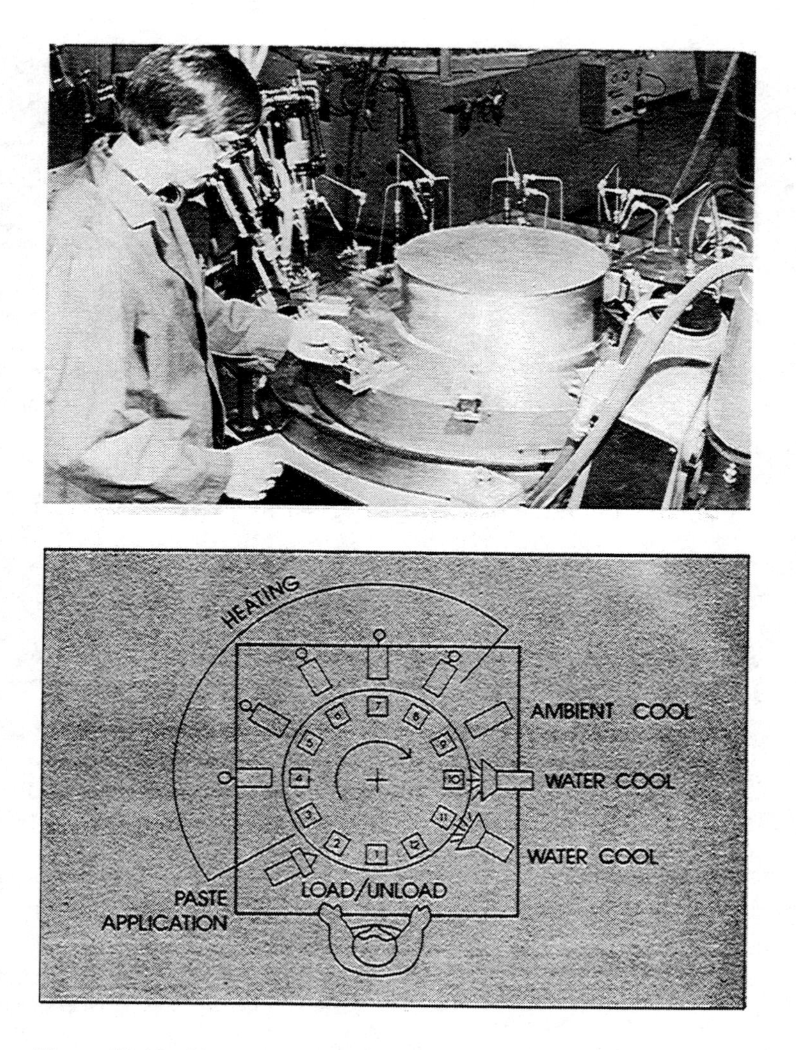

Figure 20-44 Closely-timed machine sequence includes automatic dispensing of paste alloy, heating, and cooling, for consistent part quality. (Courtesy of Fusion Incorporated)

similar to those used in brazing. Solder is usually preplaced in the joint for those methods. Dip soldering, in which the parts are dipped in a molten bath of solder, is sometimes used for joints in small parts. The automatic soldering techniques generally used in electronic assemblies are similar to automated brazing systems.

AUTOMATED WELDING SYSTEMS

When an industrial robot is used in arc welding applications, the robot guides the welding gun along a computer programmed path. Weldments of consistently high quality require accurate positioning, which can normally be achieved with the help of fixtures and proper fitting. With large workpieces and in certain other situations,

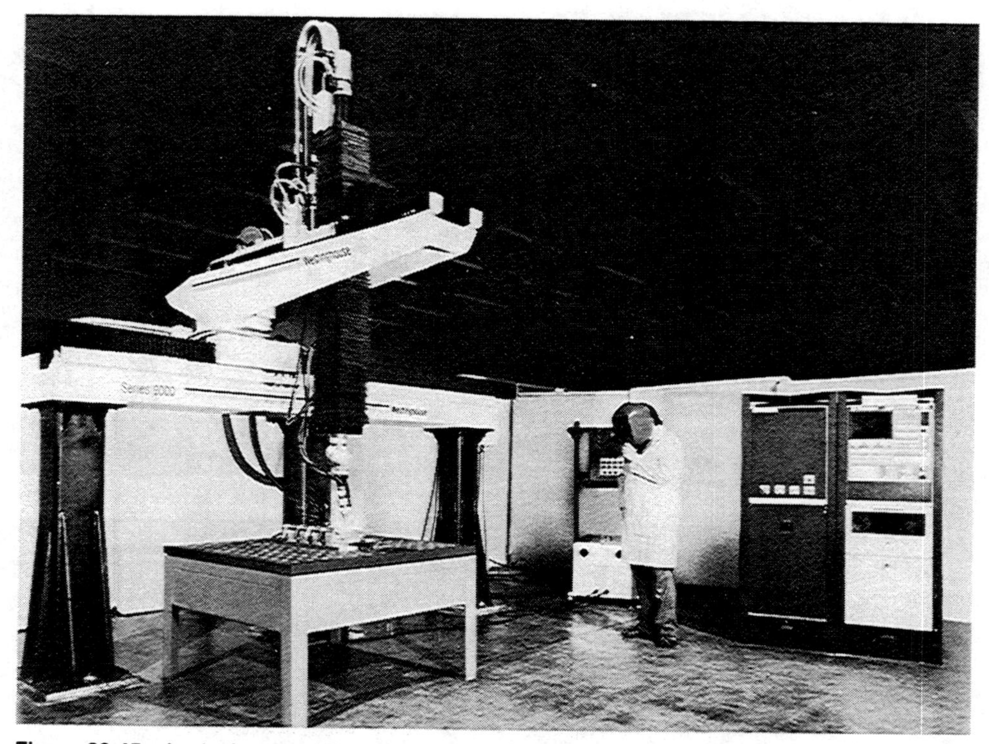

Figure 20-45 A robotic arc welding system. (Courtesy of Westinghouse Electric Corporation, Marine Division)

this is not possible or practical. These problems are resolved by using special sensor systems to permit the robot to position the welding gun properly in relation to the joint.

Such sensor systems typically are integrated with the robot, and consist of an optical sensor and microcomputer, which evaluates the sensor signals and transmits the result to the adaptive functions in the robot control system. See Figure 20-45. In the searching process, the joint is defined in three dimensions and the welding gun is positioned simultaneously. A complete search in three dimensions is often sufficient, and is performed without actuation of the arc. The robot is programmed with information relating to the material thickness and the search type, edge/overlap joint or fillet joint, and other data. Correction of the welding parameters is performed automatically.

In electron beam welding, a seam scanner unit automatically seeks the weld seam, scans its entire length for deviations from a preprogrammed path, and stores these values. See Figure 20-46. The workpiece is then automatically welded in a second pass, using corrected stored path data. See Figure 20-47. Similarly, in a seam tracker system, the scanning, correction of seam deviation, and welding are performed simultaneously using a closed-loop controller. See Figure 20-48. To accomplish this, the beam determines the measurement values by scanning ahead for an extremely short time and with a preselected frequency, to a point immediately in front of the welding position.

Figure 20-46 An electron beam welding automatic seam scanner unit. (Courtesy of Leybold-Heraeus Vacuum Systems, Inc.)

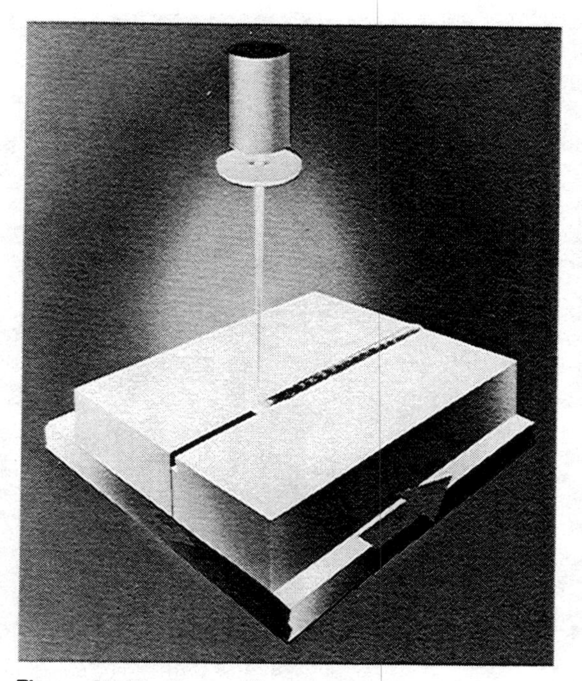

Figure 20-47 The welding process follows the seam scanner pass. (Courtesy of Leybold-Heraeus Vacuum Systems, Inc.)

Figure 20-48 The seam tracker system combines scanning and welding operations. (Courtesy of Leybold-Heraeus Vacuum Systems, Inc.)

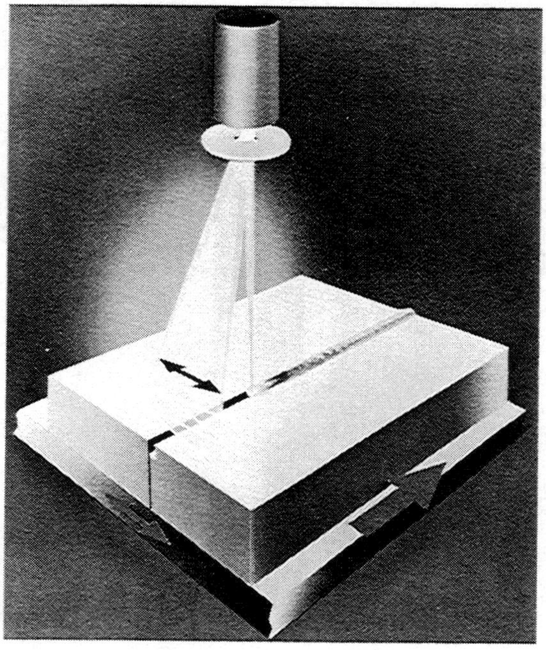

Figure 20-49 An automated electron beam welding unit. (Courtesy of Leybold-Heraeus Vacuum Systems, Inc.)

Both of these systems operate on the principle of reflected electrons, using a reduced power beam as the measuring tool. The beam is deflected transversely across the welding seam of the workpiece. The difference between the electron-reflection intensity at the seam and at the adjacent solid workpiece metal provides the measurement values for the weld-seam coordinates. These automatic EBW methods are employed to produce high quality, precision weldments efficiently and economically, in the required vacuum atmosphere. An example of an operational system and typical workpiece appears in figures 20-49 and 20-50. Similarly, laser welding can be performed on automatic systems. See Figure 20-51.

Figure 20-50 This accumulator pressure vessel is an example of a part produced with the equipment in Figure 20-49. (Courtesy of Leybold-Heraeus Vacuum Systems, Inc.)

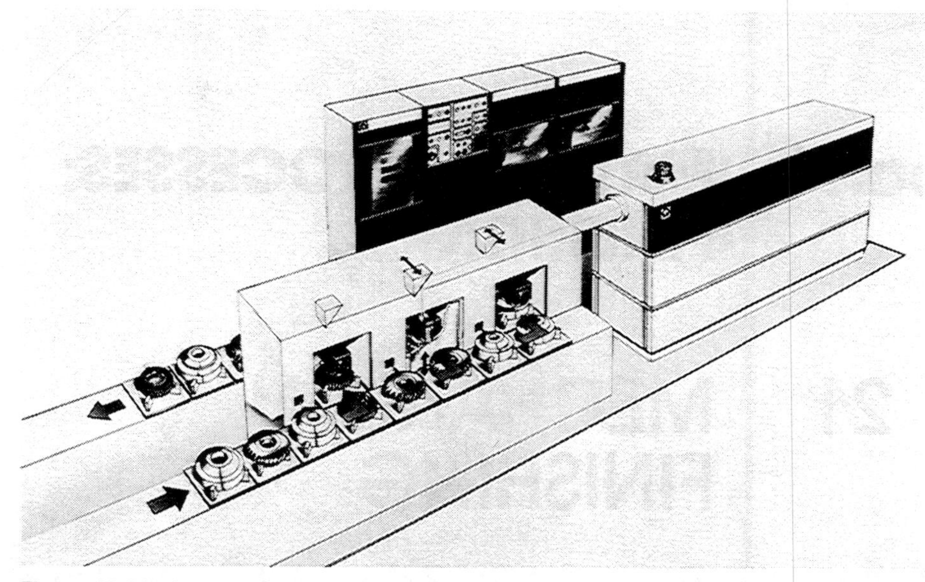

Figure 20-51 An automated laser welding system. (Courtesy of Leybold-Heraeus Vacuum Systems, Inc.)

QUESTIONS

1. Explain the difference between welding, brazing, and soldering.
2. Explain the difference between fusion welding and solid-state welding.
3. Sketch the three basic oxyfuel welding flame patterns, and explain their characteristics and uses.
4. What are the purposes of fluxes and shielding gases in welding?
5. Explain the differences in the uses of flux in welding, brazing, and soldering.
6. How does arc welding differ from oxyfuel welding? From resistance welding?
7. How does GMAW differ from FCAW? From GTAW?
8. What is the meaning of plasma arc?
9. Sketch the sequence of steps used in a typical stud welding application.
10. Explain the difference between electrogas and electroslag welding.
11. Explain how the chain of weld nuggets is produced in seam welding operations.
12. Why is flux unnecessary in resistance welding processes?
13. Explain the similarities and differences between friction welding and flash welding. Can the two processes be used interchangeably?
14. Explain the purpose of the projections in projection welding.
15. What are some of the unique applications of explosion welding?
16. Why is a vacuum necessary in EBW?
17. Explain the operation and attributes of the laser beam welding process.
18. How do LBW and EBW differ from one another?
19. Explain the automatic methods employed in welding, brazing, and soldering.
20. Prepare a library research report on current automated welding systems.

part 6 MATERIAL PROCESSES: FINISHING

21 MECHANICAL FINISHING

Mechanical finishing systems typically involve the use of abrasives to condition the surfaces of metallic and nonmetallic workpieces for the purposes of function and appearance. See Figure 21-1. The generation of smooth surfaces is a necessary step in preparing a wooden furniture part for subsequent paint or lacquer coating. Similarly, a steel spoon or knife must be smoothfinished to assure an effective silverplating operation. The appearance of both examples is therefore enhanced by mechanical finishing, and the functionality of each piece is assured because the pieces will be more lasting, due to a quality finish. Mechanical finishing is also a final manufacturing step, as pertains to abrasive blasting of a matte finish pattern on glass windows or a metallic part. The many techniques used to produce mechanical finishes are the subject of this chapter. Also included is an examination of the several important methods of cleaning workpieces as a preparatory phase of both mechanical and surface coating finishing processes.

SURFACE CLEANING

Cleaning is the process of removing unwanted material from workpieces, generally in preparation for further processing. This undesirable matter may take the form of soil, grime, grease, scale, stains, blemishes, or corrosive deposits, and is removed by a range of special techniques. Parts that are caked with hard dirt deposits are effectively cleaned by abrasive blasting or pressure hose spraying. This coarse treat-

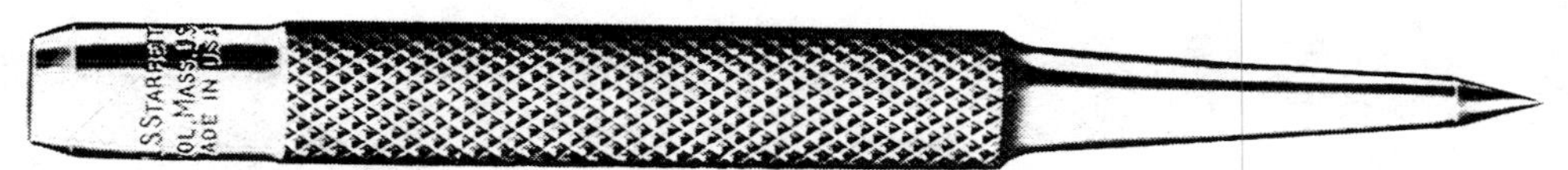

Figure 21-1 This prick punch is typical of hand tools which must undergo mechanical finishing as an element of their manufacture. The surface must be polished and buffed prior to coating or plating. The knurling provides a sure finger grip when the tool is used. (Courtesy of the L. S. Starrett Company)

ment is followed by washing and decreasing as required by subsequent finishing treatments. Cleaning may be done manually, or by using automated system for high-volume production. Some of the more common and effective industrial cleaning techniques are described here.

Alkaline cleaning employs solutions of alkaline salts, phosphates, and silicates, which soften the liquid and hold dirt particles in suspension. Organic compounds provide detergent, emulsification, and wetting action to promote cleaning. Workpieces can be immersed in such hot solutions, or can be sprayed or brushed with it. Agitation or barrel tumbling methods may be used for difficult cleaning tasks. This alkaline technique is commonly used for general cleaning, and is achieved by one or more of four major mechanisms. *Wetting* is the action of cleaning agents which weaken the metal-dirt bond by lowering the surface tension of the deposits, and is the first step in metal cleaning. In *saponification*, fatty acids react with alkaline salts to form water soluble soap, and are subsequently washed away. *Emulsification* is the process of joining two mutually immiscible or insoluble liquids, such as oil and water, through the use of a detergent, or surficant. The milky or cloudy emulsion is drained off and the part is rinsed; therefore the use of emulsifiers may be limited due to environmental concerns. In *dispersion*, a detergent lowers the surface tension of the cleaner as it reaches the workpiece, allowing it to completely and uniformly cover the part to be cleaned. The cleaner penetrates the dirt layer and disperses any oil present into small droplets, which float to the surface of the liquid cleaning solution. Each of these major mechanisms make possible the effective cleaning of workpiece surfaces.

Electrolytic cleaning involves the placement of a metal workpiece in an alkaline solution and passing a direct electric current through the circuit. This causes the liberation of gaseous oxygen bubbles at the workpiece surface, which serves to agitate the solution and free the dirt from the part. This method is generally used in ferrous metal cleaning. The workpiece can also be the cathode, where a higher scrubbing level is reached for the cleaning of nonferrous pieces.

Solvent cleaning is used to remove organic compounds, such as grease and oil, from a metal workpiece. Such dirt is readily dissolved by the solvent and then rinsed from the parts to be cleaned. Typical solvents include alcohol mineral spirits, acetone, kerosene, and chlorinated hydrocarbons, and are generally used at room temperatures. Small parts can be immersed, while larger workpieces are wiped or sprayed, with or without agitation.

Vapor degreasing is a solvent cleaning method whereby vapors from a boiling, nonflammable solvent such as trichloroethylene condense on a cool workpiece and dissolve grease or soil. This action continues until workpiece temperature reaches that of the vapor, at which point the condensation halts, leaving a clean and dry workpiece. A typical vapor degreaser unit consists of an open steel tank, with a heated solvent reservoir at the bottom, and a cooler work zone near the top. The

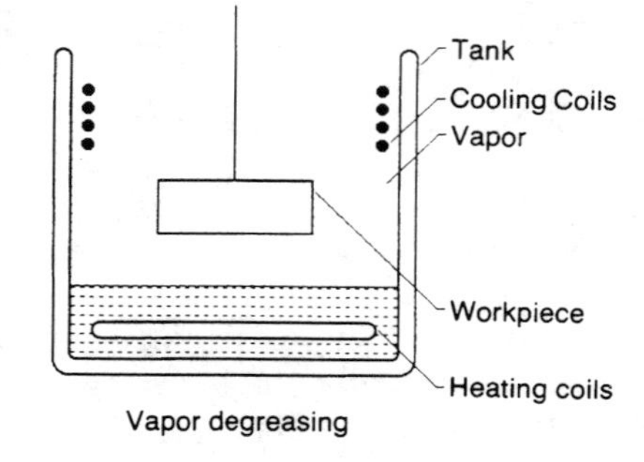

Figure 21-2 The vapor degreasing process.

simplest system features a vapor phase only. See Figure 21-2. Here, the workpiece is lowered into the vapor zone where the vapor cleaning occurs, and is especially effective on flat or simple parts. Other methods include vapor sprays, and warm and boiling liquid immersion for heavy dirt deposits on complex workpieces.

Acid cleaning employs solutions of mineral or organic acids, or acid salts to dislodge solid dirt deposits from metal surfaces. These solutions may be used in conjunction with detergents, with or without the application of heat. Acid cleaning is used on metal wires, sheets, and shapes as a final cleaning technique prior to storage or shipping. *Acid pickling* is a more severe form of chemical cleaning, commonly used to remove oxides or scale and other contaminants from forgings and castings. Pickling solutions generally include sulfuric and hydrochloric acids, volatile chemicals which are very toxic and must be used with caution to protect the worker, the product, and the environment. An accepted procedure is to subject workpieces to solvent or alkaline cleaning prior to using acidic treatments to assure uniform wetting.

Abrasive blast cleaning involves the action of abrasive particles propelled by compressed air, and sometimes water, through a nozzle, and directed at workpiece surfaces. See Figure 21-3. As such, it has a range of surface preparation uses, such as the removal of rust, scale, and paint; roughing surfaces prior to painting or coating to enhance their adhesion; and removing flash and burrs. In addition to the pressure blast system, *airless* or mechanical blast methods are used, whereby the abrasive is propelled to the workpiece by the centrifugal force of a bladed wheel.

Figure 21-3 The action of a pressure blast system gun, employing air, abrasive particles, and water. Waterless systems also are used. (Courtesy of Clemco Industries)

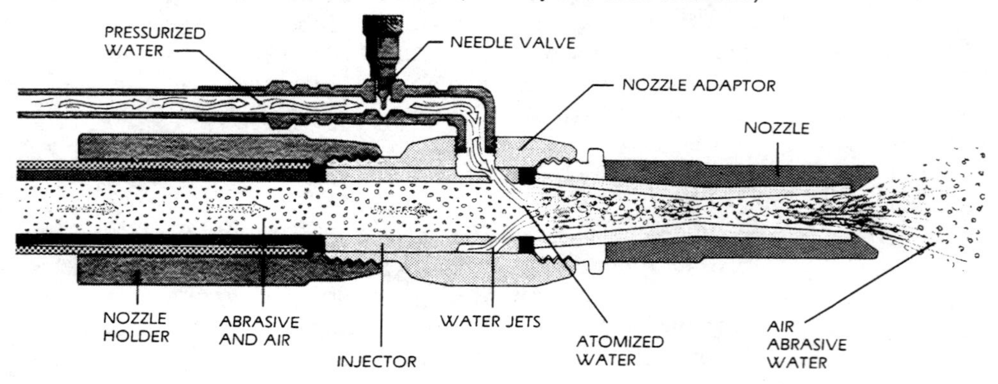

Figure 21-4 Suction blasting cabinet features an ergonomically designed seating/operating arrangement. (Courtesy of Pressure Blast Manufacturing Company, Inc.)

Suction or induction blasting cabinets restrict the size of workpieces, but are easier to use because of the confinement of the part. See Figure 21-4. Here the grains are sucked up from a hopper, forced against the workpiece by an air system, and recycled. Regardless of method, abrasive blasting is an effective way to clean workpieces. See Figure 21-5. The media used in blast cleaning include glass beads, crushed glass, garnet, aluminum oxide, steel shot and grit, plastic granules, and crushed nutshells. Blasting with nonabrasive steel shot or grit can also impart high compressive stresses on a workpiece surface, to improve the fatigue strength of parts such as springs and gears. This is known as *shot peening*. Abrasive blasting also can be a

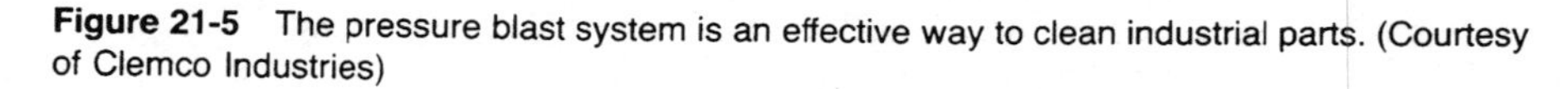

Figure 21-5 The pressure blast system is an effective way to clean industrial parts. (Courtesy of Clemco Industries)

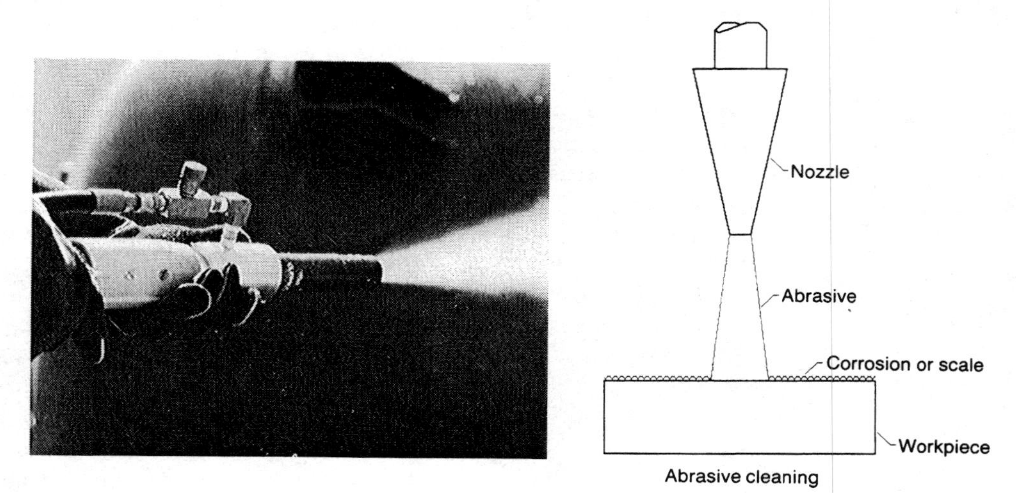

Figure 21-6 Coated abrasive flap wheel used to bore clean tubes. (Courtesy of Merit Abrasive Products, Inc.)

finishing process to produce a matte surface on glass or metal, applications which will be described later in this chapter.

Coated abrasive flap wheels are made up of tabs of abrasive-covered cloth and secured in special mandrels. These are available in many styles, and are used to descale areas difficult to manage by other means. See Figure 21-6.

When workpieces have been thoroughly cleaned by one of the previously described processes, they are ready for further finishing processes, such as the mechanical methods described in this chapter, and surface coating and coloring, which is the subject of the following chapter. The primary mechanical methods include polishing, buffing, brushing, mass finishing, and abrasive blasting.

POLISHING

As a finishing process, polishing is an intermediate abrading operation which follows grinding and precedes buffing. Its purpose is to remove deep scratchs, nicks, discolorations, and similar surface imperfections which naturally occur as a result of grinding and other machining operations. Consequently, a considerable amount of surface material is removed in the act of polishing. Conversely, there is very little material lost in buffing, whose prime purpose is to improve appearance. Inasmuch as there is resiliency in polishing and buffing heads, these are not precision processes, as are grinding, lapping, and honing.

The polishing process uses abrasive grains which are firmly attached to flexible backing such as sleeves, discs, flaps, and belts (called coated abrasives; see Chapter 13), or by adhering such grains to flexible felt or fiber wheels. This is to be contrasted to grinding, which employs abrasive grains packed and bonded to form a solid grinding wheel. Polishing, therefore, is an abrading process, and involves the use of several types of equipment.

The *flat finisher* is a machine used to smooth-finish disc clutch and brake facings, electric iron soles, door kick plates, slate tile, wooden furniture parts, chain saw guide bars, plastic tile, and skate blades. See Figure 21-7. Such machines have work heads which carry abrasive belts stretched over drive and idler drums, as shown in Figure 21-8, or abrasive drums, polishing wheels, or brushes. Typically, belt heads come in widths of 150 mm (6 inches) to 300 mm (12 inches) to handle appropriate workpieces. Larger panel sanders, up to 2.5 m (8 feet) are used to

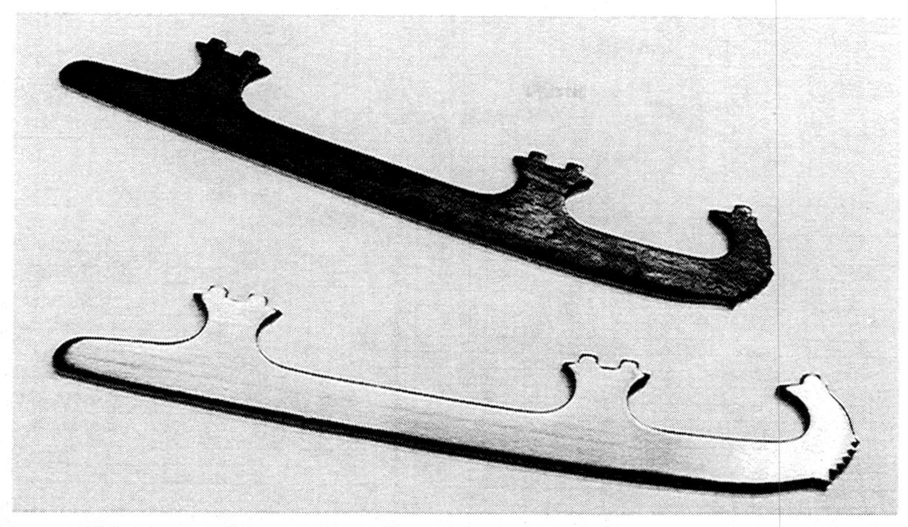

Figure 21-7 Flat finished, heat-treated skate blades, before (above) and after (below) finishing. This smoothing and cleaning operation precedes plating. A four-head wet finisher was used at the rate of 1,000 slides or 500 blades per hour. A coolant system keeps the blades cool and the abrasive belts clean. (Courtesy of Hammond Machinery, Inc.)

smooth-sand plywood sheets and furniture components. See Figure 21-9. The efficacy of an abrasive belt in such operations is that a fresh finishing surface is presented to the workpiece due to the belt action, and the belt length provides for self-cleaning and cooling. Furthermore, several heads can be joined to create a one-pass complete finishing system. For example, the skate blades shown in Figure 21-7 were polished in a four-head system, with a belt sequence of 180, 240, 320, and 400 grit. This facilitates a high production rate for this and similar products, and lends itself to automatic loading and unloading of parts. See Figure 21-10.

These versatile tools are used for moderate stock removal, fine-finishing, deburring, coil strip finishing, and oxide removing on a variety of material applications, and can be done wet or dry.

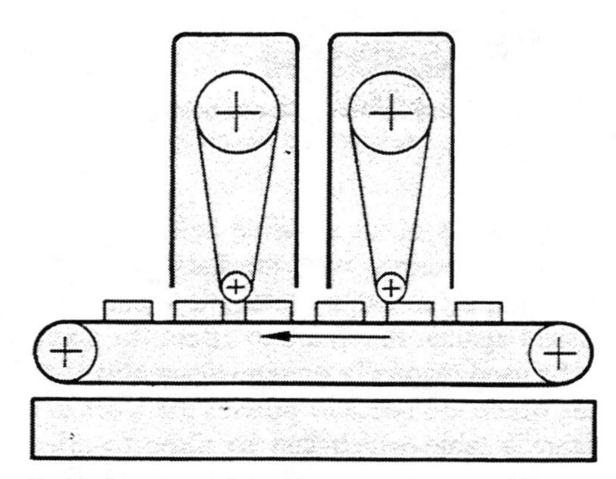

Figure 21-8 This two-head flat finisher has a continuous throughfeed conveyor to carry the workpieces to the abrasive belts at the heads. A number of heads can be added to the system according to the operational requirements. (Courtesy of Hammond Machinery, Inc.)

Figure 21-9 This table top is made of glued oak members sanded smooth by automatic machines to produce a smooth, burr-free flat surface prior to the application of protective varnish or lacquer coat. The shaped seats are smoothed with profile sanders. (Courtesy of Plymold Booths, Division of Foldcraft Co.)

Figure 21-10 These steel soles for electric irons are being ground smooth in one pass on this four-head flat finisher. (Courtesy of Hammond Machinery, Inc.)

The *platen finisher* consists of an abrasive belt stretched between two vertically mounted pulleys, much as obtained with the flat-finishing machine. The difference is that a steel back-up plate, called a platen, is located between the pulleys, and the workpiece is pushed against this plate to effect the abrading. See Figure 21-11. These machines can be used manually or semiautomatically. The machine illustrated is manually operated, where the operator presses the work to the plate. The semiautomatic machines feature a table which can be tilted to orient the workpiece to the platen. See Figure 21-12. A variety of different clamps and fixtures can be used. This machine facilitates the rapid abrading of work surfaces, where smoothing is

Figure 21-11 A standard wet model industrial platen grinder with a manual table. Semiautomatic fixtures can be mounted on this table. (Courtesy of Hammond Machinery, Inc.)

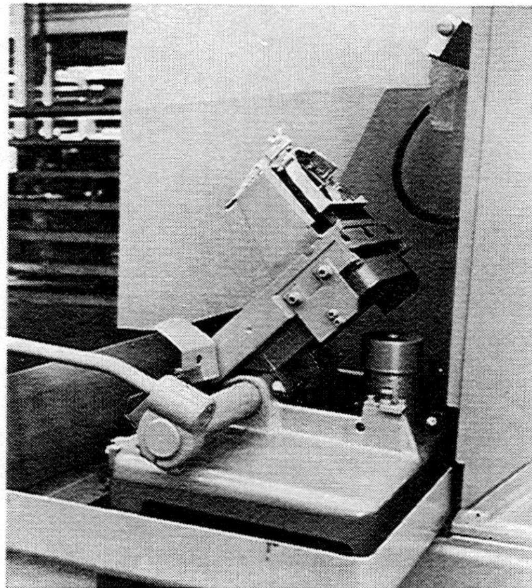

Figure 21-12 A semiautomatic tilting table can be used to accurately orient and carry the workpiece to the platen. (Courtesy of Hammond Machinery, Inc.)

more important than precise sizing. See Figure 21-13. Workpiece stops can be located on the machine to control the amount of stock removal.

The *horizontal spindle, rotary table finisher* is a flatfinishing adaptation of a grinder described in Chapter 14. Instead of a grinding wheel, wire brush wheels are used on this machine. In a typical application, gears are manually loaded on an idle station and indexed (moved) to the work station where a skiving tool removes heavy burrs as the gear is rotated. See Figure 21-14. The skiving tool (a type of

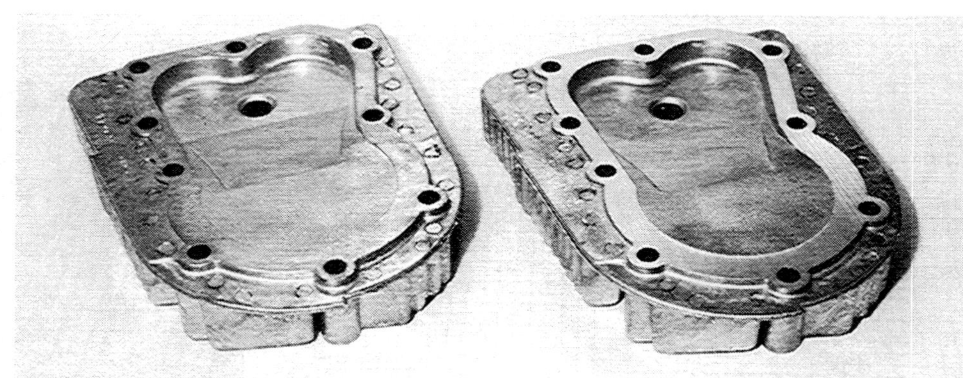

Figure 21-13 Before and after die cast aluminum cylinder head covers. The gasket surface is ground to +/− 0.005 mm (0.002 inch), at the rate of 150 pieces per hour. (Courtesy of Hammond Machinery, Inc.)

Figure 21-14 A horizontal spindle, rotary table machine with wire brush wheels instead of a grinding wheel. (Courtesy of Hammond Machinery, Inc.)

milling cutter) is then lifted after one rotation of the gear. A split wire brush—two separated wheels—then deburs the gear face at the same station, and during the same dwell cycle. See Figure 21-15.

This is a typical example of brush finishing, and the efficacy of the brush method is that it provides a flexibility necessary for parts with contours, or those requiring both flat surface and edge treatments, such as gears. Fiber brushes sprayed with

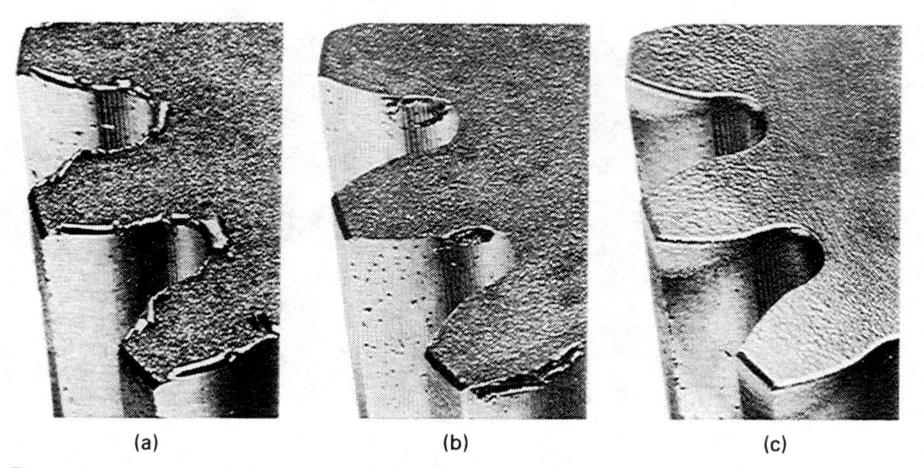

Figure 21-15 The gear finishing sequence on the above machine: (a) rough, (b) skived, (c) brushed. Note the clean, burr-free surface, produced safely, economically, and efficiently. (Courtesy of Hammond Machinery, Inc.)

liquid abrasive compounds are commonly used to polish contoured metal pieces, such as tableware and cutlery.

Edge-contour finishing poses a different set of problems than occurs with flat-finishing. Wood, plastic, and composition board parts frequently have edges which must be smoothed to remove machine marks created by cutting an edge to a desired shape. One way of smoothing such edges is to press a moving abrasive belt against the edge with a form block shaped to the profile of the workpiece edge. Other blocks can be prepared by special casting of dense plastic foams, which are then covered with abrasive to create a profile sanding block. Hard abrasive profile wheels also are used. Another more efficient method is to use a specially prepared flexible profile sanding wheel. (Sanding is a term used in wood and soft, nonmetallic material finishing. It is synonymous with the term polishing.)

The profile wheel consists of a black rubber body sandwiched between two metal disk plates. The desired profile is cut into the wheel by running it at high speed against the workpiece, which is covered with abrasive cloth. When the profiling has been completed, the rubber wheel is covered with strips of abrasive material to create an accurate sanding tool. See Figure 21-16. The construction of the tool

Figure 21-16 Profile sanding wheel with abrasive sticker strips glued in place. Note the wooden molding workpiece. The rubber wheel was profile-cut by covering the molding surface with abrasive cloth and pressing the rotating rubber wheel against it. This assures a precise profile wheel shape. (Courtesy of AC Compacting Presses)

Figure 21-17 Sanding a table top edge with a profile wheel mounted on a CNC router. (Courtesy of AC Compacting Presses)

allows the rubber to expand at operating speed to increase the outer diameter of the wheel. This outward movement of the rubber creates an air cushion gap between the rubber and the metal shaft or bushing used to attach it to a sanding machine. This feature causes the wheel to contact the workpiece with a flat footprint of about 30 mm (1.25 inches). This deformation, similar to contact of an automobile tire on the road, eliminates chattering and the resultant wavy finished surfaces, which are a usual feature of hard wheel systems. Such profile wheels can be used with common woodworking machines such as routers, table and profile sanders, shapers, and copy millers or sanders. See Figure 21-17. A variation of this system is to prepare a series of profile blocks and adhere them to a flexible belt. This subsequent profile sanding belt provides for fast speed rates and a long belt life. See Figure 21-18. Profile sanders are also used to smooth painted or lacquered edges between coats.

Figure 21-18 Here, a profile sanding belt is mounted on a rotary shaper to edge-sand a contoured wooden workpiece. (Courtesy of AC Compacting Presses)

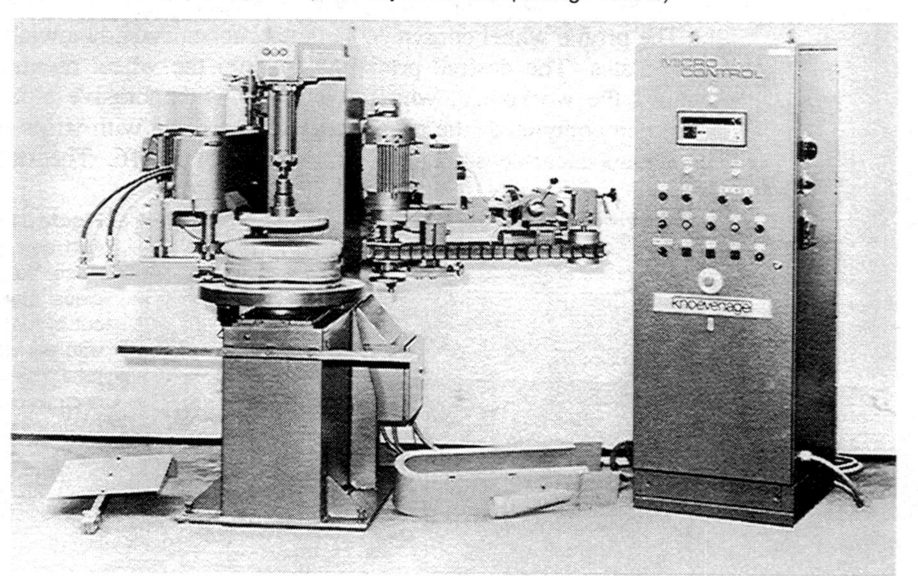

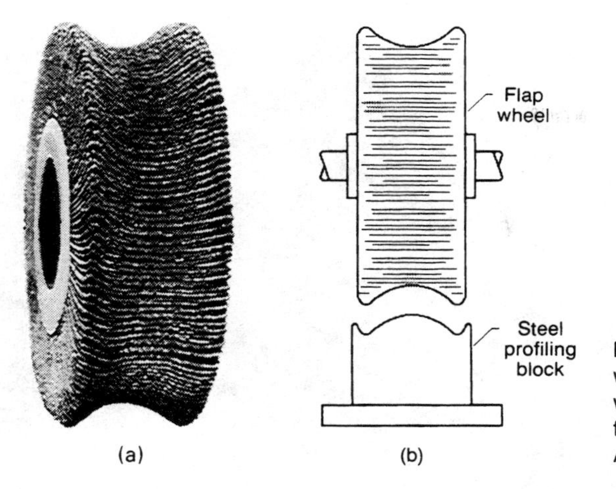

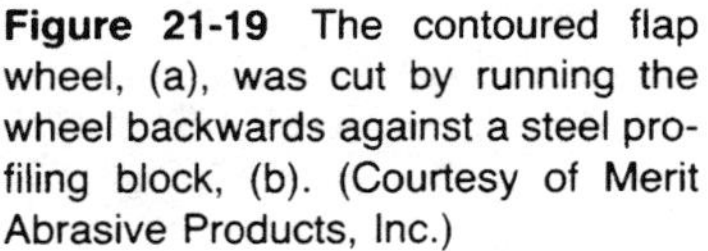

Figure 21-19 The contoured flap wheel, (a), was cut by running the wheel backwards against a steel profiling block, (b). (Courtesy of Merit Abrasive Products, Inc.)

The *flap wheel*, shown in Figure 21-6, can be preformed for edge contour finishing by using a profiling block made with carbide grit permanently bonded to a steel template. See Figure 21-19. The flaps provide a uniform finish as new cutting edges are continually being formed as the flaps wear. Noncontoured flap wheels also are used to dress and clean bottle molds, and for general polishing of almost any workpiece, of any material. See Figure 21-20. A range of woven nylon fiber discs impregnated with abrasive grains find use in many polishing applications. See Figure 21-21.

A variety of flexible polishing wheels made of felt, canvas, or leather are used both for automatic and manual smoothing operations on flat and contoured workpieces. The wheels may be coated with adhesive and rolled in abrasive grains to produce smoothing wheels to suit the shape of the part. An abrasive medium also can be sprayed onto the wheel during operation, and careful metering can assure the proper coverage to finish the workpiece without unduly loading the wheel.

Off-hand polishing operations, where the operator carefully manipulates the workpiece against a flexible wheel, are especially important in smoothing heavily contoured

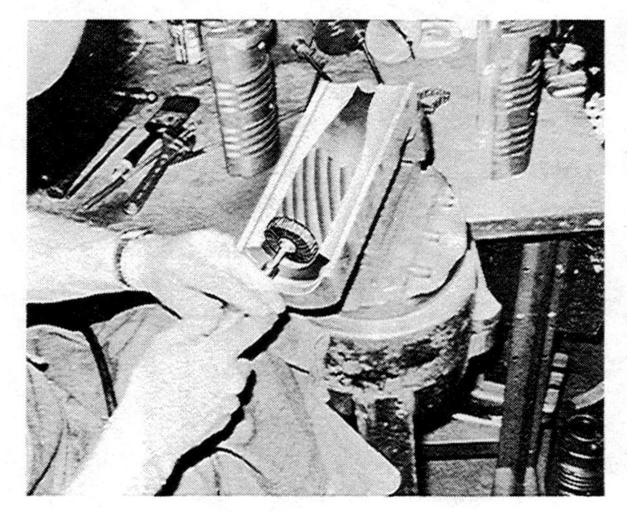

Figure 21-20 Flap wheels used to polish and clean bottle molds. (Courtesy of Merit Abrasive Products, Inc.)

Figure 21-21 Nylon fiber abrasive polishing disks. (Courtesy of Merit Abrasive Products, Inc.)

objects such as teapots, cream pitchers, plastic vases, and similar items. Here the sensitive human touch and observation is necesary to assure a quality finish. See Figure 21-22. Such operations are primarily manual, and cannot be effected by automatic systems.

Buffing is an operation used to apply a high luster to an object, and like polishing can be done manually or automatically according to the dictates of the workpiece. For example, flat tableware can be automatically buffed by passing knives and forks under a revolving buffing head on a conveyor system. See Figure 21-23. Other objects, such as a teapot, must be done by hand. Buffing follows polishing and is to be considered the final abrading operation in preparing a workpiece for the consumer market, or as a precedent to plating or other coating operations. Buffing

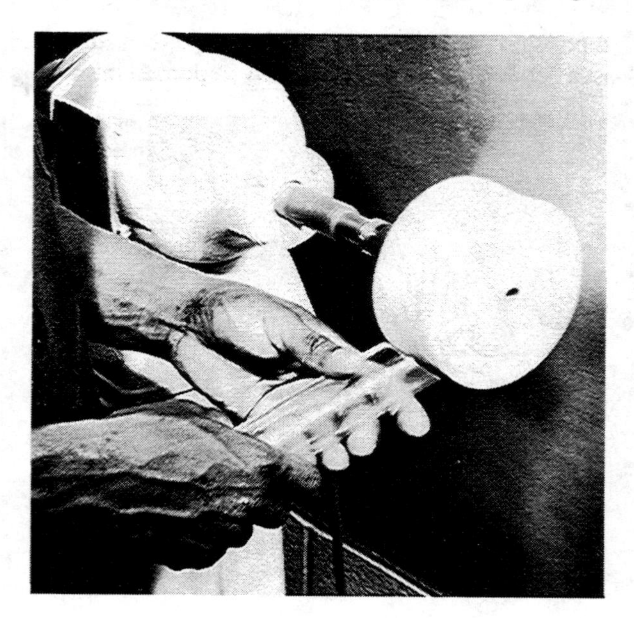

Figure 21-22 Off-hand polishing and buffing of contoured parts, such as this plastic vase, is a manual operation.

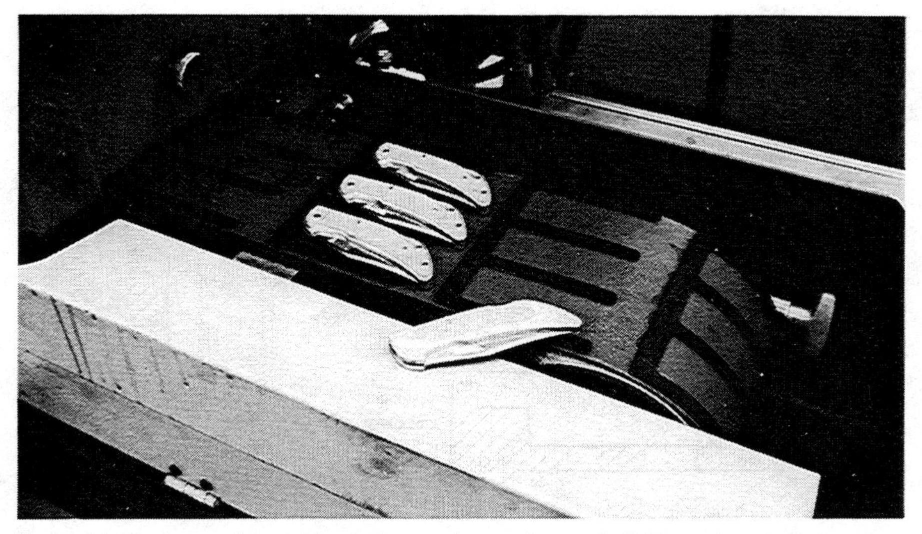

Figure 21-23 Automatic machine buffer used to apply a satin finish to these knife handles. (Courtesy of Hammond Machinery, Inc.)

also is employed after silverplating (on silver tableware for example) to impart a high luster, or a fine satin finish.

Buffing wheels are made from loosely-bound flannel discs about 25 mm (1 inch) wide, and may be assembled on a shaft to form a buffing wheel up to 300 mm (12 inches) in width. Wider wheels such as these are used to buff flat trays and plates. Buffing compounds are abrasive grains formed into a 50 × 50 × 300 mm (2 × 2 × 12 inch) solid bar, using a binder of clay, grease, or wax. In use, they are touched to a slowly-spinning wheel to coat it with just enough abrasive to do the buffing, without either glazing the wheel or discoloring the workpiece. Different grain sizes may be used to impart a satin finish to stainless steel tableware. Liquid sprays are also used in automatic systems.

Figure 21-24 This metal vase was distressed to give an attractive contrasting color and surface texture.

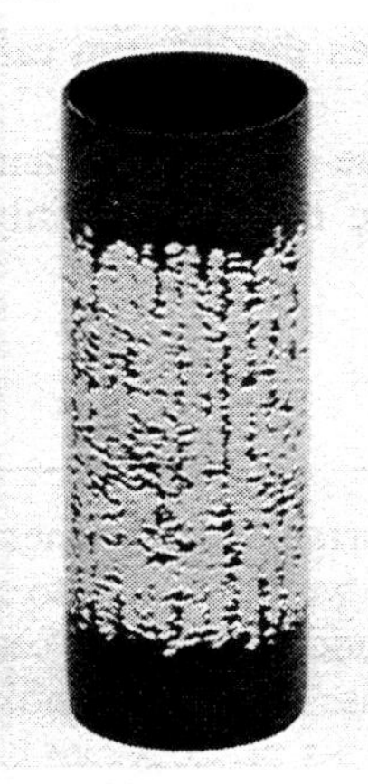

An important distinction to be made is that while polishing is an abrasive operation used to remove material, buffing and brushing primarily involve the gentle plastic deformation of a surface into a bright, or sometimes satin, smooth, and soft final result. This deformation "flows" nonferrous metals into scratch and nick areas. Generally, buffing and brushing are employed for cosmetic purposes.

Abrasive blasting was described in some detail earlier in this chapter as a method of cleaning workpieces. Impinging a surface with fine abrasive grains, either dry or in a water carrier, is also a way of conditioning a surface for plating, or for imparting a matte finish to metal, wood, glass, and plastic parts.

Abrasive distressing is done with a hand-held grinding tool to impart a decorative pattern to craft pieces. The steel vase shown in Figure 21-24 was oxide finished and then abraded with a carbide tool.

Abrasive flow machining is a special parts-finishing process whereby abrasive grains embedded in a thick slurry flow over the edges and surfaces of workpieces. Burrs can be removed, surfaces polished, and surface finish vastly improved. Radii can be applied, and material removed to clear and resize a hole or passage. Abrasive flow machining (AFM) is a versatile method, often used to advantage where hand

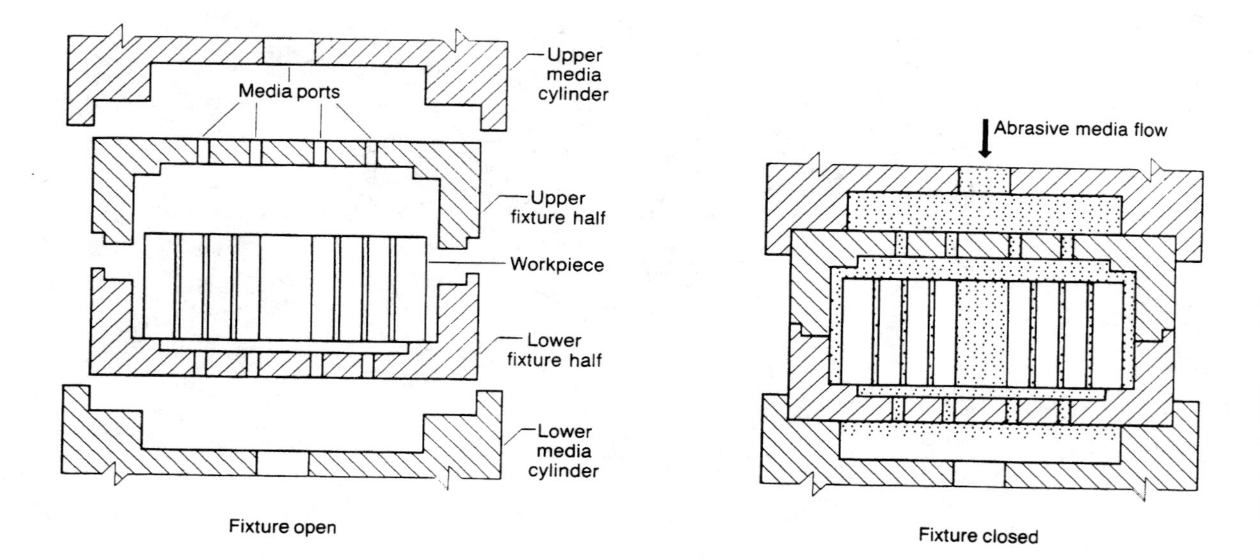

Figure 21-25 Abrasive flow machining diagram.

finishing operations have been customary. It can be applied to metals as soft as aluminum or as hard as tungsten carbide.

The equipment used in this process features two vertically opposed media cylinders, the lower one being stationary and the upper one capable of being moved up and down. See Figure 21-25. A fixture enclosing the workpiece is placed on the work table over the lower media cylinder. The upper media cylinder is brought down to clamp the fixture between the cylinders. Abrasive flow machining begins when a hydraulically driven piston pushes the media in the lower cylinder up through the fixture holding the work and into the upper cylinder. A flow cycle is completed when an opposing piston pushes the media back through the work into the lower chamber. The number of cycles depends on the hardness of the material being machined, the volume of media displaced per stroke, and the specific task being performed. Sometimes it is as few as three cycles, at other times the job can call for a hundred or more cycles.

AFM is applicable to conditions where the work requires time-consuming hand finishing, the workpiece configuration is complex, and parts are not satisfactorily or safely finished by tumbling, vibrating, or other bulk methods.

MASS FINISHING

Mass finishing is a term used to describe the process of edge and surface improvement on bulk quantities of parts. It is accomplished by loading the workpieces into a container together with an abrasive medium, water, and some additives, or it can be done dry. By applying a vibrating motion to the container, the contents tumble and rub together to clean and enhance the workpiece surfaces. This process can be used to deburr, generate edge and corner radii, remove paint, scale, rust, and other incrustation, and produce surfaces suitable for subsequent coating. Mass finishing is often the only economical method of conditioning large quantities of parts.

The media used in this finishing method include small steel, ceramic, plastic, and wooden shapes, cinders, leather cuts, nut shells, and abrasive grit particles. The compounds include detergents, lubricants, and rust inhibitors added to keep both the workpieces and media clean by preventing loading or glazing, to promote the action of the media, and to develop color or luster on the parts. There are several basic mass or bulk finishing methods.

Barrel or tumble finishing involves the use of a horizontal tumbling tank to contain the media and parts. The abrading takes place when the upper layer of the mass slides down the slope created by the rotation of the barrel, and causes pieces to rub together. The barrel is normally loaded about 50 percent full for the most efficient operation. This type of finishing is most beneficial at slow speeds for long periods of time.

Vibratory finishing occurs in an open-topped tub or bowl mounted on springs, the action being caused by a special vibratory motor, or an eccentric shaft linkage system. The movement of the medium against the workpieces occurs throughout the entire load, resulting in considerably shorter process cycles than with barrel finishing. *Tub vibrators* resemble elongated bathtubs lined with urethane, whose design permits the automatic flow-through of parts in a finishing cycle, which can vary from 2 to 45 minutes. See Figure 21-26. *Bowl vibrators* are doughnut-shaped containers, urethane lined, mounted on springs and actuated by an eccentric weight system. Parts are loaded in the machine and move spirally through the medium,

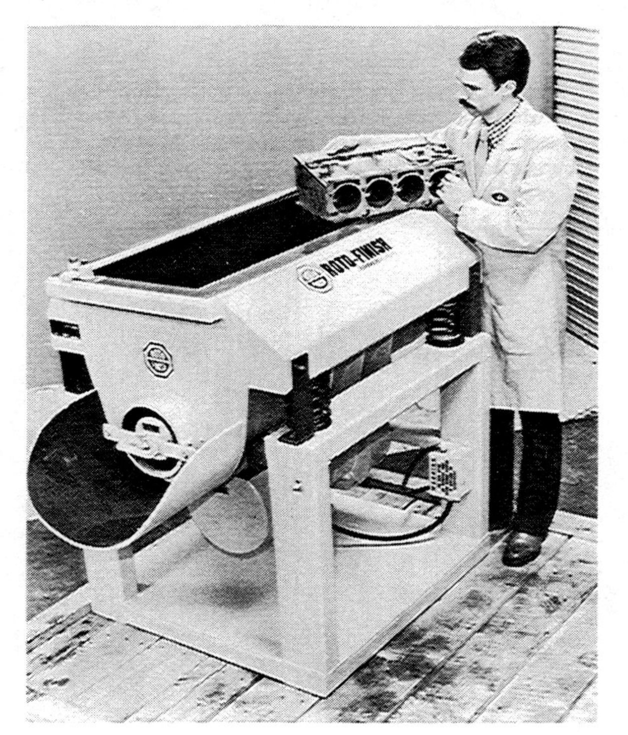

Figure 21-26 The vibratory tub finishing method. (Courtesy of ROTO-FINISH)

Figure 21-27 The bowl-type vibrator. (Courtesy of ROTO-FINISH)

and pass from the barrel onto a separating screen. These versatile batch machines can be converted to automatic continuous process systems, complete with spray rinses and magnetic separators. See Figure 21-27. Larger workpieces, or those that are made of soft materials which must be prevented from touching one another, can be placed in special fixtures to contain them during the abrading process. The shape of the bowl is important in such operations. See Figure 21-28.

Barrel and vibrating methods are used in a great number of applications, such as metallic, plastic, ceramic, composite, rubber, and wooden parts, and animal hides.

Figure 21-28 Vibratory bowl designs. Straight-wall machines can be used in any process in which the parts remain integral with the media mass. Curved-wall machines are generally used in processes in which the part density is less than or equal to the media density, to keep the parts from floating to the top of the media mass. (Courtesy of ROTO-FINISH)

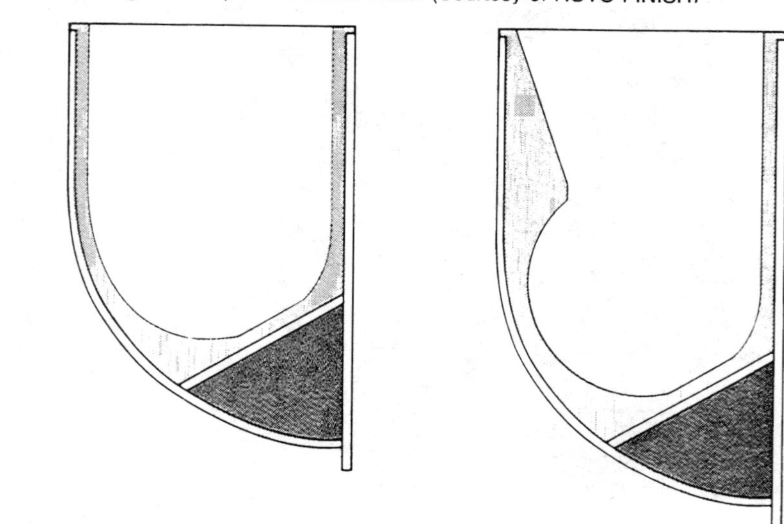

Spindle finishing is a method which requires that the workpiece be chucked on spindles which slowly rotate the parts in a fine abrasive medium. The medium container may or may not be revolved at high speeds during this process. The rate of material removal is controlled by the rotational speeds, the depth of immersion in the medium, the angle of inclination of the mounted workpiece, and the cycle time. This method is especially practical for finishing fragile parts, and for deburring machined parts.

Centrifugal finishing methods are high-energy processes, and are of two types. *Centrifugal disk finishing* employs a vertical stationary cylinder with an open top and a funnel-shaped disk bottom. The parts and medium are loaded into this container. The disk bottom rotates at high speeds, forcing the mass outward against the bowl sides which apply a braking action to the load. See Figure 21-29.

This slowing down of the load causes the parts and medium to press against each other in a scrubbing action, under forces 10 to 20 times gravity. The mass is then forced upward to the top of the load, where it falls inward and then down to the disk, where the action repeats. This provides a short cycle time and a very effective finishing process. *Centrifugal barrel finishing* is another high speed process, using a number of small drums mounted on the periphery of a turret. The turret rotates at high speed up to 100 times gravity, in one direction, while the drums rotate at a slower speed in the opposite direction. The drum loading is much like that of barrel or vibratory loading, with parts, a media, and a compound. The centrifugal turret motion compacts the drum loads into a tightly packed mass. The drum rotation causes the media to rub against the workpieces to clean and refine them. Like the centrifugal disk process, the barrel method is fast because of the high pressure abrasive action.

Figure 21-29 Centrifugal disk finishing. (Courtesy of ROTO-FINISH)

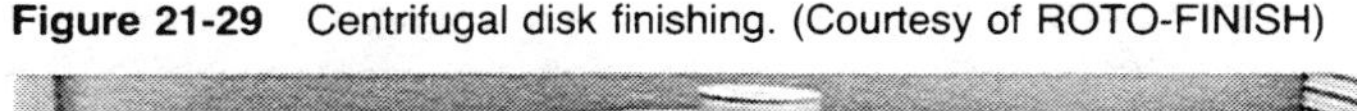

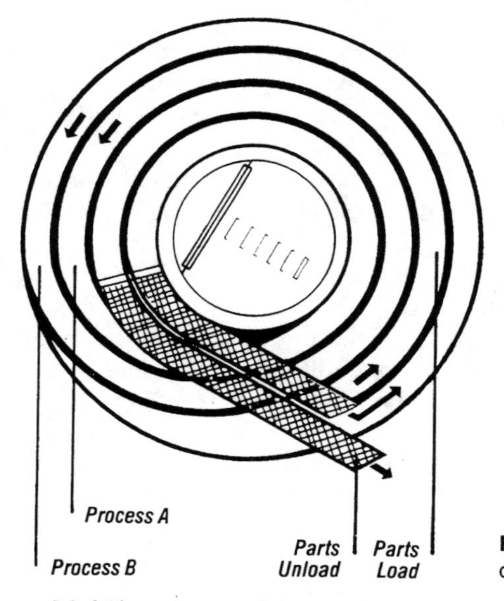

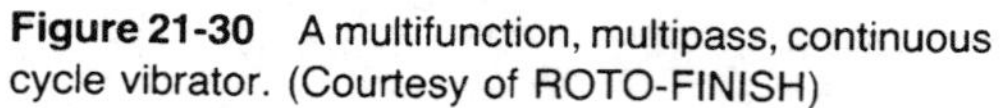
Figure 21-30 A multifunction, multipass, continuous cycle vibrator. (Courtesy of ROTO-FINISH)

Multifunction, multipass vibrators are versatile systems which permit combinations of processes and parts, and which can interface with other manufacturing methods such as grinding and casting. Parts can be fed and removed continuously, without shutting down the machine. See Figure 21-30.

Burnishing

Roller burnishing is a fast, economical way of obtaining closer tolerances and high-gloss mirror finishes on internally and externally machined surfaces. Most importantly, it eliminates secondary operations, such as grinding, and the removal of the part from the primary machine. See Figure 21-31.

Figure 21-31 Roller burnishing operations are employed to produce precision finishes on metal parts. (Courtesy of Elliott Company)

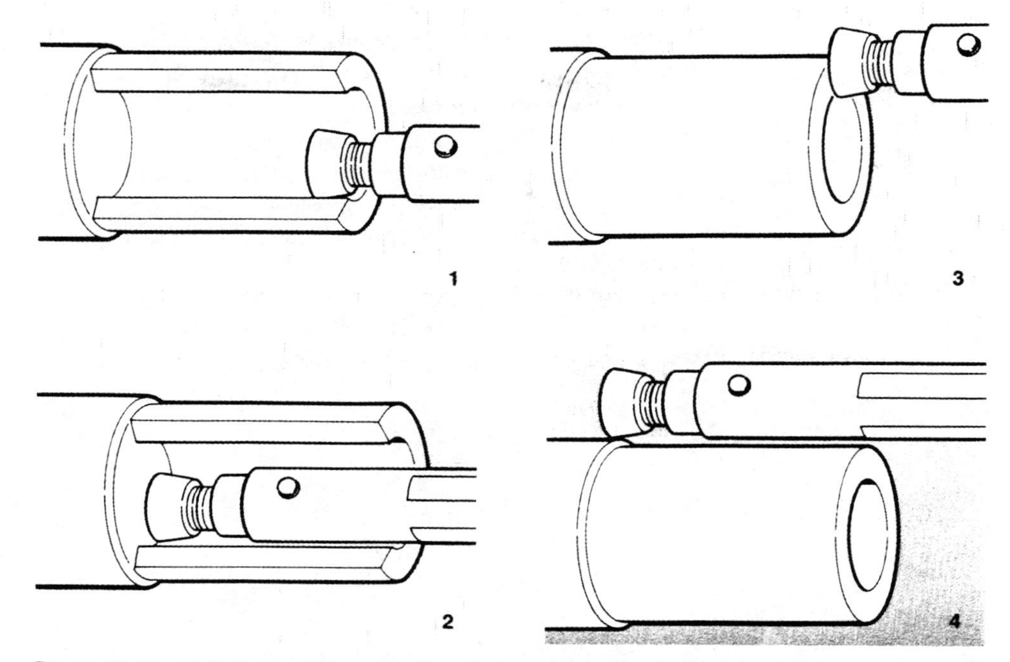

Figure 21-32 Multiple burnishing operation diagram. (Courtesy of Elliott Company)

Burnishing does not remove metal. It is a cold-working operation which simply compresses metal through plastic deformation into smooth, hard surfaces free of objectionable irregularities. This action also tends to improve the physical property of the surface by compressing its grain structure, thereby increasing its tensile strength.

This method employs a carbide roller tool for both single and multiple processes. An automatic multioperation is shown in Figure 21-32. In stage 1, the inside diameter is being burnished. In stage 2, a fillet between the inside wall and the part bottom is being burnished. In stage 3, the part end is being burnished. In stage 4, the shoulder is being burnished. The OD beyond the shoulder would be the next surface to be burnished.

The ideal applications for the carbide-roller burnishes are shafts and cylinders of nonuniform cross section. Shafts with radii, fillets, grooves, and cylinders with undercuts, tapers, and contours also are applicable. They can be finished in single machining operations, thereby saving machine time and setup costs.

QUESTIONS

1. Describe the four major mechanisms by which alkaline cleaning is achieved.
2. Prepare a sketch of the vapor degreasing process and describe how the system operates.
3. Explain the differences between the compressed air and the induction blasting techniques.
4. Describe the essential difference between grinding, polishing, and buffing.
5. Sketch a typical multiheaded flat finishing machine and explain the efficacy of using different grades of belts in a sequential finishing operation.

6. Explain the differences between a flat and a platen finisher.
7. Describe three methods of edge-contour finishing.
8. List some typical uses for flap wheels as abrasive tools.
9. Describe the reasons for using off-hand manual polishing and buffing operations.
10. Prepare a sketch of a typical AFM system.
11. Consult one of the periodicals dealing with manufacturing processes and write a short summary of a current article on mechanical finishing.
12. List some of the current books in your library which treat the subject of automated finishing methods.
13. Sketch and describe one of the mass finishing techniques.
14. Describe the burning process as a finishing technique.
15. Visit a local industry and prepare a report on the finishing methods employed there.

22 SURFACE COLORING AND COATING

Coatings and colorings are applied to clean material workpieces for the purposes of protection and/or decoration. Steel tableware is silverplated to make it more functional and visually (and tactually) appealing. A coat of lacquer is sprayed on a walnut side-table to make the surfaces more durable and attractive. Gun barrels are conversion-colored (''blued'') to resist corrosion and to make them more presentable. Other products are colored or coated to protect them during shipping, or to make their subsequent processing more effective. These are but a few of the reasons for and examples of coloring and coating methods, which is the subject of this chapter. See Figure 22-1. In short, these finishing methods are for the purposes of product preservation and presentation. The same close attention must be given to cleaning the workpieces, prior to coating, as was described in the preceding chapter. See Figure 22-2.

COLORING

Coloring a workpiece implies that its appearance has been altered for decorative or protective purposes, with no appreciable dimensionable change. By contrast, coating does result in a measurable increase (however minute) in the thickness of a surface. There are many ways of coloring metal workpieces, the most common being surface *conversion* coatings, which are inorganic films produced by chemical reactions to metal surfaces, and are to be considered *coatings* in name only. Their purpose is

Figure 22-1 Furniture is typical of a line of products to which a variety of surface finishes is applied. The tubular steel frame is coated with a polyester powder paint. The natural oak panels are finished with a durable sprayed synthetic varnish. The tabletops are covered with a high-pressure plastic laminate available in a range of colors.

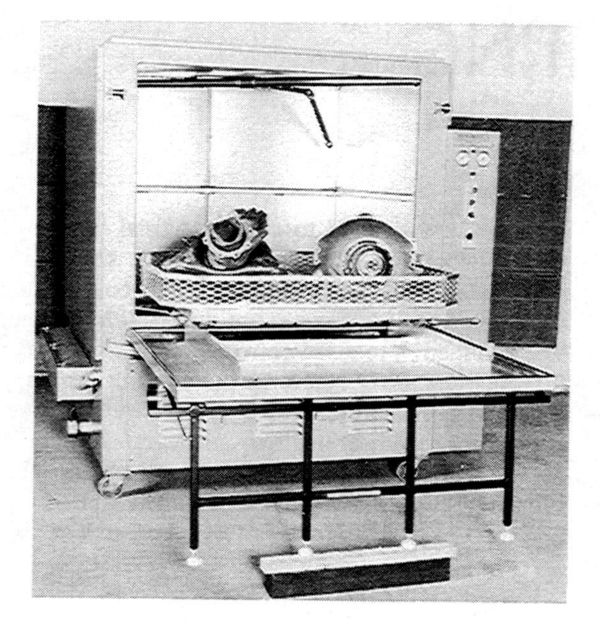

Figure 22-2 This high-speed cabinet spray washer is typical of the equipment required to clean parts before finishing. (Courtesy of Roto-Jet of America Co., Inc.)

to provide corrosion and wear resistance, to prepare a basis for subsequent organic coatings, and in some instances for appearance. These coatings are extremely thin, generally under 25 mm (0.001 inch). In fact, the resultant film is an integral part of the metal surface, and is so thin that coating *weight* (grams per square meter) of the coated area instead of thickness is now used as a means of expressing the amount of coating deposition.

Phosphate coatings are the result of the action of dilute phosphoric acid solution on the metal workpieces being treated. The surface is converted to an integral mildly protective layer of insoluble crystalline phosphate, or phosphate salts. The three principal types are zinc, iron, or manganese phosphate coatings, which are applied by spraying, brushing, or immersion. Some proprietary names of these processes are Parkerizing, Bonderizing, and Granodizing.

Chromate coatings are formed generally on nonferrous metal workpieces by applying aqueous solutions of chromic acids and salts. In the coating process, part of the surface is dissolved to form a protective film of complex chromium compounds, occasionally with electrolytic assistance. This provides an effective base for subsequent coating. Acid, due to its low electrical resistance, is often used on workpieces requiring good conductivity.

Oxide coatings are primarily employed for final decorative purposes, although they do possess some abrasion-resistant qualities, and also serve as paint bases. They are applied to metal workpieces by immersion, brushing, or spraying with a strongly oxidizing solution of caustic soda, sodium hydroxide, or other special proprietary formulations, at room or at elevated temperatures. Thorough cleaning and rinsing of properly buffed or matte-finished workpieces is a requirement for quality results. Such coloring techniques lend themselves to automated processes.

Oxide finishing solutions are used to convert iron alloys (except stainless steel) and aluminum to black colors, and copper, brass, and muntz metals to black or brown. Small parts are tumbled in the solution or carried in special inert baskets. Permanent colors are fixed by further treatment with sealers and waxes to permit handling of the objects. See Figure 22-3.

Anodizing is the process of thickening the natural oxide coat on aluminum, beryllium, magnesium, titanium, and zinc. This is an electrochemical process achieved by passing a current through an acid (usually sulfuric) electrolyte, where the workpiece is the anode. See Figure 22-4. In this fashion, the metal surface is converted to integral oxide compounds. The purpose of anodic coating is to impart to the surface the properties of this oxide, such as corrosion protection, hardness wear resistance, and electrical insulation. Additionally, the anodic film is porous and therefore can accept coloring dyes and retain lubricants. The film thickness is a factor of the type of electrolyte used and the current density, and can range from 0.0013 to 0.08 mm (0.00005 to 0.003 inch).

Figure 22-3 These metal pieces were oxide-finished using a cold conversion process. (Courtesy of Birchwood Casey)

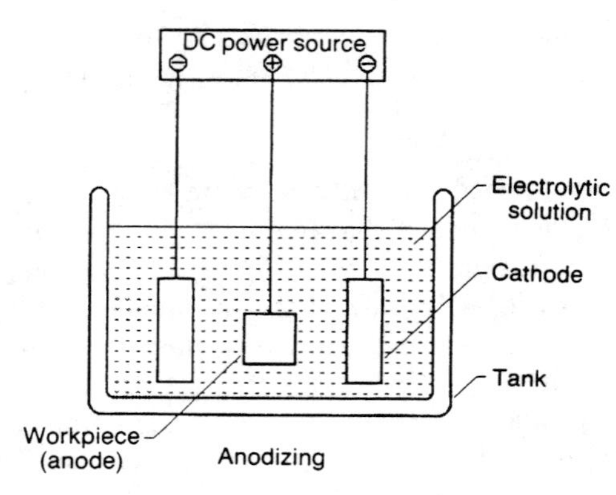

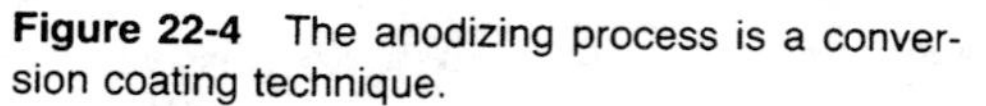
Figure 22-4 The anodizing process is a conversion coating technique.

Aluminum, for example, can be colored and sealed to provide an attractive, lasting finish. As the workpiece is removed from the anodizing tank, it is water rinsed and dipped in a bath containing an organic dye. When the desired hue is reached, the piece is washed in a nickel acetate solution and sealed in boiling deionized water to produce a clear protective film. It should be remembered that nonanodized surfaces cannot accept a dye, but must be paint coated if desired.

Heat coloring is a technique which is possible because some metals such as steel, brass, and copper achieve certain hues and coloration patterns when subjected to a flame. Workpieces must be buffed to a high luster for the best results. This coloring is a manual process done by placing a gas torch flame over the workpiece to bring out the desired color. Craft pieces such as vases and jewelry can be so colored, and should be gently solvent-cleaned and sealed with spray lacquer or some other clear finish for durability and permanence.

Staining and dyeing are terms which describe the impartation of color to porous workpieces such as wood, leather, and fabric, by using spirit or water-based colorants. Where desired, wood can be stained following proper surface preparation, such as by sanding. Application is by brushing, spraying, or immersion, and then wiping to remove excess liquid to prevent unevenness of color and raising of the grain as the solution dries. A range of colors is available, generally formulated to match a natural wood color, such as walnut, mahogany, and oak stains. Often, a natural walnut furniture piece is stained a walnut color to secure an even color in the piece. Such stains must be followed by lacquer or varnish sealing.

Fabrics are colored by dipping in a colored dye solution and rinsed in clear water. Special techniques are used to set the dyes to prevent subsequent running and fading. More information on fabric dyeing is found in Chapter 9. Some plastic materials, including synthetic fabrics, are dyed. However, color is generally imparting to solid plastic workpiece by coloring the plastic granules before forming.

COATING

Mechanical and electrolytic coatings protect workpieces by blocking off atmospheric elements which would cause corrosion or some other form of surface deterioration. Mechanical coatings, commonly called *organic* coatings, are applied by brushing, spraying, dipping, and flowing, as well as other "mechanical" methods. Paints, varnishes, lacquers, and enamels are representative of this category, and are in general use in a range of coat finishing applications. As obtained with other finishes, the coat quality is a factor of the quality of the workpiece smoothing and cleaning operations which precede it.

Paints are described as an intimate mixture of fine particles of a solid pigment dispersed in a film-forming liquid vehicle. This vehicle, such as linseed or some other drying oil and thinners, forms the coating film that gradually gels and hardens by chemically reacting with the oxygen present. Other additives include plasticizers to provide flexibility and driers to regulate the curing rate. The pigment functions to provide color, to obscure the underlying surface, and improve the durability of the film coat. These are *oil base* paints; *water base* also are used, whose efficacy is the absence of noxious fumes when drying, the convenience of cleaning painting equipment, and the elimination of fire hazards.

Varnishes and enamels are a class of finishes characterized by their ability to flow and form smooth, durable coatings on workpieces. *Varnishes* typically are clear liquids applied to a wooden substrate, and historically were produced from natural resins and other vegetable compounds. Modern varnishes are based upon synthetic resins and polymers, and set to form a tough film by polymerization. Natural varnishes harden by solvent evaporation. *Enamels* are produced by adding a pigment to the varnish, and generally are baked to secure tough finishes on such products as electric appliances. See Figure 22-5.

Lacquers, like varnishes, historically were comprised of natural resins, but more generally now are synthetic formulations. They are used to provide a clear, smooth, gloss or satin finish on a range of workpieces, and may be pigmented. They dry very quickly, are more abrasion-resistant than enamels, but require several coats to achieve satisfactory results. Typical synthetic lacquers include cellulosics, acrylics, vinyls, and polyurethanes.

Inorganic coatings are very hard, natural particle coatings which are fused to a metal substrate, and are of two classifications—porcelain enamel coatings and ceramic coatings. *Porcelain enamel* is defined as a substantially vitreous or glassy inorganic coating bonded to metal by fusion at a temperature above 425°C (800°F), according to the Porcelain Enamel Institute, Inc. These coatings are comprised of basic materials, called *frits*, which are finely milled vitreous particles produced by quenching molten

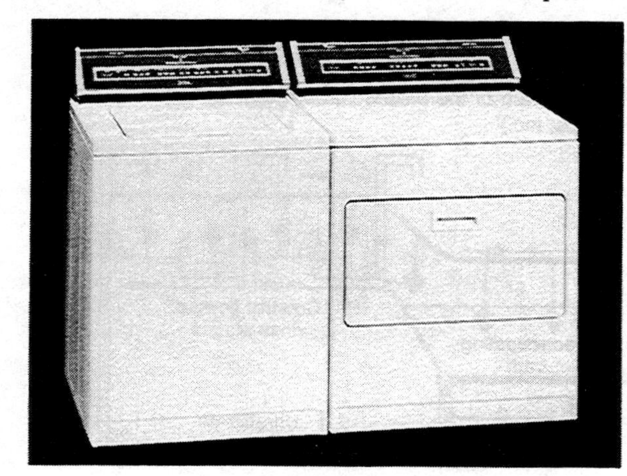

Figure 22-5 Electric appliances typically are coated with a baked on finish for protection, appearance, and ease of maintenance. (Courtesy of Whirlpool)

glass. These enamels are made of feldspar, silicates, and fluxes such as borax and soda ash. Porcelain enamels are commonly applied as a suspension of the frit in water called a slip, but may also be used in a dry form. They form a hard, durable, and easily maintained coating for sheet steel cabinets and for cast iron tubs and sinks. Application methods include flow-coating, dipping, electrodeposition, manual spray, electrostatic spray, and dry-powder spray, and are followed by firing in an enamelling furnace. The resultant coating is color permanent, corrosion- and abrasion-resistant, durable, and waterproof.

Other high density ceramic coating materials are applied to a range of metal parts for protection against corrosion and high temperatures.

Electrocoating is a method of applying organic finishes through the use of electrical energy. The workpiece to be coated is electrically charged and then immersed in a bath of paint that has been given an electrical charge of the opposite polarity. The resin and pigment migrate to the article, and a uniform film is irreversibly deposited. The part is then removed from the paint bath, rinsed to remove any excess material, and then baked to cure the finish. See Figure 22-6.

The deposition of the paint film is a result of four different processes which must occur simultaneously: electrophoresis; electrolysis; electroosmosis; and polarization.

Electrophoresis is the movement of colloidal materials dispersed in a liquid medium under the influence of potential gradient. *Electrolysis* is the dissociation and movement of ions. This refers to the deposition of solids which are in solution in the bath. *Electroosmosis*, the reverse of the potential gradient, is, in effect, wringing the water from the deposited film. *Polarization* is the ability of the deposited film to exhibit electrical resistance. It is this property of the deposited film—the ability to exhibit electrical resistance—that is responsible for the even coverage that electrocoating provides.

The initial electrical resistance between the two electrodes is relatively low. As the film is deposited, the electrical resistance increases in direct proportion to the film thickness. Since the current seeks the path of least resistance, paint is always being deposited at the fastest rate in the areas that have the thinnest deposit. Deposition ceases when the thickness of the paint film has become uniform and the applied voltage is no longer sufficient to cause additional electrophoresis, electrolysis, and electroosmosis.

Figure 22-6 A diagram of the electrocoating process. (Courtesy of George Koch Sons, Inc.)

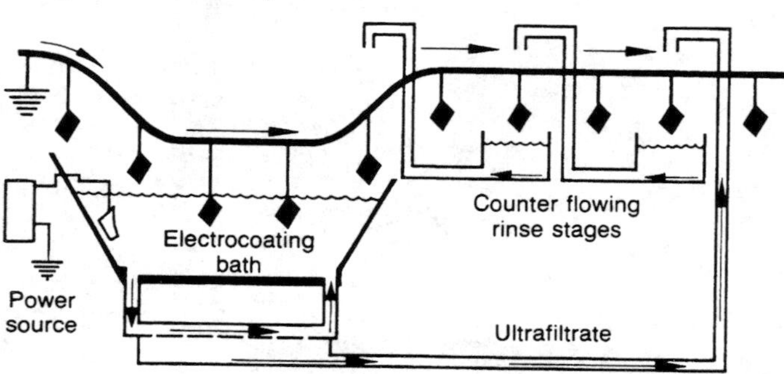

Figure 22-7 Metal panels are shown here emerging from the electrocoating bath. (Courtesy of George Koch Sons, Inc.)

The ability of the electrocoating process to form films of even thickness in intricate recessed areas is known as ''throwing power.'' This property is proportional to the voltage. For each paint, there is a limit to the voltage that can be applied. Excessive voltage will cause the film to rupture, resulting in a blemished appearance and reducing the protection against corrosion. See Figure 22-7.

Two electrocoating deposition methods are available to users of this finishing system. With the *anodic* method, the negatively charged paint particles are deposited on the positively charged workpiece. With the *cathodic* method, positively charged paint particles are deposited on the negatively charged workpiece. Although some manufacturers prefer anodic deposition, the cathodic method is generally acknowledged to provide many advantages, including better corrosion resistance, stain-free light colors, greater ''throw power,'' lower process voltages, and better coverage over weld spots, galvanized material, and dissimilar metals. Since the part is made the cathode, there is no tendency for any smut or other residual contaminants to be dissolved and included in the coating.

Powder Coating

Powder coating is a fine, dry paint which, instead of being suspended in a liquid medium, is applied in its dry form directly to the surface to be coated. Contained within each powder particle are the resin, pigments, and modifiers and, if it is a reactive system, a curing agent.

Powder is most often applied by electrostatically charging the powder particles and applying them to a grounded part. Grounding is usually done through the conveyor or fixture holding the part. See Figure 22-8.

Application usually takes place in a coating chamber, which contains the powder being sprayed and collects overspray in a reclaim system. Spray guns (along both

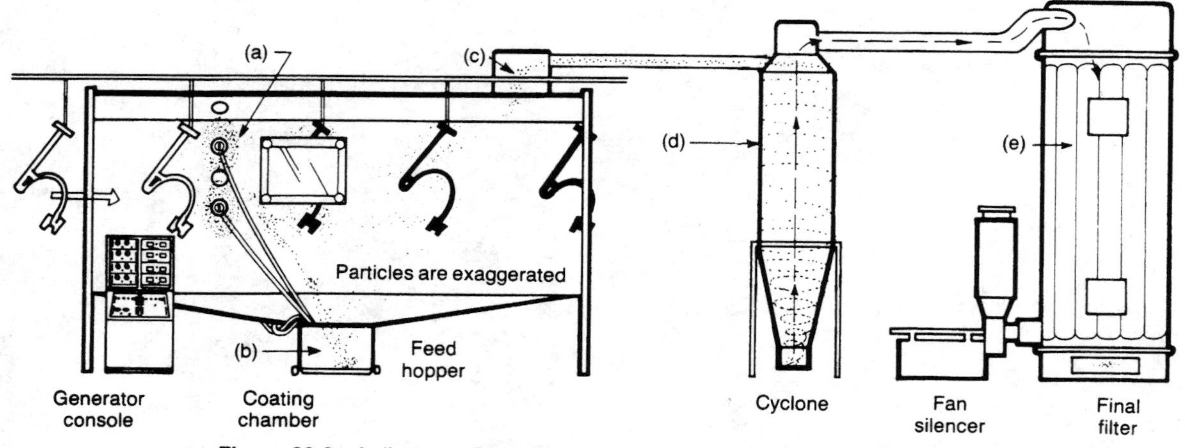

Figure 22-8 A diagram of the powder coating process. The powder is transferred at (a), the overspray falls to the feed hopper (b). Excess powders are collected at (c), and enter the reclamation system at (d). Final particle filtration occurs at (e), making this a clean and economic system. (Courtesy of Volstatic, Inc.)

sides of the parts) are controlled by a pneumatic control panel outside the coating booth. The charged powder adheres to the coated part by electrostatic charge until it enters the curing oven. Under heat, the powder melts onto the surface of the part.

If the powder is a reactive system (called a "thermoset"), curing or "crosslinking" takes place so that, once cured, the coating will not remelt. Typical thermosetting powders are epoxies, acrylics, and most polyesters.

"Thermoplastic" powders do not chemically crosslink upon application of heat, but melt and flow over the part in the oven. As the part cools after exiting the oven, the film hardens, but will remelt upon application of sufficient heat. Some typical thermoplastic powders are vinyl, nylons, and fluorocarbons.

A specialized powder application is porcelain enamel, usually for oven appliances. As porcelain is a glass, it is fired at much higher temperatures than organic powders.

Most powder coating is applied by electrostatic spray using equipment specially designed for powders. Others methods of application are fluidized bed dipping and electrostatic fluidized bed coating. See Figure 22-9. Electrostatic application of powder is made possible because powders, being generally non-conductive, will retain an electrostatic charge and will cling to a grounded substrate without dissipating that charge until melting in the cure oven. The charge may be positive or negative, but negative is the most efficient for most materials and equipment configurations, and is most widely used. Electrostatic systems are also used in liquid spray applications. The article to be painted is given a negative charge. The paint is given a positive charge and is atomized through a special round revolving nozzle. Instead of being sprayed on, the paint is actually plated on. The electrostatic force is so powerful that it actually wraps paint around the object to give a smooth even coat. Most important, there is no overspray, no fogging or drift, and no blast-through as with conventional spray guns. See Figure 22-10.

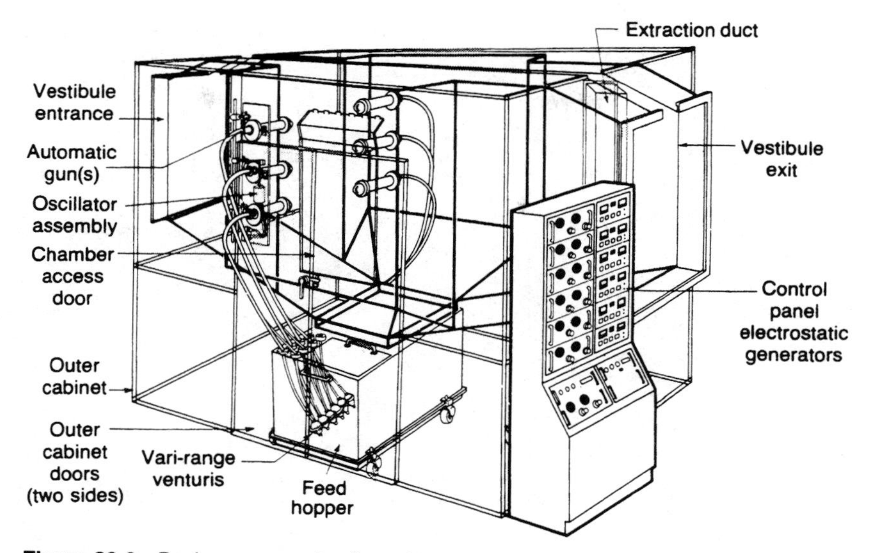

Figure 22-9 Basic components of an electrostatic powder spray coating booth. (Courtesy of Volstatic, Inc.)

Cutting tool coatings are applied by chemical vapor deposition (CVD) or physical vapor deposition (PVD) techniques. Twist drills, taps, and milling cutters often are coated with titanium nitride (TiN) to protect tools from wear abrasion, and to reduce friction and heat. CVD processes occur at temperatures in the range of 1,038°C (1,900°F). The PVD process is illustrated in Figure 22-11.

Vapor coatings involve the injection of a catalyst into a urethane coating system to accelerate the rate of cure. This process finds use in automotive trim finishing, tractor castings, truck cabs, and wooden parts.

Fluorocarbon coatings are characteristically heat-resistant to 316°C (600°F), nonsticking and nonwetting, and unaffected by most chemicals. Thermally-sprayed *metalized coatings*, usually zinc or aluminum, are used to coat steel components to provide a corrosion-resistant surface. Other methods may be sprayed by a similar process to provide hard, wear-resistant surfaces.

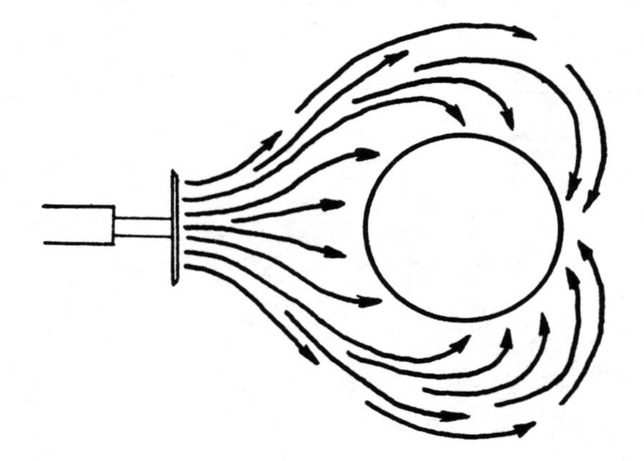

Figure 22-10 The electrostatic liquid paint spraying process. Note how the point surrounds the workpiece.

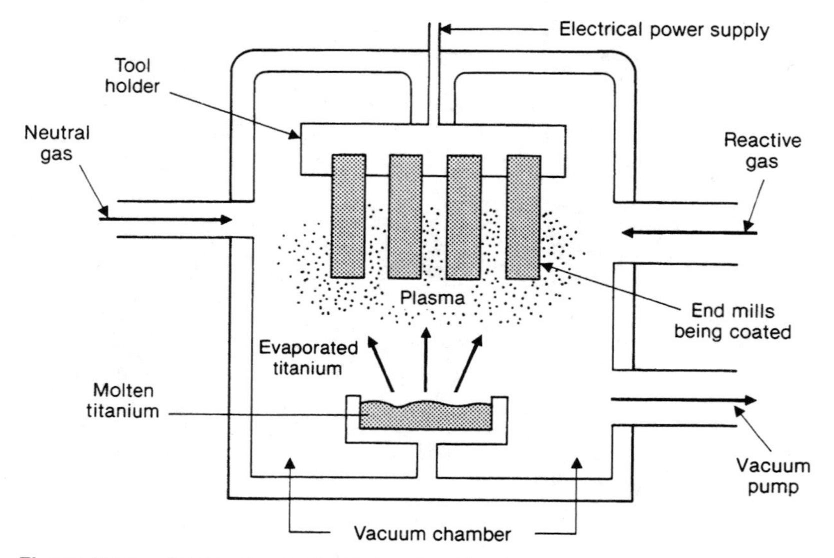

Figure 22-11 Schematic of a PVD chamber. The process operates at temperatures which reduce the risk of tool deformation and changes in tool-material hardness or metallurgy. (Courtesy of Balzers Tool Coating, Inc.)

Electroplating

The electrodeposition of a metallic coating on a metal part is accomplished under the usual electroylytic circuit theory. Very simply, *electroplating* is a method whereby a metal workpiece (the cathode) is coated with another metal (the anode) in an electrolyte solution. The process involves the placement of a thoroughly cleaned workpiece into a metallic salt solution of the metal to be applied. See Figure 22-12. When an electric current enters the circuit, it causes the ions of the metal to migrate to the workpiece, where they are deposited to form the plating coat. The anode serves to replenish the metal in the solution. The rate and quality of the deposition depend upon the workpiece distance from the anode, the density of the current, the solution temperature, and the plating duration. The efficiency of a system to evenly distribute a deposit on all surfaces and recesses of a part is commonly called *throwing power*.

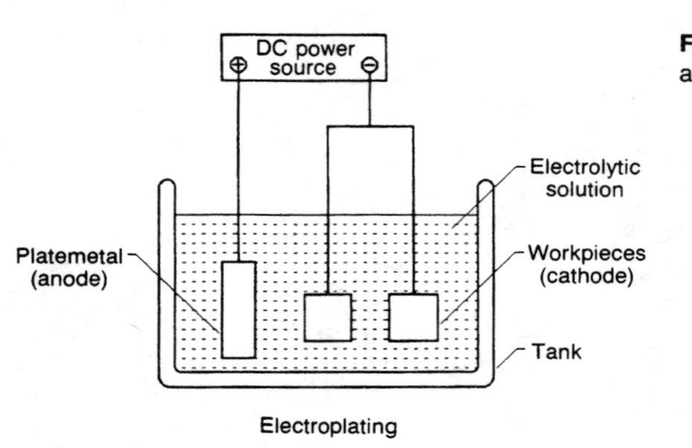

Figure 22-12 Electroplating process diagram.

Figure 22-13 Silverplated tableware. (Courtesy Reed & Barton)

The common plating metals are gold, silver, copper, nickel, cadmium, and chromium. The electrolytes employed are generally cyanide or sulphate solutions of these plating metals. The purpose of electroplating is to provide protection against corrosion and wear, to improve appearance, or to prepare a piece for further processing. Automobile bumpers and trim generally are copper plated before the chrome plating takes place. This acts to improve the surface quality before the final protective and cosmetic coat is applied. Similarly, silver-plated tableware is preplated with copper or brass for the same reason. See Figure 22-13.

Coating Equipment

Spraying is a common and versatile method of applying cleaners, adhesives, solvents, lubricants, and finishes in either low or high volume operations. Finish spraying normally occurs at pressure in the range of 200–400 kPa (30–60 psi). Three basic methods are employed. See Figure 22-14.

Figure 22-14 Typical spray gun vaporization patterns: (a) air spray. (b) air-assisted airless spray. (c) airless spray. (Courtesy of ARO Corporation)

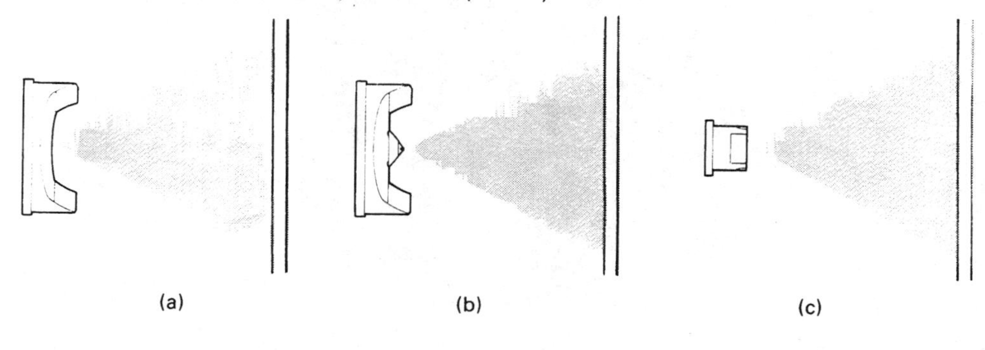

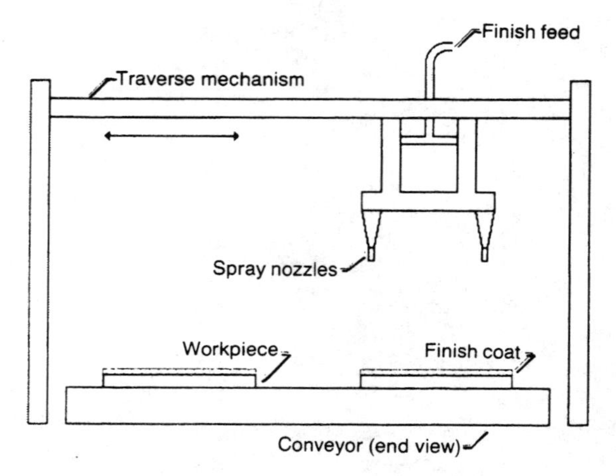

Figure 22-15 Automatic spraying systems.

Air spraying is the process of injecting a fluid stream with compressed air to achieve atomization. This method develops a very fine fluid particle mist that is excellent for high-quality finishing applications. However, the process produces a large amount of overspray, which is the presence of the fine spray in the air as well as on the surface to be coated. This may pose both environmental and health hazards, and requires adequate venting.

Airless spraying is a method of forcing a material, under high pressure, through a small orifice. The restriction created by the small orifice causes the material to atomize. This process produces very little over spray, and is an excellent method of applying finishes to steel structures, storage tanks, concrete, buildings, and heavy equipment. It is also good for spraying adhesives and lubricants.

Air-assisted airless spraying is the process of forcing material, under pressure, through a small orifice. The restriction created by the orifice sizes causes the material

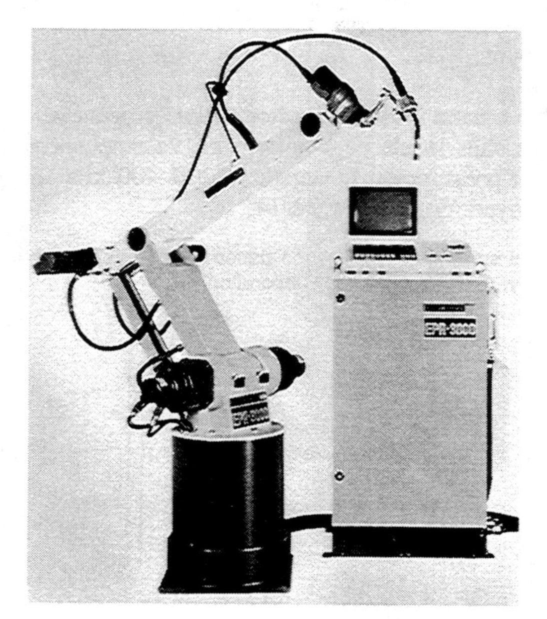

Figure 22-16 A typical computerized robotic spraying system can be programmed for a range of finishing applications. (Courtesy of the DeVilbiss Company)

to atomize. In addition, the specially designed tip introduces compressed air into the fluid stream to further atomize the material. The result is a finer quality finish than airless spray, with less overspray than with air spray. Air-assisted airless systems are used to apply finishes to lawn and patio furniture, office furniture, appliances, and farm machinery.

All spraying systems are applicable to manual and automatic applications. See Figure 22-15. Robot spray systems also are in common use in industry. See Figure 22-16.

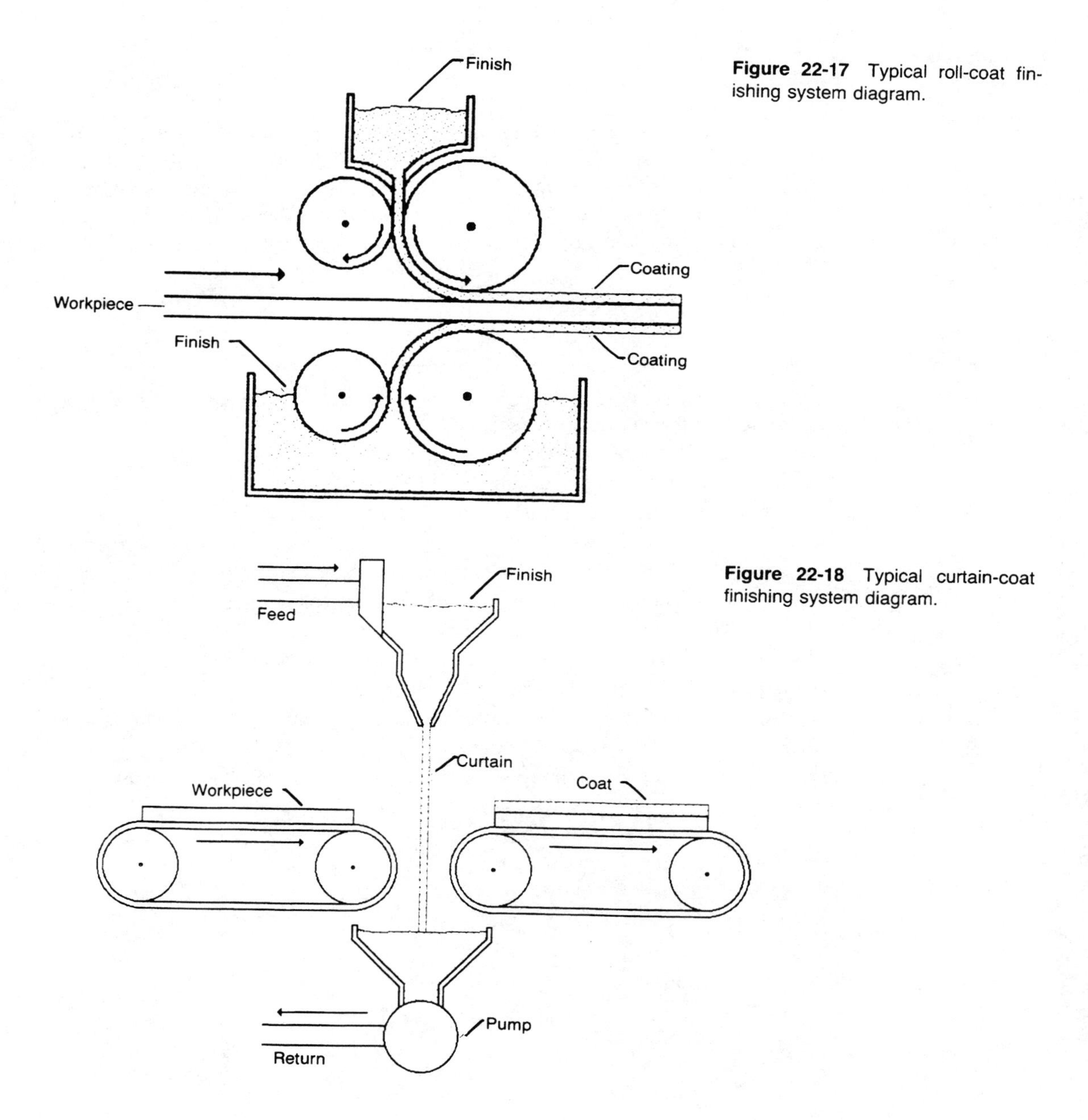

Figure 22-17 Typical roll-coat finishing system diagram.

Figure 22-18 Typical curtain-coat finishing system diagram.

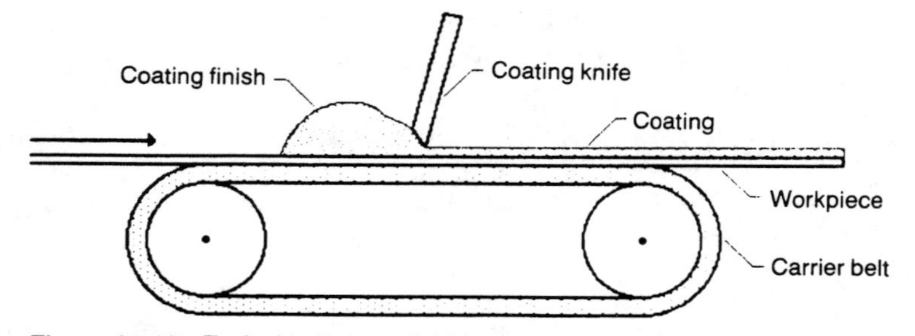

Figure 22-19 Typical knife-coat finishing system diagram.

Roll coating, *curtain coating*, and *knife coating processes*, similar to those used in the adhesive applications (see Chapter 19), are also widely used in applying surface finishes. See Figures 22-17, 22-18, and 22-19. Plastic coatings also are applied to workpieces by employing *dipping* techniques. Here, cleaned parts are carried to dipping tanks by a conveyor, and passed through ovens to cure or allowed to air dry.

QUESTIONS

1. List several *coloring* techniques to which the term *coating* is applied, and explain the reason for this apparent discrepancy.
2. Prepare a sketch of the anodizing process and explain how this operation occurs.
3. List some applications of staining and dyeing.
4. Explain the difference between mechanical and electrolytic coating processes.
5. List some examples, and explain the difference between organic and inorganic coatings.
6. Explain the four distinct phases or processes which occur with electrocoating.
7. Explain the difference between the *anodic* and *cathodic* electrocoating deposition methods.
8. Describe some of the attributes of powder coating.
9. Describe the difference between vapor, fluorocarbon, and metallized coating processes.
10. Explain the difference between electrocoating and electroplating.
11. Explain the attributes of the air, the airless, and the air-assisted spraying processes.
12. Prepare sketches of the roll, curtain, and knife coating systems.
13. Prepare a library research report on safety concerns in finishing processes.

PART 7 ADVANCED MANUFACTURING SYSTEMS

23 DESIGN FOR MANUFACTURING

Reduced to its simplest terms, design is the conscious effort directed toward giving order to the functional, material, and visual requirements of a problem. Whether the task be a dining chair or a structure to support a highway sign, the acceptable solution demands that these requirements be satisfied. *Function* requires that the chair support the human frame comfortably, and that the highway sign be contained securely. *Material* considerations must specify a sufficiently durable chair frame, and a sign construction appropriate to withstand the vagaries of time and weather. *Visual* requirements suggest that the chair be attractive to the user, pleasing to behold, and not ugly. See Figure 23-1. Appearance is less important in the design of the sign support. The main concern of the designer here is that the unit be held securely so that it can be read by the observer, and that it last for a reasonable length of time. See Figure 23-2. It is not a matter of the designer setting out to produce a visually unattractive mechanism. Indeed, some of these engineering structures are esthetic marvels. See Figure 23-3. Instead, the requirements of function and material are for this problem more important. First and foremost, the structure must work.

From the preceding facts, it follows that there is a fundamental distinction between what is commonly called industrial design and the process of engineering design. *Industrial designers*, on the one hand, are concerned with creations for the consumer which are functionally correct, materially sound, and which display a sensitivity to form and color. Such persons plan products such as furniture, tableware, brief cases, trash containers, and motorbikes. See Figure 23-4.

Figure 23-1 This dining furniture is designed to meet certain functional, material, and visual requirements. (Courtesy of Foldcraft Co.)

Figure 23-2 The prime requirements of a highway sign are functional and material—the unit must work. (Courtesy of Michigan Department of Transportation)

Figure 23-3 A bridge is first of all functional. The esthetic element occurs through the proper use of structural and material theory and application. (Courtesy of Michigan Department of Transportation)

Figure 23-4 This attractive and functional trash receptacle is the product of an industrial design team. (Courtesy of Landscape Forms)

Engineering designers employ scientific principles and techniques to make intelligent decisions regarding the materials and mechanisms necessary for devices which satisfy predetermined sets of problem parameters. These designers plan the linkages for a fork lift, design a catalytic converter, specify the employment of superplastics in an internal combustion engine, and create a system for the automatic processing of materials in a factory. See Figure 23-5. Industrial designers generally are dependent upon the results of engineering design. No hierarchal implications are intended here; each professional activity feeds the other, both contribute to the range of successful products.

The present highly competitive world marketplace demands designers with a working knowledge of materials and their processes, a dedication to the principle of function, a concern for designing artifacts that are economically producible, and an appreciation for sensitive forms. All designs are not necessarily good just because they were created by competent, free-thinking individuals. It is vital that research, experimentation, observation, and study precede any design solution.

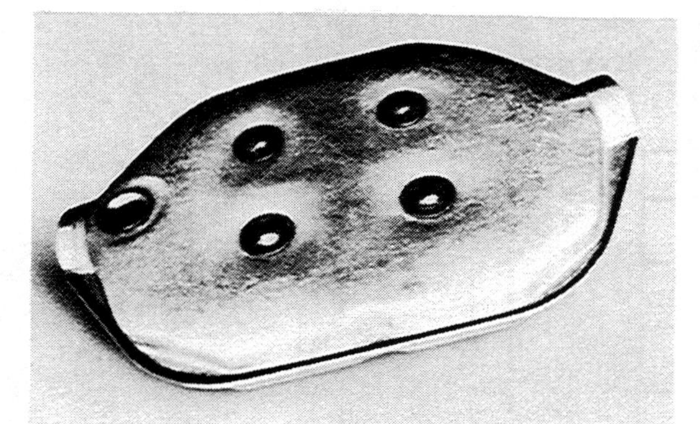

Figure 23-5 The primary requirements for this EBW catalytic converter is that it meets specified functional and material specifications to provide a safe component in an exhaust system. (Courtesy of Leybold-Heraeus Vacuum Systems, Inc.)

The subject of design for manufacturing is hardly a new concept. When Bronze Age humans first discovered that molten metal could be poured into a depression in the soil, and upon setting would assume the shape of that early sand mold, the technology of product replication was born. These ancient technologists soon learned the principles of patterns, drafts, and coring to produce modest numbers of identical bronze axe heads. Moving ahead many centuries, Eli Whitney's important development of machinery to produce American Army muskets featuring interchangeable parts occurred in 1800. In 1802, pulley blocks for the Royal Navy were being mass-produced with great success in the Portsmouth Dockyard. Each of these significant technical achievements required a reconsideration of the design of both musket and pulley block in order for them to be mass-producible. Contemporary production, with the requirements of high-speed, high-quality, low cost manufacture demands that an even closer attention be given to the theories of artifact producibility.

Design for manufacturing, then, is the rationalization of the product design process to simplify and facilitate the production scheme. The emphasis here is directed toward inventing more efficient means of cutting, forming, assembling, and finishing the various product elements. A fundamental example of this principle lies with the design of the hinges used in the manufacture of the prehung doors commonly used in both residential and commercial buildings. See Figure 23-6. One hinge is rectangular in form and must be installed in a door by marking and chiselling an opening in the door edge and frame, basically a manual or hand operation. A second hinge design features rounded corners which will better fit into a relief cut in door and frame with a power router, a machine cutting process. This latter design facilitates the automatic manufacture of door and frame assemblies. Square holes are not easily or economically cut in wood.

A traditional approach to the assurance that products be planned for efficient manufacture lies with the establishment of sets of rules for designing various components. As such, there are lists applicable to a range of process designs such as die and sand castings, extrusions, cold-headed parts, wire forms, and plastic injection moldings, among others. A typical example of some rules and considerations for the design of aluminum extrusions reads as follows:

Rule 1. Avoid unnecessary hollow shapes.
Rule 2. Avoid designs with abrupt changes in section.
Rule 3. Avoid knife edges, as they present a wavy appearance.
Rule 4. Design sections so that they have uniform wall thicknesses.
Rule 5. Use radii to eliminate all sharp corners in adjoining sections.

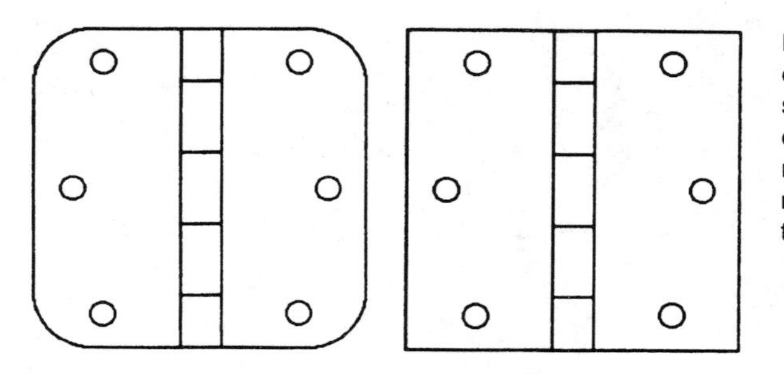

Figure 23-6 The hinge with rounded corners, left, is better adapted to installation with automated routing equipment. The square hinge, right, requires a relief cut into a wooden member by hand chiselling, a manual task.

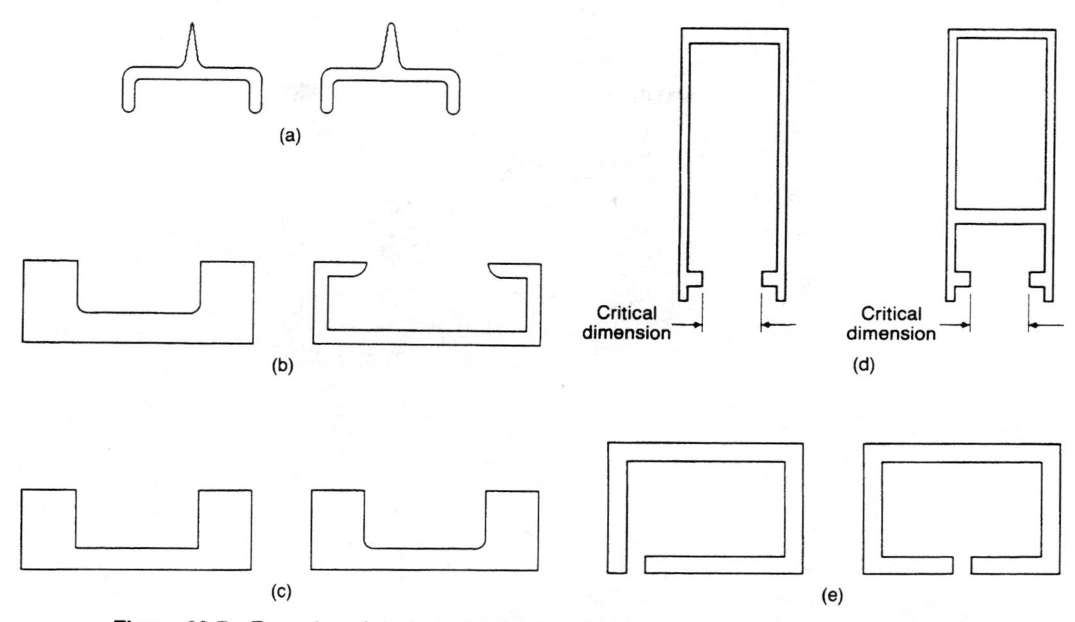

Figure 23-7 Examples of design rules for extruded members: (a) rule 3; (b) rule 4; (c) rule 5; (d) rule 6; (e) rule 7.

Rule 6. Use webs for better dimensional control.

Rule 7. Use symmetrical designs for semihollow areas.

Such rules are important for the efficient flow of molten metal into all parts of the mold. See Figure 23-7. This also results in stronger extrusions and the maintenance of dimensional control. Adhering these rules can also produce part plans which are amenable to more efficient production schemes.

DESIGN DEFINED

While there is no single accepted definition of the term ''design,'' implied in most descriptions is the concept of order and organization. Further, the term implies intention, meaning, and purpose, and directed human effort. Viewed in this light, design becomes the purposeful human effort directed toward ordering the functional, material, and visual requirements of a problem. The end result should be a workable solution to a problem having a set of predetermined specifications or parameters. Such an effort is equally applicable to both engineering and industrial designers.

Functional Requirements

First and foremost, the product must fit the purpose for which it was designed. In other words, the well-designed article works as it should. The can opener which is difficult and awkward to use, the motor mount which fails, the vehicle seat which does not adequately or comfortably support the human driver—all are examples of

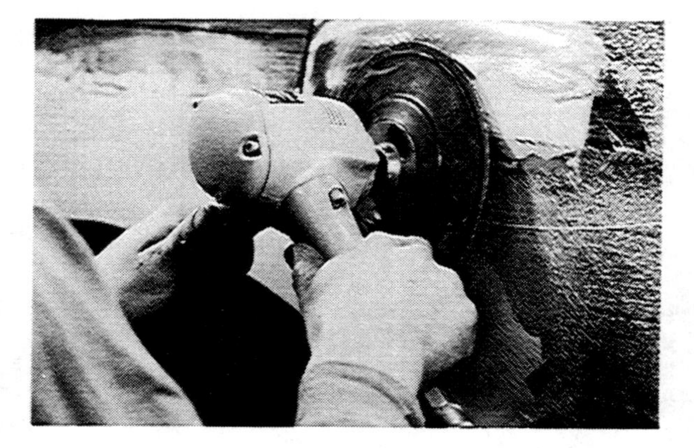

Figure 23-8 This power sander is comfortable, safe, and convenient to use. Note that the operator can hold the tool securely with both hands, and that a squeeze switch is provided for safe operation. (Courtesy of Rockwell International Corporation)

poor design. The well-designed tool is convenient and efficient to use. See Figure 23-8. The effort expended to assure this functionality requires a precise knowledge of product parameters and the skill to interpret them in design concepts. This leads to the invention of devices to support exhibit and display structures. See Figure 23-9. The problem was to create a universal connector that would join dissimilar components. The result was a unique joint which is attractive, simple and convenient to use, strong, versatile, and which works.

Figure 23-9 This universal connector is used to lock display panels and structures together to form attractive exhibit systems. (Courtesy of Nimlock Company)

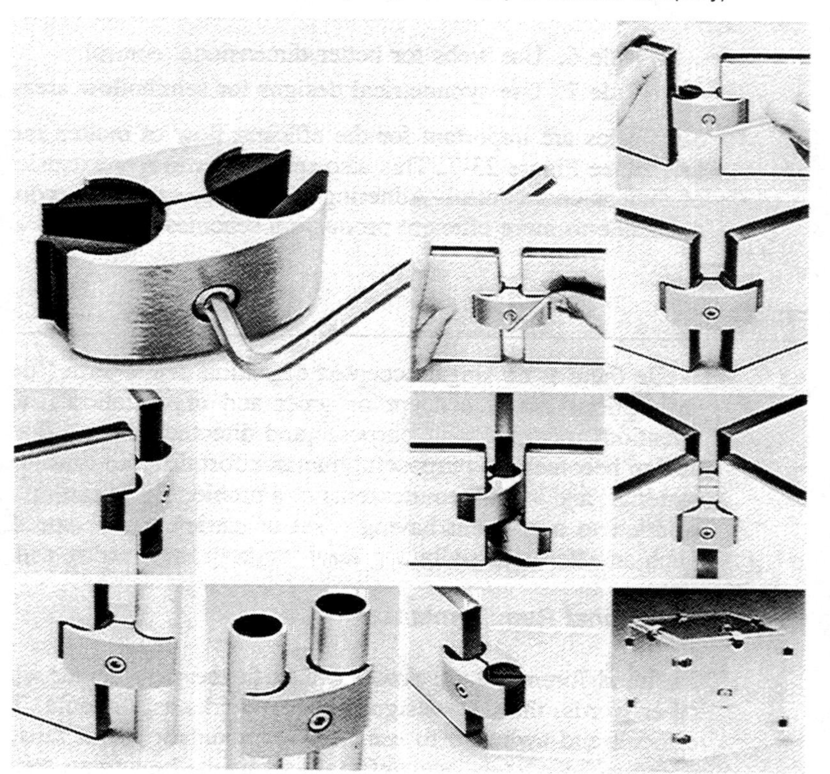

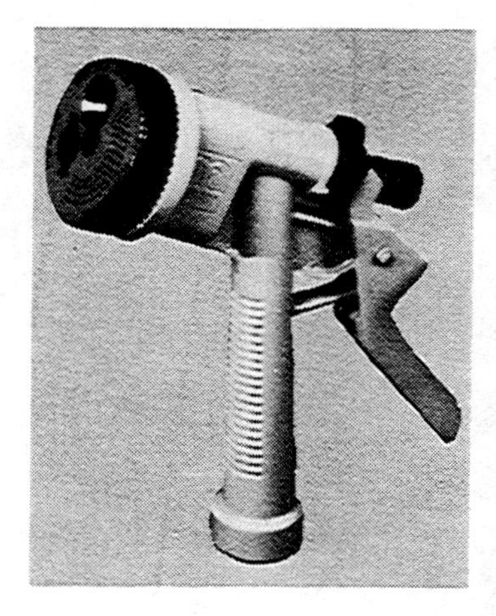

Figure 23-10 This hose nozzle gun is made of plastic for durability and maintenance-free operation. The squeeze-type trigger is easy to operate, contributing to its function. (Courtesy of Melnor Industries, Inc.)

Material Requirements

The product should reflect a single, direct, and practical use of the substance of which it is made. A maximum benefit must be derived from a minimum amount of materials. If this element of wise and proper utilization is present, the structure of the product will be sound and lasting, and it will be as strong as necessary, with no materials waste or excess bulk. The parameters of this requirement assume an extensive designer knowledge of materials and their properties, manufacturing processes, and structural systems. Materials must be selected to suit the product function. The use of plastic for the hose nozzle assures a noncorrosive, comfort insulating, easily-used garden and maintenance product. See Figure 23-10.

Visual Requirements

The product should have a pleasing appearance to the beholder or user. Simply stated, this requirement translates as a concern for the visual arrangement of the lines, the shapes, the textures, and the colors which comprise artifacts. Humans respond more positively to objects which are pleasing to the eye rather than ugly. This is perhaps the most difficult of the design requirements, for what one person views as beautiful may be visually unacceptable to another. Truly, there is no accounting for taste.

Be that as it may, designers generally are esthetically proficient, and industrial designers perhaps more so than the engineer. In many situations, form does follow function, and so the striking product can emerge purely as a result of its functional specifications. But most of the time a sensitive designer must attend to the matter of appearance. Most of the products displayed in this chapter are visually correct, display clever use of materials or joinery, have a bright, wet finish, or are otherwise interesting to look at. The ski boots in Figure 23-11 are such a product, well-constructed of durable plastic, colorful, rugged, yet sensitively conceived, and they give the impression of all these attributes. Their form necessarily follows their function, but there is an admirable attention to formal detail, an attribute of all well-designed pieces.

The definitions of design discussed earlier in this chapter alluded to a concern for order and visual organization. Such visual concerns become functional when

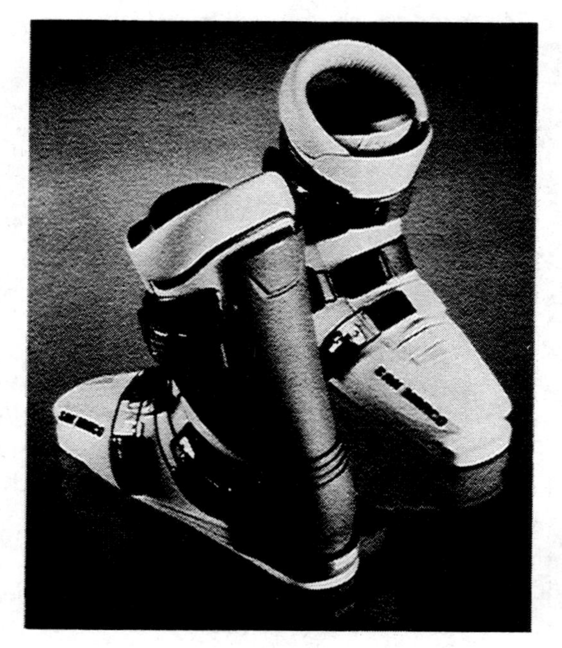

Figure 23-11 These ski boots are rugged, durable, functional, and attractive. (Courtesy of E. I. duPont de Nemours & Company, Inc.)

the issue is the placement of machine controls. See Figure 23-12. Assuming the depiction of a control panel, the orderly positions of the knobs are preferred over a random, scattered pattern. The ordered set fixes a location for an operator. The same logic holds true for the placement of an electric light switch at a standard wall height of 1.1 m (43 inches). This makes it easy to find such switches near doorways, even in the dark.

DESIGN PROCESSES

The design task rightfully is considered to be a process because, in order to achieve optimum problem solutions, a broad range of factors and subtasks must be undertaken in an orderly fashion. A schematic diagram of this process appears in Figure 23-13. A brief description of its constituent elements follows.

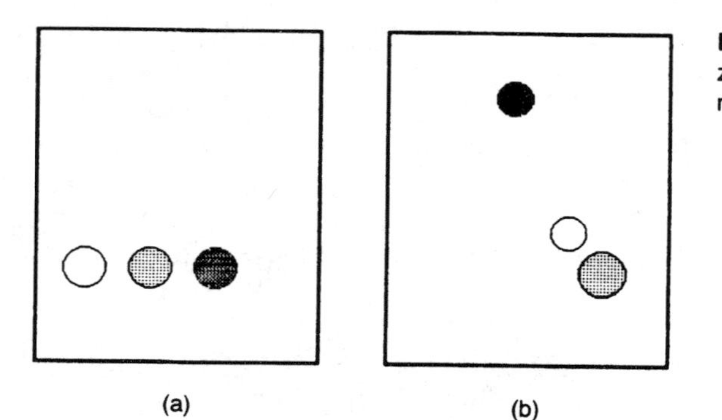

Figure 23-12 The control panel organization at (a) is preferred to the randomness of (b), because it is operator-friendly.

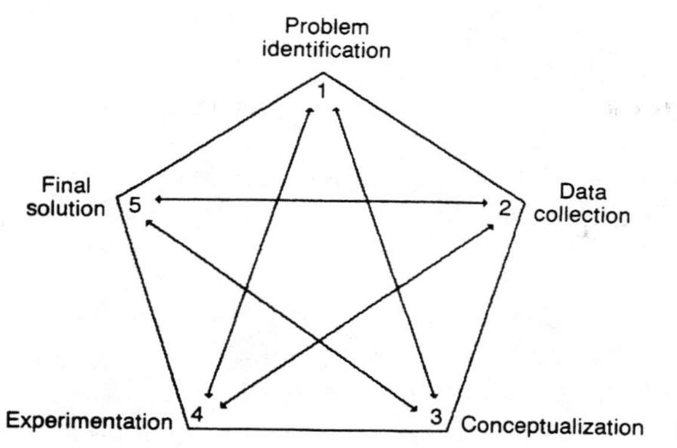

Figure 23-13 Design process diagram.

Phase 1. Problem identification. A design task may emerge as an elemental need for an existing system, such as a safety switch for a piece of production equipment, or a high performance engine, or as improvements and modifications of in-service devices. Conversely, the task may be totally innovative, exemplified by a new line of furniture or the refitting of an aircraft interior. Whatever the source, such tasks must be carefully studied and needs evaluated to ascertain their appropriateness for a specific company and the potential for satisfactory, economically feasible result. Design is costly and time-consuming. As a new, innovative, product is conceived, or an existing product improved, the primary objective is to present a commodity that will meet a real need or render a service in a manner superior to that of the former product.

Phase 2. Data collection. Once the problem has been described and given focus, pertinent data must be gathered, information retrieved, assessments made, and careful analyses undertaken. Any factors which bear upon the problem must be studied, leading to the presentation of tentative product specifications. This activity includes market analyses, assessments of the competition, and cost estimations. An evaluation of existing facilities may lead to the conclusion that such an undertaking is not cost-effective, because of possible extensive refitting or equipment acquisition. Data gathering is equally valid for its negative as well as positive conclusions.

Phase 3. Conceptualization. This activity is the hypothesis stage, where intuition and technical experience merge to produce a range of tentative problem solutions. Thinking about the task, weighing alternatives, making creative leaps, sketching possibilities, and compromising—all are elements of conceptualization. References constantly must be made to the task specifications prior to data analyses. This phase of the work is the heart of the design process, the stage at which potential design configurations emerge. This work demands introspection, discussion, debate, modification, and compromise, with a series of possible design solutions as the result.

Phase 4. Experimentation. This step in the design process involves the testing of possible solutions, building, evaluating, and modifying product prototypes, examining collected data, and evaluating marketing results. At this stage, the designer must return to the original problem statement—is the product strong enough, does it work, is it visually correct? Other appropriate questions must be asked: Is it easy and economical to produce? Can it be assembled conveniently and efficiently? Is it vandal-proof? Is it child-proof? Is it safe to use? Is it easy to repair and maintain?

Experimentation is precisely what the term implies—the opportunity to carefully check tentative problem solutions, and to detect and correct errors before the article is put into production.

Phase 5. Final solution. The logical result of the previous four phases of work is the final problem solution. At this phase of the process, considerable attention must be given to refinements resulting from methods of manufacture, simplification, use of standard components, material savings, appropriate tolerances, and others.

The total design process requires an interplay of all phases, and the necessary attention to the special design considerations which follow. An overriding factor emerges here, namely, that such concerns are attended to as the design process unfolds, and not as an afterthought. The process generally is a team effort, where a group of persons with special talents join in the design work found in industry.

DESIGN FOR MAINTENANCE

The performance and availability of industrial machinery is a function of the amount of maintenance necessary and the ease with which the service can be performed. The same holds true for any consumer product, be it a lawnmower, a tool, or a home appliance. Product usefulness is diminished considerably if it is nonoperational (deadlined) due to mechanical breakdown. This situation is further exacerbated if the malfunction was caused by improper maintenance. It is bad enough when a part breaks through normal use. It is inexcusable when cause for failure is faulty service.

There are several elements related to this issue. Assume, for example, the matter of equipment lubrication.

1. The design of the oil passage tube may be so intricate that it inhibits the flow of oil to a bearing or gear.

2. The placement of the oil hole may be so awkward that it is difficult to direct the oil to the hole, or to observe whether or not it has entered the hole.

3. The service manual may be written so poorly that people may be confused regarding proper service procedures.

4. The basic design of the components failed to consider available technology, for example, the provision of oil-impregnated bearings which require no lubrication service.

In the final analysis, these elements all are the responsibility of the designer.

The theory of design for maintenance can properly be divided into two related issues. *Maintenance*, per se, is the act of caring for or attending to. There are *preventative* maintenance procedures employed to postpone, defer, or arrest predictable failures. *Corrective* maintenance procedures are directed at restoring products to acceptable, functional operating condition after failure.

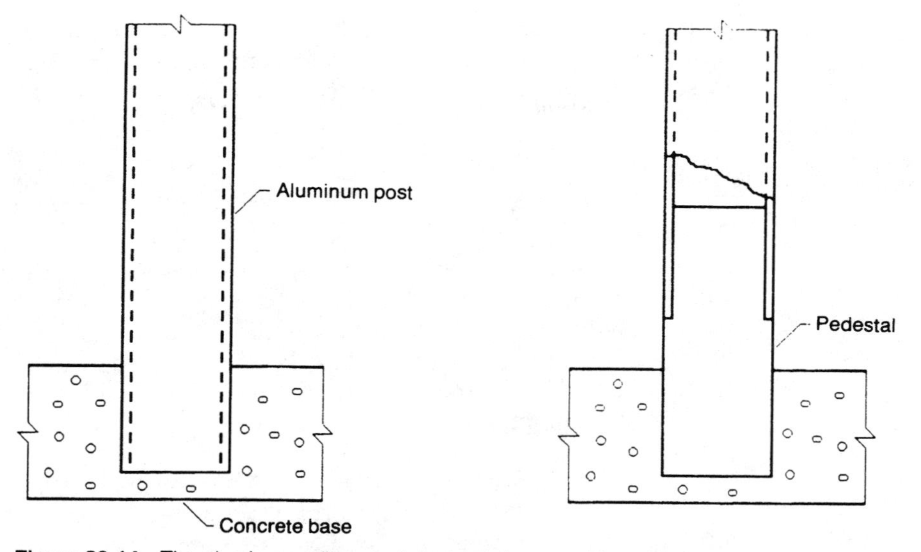

Figure 23-14 The aluminum railing post design is improved by adding a pedestal to prevent corrosion.

On the other hand, *maintainability* is a more direct and profound design issue concerned with assuring that a product perform satisfactorily for the period of its intended life with a minimum expenditure of effort and money. The principle of maintainability was violated with the design decision depicted in Figure 23-14. The exterior stair railing posts were specified to be rectangular aluminum tubing set in a concrete footing. The posts soon began to corrode from the salt broadcast on the stairway to melt the snow and ice accumulation, and eventually failed. The proper solution should have been to join the metal post to a pedestal of treated wood, plastic, or concrete, to raise them above the level of contaminants.

DESIGN FOR SAFETY

There is perhaps no design issue of greater concern to manufacturers than product safety, and product liability. Consumers purchase products assuming that they are safe to use. Designers must attend to every aspect of product planning to assure safety of use and operation. This is a subject of great range and ramification, and relates directly to the nature of the artifact. Electrical, chemical, combustible, sharp-edged, transport, and process are among the product categories each having special safety design concerns. A table saw must include adequate guards for the blade and belt, and must have a properly grounded electrical system. Further, the electrical switch must be positioned for convenient operator access and manipulation. The designer must attend to such functional requirements, in addition to adhering to any industry or governmental safety standards.

Other equipment safety elements include the "light screen" guards to provide an invisible curtain across the face of a dangerous work area, such as a punch press. The machine shuts down automatically if a hand passes through the curtain. Some machines are inoperable if both hands are not grasping control levers, again to guard against accidental hand entry.

DESIGN FOR USABILITY

The term "user-friendly" has crept into contemporary parlance, and it implies optimum convenience, definition, and ease. How many drivers have reached down to the left of an automobile steering wheel to pull the knob to release the parking brake, only to have the hood of the car open, because the hood knob is confusingly positioned next to the parking brake knob? This is a "user-unfriendly" arrangement. And the list goes on, endlessly, to include can openers, shower controls, machine adjustments, and procedures for starting a gas engine. Some are easy to use, others are not.

The reasons for violating this design requirement are in many instances justifiable. The expense of relocating or modifying components may not be cost-effective, or may lead to further problems, for example, safety, down line. Be that as it may, it is imperative that usability design considerations be attacked on the drawing board and with the prototype, and not as a reaction or response to customer complaints.

DESIGN FOR HUMAN FACTORS

An important aspect of product function relates to the interaction between product and user. The clutch, brake, and accelerator pedals are in standard locations in all automobiles, so that a driver can operate any vehicle safely without having to search for these controls. Can openers, typewriters, tennis rackets, industrial equipment, and power tools similarly are designed with a concern for the human user. See Figure 23-15. The concern therefore is to assure safe, convenient, and effective operation. The concern is also to avoid product liability claims.

Ergonomics is the discipline concerned with the influences humans impose upon the design of mechanisms, products, and environments. There are several elements which comprise the science of ergonomics, a science which is also commonly referred to as *human factors*.

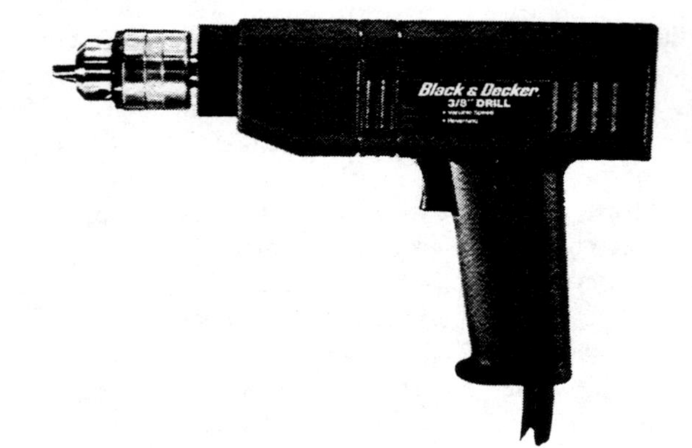

Figure 23-15 This hand drill is a lightweight, well-balanced, and rugged tool designed for human use. (Courtesy of the Black & Decker Corporation)

Anthropometric factors deal with the physical dimensions of humans, and the applications of these measures to the tools, machines, and equipment they use. Considerable information relative to these measurements has been collected, and this data bank provides designers with valuable standards to use in making appropriate design decisions. See Figure 23-16.

Physiological factors include the functions, responses, limitations, and talents of human beings. Visual, auditory, tactile, and olfactory capabilities are among the body systems which have a direct bearing upon the design and manufacture of human artifacts. An example related to home appliances appears in Figure 23-17.

Figure 23-16 Anthropometric data tables such as this are used as guides when determining critical dimensions for products used by humans.

1985 Male*

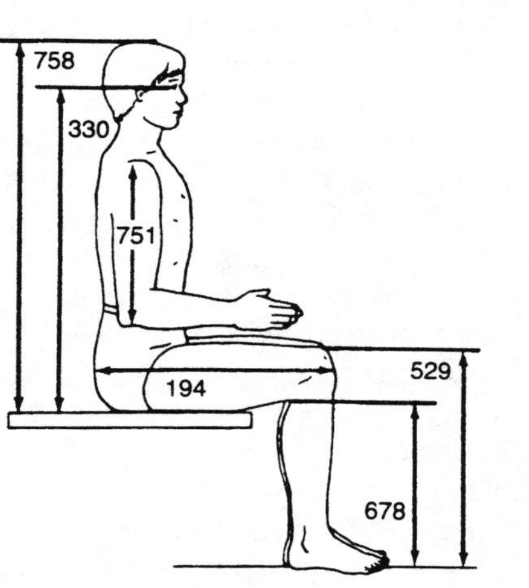

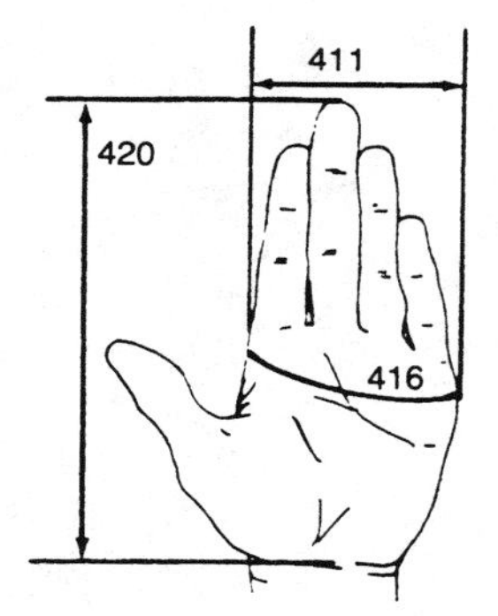

No.	Dimension	5%ile	Mean	95%ile
758	Sitting height	88..5 (34.8)	93.6 (36.9)	99.0 (39.0)
330	Eye height, sitting	76.4 (30.1)	81.3 (32.0)	86.5 (34.1)
529	Knee height, sitting	52.1 (20.5)	56.1 (22.1)	60.3 (23.7)
678	Popliteal height	40.4 (15.9)	44.0 (17.3)	47.8 (18.8)
751	Shoulder-elbow length	33.3 (13.1)	36.1 (14.2)	38.9 (15.3)
194	Buttock-knee length	56.4 (22.2)	60.8 (23.9)	65.4 (25.7)
420	Hand length	17.9 (7.0)	19.2 (7.6)	20.6 (8.1)
411	Hand breadth	8.3 (3.3)	8.9 (3.5)	9.6 (3.8)
416	Hand circumference	20.1 (7.9)	21.6 (8.5)	23.2 (9.1)

*Data given in centimeters with inches in parentheses.

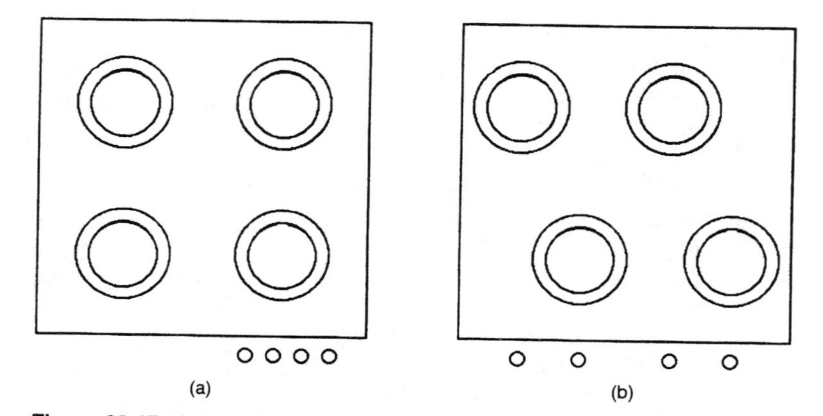

Figure 23-17 It is difficult to determine the control pattern in this stove burner arrangement (a). The alternate arrangement (b) is better because there is a visual logic in having the control lie immediately in line with the burner.

Psychological factors are those societal, cultural, mental, and emotional influences upon the things people use, how they use them, and the environments in which they live and work. One instance of human influence upon machine controls pertains to the design of membrane switches, commonly called "touch" switches. Some of these control buttons require only the passing touch of the finger of an operator, while others require a slight pressure to actuate them. Plant managers have found that some employees disliked the nonpressure switch, because of the possibility for error as the operator touches the wrong switch while searching for the correct one. There was also a feeling of need for personal control of the machine by the operator pressing the switch firmly in order to actuate it. The issue was resolved by the simple expedient of changing from touch to pressure controls. See Figure 23-18.

The aforementioned are but a few instances where ergonomic principles have made the difference between product failure or dissatisfaction, and product success.

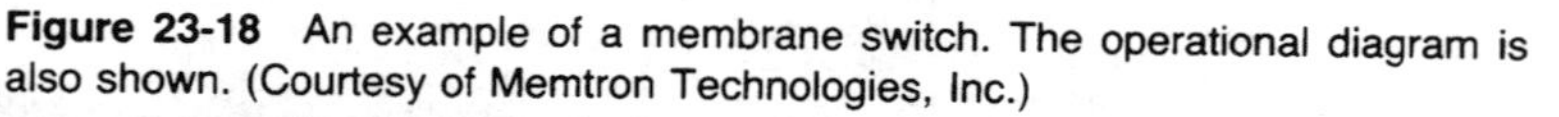

Figure 23-18 An example of a membrane switch. The operational diagram is also shown. (Courtesy of Memtron Technologies, Inc.)

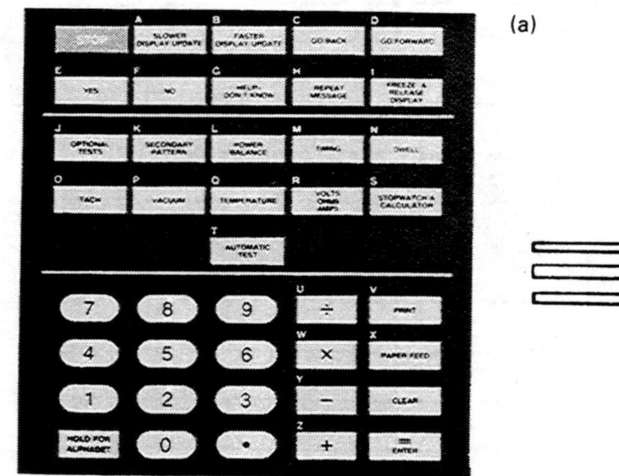

(a)

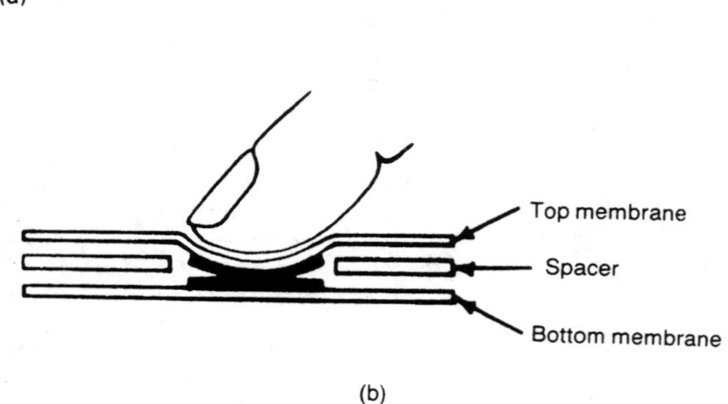

(b)

DESIGN FOR VANDALISM AVOIDANCE

Many humans have a tendency to walk away with anything they can carry, or deface attractive surroundings. Considerable expense is incurred through such wanton acts, and designers are hard-pressed to meet this challenge. Signage, or building signs, is one such problem. See Figure 23-19. Building identification lettering is architect-designed and placed to achieve a consonance with the building structure. If placed too near to the ground, it is in ready reach of a student who might wish to adorn a dormitory room with appropriate initials. By placing the lettering higher on the building, this form of vandalism is avoided. See Figure 23-20.

Telephone booths are another target for vandals. A possible solution here is to design flush-mounted, push-button switches, a built-in speaking-listening component,

Figure 23-19 This building lettering is close enough to the ground level to be removed by vandals. (Courtesy of Western Michigan University)

Figure 23-20 This building lettering is beyond the reach of vandals. (Courtesy of Western Michigan University)

and coinboxes fixed into a concrete structure. The same logic could hold for soap, towel, and tissue dispensers in comfort stations. Build these into the wall, flush-mounted, with no projecting devices. Antigrafitti walls and dividers continue to challenge the designer.

DESIGN FOR PRODUCIBILITY

Even with an apparently successful design solution, the issue of economic feasibility arises. How can the artifact be made well and at least cost? All other factors being equal, the product made from the fewest number of parts, the least intricate shape, the fewest precision adjustments, the simplest and most direct processing sequence, and the most judicious use of materials will be the most economical to manufacture. Additionally, it will be the most reliable, safest, and easiest to maintain. This is the goal of product design. While it is impossible to predict with precision the ultimate manufacturing costs at the early stages of design, this economic factor must be considered as the designing progresses. Any evidences of excessive costs should be dealt with, and appropriate design modification made. While all production aspects are significant, assembly is one of the more critical.

DESIGN FOR ASSEMBLY

The expense of product assembly is one of the highest manufacturing costs, because the process generally involves considerable part handling and special treatment. Assembly, therefore, becomes a high priority whenever productivity improvements are sought. Design for assembly is the activity directed toward *decreasing* the number of product parts, and *facilitating* the assembly of those parts remaining.

Assembly simplification properly occurs as product design begins, so that the function of each part can be scrutinized. However, many opportunities for rationalized assembly exist with the redesign of existing products. A classic instance of redesign is obtained with the familiar electric outlet box. See Figure 23-21. A number of

Figure 23-21 Design simplification leads to decreases in product assembly costs. The metal electrical box has 13 parts which must be assembled.

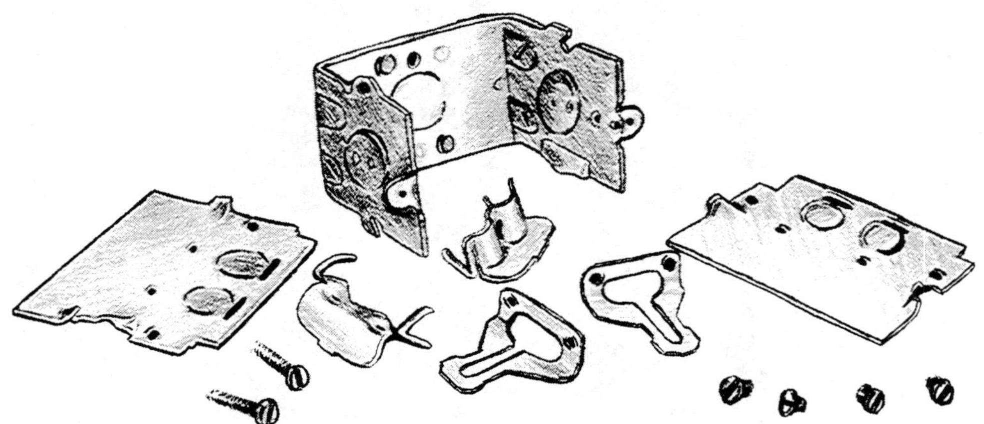

Figure 23-22 A redesigned electrical box is made of injection-molded plastic with preinserted nails, and requires no assembly.

processes are employed in the manufacture of this metal box, such as notching, lancing, bending, piercing, tapping, shearing, and screw fastening. Note that the assembled box is made of thirteen parts, including six screws. A redesigned box is shown in Figure 23-22, and this fabrication features an injection-molded polypropylene plastic shell with two nails inserted in the molding operation, for a total of three parts. The metal box retails for $1.60, the plastic model for $0.60. This product redesign resulted in a significant cost savings through the reduction of the number of total parts, and the elimination of *all* assembly operations. Parts reduction is important, because in addition to the obvious elimination of the processes required to produce the part, there is the corresponding elimination of costs associated with design, handling, storage, and inventory.

The problem of assembling wheels and axles is a common one, applicable primarily to consumer products such as lawn mowers, garden carts, seed, weed-killer, and fertilizer spreaders, and child toys. Generally, the wheel-axle systems can be permanent, with no possibility of disassembly intended. A range of options for such assemblies appears in Figure 23-23. The *spring cap* is easily assembled, and can be removed, but not without damage to the cap. *Threading* is expensive, as is drilling holes and inserting *cotter pins*, but these permit easy disassembly.

Knurling also requires additional processing, and is reasonably permanent. Automatic *riveting* is sure, automatic, and permanent. *Adhesive* caps are permanent, inexpensive, and readily automated. The selection of one of these options (and there are others) is a matter of the design specifications established for the product.

Automotive mufflers require the assembly of a number of components. See Figure 23-24. Of special significance here is that such assembly should be as "automatic" as possible to reduce overall production costs. The competition among muffler manufacturers (and installers) is fierce, and it is important to keep such costs as low as possible, while maintaining a quality product. An examination of some of these components parts provides an excellent illustration of design for producibility and assembly.

The louver tubes, shown in Figure 23-25, can be produced in several ways. A common method is to manufacture a flat, louvered blank, form it into a cylinder,

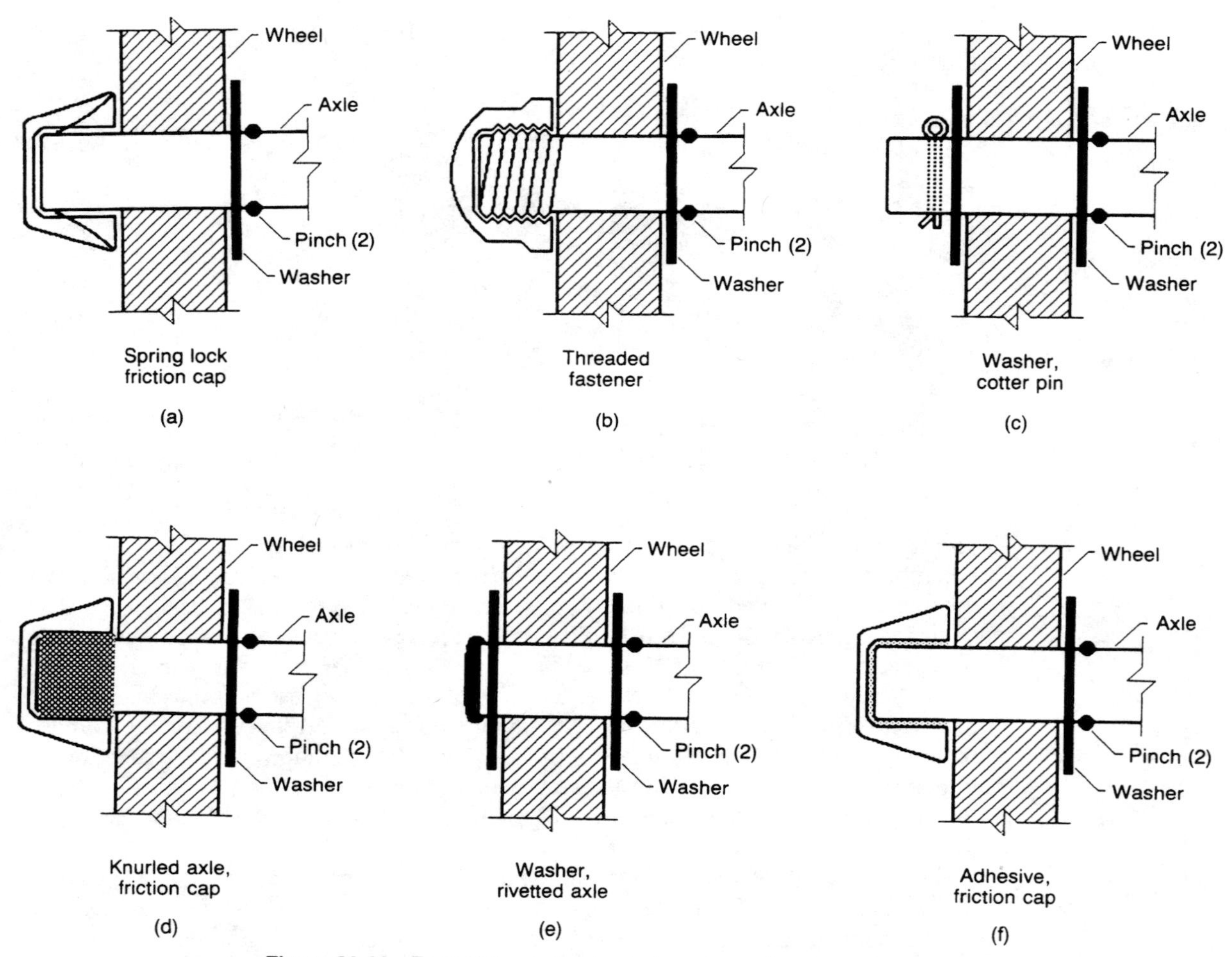

Figure 23-23 Each of these wheel-axle fastening options has advantages and disadvantages. For wheels meant to be removed, use arrangements a, b, and c. Options d, e, and f are permanent fasteners, and are easily adapted to automatic assembly. The adhesive-lock fastener is the best solution.

Figure 23-24 Typical automotive muffler components. (Courtesy of Eagle Precision Technologies, Inc.)

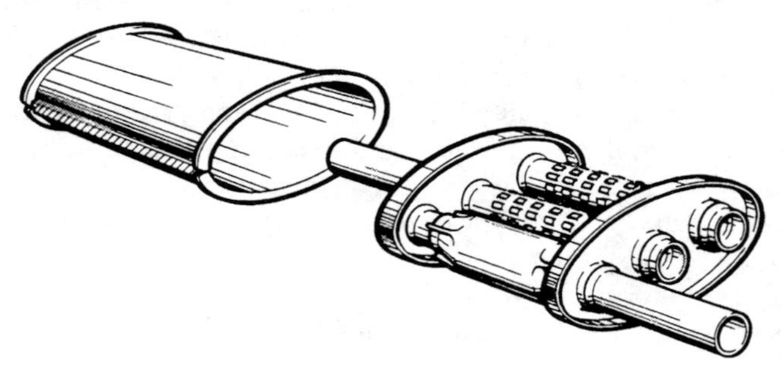

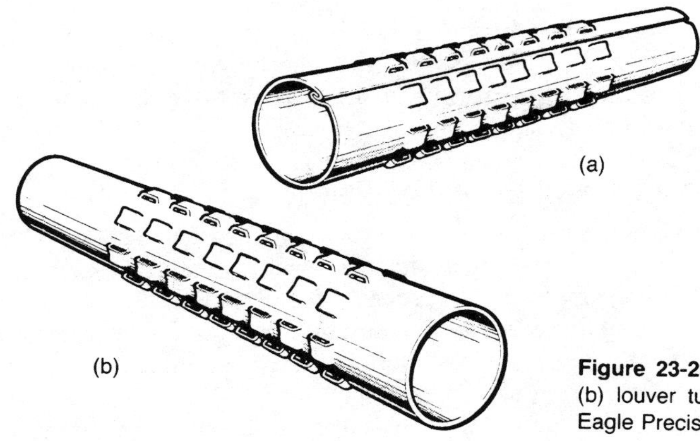

Figure 23-25 Typical (a) and redesigned (b) louver tube manufacture. (Courtesy of Eagle Precision Technologies, Inc.)

and then secure the joint with a lockseam or weldment. Another more efficient method is to form cut louvers in precut lengths of commercially available welded tubing, resulting in a substantial savings. This solution requires the development of special equipment for the process.

Baffles and endcaps must be joined to the louvered tubes or other piping in muffler assembly. A traditional method is to force the endcap over the tube to a predetermined position and spot weld it in place. A special machine design permits the cap to be slipped into position over the tube, and to expand the tube to form an annular raised ridge of metal on each side of the endcap or baffle. This provides for the quick, positive, and totally automatic assembly of the parts. See Figure 23-26.

Efforts at quantifying the assembly design process are exemplified by the work of Boothroyd and Dewhurst. This procedure involves assigning numerical values to various aspects of the assembly process. Value penalties are assessed for such

Figure 23-26 Product simplification is achieved by redesigning the assembly techniques for tubes and endcaps from a weldment (l) to a formed ridge (r). (Courtesy of Eagle Precision Technologies, Inc.)

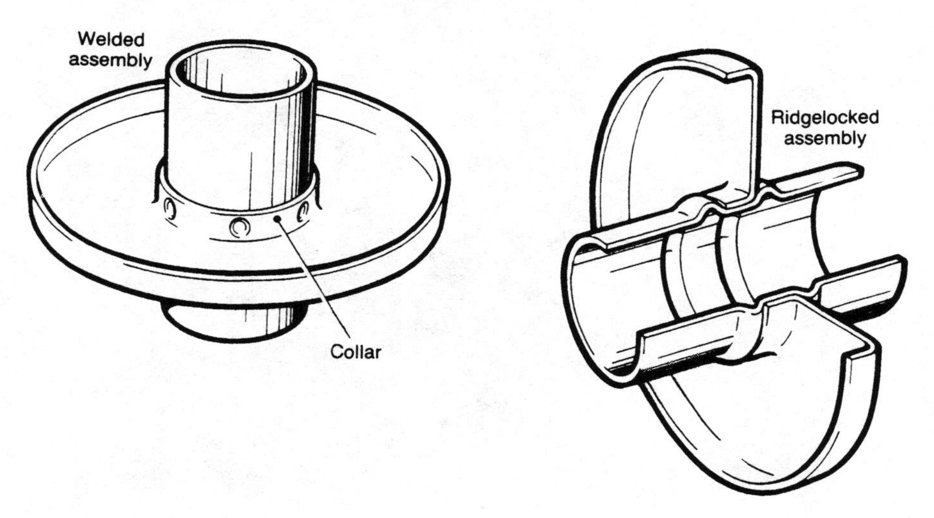

difficulties as part handling and fastening. Following the evaluation, the fabrication is redesigned by part elimination, the combination of related parts, and assembly simplification. A treatment of these assembly evaluation elements can be found in the bibliography at the end of this text.

Product Analysis—Seating Unit

Consumer products are subject to the same careful design and manufacturing analyses as are industrial components. The metal seating system for public spaces is an excellent example of this process. See Figure 23-27. The purpose of the seats is to provide comfortable, strong, durable, and attractive exterior support structures for human use. Ergonomic factors must be examined to determine proper structure dimensions for comfort and support.

Material selections were based upon an arrangement which would appear light and airy, yet be strong and lasting. Steel webbing panels are constructed of welded wire and rod, bent to shape, and welded to a structural steel tubing frame. The individual seats are positioned side by side in a variety of unit numbers, and clamp-bolted to heavy steel tubular supports. The entire unit is powder-coated, in a range of colors, to assure a weather-proof finish and a minimum of maintenance.

The visual impact of the system is exciting. There is an interplay of repetitive steel webs and tubular support, ergonomically curved, and attractively finished. The furniture is such that it can be effectively situated in parks, public waiting areas, or recreational groupings. A variety of similar and related structures have been designed as companion pieces. Cylindrical planters and trash receptacles, and banner holders are of an appearance consonant with the seating. The versatility of these units is such that they may be used in both commercial and residential settings.

Figure 23-27 These metal seating units are at once attractive functionally, substantial, and durable. (Courtesy of Landscape Forms)

Product Analysis—Mechanical Bellows Seal

A good example of an engineering design solution pertains to the extensive planning required to produce mechanical sealing systems. A typical application for such a unit is to seal the rotating shaft component of a centrifugal pump. See Figure 23-28.

The liquid to be pumped enters the suction inlet at the eye of the impeller. As the impeller rotates, at high speed, the space between the vanes is held captive by the close clearance between the front of the impeller and the pump housing. Here, the liquid is centrifugally forced to the outside of the impeller, where it exits through the discharge nozzle. At the point of discharge, the pressure of the liquid is many times higher than at the suction inlet, which accounts for the ability of the pump to push liquid to elevations far above the pump. The same discharge pressure, however, flows down behind the impeller to the drive shaft, which is connected to a driver outside the pump. This is the shaft that must be efficiently sealed if the pump is to be of any practical use. There are several ways of achieving this seal.

One obvious and common technique is to compress some resilient material, called *packing*, into an extension of the pump back-head, called the stuffing box. A typical stuffing box sealed with square rings of compression packing is shown in Figure 23-29. If this packing were to rub against the shaft, without lubrication present to prevent a build-up of frictional heat, it would soon destroy itself. Therefore, a certain amount of liquid being pumped must be allowed to flow between the packing and the shaft. Because of the surface irregularities of packing, eccentricity between the stuffing box bore and the shaft, as well as normal shaft run-out, a significant amount of packing must be used to compensate for these irregularities. This packing

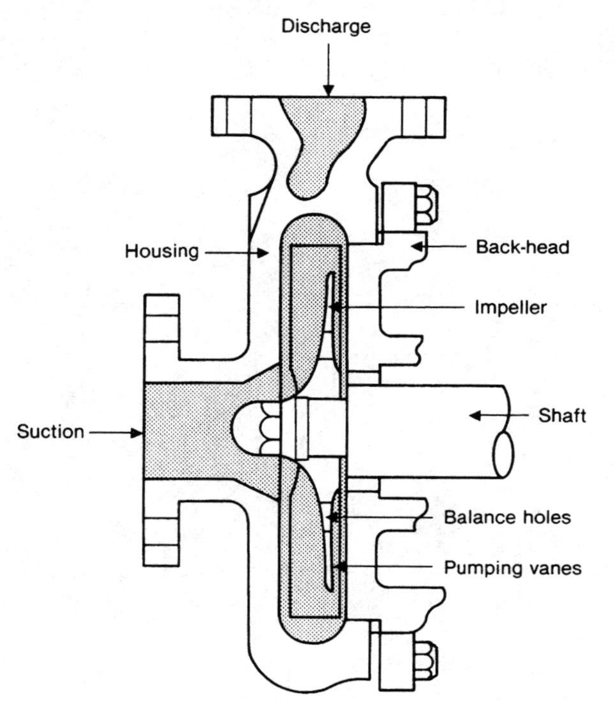

Figure 23-28 A diagram of the liquid end of a centrifugal pump. (Courtesy of Durametallic)

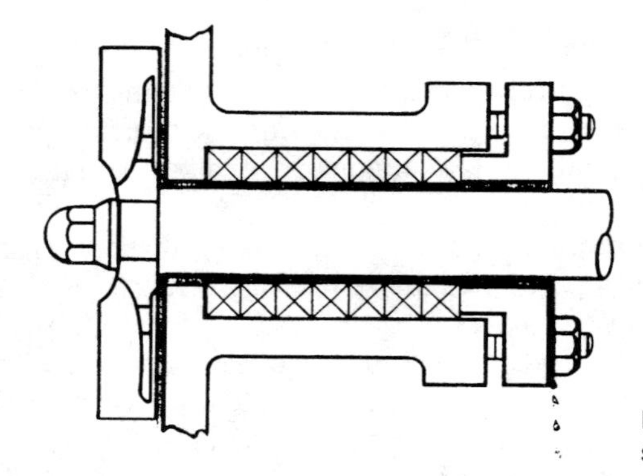

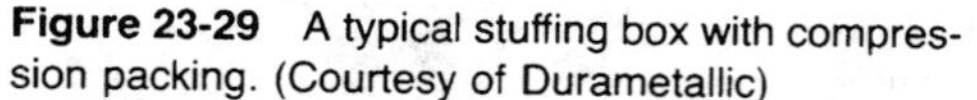

Figure 23-29 A typical stuffing box with compression packing. (Courtesy of Durametallic)

requires a generous amount of lubrication. When this flow of lubrication exits from the packing, it becomes identified as *leakage*. It is this leakage and its consequences that seals are designed to contain.

Various mechanical seals with "O" rings, gaskets, coil springs, packing, and compression rings are used to provide the required positive seal. One especially effective system is the mechanical bellows seal.

In some seal configurations, a coil spring provides the necessary elastomer bellows pressure on a seal ring system. See Figure 23-30. The metal bellows seal makes possible the sealing at higher speeds and temperatures than the elastomer bellows design. The bellows provides the necessary pressure to maintain face contact, thereby eliminating the necessity for a spring. Since the bellows is a one-piece unit supported around its entire circumference and along its total length, even pressure is applied to all points of the seal ring. It is this configuration of the bellows that allows the seal to be applied to higher shaft speeds that can be accommodated by a single coil spring over the shaft. See Figure 23-31.

Figure 23-30 A diagram of an elastomer bellows seal. (Courtesy of Durametallic)

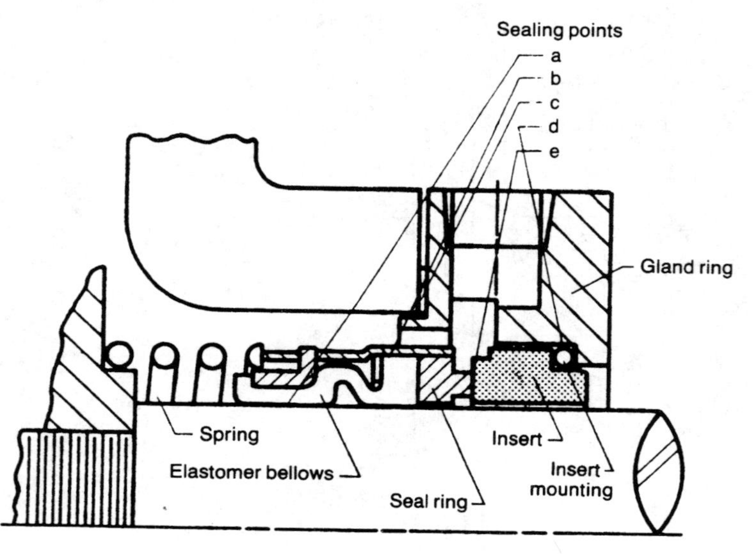

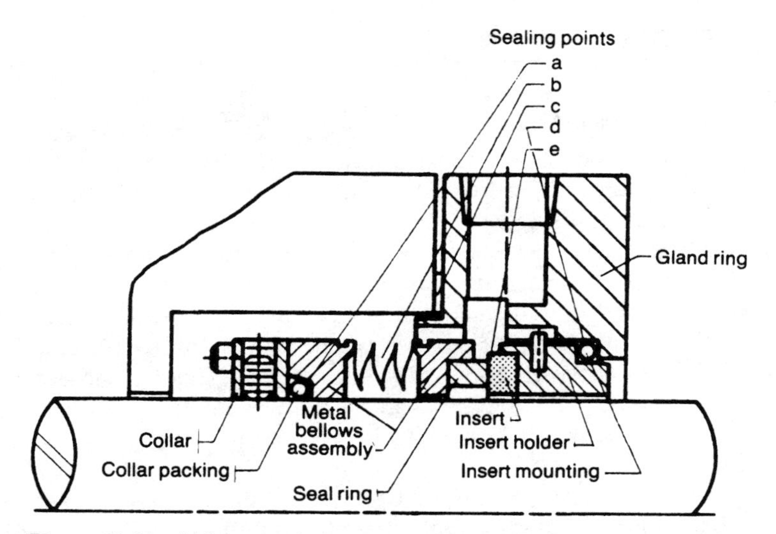

Figure 23-31 A diagram of a mechanical bellows seal. (Courtesy of Durametallic)

Aside from being used with higher shaft speeds, the metal bellows can also be applied to high and low (cryogenic) temperatures. Until recently, conventional seals had to rely on elastomers or TFE resins for secondary sealing members (shaft packing and insert mounting). These materials have a rather narrow temperature range, depending upon their composition. Even though a particular material might be suitable for use in an extreme temperature, this did not necessarily mean that it has the necessary resistance to chemical attack by the liquid that has to be sealed. Although the combined problem of temperature and chemical resistance could be generally overcome by creating an artificial compatible environment in the stuffing box, such a solution requires the use of costly "environmental controls." The metal bellows seal, therefore, is an economic solution to many high and low temperature sealing problems. See Figure 23-32.

Figure 23-32 A functional mechanical bellows seal unit. (Courtesy of Durametallic)

MISCELLANEOUS PRODUCT DESIGN

Obviously, manufacturers have a variety of process options open to them as they plan a product. The concrete mixer bowl in Figure 23-33 was originally a cast iron unit. The product redesign led to a version deep drawn from 2.65 mm (12 gauge) steel. It weights 57 percent less than the cast iron bowl, allowing a greater amount of concrete to be agitated with the same size motor. The new design also increased bowl durability and reduces wear. The horizontal flange is pierced and louvered to provide the circular ring-gear track. The dished portion of the bowl measures 600 mm (24 inches) in diameter and is 227 mm (9.25 inches) deep.

The truck frame suspension member was formerly a welded fabrication, and with redesign became a shallow-drawn piece, lighter and easier to assemble. It is made from 8 mm (0.315 inch) steel, weights 18 kg (40 pounds), and measures 975 mm (38.375 inches) long, by 100 mm (4 inches) by 100 mm (4 inches). See Figure 23-34.

While the aforementioned two formed metal parts were redesigned to simplify fabrication, this next example describes an effort directed toward ease of assembly. See Figure 23-35. The original housing unit was comprised of a cap fastened to a

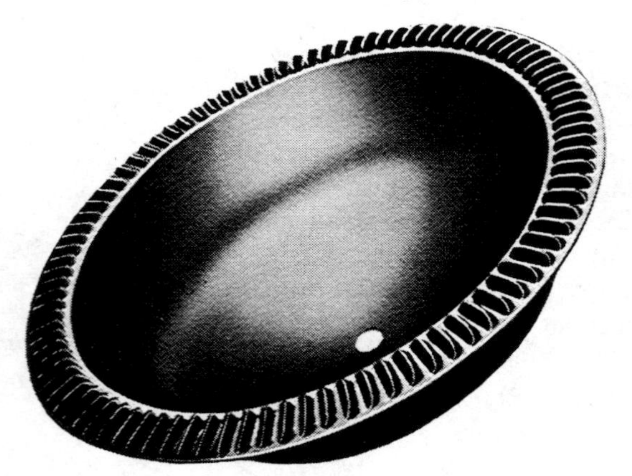

Figure 23-33 A redesigned deep-drawn steel concrete mixer bowl. (Courtesy of Commercial Shearing, Inc.)

Figure 23-34 A redesigned shallow-drawn steel truck suspension member. (Courtesy of Commercial Shearing, Inc.)

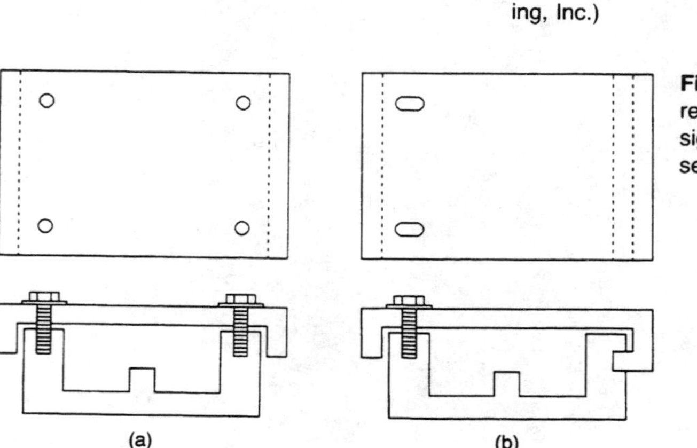

Figure 23-35 The original (a) and redesigned (b) housing unit. The redesigned part facilitates the product assembly process.

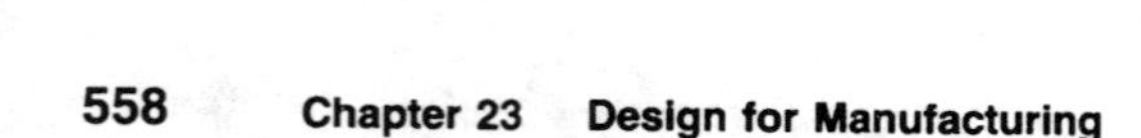

body with four machine screws. The four body holes required drilling and tapping, and the assembly was complicated by having to blind-locate the four screws into these holes. Automatic assembly was impossible. The redesigned unit featured an L-bend on one edge of the cap, two milled screw slots, and only two tapped holes. The assembly of the new unit is considerably easier, and can be at least partially if not fully automated. If the designer had originally considered the matter of component assembly during the conceptualization design phase, the better solution may well have emerged.

QUESTIONS

1. Describe the meaning and scope of the terms functional, material, and visual requirements.
2. Describe some basic differences and similarities of the work of industrial and engineering designers.
3. Write your personal definition of design, consulting references in addition to this textbook. List the references used.
4. Describe how visual requirements of the design of control panels can lead to a more functional control system.
5. Prepare a simple sketch of the design process diagram and describe the functions and interrelationships of each design phase.
6. List some examples, from your own experience or knowledge, of the following design considerations:
   a. Design for maintenance
   b. Design for usability
   c. Design for vandalism avoidance
   d. Design for producibility
   e. Design for assembly
7. Explain the difference between the terms maintenance and maintainability.
8. Prepare brief descriptions of the following human factors:
   a. Anthropometric
   b. Physiological
   c. Psychological
9. Prepare some sketches as partial solutions to the following problems:
   a. A vandal-proof restroom soap dispenser
   b. A system for easy replacement of automotive taillight bulbs
   c. A device for picking apples off a tree
   d. The control on a paint spray pressure can generally is activated by depressing a valve with the index finger, which can be tiring during prolonged operation. Design an improved spray can pressure device.
10. Disassemble one of the following items, sketch the parts, explain to the best of your ability how each was made, and describe how it was assembled. Make any design, material, or process improvements you feel are necessary.
    a. A mechanical pencil or pen
    b. A fingernail clipper
    c. A butane lighter
    d. A hand tool, such as pliers or screwdriver
    e. A light bulb (be careful!)
11. Prepare a library research report on an automatic assembly system used in producing a product from the preceding list.

24 COMPUTER INTEGRATED MANUFACTURING

Manufacturing includes all activities from the perception of product need, through the conception, design, and development of the product, production, marketing, and support of the product in use. Every action involved in these activities uses data, whether textual, graphic, or numeric. The computer, today's prime tool for manipulating and using data, offers the very real possibility of integrating the now often fragmented operations of manufacturing into a single, smoothly operating system. This approach is generally termed computer-integrated manufacturing (CIM).

The technologies that will be discussed in this chapter will be integrated to create CIM systems whose synergy will make the whole greater than the sum of its separate technologies.

COMPUTER-AIDED DRAFTING

Computer-aided drafting (CAD) is not a new technology; it has achieved acceptance and use in manufacturing design, and it has been used to replace traditional drafting techniques in areas such as architecture and cartography. It is important to understand CAD as a technology, because it interrelates with many of the other technologies described here. For example, CAD-type systems are now being used to program robots and computer numerical control (CNC) machining centers.

A CAD system is composed of a graphics terminal on which a picture of the part being designed can be displayed. Designers enter the part data by drawing on

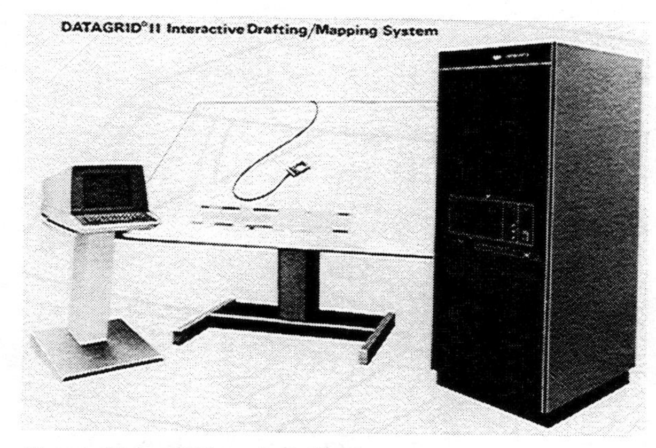

Figure 24-1 CAD work station.

a graphics tablet connected to the computer. A keyboard is used to enter dimensions and other data. The part description is then stored as one of many such part descriptions in a CAD data base. The computerized part description is not a picture, but rather a graphic representation of coordinate points and geometric shapes from which a picture can be constructed. A particularly successful standard, the Initial Graphics Exchange Standard, has been developed for transferring data from previously incompatible representations on one CAD system to another CAD system. A typical CAD work station is shown in Figure 24-1.

CAD offers many immediate benefits: Parts can be rotated (see Figure 24-2), scaled, and combined on screen in three dimensions to enable designers to better visualize them; repetitive sections can be redrawn automatically; overlays can be easily shown on screen; an engineering drawings can be easily updated and printed.

Other, more significant, benefits over the long run concern the use of the data in the CAD data base. These data—a computerized representation of the parts—can be used by the engineering and process planning functions, saving much reentering of data, eliminating sources of human error, and opening up a great avenue for cooperative design that includes feedback from engineering and manufacturing. Also, if the CAD data base is the only and therefore up-to-date source of part specifications, it eliminates a major current problem, the concurrent use of multiple versions of

Figure 24-2 Part rotation with CAD.

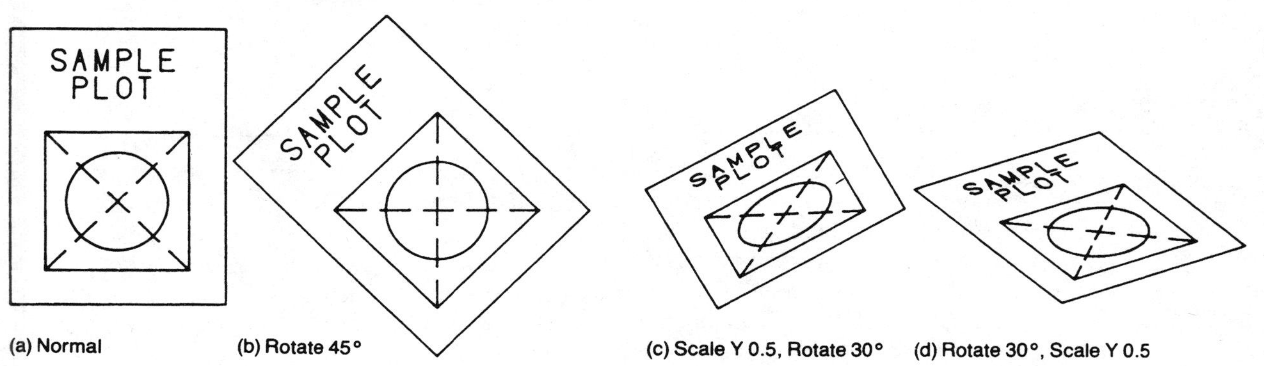

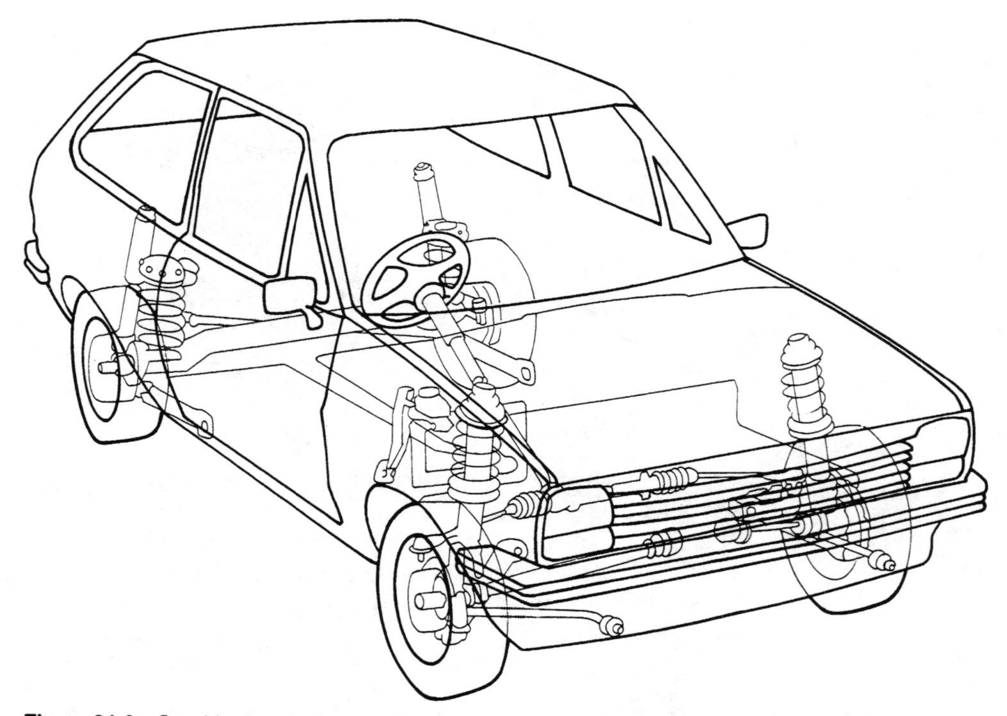

Figure 24-3 Graphics in automotive design. (Data Courtesy of Cambridge Interactive Systems, Cambridge, England)

part specifications. An example of graphics in automotive design is illustrated in Figure 24-3.

The microelectronics industry probably has the most integrated uses of CAD. A new microchip can be designed on a CAD terminal. Once the design is in the CAD data base, the chip's performance can be simulated, the design can be modified if necessary, and the masks for the chip can be made, all automatically from the data entered at the CAD terminal. Although other industries have not yet achieved this level of integration, it has become an embodiment of the computerized design-test-modify-test-fabricate model that may change the way manufacturing entities are organized. See Figure 24-4.

COMPUTER NUMERICAL CONTROL

Computer numerical control (CNC) can be defined as a form of programmable automation in which the process is controlled by numbers, letters, and symbols. In CNC, the numbers, letters, and symbols form a program of instructions designed for a particular workpart or job. When the job changes, the program of instructions is changed. This capability to change the program for each new job is what gives CNC its flexibility. It is much easier to write new programs to control movement than to make major changes in the production equipment.

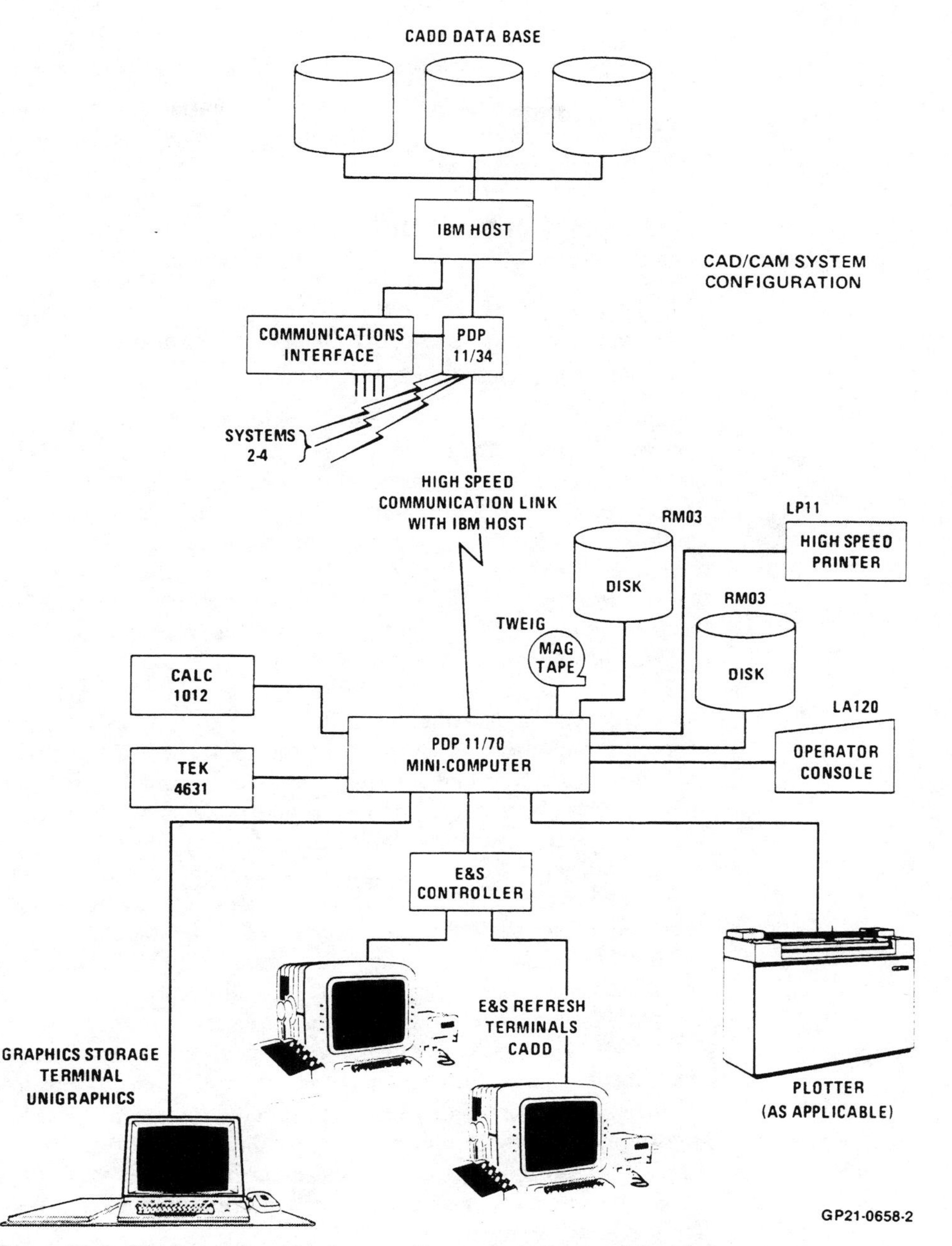

Figure 24-4 Distributed Graphics System. (Courtesy of McDonnell Douglas)

CNC equipment is used in all areas of metal parts fabrication and comprises roughly 15 percent of the modern machine tools in industry today. Equipment utilizing numerical control has been designed to perform such diverse operations as drilling, milling, turning, grinding, sheetmetal presswork, spot welding, arc welding, riveting, assembly, drafting, inspection, and parts handling. Computer numerical control should be considered as a possible mode of controlling the operations for any production situation possessing the following characteristics:

The workparts are produced in various sizes and geometries.

The workparts are produced in batches of small- to medium-size quantities.

A sequence of similar processing steps is required to complete the operation on each workpiece.

Basic components of a CNC system. An operations numerical control system consists of the following three basic components:

1. program of instructions,
2. controller unit, also called machine control unit, and
3. machine tool or other controlled process.

The general relationship among the three components is illustrated in Figure 24-5. The program of instructions serves as the input to the controller unit, which in turn commands the machine tool or other process.

Although numerical control was invented and applied some 30 years ago, it continues to change the structure of machine tools in ways that still are not fully appreciated. Computerized numerical control (CNC) has replaced the punched paper tape of the original NC tools. As machines were developed specifically for NC, the traditional lines separating machine types began to blur, and two new classes of machines began to develop.

The first class, called machining centers, generally operates with a stationary workpiece and a rotating tool. Feed of the tool in relation to the work can be handled by additional movement of the tool, the work, or both. The last method is necessary for contoured surfaces, and in complex cases may require more than three axes of movement, often five, and perhaps as many as eight. These machines primarily perform drilling, milling, and boring operations, but they also can tap, thread, and, when necessary, mill a surface that simulates work produced by turning.

The second new class of machines developed as a result of NC has rotating work and a stationary tool (except for feed). The machines are called turning centers, and they resemble lathes. They do primarily internal and external turning, drilling (of holes on the center axis), and threading. Some are equipped with powered stations that permit off-center drilling, tapping, and milling.

Both of these new classes of machines can be equipped with automatic tool-changing devices and often have automatic work loading, sensors to check on operating conditions, measuring devices, and other features than enable them to operate for long periods on different workpieces with little or no operator attention. With such versatility, a machining center and a turning center working together can perform all of the basic cutting operations on virtually any part that falls within the operating size limits of the machines.

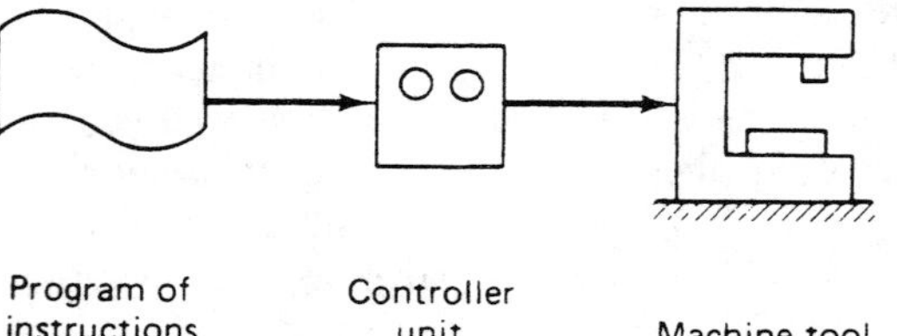

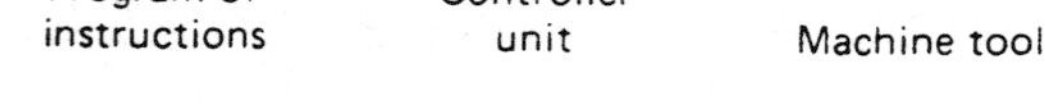

Figure 24-5 Three basic components of a numerical control system.

A number of special cutting and finishing processes supplement the basic processes performed by these machine. These include gear cutting, shearing, punching, thermal cutting, grinding, wire EDM, lasers, honing, and lapping. Although CNC was not generally applied to these operations as quickly as it was to the basic cutting operations, it is now applied to machines for each of them. (Because shearing and punching are done on presses and usually on sheet or plate material rather than on heavier workpieces used for cutting, they are usually classified as metal-forming operations. Thermal cutting also falls in this class.)

ROBOTICS

The term "robot" originates from the Czech word "robotnik," which was originated in 1921 by Karl Capek, a Czech playwright. He used the word in his drama *R.U.R.* to mean worker or serf: the machines in his play resembled people, but worked twice as hard. Since that time there have been many interpretations of what a robot is and how they are classified.

One definition of an industrial robot that has been internationally accepted was developed by the Robotics Institute of America. They define an industrial robot as "A programmable, multifunctional manipulator designed to move material, parts, tools, or specialized devices through various programmed motions for the performance of a variety of tasks."

The Japan Industrial Robot Industry Association provides a broader definition of robots by separating robots into six categories:

1. Manual manipulator—a manipulator worked by an operator.

2. Fixed sequence robot—a manipulator that repetitively performs successive steps of a given operation according to a predetermined sequence, condition, and position.

3. Variable sequence robot—a manipulator similar to that in category 2, but whose set information can be changed easily.

4. Playback robot—a manipulator which can produce, from memory, operations originally executed under human control.

5. NC robot—a manipulator that can perform a given task according to the sequence, conditions, and positions commanded via numerical data, using punched tapes, cards, or digital switches.

6. Intelligent robot—a manipulator incorporating sensory perception (visual and/or tactile), to detect changes in the work environment and, using decision-making capability, proceed accordingly.

The basic structure of a robot is similar to the human operator, with each having a body, arm, wrist, and hand or end-effector. The robot body is the foundation on which the arm is mounted. Most industrial robots have stationary bodies, but additional mobility may be provided by mounting the bodies on a track or overhead gantry.

The primary function of an arm is to move the wrist and end-effector to a desired location in the work envelope. Additional orientation of the end-effector is provided by rotation of the wrist. The end-effector consists of a hand-like device to grip parts, tools, and holding devices to enable the robot to perform the desired work task.

The classification or designation of robots may be by type of control, power source, and work envelopes.

Servo type. A nonservo robot, sometimes referred to as a pick-and-place or bang-bang robot, has a simple electromechanical control system with no feed back device. Commonly used in materials handling applications, this type of robot picks an object from one location and places it at another.

The servo robot is the most common industrial robot used in the United States. The controller continuously sends directions to the elements of the robot and then checks to make sure the components have moved as they were directed. One or more servo-mechanisms enable the arm and gripper to change direction during a continuous motion.

Power source. Robots are powered by pneumatic systems, hydraulic motors and actuators, and electric motors.

Pneumatic systems are the least expensive type of control, because of the availability of compressed air in most industrial plants. The major disadvantage of this type of power is the difficulty of providing repeated accuracy and control with compressed air. This type of power source is usually limited to nonservo, pick-and-place robots.

The first robots were powered primarily by hydraulic systems that use an electric motor to pump a fluid and a series of control valves and actuators to direct the movement of the basic components of the robot body, arm, and gripper. This type of power can handle loads with a high degree of precision.

Electric motors are being used more frequently as a power source to drive robot movement, due to their quiet and clean operation. With a servo motor, the exact

Figure 24-6 Rectangular robot.

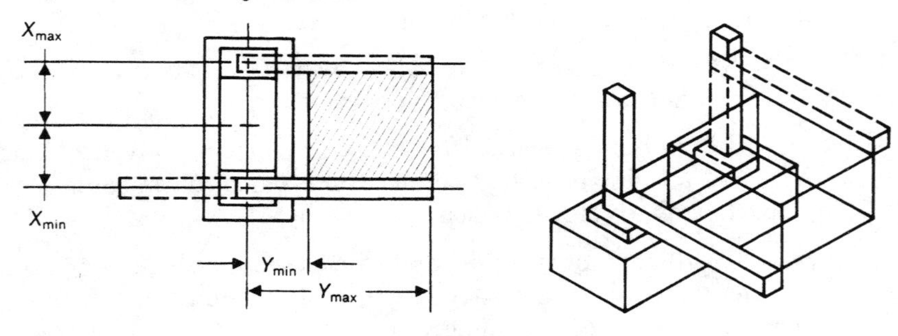

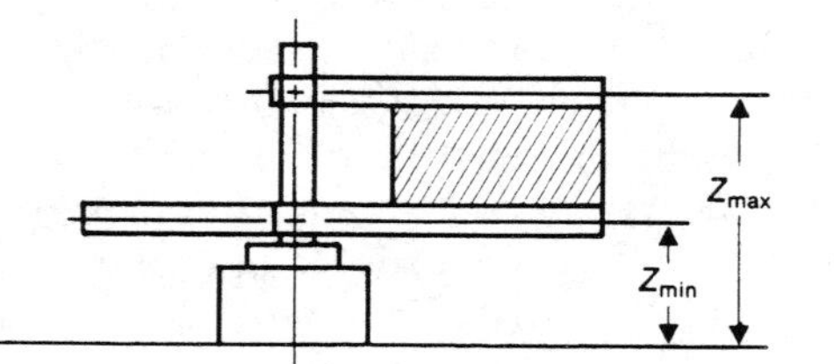

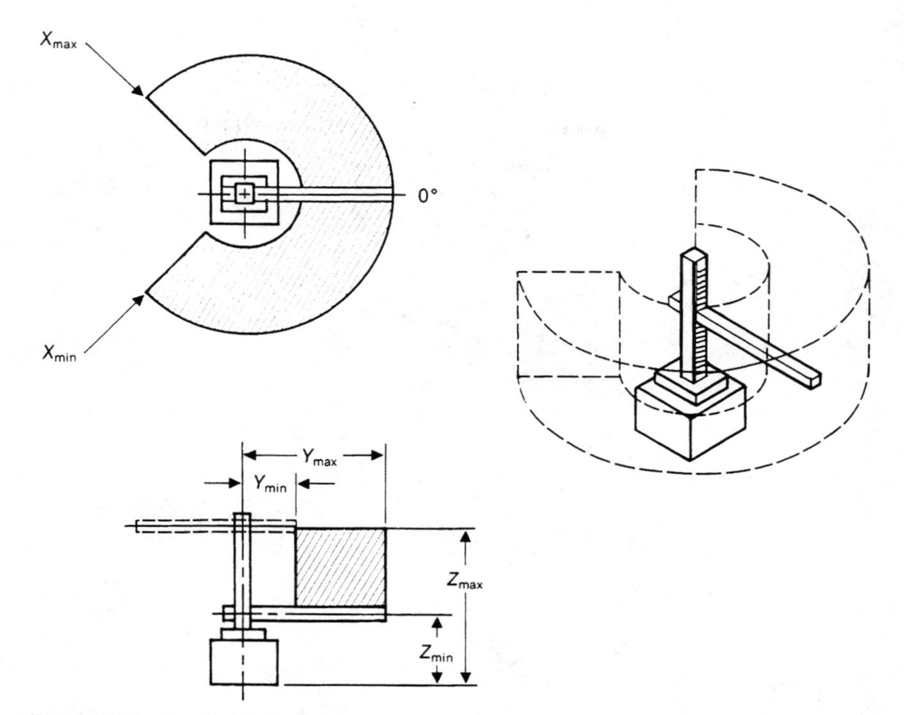

Figure 24-7 Cyclindrical robot.

position can be controlled to direct the robot to the desired position and orientation. Electric robots have limited load capacity when compared to hydraulic systems.

Work envelopes. When attempting to determine the work envelopes of robots, it becomes obvious that there are no standards that have been accepted and utilized throughout the robotics industry. Manufacturers differ on how they specify the various axes and working envelopes. To eliminate some of the confusion over terminology and dimensions, a standardized method of specifying robot motions is shown in Figures 24-6 through 24-10. The axes are designated as closely as possible to those that are used in conventional machining and NC program languages. Each of the X, Y, and Z axes have both a minimum and a maximum dimension which clearly define the work envelope in relation to the center line of the robot base. By standardizing on these dimensions, it is possible for all engineers and analysts to have a common language when evaluating or specifying a particular robot for a given application.

The axes of movement for a rectangular type robot (sometimes referred to as a Cartesian coordinate robot) are shown in Figure 24-6. This type of robot moves in the vertical and horizontal axes in simple straight line motions, forming a rectangular work space. To extend the range of these robots, they are frequently mounted on a track or overhead gantry.

The work envelope that is shown in Figure 24-7 is for a cylindrical robot. The vertical and horizontal motions are similar to the rectangular robot, with the exception of the base movement. Instead of a straight line motion, the cylindrical robot rotates about the base, forming an arc.

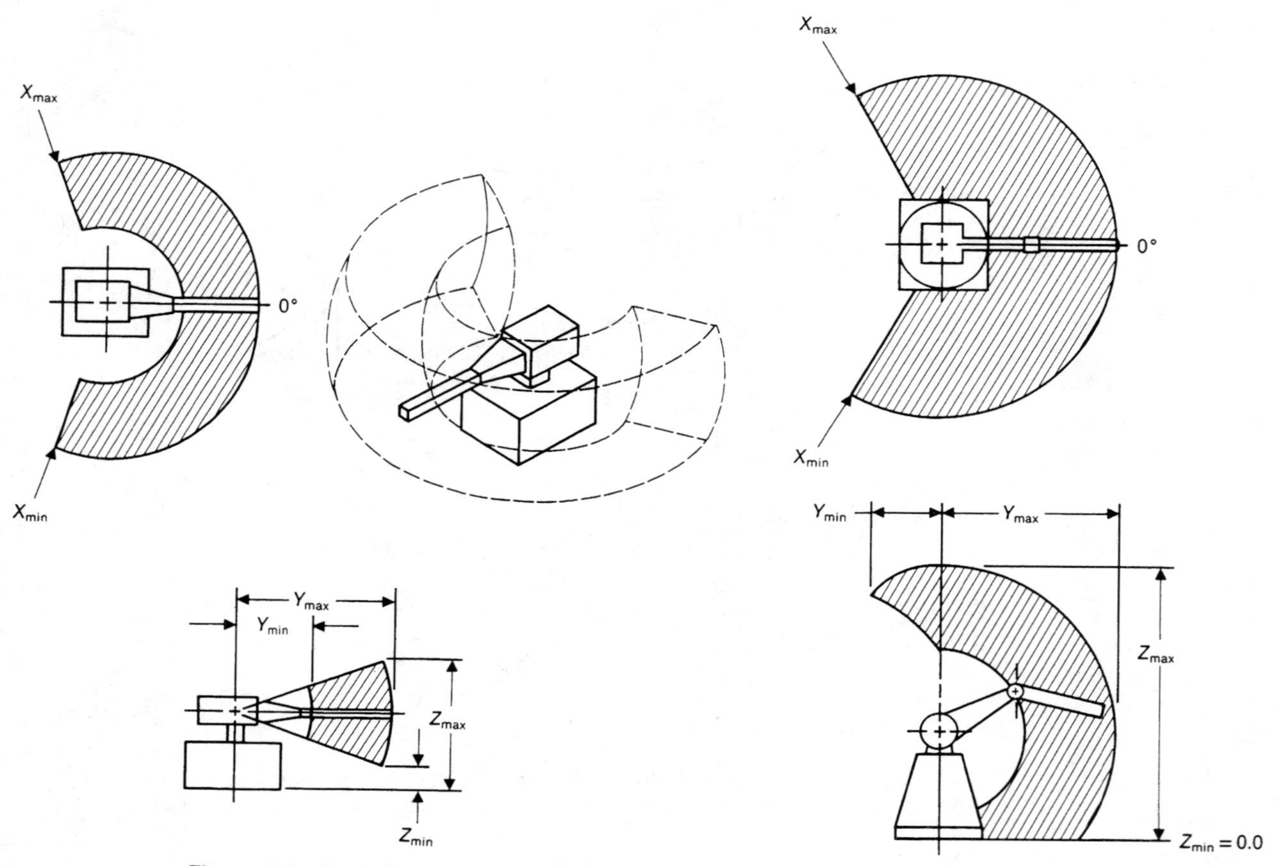

Figure 24-8 Spherical robot.

Figure 24-9 Jointed arm robot.

With a spherical robot, the reach and sweep (rotation about the base) motions are the same as with the cylindrical robot. The vertical motion results from a pivoting action of the robot arm and forms an arc in the vertical plane. The pattern of a spherical robot is indicated in Figure 24-8.

A fourth type of robot is the jointed arm robot. As shown in Figure 24-9, this type of robot forms a horizontal swing pattern similar to the cylindrical and spherical models. A unique motion in the vertical and horizontal reach planes is formed by the combination of angles that are possible between the various joints of the arm.

Additional manipulation abilities of robots may be extended beyond the limitations of the three main axes of reach (Y axis), swing (X axis), and vertical stroke (Z axis) by including wrist movements. Thirty-eight percent of all robots have provisions for controlling six axes of movement. These motions are shown in Figure 24-10.

Motion control. Three basic types of robot motion control are point-to-point, continuous path, and controlled path.

The most common motion control for robots is point-to-point control, where the path of the robot arm moves in several axes at the same time. Generally, all axes move at the maximum programmed velocity. Whichever axis has the smallest distance to move will reach its position first and then wait for the others. An example of

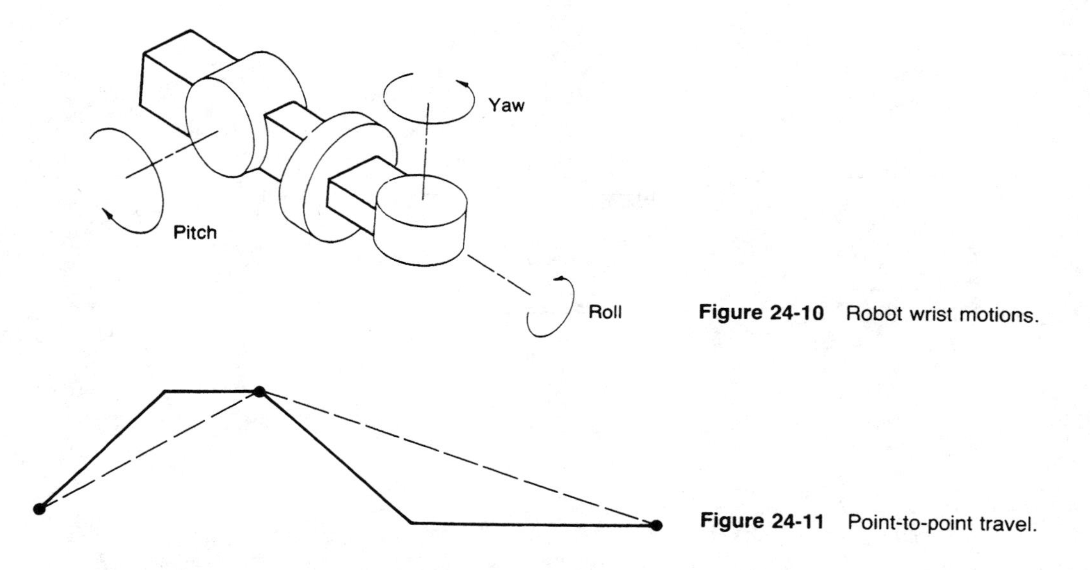

Figure 24-10 Robot wrist motions.

Figure 24-11 Point-to-point travel.

point-to-point travel is shown in Figure 24-11. Point-to-point robots are taught one point at a time with a teach pendant. The operator positions the arm at each desired point and then presses a button on the teach pendant that directs the controller to store the position data.

With continuous path robots, the operator physically moves the robot arm over the desired path of motion. The controller records the robot position at a fixed time increment. The actual path of the robot may not be a straight line, depending on the fixed time increments.

Controlled path motion follows a controlled path between discrete points. The controller computes intermediate control points by transformations between joint angles and hand position and orientation.

GROUP TECHNOLOGY

Group technology (GT) is a key philosophy in the planning and development of integrated systems. In practice, GT is defined as a disciplined approach to identifying by their attributes things such as parts, processes, equipment, tools, people, and customer needs. These attributes are then analyzed to identify similarities between and among things; the things are grouped into families according to similarities; and these similarities are used to increase the efficiency and effectiveness of managing the manufacturing process.

Although it is relatively simple to define GT, it is difficult to create and install a GT system, because of the difficulty in defining clearly how similar one part is to another. For example, parts can be categorized in terms of shape or manufacturing process requirements. These two different viewpoints require a flexible approach to the GT data base and the realization that parochial, departmental views of coding may allow some localized cost saving but miss the large corporate savings possible.

The GT concept requires that the attributes of a thing, such as a part, be identified and classified. Attributes can be visual, such as the surface finish or shape of a

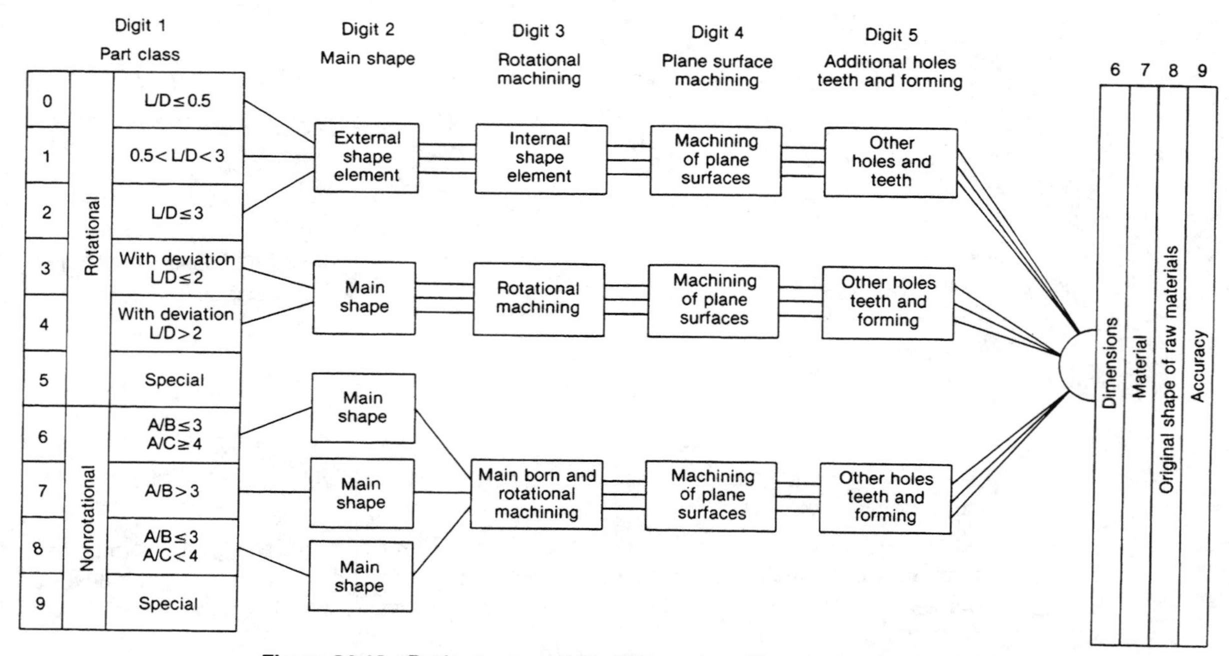

Figure 24-12 Basic structure of the Opitz system. (Source: Automation, Production Systems, and CAM by Mikell Grover, 1980)

part; mechanical, such as the strength of the material; or functional, such as the clock aspect of a printed circuit board. The attributes may also be related to the environment of the part, such as the processes or equipment necessary to make it. Because of the many possible coding strategies, it is hard to know in advance exactly what attributes will be important as the GT data base is used by more and more kinds of software. It is therefore important to guarantee that the data base structure is flexible enough to add attributes and to modify coding schemas as necessary for new applications.

One of the pioneering efforts and best known classification methods in group technology is the Opitz system, which is shown in Figures 24-12 and 24-13.

The four basic GT applications are design retrieval, FMS, purchasing support, and service depot streamlining. Design retrieval is a GT application in the design engineering area to provide the maximum potential for part standardization. Moreover, it permits greater cooperation between the design engineer and manufacturing engineering, by providing feedback about specific part attributes in the GT data base. Design retrieval will also supplement product reliability data based on the actual performance of parts with similar attributes. Finally, by determining the relationships of new parts to previously designed parts, it increases human productivity in the generation of new designs and the revisions of old designs. The software necessary to implement design retrieval is the simplest of all GT software; it involves a simple query to a GT data base for specific feasible ranges of variables.

Digit 1

		Part class
0	Rotational parts	$L/D \leq 0{\cdot}5$
1		$0{\cdot}5 < L/D < 3$
2		$L/D \geq 3$
3		
4		
5		
6	Nonrotational parts	
7		
8		
9		

Digit 2

		External shape, external shape elements
0		Smooth, no shape elements
1	Stepped to one end	No shape elements
2	Stepped to one end or smooth	Thread
3		Functional groove
4	Stepped to both ends	No shape elements
5		Thread
6		Functional groove
7		Functional cone
8		Operating thread
9		All others

Digit 3

		Internal shape, internal shape elements
0		No hole, no breakthrough
1	Smooth or stepped to one end	No shape elements
2		Thread
3		Functional groove
4	Stepped to both ends	No shape elements
5		Thread
6		Functional groove
7		Functional cone
8		Operating thread
9		All others

Digit 4

	Plane surface machining
0	No surface machining
1	Surface plane and/or curved in one direction, external
2	External plane surface related by graduation around a circle
3	External groove and/or slot
4	External spline (polygon)
5	External plane surface and/or slot, external spline
6	Internal plane surface and/or slot
7	Internal spline (polygon)
8	Internal and external polygon, groove and/or slot
9	All others

Digit 5

		Auxiliary holes and gear teeth
	No gear teeth	No auxiliary hole
1		Axial, not on pitch circle diameter
2		Axial on pitch circle diameter
3		Radial, not on pitch circle diameter
4		Axial and/or radial and/or other direction
5		Axial and/or radial on PCD and/or other directions
6	With gear teeth	Spur gear teeth
7		Bevel gear teeth
8		Other gear teeth
9		All others

Figure 24-13 Form code for rotational parts in the Opitz system. (Source: Automation, Production Systems, and CAM by Mikell Grover, 1980)

Flexible manufacturing systems exemplify a more sophisticated and more profitable GT implementation. Such an FMS can be created by identifying a cluster of machinery that can, or will be able to, service a particular family of parts. The FMS can then be streamlined to produce this part family optimally.

Purchasing support is a rather new GT application, yet almost every major manufacturer has a quasi-GT system, called a commodity code, already performing this task. A rigorous GT system may pay a tremendous financial reward by permitting all related parts, including those that do not obviously belong to the same families, to be identified throughout a factory or corporation. In one instance, a vendor of hoses offered a 50 percent reduction on hose prices if a corporation could identify all hoses and their attributes to be purchased over a given period. The corporation saved millions of dollars by rigorously following a GT system to identify all hoses to be ordered. A GT purchasing support system offers buyers a significant way to cut costs through knowledge and buyer leverage.

Service depot streamlining is a GT application which can help determine the most advantageous service parts strategy by identifying where alternate parts may be used. Standardization of parts in the service depot allows for substantial reduction in inventory and repair time, even if the standard is the most expensive item.

The premise of a GT system is that the similarity of parts and processes can be turned into substantial cost savings. The most far-reaching applications of GT will be made possible by the structuring of the parts data base itself. If the data base information is captured in attribute form and linked to applications by similarity, the ability of the data base to support manufacturing decision making will be greatly enhanced. This global application of GT to a data base design is only now gaining popularity, and research on it is still in its infancy.

One of these far-reaching applications is in the area of process planning. By performing a rigorous analysis of manufacturing processes and parts to be made, a manufacturer will improve his or her ability to move from the present method of process planning into the more highly integrated future. For example, a manufacturer may evolve from his or her present variant process planner, which uses GT to match part families to process families, to a more sophisticated generative system in which more knowledge captured in the GT data base will be used to optimize the process plan. Eventually, a highly integrated system can be achieved that delays the final process planning step until the part is to be made, optimizing not only the process but also capacity utilization. Group technology, like network technology, will be a cornerstone of CIM systems.

FLEXIBLE MANUFACTURING SYSTEMS

Flexible manufacturing systems (FMS) are expected to dominate the factory automation movement within 10 years. These FMSs will be tied to larger-scale manufacturing systems, but it is valuable to consider the FMS as a critical unit or building block in total factory integration. A FMS may be described as an integrated system of machines, equipment, and work and tool transport apparatus, using adaptive closed-loop control and a common computer architecture to manufacture parts randomly from a select family. The hardware components of an FMS may include a CNC tool, a material handling system, and an inspection station. The part family processed by the FMS is defined by GT classification. For greatest productivity, the FMS is optimized to produce only one family of parts, and conversely, the parts produced by the FMS are designed to facilitate processing by the FMS.

The concept of flexibility as used in an FMS includes:

- use of GT to achieve a part mix of related but different parts;
- batching, adding, and deleting of parts during operation;
- dynamic routing of parts to machines;
- rapid response to design changes;
- making production volume sensitive to immediate demand, and
- dynamic reallocation of production resources in case of breakdown or bottleneck.

Flexible manufacturing has been a reality in U.S. industry since its introduction in 1972, and the number of new FMS installations is doubling every two years. The number and flexibility of FMSs is expected to increase, and the cost of FMS installations is expected to drop. Although the United States was largely responsible for the technological developments of the FMS, Western Europe and Japan both have more FMS installations than this country. In fact, one of the most frequently cited FMS installations is located in the Messerschmidt-Boelkow-Blohm (MBB) plant in Augsburg, West Germany. The basic elements of this FMS are 25 CNC machining centers, multispindle gantry, and travelling-column machines; automated tool transport and tool-changing systems; an AGV workpiece transfer system; and hierarchical computer control of all these elements. The FMS is used to build wing-carrythrough boxes for Tornado fighter-bombers. A study reported by the National Academy Press shows comparisons by MBB of the performance of this FMS versus the projected performance of stand-alone NC machine tools doing the same work clearly shows the advantages of the FMS approach:

- number of machine tools decreased 52.6 percent;
- work force reduced 52.6 percent;
- tooling costs reduced 30 percent;
- throughput increased 25 percent;
- capital investment 10 percent less than for stand-alone equipment; and
- annual costs decreased 24 percent.

U.S. statistics for FMS installations are no less startling. A FMS at General Electric (GE) improves motor frame productivity 240 percent; an AVCO FMS enables 15 machines to do the work of 65; and at Mack Trucks, FMS permits 5 people to do what 20 did before. In addition to productivity enhancements, the FMS offers increased floor space capacity. GE, for example, reported that floor space capacity was increased 50 percent, with a net floor space reduction of 30 percent. A FMS can make a factory more responsive to its market—GE reported a shortening of its manufacturing cycle from 16 days to 16 hours.

An artist's rendition of a Flexible Machining Cell developed by LTV Aerospace and Defense Company is shown in Figure 24-14. The project entails machining more than 2,000 parts. Table 24-1 lists components of the cell. See also Figures 24-15 and 24-16. System highlights are described as:

Figure 24-14 Flexible Machining Cell. (Courtesy of LTV Aircraft Products Group)

complete direct numerical control (DNC);

automated machining—drilling, boring, tapping, milling, and profiling all performed in one setup;

remote automated pallet transport and machine load/unload system; and

automatic in-line part cleaning and inspection capabilities.

TABLE 24-1
FMC COMPONENTS

Elevated computer control room, which houses all but the VAPD business host computer (IBM 3081)
Eight horizontal single-spindle machining centers; a ninth machining center has been planned and the foundation installed, and will be used for part program certification
Two load/unload carrousels, each with two load/unload positions
One automated wash station
Two automatic coordinate measuring machines
Four automatic wire-guided vehicles (robo-carriers)
One central coolant distribution/chip recovery system
One manual wash station
One battery charging spur
One fixture buildup/teardown station to marry and divorce heavy fixtures and risers from machine pallets
Two material review stations for visual inspection of rejected parts
One calibration cube, which is used by the machining and inspection centers to probe ports and bores on the cube to provide a "footprint" of machine; it is used by maintenance for troubleshooting
An automatic cutter delivery system (not yet in use) from Cincinnati Milacron Incorporated (CMI) of Cincinnati, Ohio, utilizing "3 of 9 Standard" bar codes on cutter holders; once in place, human operators for cutter load/unload will be eliminated from the loop and replaced by robo-carriers and a three-axis robotic arm
Computerized cell loading, scheduling, simulation, and cutting-tool control
Palletized part load/unload at remote automatic work changer stations
Cutting tool setup in the central crib area, where electronic gauging is performed
Cutter-diameter compensation for NC programs
No assigned machine operations

Figure 24-15 Flexible Machining Cell. (Courtesy of LTV Aircraft Products Group)

Figure 24-16 Robocarrier. (Courtesy of LTV Aircraft Products Group)

COMPUTER-INTEGRATED MANUFACTURING SYSTEMS

CIM systems promise dramatic improvements in productivity, cost, quality, and cycle time. However, since full CIM has not yet been accomplished and depends on continued technological progress, the benefits are difficult to quantify accurately. Incremental gains from the implementation of individual technologies and subsystems will be substantial. These benefits are illustrated by the following data from five companies that have implemented advanced manufacturing technologies over the past 10 to 20 years:

Reduction in engineering design cost	15–30 percent
Reduction in overall lead-time	30–60 percent
Increase in product quality	2–5 times
Increase in capability of engineers	3–35 times
Increase in productivity of production operations	40–70 percent
Increase in productivity of capital equipment	2–3 times
Reduction in work-in-process	30–60 percent
Reduction in personnel costs	5–20 percent

The cumulative gains of total system integration can be expected to build on these results exponentially.

The long-range goal of CIM is the complete integration of all the elements of the manufacturing subsystems, starting with the conception and modeling of products and ending with shipment and servicing. It includes the tie-in with activities such as optimization, mathematical modeling, and scheduling.

A CIM system is created by interconnection or integration of the process of manufacturing with other processes or systems. The resultant aggregate system provides one or more of the following functions or characteristics:

1. An information communication utility that accesses data from the constituent parts of the system and serves as an information communication and retrieval system.

2. An information-sharing utility that integrates data across system elements into a unified data base.

3. An analysis utility that provides a mathematical model of a real or hypothetical manufacturing system. Employing simulation and, when possible, optimization, this utility is used to characterize the behavior of the modeled system in various configurations.

4. A resource-sharing utility that employs mathematical or heuristic algorithms to plan and control the allocation of a set of resources to meet a demand profile.

5. A higher-order entity that integrates information and processing functions into a more capable, effective processing system.

These functions and characteristics are not mutually exclusive in actual manufacturing systems; rather, they overlap significantly with all of the elements interconnected and integrated continually to form a single aggregated CIM. Perhaps the most important and least understood step in this process is the creation of an integrated system that is a higher-order entity; this is the true system-building goal.

Both horizontal and vertical growth of CIM systems can be expected as the year 2,000 approaches. State-of-the-art technology now includes small aggregates of computer-integrated tasks, often called islands of automation. Such islands of automation are found in design, where CAD workstations from different vendors share their data through a common data base and data conversion interfaces; in planning, with manufacturing resource planning (MRP) systems; and in production, where a work cell composed of a robot, machine tool, and inspection station may be coordinated by a cell controller.

In leading-edge plants, several of these islands of automation have been aggregated into larger manufacturing subsystems, termed continents of automation. At this level of integration, links exist between the design and engineering departments, with CAD terminals and data bases sharing data with CAE workstations and data bases. In planning, MRP can be linked to traditional data bases containing ordering and shipping information. On the factory floor, several work cells may be integrated with a material handling system to create an FMS.

In the factory of the future, these continents of automation will be integrated into worlds of automation that will eventually encompass not only entire factories, but also entire corporations. Because of the volume of data and complexity of decisions needed for full integration, a hierarchical structure is the only feasible way to achieve it.

A hierarchical structure has certain implications for the architecture of CIM systems. Data use and decision making must occur at the lowest levels possible. Only certain summary data will be passed upward in the hierarchy to be used in reporting the factory's state and in statistical trend analysis. Thus, an information and decision hierarchy is needed that practices management by exception. If additional information is required at upper levels, it will be requested. If local decision making cannot resolve a conflict, a decision will be requested from above. Conversely, management decisions may be communicated almost instantly throughout the system for rapid compliance.

The hierarchical structure further implies the use of distributed data based and distributed processing. A mainframe computer may be the host computer to the factory of the future. Connected to it will be an array of minicomputers, one level down in the hierarchy, each acting as host controllers to an intermediate-level manufacturing system. A mix of local and centralized data storage will be appropriate for each computer. Below the minicomputers will be microcomputers acting as cell controllers, graphics work stations, or executive work stations. Figure 24-17 shows a sheet metal data base.

The elements of this hierarchical structure can be thought of as subsystems, categorized by the role they play, although it must be remembered that categories may overlap considerably. Most subsystems of manufacturing fall into one of the following broad categories; (1) information and communication, (2) integration of processes, or (3) resource allocation. Note that each category cuts across traditional manufacturing boundaries. An information-communication subsystem, for example, could include a network that permits the geometric part data stored in the CAD data base to be transformed through GT techniques into an actual process plan and then into robot and CNC programs communicated to the factory floor. The data would then be transformed and communicated for process scheduling and material handling, right up to the delivery of the finished product.

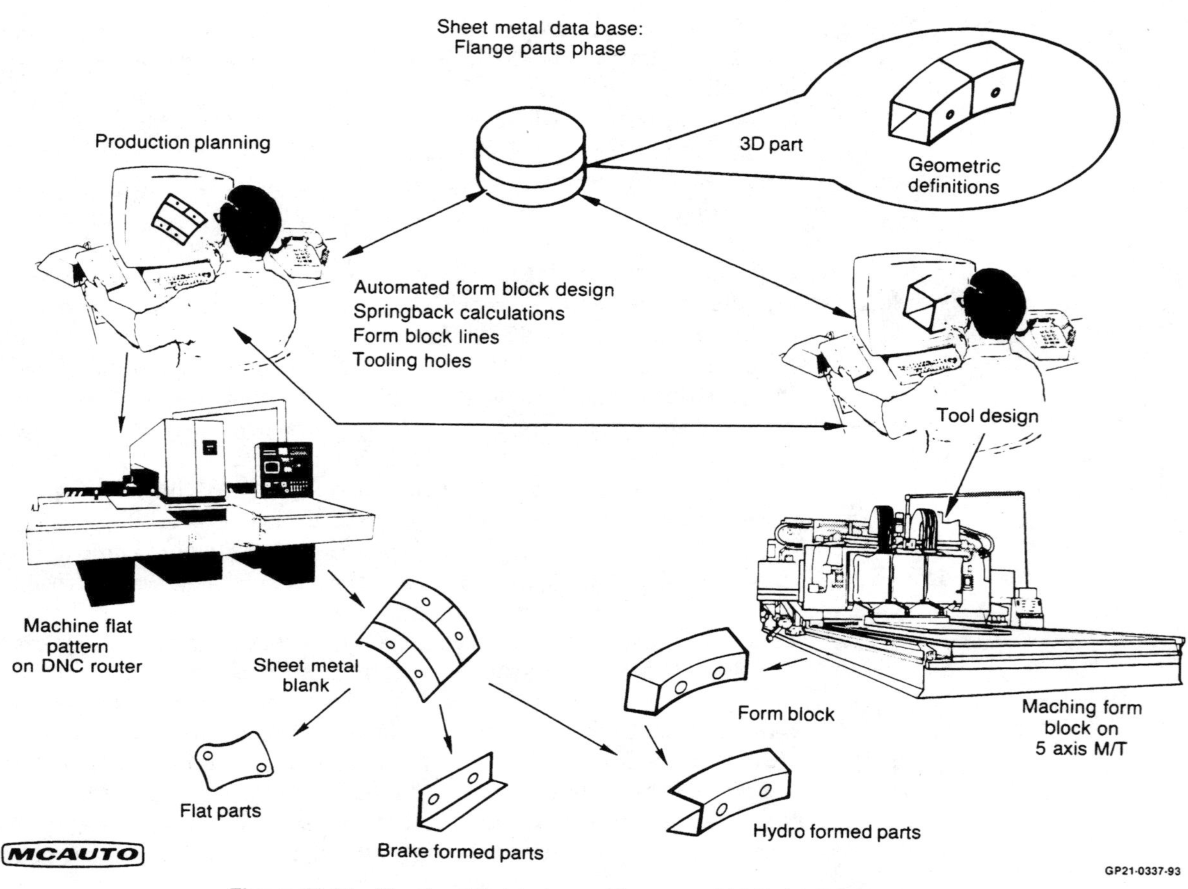

Figure 24-17 Sheet metal data base. (Courtesy of McDonnell Douglas)

Information-oriented subsystems include traditional management information system and data processing roles. These subsystems will be able to expand to include geometrical data from CAD systems, material and process data from GT coding, parts-in-process data, and order and inventory data. Information subsystems will have analytic capabilities by which the data can be massaged for quality control and trend analysis. Data retrieval will be easier for operators and decision makers through the use of new query languages or programs that will allow nonexperts access to complex data. Most, if not all, manufacturing subsystems have strong information and communication functions, even if they are primarily process-or-resource oriented.

Manufacturing subsystems on the factory floor will integrate traditional manufacturing processes by coupling and controlling previously separated processes and by carrying out computer-generated process plans. At the lowest level, this will involve data communication from sensors to a computer-controlled machine or robot. This provides the real-time adaptive control necessary to improve the work quality and throughput of individual stations. At the next level, factory floor manufacturing subsystems can integrate several processes, such as an NC machining station, an automated inspection station, and the robot which services them. In this example, the coordination is supplied by a computer which controls the work cell. The process plan is downloaded from a computer, which may be in the CAE area, to the work cell controller, which coordinates the processing by the machines in its cell. With automated inspection and data collection, the process plan may be modified to eliminate defects by responding in real time to tolerance changes. At yet a higher level, work cells are integrated into an FMS so that an automated scheduling system can assign a part in process to the next available work cell that can perform the necessary operation. This allows a system with fewer parts in process, shorter queues, fewer holding areas, and much more efficient use of floor space.

Resource allocation subsystems span a broad scope from small-scale material handling systems serving individual work cells to broadly implemented systems that monitor and control inventory, schedule work, and allocate materials to the factory floor on tight schedules. Automated material handling systems can be integrated into work cells and families of work cells to produce a powerful FMS. In turn, the FMS can be linked to production planning and capacity planning systems to form the fully computer-integrated manufacturing systems.

One of the most important reasons for implementing small-scale subsystems of manufacturing now is that they can be successively integrated into these higher

Figure 24-18 Sheet metal fabrication blanking cell. (Courtesy of Booz-Allen & Hamilton, Inc.)

SHEET METAL FABRICATION BLANKING CELL

INTEGRATED COMPUTER AIDED MANUFACTURING (ICAM)

PROJECT PRIORITY 2108

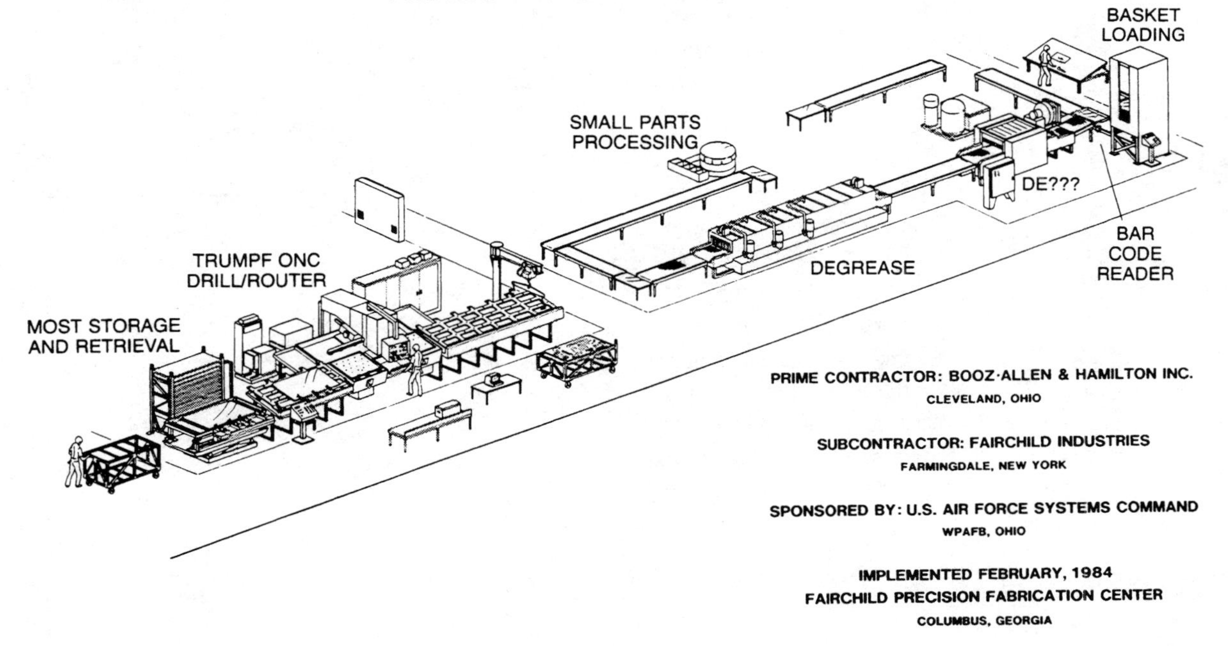

Figure 24-19 Photo of blanking cell. (Courtesy of Booz-Allen & Hamilton, Inc.)

order entities, CIM systems, that benefit from the synergy between operational programs, product data, and process data.

The following example shows a technology focused program to design, build, and implement a flexible manufacturing system for blanking and bending sheet metal. The ICAM Integrated Sheet Metal Fabrication System was developed by Booz-Allen & Hamilton Inc. and sponsored by the U.S. Air Force Systems Command. This system is shown in Figures 24-18 and 24-19.

QUESTIONS

1. What is Computer Integrated Manufacturing?
2. How can CAD reduce engineering design costs?
3. Define numerical control. How does it differ from CNC?
4. What are the basic components of a robot?
5. Describe the four types of work envelopes of robots.
6. List the basic applications of Group Technology.
7. What are the benefits of FMS?
8. Based on the benefits realized by companies that have installed CIM, why should a manufacturing company consider CIM?

25 THE TECHNOLOGY OF FUTURE MANUFACTURING*

Although all the interrelationships and long-term implications of advanced manufacturing technologies are not yet well understood, the direction of future developments is relatively clear. This chapter describes the technologies that are likely to have a major impact on manufacturing competitiveness, indicates the ways in which those technologies interact, identifies additional research needs, and discusses some of the issues that are likely to be encountered in implementation. The technologies have been divided into materials, material handling, material transformation, and data communication and systems integration. These categories are highly interdependent, and divisions are not always distinct, but this categorization provides an effective structure for a broad overview of the major technologies.

DEVELOPMENTS IN MANUFACTURING MATERIALS

Materials developments have a substantial impact on manufacturing in both product design and process engineering. New products can require different materials and materials processing, and new materials themselves often spur new products and new process development. Many of the material developments in manufacturing

* A major portion of this chapter is from *Toward a New Era in U.S. Manufacturing*, 1986, Appendix A, pp. 75–125, with the permission of the publisher, National Academy Press, Washington, D.C.

do not involve new materials, but rather substitutions, upgradings, and new concepts for conventional materials. Pressure to lower costs and raise product quality has led to some major shifts in materials selection. The consideration of doubly precoated steel for corrosion resistance in automobiles is an example. Developments in both conventional materials and new materials are equally significant to future manufacturing.

This section will focus on materials developments that are ready for manufacturing implementation, with minimal emphasis on research systems that have had little technology transfer effort. Developments in materials, polymers, ceramics, and glasses will be considered, followed by a brief discussion of some emerging issues that should be brought to the attention of policymakers.

Metals and metal-based composites. Major developments will continue in the processing of conventional metallurgical systems. For example, large tonnages of carbon and stainless steel sheet and strip will be continuously cast. Similar developments are certain in nonferrous alloy areas as well. While the continuously cast products will have some minor metallurgical variations to be considered, the major impact will be economic, allowing the basic metals industry to remain competitive in many sheet and strip applications.

Increased use of warm- and cold-formed steel parts can be expected, with emphasis on near-net-shape processes to save metal and avoid intermediate processing steps. Similar forces will continue to drive powder metallurgical processing, although it must be emphasized that advances in precision forming and powder metallurgy have been slow over many years rather than a sudden breakthrough. Powder metallurgy produced tool steels continue to offer advantages over conventional tooling stock.

Superplastic forming will continue to increase in aerospace manufacturing; some initial applications have occurred in the B-1 bomber program. See Figure 25-1. Increased market penetration will entail major changes in tooling and manufacturing

Figure 25-1 B-IB Bomber. (Courtesy of LTV Aircraft Products Group)

technology. Thermomechanically processed 7000 series aluminum alloys are available for such forming, as are aluminum-lithium alloys. However, little of these alloys are available from domestic sources; most of this material being obtained from the United Kingdom.

Many metal-processing alternatives will be examined to facilitate in-line processing systems. Improvements can be expected in the control of metal structure, with increasing awareness of the importance of grain orientation (texture), residual stress, and surface quality. In steel surface treatment, increased use of induction or laser hardening can be expected as lower cost alternatives to carburizing. Laser welding is seen as growing in the auto industry, perhaps at the expense of electron beam techniques. Power systems manufacturing may turn to welded rotor fabrication to allow increased use of attractive alloy combinations while sidestepping large forging development problems.

Many cases of metals substitution can be expected. The upgrading of alloys in small parts should increase product quality, reliability, and processing response without grossly increasing overall metal procurement cost. Aluminum can be expected to continue to replace copper in many heat exchange applications. Silicon-based switches are expected to replace iron-based magnetic devices. Also in the electronics industry, changes in plating metals are expected, with gold giving way to palladium-nickel and iron. Some observers see increased use of molybdenum-based alloys, particularly as new developments solve some of the traditional corrosion problems. Production of these alloys is energy intensive, however, and the domestic supply is limited.

In the new metals area, continued progress is expected in the development of metal matrix composites, particularly using metals such as aluminum and magnesium reinforced with silicon carbide. Three major types of reinforcement are receiving particular attention: continuous monofilament, discontinuous, and continuous multi-filament yarn. Each reinforcement requires a specific fabrication process, including diffusion bonding, hot molding, power blending, forging, casting, pultrusion, extrusion, and rolling, often in combination. Many of the material defects and anomalies that have plagued metal matrix development are attributable to the manufacturing process. As experience in these processes builds and the manufacturing technology evolves, application problems caused by material defects should decrease. Metal matrix composites are much less developed than for polymer matrixes and has not been widely disseminated for use by designers.

Progress in reducing material defects and continued application experience will result in broader applications for metal matrix composites. For example, automobile manufacturers are gaining experience with aluminum-silicon carbide in piston ring and crankshaft applications. Power systems applications are seen for nickel superalloy and stainless steel matrix composites strengthened with silicon carbide. Such systems allow increases in elastic modulus and desirable decreases in the coefficient of thermal expansion.

Much research has focused on rapidly solidified metals and amorphous metals. Introduction of such materials into the manufacturing sector is expected to be slow, with the major exception of the iron-boron-silicon-carbon system being broadly used for distribution transformers. Slow emergence is also predicted for nickel and titanium aluminides despite extensive research. Recent aluminide development has greatly increased its durability, and some jet engine applications can be expected.

Polymers and polymer-based composites. Polymers and polymer-based composites will probably continue to displace carbon steel and aluminum in a significant segment of structural and panelling applications. This trend may be most prominent in the automobile industry, but it is also likely in electronic hardware, appliance chassis applications, and home building components. Current automobiles contain about 157 pounds of plastics and polymer-based composites. By 1995, this should grow to 213 pounds as polymer materials are used increasingly in body panels. The increased use of plastic is expected to save motorists about $200 per year through fuel economy, corrosion resistance, cheaper repairs, and lower insurance rates.

The applications seen in models such as the Pontiac Fiero will spread for reasons of economy and safety. In the Fiero, the horizontal body panels are made of flexible glass fiber-filled polyester (a sheet molding compound) and the vertical panels are made of relatively stiffer glass-reinforced polyurethane (a reaction injection molded product). Beyond the body panel substitutions, further use of polymers can be expected in the automobile structure. For example, polymeric leaf springs have been used in some automobiles since 1981. While glass-polyester and glass-polyurethane materials will see major tonnage applications, future use of polymers in automobiles can be expected to rely on reaction injection molded thermoplastics as well. Beyond these applications, the expanded use of coatings (including paint) on steel can be regarded as an area where polymers will intrude further into the sheet metal markets.

The growth of polymer panels and structural shapes has necessitated the development of adhesives as a joining medium. In addition to the increased use of conventional adhesive materials, advanced work is under way on adhesive systems for higher temperature applications (epoxies, polymides), adhesives with greater strength and elastic range than epoxies, primerless adhesives (silicones perhaps), and faster-curing adhesives (cyanoacrylates, urethanes, etc.). In some instances, adhesive development overlaps sealant systems (silicones). Beyond the relatively simple polymer applications, adhesives are increasingly required for bonding dissimilar materials, particuarly when differential thermal expansion must be accommodated. For example, rivets cannot be used to join plastic liners to metal trailer bodies because of differential thermal expansion. In some metal joining developments, adhesives are being used in conjunction with spot welding to replace reveting. However, it must be emphasized that the use of adhesives for nonpolymeric joining has been slowed considerably by concerns about reliability, contamination, degradation, and consumer acceptance.

Another major shift in polymer materials use can be foreseen where flame retardation is a dominant consideration. Underwriters Laboratory interpretations of smoke, flame, and toxicity requirements are leading to shifts away from polyvinyl chloride-based systems to fluorocarbons.

High-technology ceramics. Advanced high-technology ceramics are nonmetallic materials having combinations of fine-scale microstructures, purity, complex crystal structures, and precisely controlled additives. In contrast to traditional ceramics, which are made from natural raw materials such as silica and clay, advanced ceramics are made from artificial raw materials, such as aluminum oxide, zirconia, yttria, silicon nitride, and silicon carbide, which are formed, sintered, and treated underprecisely controlled conditions. The advantage of such fine ceramics is their ability to play both functional and structural roles. Functional uses include optical devices, motors, transducers, sensors, and semiconductors; structural uses include those that require high specific strength, high wear resistance, and high corrosion resistance. Both roles will be increasingly important in manufacturing applications.

Currently, high-technology ceramics are used most often in electronics, including optical fibers, multilayer ceramic-to-metal interconnecting and mounting packages for integrated circuits, ceramic multilayer chip capacitors, piezoelectric ceramic transducers, and chemical, mechanical, and thermal sensors. Processing for these applications is generally an extension of standard ceramic technology, in which powders are pressed or formed with binders and sintered to densify the ceramics. Incremental progress has improved results, but major improvements cannot be expected until semiconductor processing techniques are applied to ceramic components. Techniques such as selective-area ion implantation and laser-induced recrystallization will greatly improve many of the properties of electronics ceramics.

High-technology structural ceramics are used as coatings and for monolithic and composite components. Major applications include tooling for metal working, wear components in a variety of abrasive environments, bioceramics for bone replacement, and military ceramics for radomes and armor. Major efforts are under way in both the United States and Japan to use structural ceramics in a variety of automotive applications, including engine wear components, turbochargers, bearings, and a variety of diesel engine components.

Significant strides have been made in the mechanical properties and reliability of monolithic structural ceramics. Understanding of strength-limiting flaws and temperature-brittle fracture behavior has improved greatly, but further work is needed to improve reliability. Improvements are needed in powder synthesis, powder properties, near-net-shape fabrication methods, microstructure control, mechanical properties, and nondestructive testing methods. Important research is also needed to identify new, more complex ceramics.

Significant advances are also being made in thermal barrier coatings and ceramic matrix composites. Ceramic matrices combined with particulates, whiskers, or fibers of a different ceramic compound or metal have yielded composites with five times the resistance to fracture of the monolithic ceramics. Recent success has been reported in the use of metal ion implantation to reduce the relative friction resistance of ceramic diesel engine parts. New research is needed to quantify the improved mechanical properties of composites, particularly fracture resistance.

In addition to the research on the composition and properties of ceramic materials, much work is needed on the processing and product design requirements of ceramics. Promising directions in ceramic processing include the use of ultrafine powders and the use of chemical routes to supplement or bypass some of the powder-processing stages. Other requirements include the processing of fine-scale layered structures, processing of ceramic composites, joining of ceramic parts, and near-net-shape processing of complex parts to minimize machining requirements.

The rate of technical progress in ceramic materials and processing will determine the pace of commercial application. Major market penetration for structural ceramics depends largely on the progress made in automobile applications. Several Japanese firms have already introduced ceramic turbocharger rotors, piston rings, swirl chambers, and camshafts. An almost totally ceramic engine is a major research objective of virtually every automobile manufacturer and should be extant by the early 1990s. As these automotive applications increase and the price falls, high-technology ceramics can be expected to see rapid growth in product and process (cutting tools) applications.

General issues related to materials. It is important to recognize the critical lack of data on and basic understanding of the physical properties of many materials. This lack is a severe handicap in manufacturing process development. Most materials handbook data have been generated for use in product design and service performance analysis rather than for process analysis. The lack of knowledge has become acute as software systems have emerged with powerful process control and process design capabilities. The requisite data inputs often involve combinations of stress, strain rate, temperature, heat transfer, friction, and so on, which have not been studied, even for classic engineering materials.

This lack of data on manufacturing materials grossly compromises the effectiveness of computer-based process models, and it tends to foster undue reliance on the few materials for which an adequate data base seems to exist. Power systems in particular are plagued by a lack of materials innovation due to the awesome data base requirements for service performance analysis, manufacturing modeling, and code adherence. With the current low return on investment in much of the primary materials industry, little supplier information is being generated. While some manufacturers develop their own data bases, other find that largely empirical trials are the least expensive approach (from a local point of view).

The scientific community has been reluctant to get involved in data-gathering efforts that involve "no new science." Federal funding agencies, reinforced by peer review systems, also have shunned this area. Progress in generating this data could have a significant impact on a variety of industries and process applications.

Another recurring concern is the frequent lack of domestic suppliers for new materials, such as some ceramics. Manufacturers are reluctant to begin using new materials systems when only one or perhaps no domestic supplier exists. Many new materials are initially imported in quantity from Japan or Europe.

A related problem is the growing tendency for manufacturers and wholesalers to limit inventory. Indeed, limiting inventories has emerged as a smart manufacturing practice, especially with high interest rates. However, this practice grossly limits the availability of new and many old materials for manufacturing trials. In fact, most of the materials in reference handbooks are not available in tryout quantities.

Lastly, there is an important interaction between the use of new materials and recycling and scrap practices. This is particularly the case in the automobile industry, in which a by-product of primarily steel construction has been the relative purity of car bodies as a source of steel scrap. The ease of recycling car bodies is being compromised by the materials substitutions now occurring, which could significantly increase material costs in a number of industries.

Although these problems slow progress, none is sufficient to prevent increased use of new materials in manufacturing if those materials are cost-effective in product, performance, and maintenance. Limited availability is probably the greatest handicap, because the machining, tooling, and processing required for many new materials can be vastly different from those for traditional metal cutting; significant research in a production environment is a prerequisite for increased use. Fortunately, enough production experience is being accumulated with many new materials, particularly polymers, to demonstrate their advantages and to encourage efforts in other areas of materials research. Despite the handicaps, significant breakthroughs can be expected so that changes in manufacturing materials will keep pace with the many other developments on the factory floor.

MATERIAL HANDLING TECHNOLOGY TRENDS

This section will assess the major trends in material handling technology. Material handling systems are used to enhance human capabilities in terms of speed of movement, weight lifted, reach distance, speed of thought, sensory abilities of touch, sight, smell, and hearing, and the ability to deal with harsh environments. In this area, it is important to distinguish between equipment technology and design technology. Equipment technology is categorized by its primary functions: transporting, storing, and controlling materials.

Transporting. The material handling function of transporting material has been affected significantly by two trends—toward smaller loads and toward asynchronous movement. The former is the result of the drive to lower inventory levels through just-in-time production. It has been manifested in the development of numerous equipment alternatives that have been downsized for transporting tote boxes and individual items rather than the traditional pallet loads. The inverted power-and-free conveyor, powered by linear induction motors for precise positioning and automatic loading and unloading, is one example of the trend to develop transport equipment for small loads. Automatic guided vehicles (AGVs) for transporting individual tote boxes are also being developed, as are specially designed conveyors and monorails for tote box movement.

The use of asynchronous movement in support of assembly has existed for many years. For example, asynchronous material handling systems were prevalent in automotive assembly before the paced assembly line was adopted at Ford in the early 1900s, and the concept has been applied recently in some European automotive assembly operations. In the early 1970s, Volvo began using AGVs to achieve asynchronous handling in support of job enlargement. Asynchronous material handling equipment is often used to allow a worker to control the pace of the process.

The trend toward asynchronous movement appears to be partially motivated by the apparent success of Japanese electronics firms in using specially designed chain conveyors that place the control of the assembly process in the hands of the assembly operators. A workpiece is mounted on a platform or small pallet, which is powered by two constantly moving chains. The platform is freed from the power chain when it reaches an operator's station. After work is completed on the workpiece, the operator connects the pallet to the chain, and it moves to the next station; if the next station has not completed its work on the previous piece, the pallets accumulate on the chain.

Asynchronous alternatives include using AGVs as assembly platforms and for general transport functions; "smart" monorails for transporting parts between work stations; transporter conveyors to control and dispatch work to individual work stations; robots to perform machine loading, case packing, palletizing, assembly, and other material handling tasks; microload automated storage and retrieval machines for material transport, storage, and control functions; cart-on-track equipment to transport material between work stations; and manual carts for low-volume material handling activities.

Storing. The major trends in material storage technology are strongly influenced by the reduction in the amount of material to be stored and the use of distributed storage. Rather than installing eight to ten aisles of automated storage and retrieval equipment, firms are now considering one- and two-aisle systems. Rather than being designed to store pallet loads of material, systems are designed to store tote boxes and individual parts. Also, rather than a centralized storage system, a decentralized approach is used to store materials at the point of use.

Among the storage technology alternatives that have emerged are storage carousel conveyors; both horizontal and vertical rotation designs are available. Furthermore, one particular carousel allows each individual storage level to rotate independently, clockwise or counterclockwise. A further enhancement of the carousel conveyor is automatic loading and unloading through the use of robots and special fixtures.

A number of microload automated storage and retrieval systems have been introduced in recent years. The equipment is used to store, move, and control individual tote boxes of material. Rather than performing pick-up and deposit operations at the end of the aisle, the microload machine typically performs such operations along the aisle, since it is used to supply material to workstations along each side of the storage aisle.

Of particular interest has been the introduction of storage equipment for product applications that previously was used for document storage in office environments. The trend toward lighter loads has resulted in a shift in technology from the ''white-collar'' environment to the ''blue-collar'' environment.

Despite the apparent need for automatic storage and retrieval of individual items, few equipment alternatives are currently available, and those that are have not gained wide acceptance. This particular void in the technology spectrum has existed for a number of years and does not appear to be of current interest to material handling equipment suppliers.

Controlling. The ability to provide real-time control of material has elevated material handling from a mundane ''lift that bale, tote that barge'' activity to a high-tech activity in many organizations. The logic control, the ability to track material and perform data input-output tasks rapidly and accurately, has had a major impact on material handling. In physical control, automatic controls have been added to a number of material handling equipment alternatives, allowing automatic transfer and assembly.

Perhaps the fastest-growing control technology today in material handling is automatic identification. Likewise, the expectation is that the greatest impact in the future will come from the application of artificial intelligence to transporting, storing, and controlling material. Among the alternative sensor technologies available to support automatic identification are a wide range of bar code technologies, optical character recognition, magnetic code readers, radio frequency and surface acoustical wave transponders, machine vision, fiber optics, voice recognition, tactile sensors, and chemical sensors.

The growth in the use of bar code technologies is due to three developments: bar code standardization, on-line printing of bar codes, and standardization labels.

The standardization of codes and labels came about through a concerted effort by the user community. The Department of Defense led the way with its LOG-MARS study; that success was followed quickly by a concerted effort by the automotive industry (the Automotive Industry Action Group). Other industries that have standardized codes and labels include the meat packing, health, and pharmaceutical industries. Others, such as the telecommunications industry, currently are involved in developing counterpart standards.

Additional developments in the control of material handling equipment include off-wire guidance of AGVs. AGV technology is one of the most prominent areas of current material handling research and implementation. Functioning as a mobile robot, the AGV is being given enhanced sensor capability to allow it to function in a path-independent mode. Through the use of artificial intelligence techniques, the AGV will be able to perform more than routine transport tasks without human intervention. Using sophisticated diagnostics it will be able to execute advanced tasks such as automatic loading and unloading of delivery trucks.

A number of European and Japanese firms are making major investments in the development of future-generation AGVs. Ranging from vehicles capable of transporting loads in excess of 200,000 pounds to those designed to transport individual printed circuit boards, a number of new entries into the U.S. market are expected within 2 to 3 years.

A related control development that will have a major impact on material handling equipment technology is interdevice communications. The manufacturing automation protocol (MAP) being developed by a number of firms led by General Motors (discussed in the section Factory Communications and Systems Technologies, under the subsection Networks) is expected to provide the common data transmission link by which many different types of manufacturing hardware, including material handling equipment, will communicate. The driving force behind this standardization effort is the desire for truly integrated manufacturing systems across the entire hierarchy of manufacturing.

Design technology. In addition to the development of new and improved material handling equipment technology, new thinking has emerged on the design of material handling systems. Specifically, computer-based analysis, including the use of simulation and color graphics-based animation, is being used increasingly to design integrated material handling systems.

Interactive optimization and heuristics also are being applied in the design of material handling systems. Considerable research has been performed in developing performance models of a variety of equipment technologies. Trade-offs between throughput and storage capacity, optimum sequencing of storages and retrievals, and the automatic routing of a vehicle in performing a series of order-picking tasks are some of the issues that have been addressed in an attempt to gain increased understanding of material handling in the future factory environment.

DEVELOPMENTS IN MATERIAL TRANSFORMATION TECHNOLOGIES

This section describes the technologies of individual computer-controlled equipment, from numerically controlled (NC) machine tools and smart robots to computer-aided design and artificial intelligence, developments whose impetus comes from rapid advances in microelectronics and computer science. Rapid developments in very large scale integration of integrated circuits have reduced the size, cost, and support requirements of information and machine intelligence while greatly increasing its capabilities. Microelectronic technology in the future will be embedded in each machine tool and robot and at every node and juncture of computer and communication networks. The capabilities provided by this embedded intelligence will revolutionize operations on the factory floor.

In addition to the application of NC to traditional metal-cutting operations, several new cutting technologies have become important in many applications. The most widely used of these is electrical discharge machining (EDM), in which the workpiece is precisely eroded or cut by electric pulses jumping between an electrode and the workpiece in the presence of a dielectric fluid. Electrodes, usually made of brass or carbon, are machined to the desired form. Although the cutting process is slow, the machines operate with little or no attention, and EDM is an efficient method of cutting many types of dies.

A major recent development in EDM is the wire cut machine, in which the electrode is replaced by a fine wire sprayed with dielectric fluid. The wire slices through the workpiece as if through cheese, making shaped cuts as the workpiece table moves by NC. The wire is constantly moving between two spools so that, in effect, fresh electrode is always being used.

Low-power lasers began to be used for precision measurement about 20 years ago. Higher-power lasers are now used for welding and for sheet and plate metal cutting. Within the past 2 years, precision machine tools that use the laser as a cutting tool have been introduced in the United States and Japan, both for drilling and for cutting contoured surfaces. Other new technologies include the use of electron beams for drilling and welding and the use of a plasma flame for cutting.

Parts can be formed from sheet or plate in a variety of presses that bend, fold, draw, punch, and trim. The average age of presses currently in use is much higher than the average age of cutting machines, and users have generally been slower to innovate, but some press-working shops have taken advantage of new technologies. For example, some shops have installed lines in which coiled sheet is unrolled, flattened, trimmed, and shaped into parts by stamping, bending, and drawing in a continuous series of operations. Others have installed transfer presses which make finished parts from strip in a continuous series of operations. Much of the progress in forming has come through better control of the material to be formed.

The only extensive use of NC in presses has been in punch presses that combine tool-changing ability with two-axis positioning of the work for punching, nibbling, trimming, and cutting with lasers or plasma flame. However, NC controls are now beginning to appear on some other types of presses.

Tooling. Cutting tools are made from a variety of materials: high-speed steel; carbides of tungsten, titanium, and boron; oxides of aluminum and silicon (ceramics); cubic boron nitride; and synthetic and natural diamonds. Major advances in cutting-tool materials sometimes cannot be fully utilized until machines designed to take advantage of their properties are generally available.

Great progress in cutting tools has been made by applying a coating of one material (in some cases, two or three coatings of different materials) onto a base material. The proliferation of tool materials and coatings has become so complicated that computer software has been developed to aid the process. The resulting tools last longer, stay sharper, and can be used to cut hard materials such as heat-treated steel and abrasive materials such as fiberglass.

As combinations of materials and coatings produce a growing list of tooling operations, the variety and volume of tooling requirements can be expected to proliferate. New product designs and performance requirements, product and process specifications, and changing lot sizes will create an ever-increasing need to match specific tooling with specific production applications. To achieve the high-quality, close-tolerance production demanded in the marketplace, manufacturers will require a large inventory of tooling to ensure that the optimal tooling is available for all production requirements. Combined with the increased expense of tooling made with rare materials and precision coatings, the costs of meeting tooling requirements will become major factors in capital budgeting decisions.

Improvements have also been made in die and mold materials. More important, however, is the change taking place in the way dies and molds are produced. Traditionally, they have been made by experienced craftspeople with a great deal of time-consuming cut and try in the finishing stages. The combination of newer EDM machines and computerized programming of die-sinking machines is removing much of the cut and try from this process.

Jigs, which serve to position the tool more accurately in relation to the work for drilling or boring, can usually be eliminated when NC machines are used. In fact, one of the major early advantages of NC was the elimination of the production and storage of jigs. Of course, if the jig also serves as a fixture to hold the work on the machine, that function is not eliminated on an NC machine.

Fixturing. Fixtures hold and locate the part being worked during machining and assembly operations. The main considerations in fixture design are positioning the part in the fixture, securing the part while the machining operation takes place, positioning the fixture relative to the machine tool, positioning the cutting tool relative to the part, and minimizing set-up times. New fixturing techniques add flexibility and programmability to minimize set-up time, maximize the flexibility of the machine, and reduce storage requirements for fixtures.

The characteristics of the fixture depend on the process being performed, the shape of the part, and the tolerances required. For example, the workpiece may be subjected to strong vibrations for torque forces during some operations such as milling, while the forces in assembly operations will be much smaller. The fixtures required for these two operations are quite different and virtually incompatible. When a variety of tasks are performed, a large number of fixtures must be developed, stored, and accessed—a very expensive undertaking.

The need for a large number of fixtures remains a problem even for flexible manufacturing systems (FMSs) that can quickly and efficiently machine a number of different parts within the same part family. The FMS can help reduce economic lot sizes and reduce the expense of keeping parts in inventory. Unfortunately, this advantage is restricted by the need to have different fixtures for different parts. The cost of multiple fixtures can account for 10 to 20 percent of the total cost of the system, and the fixtures can sometimes cost more than the rest of the system.

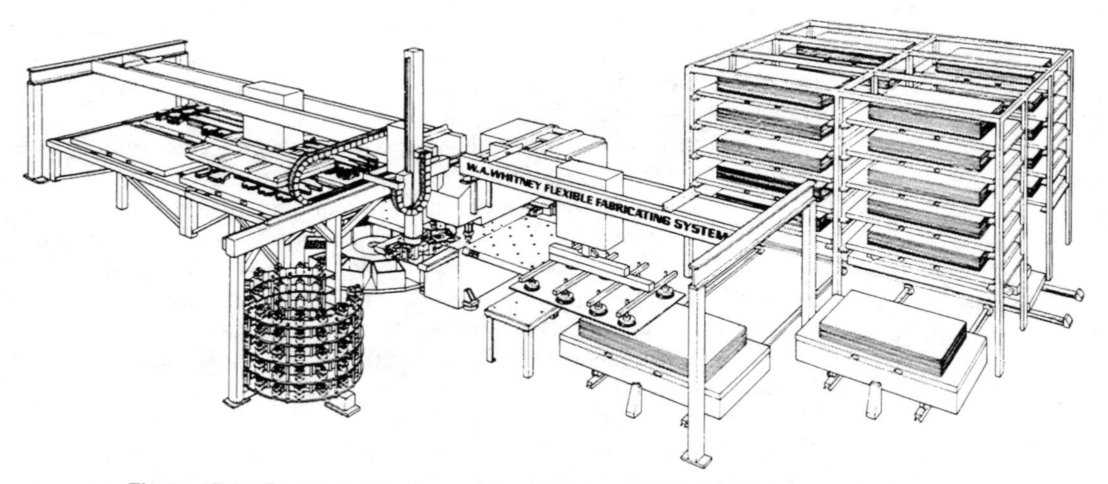

Figure 25-2 Flexible fabrication system. (Courtesy of W.A. Whitney Corp.)

Clearly, the full advantages of an FMS cannot be realized without the development of flexible fixturing that can conform to different part types and machining operations. An example of a flexible fabricating system is shown in Figure 25-2.

A number of major research efforts are focused on the problem of flexible manufacturing, and several solutions have been proposed. One approach would be to automate the current fixturing process, which uses blocks and clamps to align parts accurately. Instead of skilled toolmakers, robots could be used to assemble fixtures on coordinate measuring machines (CMMs). The fixtures would be mounted on standard pallets, permitting robots to load and unload parts easily and allowing easy alignment with machine tools. The CMM could cost $200,000 and vision-equipped robots at least $100,000; the hardware for the fixtures themselves and the software needed to control the robots would add to these amounts. Although the present cost may be prohibitive, this approach would ensure accurate location of the workpiece, and it could be used for both machining and assembly operations.

Another approach partially encapsulates the workpiece in a low-melting-point alloy prior to machining. Encapsulation has been developed specifically for milling gas turbine and compressor blades of irregular shape. The unmachined blade is precisely positioned in the encapsulation machine. Rapidly injected molten alloy surrounds the blade and provides the clamping face, protecting the blade itself. After machining, the alloy capsule is mechanically cracked open. The problem with this approach is that the blade must be positioned accurately in the very expensive encapsulation machine, which requires a different die for each workpiece. This limits flexibility and adds expense.

A third approach is programmable conformable clamps. Developed at Carnegie-Mellon University for machining turbine blades, the clamps consist of octagonal frames hinged to accept a blade. The lower half of the clamp uses plungers, activated by air pressure, that conform to the contours of the blade. A high-strength belt is folded over the top of the blade, pressing it against the plungers, which are mechanically locked in place. Accurate alignment can be done manually or automatically with sensors. Although the clamps are limited in the types and sizes of parts they can hold and their large number of moving parts may reduce reliability, they are automatic and ensure accurate alignment.

Another approach is the fluidized-bed vise, in which small spheres are held in a container with a porous floor through which a controlled air stream passes. The spheres behave like a fluid, conforming to the contours of even irregularly shaped parts; when the air flow is stopped, the spheres come together to form a solid mass that secures the part. The advantages of this approach are that a variety of part shapes can be clamped, the clamping process is automatic, and the vise is inexpensive to build and operate. However, additional research is needed to establish a predictive model for the device and to eliminate the need for an auxiliary device to determine the location and orientation of the workpiece in the vise. Research is also under way in which electrically active or thermally active polymers are used in an authentic phase change bed instead of the pseudo phase change of the air-sphere approach.

None of these approaches offers the flexibility needed in terms of variety of applications, the types and sizes of parts that can be held, and expense. They also do not address the problem of locating the workpiece. The first three approaches use mechanical stops or surfaces, and the fourth requires an additional measuring system; this problem may be overcome by combining flexible fixturing devices with sophisticated robots.

Sensors. As machine tool automation advances, the instrumentation on the machine becomes increasingly important. Most of the early problems with automation tended to be instrumentation problems. Sensors to determine what is happening and monitoring systems to evaluate the sensor information are both needed. The role of sensors in a manufacturing environment is to gather data for adaptive control system—for example, to supply guidance information to robots or to provide measurements for quality assurance and inspection systems. Sensors can provide automated equipment with vision, touch, and other senses, enabling the equipment to explore and analyze its surroundings and, therefore, behave more intelligently.

Sensors are currently used in factories to provide different types of data, such as the bipolar on-off of a limit switch, the simple numeric data of a temperature sensor, and the complex data provided by a vision sensor. Vision sensors, for example, can be used to determine part identification, orientation, and measurement data. Other sensors, such as tactile, acoustic, and laser range-finding sensors, are being used to measure force and shape, provide range data, and analyze the quality of welding processes.

Sensor technology is a very active field of research. Sensor research that shows promise for manufacturing includes micromechanics, three-dimensional vision for depth sensing, artificial skin for heat and touch sensing, and a variety of special-purpose sensing devices. Some of the special-purpose devices have no human analog. Examples are the water vapor sensors being developed for use in sophisticated adaptive-control algorithms and the optical laser spectrometry probes that monitor chemical processes in real time. The use of adaptive closed-loop control systems in manufacturing has increased the demand for a wide variety of special-purpose sensors and has stimulated the demand for sophistication in general-purpose sensors such as vision sensors.

Other research is focused on the analysis, interpretation, and use of the data provided by sensors. Through the use of VLSI techniques in IC fabrication, intelligent sensors equipped with microchips can process data even before it leaves the sensor. For example, research is under way on vision systems that can inspect IC wafer reticles. Research on this vision system is focused on the mechanical accuracy of positioning devices, on the interface to the CAD data base describing the reticle, and on modeling the fabrication process to predict what the vision system will see. The visual information itself must be interpreted to determine whether to accept the wafer under inspection or to identify the flaw and provide feedback to correct for any imbalance. This type of intelligent sensor will eventually be integrated into many elements of manufacturing.

Model-based sensor systems such as these which use process, CAD, simulation, and control algorithms are expected to provide manufacturing sensor systems of the future with very complex analysis capabilities. These analysis capabilities will far surpass the monitoring and control capabilities of human operators by being more sensitive, more precise in analysis, more rapid in feedback response, and more precise in corrective action. They will allow the factory of the future to work to very fine tolerances while maintaining consistently high quality control, approaching zero defects.

Smart robots. One of the most common uses of advanced sensor is to make robots smarter. The senses of a robot are the sensors in its work cell that provide information to the robot's central controller. The "intelligence" of the robot is determined by the combined capabilities of its controller, its sensors, and its software. Most of the robots in the world's factories today have primitive controllers and software and few, if any sensors. They mindlessly weld, paint, and pick-and-place, and some would continue to do so even if no object were present to paint, weld, or grasp. Such robots are locked into a predetermined program that does not adapt to unexpected changes in the work cell. In contrast, advanced robot systems have sensors that inform the robot of the state of its world, controllers that can interface with the advanced sensors, and software that can adapt the robot's program to reflect the changing state of its world. This is an example of adaptive behavior using a closed-loop feedback system; to a degree, it is what people do when they engage in behavior that uses the senses. It is expected that 60 percent of all robots, especially those used for inspection, assembly, and welding, will utilize vision, tactile, and other sensors within the next 10 years. Figure 25-3 shows how vision and sensor technology may be used in automatic positioning.

Smart robots have many advantages. About one-third of the cost of a robot work cell is the fixturing that holds or feeds each part in precisely the same way each time. This cost can be saved by smart robots that can find the part they need even if it is askew, upside-down, or in a bin with other parts; it is easier to change a robot program than to change the fixturing. Smart robots will be much more adaptable to product changes because they will have less fixturing to change. Smart robots will be even more adaptable to different tasks when they can easily change their end effector for a drill, deburrer, laser, or whatever tool is required.

State-of-the-art robot systems embody elements of adaptive control and are now coming into use in factories around the world. One example is arc welding robots, whose welding path is planned with the aid of a vision system that determines the

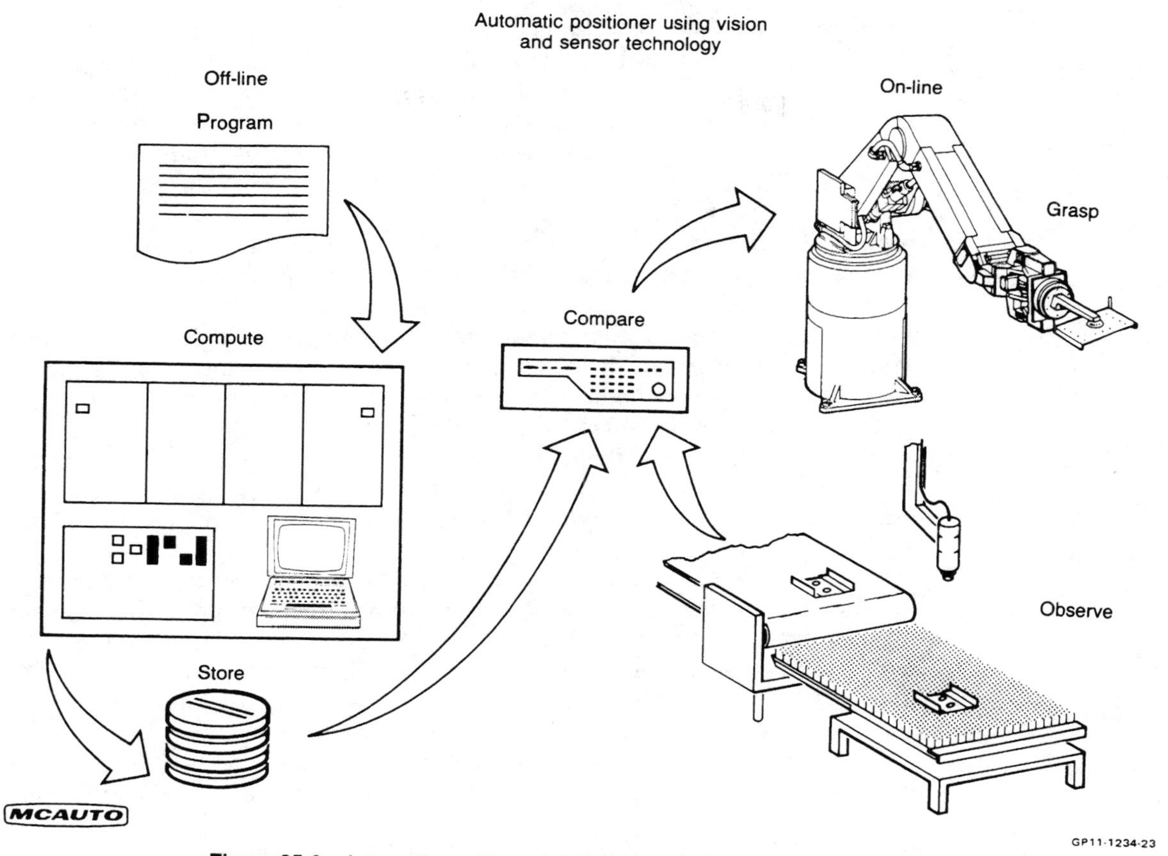

Figure 25-3 Automatic positioner using vision and sensor technology. (Courtesy of McDonnell Douglas)

location and the width of the gap to be welded. The robot software then adjusts the path and speed of the welding tool as the welding progresses. Although the welding example shows how adaptive control enables a robot to perform a task with built-in variance, the variance found in arc welding can be foreseen easily and taken into account by a human engineer or programmer. Adaptive control for robots with less structured tasks is still in the research stage.

Robots are programmed through a special-purpose computer language. State-of-the-art languages allow the robot to perform limited decision making on its own from information obtained with its sensors. However, these programming languages are limited, because they can neither interpret complex sensory data, as from a vision or tactile sensor, nor access CAD data bases to get the information they may need to identify the parts that they sense. Present languages are also robot dependent; that is, they do not allow the transfer of programs from one robot to another. This means that robots must be programmed individually by valuable, highly trained programmers.

New robot programming languages that address some of these limitations are in development in academic and commercial research laboratories. The new task level languages will allow robot programming at higher levels: The robot can be told what to accomplish or what to do with the part, and it will determine the best way to accomplish the task. The benefits expected when the new languages reach the factory floor include reducing the cost of programming, facilitating the coordination of two or more robots working cooperatively, and enabling advanced sensors to interface with the new systems.

Graphic simulation. As manufacturing systems come to include advanced systems such as smart robots, it becomes more and more important to be able to simulate their behavior. Two distinct kinds of simulation are now being used in manufacturing. One is the simulation, often graphic, of a single process, robot, or work cell. The second is the simulation, generally mathematical, of a system such as an FMS, a new or modified production line, or an entire factory. The former may be regarded as important in tactical or local decisions, the latter in strategic or system decisions. For this reason, graphic simulation will be treated here, and mathematical modeling will be covered later in the communications and systems section. The graphic simulation of lane changes by a Ford Bronco II is illustrated in Figure 25-4.

Graphic robot simulation is beginning to be used to select the most appropriate robot for a particular task or work cell and to plan the cell layout. The production engineer can use simulation to reject robots that do not visually appear to suit the task because of their arm configuration or timing constraints. Graphic simulation is also used for visual collision detection in the work cell, but this method is prone to error and is not recommended.

Some vendors of graphic robot simulators have adapted their software to generate actual robot control programs, which is termed off-line programming. It permits

Figure 25-4 Simulation of automotive performance. (Mechanical Dynamics, Inc.)

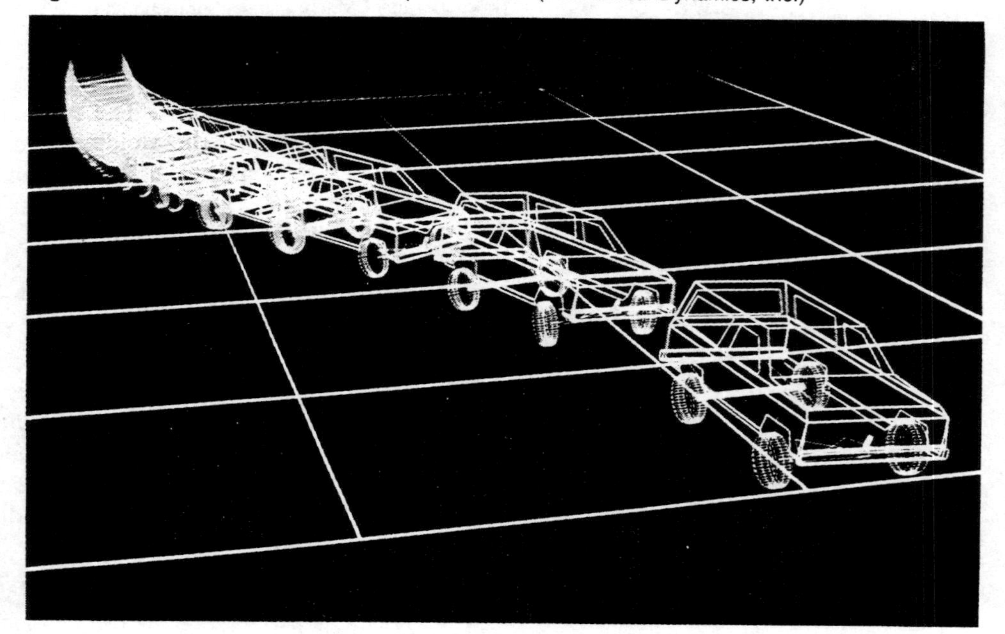

the development of robot programs without shutting down a productive work cell, thus allowing efficient, concurrent work cell design. Although programs developed off-line currently must be used on the specific robot for which the system was designed, research to include a variety of robots in the simulation system is being conducted. For example, an off-line robot programming system has been developed that can simulate any of six commonly used robots. Researchers are also working on the related problems of how to simulate a complex sensor, such as a vision sensor, and how to debug an off-line robot program that makes decisions based on advanced sensory input. Graphic simulation and off-line programming promise to provide cost, time, and personnel savings in the efficient design of programs, work cells, and processes.

Artificial intelligence. Artificial intelligence is a set of advanced computer software applicable to classes of nondeterministic problems such as natural language understanding, image understanding, expert systems, knowledge acquisition and representation, heuristic search, deductive reasoning, and planning. Artificial intelligence (AI) technology will emerge as an integral part of nearly every area of manufacturing automation and decision making. Research that will affect manufacturing is being conducted in several areas of AI, including robotics, pattern recognition, deduction and problem solving, speech recognition and output, and semantic information processing. As with simulation, AI will be used at different levels in the factory of the future. Most of the AI applications will be integrated into the software that controls automated machinery, record keeping, and decision making.

Artificial intelligence is not a new field, but the maturing fruits of 20 years of AI research are just now becoming available for commercial applications. The types of AI products that will have a significant impact on manufacturing include:

- expert systems in which the decision rules of human experts are captured and made available for automated decision making;
- planning, testing, and diagnostic systems; and
- ambiguity resolvers, which attempt to interpret complex, incomplete, or conflicting data.

The AI applications that deal with individual machines, processes, or work cells are described here; those that deal with system-level decision making will be integrated into the Factory Communications and Systems Technologies section.

Expert systems are in productive use today in isolated industries; petrochemical companies, for example, use expert systems for the analysis of drilling samples. Digital Equipment Corporation has used an expert system for a number of years, saving several million dollars annually in configuring the company's Digital Equipment Company VAX computer systems. As human experts with years of experience become scarce, the expert system provides a way in which to capture and "clone" the human expert. An interesting feature of expert systems is that they can explain the train of reasoning that led them to each conclusion. In this way, the systems also can be used to augment human decision making, in much the same way as medical expert systems have been used. Current expert systems are best suited to situations that are somewhat deterministic when the expert's rules are known. For this reason, rapid emergence of expert systems can be expected in limited areas of technical knowledge such as chip design, arc welding, painting, machining, and surface finishing. In the 1990s, expert systems are expected that will learn from experience; this means that expert systems eventually will be developed for specialties in which there are no human experts.

Although still primarily in the laboratory, one type of AI software is attempting to simplify the use and expand the applicability of programmable equipment. For example, advanced user interfaces are now being developed that use "natural language," so that a manager can type a request at his or her work-station in more-or-less plain English. The AI software will determine what he means, even if the request has been phrased conversationally or colloquially, and provide interactive assistance for decision making. By the year 2,000, managers will probably be communicating with their workstations by voice, another application of AI techniques. Artificial intelligence technology promises to make it much easier for computers and computerized equipment to be used by personnel not having computer training, such as managers, engineers, and operators on the factory floor.

FACTORY COMMUNICATIONS AND SYSTEMS TECHNOLOGIES

In contrast to the materials and process technologies just described, communications and systems technologies tend to operate at higher levels, allowing previously separate areas of manufacturing to be integrated into systems of manufacturing. A manufacturing system is defined as a system created by the interconnection and integration of processes of manufacturing with other processes or systems. This definition implies that manufacturing systems vary from a basic system, which couples a few processes, to a hierarchical system, which integrates lower-level manufacturing subsystems into the single aggregated system. Such a system is termed a computer-integrated manufacturing (CIM) system. See Figure 25-5.

This variation in complexity and level makes the concept of a manufacturing system elusive to grasp. It may be helpful to think of it as an approach, a systems approach, to incrementally integrating the functions of the manufacturing corporation.

The major characteristic of manufacturing systems is their sharing of information, their communication. Traditionally, manufacturing information has been created and communicated by humans writing on paper. This paper information was based on

Figure 25-5 Integration of functions in a computer integrated manufacturing system.

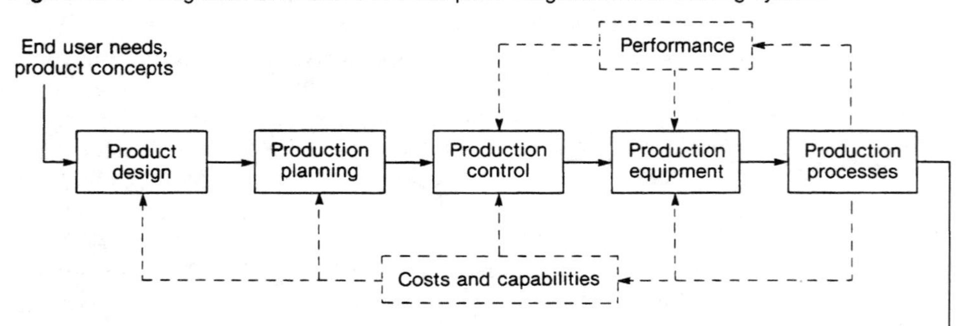

the understanding of the human expert at that moment, although often that understanding did not accurately reflect the real state of the factory at that moment. The paper method is people intensive, time-consuming to create and distribute, often inaccurate, and in frequent need of revision. As an information communication method, it virtually guarantees delay, inaccuracy, and expense.

The advent of computer technology and network communications is changing the face of the factory floor, much as office automation has changed the front office. This technology permits the system to generate its own data according to the information provided by real-time sensors built into automated machining, assembly, and inspection stations. The system gives the data to a computer, which interprets the data and takes appropriate action. This action may be to control the machining process, to replace a worn tool, or to decide whether to communicate the data, to whom, and how much.

This automated creation and sharing of information avoids the present duplication of data in several files or data bases, and it collects and communicates data at a scale and speed that will create opportunities in manufacturing never before available. For example, computer-controlled feedback permits a system to be self-diagnosing, self-maintaining, and eventually self-repairing. It allows the collection of statistical data that can be used for immediate adaptive feedback, quality control analysis, and the production of trend data. More important than any single benefit, this sharing of information makes possible the linking of systems into system aggregates. Previously disparate systems may be linked horizontally, and hierarchical adaptive control and reporting systems may be created by integrating vertically.

This information, or data, integration is the synergistic key to building manufacturing systems with broader scopes and at higher levels. The long-range goal of the manufacturing systems approach is CIM—the complete integration of the manufacturing subsystems that operate on the factory floor, the tie-in of techniques of optimization, mathematical modeling, scheduling, and data communication with the other functions (accounting, marketing, etc.) in the total manufacturing enterprise. Note that manufacturing systems are at once a means and an end. Components of the McDonnell Aircraft computer aided technology system are shown in Figure 25-6.

Systems of manufacturing are integrated through the application of several technologies: communication networks, interface development, data integration, hierarchical and adaptive closed-loop control, group technology and structured analysis and design systems, factory management and control systems, modelling and optimization techniques, and flexible manufacturing systems. Artificial intelligence techniques will be embedded in, and inseparable from, most of these technologies. Communications technologies—those associated with networks, interfaces, and data bases—may be the most critical to U.S. manufacturing progress, because they are the keys to the immediate development of manufacturing systems. On the other hand, technologies that analyze, manage, and optimize the system hold the greatest promise for improving the long-term competitiveness of U.S. manufacturing. These technologies will facilitate progress towards the goal of total integration from design to delivery. Each will be described in depth, both individually and as they relate to the full CIM concept.

Networks. The manufacturing network will be the backbone of factory communications and, therefore, of factory automation. Communications between tightly

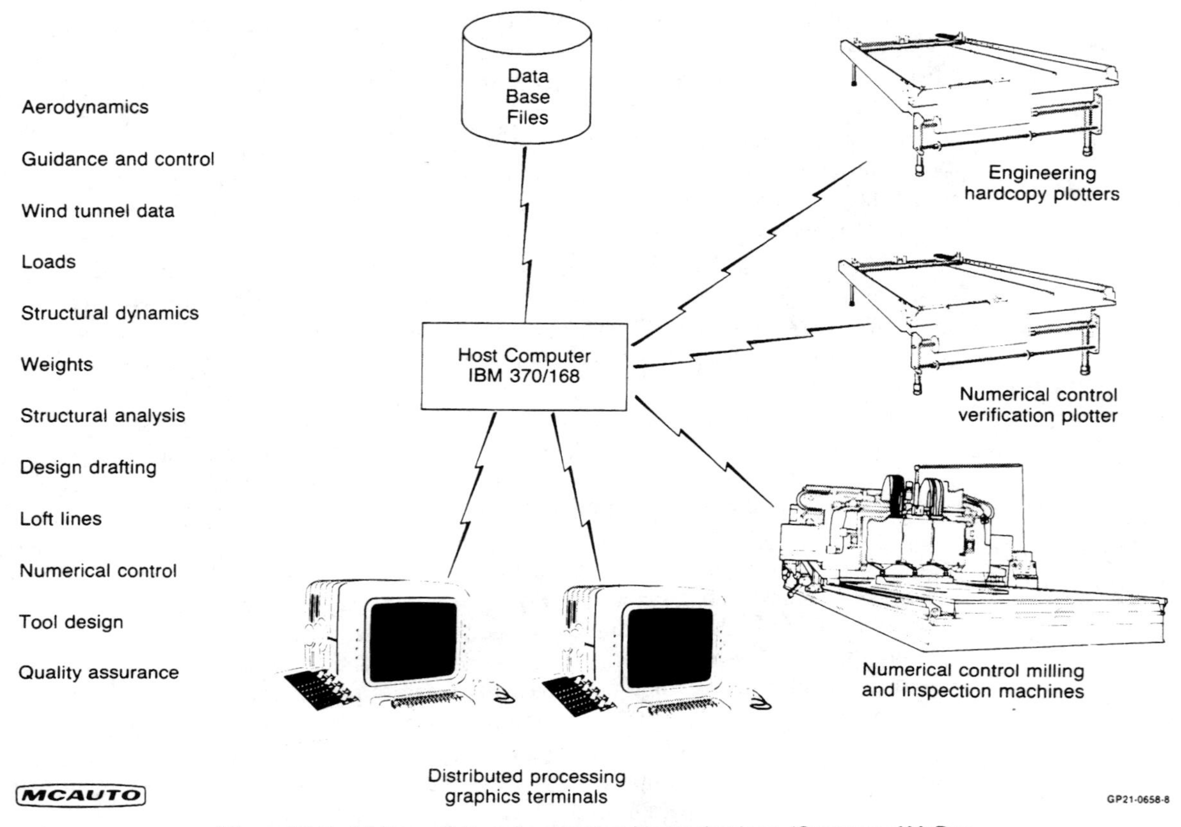

Figure 25-6 McDonnell Aircraft computer aided technology. (Courtesy of McDonnell Douglas)

coupled components, such as robots and sensors, and between elements of an FMS require that data be exchanged in real time. As the complexity of the factory system increases, including the linkage from design to planning and production, the need for factory communications will continue to expand. Networks provide the physical mechanism for this communication between heterogeneous systems. The network must not only transmit the raw data but also retain its meaning, so that a different computer, running a different program, may use it. The goal of CIM is to allow all manufacturing activities using heterogeneous hardware to communicate as though they had a common language.

Networks currently provide a protocol, or agreed-upon standard, for computer communication. Most major computer vendors, as well as manufacturing equipment vendors, have defined proprietary network protocols. Thus, a variety of incompatible networks, such as Ethernet and Modway, are in use in factories today. In addition to resolving the compatibility problem, advances in network architecture are required to meet the specific communication needs of manufacturing. The speed and traffic requirements of manufacturing communication must be taken into account, as well as provisions for interfaces between otherwise incompatible networks.

General Motors and its major vendors began work on a set of manufacturing automation protocols (MAP) for this purpose several years ago. The development of MAP has been broadened recently to include support by more than 100 major manufacturers and universities, including the National Bureau of Standards (NBS). These otherwise competitive groups realize that no single vendor can meet all the needs of a manufacturing system and that MAP may provide a solution to the communications problem between their equipment and other vendors' machines and networks.

MAP is an attempt to define the seven-level communications protocol proposed by the International Standards Organization. Although all seven levels have not yet been standardized, vendors are already selling MAP-compatible products, and far-sighted purchasers are demanding that their new hardware be MAP compatible. A recent breakthrough by Industrial Networking, Inc., has put MAP on a single microchip, which will facilitate the development of factory communication networks among heterogeneous machines. While significant challenges remain, the broad membership and participation in the MAP effort can be used as a model for specifying and solving other manufacturing system problems.

Interface standards. The network is expected to provide the physical and logical path for data communication in a factory system, but much more is required for effective communication. Networks provide the physical language and format, but do not address the semantics or effective use of the information communicated. Interface standards are needed to facilitate the effective communication of meaningful data.

The key to data integration is standardization that does not stifle innovation. Standardization of data representation within the data base is necessary to allow the full meaning of the data to be retained even when it is communicated. Current practice requires vendors of systems or modules to provide special-purpose interface definitions for each pair or family of modules that communicate. However, in some areas standards have evolved through the cooperation of users and vendors. Examples are the CLDATA file for NC machines and the Initial Graphics Exchange Standard (IGES) for CAD data base information exchange. IGES has enabled previously incompatible CAD systems, with data stored in radically different formats, to communicate that data while preserving most of the meaning. Yet these standards rapidly grow out of data as technology moves forward. CLDATA is inadequate for nondeterministic (sensor-based) machine tool programs, and IGES does not work on solid-modeling CAD systems. The IGES continues to evolve, pointing the way to wider data integration. The challenge is to define standards that will withstand the demands of continued factory innovation or to establish mechanisms to update standards as needed.

Standards are also the solution to the interface compatibility problem that arises when equipment from different vendors is used in a network. The interface connects one machine to a communications system, which is connected to other machines, computers, and communications systems. The RS-232 interface standard is a simple protocol that has allowed communication between heterogeneous microcomputers and between computers and a host of other devices. Many machines already come with limited RS-232 interface, but more progress is needed in standardizing manufacturing interfaces. MAP includes the definition of an intelligent interface which can connect previously incompatible systems to a network.

The lack of interface standards can be a major impediment to achieving CIM. If well-defined information interfaces between modules or subsystems were established for the components of manufacturing systems, components could be developed independently and enlarged as advances in technology became available. This would facilitate compatibility of the equipment of multiple vendors in the heterogeneous systems expected in the factory of the future. Interface standards of this type are the basis of research at the Automated Manufacturing Research Facility at NBS. Significant questions must be answered, however, before the information interface for the modules in the manufacturing system of tomorrow can be defined.

In addition to these interface standards for information, two other kinds of interface standards are needed. The first and most neglected is the interface between human and programmable systems. The second is the physical interface between mechanical systems.

The man-machine interface includes the commands to be given by the human to make the machine perform a task successfully and the input device or physical method—keyboard, joystick, light pen, or voice—for entering those commands. Most current programmable systems are commanded through a programming language that is proprietary to the vendor of the system. This has given rise to a Tower of Babel of control languages requiring highly trained programmers to control modern manufacturing systems. No programmer can begin to master all of the languages and input devices found in an automated factory.

Two recent trends are expected to ease the interface problem between nonprogrammers, such as engineers and technicians, and the increasingly complex programmable automation systems found in the areas of robotics, NC tools, material handling, and processing systems. Hierarchies of languages and personnel are being developed in which highly trained programmers will deal with the raw control languages and sophisticated control algorithms, less-skilled programmers will deal with a higher-level simplified language, and equipment operators will not use actual programming languages at all. This hierarchy—automation systems programmer, applications programmer, user programmer, and user—parallels the evolution of personnel in the computer field.

The second trend is the use of AI to develop task-level control languages (discussed in the Smart Robots section.) Currently under research, task programming systems will reduce programming requirements to the steps on a common process planning sheet so that programmable manufacturing systems of the future will be controlled by statements similar to those one would give to a person doing the same task. These advances will provide new generations of specialized, user-friendly manufacturing subsystems that will make the most of factory personnel.

The mechanical interface problem for the factory is solved most easily by the development and adoption of standards. The lack of standards for the newer systems is a major impediment to progress. Examples of mechanical interfaces in need of standardization include:

- robot end-of-arm and gripper attachments;
- pallets, totes, and other conveyances;

- the mechanical interface for the loading and unloading of parts and pallets at machining centers; and
- the interface between robot carts and material handling systems.

Progress with the mechanical interface problem requires the usual consortium to agree on and promulgate standards. The major roadblock has been the lack of an organized body, leadership, and focus on the problem.

Data bases. Network and information interface standards are the means of sharing data, but it is not enough to move data from one application to another. The data must be stored, and the semantics, or full meaning, of the data must remain intact when the data are retrieved. A data base provides the long-term memory, or storage facility, that contains the manufacturing data, and the information retrieval system extracts specific data from the data base.

Current practice finds a large number of information retrieval systems in place even at the same company. Each system is associated with one major function, such as accounting, shop floor information, material requirements planning, or quality control statistics. Many of these data systems are state-of-the-art information retrieval systems with complex functions to maintain and update the data base. Unfortunately, the different data bases contain redundant and conflicting data in incompatible formats. Furthermore, many of these systems are dedicated to a single computer and use proprietary data representations that are incompatible with those of other systems. Thus, the task of developing an integrated manufacturing data base management system that can include every major function in a factory is formidable.

The most serious immediate barrier to the integration of manufacturing data is the incompatibility of CAD data with information needed by computer-aided engineering (CAE) and process planning. CAD technology has become highly effective in capturing the geometry of parts, including the description of dimensions, shapes, and surfaces. Real parts, however, are made up of smaller components and may themselves be components of a larger assembly. The CAD data base currently cannot capture the relationship of the parts to the whole, but both CAE and process planning require detailed attention to the joining of separate parts, their mating surfaces and tolerances, and their overall dimensions after assembly. The CAD data base does not include knowledge or speculation of materials, but CAE needs material data for its engineering analyses, process planning needs it in the creation of NC programs, and material handling needs it to select material from inventory.

A second serious barrier to the integration of manufacturing data is the current inability to model the processing portion of the overall system (the two right-hand boxes of Figure 25-5). Such a model would allow information on production costs and capabilities to be fed back, on-line, to the product design activity as it is being performed. This capability is essential to optimizing the producibility of products at the design stage. With this capability, each decision proposed in the engineering design process would result in simultaneous information on the effect of that decision on production costs and required capabilities (relative to available capabilities) for production of the product. It also would result in major cost savings in the production activity, since it is well known that the majority of production costs are frozen at the engineering design stage. While such computer-based integration of manufacturing data is technologically feasible, many difficult problems must be solved to bring it into being.

A further problem with the data in a CAD data base is that the geometrical data cannot be searched or aggregated in the ways that have become standard for textual data. Without explicit hand coding, it is not possible, for example, to retrieve all parts which use a particular fastener. Group technology (discussed in Chapter 24) is an attempt to code and classify the geometry, function, and process data in a way that will permit the use of standard retrieval functions.

One of the keys to the data integration problem lies in the development of flexible data schemas. A schema is a method of storing data so that its meaning and accessibility are retained. Most data bases are rather fixed data schemas that restrict the new types of information that may be added and limit data retrieval capabilities. Future data bases will have more flexible schemas so that, for example, materials information can be added to the CAD data base by an engineer at a CAE station or by an expert system that contains knowledge of materials and applications.

Beyond the compatibility problem are other technical challenges to the implementation of manufacturing data base systems. For example, experts predict that future manufacturing data bases will be 20 to 50 times larger than present data bases. The size of the data base, the time tolerances for communication, and the variety of users suggest that a manufacturing system data base will be distributed across multiple heterogeneous systems, which may be in different geographical locations. This presents significant technical challenges to the achievement of a logically integrated manufacturing data base. The concepts and protocols normally used to ensure proper access, control, and update will need to be expanded to meet this sophisticated method of data base organization. Interim solutions in place today are neither geographically nor heterogeneously distributed, but progress toward these goals is being made.

A last challenge posed by the manufacturing data is the use of probabilistic or incomplete data. Current data base systems can only represent facts and cannot deal with uncertainty or conflict within their data. Manufacturing information systems of the future will depend on AI to deal intelligently with this type of information.

One of the most critical roles of people in the factory of the future will be to interact with intelligent manufacturing systems through workstations, terminals, or networked microcomputers. As expert systems and other forms of AI become embedded in systems of manufacturing, the systems will be able to perform more and more of the decision-making tasks previously performed by people. At first, these automated decisions often will have to be reviewed by people and then interactively modified, much as an architectural plan takes shape in a dialogue between client and architect. People without knowledge of the data schema will routinely query the system for information needed to make decisions. The data retrieval system will have to determine exactly what is important to the inquirer and then retrieve and massage the appropriate data. The person may even want the system's "opinion," or the system may ask for the person's opinion. The factory of the future will regard personnel and intelligent systems as partners in a dialogue that should encourage very sound decision making.

Adaptive closed-loop control. One difference between an automated system and an intelligent system is the amount and kind of feedback that is generated from an activity and passed up to a decision-making entity. This feedback allows a system to know its own state, to know when it is out of balance, and to respond to the imbalance until stability is achieved. This adaptive closed-loop control will be used at all levels of manufacturing systems from sensor-based feedback to robots or NC tools, to inspection station-based feedback to a cell controller, to a factory floor data collections system that feed back to process planning and scheduling systems. This property or adaptive feedback is the key to improving product quality, with zero defects a realistic goal (see Figure 25-5).

Adaptive feedback is also one key to better management, since only in this way can a manager know exactly the state of production, including exact costs. Systems of manufacturing have a feedforward property that will allow management to control the factory floor with an effectiveness and immediacy never before possible. (A more global discussion of hierarchical control is found in Chapter 24, Computer-Integrated Manufacturing Systems.)

Feedback and feedforward properties can also provide machines and systems with self-diagnosis features. Thus, a manufacturing system can tell if something is wrong with it and what is wrong and can suggest the remedy to a higher entity. For example, researchers at the AMRF have already demonstrated the ability to sense when a tool is about to break so that automated equipment can change the tool without the disruption caused by untimely failure; next will be limited self-maintenance and repair capabilities. When a data-intensive system breaks down, the integrity of that data is threatened. Self-diagnosis will inform the data base system of the integrity of the data and, if it is threatened, the system will take either conservative or remedial action.

Factory management and control. The factory of the future will be managed and controlled through automated process planning, scheduling, modeling, and optimization systems. The successful implementation of large-scale factory-level systems depends upon structured analysis and design systems that depend heavily on GT. Limited structured analysis systems, such as the Air Force-sponsored Integrated Computer-Aided Manufacturing Definition, have been in use for years in the analysis and design of large projects. Only through such systems can a manager know the exact state of his or her factory, and only through such exact knowledge of the present can a manager intelligently implement systems of manufacturing for the future. New systems development methodology packages, such as STRADIS, promise help in this area, but much work remains to be done before such systems are easily used by the actual decision makers. Similar work must be done on process planning and scheduling systems before they can use the feedback and feedforward properties of the hierarchical and adaptive closed-loop control systems to be found in future manufacturing systems.

Management functions will be hierarchically distributed so that "go" and "halt" decisions may be made effectively from many levels and by human or machine. Through the use of terminals on the factory floor and throughout the decision-making structure, the system can respond instantaneously to human command. At first, most of the decision making will rest in the hands of humans. Low-level manufacturing systems now work in this way. As the systems become integrated at higher and higher levels, decision rules and methods will be built into them; systems will develop plans to carry out human-specified activities. On the authorized humans' approval, the system will carry out the task, making low-level decisions on its own. If a low-level decision-making entity does not have a certain level of confidence in its decision, it may pass the decision up to the next-higher entity, be it human or computer. This new kind of man-machine interaction will allow humans to do what they do best: create, define, and communicate. The machine will do what it does best: work hard, steadily, and accurately.

Modeling and optimization systems. One tried and true method of representing an activity to a computer is through mathematical modeling. Computerized modeling tools, such as SLAM, have been used by simulation experts for years, but simulation is still more an art than a science. With experience and feedback, our ability to represent complex activities mathematically will be refined. A modeling and simulation package will be a necessary part of an intelligent structured analysis and design system. It is hard to overstate the importance of structured analysis and design systems; they will operate at high levels, with much built-in decision making.

System modeling will become a commonplace and necessary prerequisite to the successful design and implementation of large-scale manufacturing systems. This is because large projects contain too many facets to be managed effectively with only human memory and computation capability. Artificial intelligence techniques will be needed to reduce the tremendous amounts of data generated by such systems to humanly understandable terms. This intelligence must be of a higher order than the AI expert systems in existence today.

With the addition of AI, a modeling system can become an optimization system, guiding its human managers to the most productive, most cost-effective, or highest-quality utilization of resources. With such optimization capability, managers will be able to sit at their work stations and, in real time, analyze the various possibilities to determine optimal solutions and mixes. The availability of accurate information on cost, time, and quality will eliminate much of the guesswork in manufacturing decision making.

SUMMARY

All of these advanced manufacturing technologies, from materials and machine tools to the subsystems and CIM systems, provide the ability to perform traditional manufacturing tasks in a highly advantageous but nontraditional manner. Many of the individual technologies and subsystems of manufacturing can be implemented today and, in fact, must be implemented soon for a manufacturer to remain competitive. Real progress toward the factory of the future will take place through the higher-level integration of these technologies. Although a handful of domestic manufacturers continue to make progress in implementing and integrating many of the technologies described in this chapter, real barriers to full integration remain.

Specifically, standards are critically needed for the definition and communication of part data. At higher levels, the need is for proven systems of hierarchical control and feedback and usable methods of automated classification of parts and processes (GT). Required at the highest level are the evolution of structured analysis and design systems that include modelling and optimization packages, as well as intelligent user interfaces that can be used interactively by managers in real time. The technology is here now or just around the corner. U.S. manufacturing needs farsighted management and trained manufacturing engineers to put the pieces together.

QUESTIONS

1. How do materials impact manufacturing? Give an example of a new development in material and how it has affected manufacturing.
2. What are the factors that have limited the development of new materials and processes in the United States?
3. What are the primary functions of material handling equipment technology?
4. How have bar code technologies affected the storage and control of products?
5. How does fixture design influence flexible manufacturing systems?
6. Can a robot be used as a fixture? If so, how?
7. What is MAP?
8. What is the most serious barrier to the integration of manufacturing data?
9. How will factories of the future be managed?

REFERENCES

CHAPTER REFERENCES

Chapter 1

ABERNATHY, WILLIAM J., CLARK, KIM B., AND KANTROW, ALAN M., *Industrial Renaissance,* New York: Basic Books, 1983.

AYERS, ROBERT U., *The Next Industrial Revolution,* Cambridge, MA: Ballinger Publishing Co., 1984.

DENISON, EDWARD F., *The Sources of Economic Growth in the United States and the Alternatives Before Us,* New York: Committee for Economic Development, 1982.

HOUNSELL, DAVID A., *From the American System to Mass Production 1800–1932,* Baltimore: Johns Hopkins Press, 1985.

KENDRICK, JOHN W. AND GROSSMAN, ELLIOT S., *Productivity in the United States,* Baltimore: Johns Hopkins Press, 1980.

KETTERING, CHARLES F. AND ORTH, ALLEN, *American Battle for Abundance,* Detroit: General Motors Corp., 1955.

LEBERGOTT, STANLEY, *The Americans—An Economic Record,* New York: W. W. Norton, 1984.

MAYR, OTTO AND POST, ROBERT C., eds., *Yankee Enterprise,* Washington, D.C.: Smithsonian Institution Press, 1981.

WILLIAMSON, HAROLD F., ed., *The Growth of the American Economy,* Englewood Cliffs, NJ: Prentice-Hall, 1951.

Chapter 2

ADAM, EVERETT E., JR., AND EBERT, RONALD J. *Production and Operation Management.* 2nd edition. Englewood Cliffs, NJ: Prentice-Hall, 1982.

BABBAGE, CHARLES. *On the Economy of Machinery and Manufactures.* London: 1832.

CASS, EUGENE LOUIS AND ZIMMER, FREDERICK G., eds. *Man and Work in Society.* New York: Van Nostrand Reinhold, 1975.

DAVIS, KEITH AND NEWSTROM, JOHN W. *Human Behavior at Work: Organizational Behavior.* 7th edition. New York: McGraw-Hill, 1985.

FAYOL, HENRI. *General and Industrial Management.* Trans. by C. Storrs. London: Pitman, 1949.

TAYLOR, FREDERICK. *The Principles of Scientific Management.* New York: Harper & Brothers, 1911.

Chapter 3

ADAM, EVERETT E. JR. AND EBERT, RONALD J. *Production and Operations Management* 2nd ed. Englewood Cliffs, NJ: Prentice Hall, 1982.

ASFAHL, C. RAY. *Industrial Safety and Health Management,* Englewood Cliffs, NJ: Prentice Hall, 1984.

AWAD, ELIAS M. *Systems Analysis and Design,* 2nd ed. Homewood, IL: Richard D. Irwin, 1985.

FOGARTY, DONALD W. AND HOFFMAN, THOMAS R. *Production and Inventory Management,* Cincinnati, OH: South-Western Publ., 1983.

HAMMER, WILLIE. *Occupational Safety Management and Engineering,* Englewood Cliffs, NJ: Prentice-Hall, 1976.

NIEBEL, BENJAMIN W. *Motion and Time Study,* Homewood, IL: Richard D. Irwin, Inc., 1976.

Chapter 4

AMERICAN SOCIETY OF MECHANICAL ENGINEERS. *Orientation and Guide for Use of SI (Metric) Units.* 8th ed. New York: The Am. Soc. of Mechanical Engineers, 1978.

AMERICAN SOCIETY FOR TESTING AND MATERIALS. *Standard for Metric Practice.* Philadelphia: Am. Soc. for Testing and Materials, 1980.

BRAVERMAN, JEROME D., *Fundamentals of Statistical Quality Control.* Reston, VA: Reston Publishing, 1981.

BURLINGAME, LUTHER, "Pioneer Steps Toward the Attainment of Accuracy," *American Machinist,* 47 (1914): 237–43.

DOYLE, LAWRENCE E. *Manufacturing Processes and Materials for Engineers.* Englewood Cliffs, NJ: Prentice-Hall, Inc., 1985.

INSTITUTE OF ELECTRICAL AND ELECTRONIC ENGINEERS. *Standard for Metric Practice.* New York: The Inst. of Electrical and Electronic Engineers, Inc., 1982.

JURAN, J. M. AND GRYNA, FRANK M., JR., *Quality Planning and Analysis.* New York: McGraw-Hill, 1980.

LITTAUER, S. B., "Development of Statistical Quality Control in the United States," *American Statistician,* 4 (1950): 14–20.

Chapter 5

Ashby, M. F., and Jones, D. R. H., *Engineering Materials 1: An Introduction to their Properties and Applications*. New York: Pergamon, 1980.

Askeland, *The Science and Engineering of Materials*. Monterey, CA: Brooks/Cole Engineering Division, 1984.

Flinn, R. A. and Trojan, P. K., *Engineering Materials and Their Applications*. 2nd edition. Boston: Houghton Mifflin, 1981.

Guide to Engineered Materials, Metals Park, OH: *American Society for Metals,* 1987.

Guy, A. G., *Essentials of Materials Science*. New York: McGraw Hill, 1976.

John, V. B., *Introduction to Engineering Materials*. 2nd Edition. New York: Macmillan, 1983.

Ludema, K. C., Caddell, R. M., and Atkins, A. G., *Manufacturing Engineering: Economics and Processes*. Englewood Cliffs, NJ: Prentice-Hall, 1987.

Shackleford, J. F., *Introduction to Materials Science for Engineers*. New York: Macmillan, 1985.

Van Vlack, L., *Elements of Materials Science and Engineering,* 5th edition. Reading, MA: Addison Wesley, 1985.

Chapter 6

Advanced Materials & Processes, Incorporating Metal Progress, published monthly by American Society for Metals, Metals Park, OH.

ASM Metals Handbook, American Society for Metals, Metals Park, OH. This is a multivolume set of reference books with entire volumes devoted to single topics such as ferrous metallurgy, nonferrous metallurgy, corrosion, failure analysis, phase diagrams, and metallography.

Boyer, H. E. and Gall, T. L., eds., *Metals Handbook,* Desk Edition. Metals Park, OH: American Society for Metals, 1985.

John, V. B., *Introduction to Engineering Materials*. 2nd ed. New York: Macmillan, 1983.

Journal of Metals, published monthly by The Metallurgical Society, Inc., Warrendale, PA.

"The Making of Steel," American Iron and Steel Institute.

The reader is also referred to any of the textbooks listed as suggested reading for Chapter 5.

Chapter 7

Reference Books

Harper, Charles A., ed., *Handbook of Plastics and Elastomers*. New York: McGraw Hill, 1975.

Rosalind, Juran, ed., *Modern Plastics Encyclopedia 1988*. New York: McGraw Hill, 1987.

Schwartz, Seymour S. and Goodman, Sidney H., *Plastics Materials and Processes*. New York: Van Nostrand Reinhold Co., 1982.

Ulrich, Henri, *Introduction to Industrial Polymers,* Munehen, Wien, 1982.

Periodicals

Modern Plastics. New York: McGraw Hill.

Plastics Engineering. Brookfield Center, CT: The Society of Plastics Engineers.

Plastics Technology. New York: Bill Communications.

Plastics World. Boston: Cahners.

Major Plastics Organizations

The Society of Plastics Engineers. Brookfield, CT.

The Society of the Plastics Industry. New York, NY.

Chapter 8

FEIRER, JOHN L. *Cabinetmaking and Millwork*. Peoria, IL: Bennett & McKnight Publishing Company, 1977.

FEIRER, JOHN L. *Woodworking for Industry*. Third Edition. Peoria, IL: Bennett & McKnight Publishing Company, 1979.

FOREST PRODUCTS LABORATORY. *Wood Handbook: Wood As An Engineering Material*. Agriculture Handbook No. 72. Washington DC: U.S. Government Printing Office, 1987.

Furniture Design and Manufacturing. Monthly periodical.

Furniture Manufacturing Management. Monthly periodical.

Furniture Wood Digest. Monthly periodical.

JOYCE, ERNEST. *Encyclopedia of Furniture Making*. New York: Sterling Publishing Co., Inc., 1979.

KOCH, PETER. *Wood Machining Processes*. New York: The Ronald Press Company, 1964.

Mepla Manual: 32 mm System. Mepla, Inc., P.O. Box 1469, High Point, NC 27261.

Wood and Wood Products. Monthly periodical.

Chapter 9

Apparel Industry Magazine. Monthly publication.

Bobbin Magazine. Monthly publication.

Encyclopedia of Textiles. American Fabrics. 3rd Edition. Englewood Cliffs, NJ: Prentice-Hall, Inc., 1980.

"Computer-Aided Design Systems Shorten Turn Times." *Cuttings*, Vol. 6, No. 4, Sept./Oct. 1986, pp. 2–3, published by Gerber Garment Technology and Gerber Camsco.

Federal Standard No. 751a. *Stitches, Seams, and Stitching*. Washington, DC: General Services Administration.

FORTESS, FRED. "The Ultimate Objective: No Inspection Required." *Bobbin Magazine*. Vol. 27, No. 12, pp. 54–60.

FRINGS, G. S. *Fashion From Concept to Consumer*. Englewood Cliffs, NJ: Prentice-Hall, Inc., 1982.

GREENWOOD, K. M. AND M. F. MURPHY. *Fashion Innovation and Marketing*. New York: MacMillan Publishing Co., Inc., 1978.

JARNOW, J. A., M. GUERREIRO, AND B. JDELLE. *Inside the Fashion Business: Text and Readings*. 4th Edition. New York: MacMillan Publishing Company, 1987.

JOSEPH, MARJORIE L. *Introductory Textile Science*. 5th Edition. New York: Holt, Rinehart & Winston, 1986.

LOWER, JAMES. "Robotics Advance Softly." *Bobbin Magazine*. Vol. 28, No. 9, May 1987, pp. 106–113.

The Needle's Eye, c/o Union Special Corp., 400 N. Franklin Street, Chicago, Illinois 60610.

The Technology of Thread and Seams. Industrial Products Marketing, J&P Coats Limited, 155 St. Vincent Street, Glasgow G2 5PA Scotland.

TORTORA, P. G. *Understanding Textiles.* 3rd Edition. New York: MacMillan Publishing Company, 1987.

WINGATE, I. B. AND J. MOHLER. *Textile Fabrics and Their Selection.* 8th Edition. Englewood Cliffs, NJ: Prentice-Hall, Inc., 1984.

Chapter 10

ALTING, LEO. *Manufacturing Engineering Processes.* New York: Marcel Dekker, 1982.

DOYLE, LAWRENCE, E., AND OTHERS. *Manufacturing Processes and Materials for Engineers.* Third Edition. Englewood Cliffs, NJ: Prentice Hall, Inc., 1985.

KALPAKJIAN, SEROPE. *Manufacturing Processes for Engineering Materials.* Reading, MA: Addison-Wesley Publishing Company, 1984.

LINDBERG, ROY A. *Processes and Materials for Manufacture.* Third Edition. Boston: Allyn and Bacon, 1983.

Manufacturing Engineering Magazine. One SME Drive, P.O. Box 930, Dearborn, MI 48121.

Metals Handbook. 8th Edition, Volume 4, *Forming.* Metals Park, OH: American Society for Metals, 1969.

SCHEY, JOHN A. *Introduction to Manufacturing Processes.* Second Edition. New York: McGraw-Hill Book Company, 1987.

Chapter 11

AMERICAN MACHINIST. *Metalcutting.* New York: McGraw-Hill Book Company, 1982.

American Machinist. Monthly publication.

AMSTEAD, B. H., AND OTHERS. *Manufacturing Processes.* Eighth Edition. New York: John Wiley and Sons, 1987.

Cutting Tool Engineering. Monthly publication.

Gear Technology. Monthly publication.

Modern Machine Shop. Monthly publication.

SCHEY, JOHN A. *Introduction to Manufacturing Processes.* Second Edition. New York: McGraw-Hill Book Company, 1987.

TRENT, E. M. *Metal Cutting.* Second Edition. London: Butterworths, 1984.

WECK, M. *Handbook of Machine Tools.* New York: John Wiley and Sons, 1984.

WICK, C. (Ed.). *Tool and Manufacturing Engineers Handbook.* Fourth Edition. Volume 1. *Machining.* Dearborn, MI: Society of Manufacturing Engineers, 1983.

Chapter 12

ALTING, LEO. *Manufacturing Engineering Processes.* New York: Marcel Dekker, Inc., 1982.

AMSTEAD, B. H., AND OTHERS. *Manufacturing Processes.* Eighth Edition. New York: John Wiley and Sons, Inc., 1987.

DOYLE, LAWRENCE E., AND OTHERS. *Manufacturing Processes and Materials for Engineers.* Third Edition. Englewood Cliffs, NJ: Prentice-Hall, Inc., 1985.

KUTZ, MEYER. (Ed.). *Mechanical Engineers Handbook.* New York: John Wiley and Sons, 1986.

SCHEY, JOHN A. *Introduction to Manufacturing Processes.* New York: McGraw-Hill Book Company, 1987.

SHAW, MILTON C. *Metal Cutting Principles.* Oxford: Clarendon Press, 1984.

Chapter 13

American Machinist. Monthly publication.

AMERICAN SOCIETY FOR METALS. *Metals Handbook*. Eighth Edition. Volume 3. *Machining*. Metals Park, OH: American Society for Metals, 1967.

BHATEYA, C. AND R. LINDSAY. (Eds.). *Grinding: Theory, Techniques, and Troubleshooting*. Dearborn, MI: *Society of Manufacturing Engineers*, 1982.

Cutting Tool Engineering. Monthly publication.

DOYLE, LAWRENCE E. *Manufacturing Processes and Materials for Engineers*. Third Edition. Englewood Cliffs, NJ: Prentice-Hall, Inc., 1985.

GRINDING WHEEL INSTITUTE. *Abrasive Machining*. Cleveland: Grinding Wheel Institute, 1984.

Manufacturing Engineering. Monthly publication.

MCKEE, RICHARD L. *Machining With Abrasives*. New York: Van Nostrand Reinhold Company, 1982.

PINKSTONE, WILLIAM G. *The Abrasive Ages*. Lititz, PA: Sutter House, 1974.

SHAW, M. C. *Metal Cutting Principles*. Oxford: Oxford University Press, 1984.

Chapter 14

BELFORTE, DAVID, AND MORRISS LEVITT. (Eds.) *Industrial Laser Annual Handbook*. Tulsa, OK: PennWell Books, 1987.

DROZDA, T., AND C. WICK, (Eds.). *Tool and Manufacturing Engineers Handbook. Volume 1. Machining*. Dearborn, MI: Society of Manufacturing Engineers, 1983.

JOHNSON, JIM. *Laser Technology*. Benton Harbor, MI: Heath Company, 1986.

LASER INSTITUTE OF AMERICA. 5151 Monroe Street, Suite 102 W., Toledo, OH 43623.

Lasers and Optronics. Monthly Journal.

LUXTON, JAMES T., AND DAVID E. PARKER. *Industrial Lasers and Their Applications*. Englewood Cliffs, NJ: Prentice-Hall, Inc., 1985.

MALLOW, ALEX, AND LEON CHABOT. *Laser Safety Handbook*. NY: Van Nostrand Reinhold Company, 1978.

SAUNDERS, RICHARD, AND OTHERS. Coherent, Inc. *Lasers*. NY: McGraw-Hill Book Company, 1980.

WELLER, E. J. (Ed.) *Nontraditional Machining Processes*. Dearborn, MI: Society of Manufacturing Engineers, 1983.

Chapter 15

AVITUR, B. *Handbook of Metalforming Processes*. New York: Wiley-Interscience, 1983.

DOYLE, LAWRENCE E. *Manufacturing Processes and Materials for Engineers*. Englewood Cliffs, NJ: Prentice-Hall, Inc., 1985.

EARY, D. F., AND E. A. REED. *Techniques of Pressworking Sheet Metal*. Englewood Cliffs, NJ: Prentice-Hall, Inc., 1974.

LANGE, K. (Ed.) *Handbook of Metalworking*. New York: McGraw-Hill Book Company, 1985.

Manufacturing Engineering. Monthly Journal of the Society of Manufacturing Engineers.

Metals Handbook. Volume 4. *Forming*. Metals Park, OH: American Society for Metals, 1969.

WICK, C., AND OTHERS. (Eds.) *Tool and Manufacturing Engineers Handbook*. Volume 2. 4th Edition. Dearborn, MI: Society of Manufacturing Engineers, 1984.

WILHELM, D. *Understanding Presses and Press Operations*. Dearborn, MI: Society of Manufacturing Engineers, 1981.

Chapter 16

AMERICAN FOUNDRY SOCIETY. *Metal Casting and Molding Processes*. Des Plaines, IL: American Foundry Society, 1975.

AMERICAN SOCIETY FOR METALS. *Metals Handbook*. Volume 5. *Forgings and Castings*. Eighth Edition. Metals Park, OH: American Society for Metals, 1970.

DOYLE, LAWRENCE E., AND OTHERS. *Manufacturing Processes and Materials for Engineers*. Englewood Cliffs, NJ: Prentice-Hall, Inc., 1985.

GROENVELD, T. G., AND A. L. PONIKVAN. (Eds.). *Designing for Thin-Wall Zinc Die Castings*. New York: International Lead Zinc Research Organization, Inc., 1986.

HAMILTON, EDWARD. *Patternmaking Guide*. Des Plaines, IL: American Foundry Society, 1976.

METAL POWDER INDUSTRIES FEDERATION. *P/M Design Guidebook*. Princeton: Metal Powder Industries Federation, 1983.

Modern Casting. Monthly publication.

SCHEY, JOHN A. *Introduction to Manufacturing Processes*. Second Edition. New York: McGraw-Hill Book Company, 1987.

STEEL FOUNDER'S SOCIETY OF AMERICA. *Steel Castings Handbook*. Des Plaines, IL: Steel Founder's Society of America, 1980.

YANKEE, HERBERT W. *Manufacturing Processes*. Englewood Cliffs, NJ: Prentice-Hall, Inc., 1979.

Chapter 17

AMERICAN SOCIETY OF MECHANICAL ENGINEERS. *Metals Engineering Processes*. New York: McGraw-Hill Book Company, 1958.

AMERICAN SOCIETY FOR METALS. *Metals Handbook. Vol. 5. Forging and Casting*. Eighth Edition. Metals Park, OH: American Society for Metals, 1970.

BYER, THOMAS G. (Editor). *Forging Handbook*. Cleveland, OH: Forging Industry Association, 1985. (Produced in collaboration with the American Society for Metals.)

DOYLE, LAWRENCE E., AND OTHERS. *Manufacturing Processes and Materials for Engineers*. Englewood Cliffs, NJ: Prentice-Hall, Inc., 1985.

LINDBERG, ROY A. *Processes and Materials of Manufacture*. Third Edition. Boston: Allyn and Bacon, Inc., 1983.

Chapter 18

American Fastener Journal. Monthly journal.

Assembly Engineering. Monthly journal.

Fastener Industry News. Monthly journal.

Fastener Technology International. Monthly journal.

LANE, J. *Automated Assembly*. Second Edition. Dearborn, MI: Society of Manufacturing Engineers, 1986.

PARMELY, ROBERT O. (Ed.). *Standard Handbook of Fastening and Joining*. New York: McGraw-Hill Book Company, 1977.

WICK, CHARLES AND RAYMOND F. VEILLEUX. (Eds.). *Quality Control and Assembly*. Volume 4. Tool and Manufacturing Engineers Handbook. Dearborn, MI: Society of Manufacturing Engineers, 1987.

Chapter 19

Adhesives Age Magazine. Monthly journal.

BRUNO, E. J. (Ed.). *Adhesives in Modern Manufacturing*. Dearborn, MI: Society of Manufacturing Engineers, 1970.

EDITOR, *Machine Design Magazine*. *Engineering Adhesives*. Cleveland, OH: Penton/IPC, Inc., 1982.

GILLESPIE, ROBERT H. *Adhesives for Wood*. Park Ridge, NJ: Noyes Publications, 1984.

LANDROCK, ARTHUR. *Adhesives Technology Handbook*. Park Ridge, NJ: Noyes Publications, 1985.

SCHNEBERGER, GERALD L. *Adhesives in Manufacturing*. New York: Marcel Dekker, Inc., 1983.

SHIELDS, J. *Adhesives Handbook*. Third Edition. London: Butterworth and Company, Ltd., 1984.

SKEIST, IRVING (Ed.). *Handbook of Adhesives*. Second Edition. New York: Van Nostrand Reinhold Company, 1977.

WAKE, WILLIAM C. *Adhesion and the Formulation of Adhesives*. Second Edition. New York: Applied Sciences Publishers, 1982.

Chapter 20

BRUMBAUGH, JAMES E. *Welders Guide*. Indianapolis: The Bobbs-Merril Co., Inc., 1983.

DOYLE, LAWRENCE E., AND OTHERS. *Manufacturing Processes and Materials for Engineers*. Englewood Cliffs, NJ: Prentice-Hall, Inc., 1985.

SCHWARTZ, M. M. *Metals Joining Manual*. New York: McGraw-Hill Book Company, 1979.

SMITH, DAVE. *Welding Skills and Technology*. New York: McGraw-Hill Book Company, 1985.

Standard Welding Definitions and Terms. ANSI-AWS A3.0–85. Miami, FL: The American Welding Society, Inc., 1985.

Welding, Brazing, and Soldering. Volume 6, Metals Handbook, Ninth Edition. Metals Park, OH: American Society for Metals, 1983.

Welding Handbook. Seventh Edition. Miami, FL: The American Welding Society, Inc. 1984.

A. Volume 1. *Fundamentals of Welding*. 1976.
B. Volume 2. *Welding Processes—Arc and Gas Welding and Cutting, Brazing, and Soldering*. 1978.
C. Volume 3. *Welding Processes—Resistance and Solid State Welding and Other Joining Processes*. 1980.
D. Volume 4. *Metals and Their Weldability*. 1982.
E. Volume 5. *Engineering, Costs, and Quality and Safety*. 1984.

The Welding Journal. Monthly Journal of the American Welding Society.

WOODGATE, RALPH W. *Handbook of Machine Soldering*. New York: McGraw-Hill Book Company, 1983.

Chapter 21

AMERICAN SOCIETY FOR METALS. *Metals Handbook*. Ninth Edition. Volume 5. *Surface Cleaning, Finishing, and Coating*. Metals Park, OH: American Society for Metals, 1982.

AMSTEAD, B. H., AND OTHERS. *Manufacturing Processes*. New York: John Wiley and Sons, 1987.

DAVIES, M. A. S. (Ed.). *Finishing Handbook and Directory*. London: Sawell Publications, Ltd., 1985.

DOYLE, LAWRENCE E., AND OTHERS. *Manufacturing Processes and Materials for Engineers*. Englewood Cliffs, NJ: Prentice-Hall, Inc., 1985.

Industrial Finishing. Monthly journal.

Metal Finishing Guidebook and Directory. Hackensack, NJ: Metals and Plastics Publications, Inc., 1986.

Chapter 22

AMERICAN SOCIETY FOR METALS. *Metals Handbook*. Ninth Edition. Volume 5. *Surface Cleaning, Finishing, and Coating*. Metals Park, OH: American Society for Metals, 1982.

AMSTEAD, B. H., AND OTHERS. *Manufacturing Processes*. New York: John Wiley and Sons, 1987.

DAVIES, M. A. S. (Ed.). *Finishing Handbook and Directory*. London: Sawell Publications, Ltd., 1985.

DOYLE, LAWRENCE E., AND OTHERS. *Manufacturing Processes and Materials for Engineers*. Englewood Cliffs, NJ: Prentice-Hall, Inc., 1985.

Industrial Finishing. Monthly journal.

Metal Finishing Guidebook and Directory. Hackensack, NJ: Metals and Plastics Publications, Inc., 1986.

Products Finishing. Monthly publication.

Chapter 23

ALEXANDER, DAVID C. *The Practice and Management of Industrial Ergonomics*. Englewood Cliffs, NJ: Prentice-Hall, Inc., 1986.

ANDREASEN, N. MYRUP, S. KAHLER, AND T. LUND. *Design for Assembly*. New York: Springer Verlag, 1983.

BOOTHROYD G., AND P. DEWHURST. *Design for Assembly*. Wakefield, RI: Boothroyd Dewhurst, Inc., 1983.

BRALLA, JAMES B. (Ed.). *Handbook of Product Design and Manufacture*. New York: McGraw-Hill Book Company, 1986.

BRONIKOWSKI, RAYMOND J. *Managing the Engineering Design Function*. New York: Van Nostrand Reinhold Company, 1986.

BUSCH, AKIKO. (Ed.). *Product Design*. New York: PBC International, Inc., 1984.

CROSS, NIGEL. *Developments in Design Methodology*. New York: John Wiley and Sons, 1984.

HEGINBOTHAM, W. B. *Programmable Assembly*. New York: Springer-Verlag, 1984.

HUBKA, V. *Principles of Engineering Design*. London: Butterworth Scientific, 1982.

KONZ, STEPHAN. *Facility Design*. New York: John Wiley and Sons, 1985.

LEECH, D. J., AND B. T. TURNER. *Engineering Design for Profit*. New York: John Wiley and Sons, 1985.

LINDBECK, JOHN R. *Designing Today's Manufacturing Products*. Peoria, IL: Bennett and McKnight, 1972.

LORENZ, CHRISTOPHER. *The Design Dimension*. New York: Basil Blackwell, Ltd., 1986.

LUCIE-SMITH, EDWARD. *A History of American Design*. New York: Van Nostrand Reinhold Company, 1983.

PAPANEK, VICTOR. *Design for the Real World.* New York: Van Nostrand Reinhold Company, 1984.

PULOS, ARTHUR J. *American Design Ethic.* Cambridge, MA: MIT Press, 1983.

RAY, MARTYN S. *Elements of Engineering Design.* Englewood Cliffs, NJ: Prentice-Hall, International, 1985.

TRUCKS, H. E. *Designing for Economical Production.* Dearborn, MI: Society of Manufacturing Engineers, 1987.

VAN TYNE, C. J., AND B. AVITZUR (Eds.). *Production to Near Net Shape.* Metals Park, OH: American Society for Metals, 1983.

Chapter 24

BERTOLINE, GARY R. *Fundamentals of CAD.* Albany, NY: Delmar Publishers, 1985.

GROOVER, MIKELL P. *Automation, Production Systems, and Computer-Aided Manufacturing.* Englewood Cliffs, NJ: Prentice-Hall, 1980.

GROOVER, MIKELL P. AND ZIMMERS, EMORY W. JR. *CAD/CAM.* Englewood Cliffs, NJ: Prentice-Hall, 1984.

GUPTON, JAMES A. JR. *Computer-Controlled Industrial Machines, Processes, and Robots.* Englewood Cliffs, NJ: Prentice-Hall, 1986.

PUSZTAI, JOSEPH AND SAVA, MICHAEL. *Computer Numerical Control.* Reston, VA: Reston Publishing Company, 1983.

TIJUNELIS, DONATAS AND MCKEE, KEITH E. *Manufacturing High Technology Handbook.* New York: Marcel Dekker, 1987.

Chapter 25

COMMITTEE ON THE STATUS OF HIGH-TECHNOLOGY CERAMICS IN JAPAN. *High-Technology Ceramics in Japan,* p. 30. Washington, D.C.: National Academy Press, 1984.

GROOVER, MIKELL P. *Automation, Production Systems, and Computer-Aided Manufacturing.* Englewood Cliffs, NJ: Prentice-Hall, 1980.

"HIGH TECHNOLOGY PRODUCTION: STRATEGIES AND APPLICATION." IEEE VIDEOCONFERENCE, 1986.

MANUFACTURING STUDIES BOARD. *Toward a New Era in U.S. Manufacturing.* Washington, DC: National Academy Press, 1986.

THOMPSON, BRIAN. "Fixturing: The Next Frontier in the Evolution of Flexible Manufacturing Cells." *CIM,* Mar/Apr, 10–13, 1985.

WINSTON, PATRICK H. AND KAREN A. PRENDERGAST. *The AI Business.* Cambridge, MA: The MIT Press, 1984.

INDEX

D

E

F

G

N

O

P

Q

R

S

W

X

Y

Z